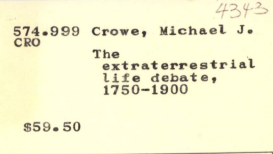

THE EXTRATERRESTRIAL LIFE DEBATE 1750–1900

*The idea of a plurality of worlds
from Kant to Lowell*

MICHAEL J. CROWE
University of Notre Dame

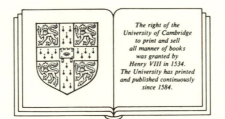

The right of the
University of Cambridge
to print and sell
all manner of books
was granted by
Henry VIII in 1534.
The University has printed
and published continuously
since 1584.

Cambridge University Press 4343

Cambridge
London   New York   New Rochelle
Melbourne   Sydney

Published by the Press Syndicate of the University of Cambridge
The Pitt Building, Trumpington Street, Cambridge CB2 1RP
32 East 57th Street, New York, NY 10022, USA
10 Stamford Road, Oakleigh, Melbourne 3166, Australia

First published 1986

Printed in the United States of America

Library of Congress Cataloging in Publication Data
Crowe, Michael J.
The extraterrestrial life debate 1750–1900.
Bibliography: p.
Includes index.
1. Plurality of worlds – History. 2. Life on other
planets – History. I. Title.
QB54.C76   1985   574.999   85-7842
ISBN 0 521 26305 0

British Library Cataloging-in-Publication applied for

In loving memory of my parents,
Claire Huesman Crowe (1900–1962)
and Michael P. Crowe (1899–1976)

Contents

Part II From 1800 to 1860

Illustrations

Preface

Possibly tomorrow, or perhaps a year or a century from now, astronomers may make one of the most important discoveries that scientists have ever sought: the detection of an extraterrestrial civilization. Establishing contact with such a civilization could produce extraordinary effects, possibly including solutions for our most pressing scientific and technological problems.[1] As has often been noted, even the discovery of conclusive evidence for extraterrestrial life would have far-reaching implications for our philosophical, religious, and social thought.

This book contains a more modest discovery than either of those. Although appearing on nearly every page, it can be summarized in a sentence: The question of extraterrestrial life, rather than having arisen in the twentieth century, has been debated almost from the beginning of recorded history. Between the fifth-century B.C. flowering of Greek civilization and 1917, more than 140 books and thousands of essays, reviews, and other writings had been devoted to discussing whether or not other inhabited worlds exist in the universe. Moreover, as documented in this book, the majority of educated persons since around 1700 have accepted the theory of extraterrestrial life and in numerous instances have formulated their philosophical and religious positions in relation to it. To put this point differently, even if no UFOs hover in our heavens, belief in extraterrestrial beings has hovered in the human consciousness for dozens of decades. Although the moon and Mars are as barren as giant bricks, moonlings and Martians long ago began to invade our culture and influence our thought, and they now occupy increasing roles in our cinematic and literary creations. Our extraterrestrials may no more exist than the gods of the Greeks, but their effects are no less indisputable. Just as paranoia destroys real lives, just as atheists admit the influence of belief in God, so should we see the invasion of the extraterrestrials as long since under way. This book has, in short, been written in the conviction that even if no extraterrestrials exist, their influence on terrestrials has been immense.

The particular focus of the present study is the debate that developed between 1750 and about 1900 on one planet concerning the question whether or not life exists on any other body in the universe. But this book

xiii

has been envisioned as more than a dispassionate description of the past; it has also been written in the hope of making a modest contribution to the present phase of this long debate. As Pierre Duhem remarked in an analogous context, to give the history of a theory can also be to provide a critique of it. Consequently, while striving to record accurately the ideas espoused by the persons discussed, I have not hesitated to analyze and to assess their positions. Moreover, larger patterns and principles have been sought, some of which seem suggestive concerning current issues in the debate.

The decision to concentrate on the period from the middle of the eighteenth century to the early twentieth century was made in light of a number of factors. The developments from antiquity to the first half of the eighteenth century have recently been capably discussed by Steven J. Dick in his *Plurality of Worlds: The Origins of the Extraterrestrial Life Debate from Democritus to Kant*. This has made it possible to skim over these earlier materials, devoting only the first chapter to them. The decision to terminate this study in the early twentieth century derived both from the vast quantity of materials that had appeared by then and from my determination to base this study on first-hand knowledge of the great majority of these items, many exceedingly rare. Although the general coverage of these materials terminates at 1900, the analysis of the canals of Mars controversy is carried through to its culmination around 1910.

A vexing problem has been the choice as to what materials should be analyzed. Essentially all works of science fiction, a genre with a long and in many cases distinguished history, have been excluded. The vastness of this literature and its tendency to tell more about what humans can imagine than what they in fact believe have necessitated this resolve, which has been broken in a few cases in which the continuity of the story has required such inclusion. However, it is hoped that this research may prove of value to historians of science fiction, whose studies have frequently illuminated topics treated in this book.

Astronomical writings have been a special focus of this study, and initially they were the sole focus. However, it soon became clear that no sharp demarcations among scientific, philosophical, and religious approaches were either desirable or possible in the study of an idea that in many instances has seemed more like a metaphysical postulate or religious principle than a scientific hypothesis. In fact, the absence of clear boundaries separating astronomical, literary, philosophical, and religious discussions of the idea of a plurality of worlds is one of the themes of this book.

Geographically, the concentration has been on the nations of Europe and North America. Even if the mutual interactions among these nations and the scarcity of such interactions with oriental, African, and South

American conceptions were not to justify this choice, the author's lack of knowledge of oriental and African languages would have necessitated this exclusion in a study in which extensive efforts have been made to trace ideas to their original sources.

The audience for whom this book has been written is large, including all who have an interest in extraterrestrials, but especially those whose concern extends to how we terrestrials have responded to one of our most challenging quests. The general reader must be prepared to forgive a flood of references, for this book has been written in the hope of withstanding and repaying the scrutiny of scholars, especially my colleagues in the community of historians of scientific ideas. Historians of religion and theology will, it is hoped, find materials of relevance here, perhaps more than they expect. So also, those concerned with the development of literature and of philosophy may find that this research will illuminate issues of direct interest to them. Even those categories are not capacious enough to include the array of individuals who are part of this story, but mention of such names as Arago and Aristotle, Balzac and Beethoven, Cavour and Chalmers, Darwin and Dostoevsky, Emerson and Engels, Flammarion and Franklin, Galileo and Grandville, Hegel and the Herschels as participants in the debate may suggest that interest in the idea of other worlds has been nearly as broad as learning itself.

Readers of this book may wish to keep in mind that this is the first detailed scholarly study in English analyzing these materials. I have, however, profited greatly from the publications of a host of authors, among whom the ten whose researches contributed most directly to this study are R. V. Chamberlin, S. J. Dick, S. L. Engdahl, C. Flammarion, K. S. Guthke, W. G. Hoyt, S. L. Jaki, A. O. Lovejoy, G. McColley, and M. H. Nicolson.[2] References to their writings recur throughout the text, but no doubt inadequately express the debt I owe to their erudition. Readers may also wish to be aware of my conviction that large as the present volume may be, the story contained in it is far from complete. Certainly the unconcluded debate on extraterrestrials will continue into the future, and so, it is hoped, will research into its past.

Acknowledgments

The research and writing of this book have been greatly assisted by many persons and institutions. The National Science Foundation has supported this research under grants SOC-7709638 and SES-8007886. To that institution and to Dr. Ronald J. Overman, Assistant Program Director for History and Philosophy of Science, warmest expressions of gratitude are due. The University of Notre Dame supported this work by a one and one-half semester leave of absence, a travel grant, and three summer support grants, the most recent from Notre Dame's Institute for Scholarship in the Liberal Arts.

Approximately forty research libraries in six countries have been consulted. Deep appreciation is due the librarians of the University of Notre Dame, including James T. Deffenbaugh, Maureen Gleason, Robert J. Havlik, Joseph Lauck, Marie K. Lawrence, Anton C. Masin, Pamela Paidle, and Joseph Ross. For especially extensive assistance I am indebted to the staffs of the following libraries: the Regenstein Library and the Eckhart Library of the University of Chicago; the John Crerar Library; the University of Cambridge libraries, particularly those of Trinity College and the Institute of Astronomy; the Library of Congress; and the British Library. Very valuable assistance came from Brenda Corbin, Librarian of the U.S. Naval Observatory, and from Enid Lake, formerly Librarian of the Royal Astronomical Society.

Numerous professional colleagues have contributed to this book by generously giving advice and/or criticisms. Among Notre Dame colleagues, Thomas Kselman, Ernan McMullin, Thomas Werge, and Phillip R. Sloan have all read and commented on sections relevant to their expertise, the last having been a continuous source of encouragement and sound advice. Steven J. Dick of the U.S. Naval Observatory has read the entire manuscript, supplying detailed comments from his extensive knowledge of the area. His suggestions have immeasurably improved the book. The late William G. Hoyt of the Lowell Observatory read a number of chapters, contributing greatly to their improvement. Richard Waldron of the New Jersey Historical Society shared with me his unpublished Freneau materials and advised on the second section of Chapter 3. Karl S. Guthke of Harvard University and Walter Schatzberg of Clark

University read and commented on the fourth section of Chapter 3, which is partly based on their writings. E. Robert Paul of Dickinson College provided very helpful suggestions on Chapters 4 and 5, and John Hedley Brooke of the University of Lancaster drew on his research on the Whewell debate to make excellent recommendations on Chapters 6 and 7. John G. Burke of UCLA, who is preparing a history of the study of meteorites, supplied commentary on the sixth section of Chapter 8. Both Norriss S. Hetherington and William Sheehan provided rich and detailed commentary on Chapter 10. A list of the other scholars who have contributed to aspects of this research would be extremely long, but special thanks are due to Donald Beaver, Andrew Burgess, John Burnham, Frederic B. Burnham, E. Gerard Carroll, I. Bernard Cohen, W. Paul Fayter, André Goddu, Michael A. Hoskin, Stanley L. Jaki, Timothy Lenoir, Sydney Ross, and Richard S. Westfall. I take full responsibility for mistakes remaining in this book.

Research assistance has been provided over the years by a number of Notre Dame undergraduate and graduate students, some of whom are now promising young scholars; these include Thomas Berry, Otto (Barry) Bird, Orville R. Butler, Mary Kane, William Kane, Shane Little, Therese Anne Brown Matthews, Daniel Meuleman, Mark Moes, Thomas Pearson, John Roda, Kenneth Taylor, and Margaret Humphreys Warner.

I am indebted to the following publishers and persons for permission to quote materials from the items indicated: Cambridge University Press, for passages from G. W. Leibniz, *New Essays on Human Understanding,* translated by Peter Remnant and Jonathan Bennett (Cambridge, England, 1981); Harvard University Archives, for passages from letters of C. W. Eliot, E. C. Pickering, and W. H. Pickering, as quoted in Bessie Zaban Jones and Lyle Gifford Boyd, *The Harvard College Observatory: The First Four Directorships, 1839–1919* (Cambridge, Mass.: Harvard University Press, 1971); Hermann éditeurs des sciences et des arts, for passages from E. M. Antoniadi, *La planète Mars* (Paris, 1930); Houghton Mifflin Company, for passages from *Young Emerson Speaks,* edited by Arthur C. McGiffert, Jr., copyright 1935 by the Ralph Waldo Emerson Memorial Association, copyright renewed 1966 by Arthur Cushman McGiffert, Jr., and the Ralph Waldo Emerson Memorial Association; Stanley L. Jaki, for passages from his translation of Immanuel Kant, *Universal Natural History and Theory of the Heavens* (Edinburgh: Scottish Academic Press, 1981), and from his translation of J. H. Lambert, *Cosmological Letters on the Arrangement of the World-Edifice* (New York: Science History Publications, 1976); Oxford University Press, for passages from G. F. W. Hegel, *Hegel's Philosophy of Nature,* translated by A. V. Miller (Oxford, 1970); Routledge & Kegan Paul, for passages from Nicolas Cusanus, *Of Learned Ignorance,* translated by Germain Heron (London, 1954); the Royal Astronomical Society, for quotations from unpublished manuscripts of Sir William Herschel; The Master and Fellows of Trinity College, Cambridge, for passages from unpublished writings of William Whewell; University of Arizona Press, for passages from William Graves Hoyt, *Lowell and Mars* (Tucson, 1976).

The typing of this manuscript in its various drafts has been the work of persons too numerous to mention. However, the care, patience, and professionalism shown by Susan Curtis and Cheryl A. Reed in preparing its final form were crucial to the completion of this book.

Helen Wheeler and Richard L. Ziemacki of Cambridge University Press have overseen the publication of this book, invariably supplying assistance and encouragement, and to them I am deeply indebted.

Finally, warm thanks to my wife, Mary Ellen, who in numerous ways has contributed to this project, most recently by helping proofread the entire manuscript. While sharing and supporting my conviction in the importance of this project, she has repeatedly reminded me that there are more important matters.

Introduction: before 1750

1

The plurality of worlds debate before 1750: a background survey

1. The debate in ancient and medieval science and philosophy

The roots of the plurality of worlds debate as it existed in 1750 extend back to antiquity. The present chapter sets the stage in a summary fashion for the post-1750 developments by drawing on Steven J. Dick's recent history of the debate up to the time of Kant, as well as other relevant studies.[1]

Regarding the existence of other worlds, the ancients of both Greece and Rome were deeply divided. Arguing the affirmative were the Epicureans, so called after Epicurus (341–270 B.C.), who developed certain ideas that had originated with Democritus and Leucippus two centuries earlier. Among the theories that we today consider most modern are (1) that matter is composed of atoms, (2) that the present state of nature is the result of a long evolutionary process, (3) that life exists elsewhere in the universe, and (4) that there is no God, or at least no personal God. Modern though these ideas may seem to us, they all indisputably date from antiquity, where they can be seen in Epicurus's "Letter to Herodotus," in a passage in which the philosopher's atheism is implicit:

. . . there are infinite worlds both like and unlike this world of ours. For the atoms being infinite in number . . . are borne far out into space. For those atoms . . . have not been used up either on one world or on a limited number of worlds, nor on all the worlds which are alike, or on those which are different from these. So that there nowhere exists an obstacle to the infinite number of worlds.[2]

Later in the letter he adds that "we must believe that in all worlds there are living creatures and plants and other things we see in this world. . . ."[3]

In reading these passages it is imperative to recognize that in them, as well as in most other ancient discussions, the term *world* had a meaning very different from the meaning we now give it. The Epicurean "innumerable worlds" were not the solar systems of remote stars; in fact, in Greek astronomy, the stars were typically thought of as located in a starry vault not greatly distant from the orbit of the outermost planet of our system. Rather, the postulated worlds of Epicurus were separate systems unseen by humans, each with its own earth, sun, planets, and stars.[4] This fact shows that the basis of Epicurean pluralism lay not in direct observation but in the metaphysical materialism and atomism of its philosophy. Other worlds must exist because some of the chance conglomerations of infinite atoms in an infinite universe must form worlds, all things being possible. As Metrodorus of Chios, Epicurus's leading disciple among his contemporaries, put it, "It would be strange if a single ear of corn grew in a large plain or were there only one world in the infinite. And that worlds are infinite in number follows from the causes [i.e., atoms] being infinite."[5] The passages from Epicurus and Metrodorus illustrate the Epicurean endorsement of an idea that Arthur Lovejoy called "the principle of plenitude," the doctrine that "no genuine potentiality of being can remain unfulfilled, that the extent and abundance of the creation must be as great as the possibility of existence and commensurate with the productive capacity of a 'perfect' and inexhaustible Source, and that the world is better, the more things it contains."[6] For the Epicureans, this "Source" was infinite nature, whereas in later centuries some religious authors identified it with the omnipotent Creator-God.

The most influential proponent of Epicurean philosophy was the Roman poet Lucretius (ca. 99–55 B.C.), who in his *De rerum natura* blended elegant Latin verse with Epicurean ideas to explain topics ranging from the origin of language to optical illusion, from the sweetness of wines to the evolution and structure of the universe. Concerning this last topic, the poet asserts:

Granted, then, that empty space extends without limit in every direction and that seeds innumerable are rushing on countless courses through an unfathomable universe . . . , *it is in the highest degree unlikely that this earth and sky is the only one to have been created* and that all those particles are accomplishing nothing. This follows from the fact that our world has been made by the spontaneous and casual collision and the multifarious, accidental, random and purposeless congregation and coalescence of atoms whose suddenly formed combinations could serve [to produce] . . . earth and sea and sky and the races of living creatures.[7]

The theological conclusion he draws from this doctrine is that "*nature is free and uncontrolled by proud masters* and runs the universe by herself without the aid of gods."[8] After the rediscovery in the fifteenth century of Lucretius's poem, a host of authors from Gassendi to Newton and Kant investigated whether or not Epicurean atomism, evolutionism, pluralism, and atheism were detachable from each other.

Attacks on the materialist, atomist, and pluralist positions date at least from the period of Plato and Aristotle, both of whom opposed the claims of Democritus and Leucippus for the existence of innumerable worlds.[9] In his *Timaeus,* Plato (428–348 B.C.) asserts that "there is and ever will be one only-begotten and created heaven," basing his claim on the arguments (1) that the uniqueness of the Creator implies the uniqueness of the creation and (2) that were the creation a composite, it would be subject to dissolution and decay.[10] The writings of Aristotle (384–322 B.C.) contain an array of arguments against a plurality of worlds. In his *On the Heavens* he expounds his doctrine of natural place. Earthly and watery substances move downward, he explains, because they seek their natural place, the center of the earth, whereas air and fire move upward, again tending to their natural place. A piece of earth can, of course, be forced upward, but that is a forced, "violent" motion. The relevance of this doctrine is that Aristotle notes that were other worlds to exist, they would have to be composed, as is our own, of earth, air, fire, and water. Consequently, a portion of earth could be moving in natural motion in our world, whereas it would be in violent motion with respect to another world, an obvious contradiction. Another argument appears prominently in Aristotle's *Metaphysics,* where he explains the motion in our planetary system as due to the Prime Mover acting at its periphery. Were there more worlds than one, a plurality of Prime Movers would be necessary, an idea he rejects as philosophically and religiously unacceptable.[11]

Although the plurality of worlds debate was most intense between the Epicureans and Aristotelians, other schools and individuals were involved. The Pythagoreans, for example, were reported to have believed that "the moon is terraneous, is inhabited as our earth is, and contains animals of a larger size and plants of a rarer beauty than our globe affords. The animals in their virtue and energy are fifteen degrees superior to ours, emit nothing excrementitious, and the days are fifteen times longer."[12] Plutarch (ca. 46–120 A.D.) also speculated on lunar life in his *De facie in orbe lunae,* and Lucian of Samosata (ca. 120–200 A.D.) composed two fictional moon voyages.[13] "Plurality of worlds" is in one sense ambiguous: It can mean either a number of simultaneously existing worlds or a succession of worlds in time. The Stoics favored the latter doctrine, the Roman statesman and orator Cicero (106–43 B.C.), for example, accepting it from them, while rejecting as absurd the idea of coexisting worlds, although he left open the possibility of life on the moon.[14]

Early Christian scholars faced the formidable task of forging a viable intellectual tradition responsive to the sophisticated questions raised by

Greek and Roman authors. To the idea of a plurality of worlds, their response was initially negative; in the third century, Hippolytus, for example, rejected it, as did Eusebius, bishop of Caesarea, in the fourth century and Theodoret, bishop of Cyprus, in the fifth century.[15] Augustine of Hippo (354–430) did likewise, although he was more concerned to refute the Stoic doctrine of successive worlds. His opposition to both successive and simultaneous worlds, as well as to the principle of plenitude, is evident in his *City of God:*

For if they imagine infinite spaces of time before the world during which God could not have been idle, in like manner they may conceive outside the world infinite realms of space, in which, if any one says that the Omnipotent cannot hold His hand from working, will it not follow that they must adopt Epicurus' dream of innumerable worlds?[16]

The possibility of a plurality of worlds was discussed in the thirteenth century as the Christian scholars of the West gradually gained access to the writings of antiquity. One of the most important of these scholars, Albertus Magnus (1193–1280), commented: "Since one of the most wondrous and noble questions in Nature is whether there is one world or many, . . . it seems desirable for us to inquire about it."[17] Such inquiry occurred and can be found in the writings of Michael Scot (d. ca. 1240) in Spain, William of Auvergne (ca. 1180–1249) in Paris, and Roger Bacon (1214–92) at Oxford. Albertus Magnus and his pupil Thomas Aquinas (1224–74) also added to the literature on the subject. All rejected a plurality of worlds – not a surprising result, given the enthusiasm of the period for the Aristotelian writings, from which, directly or indirectly, most of their antipluralist arguments came. This was certainly true of Thomas Aquinas, but with one important difference: As a Christian, he felt compelled to urge that the singularity of our world in no way contradicts belief in an omnipotent God. However, the finely wrought distinctions he employed to this purpose did not satisfy some of his contemporaries. In 1277 there occurred an ironic event that Pierre Duhem and others have claimed to be a major cause of modern science. The bishop of Paris, Etienne Tempier, pressed by theologians concerned that philosophers of Aristotelian inclination were championing doctrines that could be construed as limiting God's power, issued a condemnation of 219 such propositions. Included in the condemnation was proposition 34: "that the First Cause cannot make many worlds."[18]

Abruptly after 1277 the milieu changed, with many authors formulating analyses aimed at showing that God could create multiple worlds. Although few urged that God did in fact do so, this process led to a valuable reexamination and critique of Aristotle's antipluralist arguments. Examples of this trend are Jean Buridan (ca. 1295–1358), rector of the University of Paris, and the Oxford-educated Franciscan, William of Ockham (ca. 1280–1347), both of whom called into question Aristotle's argument based on his doctrine of natural place. The former urged that God could create worlds composed of other elements having their own natural places, whereas his more radical English contemporary relativized the notion of natural place by claiming that in different locales, the same elements would have correspondingly different natural places.[19] An especially fascinating case is Nicole Oresme (1325–82), tutor to the future Charles V of France and eventually bishop of Lisieux, who in his translation of and commentary on Aristotle's De Caelo provided a brilliant critique of some of Aristotle's ideas. Admitting that no philosophical or scientific reasons would bar the possibility of a succession of worlds in time or a nested series of concentric worlds, he went on in considering the case of spatially separate worlds to urge that the motions of bodies were governed by their surroundings: For example, wood in water rises. As Dr. Dick puts it, "In a single stroke, with this definition Oresme transferred significance from the earth—outer sphere relation to a heavy body—light body relation independent of where the bodies were situated."[20] Despite this and other criticisms of Aristotle's arguments, Oresme concludes that ". . . God can and could in his omnipotence make another world besides this one or several like or unlike it. Nor will Aristotle or anyone else be able to prove completely the contrary. But, of course, there has never been nor will there be more than one corporeal world. . . ."[21] The contrast, so striking in Oresme but present in other authors as well, between their openness to the possibility of God creating other worlds and their denial that he did so cries out for explanation. Was it their reading of Scripture, fear of church authorities, doubts about whether or not pluralism could be reconciled with the Christian doctrine of the atonement, or other factors that caused Oresme's denial of pluralism? The problem is exacerbated by the fact that these late-thirteenth and fourteenth-century authors examined only the question of other worlds, not that of other inhabited worlds.[22] There is little direct evidence that Scripture was a major factor. Aquinas had cited John 1:1 – "The world was made by Him" – as a reason for believing that only one world exists,[23] but this appears in his writings more as an undeveloped aside than as a substantive argument.

The cases of Nicholas of Cusa and William Vorilong shed light on the question whether or not fear of church authorities or tensions with the doctrine of the atonement influenced the debate. In 1440, Nikolaus Krebs (1401–64), commonly known as Nicholas of Cusa or Cusanus, published De docta ignorantia, a remarkable if also enigmatic masterpiece of the Middle Ages. Therein Cusanus endorses the idea of other inhabited worlds:

Life, as it exists on earth in the form of men, animals and plants, is to be found, let us suppose, in a higher form in the solar and stellar regions. Rather than think that so many stars and parts of the heavens are uninhabited and that this earth of ours alone is peopled – and that with beings, perhaps of an inferior type – we will suppose that in every region there are inhabitants, differing in nature by rank and all owing their origin to God, who is the centre and circumference of all stellar regions.[24]

He even speculates on the nature of his extraterrestrials, prefacing his statement by the admission that such speculation is groundless:

Of the inhabitants then of worlds other than our own we can know still less, having no standards by which to appraise them. It may be conjectured that in the area of the sun there exist solar beings, bright and enlightened denizens, and by nature more spiritual than such as may inhabit the moon – who are possibly lunatics – whilst those on earth are more gross and material. (pp. 115–16)

Having populated the sun and moon, he adds: "And we may make parallel surmise of other stellar areas that none of them lack inhabitants, as being each, like the world we live in, a particular area of one universe which contains as many such areas as there are uncountable stars." (p. 116)

A superficial knowledge of the extraterrestrial life debate, including belief in the myth that Giordano Bruno was martyred for his pluralistic convictions, might lead one to suspect that these claims of Cusanus reveal a person with little sense of the politically acceptable, if not a man destined for imprisonment or burning at the stake. Nonetheless, as Pierre Duhem stressed:

When for the first time in Latin Christianity, one hears a person speak of the plurality of inhabited worlds, it is proposed by a theologian who a few years earlier had spoken at an ecumenical council; a person who in a very celebrated book sought to divine the characteristics of the inhabitants of the sun and of the moon, went on to be honored by the confidence of popes [and by] the most elevated ecclesiastical honors. . . .[25]

All this is correct: Cusanus's political sensitivities were such that in 1437 he had been sent to Constantinople for the Council of Basel; moreover, eight years after his *Of Learned Ignorance* he was made a cardinal of the Catholic church. Although the issue of the reconcilability of pluralism with Christian conceptions of a divine incarnation and redemption was not treated by Cusanus nor, so far as is known, by the other authors who in the period after 1200 examined the idea of other worlds, this was done by the French theologian William Vorilong (d. 1463), who, after giving reason for believing that God could create another inhabited world, added:

If it be inquired whether men exist on that world, and whether they have sinned as Adam sinned, I answer no, for they would not exist in sin and did not spring from Adam. . . . As to the question whether Christ by dying on this earth could redeem the inhabitants of another world, I answer that he is able to do this even if the worlds were infinite, but it would not be fitting for Him to go unto another world that he must die again.[26]

By the end of the Middle Ages, the challenge of modifying the Aristotelian system so as to accommodate a God who could create multiple worlds had in some measure been met. Further challenges were soon forthcoming. One of these came from the publication in 1473 of the recently discovered *De rerum natura* of Lucretius. With this, Western man was forced to confront another of the powerful philosophical systems forged in antiquity, one far less easily reconciled to Christianity. A still more formidable challenge appeared exactly seventy years later, when Copernicus claimed that our earth, rather than being the center of the universe, is but one of the planets. That claim, which before long some saw as giving legitimacy to the idea of a plurality of worlds, was nothing less than revolutionary.

2. *From Copernicus and Bruno to Fontenelle and the Newtonians*

In surveying the extraterrestrial life debate between 1500 and 1750 it is important to ask how during that period pluralism went from being a belief of a few to a dogma taught in scientific textbooks and preached from pulpits. Two broad explanations have been proposed: (1) that astronomical developments were decisive, especially the heliocentric astronomy of Copernicus, Kepler, and Galileo, as well as the observations made possible by the invention around 1608 of the telescope, or (2) that the crucial factor was a supportive shift in the evolving philosophical and religious mentality of the period. In his brilliant *Great Chain of Being*, Arthur Lovejoy advocated the latter view, urging that the features that distinguish the modern from the medieval cosmos

. . . owed their introduction, and for the most part, their eventual general acceptance, not to the actual discoveries or to the technical reasonings of astronomers, but to those originally Platonic metaphysical preconceptions which . . . had . . . been always repressed and abortive in medieval thought. . . . These features [were chiefly] derivative from philosophical and theological premises. They were, in short, manifest corollaries of the principle of plenitude.[27]

Although Lovejoy qualified his thesis by admitting that Bruno's plenitude-based presentations of the pluralist position contributed little to the acceptance of pluralism, far less than the Cartesian cosmology that Fontenelle used in advocating extraterrestrials,[28] the validity of his position seems to be called into question by Steven J. Dick's analysis of the period in his recent and more detailed *Plurality of Worlds*. Without explicitly challenging Lovejoy's thesis, Dick declared his intention "to follow the thread [of the development of the extraterrestrial life debate] from the point of view of the history of science."[29] In this sketch of pre-1750 developments, no attempt can be made to resolve this complex question. However, the reader should be aware of its prominent place in the historiography of the extraterrestrial life debate and its significance even for contemporary discussions of the question of life elsewhere. A number of authors have, for example, suggested that the diversity of views characteristic of even the most recent writings in exobiology derives from the different metaphysical assumptions underlying the presently held positions.[30]

In studying the period from 1500 to 1750, one is struck by the hesitancy, indeed resistance, shown by some of the founders of the scientific revolution to a full-blown pluralism. In fact, if pluralism be defined as the doctrine that the earth is but one of the inhabited planets of our solar system and the stars are suns surrounded by inhabited planets, then Copernicus, Kepler, Brahe, and Galileo cannot be called pluralists.[31] That Nicolaus Copernicus (1473–1543) never in his writings discussed the question of other worlds is partly explained by his highly mathematical approach, but it is also due to the cautious, in many ways conservative, character of his thought. In this and in other ways he contrasts sharply with Giordano Bruno (1548–1600), whose passion for the new and daring was scarcely more limited than the infinite universe he championed in such books as *La cena de le ceneri* and *De l'infinito universo et mondi,* both published in 1584, and *De immenso,* of 1591. Bruno was passionately pluralist, populating not only planets but also stars, and even attributing souls to the planets, stars, meteors, and the universe as a whole. His sources, including Lucretius, Cusanus, Palingenius, Paracelsus, Copernicus, and the Hermetic writings, seem to have been more numerous than his followers, at least until the eighteenth- and nineteenth-century revival of interest in Bruno as a supposed "martyr for science." It is true that he was burned at the stake in Rome in 1600, but the church authorities guilty of this action were almost certainly more distressed at his denial of Christ's divinity and alleged diabolism than at his cosmological doctrines.[32]

The struggles of Johannes Kepler (1571–1630) with pluralism make one of the most dramatic stories in the entire history of the extraterrestrial life debate. With deep interest and a sense of foreboding, Kepler awaited in 1610 the arrival of a copy of Galileo's recently published *Siderius nuncius,* in which the Italian astronomer announced his telescopic discoveries of mountains on the moon, four satellites of Jupiter, and the vast number of stars beyond the reach of unaided sight. Kepler had heard rumors of its revelations, fearing that it supported a Brunonian rather than a Keplerian universe. At least that is the inference that emerges from his published response to Galileo's *Sidereal Messenger*, where he expressed his relief in such statements as the following:

. . . I rejoice that I am to some extent restored to life by your work. If you had discovered any planets revolving around one of the fixed stars, there would now be waiting for me chains and a prison among Bruno's innumerabilities. I should rather say, exile to his infinite space. Therefore, by reporting these four planets revolve, not around one of the fixed stars, but around the planet Jupiter, you have for the present freed me from the great fear which gripped me as soon as I had heard about your book. . . .[33]

Despite his abhorrence of Bruno's "innumerabilities," Kepler was not opposed to life on at least some planets or on the moon. His reason for favoring life on Jupiter is revealed in this remark:

Those four little moons exist for Jupiter, not for us. Each planet in turn, together with its occupants, is served by its own satellites. From this line of reasoning we deduce with the highest degree of probability that Jupiter is inhabited. Tycho Brahe likewise drew the same inference, based exclusively on a consideration of the hugeness of these globes.[34]

Kepler's claim here and elsewhere that Brahe believed in extraterrestrial life is probably incorrect; this has been shown by Dick, who specifically analyzed a letter in which Kepler attributed "living creatures" to the stars, adding that "not only that unfortunate Bruno, . . . but also my Brahe held this opinion, that there were stellar inhabitants." Dick's analysis demonstrates that Kepler's enthusiasm for extraterrestrials led him to misconstrue, in fact to invert, a reduction to absurdity argument formulated by Brahe. Brahe's argument was that if the Copernican theory were true, then the stars would be at immense distances, and this in turn would entail a vast wastage of space, unless there were stellar inhabitants. But the idea of such inhabitants being to Brahe absurd, the Danish astronomer urged rejection of the Copernican system itself.[35]

   Kepler was very pleased by Galileo's detection of earthlike features on the moon; in fact, Kepler had a few years earlier published his own pretelescopic arguments for a moon surface similar to the earth's.[36] Moreover, in his fictional lunar voyage *Somnium*, drafted in 1609 but published posthumously in 1634, Kepler presented further arguments for life on the moon. Nonetheless, Kepler remained convinced that "the center of the world is the sun," that "there is no globe nobler or more suitable for man than the earth," and that man is the "predominant creature" in all creation.[37] These points he defends in his response to Galileo and at greater length in his *Epitome astronomiae Copernicanae* (1618–21), where Kepler states that "the Earth was going to be the home of the speculative creature, and for his sake the universe and world have been made."[38] Anthropocentrism, the case of Kepler dramatically demonstrates, died hard.

Galileo Galilei (1564–1642), who never in his writings or letters mentioned Bruno,[39] did in at least six passages comment on the question of extraterrestrial life. These range from rejection to cautious reserve concerning it, both features being present in the following statement in his 1613 "Letter on Sunspots":

I agree with Apelles [Christopher Scheiner] in regarding as false and damnable the view of those who would put inhabitants on Jupiter, Venus, Saturn, and the moon, meaning by "inhabitants" animals like ours, and men in particular. Moreover, I think I can prove this. If we could believe with any probability that there were living beings and vegetables on the moon or any planet, different not only from terrestrial ones but remote from our wildest imaginings, I should for my part neither affirm nor deny it, but should leave the decision to wiser men than I.[40]

Thus, Galileo's writings, like those of Copernicus, Brahe, and Kepler, gave little direct support to a full-blown Brunonian pluralism; however, indirect support was a different matter. As early as 1550, only seven years after Copernicus's treatise, the Lutheran scholar Philip Melanchthon (1497–1560) spoke out in opposition to the new cosmology, including the tendency he saw in it to support a plurality of worlds:

. . . the Son of God is One; our master Jesus Christ was born, died, and resurrected in this world. Nor does he manifest Himself elsewhere, nor elsewhere has He died or resurrected. Therefore it must not be imagined that Christ died and was resurrected more often, nor must it be thought that in any other world without the knowledge of the Son of God, that men would be restored to eternal life.[41]

In the same period, similar antipluralist arguments were made by the Protestant theologian Lambert Daneau (ca. 1530–95), who especially attacked the Epicureans.[42] On the other hand, in Italy, G. B. Benedetti (1530–90), an early convert to Copernicanism, suggested in 1585 that the planets might be inhabited.[43]

Galileo's caution was by no means imitated, even among those most directly linked with him. From a Naples prison, Tommaso Campanella (1568–1634), in whose writings Frances Yates has detected Brunonian influences,[44] wrote his *Apologia pro Galileo* (1622), which contains a defense not only of Galileo and the Copernican system but also of the idea of a plurality of worlds. Citing the condemnation of 1277 and Cusanus, as well as making skillful use of Galileo's observations and his own gifts for reinterpreting Scripture, Campanella includes such statements as the following, which was surely of dubious value in saving Galileo from the charge that he supported pluralism:

If the inhabitants which may be in other stars are men, they did not originate from Adam and are not infected by his sin. Nor do these inhabitants need redemption, unless they have committed some other sin. I am constrained to set forth those passages in Ephesians 1, and Colossians 1, "And through him to reconcile all things unto himself, making peace through the blood of his cross, both as to the things that are on earth, and the things that are in heaven." In his *Epistle* on the solar spots Galileo expressly denies that men can exist in other stars . . . , but affirms that beings of a higher nature can exist there. Their nature is similar to ours, but is not the same, despite whatever sportful and jocose things Kepler says at such length in his prefatory dissertation to the *Starry Messenger*.[45]

Such was the defense of Galileo made by a man who in 1611, shortly after reading Galileo's *Starry Messnger,* had written to urge him to speculate on planetary dwellers.[46]

Campanella's *Apology,* Galileo's *Starry Messenger,* and Kepler's *Somnium* were all in the background of *The Discovery of a World in the Moone, or, A Discourse Tending to Prove That 'Tis Probable There May Be Another Habitable World in That Planet,* which appeared anonymously in England in 1638. Its author was the young Rev. John Wilkins (1614–72), an Oxford graduate who in time became master of Trinity College, Cambridge, and later bishop of Chester. In the process of advocating the Copernican theory and lunar life, Wilkins makes a number of arguments supportive of pluralism in general, his efforts to answer religious objections falling into this category. Frequently stressing that he can "promise only probable arguments," he states his preference for "Campanella's . . . conjecture . . . that the inhabitants [of the moon] are not men as we are. . . ." Employing the principle of plenitude, he adds: "there is a great chasm betwixt the nature of men and angels: it may be the inhabitants of the planets are of a middle nature between both these. It is not impossible that God might create some of all kinds, that so he might more completely glorify himself. . . ."[47] The merits of his astronomical arguments were and continue to be disputed,[48] but the impact of Wilkins's *Discovery* is undeniable, Professor Nicolson calling it "the most influential . . . [o]f all the works of 'popular astronomy' in England during the seventeenth century. . . ."[49]

In 1640, Wilkins published a second book supporting the Copernican theory, his *Discourse Concerning a New Planet,* the new planet being the earth. One paragraph in that book has been used to call into question the widespread belief that opposition to Copernicanism arose from man's pride in the central place in the cosmos assigned him in the Aristotelian-Ptolemaic cosmos. In that paragraph, Wilkins cites as a reason urged against Copernicanism the "vileness of our earth," which, it is supposed, "consists of a more sordid and base matter than any other part of the world; and therefore must be situated in the centre, which is the worst place, and at the greatest distance from those purer incorruptible bodies, the heavens."[50] In 1638, just as Wilkins's *Discovery* was being published, a lunar voyage fantasy written by Bishop Francis Godwin (1552–1633) also appeared. This encouraged Wilkins to embellish the third (1640) edition of his *Discovery* with a favorable discussion of the possibility of travel to the moon. Such ideas increased the opportunities, already ample, for wits to poke fun at Wilkins. Were one to compile an anthology of extraterrestrial humor, one would certainly want to include Robert South's jest about Wilkins wishing to obtain a bishopric on the moon, the Duchess of Newcastle's jibe to him about the need for lunar lodgings, and especially Samuel Butler's comic poem "The Elephant in the Moon," in which astronomers report sighting such a beast in their telescope, only to find that a mouse has invaded their instrument.[51]

Especially after Wilkins's book, the moon became the battleground for a war, lasting hundreds of years, concerning lunar life. Observations were certainly relevant; rarely, however, were they decisive. For example, the best moon maps of the middle seventeenth century were those in the *Selenographia* (1647) of Johannes Hevelius (1611–87) and the *Almagestum novum* (1651) of Giovanni Battista Riccioli (1598–1671), the latter map having been drawn by Riccioli's fellow Jesuit F. M. Grimaldi (1618–63). Whereas Hevelius drew seas on the moon and advocated "selenites," as he called his lunar inhabitants, Riccioli denied that water exists on the moon and above his map proclaimed that "No Men Dwell on the Moon" and "No Souls Migrate to the Moon" (Figure 1.1).

Many intellectuals in the first half-dozen decades of the seventeenth century seem to have been so confused, so undecided, or so cautious about the question of a plurality of inhabited worlds that scholars have had difficulty in specifying the actual views they held. For example, the position taken by Robert Burton (1577–1640) in his *Anatomy of Melancholy,* first published in 1621 but repeatedly revised and updated, has been described by some as antipluralist and by others as propluralist.[52] In France, the scientifically astute Minim friar Marin Mersenne (1588–1648) wavered on that question as well as the Copernican theory, but ultimately opted for the single-world view.[53] Because Blaise Pascal (1623–62) possessed one of the finest scientific minds of the century as well as a profound religious sensitivity, it would be especially fascinating to know his conclusion concerning the question of life elsewhere. That he pondered this problem is clear from his *Pensées,* which contains such enigmatic remarks as "The eternal silence of these infinite spaces frightens me." One characterization of Pascal's position is that presented by Lovejoy, who found in Pascal "the curious combination of a refusal to accept the Copernican hypothesis with the unequivocal assertion of the Brunonian."[54] Leaving aside the disputed question whether or not Pascal was anti-Copernican, it is true that in the Brunschveicg ordering of the *Pensées,* Pascal's "infinite spaces" sentence is followed by the apparently Brunonian exclamation: "How many kingdoms know us not!"[55] How-

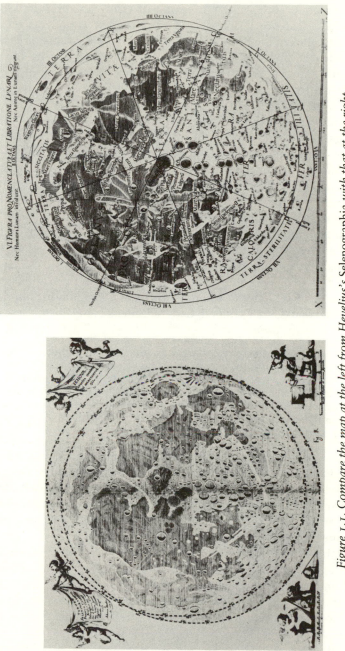

Figure 1.1. Compare the map at the left from Hevelius's Selenographia with that at the right, drawn by Grimaldi for Riccioli's Almagestum novum. Whereas Hevelius's moon contained seas and selenites, Riccioli's lacked both water and inhabitants.

ever, in the more authoritative ordering of Louis Lafuma, the second ("kingdoms") sentence is transferred to a context entirely terrestrial, rescuing Pascal from the Brunonians and preserving the sense of Pascal's "eternal silence" phrasing.[56] Elsewhere, and more accurately, Lovejoy wrote that "Pascal seems to conceive of mankind as alone in a dead infinity of matter. . . ."[57] And this description fits well with the other passages in the *Pensées* that touch on such matters.[58]

An examination of the writings of René Descartes (1596–1650) reveals his paradoxical position in the pluralist debate; as Dick, referring to the vortex cosmology Descartes expounded in his *Principia philosophia* (1644), puts it: "Compelling though the inference from vortices to an infinity of inhabited planets might be, it was not one that Descartes made in the *Principles* itself or in his private correspondence."[59] In fact, Descartes's discussions of pluralism lead one to wonder, had he lived long enough to encounter the pluralistic universe constructed in 1686 on Cartesian cosmological principles by Fontenelle, if his reaction might have been dismay rather than delight. The cautious character of Descartes's published and private pronouncements relevant to pluralism is illustrated by his June 6, 1647, letter to Chanut, who had asked him for his views. After noting that Christ's blood had redeemed many men, Descartes adds:

I do not see at all that the mystery of the Incarnation, and all the other advantages that God has brought forth for man obstruct him from having brought forth an infinity of other very great advantages for an infinity of other creatures. And although I do not at all infer from this that there would be intelligent creatures in the stars or elsewhere, I also do not see that there would be any reason by which to prove that there were not; but I always leave undecided questions of this kind rather than denying or affirming anything.[60]

Nonetheless, by putting forward a physical system in which each star was considered a sun and, as such, possibly surrounded by planets, Descartes had opened a door through which many would pass to pluralism. In fact, as both Dick and Lovejoy conclude, the Cartesian system was the single most important source of pluralist cosmologies in the latter half of the seventeenth century.[61]

Not all, however, construed the Cartesian system in the same way. For example, among the earliest Cartesians, Henry Regius (1598–1679) of Holland, in his *Fundamenta physices* (1646), avoided assigning planets to the stars, as did Jacques Rohault (1620–75) in his *Traité de physique* (1671). Nor did either advocate extraterrestrials.[62] Both of these positions were, on the other hand, endorsed by Henry More (1614–87) in his 1646 poem *Democritus Platonissans, or, An Essay upon the Infinity of Worlds out of Platonick Principles*. More's role in the debate is especially interesting, because whereas in his *Psychothasia Platonica* (1642) he had rejected the Epicurean idea of an infinity of worlds, in his 1646 work he endorsed it, citing on its behalf Descartes, "that sublime and subtil Mechanick," who, More adds, "though he seem to mince it must hold infinitude of worlds, or which is as harsh one infinite one."[63] The Cambridge Platonist More, whose Platonism was very selective, departed also from Descartes by his pluralist assertions in *Democritus Platonissans* that

> And what is done in this Terrestiall starre
> The same is done in every orb beside.
>
>   (Stanza 13)
>
> . . . long ago there Earths have been
> Peopled with men and beasts before this Earth,
> And after this shall others be again
> And other beasts and other humane birth. . . .
> Another Adam once received breath
> And still another in endless repedation
> And this must perish once in finall conflagration.
>
>   (Stanza 76)

Behind this stood More's vision of a God who did "his endlesse overflowing goodness spill/In every place; which streight he did contrive/Int' infinite severall worlds. . . ." (Stanza 50) More, whose pluralism had its origin in Epicurean and Cartesian systems, eventually rejected both philosophies. In his *Antidote against Atheism* (1653) he vigorously attacked atomism, and in his *Divine Dialogues* (1668) he broke completely with Descartes. More's pluralism, however, persisted; in the latter work he included a speculation that men on other planets were saved by God revealing to them the mystery of Christ's incarnation and redemption.[64] More's pluralism was not without influence; for example, it led his fellow Cambridge Platonist Ralph Cudworth (1617–88) to endorse pluralism.[65]

As the case of More illustrates, seventeenth-century pluralism was linked in important but complex ways to Epicurean philosophy. Some saw the Epicurean endorsement of that doctrine as but another reason for rejecting an atheistic system; others, however, realizing the merits of some features of that system, undertook the delicate task of deciding which aspects could be reconciled with their Christian commitments. In the first half of the century, few figures faced that task with deeper concern or broader background than the French priest and scientist Pierre Gassendi (1592–1655). The drama of the situation is suggested by a letter Gassendi, an active astronomical observer and enthusiast for atomism, received in 1642 from Pierre de Cazre, rector of the College of Dijon:

Ponder less on what you yourself perhaps think than on what will be the thoughts of the majority of others who carried away by your authority or your reasons, become persuaded that the terrestrial globe moves among the planets. They will conclude at first that, if the earth is doubtless one of the planets and also has inhabitants, then it is well to believe that inhabitants exist on the other planets and are not lacking in the fixed stars, that they are even of a superior nature and in proportion as the other stars surpass the earth in size and perfection. This will raise doubts about Genesis which says that the earth was made before the stars and that they were created on the fourth day to illuminate the earth and measure the seasons and years. Then in turn *the entire economy of the Word incarnate and of scriptural truth will be rendered suspect.*[66]

Gassendi faced a further tension in regard to pluralism: As a Christian convinced of a Creator-God who could act freely, he was careful to stress that God could create a multiplicity of worlds. However, in his 1649 and 1658 expositions of Epicureanism, he concluded against God having done so. He did distinguish between the idea of a plurality of worlds in the classical sense of isolated universes and the more modern stars-as-suns formulation, but in both cases he jettisoned this feature of atomism.[67] In doing this, however, the question of other worlds had been raised, and it was raised again in England when in 1654 Walter Charleton (1619–1707) published a defense of atomism derived largely from Gassendi's 1649 book and including a similar discussion of the plurality of worlds question.[68]

In 1657, Pierre Borel (1620?–71), a Protestant French physician, published a book arguing for life throughout the universe. Entitled in its English translation of 1658 *A New Treatise Proving a Multiplicity of Worlds. That the Planets Are Regions Inhabited, and the Earth a Star, and that It Is out of the Center of the World in the Third Heaven, and Turns Round before the Sun Which Is Fixed. And Other Most Rare and Curious Things,* it is itself a "curious thing" consisting, as one Borel scholar remarked, of "a melange of erudition, of the most recent scientific knowledge, and of naiveté."[69] Among the sources of Borel's treatise are Descartes, on whom Borel wrote a book, Lucretius, from whom Borel selects many passages for quotation, and his own lunar observations.[70] Borel's naiveté is evident in his argument for lunar life that the Paradise bird, being found on the earth only when dead, must come from our satellite. His claim that the sun is both the center of the universe and the source of light for the inhabited stars also does him little credit. The principle of plenitude, which Borel frequently invokes, as well as his stress that pluralism helps man attain a better perspective on his place in the universe, show his concern for philosophical issues. Overall, it is not unfair to describe Borel's book as revealing an author of substantial erudition but a rather limited sense of the scientific.

The pluralist potential of Cartesianism cosmology was exploited far more effectively by Bernard le Bovier de Fontenelle (1657–1757), who in 1686 created a sensation by his *Entretiens sur la pluralité des mondes.* From its January 1686 preview in the *Mercur galant* as a work of science "fashioned so gallantly that there is nothing savage about it," and even readable by women, few could have inferred its eventual influence.[71] Presented anonymously in the form of five dialogues between a philosopher and a charming marquise, it had by May 1686 acquired an author, Pierre Bayle ascribing it to Fontenelle,[72] and by 1687 a sixth dialogue as well as a place on the Catholic Index. While Rome pronounced it dangerous, relenting in 1825 but reinstating it on the Index in 1900,[73] the public proclaimed it a delight. By 1688, English readers had access to it in three different translations, with four more eventually following.[74] Other Europeans had to wait longer, but by 1800 it had been rendered into Danish, Dutch, German, Greek, Italian, Polish, Russian, Spanish, and Swedish.

Why such success? Its popularity was partly due to the elegance and charm of Fontenelle's prose, as well as to the wit of a man who even in his nineties could on meeting a young lady exclaim: "Ah! If I were only eighty now!"[75] The topic and its timeliness added as well to its appeal. Drawing on Descartes's vortex cosmology and the decades of telescopic observation by then available, Fontenelle constructed a seemingly plausible scientific case while incorporating a number of the delightful devices developed by the seventeenth-century pioneers of science fiction. Surely it is more in fiction than in physics that the origin of Fontenelle's Venusians and Jupiterians, for example, is to be sought. The former, he suggests, are "little black people, scorch'd with the Sun, witty, full of Fire, very Amorous, . . . ever inventing Masques and turnaments in honour of their Mistresses." His denizens of distant Jupiter are "Flegmatik: They are people who know not what 'tis to laugh, they take a day's time to answer the least question . . . ; and are so very grave, that were *Cato* living among 'em, they would think him a merry Andrew."[76] Earthlings fare hardly better, being described as having "no fix'd or determin'd character; some are made like the Inhabitants of Mercury, some like those of Saturn . . . ," and elsewhere they are compared in their resistance to Copernicanism to that "Mad man at Athens, who fancy'd all the Ships were his that came into the Port Pyraeum."[77] Fontenelle's planetarians show how readily he forsook his prefatory statement designed to avoid scriptural skirmishes, that his extraterrestrials were not to be considered men. Borrowing this reservation from Wilkins, who was probably Fontenelle's chief inspiration,[78] Fontenelle, as perhaps the first of the *philosophes,* seems to have felt few sympathies with those "excessively tender . . . in Religious matters." Borel was also in the background, providing Fontenelle with a Venus forty times smaller than the earth. In his 1708 revision of his *Conversations on the Plurality of Worlds,* Fontenelle enlarged Venus to one and one-half times the size of the earth, whereas by 1742 he had shrunk it to earth size, which is approximately correct.[79]

Fontenelle was not uninformed scientifically or uncritical; his Copernican first dialogue is followed by two dialogues debating lunar life. These indicate an author aware of the conflicting claims and wavering in his conclusions. Surprisingly, his scruples about selenites disappear for other planetary satellites. Cartesian in cosmology, but in little else,[80] Fontenelle proceeds in his fifth dialogue to go beyond Descartes by encircling the stars with planets. These, Fontenelle populated, as well as comets, but not the sun or stars. His sixth dialogue includes a summary of five main arguments employed in his book:

[1] the similarities of the planets to the earth which is inhabited; [2] the impossibility of imagining any other use for which they were made; [3] the fecundity and magnificence of nature; [4] the consideration she seems to show for the needs of their inhabitants as having given moons to planets distant from the sun, and more moons to those more remote; and [5] that which is very important – all that which can be said on one side and nothing on the other. . . .[81]

None of these arguments, except the fifth, was new; what was original was the captivating format Fontenelle furnished for them. What was needed in the period after Fontenelle's book was a more scientifically developed presentation of pluralism.

This was soon supplied by one of Fontenelle's readers, the distinguished Dutch scientist Christiaan Huygens (1629–95), whose so-called *Cosmotheoros* appeared posthumously in Latin in 1698 and by year's end was available in English as *Celestial Worlds Discover'd: or, Conjectures Concerning the Inhabitants, Plants and Productions of the Worlds in the Planets,* with Dutch, French, German, and Russian renderings following within two decades. Although its influence is indisputable, it may be asked whether or not Alexander von Humboldt's derisive characterization of it as consisting "of the dreams and fancies of a great man"[82] was justified. Is it a work of scientific merit, comparable to the other productions of Huygens's genius? Or should it be classified with Fontenelle's treatise, to which it bears some similarities? For example, the five main arguments cited previously as employed by Fontenelle are all in the *Cosmotheoros*.[83] So also are speculations that go far beyond the facts; as an illustration, consider the long series of inferences Huygens founds on his "Principle . . . that the other Planets are not inferior in dignity to ours." (p. 61) From this he concludes that the inhabitants of the planets must "have made as great advances [in astronomy] as we have." (p. 61) This leads him to attribute to them instruments as well as such organs as hands necessary for their use. Geometrical and optical attainments, as well as the art of writing and an advanced social structure, also follow (pp. 61–81). One wonders whether an excess or deficiency of imagination is shown when Huygens argues that their geometry and music must be nearly identical with ours (pp. 84–91). To Fontenelle's question whether or not planetarians have other senses, a conjecture having its origin in the Pyrrhonism of Montaigne, Huygens's response is largely negative (pp. 50–1). This approach is characteristic: Whereas Fontenelle delighted in the diversity among his extraterrestrials, Huygens stressed their similarities to terrestrial forms.[84]

Huygens is not, however, uncritical: Kepler, who created "pretty Fairy Stories of the men in the Moon" (p. 3), Kircher, who wrote "idle unreasonable stuff" (p. 101), and Fontenelle, who failed to "carry . . . the business any farther" than Cusanus and Bruno (p. 3), are all criticized, as are the Epicureans and Cartesians, who conceive of living creatures as "haply jumbled together by a chance Motion of I don't know what little Particles." (p. 21) Huygens, whose pluralism was already evident in his *Systema Saturnium* (1659) and who had begun to compile notes for his *Cosmotheoros* by 1686,[85] took his work very seriously. He asserts that the quest for evidences of extraterrestrials had by the end of the seventeenth century become "not impracticable" and with "very good room for probable Conjectures." (pp. 3–4) Moreover, especially in the second of the two books making up his treatise, he scrutinizes astronomical observations related to life elsewhere. An analysis of his arguments reveals, however, that whereas his evidences against life on the moon are detailed and nearly decisive, the empirical support he finds for life on the planets is slim. For example, he infers that spots seen on Jupiter (pp. 25–6) are clouds and, as such, evidences of water, although he departs from his stress on uniformity by suggesting that different forms of water may exist on the various planets (p. 27). His empirically based attribution of a thick "Atmosphere surrounding *Venus*" (p. 110) is one of the anticipations of modern results credited to Huygens, but he leaves unanswered the problem of how his Venusian astronomers would overcome this obstacle to observation. He recognizes the excessive heat brought by the sun to the inner planets and the deficiency in heat for the more remote planets, but he says little in explanation of how their inhabitants adjust to these conditions. In short, much of his commendable presentation of observation is contrary or irrelevant to planetary life. He fares better in discussing the possibility of life beyond our solar system. One cannot but praise his attempt to estimate the distance of the stars from an analysis of how far our sun would have to be removed to make it comparable in brightness to Sirius; this he capably uses to urge that our failure to detect planets around stars is to be expected (pp. 149–55). Huygens falls somewhat into overstatement in his claim that "all the greatest Philosophers of our Age . . . have the Sun of the same nature with the fix'd Stars," and he uses a weak form of argument in asking "why may not every one of these Stars or Suns have as great a retinue as our Sun, of Planets, with their Moons . . . ?" (p. 149) Nonetheless, the first of these ideas was gaining ever increasing acceptance, and the second was not far behind.

In assessing the scientific character of Huygens's *Cosmotheoros,* a crucial consideration is how he presented his conclusions. In this regard, he deserves high praise; conjectures abound, but conjectures they are labeled and their relative probabilities assessed. This feature of his book goes far to justify it as a scientific treatise, especially in the context of the hypothetico-deductive methodology of science championed by Huygens in the preface to his classic *Treatise on Light.*

One of the least known but most important influences of Huygens's *Cosmotheoros* involved England's Royal Astronomer, John Flamsteed, who recommended the *Cosmotheoros* to the vicar of Greenwich, Archdeacon Thomas Plume (1630–1704). Plume's fascination with it led him to bequeath 1,902 pounds to Cambridge University to "erect an observatory and to maintain a professor of astronomy and experimental philosophy, and to buy or build a house with or near the same."[86] Thus came to be established the Plumian professorship of astronomy at Cambridge, a position held by some of the most distinguished astronomers of modern times.

Because Sir Isaac Newton (1642–1727) was the most gifted scientist of his day, his contemporaries and successors concerned about the question of a plurality of worlds searched his writings for indications of his position on this issue. Only a few were to be found, even in his unpublished writings. His *Principia* of 1687 nonetheless exerted an influence on the debate by discrediting Descartes's cosmology and cosmogony and by providing the framework within which most subsequent discussions would be conducted. Appreciation of the merits of the Newtonian synthesis, however, took time. Few recognized his genius as rapidly as Rev. Richard Bentley (1662–1742), a Cambridge-educated divine who by 1691 had begun to study the *Principia.* This background Bentley employed when he was chosen to inaugurate in 1692 the lecture series funded by Robert Boyle, who specified that the lectures be aimed at "proving the Christian religion."[87] Before publishing the last two of his eight Boyle lectures, Bentley wrote to Newton to ask how he conceived that his system related to religion. A pleased Newton promptly responded in four letters that were themselves published in 1756. They touch on many topics, including Bentley's mention of the idea of a succession of worlds. Newton's response to this is that "the Growth of new Systems out of old ones, without the Mediation of a divine Power, seems to me apparently absurd."[88] This and Newton's rejection of any purely mechanical explanation of the formation of the solar system because of the problems of explaining how the "shining" matter of the sun could have been separated from the "opaque" matter of the planets[89] have been taken to be evidence of his hostility not only to the Epicurean cosmogony but also to any evolutionary approach to the cosmos.

Bentley, equally opposed to Epicureanism, treated a number of themes of that philosophy in his seventh and eighth lectures. Because Bentley's seventh lecture includes an endorsement of the doctrine "supposed by Astronomers" that "every Fixt Star [is] of the same Nature with our Sun; and each may very possibly have Planets about them . . . ,"[90] one is not surprised to find him in his eighth lecture discoursing on other worlds. Bentley faced the serious problem of trying to prove the existence of God from the beneficent design evident in nature, doing this in the enlarged cosmos of the late seventeenth century, wherein, as Bentley puts it, "we need not nor do not confine and determine the purposes of God in creating all Mundane Bodies, merely to Human Ends and Uses."[91] Bentley, however, hastens to add that God could have created the universe for man because "the Soul of one vertuous and religious Man is of greater worth and excellency than the Sun and his Planets and all the Starrs in the World."[92] Bentley also states that although he "dare" not argue that the stars provide an advantage to us, he does not believe that they can be explained as existing for themselves, that is, as clumps of matter bereft of inhabitants.[93] From this he draws the pluralist conclusion that the heavenly bodies "were formed for the sake of Intelligent Minds. . . ." Just as the earth was made for man, "why may not all other Planets be created . . . for their own Inhabitants which have Life and Understanding?"[94] The question of the relation of extraterrestrials to Adam's fall and Christ's incarnation he dismisses by urging that they need not be men, adding that God "may have made innumerable Orders and Classes of Rational Minds; some higher in natural perfections, others inferior to Human Souls. [These] would constitute a different Species. . . ."[95]

Later in his eighth lecture, Bentley praises God's wisdom in placing our planet at just such a distance from the sun that we are neither too cold nor too hot. This pious thought produces a problem: How are extraterrestrials to survive on Mercury and Saturn, distanced so differently from the sun? Bentley's answer is that

. . . the Matter of each Planet may have a different density and texture and form, which will dispose and qualifie it to be acted on by greater and less degrees of Heat according to their several Situations; and that the Laws of Vegetation and Life and Sustenance and Propagation are the arbitrary pleasure of God, and may vary in all Planets . . . , in manners incomprehensible to our Imaginations.[96]

It would be fascinating to know how Newton viewed this radical claim made by Bentley: not only other worlds, but worlds made of matter with different dispositions and qualities and life forms governed by different laws, these being set at God's "arbitrary pleasure." This seems a much more voluntarist view than the Newton of the *Principia* would have permitted. Leaving that question aside for the moment, we should above all note that in Bentley's *Confutation of Atheism* the public encountered for the first time the joining of Newtonianism, Christianity, and pluralism. This combination would within a few decades become commonplace.

Four passages from Newton's post-1700 writings relate to pluralism. The first occurs in one of the queries that Newton added to the 1706 Latin edition of his *Opticks*. After attempting to free atomism from its Epicurean associations by stating that "God in the Beginning form'd Matter in solid, massy, hard, impenetrable, moveable Particles,"[97] Newton adds that "God is able to create Particles of Matter of several Sizes and Figures, and in several Proportions to Space, and perhaps of differing Densities and Forces, and thereby to vary the Laws of Nature and make Worlds of several sorts in several Parts of the Universe."[98] The striking closeness of the latter portion of this passage to the claims made earlier by Bentley in his eighth lecture, combined with the radical departure of this position from the assumption of the uniformity of nature, suggests that Newton may have been influenced by Bentley or, alternatively, that Newton communicated more to his Cambridge colleague than his four letters to Bentley reveal.

The second passage comes from the "General Scholium" that Newton added to the 1713 edition of his *Principia*, which edition he published partly at the urging of Bentley, by then master of Trinity College, Cambridge.[99] That scholium contains the statement that "if the fixed stars are the centres of other like systems, these, being formed by the like wise counsel, must be all subject to the dominion of One . . . ," that is, of God.[100] Although hypothetical in form, this statement certainly suggests that in Newton's universe, the stars were encircled by planets, as Bentley had also averred.

The third passage, which dates from after 1710 and exists in a number of variants,[101] was not published until David Brewster unearthed it for his 1855 biography of Newton to show that Newton believed in extraterrestrials. In the passage as given by Brewster, Newton states that at the final judgment, Christ

. . . will give up his kingdom to the Father, and carry the blessed to the place he is now preparing for them, and send the rest to other places suitable to their merits. *For in God's house (which is the universe,) are many mansions, and he governs them by agents which can pass through the heavens from one mansion to another. For if all places to which we have access are filled with living creatures, why should all these immense spaces of the heavens above the clouds be incapable of inhabitants?*[102]

Leaving aside a dozen differences in punctuation, spelling, and so forth, between Brewster's version and the original manuscript, it is interesting to note, as Brewster did not, (1) that the portion of the statement following the words *"many mansions"* is, in Newton's manuscript, not only not italicized but also crossed out, and (2) that added at the end is this nonitalicized, non-crossed-out sentence: "We are also to enter into societies by Baptism & laying on of hands & to commemorate the death of X in our assemblies by breaking of bread."[103]

The fourth and final passage comes from the record made by John Conduitt, the husband of Newton's niece, of a conversation he had with Newton on March 7, 1725, two years before Newton died.[104] In that conversation Newton revealed to Conduitt his "conjecture" that "revolutions" occur in the solar systems of stars "all which he took to be suns enlightening other planets, as our sun does ours." These revolutions, according to Newton, result from solar emissions forming into satellites of planets and when sufficiently large flying off as comets and eventually falling into the suns. Although thereby replenishing these suns, the excessive heat generated has the unfortunate side effect of destroying life on the earths served by those suns. Conduitt also learned that "He seemed to doubt whether there were not intelligent beings superior to us, who superintended these revolutions of the heavenly bodies, by the direction of the Supreme being." Concerning the question of how the inhabited earths were repopulated after such revolutions, Newton's statement was that "that required the power of a creator." If Conduitt's record is correct, it shows that Newton definitely believed in planetary systems around all stars, a succession of "worlds" on this earth, and almost certainly life in many other locations. His "intelligent beings superior to us" may be demigods, angels, or some special form of extraterrestrial life.

The linking of pluralism with Newtonianism and natural theology, begun by Bentley, was continued by many authors, few of whom did it as successfully as Rev. William Derham (1657–1735). While serving in 1714 as chaplain to the future King George II, Derham published his *Astro-Theology, or A Demonstration of the Being and Attributes of God from a Survey of the Heavens.*[105] Written as a companion piece to his earlier *Physico-Theology,* Derham's *Astro-Theology* was devoted to delineating the design of the Deity in the celestial realm as his earlier book had done for the terrestrial. Both books were extremely popular: By 1777, his *Astro-Theology* had attained fourteen English and six German editions. Elementary enough to appeal to the literate populace, to whom it taught much astronomy, Derham's *Astro-Theology* also contained riches for the astronomically refined. Especially in Germany, it served to spread knowledge of the Newtonian synthesis from which it took its overall orientation.

Pluralism permeates the book, adding to its attractiveness and religious appeal. Derham's indebtedness to Huygens's *Cosmotheoros* did not preclude his arguing against the Dutchman's waterless and lifeless moon. The importance of Derham's book in the extraterrestrial life debate derives not from such detail but from its striking and timely formulation of the question of world systems. Three of these are delineated in Derham's "Preliminary Discourse," these being the "Ptolemaick," which he rejects, the "Copernican," which he accepts, but only as a precursor to his "New Systeme." This third system includes the Copernican, but goes beyond it by the supposition that "there are many other Systemes of *Suns* and *Planets,* besides that in which we have our residence; namely, that every Fixt Star is a sun and encompassed with a Systeme of Planets, both Primary and Secondary as well as ours."[106] One of his arguments for this system is that it "is far the most magnificent of any; and worthy of an infinite CREATOR. . . ." (p. xliv)

That Derham should be seen more as symbol than as the source of the increasing acceptance of pluralism in early eighteenth-century England is shown by mention of Thomas Burnet (1635?–1715), John Ray (1628–1705), and Nehemiah Grew (1641–1712), all of whom in the three decades before Derham's *Astro-Theology* had endorsed pluralism in physicotheological treatises. The approach taken in these books is suggested by the title of Ray's 1691 *The Wisdom of God Manifested in the Works of Creation.*[107] Important and widely read as these authors were, Derham's conceptualization of the state of astronomy as involving a "New Systeme" reveals most dramatically the distance that astronomy had come since the days of Copernicus. Logically considered, the cosmos of Copernicus may not appear far removed from Derham's "New Systeme," but historically the two systems were separated by an interval capacious enough to include Kepler, Galileo, Descartes, Gassendi, and a host of others boldly embracing the heliocentrism of Copernicus, but hesitating at the pluralist universe championed by Derham.

3. Pluralism in the early eighteenth century: "this is the best possible world" or "this Earth is Hell"

In the first half of the eighteenth century, numerous authors endorsed the pluralist position. Derham was among them; so also were Brockes, Franklin, Haller, Mather, Maupertuis, Voltaire, and Young, whose writings, for reasons of conceptual unity, are most conveniently discussed in subsequent sections. The paths to pluralism taken by these authors were varied: for some, religion; for others, astronomy; for most, a combination of these two and other factors helped them on their way. Moreover, they in turn pushed pluralism in diverse directions; some used it to embellish poetry, others to embarrass traditional piety; some to expand philosophy, others to enrich astronomy. However they came to it, whatever they drew from it, pluralism was increasingly appearing in their writings. Moreover, its image was shifting from that of a radical conjecture requiring careful justification to that of an accepted doctrine in need of amplification and integration into other systems of thought.

A perennial source of pluralism's appeal has been its adaptability to diverse philosophical systems. This point can be illustrated by examining the uses found for it by four leading philosophers of the period: Locke, Berkeley, Leibniz, and Wolff. In his *Essay Concerning Human Understanding* (1689), John Locke (1632–1704) stressed that man's capacity for attaining ideas is limited by his senses, whereas beings on other worlds, assisted by "senses and faculties more or perfecter than we have," may develop ideas unavailable to us. The famous empiricist supported the existence of such beings by stating: "He that will consider the infinite power, wisdom, and goodness of the Creator of all things will find reason to think it was not all laid out upon so inconsiderable, mean and impotent a creature as he will find man to be; who in all probability is one of the lowest of all intellectual beings."[108] Similar theological reasons, as well as the implausibility of believing that the stars were made solely for us, are cited on behalf of pluralism in Locke's later *Elements of Natural Philosophy*, for which Huygens's *Cosmotheoros* was a major source.[109]

The Irish idealist Bishop George Berkeley (1685–1753), like Locke, found pluralism useful, but for quite different philosophical purposes. In the second dialogue of his *Three Dialogues between Hylas and Philonous* (1713), Berkeley has his spokesman Philonous ("mind-lover") argue against the materialist Hylas that the intricate ordering of the universe ill accords with skepticism. Berkeley's universe includes "innumerable worlds [in which] the energy of an all-perfect mind [is] displayed in endless forms."[110] Pluralism is used also in Berkeley's *Alciphron, or the Minute Philosopher* (1732), written, according to its subtitle, as *An Apology for the Christian Religion, against Those Who Are Called Free-Thinkers*. To combat the objection to Christianity deriving from the presence of "so much vice and so little virtue upon earth," Berkeley's Euphanor asserts:

And, for aught we know, this spot, with the few sinners on it, bears no greater proportion to the universe of intelligences than a dungeon doth to a kingdom. It seems we are led not only by revelation, but by common sense, observing and inferring from the analogy of visible things, to conclude there are innumerable orders of intelligent beings more happy and more perfect than man. . . .[111]

The theme of the lowness of man, present in the pluralism of both Locke and Berkeley, is not usually associated with Gottfried Wilhelm

Leibniz (1646–1716), who, it is frequently said, championed the doctrine that ours is the best of all worlds. Leibniz's position on this issue is more plausible than it may seem to those who know it only from its satirization in Voltaire's *Candide*. In what follows, this point will be clarified, but no more than a beginning can be made on the more complex task of determining the position of Leibniz in the pluralist debate.[112] The difficulty of this task, which only a Leibniz scholar could carry to completion, is complicated by the fact that although passages on "possible worlds" abound in Leibniz's writings, many of these refer less to the existing universe than to the question of the types of universes God could have created.

Although Leibniz's interest in the idea of possible worlds can be traced to a manuscript written as early as 1676 and to his 1698 correspondence with Johann Bernoulli,[113] his first extended published presentation of his views came in his *Théodicée* (1710). Early in that work he portrays God in creating our universe as selecting it from among all possible universes: "for this existing world being contingent and an infinity of other worlds being equally possible, and holding, so to say, equal claim to existence with it, the cause of the world must needs have had regard or reference to all these possible worlds in order to fix upon one of them."[114] Moreover, Leibniz adds: "this supreme wisdom, united to a goodness that is no less infinite, cannot but have chosen the best." (p. 128)

Leibniz realized that his claim that our universe is the best possible is open to a host of objections. He responds to a number of these in his *Théodicée*. For example, he writes:

It is true that one may imagine possible worlds without sin and without unhappiness, and one could make some like Utopian and Severambian romances: but these same worlds would be very inferior to ours in goodness. I cannot show you this in detail. For can I know and can I present infinities to you and compare them together? But you must judge with me *ab effectu*, since God has chosen this world as it is. We know, moreover, that often an evil brings forth a good whereto one would not have attained without that evil. (p. 129)

Among his various responses to the problem of the presence of sin and evil in the world, Leibniz proposes one that was unavailable to pre-Copernican Christians who had not realized that the universe contains "an infinite number of globes, as great and greater than ours, which have as much right as it to hold rational inhabitants, though it follows not at all that they are human." Within the universe, "It may be that all suns are peopled only by blessed creatures" and our earth "is almost lost in nothingness. . . ." This perspective leads Leibniz to infer that "since all the evils that may be raised in objection before us are in this near nothingness [our earth], haply it may be that all evils are almost nothingness in comparison with the good things which are in the universe." (p. 135)

Such passages suggest that one could paraphrase Leibniz's overall position as: "This is the worst world (planet) in the best possible universe." Leibniz also seeks to give new signficance to the traditional claim that Adam's fall was in one sense "a fortunate sin" because it led to Christ's incarnation and redemption. He endorses this idea, but sets it within the context of his pluralist cosmos, to the whole of which Christ gave "something more noble than anything there would otherwise have been amongst created beings." (p. 378)

In the years just before the publication of his *Théodicée*, Leibniz was composing his *Nouveaux essais sur l'entendement humain*, written in reaction to Locke's *Essay Concerning Human Understanding*. This long but uncompleted work was first published only in 1765. Although disagreeing with Locke's fundamental philosophical position, Leibniz in his *Nouveaux essais* reveals his agreement with Locke on the ideas of a chain of being and a plurality of worlds. Leibniz's spokesman states: ". . . I believe that the universe contains everything that its perfect harmony could admit. It is agreeable to this harmony that between creatures which are far removed from one another there should be intermediate creatures, though not always on a single planet or in a single [planetary] system. . . ."[115] This statement is one of many in which Leibniz conveys his view, important for pluralism, that all forms of being capable of existing in the universe will attain actuality, thereby contributing to its excellence. Shortly after it occurs a passage that shows not only Leibniz's broad reading in the pluralist literature and his deep concern for theological issues but also his sophisticated wit:

If someone . . . came from the moon . . . , like Gonsales [Godwin's cosmic voyager] . . . , we would take him to be a lunarian; and yet we might grant him . . . the title *man* . . . ; but if he asked to be baptized, and to be regarded as a convert to our faith, I believe that we would see great disputes arising among the theologians. And if relations were opened up between ourselves and these planetary men – whom M. Huygens says are not much different from men here – the problem would warrant calling an Ecumenical Council to determine whether we should undertake the propagation of the faith in regions beyond our globe. No doubt some would maintain that rational animals from those lands, not being descended from Adam, do not partake of redemption by Jesus Christ. . . . Perhaps there would be a majority decision in favour of the safest course, which would be to baptize these suspect humans conditionally. . . . But I doubt they would ever be found acceptable as priests of the Roman Church, because until there was some revelation their consecrations would always be suspect. . . . Fortunately we are spared these perplexities by the nature of things; but still these bizarre fictions have their uses in abstract studies as aids to a better grasp of the nature of our ideas. (p. 314)

In another passage from his *Nouveaux essais*, Leibniz expresses pessimism concerning the astronomical search for extraterrestrials:

Until we discover telescopes like those of which M. Descartes held out hope, which would let us pick out things no bigger than our houses on the lunar surface, we shall be unable to settle what there is on any globe other than ours. . . . Thus, although in some other world there may be species intermediate between man and beast . . . , and although in all likelihood there are rational animals somewhere, which surpass us, nature has seen fit to keep these at a distance from us so that there will be no challenge to our superiority on our own globe. (pp. 472–3)

Thus, we find in Leibniz a philosopher whose confidence in the existence of extraterrestrials was by no means shaken by his conviction that empirically we can know nothing of them.

One of the first champions of the system of Leibniz was Christian Wolff (1674–1754), who taught it at Halle and Marburg and expounded it in his voluminous writings. Wolff, who also taught mathematics and science, was especially interested in using Leibnizian rationalism to reconcile religion and science. Such a concern pointed to pluralism, which is endorsed by Wolff in a number of his writings, frequently on physicotheological grounds. Of all Wolff's pluralist passages, none attracted more attention than one from the third volume (1735) of his *Elementa matheseos universae,* where he sets out to calculate the height of Jupiterians. His method is based on the assumptions (1) that bodily size is proportional to the size of the eye and (2) that the square of the diameter of the pupil is inversely proportional to the intensity of the available light. He notes that Jupiter is 26/5 times farther from the sun than the earth and consequently receives light of $(5/26)^2$ the intensity on earth. This would indicate that the eyes and hence the bodies of Jupiterians should be 26/5 times larger than those of earthlings, but on the principle(?) that eye dilation is greater in more intense light, he changes this fraction to 26/10 or 13/5. Taking the average human height to be 5 7/32 Paris feet, Wolff assigns his Jovians the height of 13 819/1,440 feet![116] For a century and a half or more, persons in the pluralist debate called attention to this calculation. A few praised it, but most, including Voltaire, d'Alembert, and Proctor, cited it as a silliness to be shunned.

Early eighteenth-century intellectuals interested in learning about pluralism from astronomical authors rather than philosophers could turn not only to William Derham but also to other writers. Frenchmen, for example, could consult Pierre Julien Brodeau de Moncharville (d. 1711), who in 1702 published his *Preuves des existences, et nouveau système de l'univers* (Paris), urging that millions of inhabited planets may exist.[117] At Oxford, students learned of pluralism from David Gregory (1661–1708), the Savilian professor of astronomy, as well as from his eventual successor in that chair, John Keill (1671–1721). Whereas Gregory was noncommittal in his *Astronomiae physicae et geometricae elementa* (1702), Keill was outspoken, claiming in his *Introductio ad veram as-*

tronomiam (1718) that it is "no ways probable" that God would create stars without placing planets around them "to be nourished, animated, and refreshed with the Heat and Light of these Suns. . . ."[118] Oxford's Savilian professor of geometry at that time was Edmond Halley (1656–1742), who in 1720 became Astronomer Royal. Not only did Halley endorse pluralism, he argued from his statement that all planets "are with Reason suppos'd habitable" to the possibility that habitable globes exist beneath the earth's surface. He had proposed subterranean globes to explain apparent shifts in the earth's magnetic poles, but was delighted to add concerning their habitation: "Thus I have shew'd a possibility of a much more ample Creation, than has hitherto been imagin'd. . . ."[119] At Cambridge, William Whiston (1667–1752), who in 1702 succeeded Newton as Lucasian professor of mathematics, repeatedly advocated pluralism. As early as his *New Theory of the Earth* (1696), Whiston urged that other planets and planetary systems have inhabitants subject to moral trials.[120] Two decades later, in his *Astronomical Principles of Religion,* Whiston extended his pluralism by proposing denizens dwelling in the interiors of the earth, sun, planets, and comets; moreover, he posited "Not wholly Incorporeal, but Invisible Beings" living in planetary atmospheres.[121] In the intervening years, Whiston advanced the Newtonian system by publishing his Cambridge astronomical lectures wherein he identified stars as suns,[122] but in 1710 he lost the Lucasian chair because of charges of religious heterodoxy, in particular, of Arianism. Nonetheless, he continued to lecture at various locations on astronomy and to spread the pluralist doctrine.

One such location was Button's coffeehouse in London. Recent researches indicate that Whiston's lectures at Button's in 1715 introduced Alexander Pope (1688–1744) to the new pluralist universe.[123] Two decades were to pass before Pope presented pluralism in his *Essay on Man,* but from the time of its publication, wherever or in whatever language that poem was read,[124] Pope's message that the "proper study of mankind" must include consideration of extraterrestrials was brought before the public. That point Pope's poem sets out in the following lines, which had such appeal to Thomas Wright and Immanuel Kant that they quoted them when in the 1750s they published their theories of the universe:

> He, who thro' vast immensity can pierce,
> See worlds on worlds compose one universe,
> Observe how system into system runs,
> What other planets circle other suns,
> What vary'd Being peoples ev'ry star,
> May tell why Heav'n has made us as we are.[125]

Other pluralist passages quoted by Kant include these:

Who sees with equal eye, as God of all,
A hero perish, or a sparrow fall,
Atoms or system into ruin hurl'd
And now a bubble burst, and now a world.
(I, lines 87–90)

Superior beings, when of late they saw
A mortal man unfold all Nature's law,
Admir'd such wisdom in an earthly shape,
And shew'd a Newton as we shew an Ape.
(II, lines 31–4)

It is suggestive of the transformation that Western man was undergoing that whereas John Donne a century earlier had taken the vision of a vast pluralistic universe to signify cosmic chaos, Pope rejoiced in it as a symbol of order.[126]

Pluralism appealed to many of Pope's literary contemporaries, including two poets who also imbibed at Button's. Around 1713, both John Gay (1685–1732) and John Hughes (1677–1720) published poems with pluralist themes.[127] Joseph Addison (1672–1719), who helped arrange for Whiston's lectures, as well as Sir Richard Steele (1672–1729), with whom Addison conducted the *Spectator,* contributed essays to that literary daily endorsing the pluralist position. For example, in *Spectator* 519 of October 25, 1712, Addison, after noting that an "Infinity of Animals" exists on the earth, adds:

The author of the *Plurality of Worlds* draws a very good Argument from this Consideration, for the *peopling* of every Planet as indeed it seems very probable from the Analogy of Reason, that if no part of Matter, which we are aquainted with, lies waste and useless, those great Bodies which are at such a Distance from us should not be desart and unpeopled, but rather that they should be furnished with Beings adapted to their respective Situations.[128]

The principle of plenitude in the form of the "exhuberant and overflowing Goodness of the Supreme Being" as well as the notion of a "Scale of Being" are cited in support of pluralism in this essay. In *Spectator* 565 (July 9, 1714), the influence of Pascal is apparent as Addison attempts to deal with the place of the individual in this vast cosmos. To relieve this tension he stresses God as "Omnipresent" and "Omniscient."[129] Steele's views appear in *Spectator* 472 (September 1, 1712), where he asserts that each star "is a *Sun* . . . performing the same offices to its dependant Planets, that our glorious Sun does to this." And he adds that "Enquiries of the *Sight* . . . make their Progress through the immense Expanse to the *Milky Way,* and there divide the blended Fires of the *Galaxy* into infinite and different worlds, made of distinct Suns, and their peculiar Equipages of Planets. . . ."

In the *Spectator* of March 29, 1912 (#339), Addison welcomed the appearance of *The Creation,* a long didactic poem by the physician-poet

Richard Blackmore (1654–1729). *The Creation,* Addison advised, "is executed with so great a Mastery, that it deserves to be looked upon as one of the most useful and noble Productions in our *English* verse." Even Blackmore's adversary John Dennis described the poem as having "equall'd that of LUCRETIUS in the Beauty of its Versification, and infinitely surpass'd it, in the Solidity and Strength of its Reasoning."[130] The superiority seen by Dennis in Blackmore's reasoning no doubt derived from Blackmore's contra-Lucretian concern to "demonstrate the existence of a Divine Eternal Mind." Although separated by eighteen centuries and by very different theological views, Blackmore and his Roman predecessor both espoused pluralism, albeit in signficantly different forms. This is evident in the astronomical second book of *The Creation,* where, despite hesitating to endorse Copernicanism, Blackmore writes of the solar system:

> Yet in this mighty system, which contains
> So many worlds, such vast etherial plains,
> But one of thousands, which compose the whole,
> Perhaps as glorious, and of worlds as full.
>
> All these illustrious worlds, and many more,
> Which by the tube astronomers explore;
> And millions which the glass can n'er descry,
> Lost in the wilds of vast immensity;
> Are suns, are centres, whose superior sway
> Planets of various magnitude obey.[131]

In his third book he states:

> We may pronounce each orb sustains a race
> Of living things, adapted to the place.
> Were the refulgent parts, and most refin'd
> Only to serve the dark and base design'd?

As these lines suggest, the earth in Blackmore's conception was a lesser body, an "Inferior habitable seat."

> Witness, ye stars, which beautify the skies,
> How much do your vast globes, in height and size,
> In beauty and magnificence, outgo
> Our ball of Earth, that hangs in the clouds below!
> Between yourselves, too, is distinction found,
> Of different bulk, with different glory croun'd. . . .

Matthew Prior (1664–1721) also recognized the poetic, religious, and moral uses of pluralism; in his *Solomon on the Vanity of the World* (1718) he has the biblical prophet assert sentiments similar to Blackmore's:

> And in that Space, which we call air and Sky
> Myriads of Earths, and Moons and Suns may lye,
> Unmeasur'd and unknown by human eye.

Unknown they may be, but they teach man a lesson:

> But do these Worlds display their Beams, or guide
> Their orbs to serve thy use, to please thy Pride?
> Thy self but Dust, thy stature but a Span
> A moment thy duration; foolish man![132]

Nothing more dramatically demonstrates the popularity of pluralism in the period from 1710 to 1750 than the publication during those decades of a dozen or more books devoted to that subject.[133] A number of these took the form of dissertations, for example, that presented at Utrecht University in 1726 by William Arntzen (1704–35), who argued for lunar life, citing a host of earlier authors and an array of observations. Nevertheless, as Dick puts it, Arntzen was "the victim of a metaphysical framework that caused him overzealously to transfer terrestrial experience to his interpretations of lunar phenomena. . . ."[134] Whereas Arntzen called the moon "another earth," Eric Engman, in his 1740 Upsala dissertation directed by Anders Celsius (1701–44), offered alternative interpretations of observations cited in favor of lunar life, concluding against a lunar atmosphere and the moon's habitability.[135] Three years later, Celsius directed the dissertation of Isacus Svanstedt, who used both scientific and theological arguments to champion a richly populated universe, including even the possibility of lunar life. A trio of German religious writers also discussed the pluralist position, all finding for the affirmative. Both the clergyman Andreas Ehrenberg (d. 1726), in his 1711 book, and the school rector Johann Schudt (1664–1722), writing a decade later, speculated that the inhabitants of Mars are served by one or two moons.[136] Both also mentioned the problem pluralism presents for the Christian doctrine of the atonement, but they were above all intent on integrating pluralist thought with physicotheology.[137] Ehrenberg, in response to an attack on his book by Georg Pertsch (1651–1718), published a second pluralist volume.[138] The hymnologist David Schöber (1696–1778) discussed the moon and planets in a 1748 book, arguing for their habitability and even calculating, by means of the method developed by Wolff, the size of each planet's inhabitants. Schöber's chief concern, however, was to reconcile pluralism with the Christian scheme of redemption, a task taking up half his volume. Two German scientists, Johann Herttenstein (1676–1741) and Johann Hennings (1708–64), also advocated pluralism. Herttenstein's book, rich in both philosophical and astronomical references, especially stresses similarities among the planets, whereas Hennings's more scientific treatise attributes life to the planets but not their satellites.

The last two books in this cluster were by Englishmen. In 1711, Daniel Sturmy, rector of East-Hatley, published *A Theological Theory of a Plurality of Worlds,* devoted to advancing the pluralist cause by three lines of argumentation. First he cites such ideas as the plenitude of God's powers and the scale of being as evidence that all planets and satellites are inhabited. In that section he stresses the diversity among planetarians: "Upon the Bodies of some Creatures . . . , Fire may make as agreeable an Impression, as fine Aire, and serene Gales of Wind, or the moderate Heat of the Sun, do make upon the Bodies of Men."[139] In his next section Sturmy seeks scriptural indications of extraterrestrials, and in his final section he presents pluralism as supporting "Practical Divinity" by, for example, promoting heavenly-mindedness and encouraging "a solid and unaffected Contempt of the World." Twenty-five years later, the physician John Peter Biester, in his *Enquiry into the Probability of the Planets Being Inhabited,* advocates life on all planets, but not their satellites. Biester's chief concern is with the problem that the sun's rays fall with very different intensities on the various planets. He attempts to overcome this difficulty in various ways, for example, by having Mercurians populate their planet's pole and by confining Jovians to Jupiter's equator. His most valiant efforts are expended on the Saturnians, whom he has move with the seasons and derive benefit from the rays reflected from Saturn's ring. The powers of the Divinity are also drawn upon to supply creatures capable of surviving the thermal hardships he has been unable otherwise to prevent, but overall Biester uses omnipotence more sparingly than many of his contemporaries.

Some authors, both before and after Derham, were skeptical of or opposed to pluralism. One such author was Thomas Baker (1656–1740), a learned antiquarian, who for a number of decades was a fellow of St. John's College, Cambridge. A chapter of his *Reflections upon Learning* (1699) criticized the claims of the pluralists, especially Huygens. In its concluding paragraph, which was quoted by such later authors as Catcott, Nares, and Maxwell, he objects to those "World-Mongers" who, judging things only by their bulk, assert that God would waste the vast heavenly bodies by leaving them lifeless. Such persons should realize, Baker states, that

. . . there is more Beauty and contrivance in the Structure of the Human Body, than there is in the Glorious Body of the Sun, and more perfection in one Rational immaterial Soul, than in the whole Mass of Matter. . . . There cannot then be any absurdity in saying, that all things were created for the sake of this inferior World, and the Inhabitants thereof, and they that have such thoughts of it, seem not to have consider'd, who it was that died to redeem it.[140]

At about the same time, Rev. Robert Jenkin (1656–1717), who in 1711 became master of St. John's, Cambridge, included an antipluralist chapter in his *Reasonableness and Certainty of the Christian Religion.* Among his seven arguments, one is that although the planets could have been designed for habitation, it seems more probable that they were intended to serve as "Mansions of the Righteous, or Places of Punishment for the wicked, after the Resurrection. . . ."[141] The dissenting English clergyman Isaac Watts (1674–1748), best known as a hymn writer, has in two recent and excellent books been cited as an opponent of pluralism on the basis of four lines from his 1709 poem "The Creator and Creatures":

> Thy voice produc'd the seas and spheres,
> Bid the waves roar, and planets shine;
> But nothing like thy Self appears,
> Through all these spacious works of thine.[142]

Possibly these lines were written as opposing pluralism, but they can also be construed as claiming that Christ came only to this earth. The latter interpretation gains support from the fact that in his *First Principles of Astronomy and Geography,* Watts asserts that "'tis probable that [the planets] are all Habitable Worlds furnished with rich Variety of Inhabitants to the Praise of their great Creator."[143] Among German religious writers and theologians from this period whose views on pluralism have been examined, nine favored it, whereas three were opposed.[144]

The place in the pluralist debate of Pierre Bonamy (1694–1770), a Parisian *literateur,* is rather curious. On the one hand, in 1736 he contributed an essay, rich in bibliography, surveying the ancient writings on the question of a plurality of worlds.[145] On the other hand, Bonamy's own position seems to have been antipluralist; this is suggested by his reference to such speculations as "folie," by his description of Fontenelle's pluralism as "an ingenious badinage ventured to enliven a conversation," by his comment that the support given to the doctrine by astronomers justifies one in suspecting that "it cannot be absolutely false," and by his stress that his essay should be seen as concerned not with the truth of the pluralist "sentiment" but only with making clear that the ancients taught the doctrine.[146] However, the opposition to pluralism of Baker and Jenkin, as well as cautions expressed by Watts and Bonamy, were uncharacteristic of the age and had little effect in stemming the ever increasing enthusiasm for the pluralist message.

In concluding this chapter, mention should be made of two groups to whom pluralism had a special appeal; these were the deists and those who sought within the new cosmology to determine the position of hell. One of the earliest deists was Charles Blount (1654–93), who in his *Oracles of Reason* (1693) endorsed the pluralist position.[147] The later and more prominent deist John Toland (1670–1722) contributed to the pluralist debate by his *Account of Jordano Bruno's Book of the Infinite Universe and Innumerable Worlds* (1726), which is essentially a translation of part of Bruno's volume. The effect of these publications on most traditional Christians must have been to cast Blount, Bruno, and Toland, and with them pluralism, into disrepute; however, distant as most of

Bruno's doctrines remained from orthodoxy, his pluralism was beginning to appear prophetic.

Of a number of eighteenth-century books devoted to designating the position of hell in the physical universe, the most famous was written by Tobias Swinden (1659–1719) and the most strange by Jacob Ilive (1705–83). The Cambridge-educated Swinden, who was vicar of Shorne, created a stir with his *Enquiry into the Nature and Place of Hell* (1714), which attained both French and German translations. The sun, being both fiery and conveniently capacious, is, Swinden asserts, the most probable place of hell. The stars he accepts as possibly suns and perhaps surrounded by inhabited planets, but hells he hesitates to associate with them, their inhabitants probably not having fallen, and scripture in any case mentioning but a single hell.[148] Ilive, a London printer, proposed a different location for hell in a bizarre 1733 book that also endorses pluralism, primarily on the basis of the "many mansions" passage in Scripture. Ilive places hell not in the sun as Swinden said nor in the earth as tradition taught; rather, as he asserts in the subtitle of his book, "this Earth is Hell" and "the Souls of Men are the Apostate Angels."[149] Ilive's near contemporary John Nichols, in his *Literary Anecdotes of the Eighteenth Century,* described Ilive as "somewhat disordered in his mind." Ilive's book contains little evidence to the contrary, but it does suggest how tantalizing pluralism could appear to persons of excessive imagination.

By 1750, pluralism had been championed by an array of authors, including some of the most prominent figures of the age. Presented with exceptional appeal by Fontenelle, given legitimacy in scientific circles by Huygens and Newton, reconciled to religion by Bentley and Derham, set to poetry by Pope and Blackmore, integrated into philosophical systems by Berkeley and Leibniz, taught in textbooks by Wolff and in taverns by Whiston, the idea of a plurality of worlds was winning international acceptance. Revolutionary this transformation was, although from the perspective of the present its foundation was frail, resting as it did on analogical arguments of dubious force, on metaphysical principles such as the principle of plenitude, and on a scattering of astronomical observations. However that may be, the era of the extraterrestrial had begun, and it would continue even unto the present.

PART I
From 1750 to 1800

2

Astronomers and extraterrestrials

1. Wright, Kant, and Lambert: pioneer sidereal astronomers and proponents of a plurality of worlds

During the second half of the eighteenth century, the idea of a plurality of worlds won the attention of many if not most European astronomers. This chapter illustrates this point while developing three specific theses: (1) that the astronomical writings of the pioneers (Wright, Kant, and Lambert) of the most important astronomical advance of that half-century, the creation of sidereal astronomy, were permeated by pluralist concerns, (2) that the career of the premier astronomer of that period (Sir William Herschel) was deeply influenced by ideas of extraterrestrial life, and (3) that four of Herschel's leading Continental contemporaries (Schröter, Bode, Laplace, and Lalande) were scarcely less involved in the pluralist quest than Herschel himself.

In 1750, two centuries after the commencement of the Copernican revolution, a second astronomical revolution began. Thomas Wright, Immanuel Kant, and Johann Lambert, in books published between 1750 and 1761, proposed daring new theories of the sidereal region, the realm of stars and nebulae. This second or sidereal revolution, which culminated only in the 1920s, encompassed three conclusions: (1) that the Milky Way is an optical effect arising from the scattering of millions of stars over a roughly planar area, (2) that these stars form a giant disk-shaped structure many light-years in diameter, and (3) that many of the nebulous patches seen in the heavens are galaxies comparable in magnitude to our own Milky Way galaxy. In discussing the writings of Wright, Kant, and Lambert it will be important to ask not only what contribution each made to this revolution but also how he came to make it. In doing this we shall see that this revolution was intimately related to the extraterrestrial life debate, a point too rarely recognized.

Thomas Wright of Durham, born in 1711, died in obscurity in 1786, despite having published a number of books, most notably his *Original*

Theory of the Universe. The fluctuations in Wright's reputation over the past two centuries furnish an instructive study of the difficulties in interpreting his astronomical writings.[1] Although the successes and reversals in the sidereal revolution influenced Wright's standing, a central problem in his historiography was already evident in the efforts made in the 1830s by two Americans, C. S. Rafinesque and C. Wetherill, and in the 1840s by an Englishman, Augustus De Morgan, to rescue his reputation, for these authors championed quite different Wrights. De Morgan portrayed him as a "discoverer" who was "entitled to have his speculations considered, not as the accident of a mind which must give the rein to imagination . . . , but as the justifiable research and successful conclusion of thought founded on both knowledge and observation."[2] On the other hand, Wetherill and Rafinesque praised him for "disdaining the narrow constructed algebraic Newtonian standard," for possessing the "wisdom of an Ancient Sage," and for promoting "the most expansive, sublime, and religious conception of the Universe or INFINITE CREATION."[3] Celestial scientist or religious sage? A century after De Morgan, F. A. Paneth repeatedly proclaimed Wright a pioneer of sidereal astronomy, even proposing the erection on Wright's house of a marble plaque inscribed: "He was the first to explain the Milky Way."[4] Paneth's position was shown to be unacceptable in the early 1970s, when Dr. Michael Hoskin, drawing on previously unpublished Wright manuscripts, revealed that Wright's religious goals were paramount and that he had not one but three theories of the Milky Way, all wrong and increasingly so.[5] It is Hoskin's Wright that will concern us, although it will be suggested that Wright's *Original Theory* as well as the pioneering books of Kant and Lambert can be properly understood only if they are read not least as tracts on extraterrestrials.

Although he lacked university training, Wright possessed a passion for learning so intense that in 1729, as he recorded in his journal, "Father, by ill advice, think him mad. Burn all the books he can get and endeavor to prevent Study."[6] As becomes clear from a Wright manuscript of 1734, this son of a carpenter had by then set for himself the colossal task not simply of constructing a new conception of the physical cosmos but also of integrating into it the domain of the spiritual. Heaven and hell, comets and the chaos, the stars and the "Sedes Beatorum," planets and patriarchs, all these and "an infinite number of worlds" Wright sought to arrange and even to display on a giant chart (now lost) for which his manuscript was to provide the explanation. At the center of creation Wright placed the "Paradise of mortal spirits . . . surrounding the Sacred Throne of Omnipotence." Beyond this was the "Region of Mortality, in which all sensible being such as ye planetary bodies are imagined to circumvolve in all manner of direction round the Devine Presence. . . ."

Finally, outside this shell of stars, he placed "the shades of Darkness & Dispare supposed to be the Desolate Regions of ye Damn^d."[7] Although scarcely mentioning the Milky Way in this work, Wright later claimed that this sketch "prov'd afterwards the foundation of his Theory of ye Univers a much more perfect work." (p. 15) In fact, this manuscript reveals a speculative mind less in the tradition of Isaac Newton than in that of Tobias Swinden. Wright's level of learning had advanced by 1742, when he published a magnificently illustrated text on the solar system entitled *Clavis coelestis* and enriched by references to the astronomical works of Bradley, Halley, Huygens, Newton, and Whiston. The expository character of the book makes pluralist references rare, but in his short section on stars he labels them "great Globes of Fire like the Sun [which] may very possibly be the Centers of other Systems of Planets. . . ."[8]

The title page of Wright's *An Original Theory or New Hypothesis of the Universe* (1750) not only carries the promise of "Solving . . . the General Phenomena of the Visible Creation; and Particularly the Via Lactea" but also suggests its pluralist and religious orientation by quoting Edward Young's lines:

> One Sun by Day, by Night ten Thousand shine,
> And light us deep into the Deity.[9]

His "Preface" conveys the same message by a quotation from Huygens's *Cosmotheoros* on the value of speculation and one from Pope's *Essay on Man* in support of a plurality of worlds. Wright's universe consisted of a "vast Infinity of Worlds . . . crouded full of Beings, all tending through their various States to a final Perfection. . . ." (p. viii) Of the nine letters that make up Wright's volume, the first is a sustained argument for the pluralist position, which Wright claims "has ever been the concurrent Notion of the Learned of all Nations . . ." (p. 3), and which he supports by quotations from such authors as Bruno, Milton, Huygens, Newton, Derham, Pope, and Young. His second letter is largely aimed at justifying analogical reasoning, so important to pluralists. Wright states: "All then that I pretend to argue for, is a Universality of rational Creatures to people Infinity, or rather such Parts of the Creation, as from the Analogy and Nature of Things, we judge to be habitable Seats for Beings, not unlike the mortal human." (p. 12) His third letter supplies data on the planets, marshaled on behalf of "the Thing I have chiefly in this letter attempted to demonstrate [i.e.] that the planetary Bodies in general, are meer terrestrial, if not terraqueous Bodies, such as this we live on. . . ." (p. 26) His arguments for this claim are hardly impressive, consisting of little more than evidences that the planets shine by reflected light. The case he makes in his next letter for the stars being suns is stronger, but to surround them with planets he has recourse not to the telescope but to teleology: "If the Stars were ordained merely for the Use of Us, why so much Extravagance and Ostentation in their Number, Nature, and Make?" (p. 33) He concludes the letter by quoting Joseph Addison:

When I consider that infinite Host of Stars . . . with those innumerable Sets of Planets or Worlds, which were then moving round their respective Suns; when I still enlarge the Idea, and supposed another Heaven of Suns and Worlds rising still above this which we discovered; and these still enlightened by a superior Firmament of Luminaries, which are planted at so great a Distance, that they may appear to the Inhabitants of the former as the Stars do to us; in short, whilst I pursued this Thought, I could not but reflect on that little insignificant Figure which I myself bore amongst the Immensity of God's Works. . . .[10]

This passage is noteworthy because of its suggestion that higher-order universes may exist, which may, however, be so distant that they appear to us as stars. Later in his book Wright advocates a similar idea.

Wright's fifth and sixth letters deal primarily with the stars, attention being given to estimating their distances, to speculating on their motions, and in general to raising the question of their spatial arrangement. Researches of such astronomers as Huygens, Halley, and Bradley are capably used, as are some of Wright's own observations. Nebulae are noted, and a crude calculation leads him to suggest that nearly four million stars are to be seen in the heavens. All of this is blended with much piety and pluralism, nothing catching the spirit of this better than his quoting of Young's lines:

Devotion! Daughter of Astronomy!
An indevout Astronomer is mad.[11]

In a broad sense, deism more than devotion may have been the daughter of post-Newtonian astronomy, but in any case the astronomy of Wright was enriched by the principle of plenitude and by pluralist notions only loosely connected with technicalities of the science. The appeal of this astronomy was such that it stimulated the Durham astronomer to exclaim after calculating the number of visible stars: "What! a vast Idea of endless Beings must this produce and generate in our Minds; and when we consider them all as flaming Suns, Progenitors, and *Primum Mobiles* of a still much greater Number of peopled Worlds, what less than an Infinity can circumscribe them, less than an Eternity comprehend them, or less than Omnipotence produce and support them . . . ?" (pp. 42–3) His enthusiasm for the notion of the plenitude of God's action is evident in his statements:

Suns crowding upon Suns, to our weak Sense, indefinitely distant from each other; and Miriads of Miriads of Mansions, like our own, peopling Infinity, all subject to the same Creator's Will; a Universe of Worlds, all deck'd with Mountains, Lakes, and Seas, Herbs, Animals, and Rivers, Rocks, Caves, and Trees; and all the Produce of indulgent Wisdom, to chear Infinity with endless Beings, to whom his Omnipotence may give a variegated eternal Life. (p. 46)

Wright's expository skills are evident everywhere in his book, but his originality is apparent primarily in his seventh letter, where he expounds his theories of the Milky Way. Wright shows that if we were located in an extended planar array of stars, then we would see the stars as they appear in the Via Lactea (p. 62). Thus, Wright can be credited with recognizing the optical character of the Milky Way, but he does not propose that the stars form a disk; rather, he suggests two other theories. The first is that the stars of our system are located in the wall of a giant hollow sphere. On this model, the Milky Way is explained as the appearance seen when we look in a direction parallel to a plane tangent to the sphere in the region where we are located. Wright's other theory places the sun in a flattened ring of stars comparable in shape to the ring of Saturn. Just as a denizen of the Saturnian ring system would see a higher density of ring particles when looking in any direction parallel rather than perpendicular to the plane of the ring, so also would we on this model see ourselves encircled by a diffuse glow. Ingenious though these two theories may be, Wright (as discussed later) eventually forsook them.

Wright's religious concerns are in the forefront in his final two letters, as he places God at the center of creation, associated with a giant globe where "the Vertues of the meritorious are at last rewarded. . . ." (p. 81) Although showing some restraint in speculating about the forms of his extraterrestrials, he suggests: "Man may be of a very inferior Class; the second, third, or fourth perhaps, and scarce allowed to be a rational Creature." (p. 81) His pious purposes also appear when his calculation that possibly 170 million inhabited globes exist within "our finite view" spurs him to reflect: "In this great Celestial creation, the Catastrophy of a World, such as ours, or even the total Dissolution of a System of Worlds, may possibly be no more to the great Author of Nature, than the most common Accident in Life with us. . . ." (p. 76) This deistic remark, made in a book containing no mention of Christ, stimulates Wright to comment: "This Idea has something so chearful [!] in it, that I own I can never look upon the Stars without wondering why the whole World does not become Astronomers. . . ." (p. 76) Wright's blending of the spiritual and material is evident even in some of his elaborate diagrams; for example, in his Plate XXXII (based on his first model) he represents numerous shells of stars as each centered on the "Eye of Providence" (Figure 2.1). This diagram is historically significant because it shows pictorially what his ninth letter presents verbally: his belief in a plurality not only of worlds but of universes as well. These, in a passage of particular importance, he identifies as nebulae, that is, "the many cloudy Spots, just perceivable by us, as far without our starry Regions, in which tho'

Figure 2.1. Wright's representation of infinity, showing sections of shells of stars, each having at its center the "Eye of Providence."

visibly luminous Spaces, no one Star or particular constituent Body can possibly be distinguished. . . ." (p. 83) Thus, although Wright fails to attain the true theory of the Milky Way, he does put forth both the first and third of the main points of the sidereal revolution.[12]

As was learned in the 1960s with the publication of Wright's manuscript entitled *Second or Singular Thoughts upon the Theory of the Universe,* he substantially altered during his later years his views on the structure of the universe, but unfortunately for his reputation these alterations were largely retrogressive. For example, in this manuscript the stars appear as volcanoes in a solid firmament, while the Milky Way

. . . is looked upon as no other than a vast chain of burning mountains forming a flood of fire surrounding the whole starry regions and no how different from other luminous spaces [nebulae], but in ye number of stars that compose them, or where there are none, in the vast floods of celestial lava that form it.[13]

His pluralist inclinations, however, persisted, with discussions of extraterrestrials being as abundant in this manuscript as in his 1750 book. This was possible partly because Wright retained his spheres (pp. 63–4), making them concentric, the firmament of one sphere being the interior of the central sun of the next sphere!

Celestial scientist or religious sage? That question can now be asked again, and Rafinesque's term for Wright – "Celestial Sage" – accepted, as well as his description of Wright's nine letters in his 1750 book as "quite equal to those of Fontenelle on a plurality of worlds. . . ."[14] Wright's theories of the universe become most easily understandable when he is viewed not primarily as a pioneer sidereal astronomer but rather as a man possessing modest learning, a strong inclination toward speculations linking the spiritual with the astronomical, and an extreme passion for pluralism. More than anyone before his time, he carried pluralism into the sidereal realm and in doing so advocated some ideas of lasting importance. Just as the Greeks became more concerned with the planets when they viewed them as demigods, so also Wright, by attributing intelligent life to the regions around the stars, was disposed to view them with increased interest. Thus, his blending of the spiritual and the material, although methodologically retrogressive, was in some ways scientifically productive. This striking fact is also illustrated by a book published in 1755 by Immanuel Kant, whose relationship to Wright is among the most curious in the entire history of astronomy.

Born in Königsberg and never venturing more than sixty miles from that provincial town, standing in maturity barely five feet two inches, and never weighing more than a hundred pounds, Immanuel Kant (1724–1804) is now recognized as a giant of thought who boldly explored the depths of the human mind and the heights of the heavens. This son of a saddler created a "Copernican revolution" (as he aptly called it) in philosophy by publishing when he was fifty-seven his *Critique of Pure Reason*. However, it is the young Kant of the 1750s and of the sidereal revolution who is of present concern. About this Kant less is known; in fact, a long-standing dispute continues as to whether or not the young Kant possessed scientific credentials of significance. Crucial to this controversy is the assessment of Kant's *Allgemeine Naturgeschichte und Theorie des Himmels* (1755). William Hastie, the Glasgow theologian who in 1900 became its first English translator, cited a number of nineteenth-century scientists, including Helmholtz and Kelvin, to support his description of Kant as a "thorough scientist" entitled to be called "the true founder of Physical Astronomy in its widest range" and "the great founder of the modern scientific conception of Evolution." Hastie also praised Kant's *Universal Natural History and Theory of the Heavens* (as he entitled his translation) "as the most wonderful and enduring product of [Kant's] genius."[15] Not all scholars accepted Hastie's views. In 1908, Svante Arrhenius characterized Kant's book as filled with scientific errors from which an understanding of mechanics would have saved him. In a 1924 lecture, the astronomer C. V. L. Charlier described Kant's book as "scientifically, of very small value" and, when assessed as a popular work, "unsuitable and even dangerous as inviting feeble minds . . . to vain and fruitless speculation."[16] More recently, Irving I. Polonoff, in a detailed study of the young Kant, concluded that he possessed only a quite limited knowledge of science, and Stanley L. Jaki, taking a very different approach, reached the same conclusion.[17]

Trained at Königsberg in Leibnizian and Wolffian thought, exposed at the same time by his teacher Martin Knutzen to Newtonianism, so devoted to Lucretius's *De rerum natura* that he knew long passages by heart, the young Kant of the 1750s was in a number of ways prepared to discuss cosmological questions – and to be enthusiastic about extraterrestrial life ideas. Both these features are found in his *Universal Natural History and Theory of the Heavens*. Of special importance for that book was his chance reading of a 1751 review of Wright's volume, which review appeared in the Hamburg *Freye Urtheile und Nachrichten*.[18] Although this review gave more attention to Wright's religious and pluralist speculations than to his ideas on the Milky Way, it contained enough on the latter to lead Kant to the disk theory. It did not, however, mention Wright's speculation that nebulae may be other universes, a conceptual leap Kant himself attained. Thus, the young Kant presented in his 1755 book all three fundamental principles of sidereal astronomy: the Milky Way as optical effect and as disk, and nebulae as other Milky Ways. These speculations, along with Kant's proposal in the same volume of a form of the nebular hypothesis of planetary formation, are what have given it such scientific celebrity. However, it is important to examine Kant's book in terms of the totality of its approach and contents.

In the preface to his book, Kant attempts to resolve a number of problems faced in his inquiry. Some of these are associated with his espousal of an Epicurean and evolutionary approach to the universe. Writing at a time when his contemporaries were attempting to prove God's existence and wise design in such books as *Ichthyotheologie, Insectotheologie,* and *Petinotheologie* (birds),[19] Kant expresses his desire to avoid both the excesses of such efforts and the atheistic implications of rash attempts to "assign to a nature *left to itself* such processes in which one rightly perceived the immediate hand of the Supreme Being. . . ."[20] The brilliance and boldness of the author who later composed the *Critique of Pure Reason* are evident in his criticism of physicotheologians, against whom he urges that God's designing action is not primarily to be sought in such particulars of nature as an insect's wing, but rather in the laws that God providentially imposed on matter at its creation. In opposition to the Epicurean explanation of natural objects as resulting from blind chance, Kant asks:

How would it be possible that things of different natures should have, *in connection with one another,* seemingly worked for such excellent co-ordination and beauties, nay for the purpose of such beings [as] . . . men and animals, if they did not bespeak a common origin, namely an infinite intellect in which all things were designed in respect of essential properties?[21]

For Kant, this idea provided an answer to the Epicureans and physicotheologians, as well as a compelling proof of God's existence: *"there is a God precisely because nature can proceed even in chaos in no other way than regularly and orderly."* (p. 86)

Kant's book, unlike Wright's, contains not just a cosmology; it offers also a cosmogony, an account of the origin and development of the universe. Kant attempts to set aside the religious difficulties associated with the latter partly by modifying, partly by reinterpreting the classical Epicurean approach; the scientific and methodological problems he counters by claiming, as Newton had not, that Newtonian principles provide the basis for a cosmogony. The sphericity of the heavenly bodies, the simplicity of the law of attraction, and the emptiness of the intervening spaces embolden Kant to paraphrase Descartes: *"Give me matter, I will build a world out of it!"* (p. 88) He does make the admission, presumably forgotten in his detailed speculations about living beings on other planets at the end of his book, "that the formation of all celestial bodies . . . will sooner be understood than the production of a single herb or of a caterpillar will become evidently and completely clarified from mechanical reasons." (p. 88) Concerning his indebtedness to Wright, whose book Kant knew only from the *Freye Urtheile* review,

Kant states: "I cannot exactly determine the borderline which lies between Mr. *Wright's* system and my own and in what pieces have I simply
copied or further developed his sketch."[22] Finally, Kant stresses his reliance on the method of analogy, but for its adequacy he provides scant
justification.

The power of the method of analogy, at least as a method of discovery,
has rarely been more effectively exhibited than in the first of the three
parts of Kant's book, where his most famous sidereal speculations appear. Kant begins by stating: "the fixed stars, as so many suns, are centers
of similar systems in which all may be arranged just as greatly . . . as in
ours. . . ." (p. 101) In support of these "similar systems" of planets,
Kant does not offer, nor could he, any direct observational evidence.
Analogy was also important for Kant's theory of the Milky Way as a
circular system of stars; in particular, he labels it a "zodiac" of stars.
Nonetheless, he was enough of a Newtonian to note that an orbital
motion in the system of stars would prevent its gravitational collapse.
Concerning nebulae, he remarks that their frequently ellipitcal form as
well as their remoteness, indicated by the feebleness of their light, suggest
that they are other "Milky Ways" (p. 107). In 1971, K. G. Jones examined the observational basis of Kant's claim, showing that his support
derived from a list of nebulae compiled by Derham. Of the small number
of nebulae described by Derham as elliptical, in fact only one – Andromeda – is elliptical. As Jones summarizes his analysis, Kant's observational
basis for his identification of nebulae with other sidereal systems was
"quite nugatory."[23] Kant's own comment on the conclusiveness of his
theory is very different: "if conjectures, in which analogy and observation agree perfectly . . . , have the same dignity as formal proofs, one
must hold the certainty of these systems to be demonstrated." (p. 107)
Kant's theory that nebulae are other universes was debated until its basic
correctness was established in the 1920s.

The cosmological approach of the first part of Kant's book gives way
to cosmogonical concerns in its second part, where Kant presents his
version of what has come to be known as the nebular hypothesis. Basically, his theory is that in an infinite universe, matter gradually condenses
as a result of gravitational forces around certain especially dense masses,
and that as it does so, some of the matter begins to revolve around these
original centers, this matter finally forming itself first into rings and
eventually into individual spheres. Stated in this way, it becomes clear
that Kant's theory was devised to apply not only to planetary systems, as
was Laplace's later theory, but also to galaxies and systems of galaxies as
well. Thus, it is natural that Kant suggests that as the solar system is
centered on the sun, so also the Milky Way must possess a single massive
center, which he identifies with the bright star Sirius. Moreover, Kant

urges, notwithstanding the infinitude of his cosmos, that it too must possess a single center, the spiritual function of which he leaves to Wright's assigning. The energetic young Kant develops this theory in great detail, although without writing a single equation. Elements of Newtonian mechanics are invoked in this evolutionary project that Newton had eschewed and that in contemporary astronomy largely defies resolution. Too frequently, however, Kant's Newtonian approaches are distressingly vague or, when specific, are wrong. The latter is the case when Kant argues that certain repulsive forces act between particles, causing a sort of Lucretian swerve and producing the circular motion of matter around the primitive centers of gravitational force. All in all, as Jaki's detailed analysis of Kant's cosmogonical theory reveals, it was erected on a frail scientific foundation.[24]

Analogy again played its role in this theory, but as becomes clear in the seventh chapter of this part of Kant's *Theory,* the principle of plenitude was even more fundamental:

Now, it would be senseless to set Godhead in motion with an infinitely small part of his creative ability and to imagine his infinite force, the wealth of a true inexhaustibility of nature and worlds to be inactive and locked up in an eternal absence of exercise. Is it not much more proper, or to say better, is it not necessary to represent the very essence of creation as it ought to be, to be a witness of that power which can be measured by no yardstick? For this reason the field of the manifestation of divine attributes is just as infinite as these themselves are. (p. 151)

Kant uses this principle to urge that in a spatially infinite universe, "the cosmic space will be enlivened with worlds without number and without end." (p. 152) This process takes place over time, the first worlds being formed near the "center" of Kant's infinite cosmos, with chaos gradually giving way to order in the more remote regions. He states that the formation of worlds

. . . is not the work of a moment. . . . it is efficient throughout an entire sequence of eternity with an ever growing degree of fruitfulness. Millions and entire mountains of millions of centuries will flow by, within which always new worlds and world-orders form themselves one after another in those reaches distant from the center of nature and reach perfection. . . . (p. 154)

Kant also claims that worlds inevitably decay and are destroyed, this process likewise spreading from the center. The geometry of this development is such that the rate of world formation is more rapid than that of destruction, and, moreover, the "Phoenix of Nature" is such that the destruction of worlds generates materials out of which future worlds will form. Kant also urges that the inhabitants of worlds near the center, being made of the cruder matter common there, are at a low intellectual level, whereas "the most perfect classes of rational beings [are] farther

from that center than near it." (p. 167) This hierarchy of beings forms for
Kant a great chain of rational beings in which they progress through "all
infinity of time and space with degrees, growing into infinity, of perfection of the ability of thinking, and bring themselves gradually closer to
the goal of the highest excellence, namely, to divinity. . . ." (p. 168) In
the eighth chapter of the second part of his book Kant further develops
his thesis that God's design for the cosmos is to be seen in general laws
rather than in individual adaptations and that consequently some specific
defects in nature are reconcilable with God's overall design. In summarizing his position he states that nature's "unlimited fruitfulness . . . has
brought forth the inhabited celestial globes as well as the [uninhabited]
comets, the useful mountains and the harmful crags, the habitable lands
and empty deserts, the virtue and the vice." (p. 181)

The sidereal perspective prevalent in most of Kant's book is replaced
by a concentration on the solar system in his concluding third part, which
he describes as "an essay on a comparison, based on the analogies of
nature, between the inhabitants of the various planets." (p. 182) Using a
passage from Pope's *Essay on Man* to introduce this, as he had the two
earlier parts of his book, Kant also embellishes his argument with a
quotation from Albrecht von Haller's pluralist poetry. Concerning the
probabilities of planetary life, Kant remarks: "it is not even necessary to
assert that all planets must be inhabited, although it would be sheer
madness to deny this in respect to all, or even to most of them." (p. 184)
Just as deserts exist on the earth, so may Jupiter presently be without life,
although there will come a time "when the period of its development is
completed." (p. 184) Kant's analysis rests on a number of assumptions;
among these are (1) the great chain of being and (2) the idea that from the
type of matter dominant on a planet, one can make inferences about its
inhabitants. The first belief underlies Kant's comment that "the infinity
of nature includes within herself . . . all beings which display her overwhelming richness. From the highest class of thinking beings to the most
abject insect, no member is indifferent to nature; and nothing can be
missing without breaking up the beauty of the whole, which consists in
interconnectedness." (p. 185) The second assumption, no less fantastic to
modern readers, is formulated in this statement:

*The excellence of thinking natures, the promptness in their reflections, the clarity
and vivacity of the notions that come to them through external impression,
together with the ability to put them together . . . in short the whole range of
their perfection, stands under a certain rule, according to which these become
more excellent and perfect in proportion to the distance of their habitats from the
sun.* (p. 189)

Kant's Mercurians and Venusians are consequently dullards, whereas his
earthlings occupy "exactly the middle rung . . . on the ladder of beings,"

and his Jovians and Saturnians are greatly superior beings. As he states: "From one side we saw thinking creatures among whom a man from Greenland or a Hottentot would be a Newton, and on the other side some others would admire him as [if he were] an ape." (p. 190) Not the least curious feature of this claim is that having doubted the existence of Jupiterians six pages previously, he now bestows superbeings on that planet. Not only brilliance of intellect but also a measure of freedom from decay and death are among the advantages of life on the outer planets. In fact, Kant speculates that our souls may find a future habitation on them: "Who knows, whether the moons do not orbit around Jupiter to shine finally on us?" (p. 196) Hints that Kant may have been concerned with reconciling Christianity and his pluralist position are present in this part, for example, in what he calls "one more escapade . . . into the field of phantasy." He asks whether the denizens of the giant planets may be "too noble and wise" to sin, and whether those "who inhabit the lower planets are grafted too fast to matter . . . to carry the responsibility of their actions before the judgment seat of justice?" (p. 195) Lest earthlings be the sole sinners in the solar system, Kant offers the "miserable comfort" that perhaps Martians suffer from the same affliction. In his concluding paragraph, Kant urges that "when man has filled his soul with such consideration . . . , then the spectacle of a starry heaven in a clear night gives a kind of pleasure which only noble souls can absorb." (p. 196)

What is to be said of Kant's 1755 book? On the informational level, it is noteworthy that at first it attracted almost no attention, his publisher having gone bankrupt in 1755, with most copies becoming unavailable until the mid-1760s. Considered as a work of science, it contains some brilliant flashes of insight, but little to show that Kant had a command of the advanced science of his day. As an analysis of physicotheology, it provided a point of view that might have done much to mitigate the clash occasioned a century later by Darwin's evolutionary arguments. As a treatise on extraterrestrials, it is in some ways remarkable: Almost never before and almost never subsequently did an author advocate such widespread life in both the solar and sidereal systems. The book reveals that the philosopher, later so famous for his critiques of speculative systems, wrote not without having experienced their siren call. That Kant later realized that he had let his imagination roam too far in the 1750s is indicated by the fact that when in 1791 J. F. Gensichen prepared an edition of Kant's book to accompany the German translation of three astronomical papers by William Herschel, he noted that Kant had forbidden him to include any materials beyond his fifth chapter because of their excessively hypothetical character.[25]

Kant may have excised extraterrestrials from the edition, but there is no reason to believe that the maturing Kant forsook the pluralist position, which he in fact advocated in many later writings. For example, his *The One Possible Basis for a Demonstration of the Existence of God* (1763) contains not only a section devoted to summarizing the main astronomical doctrines of his 1755 book but also an expression of the feeling of wonder he experienced as he turned from the revelations of the microscope to those of the telescope:

. . . when I see the intrigue, power and a scene of turmoil in a bit of matter and then turn my attention from that to boundless space, swarming with worlds like bits of dust, no human language can express the feeling such reflection excites and all subtle metaphysical analyses fall far indeed from the grandeur and value that are part and parcel of such a vision.[26]

In his *Observations on the Feeling of the Beautiful and Sublime* (1764), Kant remarks: "it will not be necessary for women to know more of the cosmos than is necessary to make the aspect of the sky touching to them on a fine night, after they have grasped, to a certain extent, that there are more worlds, and on them more creatures of beauty to be found."[27]

The degree of confidence Kant retained into the 1780s in the pluralist position is revealed in a section from his *Critique of Pure Reason* (1781) in which opinion, knowledge, and belief are compared. Kant asserts:

I should not hesitate to stake my all on the truth of the proposition – if there were any possibility of bringing it to the test of experience – that, at least, some one of the planets, which we see, is inhabited. Hence I say that I have not merely the opinion, but the strong belief, on the correctness of which I would stake even many of the advantages of life, that there are inhabitants in other worlds.[28]

In 1784, Kant published both a review of Herder's *Ideen zur Philosophie der Geschichte der Menschheit* and his own essay entitled "Idea for a Universal History from a Cosmopolitan Point of View." In the review, while agreeing with Herder's pluralism, he objects to his former student's advocacy of the doctrine of transmigration of souls. In his essay, Kant, in the context of a discussion of the lowly nature of man, suggests:

If . . . we carry out well the mandate given us by Nature, we can perhaps flatter ourselves that we may claim among our neighbors in the cosmos no mean rank. Maybe among them each individual can perfectly attain his destiny in his own life. Among us, it is different; only the race can hope to attain it.[29]

Despite Kant's stress on the difference between men and extraterrestrials in his review of Herder, an opposite approach is evident in his *Fundamental Principles of the Metaphysic of Morals* (1785). In that book, he repeatedly describes his goal as being to formulate an ethics applicable not just to mankind, but also to all "rational beings." One reason for his emphasis on "rational beings" was no doubt his conviction that extraterrestrials fitting that category exist. That Kant formulated his philosophy in such a way as to include extraterrestrials is also suggested by a famous passage from the conclusion of his *Critique of Practical Reason* (1788):

Two things fill the mind with ever new and increasing admiration and awe, the oftener and more steadily they are reflected on: the starry heavens above me and the moral law within me. . . . The former . . . broadens the connection in which I stand into an unbounded magnitude of worlds beyond worlds and systems of systems. . . . The former view of a countless multitude of worlds annihilates, as it were, my importance as an animal creature, which must give back to the planet (a mere speck in the universe) the matter from which it came. . . .[30]

In 1790, Kant's *Critique of Judgement* appeared, containing a number of comments on extraterrestrial life, which he describes as "a matter of opinion; for if we could get nearer the planets, which is intrinsically possible, experience would decide whether such inhabitants are there or not; but as we never shall get so near to them, the matter remains one of opinion."[31]

As Kant's fame spread ever more widely, efforts were made to publish his lectures, for example, those he gave on anthropology and those on philosophical theology. A passage from his *Anthropology from a Pragmatic Point of View* (1798) consists of a warning against efforts to speculate in detail on the form of extraterrestrial beings, whom authors invariably describe as similar to men.[32] Kant's *Lectures on Philosophical Theology,* first published in 1817, were capacious enough to contain not only a remark supporting the probability of life on the moon but also one on the relevance of pluralism to the Leibnizian doctrine that this is the best of all possible worlds: "For surely if our terrestrial globe were the whole world, it would be difficult to know it for the best, and to hold by this with conviction. For to speak honestly, on this earth the sum of sorrow and the sum of good might very well just about balance each other."[33]

These passages show what a large role the idea of a plurality of worlds played in the thought of the most eminent philosopher of modern times. Moreover, they suggest that the "starry heavens" that so filled Kant with awe were not the heavens of traditional astronomy, but rather a densely denizened domain wherein millions of inhabited planets orbited suns clustered in endless hierarchies of systems. This was an extraordinarily imaginative conception of the cosmos, but as Kant must eventually have realized, it was at most only partially grounded in science.

Included with the summary of his astronomical system that Kant published in 1763 is the lament that his 1755 system was unknown even to "the celebrated *J. H. Lambert,* who six years later in his *Cosmological Letters* (1761) presented exactly the same theory of the systematic composition of the universe as a whole, of the Milky Way, and of the nebulae. . . ."[34] Kant attributes more to Lambert in this passage than is justified, but about the brilliance of Johann Heinrich Lambert (1728–77), whom Kant in 1765 described as "the greatest genius in Germany,"[35] there can be no doubt. A philosopher of such importance that Kant considered dedicating the *Critique of Pure Reason* to him, Lambert is also ranked among the leading astronomers, mathematicians, and physicists of his century. The majority of Lambert's most significant writings came after 1765, when he was elected to the Prussian Academy of Sciences in Berlin, but of present concern is his *Cosmologische Briefe über die Einrichtung des Weltbaues* (1761).

As Lambert recounted to Kant in 1765, this book had its origin one evening in 1749 when "from my window I looked at the starry sky, especially the Milky Way. I wrote down on a quarto sheet the idea that occurred to me then, that the Milky Way could be viewed as an ecliptic of the fixed stars, and it was that note I had before me when I wrote the *Letters* in 1760. In 1761 I heard in Nürnberg that an Englishman had had similar thoughts a few years before."[36] As this letter makes clear, Lambert arrived independent of Wright and Kant at the conception of the Milky Way as the optical effect of an extended planar distribution of stars. Moreover, he shared with Kant the disk theory of the Milky Way,[37] and possibly the view that nebulae are independent systems comparable to the Milky Way.[38] These conceptions made Lambert's book a classic of cosmology, but they neither exhaust its contents nor adequately reflect his overall approach.

In comparing Lambert's book with those of Wright and Kant, one finds that, unlike Kant, Lambert was concerned only with cosmology, not·cosmogony.[39] In all three books, however, and most pervasively in Lambert's volume, discussions of extraterrestrial life abound. So he wrote it, and so was it read by his contemporaries. Lambert in 1765 stated to Frederick the Great that the public considered his *Letters* as a second volume "to the one on plurality of worlds by Fontenelle," and an Italian physics professor, Rev. Giuseppe Toaldo, praised it as the most beautiful of pluralist books.[40] Moreover, in his preface, Lambert explicitly states:

I would have wished to make the considerations in these letters into a second part of Fontenelle's dialogues . . . if I could have written them with [his] so admirable liveliness, and had clever ideas been to me so flowing and rich. How far that which I say on the world edifice can serve basically as a continuation of Fontenelle's thoughts each reader will be able to judge if he will omit the vortices which Fontenelle so willingly uses.[41]

Although Lambert lacked Fontenelle's breezy style, his more extensive astronomical knowledge spurred him to spread life far beyond the solar system that had been the Frenchman's chief concern. Another prominent feature of Lambert's *Letters* is his intense interest in comets. Roger Jaquel, a leading Lambert scholar, even describes his cosmology as "la plus cométoplethorique" in the entire history of cosmologies.[42] Lambert's passion for populating comets as well as his overall pluralism derived from the teleological approach everywhere evident in his volume. As his interlocutor comments in the seventh of the twenty letters that make up Lambert's treatise, "From teleology you derive the purposes, and from experience you find that the means for such purposes are available."[43] Lambert's teleological methodology appears not only in his scientifically well formulated efforts to show, contra Whiston and others, that earthlings need not fear cometary collisions but also in his claim – far less satisfactorily supported – that thousands or possibly millions of comets circulate in our solar system and, by analogy, in others as well (p. 72). A knowledgeable Newtonian, Lambert was aware that comets could move either in elliptical orbits or in nonrecurring parabolic or hyperbolic paths. The latter paths he incorporates into his system by proposing that such comets pass from one system to another and are populated by astronomically interested extraterrestrials blessed with an atmosphere that by expanding at certain times preserves livable conditions on these comets. In one of the most fantastic passages from a book that William Herschel described as "full of the most fantastic imaginations,"[44] Lambert uses his interlocutor to describe in more detail these travelers from system to system:

I rank highest the astronomers which you, Sir, place on such celestial bodies. Their route proceeds from suns as we go from city to city on earth, and when in our case a few days go by they count in myriads of our years. . . . Our greatest measures are their differentials, and our millions can hardly suffice as their table of multiplication. They know the warmth and brightness of each sun, and with a single conclusion they determine the general characteristics of the inhabitants of each planet which orbits around it at a given distance. Their year is the time from one sun to another. Their winter falls in the middle of the intervening space, or of the journey which they make to another sun, and they celebrate the moment when their former course turns into a new one. The perihelion of each course is their summer. (p. 79)

Lambert may have been less than serious when he penned that passage, but this scientist-philosopher, known for the traditional character of his religious orientation and for his Leibnizian-Wolffian philosophical background, meant to be taken literally in his assertions that "The Creator . . . is much too efficient not to imprint life, forces, and activity on each speck of dust. . . . [I]f one is to form a correct notion of the world, one should set as a basis God's intention in its true extent to make the whole world inhabited. . . ." (p. 102) Not only does Lambert make all heavenly bodies inhabitable, he places on each "innumerable inhabitants of all possible kind and form." (p. 55) This must be so because "all possible varieties which are permitted by general laws ought to be realized, because perfection becomes thereby greater." (p. 89) Although sharing the principle of plenitude with Wright and Kant, Lambert specifically rejects the spatial infinitude of the universe (p. 47). Thus, the system of hierarchically arranged Milky Ways that he proposes was meant to be taken as ceasing at some point, perhaps at the thousandth order (p. 169).

Fantastic some of these ideas certainly were, but far more frequently than Kant, Lambert labels particular claims as speculative and attempts to specify the degree of credibility each deserves.[45] In his concluding letter, for example, Lambert admits: "And truly, have I not already rambled somewhat beyond the limits of what is credible? I drew conclusions, and surely enough without having in each case appropriate observational evidence on hand and without knowing how far they would lead." (p. 184) Lambert, who was very aware of the need for quantitative laws and empirical verification, urges in that letter that simplicity, harmony, and connectedness are also factors to be considered in evaluating hypotheses. It was his sensitivity to these features of scientific argumentation that made him anxious to "step before the judgment seat of reason." (p. 185)

Lambert's volume secured a more international readership during the eighteenth century than did those of Wright and Kant. Whereas Wright's volume was first republished only in the 1830s, and Kant's book, despite some late-eighteenth-century German editions, was translated into French and English only after more than a century, Lambert's treatise was made available to the French public in 1770 by means of a condensed translation prepared by his Swiss friend J. B. Merian. In preparing that publication, Merian left out much of Lambert's science and most of his methodological qualifications, but did not dilute Lambert's pluralist doctrines. French readers were far better served by the 1801 full translation prepared by the Toulouse astronomer Antoine Darquier (1718–1802), to which extensive notes were added by the Dutch astronomer J. M. C. van Utenhove (1773–1836). Russian readers gained access to Lambert's ideas in Makhail Rozin's 1797 translation of the Merian condensation, which was also the text used when James Jacque published in 1800 his English translation. It would be a revealing research project to determine whether it was the pioneering features of Lambert's sidereal hypotheses, so much in advance of his age, or his obviously exciting and clearly contemporary pluralist speculations that accounted for these various editions of his book. One may suspect the latter.

The degree to which pluralist ideas pervade the books of Wright, Kant, and Lambert is very striking. This feature is partly due to the desire they shared for investigating broad cosmological questions. In this they were strongly influenced by their faith in the principle of plenitude. Believing God active to the full extent of his omnipotence and convinced that this entailed his populating the universe with a great chain of beings, they sought to determine God's design for their domiciles. Ironically, these three pioneers of stellar astronomy may have had no more concern about stars than their contemporaries; rather, it was inhabited planetary systems that interested them and that they sought to arrange into systems. The Milky Way for them was not primarily a giant array of glowing globes, but rather a visible collection of sources of heat and light serving the myriads of beings living on the admittedly unobservable planets of the stellar systems. If it is correct that to some extent their pluralism motivated and aided them in their search for those ideas that have made their books famous, this would illustrate Lambert's statement in his *Cosmological Letters* (p. 46) that a teleological approach frequently has heuristic benefits.

Wright, Kant, and Lambert wrote their books in the middle years of the Enlightenment, which R. G. Collingwood characterized as the age of the endeavor to "secularize every department of human life and thought."[46] Their books, which can be seen as extended endorsements of Young's "An undevout astronomer is mad," at first glance fit Collingwood's characterization no better than they harmonize with the Enlightenment as an age of empiricism, Newtonianism, and this-worldliness. Wright, Kant, and Lambert were interested in, and to differing degrees informed about, the telescopic observations of their contemporaries, but analogical argumentation and the principle of plenitude were more influential factors in their thought. Newton was known to them to various degrees, but they went far beyond and in some cases contrary to his scientific and methodological principles. Wright's spherical shell universes, Kant's dull Mercurians and super Saturnians, and Lambert's cometary astronomers are now seen as bizarre companions to the more durable doctrines developed in their books. In a deeper sense, however, support for Collingwood's characterization can be found in Kant's and Lambert's angelic superbeings and in Wright's God-centered shells when it is suggested that these three authors were in fact attempting to transform traditional religious notions by interpreting them in physical terms. This blending of the spiritual and the celestial was, of course, not confined to these three authors and will be encountered again and again in this book, not least in the writings of the greatest of sidereal astronomers, William Herschel, who without knowing of their books embarked in the 1770s and 1780s on much the same quest and carried it farther.

2. Sir William Herschel: "promise not to call me a Lunatic"

Before discussing Herschel, it will be convenient to consider James Ferguson (1710–76), who, it will be suggested, importantly influenced Herschel. Although Ferguson's formal education consisted of only three months in a Scottish grammar school, this shepherd turned instrument maker and astronomical author published a number of popular scientific treatises, most notably his *Astronomy Explained upon Sir Isaac Newton's Principles* (1756), which attained seventeen editions. Its appeal was due not only to its clear if elementary explanations but also to its pious pluralism. Describing astronomy as "the most sublime, the most interesting, and the most useful" of the sciences, Ferguson urges that through it, our understanding becomes "clearly convinced, and affected with the conviction, of the existence, wisdom, power, goodness, and superintendency of the SUPREME BEING!"[47] These characteristics of the Creator assure Ferguson that God made the stars not primarily to illuminate our earth, which our moon does far better, but to serve their own systems of planets. As he states, astronomy "discovers to us . . . an inconceivable number of suns, systems, and Worlds, dispersed through boundless space. . . ." And he adds: "From what we know of our own System, it may be reasonably concluded that all the rest are with equal wisdom contrived, situated, and provided with accommodations for rational inhabitants." In short, Ferguson's universe contains: "Thousands of thousands of Suns . . . attended by ten thousand times ten thousand Worlds . . . peopled with myriads of intelligent beings, formed for endless progression in perfection and felicity." (pp. 3–5)

Ferguson focuses chiefly on the solar system, concerning which he notes that the outer planets are provided with extra moons, and in the case of Saturn with a ring, to illuminate the nights of their inhabitants. He states that "by the assistance of telescopes we observe the Moon to be full of high mountains, large valleys, and deep cavities. These similarities leave us no room to doubt but that all the Planets and Moons in the System are designed as commodious habitations for creatures endowed with capacities of knowing and adoring their beneficent Creator." (p. 4) So confident was he of lunar life that he remarks that the earth not only is "a moon to the moon" but also, by having a more or less fixed position in the moon's sky, allows lunarians to determine lunar longitude (pp. 16–18). Concerning comets, he admits: "The extreme heat, the dense atmosphere, the chaotic state of the comets, seem at first sight to indicate them altogether unfit . . . for rational beings. . . ." Nonetheless he populates them, urging that God's "infinite power and goodness" enable him to "make creatures suited to all states and circumstances." That God so acts, Ferguson asserts, is shown by the fact that "matter exists only for the sake of intelligence [and is] always . . . pregnant with life, or necessarily subservient thereto. . . ." (pp. 27–8)

That teleology more than the telescope, that religious conviction more than rigorous calculation inspired Ferguson to such fancies goes without saying. Nonetheless, his book was sufficiently rich in astronomical information that when the *Encyclopaedia Britannica* was launched in 1771, the editors presented as the long, unsigned article "Astronomy" an abbreviated version of Ferguson's *Astonomy Explained*. This energetic author carried his enthusiasm for pluralist astronomy to the English public in other ways as well, for example, by popular lectures on astronomy in such cities as Bath, Bristol, and London. Although he was not the first to present pluralism to British children, this having been done in 1758 in a book that has been attributed by some to John Newbery and by others to Oliver Goldsmith,[48] Ferguson's *Easy Introduction to Astronomy for Young Gentlemen and Ladies* (1768) was one of the most successful works of its kind, attaining a dozen editions. Using a dialogue form in this book, he has his Eudosia exclaim: "I cannot imagine the inhabitants of our earth to be better than those of other planets. On the contrary, I would fain hope that they have not acted so absurdly with respect to [God], as we have done."[49] Ferguson's talents were recognized by the Royal Society, which elected him to membership, and by King George III, who bestowed fifty pounds per year on him. Of more immediate relevance are the indications that his pluralist claims influenced three authors who became central figures in the pluralist debate: Tom Paine, David Brewster, and William Herschel.

The eminence of Sir William Herschel (1738–1822) among astronomers is indisputable, but recent researches may require a revision in the portrayal of the career of this Bath musician who became a full-time astronomer only after his 1781 discovery of Uranus when King George III rescued him from "crochets and quavers" by funding his investigations. Traditionally Herschel has been presented as a tireless telescopic technician, as a model empiricist who, as Edwin Hubble put it, did not, like Kant, speculate about the nebulae but observed them by the thousands.[50] Herschel has also been seen as eschewing the philosophical controversies of the Enlightenment to concentrate on what his telescopes would reveal about the timeless night sky, which he studied with a detachment less easily detected in the physicotheological excursions of Wright, Lambert, and Ferguson.[51] Attractive as this image of him may be, it ill accords with some of the recent researches on Herschel published from Cambridge University by Michael Hoskin and some of his former students or with materials I have uncovered in Herschel's unpublished manuscripts.[52] Considered conjointly, these studies suggest (1) that Herschel was less an isolated empiricist than a speculatively inclined celestial naturalist, quixotically caught up in a quest for evidence of extraterrestrials, (2) that many of his efforts make most sense when seen as attempts to transform pluralism from being a delight of poets, a doctrine of metaphysicians, and a dogma of physicotheologians into a demonstration of astronomers, and (3) that pluralism was a core component in Herschel's research program and as such influenced many of his astronomical endeavors, especially (if far from exclusively) in his formative years.

About Herschel's development from 1753, when he joined his father's Hanoverian regimental band as an oboist, to the mid-1770s, when astronomy became the consuming avocation of the by then Bath organist, less is known than one might hope. That his father had instilled in him a love of learning is indicated by his determination in 1756, while stationed with the band in England, to read Locke's *Essary Concerning Human Understanding*. That this zeal did not diminish after 1757, when he forsook the band to work as a musician in England, is shown by the fact that by 1761 he had read Leibniz's *Théodicée*. His growing attachment to astronomy is reflected in his 1773 purchase of Ferguson's *Astronomy*. Herschel's interest in these three books has been documented by his biographers;[53] what has not been noted is that all three contain endorsements of pluralism.[54] That Herschel's involvement with Ferguson may have gone beyond the *Astronomy,* which he took "to bed with a bason [sic] of milk or a glass of water" for a number of months,[55] is suggested by the fact that Ferguson lectured in Bath in 1767 and again in 1774. Given that by 1773 Herschel was sufficiently committed to astronomy that he was constructing his own telescopes, it seems probable that he attended one or both of Ferguson's lecture series.[56] Moreover, internal evidences from Herschel's writings point to the conclusion that Herschel was powerfully influenced by Ferguson's conception of astronomy, including its pluralist component.

Herschel's scientific debut should be dated not from his 1781 discovery of Uranus but from May 1780, when two of his papers were read to the Royal Society, the longer of these being his "Astronomical Observations Relating to the Mountains of the Moon." Behind this paper lies a fascinating story, recoverable from Herschel's unpublished manuscripts, which provide evidence that at that time he believed that he was already on the verge of a discovery more revolutionary than his detection of Uranus. Herschel's lunar mountain paper led Neville Maskelyne, the Astronomer Royal, to request details from Herschel on his methods of measurement and to ask about his statement in that paper that "the knowledge of the construction of the Moon leads us insensibly to several consequences . . . such as the great probability, not to say almost absolute certainty, of her being inhabited."[57] No doubt Maskelyne's query about Herschel's lunar life claim was intended to suggest the impropriety of its inclusion in a formal scientific paper. The hint was hardly taken; instead, Herschel responded with a disquisition that Maskelyne chose not to expose to the public when he appended the measurement section of Herschel's letter to the eventually published paper. Only with the 1912 publication of Herschel's collected papers did his full letter become known. Although Herschel was aware in 1780 of the evidence against a lunar atmosphere, based on the sharpness with which the moon occults stars, and had himself made observations supportive of this conclusion (I, pp. xci–xcii), he nonetheless argues for lunar life in this letter, suggesting that his belief in it "may perhaps be ascribed to a certain Enthusiasm which an observer, but young in the Science of Astronomy can hardly divest himself of when he sees such wonders. . . ." After requesting Maskelyne to "promise not to call me a Lunatic," he quotes extensively from a document drafted eighteen months earlier in which he uses arguments from analogy as applied to the similarities between the earth and the moon as a basis for asking:

. . . who can say that it is not extremely probable, nay beyond doubt, that there must be inhabitants on the Moon of some kind or other? Moreover it is perhaps not altogether so certain that the moon *is* out of the reach of observation in this respect. I hope, and am convinced, that some time or other very evident signs of life will be discovered on the moon. (I, p. xc)

Maskelyne's dismay cannot have diminished when later in the letter Herschel explains: "The earth acts the part of a Carriage, a heavenly waggon to carry about the more delicate moon, to whom it is destined to give a glorious light. . . . For my part, were I to chuse between the Earth and Moon I should not hesitate to fix upon the moon for my habitation." (I, p. xc)

As unrestrained as Herschel's letter to Maskelyne appears, it may have taken some reserve on Herschel's part to resist making even stronger statements. This is evident from Herschel's unpublished compilation of his lunar observations, which reveals that in 1780 Herschel believed he already possessed substantial observational evidences of life on the moon. For example, among the earliest of his lunar observations is that dated May 28, 1776, when with a newly acquired telescope, he saw

. . . something I had never observed before, which I ascribed to the power and distinctness of my Instrument, but which may perhaps be an optical fallacy. . . . I believed to perceive something which I immediately took to be *growing substances*. I will not call them Trees as from their size they can hardly come under that denomination, or if I do, it must be understood in that extended signification so as to take in any size how great soever. . . . My attention was chiefly directed to Mare humorum, and this I now believe to be a forest, this word being also taken in its proper extended signification as consisting of such large *growing substances*.[58]

Herschel proceeds to sketch the forest (Figure 2.2) and to analyze whether or not a lunar forest would be visible from the earth:

Our tallest trees would vanish at that distance. It is not impossible but that the vegetable Creation (and indeed the animal too) may be of a much larger size on the Moon than it is here; tho' perhaps not very likely. And I suppose that the

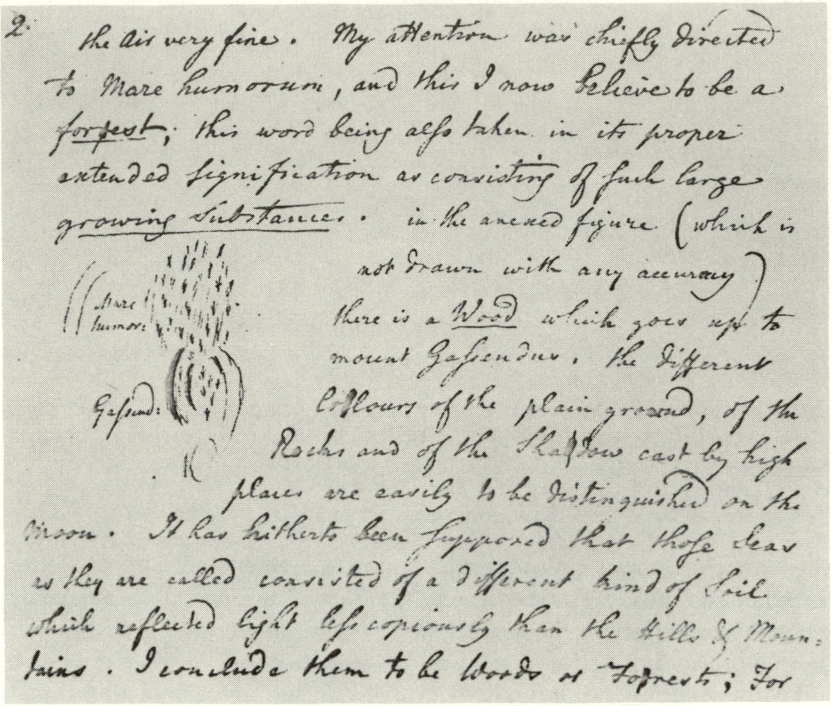

Figure 2.2. A portion of Herschel's unpublished notes on his lunar observations, including his drawing of the lunar forest he believed he had observed. (Courtesy of the Royal Astronomical Society.)

borders of forests, to be visible, would require Trees at least 4, 5 or 6 times the height of ours.

But the thought of Forests or Lawns and Pastures still remains exceedingly probable with me, as that will much better account for the different Colour, than different coloured soils can do.[59]

Herschel's ambivalence about these observations led him in late 1778 to compose a new analysis, portions of which he quoted in his Maskelyne letter, but the following passages, which show that Herschel believed he had evidence not only of forests but also of lunar towns, were not included:

As upon the Earth several Alterations have been, and are daily, made of a size sufficient to be seen by the Inhabitants of the Moon, such as building Towns, cutting canals for Navigation, making turnpike roads &c: may we not expect something of a similar Nature on the Moon? – There is a reason to be assigned for circular-Buildings on the Moon, which is that, as the Atmosphere there is much rarer than ours and of consequence not so capable of refracting and (by means of clouds shining therein) reflecting the light of the Sun, it is natural enough to suppose that a Circus will remedy this deficiency. For in that shape of Building one half will have the direct and the other half the reflected light of the Sun. Perhaps, then on the Moon every town is one very large Circus? . . . Should this be true ought we not to watch the erection of any new small Circus as the Lunarians may the Building of a new Town on the Earth. Our telescopes will do this. . . . By reflecting a little on this subject I am almost convinced that those numberless small Circuses we see on the Moon are the works of the Lunarians and may be called their Towns. . . . Now if we could discover any new erection it is evident an exact list of those Towns that are already built will be necessary. But this is no easy undertaking to make out, and will require the observation of many a careful Astronomer and the most capital Instruments that can be had. However this is what I will begin.[60]

Having adopted this remarkable research program, which probably was no small factor in his efforts to build better telescopes, Herschel set about making numerous lunar observations. His lunar observation book shows that to classify the lunar "circuses" he chose at first the labels "Metropolis, Cities, Villages" but finally satisfied himself with the more prosaic terms "Large places, Middling places, Small places."[61] His June 17, 1979, entry records his observation of "a Cut or Canal that seems evidently to be the effect of Art rather than of Nature," and a month later, seeing a new spot in the Mare Crisium region, he writes: ". . . I find it is a city."[62] Extensive lunar observations from 1780 and 1781 are recorded, many from the earlier year being height determinations of lunar mountains. The latter year produced richer results, his observations of late June yielding numerous patches of "vegetation," "turnpike roads," and "circuses."[63] On another evening he reports regions "tinged with green."[64] In 1783 he records that a star passing behind the moon disappeared slowly, indicating a lunar atmosphere, and also in 1783 he espies "two small pyramids."[65] Herschel's lunar observations seem to be far less frequent after 1783, even though in that year he confided to Alexander Wilson, a Scottish astronomer, what he had done and hoped to do toward establishing the existence of life on our moon:

The attempt of finding traces of animation in the Moon has now been 5 or 6 years one of those I have endeavoured to render practicable, and tho' I have met with no self evident or occular demonstration of the moons being inhabited, yet do I still hope that a good many of my observations will at least render the reasons we may alledge from analogy more forcible. The highest power I have hitherto been able *conveniently* to use in viewing the *Moon* is 932. Hence it is easy to calculate what sort of Objects we may expect to see. However the many interruptions [e.g., his discovery of Uranus!] I have within these last two years met with, have prevented my Observations on this subject to be so frequent as I now, with improved instruments, hope to make them. Very happily our gracious sovereign has enabled me to follow a study, which from my excessive attachment to it I formerly followed in spite of other employments.[66]

It is unknown how many other astronomers were aware of Herschel's hopes for detecting lunar life; one suspects that part of the process of his professionalization was to learn that discoursing on such matters gave support to those who thought him "fit for bedlam."[67] Whatever the case may be, it is a mark of his professionalism that never in his published writings did he claim observational evidence of lunar life. Probably he dismissed his observations of "forests," "cities," "turnpike roads," and such as among those tricks of the telescope that he lamented in a 1782 letter to Alexander Aubert:

These instruments have played me so many tricks that I have at last found them out in many of their humors. . . . I have tortured them with powers, flattered them with attendance to find out the critical moments when they would act, tried them with specula of a short and of a long focus, a large aperture and a narrow one; it would be hard if they had not been kind to me at last.[68]

Herschel was convinced that life exists not only on the moon but also on the planets and their satellites. The similarities between the earth and Mars are stressed in his 1784 paper on that planet to support his repeated references to the inhabitants of Mars (I, pp. 138, 148, 156). In fact, his concluding comment is that Mars "has a considerable but moderate atmosphere, so that its inhabitants probably enjoy a situation in many respects similar to ours." (I, p. 156) In other papers he casually refers to "the inhabitants of Saturn or the Georgian planet [Uranus]" (I, p. 422) and to "the inhabitants of the satellites of Jupiter, Saturn, and the Georgian planet." (I, p. 481) Shortly after the discoveries of Ceres and Pallas, Herschel reported his observation that they "have an atmosphere of considerable extent." (II, p. 194) This spurious observation, one suspects, may have had its origin in his pluralist proclivities.

Herschel's writings show that he also believed that inhabited planets orbit other stars. In a 1783 paper, he justifies observations of the variable star Algol by claiming that such observations could verify the existence of "a plurality of solar and planetary systems." (I, p. cvii) Although admitting such extra-solar-system planets "can never be perceived by us" (I, pp. 416–17), he states (1789) that every star "is probably of as much consequence to a system of planets, satellites, and comets, as our own sun. . . ." (I, p. 330) The point is most explicit in his 1795 statement that analogy supports the conclusion that "since stars appear to be suns, and suns, according to the common opinion, are bodies that serve to enlighten, warm, and sustain a system of planets, we may have an idea of numberless globes that serve for the habitation of living creatures." (I, p. 482) The far-reaching significance of this claim is seen when it is recalled that during the 1780s Herschel had come to believe that most of the hundreds of nebulae he had detected with his telescopes were entire universes comparable to the Milky Way. It is within this context that one should view the exclamation of the poetess Fanny Burney, who in 1786 visited Herschel: "he has discovered fifteen hundred universes! How many more he may find who can conjecture?"[69]

The continuing intensity of Herschel's pluralist convictions is clearly revealed in a 1795 paper, especially when this paper is seen in relation to an event reported in the *Gentleman's Magazine* for 1787. A certain Dr. Elliot was brought to trial in London for having set fire to a lady's cloak by firing a pair of pistols near it. Insanity was the plea made for Elliot, in support of which a Dr. Simmons recounted examples of Elliot's bizarre behavior, especially his having prepared a paper for submission to the Royal Society in which he maintained the sun to be inhabited.[70] This incident leads one to wonder what may have been the reaction among readers of the Royal Society's *Philosophical Transactions* when in 1795 they encountered a paper in which Herschel theorized that the sun consists of a cool, solid, spherical interior above which floats an opaque layer of clouds that simultaneously reflects the rays of the glowing exterior region and shields the interior region from excessive heat and light. As he states: "The sun . . . appears to be nothing else than a very eminent, large, and lucid planet, evidently the first, or in strictness of speaking, the only primary one of our system. . . . Its similarity to the other globes of the solar system . . . leads us to suppose that it is most probably . . . inhabited . . . by beings whose organs are adapted to the peculiar circumstances of that vast globe."[71] Herschel contrasts his theory with that of "fanciful poets" who portray the sun "as a fit place for the punishment of the wicked," urging that his claim rests *"upon astronomical principles."* (I, p. 479) He argues his case by means of analogy, claiming that telescopes reveal that the moon has numerous similarities to the earth. The obvious differences he dismisses by noting that terrestrial beings flourish in a variety of circumstances:

While man walks upon the ground, the birds fly in the air, and fishes swim in water; we can certainly not object to the conveniences afforded by the moon, if those that are to inhabit its regions are fitted to their conditions as well as we on this globe are to ours. An absolute, or total sameness, seems rather to denote imperfections, such as nature never exposes to our view. . . . (I, p. 481)

Finally, he urges that terrestrials who deny life to the sun have no more logic on their side than inhabitants of a planetary satellite who deny life to the primary around which they revolve. Such arguments support E. S. Holden's statement that Herschel's views on solar and lunar life "rest more on a metaphysical than a scientific basis. . . ."[72]

Holden's conclusion, however, needs to be qualified in one important way that helps explain why the premier astronomer of that day adopted such a strange theory. Although as early as 1780 Herschel had considered a form of this solar model (I, p. xcvi), he had between then and 1795 accumulated astronomical evidences that, when viewed in terms of his pluralist metaphysics, substantially increased the attractiveness of that model. In particular, during this period Herschel's sidereal researches had led him to observe what he describes in his 1795 solar paper as "very compressed clusters of stars." He goes on to state concerning stars in such cluster that

. . . it will hardly be possible to assign any sufficient mutual distance [to them] to leave room for crowding in those planets, for whose support those stars have been, or might be, supposed to exist. It would seem, therefore, highly probable that they exist for themselves; and are, in fact, only very capital, *lucid,* primary planets, connected together in one great system of mutual support. (I, pp. 482–3)

Thus, Herschel had found a way to save these stars from being "mere useless brilliant points" (I, p. 484), or, put differently, to rescue his teleologically based pluralist metaphysics from a serious difficulty. That Herschel's solar theory was no passing fancy in his thought is shown by his having elaborated it further in an 1801 paper in which he refers to the sun as "a most magnificent habitable globe" (II, p. 147) and by his 1814 description of stars as "so many opaque, habitable, planetary globes." (II, p. 529) Moreover, although Agnes Clerke, writing in 1885, described Herschel's solar model as more primitive than that of Anaxagoras, it persisted as the preferred theory of the sun until the 1850s.[73]

Herschel's pluralist convictions did not extend to comets. This and other interesting points are evident in extensive unpublished notes he made in 1799 when he first encountered Lambert's *Cosmologische Briefe,* reading it in J. B. Merian's French condensation. His published description of Lambert's book as "full of the most fantastic imaginations" (II, p. 318) is expressed in more detail in these notes, a selection from which follows, the numbers cited referring to the pages of Merian's text:

24. Worlds in grains of sand, and inhabitan[t]s! This paragraph is too poetical to be philosophical.
26. "We cannot make a step without destroying worlds and without creating new ones!" What an abuse of words is this kind of language!
38. The author supposes "that fire may have its inhabitants." Very poetical!
60. The author seems to be perfectly in the secrets of the Creator. He makes as many Celestial bodies as *he* can find room for . . . He tells us which is *the most perfect plan.*
64. The author now is so fond of his comets that he finds it necessary to apologize for the existence of planets.
79. Now we also have traveling globes for astronomers and the authors [sic] pity for not being upon one of them.
140. The *auther* [sic] now uses all the licence [sic] of the poets in the flights of fancy. He confesses that it makes his head giddy, and that he does not know where to stop. I do not call this Astronomy, but wild imagination.[74]

This selection, which includes all Herschel's notes on Lambert's pluralism, is interesting also for what it does not contain, that is, any explicit criticism of Lambert's extreme teleological approach. This suggests that Herschel was not put off by Lambert's pluralism, but only by the form he gave it. Although Herschel had observational reasons for rejecting Lambert's cometary inhabitants, his negative reaction to this idea was probably primarily due to the fact that in his own evolutionary cosmology he had found another teleological justification for comets: as mechanisms for the rejuvenation of stars (I, p. 478).

If one looks broadly at Herschel's career, at least three questions come to mind that Herschel scholars have perhaps not fully answered. First, what inspired him to forsake music for astronomy? It cannot have been a desire to discover (as he did) a new planet; for weeks after detecting Uranus he identified it as a comet. Nor was it a determination to become (as he did) the founder of observational stellar astronomy; this field scarcely existed in the early 1780s.[75] Second, what drove him to unparalleled efforts to build giant telescopes, the usefulness of which was scarcely obvious in an age dominated by positional astronomy? As recent research has shown, Herschel's scientific contemporaries at first viewed both this stress on large telescopes and his overall conception of astronomy as peculiar.[76] Third, what led King George III not only to bestow an annual salary on Herschel but also to lavish 4,000 pounds on the construction of a telescope of unprecedented size? The present analysis of Herschel's extraterrestrial life ideas suggests a partly conjectural reconstruction of his career that may shed light on these questions. Herschel's entry into astronomy can be attributed in good part to the appeal of the pious pluralism present in the works of such authors as Ferguson. Captivated by Ferguson's claims for lunar life, Herschel boldly if naively sought to detect it directly, being encouraged by his early, albeit ambiguous, observations. This hope fired his passion for improved telescopes, for which he also found other uses. As the circle of his astronomical associates widened, he learned that they looked askance at his extraterrestrial endeavors. Whether or not Herschel, after Maskelyne's rebuke, shared his hope to detect lunar life with others besides Wilson is unclear, but one can conjecture that the munificence of the monarch may have been motivated by Herschel confiding to the king that his discovery of the Georgian planet was only a prelude to a more dramatic discovery toward which his progress had been halted by the inadequacies of his instruments. Such a secret conversation may have contributed to the king's delusion, when beset in 1788 by mental illness, that he could see Hanover through Herschel's telescopes.[77]

This reconstruction is admittedly conjectural; what is certain is Herschel's lifelong commitment to ideas of extraterrestrial life. This first led to observational efforts but gradually gave way to what Simon Schaffer has

described as "his relentless pursuit of the material conditions for extraterrestrial life. . . ."[78] The theories Herschel developed in this later
quest, such as that of life on the sun, may ultimately have been no more
successful than his earlier endeavors. Nonetheless, they supplied encouragement to a generation or more of pluralists, including (as will be
shown) Herschel's brilliant son John. Lest this analysis of Herschel's
career be misunderstood, it is important to ask whether Herschel's extraterrestrial life ideas justify labeling him a "lunatic," as he feared Maskelyne might do. At a time when historians of science have demonstrated
the depth of Kepler's commitment to Pythagoreanism and Newton's
commitment to alchemy, such a label is certainly unsuitable. Herschel,
like Kepler and Newton, was a man of remarkable genius and *a man of
his time*. The legitimacy of the latter point will become more evident as
we encounter many other eighteenth-century intellectuals scarcely less
committed to pluralism than Herschel.

3. Herschel's continental contemporaries: Schröter and Bode, Laplace and Lalande

Had a late-eighteenth-century earthling sought to assemble his planet's
five most prominent astronomers to debate the extraterrestrial life question, he might well have chosen Herschel, Schröter, Bode, Laplace, and
Lalande. Although flawed by the fact of the group having five for the
affirmative, this plan would have resulted in a lively controversy, because, as this section shows, the latter four developed the pluralist doctrine in directions that diverged not only from Herschel but also from
each other.

Johann Heironymous Schröter (1745–1816) has sometimes been
called "the Herschel of Germany,"[79] because in many ways his career
paralleled that of Herschel, from whom he had secured some of his first
telescopes. By the mid-1790s, Schröter, the chief magistrate of Lilienthal
in northern Germany, had overseen the construction of his own 18.5-
inch-aperture reflector, the largest then on the Continent. His decades of
diligent observation ceased only in 1813 when invading French forces
pillaged his observatory, destroying his records. Planetary and lunar observations were his passion, the latter leading to his two-volume *Selenotopographische Fragmente* (1791, 1802), the most detailed study of the
moon up to that time. More than a mapping of the moon, it is filled with
discussions of lunarians, or "selenites," as Schröter, following Hevelius,
called them. In the first volume he sets the stage for selenites by frequent
claims for a lunar atmosphere, saving his main arguments for its final
three sections, where Schröter confesses to being

. . . fully convinced *that every celestial body may be so arranged physically by the Almighty as to be filled with living creatures organized conformably to its physical plan and praising the power and goodness of God, and that the infinite grandeur of the Creator ought to be glorified in the analogous multiplicity of the physical arrangement of the celestial bodies as it is also certainly revealed in the infinite variety of their living creatures.*[80]

In support of this claim he cites similarities among the planets, suggesting that to assert that life is confined to the earth would be comparable to proclaiming that in a forest only one of numerous similar trees is fruitful.

All this, as well as his argument from ancient and modern authorities (p. 671), is thoroughly traditional; what appear remarkable are his subsequent observational claims. After suggesting that were we transported to the summit of a lunar mountain, we would encounter sights showing the solicitude of the Almighty for the selenites, he adds:

I at least imagine . . . the [lunar] landscapes of *Plato* and *Newton* together with the adjacent gray surface of Maris Imbrium to be just as fruitful as the Campanian plain [of Italy]. Here nature has ceased to rage, there is a mild and beneficial tract given over to the calm culture of rational creatures, who . . . give thanks for the fruit of the field and perhaps only fear that Mount Blanc and [certain] cratermountains may cause new disorders through new eruptions and overflow many moon cottages. At least the southern region of Mount Blanc has many similarities in general with the Phlegrean plains, and the moon Alps terminate frequently, with the new small crater just as the Appenines of our Italy with Vesuvius. (p. 693)

Lest such ideas appear "far fetched fantasies," he reminds his reader of their empirical support, suggesting that the color changes he had reported observing on the moon may, "not improbably," be ascribed to alterations in the areas of the moon under cultivation. So also, it is perhaps the case that

. . . many a small spot . . . is a constructed habitation of the rational inhabitants of the moon; and perhaps in that and in [lunar] industries . . . lies the explanation why many such an object is frequently invisible under identical angles of illumination, but then, when visible, appears sometimes bright, sometimes dark. . . . Just such a changing appearance would be given by many a populous terrestrial city frequently covered by fog, were it observed by an eye on the moon. (p. 674)

Eleven years later, in the second volume of his *Selenotopographische Fragmente,* Schröter continued to argue for his "selenites"; in fact, under this term its index refers to no less than fifteen sections relevant to lunar life.

In 1792, Schröter published a long paper in the *Royal Society Transactions* in which he puts forward claims for a lunar atmosphere based on a February 24, 1792, observation of a lunar twilight, as well as observations indicative of Venusian mountains "of such enormous height, as to exceed 4, 5, and even 6 times the perpendicular elevation of Cimboraco, the highest of our mountains."[81] Schröter's observation of a lunar twilight is now known to have been an illusion, and even in 1793 his Venusian observations were challenged by Herschel, who reobserved Venus, agreeing that it has an atmosphere, but asserting concerning Schröter's Venusian mountains "that neither want of attention, nor a deficiency of instruments, could occasion my not perceiving *these mountains of more than 23 miles in height. . . .*"[82] In this paper Herschel advocates an opaque Venusian atmosphere, making the surface features of Venus largely invisible. Schröter's 1795 rejoinder contains arguments for a "generally clear and transparent" atmosphere for Venus. Among these is one in which Schröter's pluralist convictions, never far beneath the surface, come to the forefront. He states that Venus must have a transparent atmosphere because

I cannot think . . . that Providence would bless the inhabitants of Venus, incomparably less than with the happiness of seeing the works of almighty power, and of discovering, like a Herschel, still more and more distant regions of the universe. We must . . . adhere to this analogy, till indisputable experiments convince us of the contrary. . . .[83]

This passage reveals an author whose excessive readiness to interpret planetary and lunar detail was rivaled by his rashness to specifying the ways of the Divine.

In 1796, Schröter compiled his Venusian researches into the first book ever written about that planet. Here his pluralist convictions also appear, for example, in his repetition in essentially identical wording of the first passage quoted from his book on the moon.[84] Schröter's *Kronographische Fragmente* (1808) displayed once again his tendency to attribute earthlike features to the planets. Joseph Ashbrook, who recently examined this rare work on Saturn, described it as "an odd blend of honest observing records and extravagantly wrong conclusions from them. Its author maintains stoutly that the ring system is a solid body, studded here and there with mountains, and possessing its own atmosphere."[85] Next came Schröter's *Hermographische Fragmente* (1815–16), which bestows on Mercury an atmosphere, large mountains, and a terrestrial period of rotation as erroneous as that he had earlier found for Venus.[86] In 1881, Schröter's observations and analyses of Mars were located and published. It is ironic that in the case of Mars, which has permanent features detectable by his telescopes, Schröter attributed what features he saw to windblown cloud formations that he believed hid the true features of the planet.[87]

Schröter's large telescopes and energetic observation of planetary and lunar surfaces may justify the description of him as "the Herschel of Germany," but Ashbrook proposed a more suitable sobriquet for Schröter: "the Percival Lowell of his age."[88] Moreover, it may be suggested that the tragedy of Schröter's life was less the overrunning of his observatory by French forces than the insidious influence of an insufficiently developed critical sense. Like Herschel, he was a pluralist with much imagination; unlike his more famous contemporary, Schröter never learned that large telescopes and diligent observation are not of themselves sufficient to transform an amateur into a professional astronomer.

Among late-eighteenth-century advocates of extraterrestrials, few, if any, championed them with more frequency, fervor, or influence than Johann Elert Bode (1747–1826), whose expository skills were already evident when in 1768 he published an introduction to astronomy entitled *Deutliche Anleitung zur Kenntniss des gestirnten Himmels*. His calculatory capabilities were recognized by Lambert, who in 1772 brought him to the Berlin Academy. This led to his becoming editor of the *Astronomisches Jahrbuch* for over five decades, as well as director from 1786 of the Berlin Observatory. Such credentials gave credibility to claims he made for the pluralist position, which he adopted at least as early as the 1772 second edition of his *Anleitung*, to which he added a pluralist chapter (discussed later) as well as the numerical approximation of planetary distances now known as Bode's law. In 1776 he proposed a model for the sun similar to that put forward two decades later by Herschel. After attributing a protective layer to the sun and inhabitants to its supposedly cool core, Bode describes the sun as "a dark planetary body which as our earth consists of land and water and exhibiting on its surface all the unevenness of mountains and valleys and also surrounded up to a certain height by a thick atmosphere."[89] Concerning solarians, he asks:

Who would doubt their existence? The most wise author of the world assigns an insect lodging on a grain of sand and will certainly not permit . . . the great ball of the sun to be empty of creatures and still less of rational inhabitants who are ready gratefully to praise the author of their life.

Its fortunate inhabitants, say I, are illuminated by an unceasing light, the blinding brightness of which they view without injury and which, in accordance with the most wise design of the all-Good, communicates to them the necessary warmth by means of its thick atmosphere. (p. 246)

In 1778, Bode published his *Kurzgefasste Erläuterung der Sternkunde und den dazu gehörigen Wissenschaften*, a more advanced presentation of astronomy than his *Anleitung*. Together, these books "dominated German astronomical literature for the next half-century"[90] and simultaneously championed such an extreme form of pluralism that Bode's position has been labeled "un panpopulationnisme cosmique."[91] Essentially every celestial body — sun or star, planet or satellite, and even comets — Bode populates with rational beings. For example, in his *Erläuterung*, after stressing the similarities of planets to the earth, Bode asks: "should not this correspondence extend also to habitability which is the most important goal of creation? With what spurious reasons can this still be contested, given that the ancient philosophers and astronomers who lacked as many proofs as we have, believed in the plurality of inhabited worlds?"[92] In response to those who deny life on the planets because of the greatly different intensity with which solar rays fall on them as compared with the earth, Bode urges that atmospheric conditions strongly influence the surface temperature of each planet, making life possible. He also asserts that the variety in terrestrial life indicates "that the rational inhabitants, and even the animals, plants, etc. of the other planetary bodies are characterized by forms different from those which occur on our earth." Nonetheless, he assures his readers that "rational beings" on the planets "are ready to know the author of their existence and to praise his goodness." (pp. 375–6) In support of life on the planetary satellites, he stresses the similarity of the moon to the earth and also that each primary reflects more light to its satellite than it receives from it. The teleological basis of his position is obvious when he urges that, were "the immensely large ball of the sun desolate and bereft of rational and living creatures, the intentions of the Eternal would be restricted. . . ." (p. 376) Elsewhere in his *Erläuterung*, this protégé of Lambert argues for cometarians, who, he suggests, "wander with their dwelling place from the sun to the farthest limits of its province, and can consequently observe it from far distant points and from different sides. . . . Who can conceive what special arrangements of the wise Creator in regard to the climate, zones, dwelling places, sectioning of creatures, natural products, may not be expected for all those on a cometary body?" (p. 491) Finally, he stresses that intelligent life, so widespread in our solar system, is similarly distributed in the other systems scattered throughout space (p. 509). The pluralist impact of his *Anleitung* and *Erläuterung* was enhanced during this period, when Bode added such extensive notes to the German translation (1780) of Fontenelle's *Entretiens* made by W. C. Mylius that he doubled the size of that volume.

Bode's pluralist efforts continued into the nineteenth century. For example, his *Allgemeine Betrachtungen über das Weltgebäude* (1801) is a republication as a short book of the pluralist chapter he had added to the second edition of his *Anleitung*. His attachment to that essay, as well as its popularity with the public, led Bode to revise it for publication as the third and concluding part of his *Betrachtung der Gestirne und des Weltgebaudes* (1816). Bode begins the 1816 version of this long essay by stressing God's beneficence in bestowing a variety of forms of life on the earth, following this by a survey of the solar system. Concerning our moon he reveals that certain parts "can be taken for cultivated lands, forests, etc"; moreover, in some areas, "Schröter has discovered traces of incidental physical alterations, which may indicate natural upheavals and a culture perhaps organized by its inhabitants."[93] Bode's acceptance of Schröter's claims is also evident elsewhere, as in his discussion of the supposed mountains and valleys of Mercury and Venus (pp. 356, 360). Moreover, Bode had published some of Schröter's papers in his *Astronomisches Jahrbuch* and would soon publish even less reliable pluralist observational claims made by Gruithuisen. Schröter's assertions about Mercury and Venus contribute to Bode's statement that if on the planets "lands and seas, mountains and valleys are present and clearings and condensations occur in their atmospheres . . . ; if they have several moons as companions, etc; so it is well established that they are entirely similar to our earth and consequently likewise habitable." (p. 365) That teleological considerations contributed no less than telescopic observations to this conclusion is suggested by Bode's assertion that

If they had no inhabitants . . . , what would their purpose and destiny be, and what else could we take to be the intentions of the Creator for all these great and wise plans and arrangements? Perhaps it was [in order] that they might decorate the starry heavens here and there with bright points? Certainly not. . . . Never! How would this be reconciled with his wisdom, which always and exactly selects means in accord with ends. . . . (p. 365)

Although stressing the empirical character of his own proofs for pluralism, Bode recommends that his readers also examine the more obviously metaphysical justifications for pluralism put forth by E. G. Fischer in a 1792 paper (discussed later) that Bode had published in his astronomical journal (p. 365).

After attributing life to the planetary satellites, Bode argues for cometary inhabitants, on whom, he suggests, "either the very unequal actions of the sun make no impression, or the goodness of the Creator has affected arrangements to provide them with security against these extraordinary alterations." (p. 371) Nor does he neglect the sun:

The sun itself can be populated. Suppose it an actual fireball; still its habitability remains possible according to the inexhaustible design of eternal omnipotence and wisdom. Or suppose, in accordance with the probably more correct opinion, it to be an electrified, fireless sphere enveloped in a condensed etherial mass of light; then will the habitability of the amazingly broad area of its surface be conceivable to us, and it cannot be lacking in inhabitants. These fortunate solar citizens [although] . . . illuminated almost incessantly by the light material, will remain undazzled, cool, and secure in the midst of the sun's glare [protected] under the shadow of the Almighty. (p. 372)

Bode even bestows windows on his "Sonnenbürgern" by suggesting that sunspots may provide them a view of the rest of the universe (p. 373).

Bode's "panpopulationnisme cosmique" extends to the stars, among which our sun is "one of the smallest." (p. 379) Bode, who was one of the earliest advocates of the multiple universes of Kant, Lambert, and Herschel, selects from their writings the most pluralistic features, opting for Kant's infinite universe over Lambert's finite universe, but retaining Lambert's inhabited comets. Kant's cosmology was especially in the background of Bode's statement that

Perhaps there are celestial bodies which are inhabited by imperfect beings such as we are, yet others, and presumably the majority, can be occupied by inhabitants of higher faculties of soul and greater agility of body. Does the idea, which Lambert, Kant, Bonnet and other philosophers assert, appear unreasonable that the mental energies [Seelenkräfte] of rational creatures can undergo not imperceptible alterations in accordance with the different degree of the fineness of the corporeal matter which surrounds their thinking natures, that these depend upon the different distances of the planetary spheres from the central point of their system, and improves with increasing distances from it, so that there is an orderly gradation of perfections of organic and living creatures on the planets of our own and all other solar systems[?] (pp. 392–3)

The evolutionary cosmogony of Kant rather than the static cosmology of Lambert was surely the source of Bode's admission that perhaps not all worlds are presently habitable, some being in the process of formation or destruction (p. 402). But he goes beyond both Kant and Lambert and indeed beyond astronomy itself when he speculates on the center of the universe:

Who knows but there glows at this middle point a more than terrestrial sun and that there the omnipotence of God shines in highest brilliance? . . . From here the hand of the Eternal at the beginning of all things formed the suns with their spheres. . . . From here, all suns, world systems, and milky ways receive their most excellent systematic arrangement. . . . From here . . . the omnipresence of the universal monarch of worlds governs and cares for person and angel and for the worm as well. . . . (pp. 399–400)

The richness of his imagery and the piety of his pluralism show through when he adds:

Struck through with a holy and reverential shudder, I think back to that time before time existed, to when nothing except God, the All-Sufficient and Necessary One, was, to the time when the visible began. – The prime matter of nature still chaotically slumbered. – It pleased the infinite Creator to present a glimpse of his splendor and greatness, and the universe resulted. His wisdom chose among all possible worlds the best, and the breath of his mouth brought it to reality. The Eternal made around the foot of his throne suns without number, alloted and counted out for each its spheres, and millions of spirits of high lineage were witnesses and admirers of these splendid creations. (p. 400)

What explains the remarkable appeal of the writings of Bode, whose *Anleitung* has been described as "the most widely read astronomical work in Germany until the middle of the nineteenth century"?[94] One of the sources of his success is that he embellished his treatises on the technicalities of astronomy with a poetically pleasing and religiously relevant pluralist message drawn from the most extreme extraterrestrial life ideas of Kant, Lambert, Herschel, and Schröter. When the history of popularization of astronomy is written, Bode will occupy a prominent place in it; moreover, as is illustrated in this history, Bode's practice of boldly blending pluralist themes with astronomical detail will be seen not as an isolated strategy but as a recurrent pattern for astronomical authors who have attained widespread public appeal.

The mathematical dexterity displayed by Pierre Simon Laplace (1749–1827), most notably in the five volumes of his *Mécanique céleste,* won him accolades from his former pupil Napoleon, as well as the title "the Newton of France." In fact, in his celestial mechanics, Laplace had gone beyond Newton by demonstrating what Newton had doubted: the long-term stability of the solar system. Moreover, in the various editions of his *Exposition du système du monde,* first published in 1796, Laplace developed what Newton had specifically denied as a possibility: a theory of the formation of the solar system. Known as the "nebular hypothesis," Laplace's theory explained the solar system as resulting from the rotation, contraction, and condensation of a primitive solar fluid.[95] Because the nebular hypothesis implied that planets are an expected result of the evolution of stars, it provided powerful support for the pluralist contention that stars are in most cases encircled by planets.

Laplace presented the nebular hypothesis in the concluding chapter of his *Exposition,* endorsing pluralism in the same chapter. His arguments for planetarians are quite traditional, depending primarily on analogies between the earth and planets. He also states: "it is not natural to suppose that matter . . . should be sterile upon a planet so large as Jupiter. . . ."[96] He shows more restraint than such astronomers as Kant, Lambert, and Bode; for example, in regard to the physical form of planetarians, he restricts himself to suggesting that they ought to exhibit "a diversity of organization suited to the various temperatures" of the planets (II, p. 355). Laplace then cautiously presents his nebular hypothesis, drawing from it the conclusion that the stars are "suns, which may be the foci of many planetary systems. . . ." (II, p. 366) He concludes his book by stating that although astronomy has shown man to be "almost imperceptible in the vast extent of the solar system, itself only an insensible point in the immensity of space," man should be aware that the "sublime results to which [astronomical] discovery has led, may console him for the limited space assigned him in the universe." (II, p. 374) The similarity of Laplace's suggestion to the paradox, presented more than a century earlier by Pascal, that man, despite his physical insignificance, can comprehend the vast universe, should not obscure the radical difference between the cosmic conceptions of these two French authors. Diminished if not destroyed is Pascal's contrast between the minuteness of man and the magnitude of mindless matter, for in the immense spaces of Laplace's universe, living beings are profusely scattered.

Deeper changes were also under way, as can be illustrated from a discussion of Laplace's religious views. Laplace and Napoleon periodically conversed on astronomical topics; for example, while sailing to Egypt, Napoleon asked Laplace and other scientists on board for their views on whether or not the planets were inhabited and how the universe was formed. Although by no means an orthodox Christian, Bonaparte's distress at some of the antitheist remarks of the group provoked him to point to the stars and exclaim: "You may say what you like, but who made all those?"[97] The story is also told that after Napoleon had perused Laplace's *Exposition du système du monde,* he commented to Laplace: "Newton has spoken of God in his book. I have already gone over yours and I have not found this name a single time." To this, Laplace responded: "Citizen First Consul, I have no need for that hypothesis."[98] Laplace's assertion raises the question whether or not he had forsaken the Catholicism in which he had been raised. George Sarton, in 1941, suggested that Laplace's response should be seen only as a statement that in his book he had been able to explain the leading features of the universe without recourse to God operating as advocated by Newton — acting periodically to restore order to the supposedly unstable solar system.[99] Such an interpretation seems difficult to sustain in light of a document in Laplace's handwriting uncovered in 1955 by Roger Hahn, who concluded that Laplace had become a severe critic of revealed religion, that he had come to base "his faith not upon revelation of the Holy Book, but rather on his creed of the invariability of natural laws."[100] Hahn also asserted that Laplace's views of Christianity moderated in the years before his death, but overall the evidence points to the probability of an agnostic element in the astronomer's thought. This is supported by the account of another exchange between Napoleon and Laplace preserved by William Herschel, who met with them during an 1802 visit to France. After asking Herschel about the construction of the heavens, Napoleon became involved in an argument with Laplace. According to Herschel:

The difference was occasioned by an exclamation of the first Consul, who asked in a tone of exclamation or admiration (when we were speaking of the extent of the sidereal heavens): "And who is the author of all this!" Mons. De la Place wished to shew that a chain of natural causes would account for the construction and preservation of the wonderful system. This the first Consul rather opposed. Much may be said on the subject; by joining the arguments of both we shall be led to "Nature and nature's God."[101]

Tensions between astronomy and religion were also felt by Jérôme Lalande (1732–1807), whose early training came from the Jesuits of Lyon. In 1760 he became professor of astronomy at the Collège royale and editor of the *Connaissance des temps,* which he edited until 1776, and again from 1794 until his death. In addition to presenting more than 150 papers to the Paris Academy, he authored many books, including his *Traité d'astronomie,* first published in two volumes in 1764, and later republished in expanded form. His numerous writings and stimulating lectures greatly expanded interest in astronomy among the French. Indeed, it is for this, more than for his research contributions, that he is remembered.

A Frenchman of the 1790s, wishing to learn something of astronomy, might well turn to the third edition of Lalande's massive *Traité d'astronomie,* published in 1792 in three volumes under the title *Astronomie.* There he would find not only much technical astronomy but also a section on the plurality of worlds. This begins with a historical discussion in which Lalande claims that the "greatest philosophers" have espoused this doctrine and that "the resemblance is so perfect between the earth and the other planets that if one admits that the earth was made to be inhabited, one cannot refuse to admit that the planets were made for the same purpose. . . ."[102] Given all the similarities of the planets to the earth, "Is it," he asks, "possible to suppose that the existence of living and thinking beings is limited to the earth? On what would this privilege be founded . . . ?" (III, p. 353) Not only does Lalande urge that the planets are inhabited, he also notes that Gowin Knight had suggested that the sun and stars could themselves sustain inhabitants (III, p. 353). Turning to the religious tensions associated with pluralism, Lalande states:

> There have been some writers, as timid as they are religious, who have condemned this system [pluralism] as contrary to religion; this was a bad way to promote the glory of the creator. If the extent of his works announces his power, can one supply any idea more magnificent and more sublime? We see with the naked eye many thousands of stars; an ordinary telescope reveals many more in every region of the sky. . . . [I]magination pierces beyond the telescope; it sees a new multitude of worlds infinitely larger. . . . (III, p. 354)

These apparently pious sentiments are followed by a caution concerning the argument for pluralism from final causes, Lalande suggesting not that pluralism is questionable, but only the inference that the planets were designed to be inhabited (III, p. 354).

Lalande's efforts to popularize astronomy led him to publish in 1800 an edition of Fontenelle's *Entretiens,* complete with a biographical introduction on Fontenelle and astronomical annotations of the test. In praising Fontenelle's achievements, especially his efforts to make science better known, Lalande reveals his special debt to that author: ". . . I am indebted to him for the germ of that insatiable activity of mind I have experienced ever since the age of sixteen."[103] The remainder of Lalande's

introduction consists essentially of a republication of the pluralist section
of his *Astronomie.*

About the orthodoxy of Laplace's religious beliefs, doubts remain, as
has been noted; concerning Lalande's convictions, the evidence is conclusive. At least as early as 1800 he publicly proclaimed his atheism in the
preface he wrote for Sylvain Maréchal's *Dictionnaire des athées.* Moreover, in that period he took to signing his letters "Lalande doyen des
athées."[104] In the 1806 supplement to the *Dictionnaire des athées,* which
Lalande himself edited, the astronomer revealed: "The spectacle of the
heavens appears to all the world a proof of the existence of God. This I
believed for nineteen years; today I see only matter and motion."[105]

Lalande, and possibly Laplace, bring this chapter to a dramatic conclusion. Whereas the six authors – Wright, Kant, Lambert, Herschel, Schröter, and Bode – analyzed earlier in this chapter proclaimed their delight in
seeing the action of the Almighty in a universe rich in extraterrestrials,
Lalande, no less a pluralist, stated: "I have searched through the heavens,
and nowhere have I found a trace of God."[106] Lest Lalande, the self-
proclaimed "dean of atheists," be dismissed too soon from the side of the
angels, it is interesting to ask whether any of the eight, except Lalande
and possibly Laplace, would have endorsed an earlier author's critique of
physicotheological proofs for God's existence. Writing in an ironic vein,
this author stated: "I marvel at the daring with which these people undertake to speak of God. In addressing their argument to unbelievers, their
first chapter is to prove the existence of God from the work of
nature. . . . [Finally] they give them, as a complete proof [of God], the
course of the moon and planets. . . ." And he added: "Nothing is more
calculated to arouse their contempt."[107] What gives point to this passage
is that it is from Pascal's *Pensées,* possibly the most profound defense of
Christianity ever written. Although Pascal's attack was aimed only at
those claiming that astronomy supplies proof of God, not at believers
perceiving the presence of the Divine in nature, his remarks raise the
question whether or not the confidence with which, for example, Bode
wrote of "the intentions of the Creator" may have acted to discredit the
religious position he sought to support.

3

Extraterrestrials and the Enlightenment

1. The idea of a plurality of worlds in Britain: "One sun by day, by night ten thousand shine"

Some remarks on the eighteenth century as the Age of Enlightenment will prepare the way for the discussion in this chapter of the pluralist debate in the second half of the eighteenth century. F. L. Baumer has stated: "The Enlightenment . . . , as Ernst Troeltsch and many others were later to say, was the hinge on which the European nations turned from the Middle Ages to 'modern' times, marking the passage from a supernaturalistic-mythical-authoritative to a naturalistic-scientific-individualistic type of thinking."[1] The claim that man's view of himself and the universe changed dramatically in the Enlightenment fits with the fact that pluralism evolved during the eighteenth century from being seen as a bizarre and irreligious speculation to becoming an accepted doctrine of astronomy and natural theology. Nonetheless, this chapter will demonstrate that this transformation took place at different times and in diverse ways in various countries. Moreover, support will be found for those who have stressed that Christian conceptions continued to play a powerful role in the eighteenth century, as well as for those such as Carl Becker who have suggested that a number of the differences between medieval and eighteenth-century thought are more apparent than real.[2]

Literate Londoners living in the 1750s, had they chanced upon Thomas Wright's book, would have found nothing out of the ordinary in his advocacy of extraterrestrials. They would already have encountered such creatures in Pope's *Essay on Man*, which was but one of many poems promoting pluralism during the preceding decades. In the 1720s, for example, two Scottish poets employed pluralist themes. David Mallet (1705?–65) included a celestial second canto in his "Excursion" (1728), describing the heavens in which

> Ten thousand suns blaze forth, each with his train
> Of worlds dependent, all beneath the eye
> And equal rule of one eternal Lord.[3]

The great chain of being as well as man's difficulty in comprehending this cosmos are given expression as Mallet asks, concerning these worlds,

> What search shall find
> Their times and seasons! their appointed laws
> Peculiar! their inhabitants of life,
> And of intelligence, from scale to scale
> Harmonious rising and in fix'd degree;
> Numberless orders, each resembling each,
> Yet all diverse!

Moreover, Mallet's universe is changing, as one or another sun, "grown dim with age," dies, an event that

> Millions of lives, that live but in his light
> With horror see, from distant spheres around
> The source of day expire, and all his worlds
> At once involv'd in everlasting night!

James Thomson (1700–48) was nearly simultaneously composing *The Seasons,* one of the most widely read physicotheological poems of the century. Therein Thomson describes the sun as "Soul of surrounding worlds!" and as

> Informer of the planetary train!
> Without whose quickening glance their cumbrous orbs
> Were brute unlovely mass, inert and dead,
> And not, as now, the green abodes of life!
> How many forms of being wait on thee . . . ![4]

Although Thomson's influence on Mallet is widely recognized, it was from Mallet (if ultimately from Newton) that Thomson derived his reference to comets as serving "To lend new fuel to declining suns,/To light up worlds, and feed the eternal fire."[5]

During the 1730s, the array of pluralist poems further increased. Robert Gambol, in his anonymous *Beauties of the Universe* (1732), wrote that after death the soul "Will clearly view what here we darkly see:/Those planetary worlds, and thousands more,/Now veil'd from human sight, it shall explore."[6] Two years later, Henry Baker (1698–1774), best known as a microscopist, published *The Universe: A Poem Intended to Restrain the Pride of Man.* An enthusiast for the great chain of being, he compares the revelations of the microscope and telescope, suggesting that both reveal the abundance of life. The latter instrument shows that

> Nor can those other worlds, unknown to this,
> Lest stor'd with Creatures, or with Beauty be,
> For God is uniform in all his Ways,
> And everywhere his boundless Pow'r displays.[7]

The teachings of the telescope and microscope also inspired the Irish poet Henry Brooke (1703–83), who in his *Universal Beauty* (1735) described a cosmos characterized by both unity and variety, a universe

> Where countless orbs through countless systems shine; . . .
> Yet boundless all those worlds that roll within:
> Each world as boundless as its native race,
> That range and wanton through its ample space. . . .[8]

Lunar life was not forgotten; no less a poet than Thomas Gray (1716–71), while studying at Cambridge in 1737, wrote of it in Latin verse, predicting in his "Luna habitabilis" that our satellite would in time become a British colony.[9] Moses Browne (1704–87), who embellished his "Essay on the Universe" (1739) with technical footnotes, thought travel to the moon absurd, but of pluralism he approved, populating the planets

> With creatures, suited to their various seat,
> Intense degrees of cold or heat to bear
> Of light, or gloom, a pleasing, proper share,
> To them agreeable, by nature blest,
> Painful, howe'er, imagin'd to the rest.[10]

Browne was aware of both the scientific problems and religious difficulties of pluralism; these he suggests by asking whether other planetary systems were

> Form'd when the world at first existence gain'd?
> And to one final period all ordain'd?
> Or, since wide space they independent fill,
> Apart created, and creating still?
> Do scriptures clear, the aw'd assent oppose?
> They chiefly *our* original disclose.
> Do they assert, ere we in being came,
> God ne'er was own'd by the Creator's name?[11]

As an Anglican of evangelical inclination, Browne no doubt felt certain that such questions posed no irresoluble problems for Christianity; others suspected the reverse. In fact, the irony of such physicotheological poetry is that in spirit it was frequently deist, even if this was far from the design of those who wrote it. Whereas God the Almighty was celebrated in numerous verses, Christ the Redeemer scarcely appeared. This was the case, and probably the intention, in Alexander Pope's 1738 paraphrase of the "Lord's Prayer," which includes the caution:

> Yet not to Earth's contracted Span
> Thy goodness let me bound,
> Or think Thee Lord alone of man,
> When thousand worlds are round. . . .

Whereas Pope called this "The Universal Prayer," some of his contemporaries dubbed it "The Deist's Prayer."[12] This serves to suggest what later discussions will document: that the rapid assimilation of pluralism to religion in the eighteenth century was not unrelated to the spread of deism.

Of all eighteenth-century English poems using scientific materials, only one rivaled and possibly surpassed Pope's *Essay on Man* in popularity. This was *Night Thoughts,* published between 1742 and 1745 by the already sixtyish rector of Welwyn, Edward Young (1683–1765). Translated into French, German, Italian, Magyar, Portuguese, Spanish, and Swedish, it was, according to one of Young's biographers, "for more than a hundred years . . . more frequently reprinted than probably any other book of the eighteenth century."[13]

Young's *Night Thoughts* is divided into nine "Nights," in the last of which the author makes a final attempt to reform the libertine Lorenzo by means of "A Moral Survey of the Nocturnal Heavens." Whereas other physicotheological poets were finding in flora and fauna the finest field for proving the Deity, Young above all seeks Him in the celestial:

> Devotion! daughter of Astronomy!
> An undevout astronomer is mad.
> True; all things speak a God; but in the small,
> Men trace out Him; in great, He seizes man. . . .[14]

Young's universe was thoroughly pluralistic, being constructed, if not from Derham, at least on Derham's "New Systeme":

> One sun by day, by night ten thousand shine,
> And light us deep into the Deity. . . .
> (IX, 748–9)

It was also Newtonian; he tells Lorenzo:

> Then mark
> The mathematic glories of the skies,
> In number, weight, and measure, all ordain'd.
> (IX, 1079–81)

But it was not Lucretian:

> Know'st thou th' importance of a soul immortal?
> Behold this midnight glory: worlds on worlds!
> Amazing Pomp! redouble this amaze;
> Ten thousand add; add twice ten thousand more;
> Then weigh the whole; one soul outweighs them all. . . .
> (VII, 993–7)

Nor, at first sight, seems earth to hold a privileged place:

>. . . what swarms
>Of worlds that laugh at earth! immensely great!
>Immensely distant from each other's spheres!
>What, then, the wondrous space through which they roll?
>At once it quite engulfs all human thought;
>'Tis comprehension's absolute defeat.
>
>> (IX, 1102–7)

And the poet asks in regard to our planet:

>Canst thou not figure it, an isle, almost
>Too small for notice in the vast of being;
>Sever'd by mighty seas of unbuilt space
>From other realms; from ample continents
>Of higher life, where nobler natives dwell. . . .
>
>> (IX, 1603–7)

Such ideas raised a host of questions, scientific and religious, concerning the extraterrestrials encountered late in "Night Ninth":

>What 'er your nature, this is past dispute,
>Far other life you live, far other tongue
>You talk, far other thought, perhaps, you think,
>Than man. How various are the works of God?
>But say, what thought? Is Reason here Inthroned,
>And absolute? Or Sense in arms against her?
>Have you two lights? or need you no reveal'd? . . .
>And had your Eden an abstemious Eve? . . .
>Or if your mother fell, are you redeemed? . . .
>Is this your final residence? If not,
>Change you your scene, translated? or by death?
>And if by death; what death? Know you disease?
>
>> (IX, 1766–81)

Young leaves many of these questions unanswered, but not all. In fact, whereas other pluralists of the period usually eschew the delicate dilemmas raised by pluralism for Christianity, Young commits himself. His extraterrestrials, undefined physically, are specified spiritually:

>Each of these stars is a religious house;
>I saw their altars smoke, their incense rise;
>And heard hosannas ring through every sphere. . . .
>
>> (IX, 1881–3)

How far, one may wonder with Lorenzo, does such cosmic piety extend?

>Ten thousand worlds, ten thousand ways devout,
>All nature sending incense to the Throne,
>Except the bold Lorenzos of our sphere?
>
>> (IX, 1898–900)

Young, definitely no deist, answers this query in his concluding apostrophe to the Holy Trinity. God the Son is introduced with the words

> And Thou the next! yet equal! Thou, by whom
> That blessing was convey'd; far more! was bought;
> Ineffable the price! by whom all worlds
> Were made; and one redeem'd! . . .
> Thou God and mortal! thence more God to man! . . .
> Who disembosom'd from the Father, bows
> The heaven of heavens, to kiss the distant earth!
> Breathes out in agonies a sinless soul!
> Against the cross, Death's iron sceptre breaks!
>   (IX, 2262–5, 2348, 2352–5)

Young's poem came to be cherished by readers as diverse as Napoleon, John Wesley, who edited an edition of it, William Blake, who prepared illustrations for it, and countless cottagers, who placed it beside their Bible and Bunyan. Although Young's fame scarcely survived his century, his Christian vision of a fallen earth rolling among a host of inhabited but sinless orbs was given new life in the nineteenth century by Thomas Chalmers and in the twentieth by C. S. Lewis. Pluralism, a delight to deists, was, as the instances of Young, Chalmers, and Lewis show, not without appeal to traditional Christians.

One year after the publication in 1750 of the first complete edition of Young's *Night Thoughts*, William Hay (1695–1755), a Whig member of Parliament, presented a strikingly different attempt to formulate a Christian conception of the cosmos. Entitling his treatise *Religio Philosophi: or, The Principles of Morality and Christianity Illustrated from a View of the Universe, and Man's Situation in It* (London), Hay makes clear from his first page his agreement with the Copernicans, who "with the greatest Probability (almost Certainty) imagine each Fixed Star to be a Sun with Planets . . . surrounding it . . . ; and all such Planets to be inhabited as well as the Earth. . . ." Moreover, Hay claims that ". . . Praise and Thanksgiving are continually ascending to [God's] Throne . . . from every Quarter of the Universe," forming thereby "a general Religion, a joint Communion, a Universal Church." This thought, he suggests, "bids us love them, not as our own Species, but as our Fellow-creatures, and as Members of the same Church and Communion. . . ." (pp. 14–15) Although Hay leaves unspecified how we are to love these extraterrestrials with whom we can have no contact and whose very existence is problematic, he does assert it as "most probable" that they, like us, violate "The Universal Principle[s] of Morality," necessitating that God recall them "to a Sense of their Duty [by means] which may be as various as the Globes themselves. . . ." (pp. 15–17) In particular, he suggests that God acts on some planets

. . . as a tremendous Judge, . . . in others as an indulgent Parent. In some only alarming, in others correcting; in some destroying, in others rebuilding; in some changing only individuals, in others entire Species; and exalting the Rational Creatures of each Globe from a Material to a Spiritual, from a Mortal to an Immortal State; transforming them into Angels; and from those Seminaries perpetually increasing the Host of Heaven. (pp. 34–5)

Although Hay states that "the Moral Transactions in one Globe are distinct from those of another . . ." (p. 17), he speculates that some of the changes wrought by God on a planet may be visible elsewhere: "when our own Globe was overflowed . . . in the days of Noah, Astronomers in Mars, Venus, or the Moon might observe it darker . . . ; and perhaps they will hereafter observe it of a more ruddy Complection at the time of the [earth's] general Conflagration. . . ." (pp. 20–1) Traditional Christians, reading this far in any of the four editions that Hay's book attained, probably found much that was new but little that was objectionable. This soon changed.

Hay's book becomes daringly different and heterodox when, in discussing the redemption, he states that the second person of the Trinity, concerned "to instruct, redeem, judge and govern the Race of Adam . . . united to himself Jesus: and for the same or similar Ends he may have, and probably hath, united to himself other rational Creatures in other Planets. Jesus is not their Saviour, nor they the Saviours of Mankind. . . ." (p. 139) Hay's claim that the idea of a multiply incarnated God had been little discussed before was correct, but he was surely wrong in explaining this as simply due to the dominance of the Ptolemaic system and to fear of repression (pp. 141–4). The idea of turning Christ into a cosmic Krishna was then, and remains, a notion that Christian theologians, as well as such critics of Christianity as Tom Paine, have judged to be irreconcilable with that religion.[15] However this may be, Hay's book illustrates the fact that the idea of a plurality of worlds was in the air, available to those who would use it. And a number of authors employed it in even more radical ways than he had done.

The demise of Henry St. John, Viscount Bolingbroke (1678–1751) silenced, it seemed, a man whose eloquence had led his Tory associates to select him as secretary of state and whose brilliance had disposed Pope to describe Bolingbroke as his "Guide, Philosopher, and Friend." However, the publication in 1754 of the five volumes of Bolingbroke's *Philosophical Works,* consisting largely of pieces not published previously, provoked Samuel Johnson to label him "a scoundrel, and a coward: a scoundrel, for charging a blunderbuss against religion and morality; a coward, because he had not the resolution to fire it off himself, but left half a crown to a beggarly Scotchman, to draw the trigger after his death!"[16]

Bolingbroke's "blunderbuss" was fired partly by pluralist powder. For example, in a postscript to the second of his "Essays Addressed to Alexander Pope," Bolingbroke advocates the pluralist "hypothesis . . . which we find in the Cosmotheoros of Mr. Huygens, and from which Fontenelle has borrowed the materials of his pretty book of the Plurality of worlds."[17] The historical insensitivity shown by Bolingbroke in attributing an influence on Fontenelle's book to a volume published a dozen years after it is partially redeemed by his correctly noting that the pluralist doctrine was of ancient origin. The enthusiasm of the statesman who had negotiated the Treaty of Utrecht for the hypothesis that there exist "numberless worlds [which are] habitations of ten thousand millions of intellectual corporeal beings" (II, p. 145) stems, he states, from its potential for fostering "a sort of truce" between the philosopher and the divine (II, pp. 153–4). Both can agree that the range of "sense and intelligence" from animal to man on this planet implies that "this evidently unfinished intellectual system [must be] continued upwards" to beings bridging the gap between man and God (II, p. 145). The theologian may thus accept the great chain of being, but he must forsake his "legions of angels, of demons and genii, and of pure and impure spirits which pagan theology invented, and Jews and Christian adopted." (II, p. 145) To encourage abandonment of the angelic orders, Bolingbroke urges that it is easy to conceive of corporeal creatures "tempered in a finer clay, cast in nobler moulds, than the human, and animated by spirits more subtle and volatile than ours. . . ." (II, p. 147) Admitting that his pluralism is "purely hypothetical," he claims that it is "less liable to have . . . absurd notions and practices grafted upon it, as have been grafted on the other." (II, p. 147) And he adds that the pluralist position, unlike the "pneumatical" (spiritual), is supported by "a plain, direct, and unforced analogy. We know that there are habitations: and we assume that they are inhabited." (II, p. 149)

The fertility of pluralism in providing utopias is illustrated when in a later essay Bolingbroke stresses man's difficulty in "uniting under one form of government, or submitting to one rule of life" by suggesting in contrast that "All the inhabitants of some other planet may have been, perhaps, from their creation united in one great society, speaking the same language, and living under the same government; or too perfect by their nature to need the restraint of any." (IV, pp. 45–6) In a more theologically oriented section, Bolingbroke attacks theologians who assert that "man and the happiness of man, were the final causes of the creation. . . ." (IV, pp. 317–18) This view must be surrendered, because modern astronomy shows that we "cannot doubt that numberless worlds and systems of worlds . . . inhabited by living creatures . . . compose this amazing whole, the universe. . . ." (IV, p. 320) To the divine's argument that the rationality of man makes him a sufficiently elevated entity to be worthy of God having created the universe for him, Bolingbroke responds by returning to his claim, based on the great chain of being, that higher orders probably exist elsewhere within the *material* universe (IV, p. 320). Bolingbroke's deism and materialism are evident in his apparent rejection even of the notion that the universe was created for intelligent beings. What he suggests concerning the earth, he would presumably apply elsewhere as well; after noting that man is fit to inhabit the earth, he asks: "But will it follow, that the planet was made for him, not he for the planet? The ass would be scorched in Venus or Mercury, and be frozen in Jupiter or Saturn. Will it follow, that this temperate planet was made for him, to bray and eat thistles in it?" (IV, p. 319) Later in the same essay the Tory statesman recommends a sort of cosmic resignation or optimism, based partly on pluralism and the great chain of being. His argument is that because every creature in the universe "is adapted . . . to the place he is to inhabit . . . ," it is foolish for an individual to aspire to excellences of sense or intellect beyond those appropriate to his position in that design or chain (IV, pp. 379–80). Indeed, "the general state of mankind in the present scheme of providence is a state not only tolerable, but happy." (IV, p. 386) A "general providence" Bolingbroke accepts, but a "particular providence," especially as carried out by angels, he rejects, urging once again that the chasm between man and God is bridged not by angels but by extraterrestrials (V, pp. 83–93).

Perhaps Dr. Johnson need not have been so apprehensive; Bolingbroke's "blunderbuss" seems to have discharged with less force than feared. David Hume, far more friendly to the viscount's views than Johnson, remarked: "The Clergy are all enrag'd against him; but they have no Reason. Were they never attack'd by more forcible Weapons than his, they might forever keep Possession of their Authority." Moreover, by the 1790s Edmund Burke could ask derisively, "Who now reads Bolingbroke? Who ever read him through?"[18]

To Burke's second query, these researchers have uncovered an answer: Samuel Pye, M.D. The evidence for this comes from the curious if ingenious volume that Pye published in 1765 with the title *Moses and Bolingbroke,* and with the intent of employing pluralism against Bolingbroke.[19] Determined to save Moses from Bolingbroke's barbs in his *Philosophical Works,* Pye contrasts the views of "Moses, in *his* Pentateuch, and Bolingbroke, in *his* . . ." (p. vi), using for this is a dialogue between Bolingbroke befuddled to find himself in an afterlife of which he "had not the least notion" (p. 1) and a Moses maddened at the measure to which Genesis "has been *metaphrased, paraphrased, explained, obscured, allegorized, spiritualized.* . . ." (p. 40) Pye, who provides Bolingbroke in most cases with lines from his *Works* and other writings, has his Moses interpret Genesis as an account applicable to the formation of our entire solar system, not simply of the earth. He also adds that "the inhabitants of Jupiter, Saturn, and every other planet of the finite system, may . . . be supposed to have the foundation of their subjection to the same Lord, discovered to them in a similar manner . . . ; that is, by a succinct account of the creation of the same system, and the particulars of the formation of their respective planets." (pp. 58–9) To illustrate this point, Pye provides some sections of a Jupiterian Genesis, a few lines of which may be cited:

GENESIS Chap. I
1. In the beginning God created the heaven and Jupiter.
16. And God made five great lights; the greater light to rule the day, and the lesser lights to rule the night: and stars also.
Chap. II
2. And on the fifteenth day God ended his work which he had made: and he rested on the fifteenth day from all his work, which he had made.
3. And God blessed the fifteenth day, and sanctified it. . . .[20]

At the end of Pye's publication, a duly impressed and much mortified Bolingbroke recants.

Pye's playful yet seriously intended volume secured only a single printing, but Pye supplemented it in 1766 by his *Mosaic Theory of the Solar, or Planetary, System,* designed to show that the Mosaic account of Creation can be reconciled with a scientific theory of the formation of the solar system. Both of these works seem to be almost unknown to historians of astronomy, Bolingbroke scholars, and later participants in the extraterrestrial life debate. The loss, however, has not been great.[21]

A recent brief survey of the history of the extraterrestrial intelligence concept raises the question whether or not any eighteenth-century figures rejected the pluralist position.[22] The answer to this query is that a substantial number of authors from that century either doubted or rejected the doctrine. To cite one example, noteworthy because it is a group rather a single person, mention may be made of the Hutchinsonians. These were the followers of John Hutchinson (1674–1737), who in his *Moses's Principia* (1724–7) called into question Newtonian science and advocated a system of his own, based partly on his belief that Old Testament Hebrew, if read without points, would reveal the ways of the universe and of God. In fact, as one of the scholars who have recently studied this little known group put it, "The hands may have been those of Esau, but the voice was that of Descartes."[23] Their anti-Newtonianism extended to rejection of Sir Isaac's active principles, action at a distance, and void interstellar space. Religiously they tended to be Augustinian, scriptural literalists, and High Church. The movement attracted some figures of prominence, including Alexander Catcott, Duncan Forbes, Bishop George Horne, William Jones of Nayland, and John Parkhurst.

Duncan Forbes (1685–1747), as lord president of Scotland's Court of Session, was perhaps the most distinguished and also among the earliest Hutchinsonians. His antipluralist position is evident even from the opening pages of his *Reflexions on the Sources of Incredulity with Regard to Religion* (1753), where he states:

> It is rash to affirm, that the universe, or even the solar system, was made principally for the sake of the earth, or of man. . . . It is at the same time not certain . . . that there was any other use for creating these immense heavenly bodies, but to regulate the motion of the earth; to produce the other effects which some of them evidently have, and all of them in a greater or smaller proportion may have on the earth; and to raise in man that idea of the magnificence, power, and skill of the Creator. . . .[24]

Despite this statement, Forbes adds: "But we cannot deem it impossible, that beings may have been made, fit to reside, to act, and to think, that in the very centre, as well as on the surface of the sun." (p. 2) John Parkhurst (1728–97) was a Cambridge-educated biblical scholar, best known for his *English and Hebrew Lexicon* (1762). The format of this book was capacious enough that in his discussion of a certain Hebrew word he described Kepler's *Somnium,* commenting: "What Kepler proposed as a dream, Huygens, and a long list of Kepler's Newtonian followers, have treated as a reality, or at least as a high probability!"[25] At another point, Parkhurst suggests: "The modern philosopher, who imagines the moon and planets to be inhabited worlds, and the fixed stars suns to other systems, may perhaps find enough to awaken him from this amusing and delusive dream, in the excellent Mr. Baker's *Reflections on Learning,* chap. viii . . . or in *Catcott on the Creation*, p. 20."[26] The second of the authors referred to by Parkhurst is Alexander Catcott (1725–79), who learned his Hutchinsonianism from his father and then fostered it at Oxford while studying there in the 1740s. It was possibly through the younger Catcott that his fellow Oxford students George Horne (1730–92) and William Jones (1726–1801) became part of the movement. Horne later became president of Magdelen College, vice-chancellor of Oxford, and eventually bishop of Norwich. Jones became curate of Nayland and extended the scientific side of the Hutchinsonian cause in a number of books. The latter task was also taken up by Catcott, who became a scriptural geologist very active in field work.[27] Catcott came to write the 1756 book referred to by Parkhurst because he was disturbed by claims made by the bishop of Clogher, Robert Clayton (1695–1758), in a volume aimed at answering Bolingbroke.[28] Clayton having endorsed pluralism in his book, Catcott included in his response a section attacking that doctrine, which he felt had been used against

Christianity by deists. Drawing on Scripture, especially Genesis and Revelation, Catcott asserts that the visible universe "was made for the service of man alone; and therefore that the planets and fixed stars have no peculiar inhabitants of their own. . . ." (p. 24) Then, responding directly to the propluralist arguments of Clayton, Catcott discusses the determinations of solar parallax present in the astronomical literature of that period. Finding these very discordant, he urges that Clayton's assertion that the stars are immensely distant and hence suns is dubious at best, Catcott favoring the view that "the fixed stars are only so many *specula* [mirrors]. . . ." (p. 30) Catcott even more vigorously objects to Clayton's claims that pluralism "may be of great use, in *abating* our *pride,* and *exalting* our *notions* of the *great Creator.* . . ."[29] To the first claim, he responds that pluralism in fact fosters pride "by giving room to the wildest genius to indulge his extravagant fancy in acting the god and making (out of his own little head) an *infinity of worlds.*" (p. 31) To Clayton's second claim, Catcott urges that pluralism, rather than exalting our notion of God, will "tend in weak minds to lessen the idea of his *goodness* and *concern* for *man;* and so introduce infidelity and atheism in the world. And I am sorry to say it, that several of our modern philosophers have been these weak men, and have argued against christianity from this very circumstance." (p. 32)

It is a curious fact of history that it was only in 1817, after the Hutchinsonian movement had largely died out, that the antipluralist portion of its program was given to its fullest expression. In that year a London bookseller, Alexander Maxwell, devoted a long book to arguing against pluralism, drawing heavily on Hutchinsonian sources. In his book, Maxwell cites Catcott as typical of the group, adding, however, that "Some few among them, I believe, consider the planets as inhabited; but this is not generally the case."[30] The scientifically unsophisticated character of Maxwell's book (discussed subsequently) prevented it both from wounding the pluralist cause and from resuscitating the Hutchinsonian system. However, the effect of the Hutchinsonians was not limited to those who adopted their entire position. Some selectively borrowed from the movement, as the next figure, involved in creating his own movement, illustrates.

Little of signficance for history would seem discernible in the sight of a smallish man in his mid-fifties riding to London in 1759 with his horse at slack rein that he might read. Yet when this journey is part of the quarter of a million miles that the man was riding to foster a revolution, when the book is Huygens's *Celestial Worlds Discover'd,* and the man is John Wesley (1703–91), then the historian may take notice.[31] At that time Wesley was three decades distant from his ordination in the Church of England and two decades removed from that May 24, 1738, meeting on

Aldersgate Street when he felt his heart "strangely warmed," an event that Lecky labeled the beginning of an "epoch in English history." So energetic was Wesley that he preached forty thousand sermons; so eloquent were these that from the "holy Club" of his Oxford days there arose the Methodist church, which eight decades after his death numbered twelve million members worldwide. Wesley's *"Methodist Revolution,"* as Bernard Semmel called it in his recent book of that title, was undoubtedly of great signficance, but historians still dispute whether it was "repressive and regressive" or "liberal and progressive."[32] Not unrelated to this question is that of Wesley's attitude to the sciences[33] and specifically to the pluralist position.

Concerned to improve the level of learning among his ever increasing flock, Wesley in 1758 started to compose the two-volume work that appeared in 1763 as *A Survey of the Wisdom of God in the Creation, or a Compendium of Natural Philosophy.* Although basing it on a Latin treatise by Johann Franz Buddeus (1667–1729), Wesley made it substantially his own by including materials from Derham, Hutchinson, Ray, and others. It was probably in preparation for his *Survey* that Wesley in 1759 read Huygens's volume, his curious reaction to which he recorded in his *Journal:*

He surprised me. I think he clearly proves that the moon is not habitable; That there are neither:

Rivers nor mountains on her spotty globe:

That there is no sea, no water on her surface, nor any atmosphere. And hence he rationally infers that "neither are any of the secondary planets inhabited." And who can prove that the primary are? I know the earth is. Of the rest I know nothing. (II, p. 515)

That Wesley, who was originally a pluralist (XIII, p. 396), should see Huygens's rejection of lunar life as reason for jettisoning the entire pluralist position may in part be explained by the supposition that Wesley's readings prior to 1759 in the Hutchinsonians predisposed him to the reversal of mind recorded in the foregoing quotation.[34]

Wesley's *Survey* advocates, albeit cautiously, an antipluralist position. His main argument is largely an elaboration of the previously cited statement from his *Journal,* bolstered by the quotation of the passages from Huygens's book wherein he explained the evidence against lunar life. Because he believed that Huygens had refuted the possibility of life on our moon as well as on the moons of other planets, Wesley concluded: "I doubt that we shall never [*sic*] prove that the Primary are: And so the whole ingenious Hypothesis, of innumerable Suns and Worlds moving round them, vanished into Air."[35] Wesley's antipluralist position quickly provoked a controversy in the *London Magazine.* The long 1765 letter

that Wesley published in response to his critic begins with an admission
of some factual errors and an explication of some unclarities, but soon
Wesley sets about rebutting his opponent's claims. One of the most
important of these concerns whether or not astronomers had proved that
the stars are so distant that they must be solar sized and hence suns.
Wesley's response is that no judgment can be made on this until "parallaxes can be taken with greater certainty than they are at present." (XII,
p. 397) Wesley also opposes the claim that the moons of Jupiter and of
Saturn and Saturn's ring count as evidence of the habitability of those
two outer planets (XIII, pp. 397–8). The major thrust of Wesley's response to this attack on his work of natural theology is ironically to
encourage caution in making assumptions about God's design in the
universe. This is typified in the following passage, wherein Wesley first
quotes, then answers, his opponent:

"They who affirm, that God created those bodies, the fixed stars, only to give
us a small, dim light, must have a very mean opinion of the divine wisdom." I do
not affirm this; neither can I tell for what other end He created them: He that
created them knows. But I have so high an opinion of the divine wisdom, that I
believe no child of man can fathom it. It is our wisdom to be very wary how we
pronounce concerning things which we have not seen. (XIII, p. 398)

In short, Wesley admonishes his adversary: "Be not so positive. . . ."
(XIII, p. 399) Wesley's confidence in the arguments in this letter led him
to republish it in a number of subsequent editions of his *Survey*.

Wesley's fullest statement on the question of a plurality of worlds
appears in a sermon composed late in his life, probably in 1788, entitled
"What Is Man?"[36] The text for this sermon Wesley gives as: " 'When I
consider thy heavens, the works of thy fingers, the moon and the stars,
which thou has ordained; what is man?' Psalm viii 3,4." (VII, p. 167)
After elaborating on the minuteness of man within the vastness of space
and the eternity of time, he urges first that we should consider "what
David does not appear to have taken into his account; namely, that the
body is not the man; that man is not only a house of clay, but an
immortal spirit which is of infinitely more value . . . than the whole
material creation." (VII, p. 171) Second, Wesley characteristically recalls
God's injunction: "As the heavens are higher than the earth, so are my
thoughts higher than your thoughts. . . ." (VII, p. 171) Then he stresses
that God's concern for man was so great that "God gave his Son." (VII,
p. 172) This third response leads Wesley to discuss what was for Christianity the most delicate problem arising from pluralism:

"Nay," says the philosopher, "if God so loved the world, did he not love a
thousand other worlds, as well as he did this? It is now allowed that there are
thousands, if not millions, of worlds, besides this in which we live. And can any
reasonable man believe that the Creator of all these, many of which are probably

as large, yea, far larger than ours, would show such astonishingly greater regard to one than to all the rest?" (VII, p. 172)

To combat this unidentified philosopher, Wesley formulates two arguments. The first is that even if millions of inhabited worlds exist, God in his "infinite wisdom" presumably had good reasons "to show his mercy to ours, in preference to thousands or millions of other worlds." (VII, p. 172) Second, Wesley repeats in abbreviated form his Huygens-based antipluralist arguments (VII, pp. 172–3).

In describing Wesley's attitude to pluralism as negative or at least cautious, a number of qualifications are necessary. For example, his position did not preclude him from including in his *Survey* three pages of pluralist religious prose by his Calvinist friend Rev. James Hervey (1714–58). These were from Hervey's "Contemplations on the Starry Heavens," which appeared in his *Meditations and Contemplations* (1745–7), an extremely popular work. Although Wesley altered these passages so as to simplify their ornate prose and to soften their pluralist proclamations, he preserved their main message.[37] Further pluralist passages were introduced into his *Survey* when, for its third edition (1777), Wesley included book-length extracts from both Bonnet's *Contemplation de la nature* and Dutens's *Inquiry into the Origin of the Discoveries Attributed to the Ancients*. Bonnet's pluralism (discussed later) was linked to the great chain of being idea, the latter being so appealing to Wesley that he himself made the translation of the extracts included. Dutens's thesis that the ancients possessed many modern scientific notions was carried too far, but he was correct in asserting that the idea of a plurality of worlds had ancient roots.[38] A final example of Wesley's involvement with pluralist writings is that in 1770 he edited an abridged edition of Young's *Night Thoughts*.[39]

Was Wesley's view of astronomy "regressive" or "progressive"? Certainly this religious leader deserves credit for encouraging the study of science by writing his *Survey* and two other scientific books. However, when he warns in the concluding section of his *Survey* that "there is reason to fear that even the Newtonian, yea, and Hutchinsonian system, however plausible and ingenious . . . are yet no more capable of solid, convincing proof, than the Ptolemaic or Cartesian" (XIII, pp. 490–1), he reveals his limited capability for judging the science of his day. His disdain for theory and for mathematics was also regressive. However, if the concluding section of his *Survey,* which he titled "Remarks on the Limits of Human Knowledge,"[40] be looked at broadly, it is possible to see it as a justified and timely call to scientists and especially natural theologians to "Be not so positive. . . ." The astronomical excesses of Kant, Lambert, and especially Wright are reminders of the dangers of dogmatism in science, and John Dillenberger's statement in his historical *Protestant*

Thought and Natural Science that "natural theology was mainly responsible for the demise of Christianity in many areas"[41] suggests that Wesley's wariness of an excessive rationalism in natural theology was not
misplaced, especially in a nation that would soon face the Huttonian and
Darwinian revolutions. In this regard, it is interesting and ironic to note
that for the 1815 edition of Wesley's *Survey,* B. Mayo of Philadelphia
carried out a revision and updating. Mayo's emendations were most
extensive in the astronomy section, which he found "abounding" in
"obsolete . . . ancient opinions." Mayo's updatings include not only the
removal of Wesley's *London Magazine* letter and his quotations from
Huygens but also the rewriting of the section so as to advocate the
doctrine of a plurality of worlds and, what was surely regressive, life on
the moon.[42]

   Whereas some called him the "Socrates of Edinburgh" or even "St.
David" (Voltaire), the majority probably preferred Boswell's "Great Infidel" or even John Wesley's description of him as "the most insolent
despiser of truth and virtue that ever appeared in the world."[43] The man
who appeared so differently to his contemporaries was David Hume
(1711–76), whose 1742 essay "The Sceptic" contains two remarks that
if not directly opposed to pluralism, at least convey his concern about an
implication frequently drawn from it. Hume's first statement, the second
being similar, is:

> *Nothing can be more destructive,* says FONTENELLE, *to ambition, and the
> passion for conquest, than the true system of astronomy. What a poor thing is
> even the whole globe in comparison of the infinite extent of nature?* This consid-
> eration is evidently too distant ever to have any effect. Or, if it had any, would it
> not destroy patriotism as well as ambition?[44]

The depth of Hume's opposition to many of the teachings of traditional
religion became clear with the publication three years after his death of
his *Dialogues Concerning Natural Religion,* which contains a powerful
attack on the argument for God's existence from the design in nature.
Among the religious leaders upset at its contents was John Wesley, who
criticized Hume in one of his last sermons.[45] It is a curious feature of
Hume's *Dialogues,* however, that the position adopted in it on the pluralist question is closer to that of Wesley than to that of Bolingbroke. In an
attempt to show that we cannot legitimately infer an intelligent designer
of the universe from the supposed presence of intelligent beings throughout the universe, Hume has Philo urge:

> Is there any reasonable ground to conclude, that the inhabitants of other planets
> possess thought, intelligence, reason, or anything similar to these faculties in
> men? When nature has so extremely diversified her manner of operation in this
> small globe; can we imagine that she incessantly copies herself throughout so
> immense a universe?[46]

Shortly thereafter, Philo admits that similarities have been observed between the earth and other planets, but he urges that what is required to show the presence of intelligence in them is not these similarities, but rather similarities in their processes of formation (pp. 150–1). In a parallel argument, Philo claims that despite the presence of intelligence on earth, for all we can know it is possible that "Many worlds might have been botched and bungled, throughout an eternity, ere this system was struck out. . . ." (p. 167) Although the dialogue form of this book as well as Hume's tendency to employ irony make interpretation of his writings open to question, these passages suggest that on the question of intelligent life elsewhere, David Hume and John Wesley, who differed so sharply on many matters, were nearly at one. But the Augustinian Wesley and the analytical Hume shared more than that: Both were skeptical of the natural theology then so dominant. Perhaps that is why both doubted the pluralist doctrine that had come to form one of the arches in the impressive but frail edifice that eighteenth-century physicotheologians were erecting.

In 1747, as Wesley made the first of his forty-two journeys to preach in Ireland, a reclusive Swedish nobleman settled in London. Both, although unaware of it at the time, were in the process of preparing the way for major new religious denominations. When Wesley learned of his fellow Londoner's ideas, he labeled him a "madman," whereas Kant called him the "the archvisionary of all visionaries." Nonetheless, both Wesley and Kant took the views of Baron Emanuel Swedenborg (1688–1772) with sufficient seriousness that each composed a critique of his remarkable claims.[47] Their attacks, although not without effect, did not prevent a group of Swedenborg's disciples from founding in London in the 1780s the Church of the New Jerusalem, which now has a worldwide membership of about thirty thousand, nor did their criticism deter such intellectuals as Blake, Goethe, Emerson, Hugo, Henry James, Sr., and Yeats from taking a strong interest in Swedenborg's system.

Born in Stockholm, Swedenborg attended Upsala University, where his father had taught Lutheran theology before becoming bishop of Skara. After an extended European tour that included three years in England, where the scientifically inclined Swedenborg met Flamsteed and Halley and studied the works of Newton, he was appointed to a position on the Swedish board of mines. This he held between 1716 and 1747, publishing during that period an array of scientific books, the merits of which have been rather differently assessed by subsequent scholars, but the influence of which most agree was small.[48] In 1734, Swedenborg published his *Opera philosophica et mineralia,* the first volume of which contained his *Principia rerum naturalium,* a work more in the Cartesian than in the Newtonian tradition. Ever since 1879, when the Swedish astronomer Magnus Nyren claimed that Swedenborg in his *Principia* had proposed early forms of both the nebular hypothesis and the disk theory of the Milky Way, historians of astronomy have debated these claims, generally concluding against Swedenborg.[49] The value of his physiological and psychological doctrines as expounded in his two-volume *Oeconomia regni animalis* and in the three volumes of his *Regnum animale* is also a subject of controversy, but these works do show his great energy, extensive knowledge, and increasing concern with the more spiritual and philosophical aspects of nature.

The transition from Swedenborg-scientist to Swedenborg-seer came in the 1740s. During that decade he not only published his *De cultu et amore Dei,* an account of the Creation derived from biblical and scientific sources, but also began to undergo extraordinary religious experiences. Whereas Wesley in 1738 felt his heart "strangely warmed," Swedenborg in 1743 and thereafter experienced almost unparalleled visions. In his 1769 autobiographical letter to Rev. Thomas Hartley, Swedenborg described the "circumstance, that I have been called to a holy office by the Lord Himself, who most mercifully appeared before me, His servant, in the year 1743; when He opened my sight into the spiritual world and enabled me to converse with spirits and angels, in which state I have continued up to the present day."[50] Among the earliest of the numerous theological books that flowed from the mind of the thereby transformed Swedenborg was his *Arcana coelestia* (1749–56), the eight volumes and 3 million words of which were devoted to revealing the hidden allegorical meaning of Genesis and Exodus. Written in Latin and published anonymously in London, it secured few readers, and even they must have been shocked at its author's prefatory remark that "it has been granted me now for some years to be constantly and uninterruptedly in company with spirits and angels, hearing them speak and in turn speaking with them."[51]

The relevance of Swedenborg's *Arcana coelestia* to the extraterrestrial life debate is that scattered throughout it are sections recounting his conversations with the spirits and angels of each of the planets of our system, as well as with those of our moon and of five "earths" in other solar systems. These sections he gathered together into a small 1758 volume entitled *De telluribus in mundo nostro solari.* . . . When in 1787 Rev. John Clowes (1743–1831), one of the founders of the Church of the New Jerusalem, published his English translation of this work, he gave it the title *Concerning the Earths in Our Solar Systems Which Are Called Planets; and Concerning the Earths in the Starry Heavens; Together with an Account of Their Inhabitants, and Also of the Spirits and Angels There; From What Has Been Seen and Heard.*[52] Although pluralist passages occur in a number of Swedenborg's other books, this work is his main contribution to pluralist thought.

Swedenborg's *Earths* begins with its author's statement that "since I had an ardent desire to know whether there were other earths . . . , it has been granted me by the Lord to speak . . . with spirits and angels who are from other earths, with some for a day, with some for a week, with some for months. . . . It has now been granted me to enjoy this privilege for twelve years." (#1) These spirits and angels, he reveals, are all "from the human race," meaning that those associated with each planet are the departed spirits of the "humans" who lived on that planet (#1). Not only are no angels or spirits directly created, but also the "humans" from which they come are, at least physically, similar to earthlings, some, however, being smaller or larger or of different colors or manners of walking. Teleological considerations are used in conjunction with the observed similarities of planets to the earth and of the stars to our sun to urge the likelihood of life elsewhere (#3–4). The similarity of his extraterrestrials to the inhabitants of our planet combines with the scantiness of his descriptions of the physical features of other planets to suggest that Swedenborg's scientific erudition entered but slightly into his composition. An apparent exception may be seen in his lunarians, who speak very loudly and "from the abdomen, and thus from some collection of air therein; the reason of which is, that the Moon is not surrounded with an atmosphere of the same kind as that of other earths." (#111) Repeatedly Swedenborg expounds on his microcosm-macrocosm doctrine of the "Grand Man" making up his entire finite universe. This "Grand Man" in turn corresponds to the Lord. The inhabitants of Mercury, for example, correspond to memory, a faculty in which they excel (#10), whereas his lunarians correspond to the "ensiform or xiphoid cartilage." (#111) These characterizations seem to be remnants of earlier astrological associations.[53]

Although Swedenborg frequently includes references to the horses, cows, and goats of other planets and to the houses of his extraterrestrials, he is more concerned to recount their moral and mental characteristics as well as their forms of worship and social organization. The beings he encounters elsewhere are generally morally and spiritually superior to earthlings, although he makes clear that we are not alone in having harlotry, robbery, idolatry, and savagery. However, in Swedenborg's vision, our earth is unique in one respect:

. . . it pleased the Lord to be born, and to assume the Human, on our Earth, and not on any other. THE PRINCIPLE REASON *was for the sake of the Word, that it might be written on our Earth; and when written might afterwards be published throughout the whole Earth; and that, once published, it might be pre-*

served for all posterity; and thus it might be made manifest, even to all in the other life, that God did become Man. (#113)

A striking feature of this is that Christ's communicative function in his terrestrial incarnation seems to be given primacy over his redemptive role. The communication of God's Word beyond our earth is made possible through angels and spirits of our planet convening with their counterparts of other planets, who in turn, under special conditions, talk with the living inhabitants of their planets. A curious concomitant of the coming of Christ as human only to our planet is that writing and hence science are to be found nowhere else in the universe (#136, 155).

If Swedenborg's book is compared to Dante's *Divine Comedy*, one cannot but be struck by the detachment evident in the former's narrative: Nary an exclamation escapes Swedenborg's lips as he recounts his extraordinary visions. Although on his journey he encounters some individuals (e.g., Christian Wolff) and groups (e.g., Jesuits) of whom it is clear that he does not approve, there is far less politics – and also less poetry – than in the Italian's epic. The messages conveyed in Swedenborg's *Earths* are characteristically moral – be like the deeply spiritual inhabitants of Mars or those of the third of the five "earths" he sees beyond our solar system – or doctrinal. In the latter regard he repeatedly expounds his less than orthodox doctrine of the Trinity. In considering this contrast between Dante and Swedenborg, one must not lose sight of the fact that the latter meant his tale to be taken completely literally. As Svante Arrhenius commented in discussing these "strangest fantasies," "Swedenborg was an extraordinarily conscientious man, and there can be no doubt, that he believed what he maintained."[54] Because of this, Swedenborg would not be pleased by Marjorie Hope Nicholson's statement that "Swedenborg's religious writings are permeated with his reading in cosmic voyages and in the science and philosophy out of which they sprang."[55]

Assessing the significance of Swedenborg's *Earths* in his overall religious doctrine and its dissemination is difficult. On the one hand, his followers relatively rapidly translated this brief book into other languages. By 1770 it was available in the German of F. C. Oettinger, by 1787 in English by John Clowes, and French and Italian translations followed in the nineteenth century. Moreover, the extraordinary character of his pluralist claims helps explain why one of his late-nineteenth-century disciples lamented: "Much ridicule has been heaped upon the New Church because its members believe the other planets to be intended, like this globe, for human habitation. Perhaps there is no point to which our people are so sensitive as this, for they continually hear scornful remarks, as if the single fact of our holding this belief were enough to discredit all else."[56] On the other hand, it seems to have been those cases in which Swedenborg made apparently correct predictions of terrestrial events – the Stockholm fire, Queen Ulrika Eleonora's secret, or his own death – that attracted attention to this reclusive author. His advocacy of extraterrestrials, admittedly extraordinary in its form, was nothing new to the eighteenth century. However, in the long run, as the evidence against life on our moon and on the planets of our systems has mounted, the credibility of the visions of the Swedish scientist-turned-seer has not been increased.

Swedenborg was, of course, on the periphery of British religious and social life; that was scarcely true of Soame Jenyns (1704–87), a longtime member of Parliament, a religious writer, and a minor poet. In "An Essay on Virtue" (1752), Jenyns employs pluralism to urge that God's goodness extends throughout the universe:

> Hence soul and sense, diffus'd through ev'ry place,
> Make happiness as infinite as space;
> Thousands of suns beyond each other blaze,
> Orbs roll o'er orbs, and glow with mutual rays;
> Each is a world, where, form'd with wondrous art,
> Unnumber'd species live through ev'ry part. . . .

To strengthen his point, he adds that plants, animals, and higher beings everywhere are governed by God's eternal law, which "is only this/That all contribute to the general bliss."[57] In a prose "Disquisition" of 1782 entitled "On the Chain of Universal Being," Jenyns draws on pluralism to fill out the upper reaches of that chain.[58]

That not only poets but also preachers could espy in the heavens of pluralist astronomy rich resources from which to embellish their writings is well illustrated by "On the Glory of God, as Displayed by the Heavenly Luminaries," a sermon by the Unitarian divine Newcome Cappe (1733–1800). A reading of this sermon suggests that it was untypical of its time only in the elegance of its prose. Not only Cappe's York congregation but also those of many other areas of Britain must have been urged to an appreciation of God's grandeur by allusions to the order and harmony of the planetary motions and the vastness of the universe. Many of his contemporary clergymen no doubt also revealed that

> With a very few exceptions, every star . . . is another sun unto another system; placed in the centre of many worlds, and affording unto each . . . their proper measure both of light and heat. . . . If so many suns, how many worlds? If so many worlds, what numbers can express the inconceivable multitude of their inhabitants? all of them the creatures of divine power, the monuments of divine wisdom, the objects of divine love! – Think then . . . how many myriads of unnumbered worlds are at that moment rejoicing in the goodness of their Maker. . . . Of all these innumerable worlds, that one on which we live, vast as we conceive it, is among the least that we behold.[59]

The better educated preachers of this period, in elaborating the pluralist theme, may have shown more restraint than the dissenting divine whose discussion of the benefits we receive from the moon concludes with the reminder that our planet more than repays those benefits by its services to the inhabitants of the moon (p. 342). Nevertheless, few of even the most scientifically severe of the day would have objected to Cappe's praises for God's positioning of our planet at just that distance from the sun that provides our mild climates (pp. 346–7). Another feature of Cappe's sermon was his eschewal of efforts to answer the pluralist objection to Christianity. This gives his sermon a simplicity that no doubt added to its effectiveness in inspiring a feeling for God's glory. By the time of Cappe's death, if not before, these difficulties were becoming increasingly hard to avoid.

James Beattie (1735–1803) is now primarily known as one of Scotland's more prominent eighteenth-century poets, but his profession was teaching moral philosophy and logic at Marischal College, Aberdeen. In fact, it was through two of his philosophical and apologetic works that, in the eyes of his contemporaries, he achieved his greatest triumphs. His *Essay on the Immutability of Truth* (1770) employs the Scottish Common Sense philosophy of Thomas Reid (1710–96)[60] to refute the skepticism of Hume, whereas his *Evidences of the Christian Religion* (1786) assails those who doubt that Christianity can be reconciled with reason and common sense. In the concluding section of the latter book, Beattie responds to those who cite the smallness of man's planet in the vast universe as reason to ask "is it possible to imagine that such creatures as we are can be of so great importance, as that the Deity should send his Son . . . into this little world. . . ?"[61] The first of Beattie's replies is to stress God's ability to create "this boundless universe, with all the variety of beings it contains." (p. 179) Beattie's second rejoinder centers on and cites Richard Bentley's statement that "the soul of one virtuous man is of greater worth and excellency, than the sun and his planets, and all the stars in the world" (p. 181), whereas his final reply posits extended effects from the redemption. He states that extraterrestrials "will not suffer for our guilt, nor be rewarded for our obedience. But it is not absurd to imagine, that our fall and recovery may be useful to them as an example; and that the divine grace manifested in our redemption may raise their adoration and gratitude into higher raptures and quicken their ardour to inquire . . . into the dispensations of infinite wisdom." (p. 184) Moreover, he suggests that this view is "not mere conjecture [but] derives plausibility from many analogies in nature; as well as from holy writ, which represents the mystery of our redemption as an object of curiosity to superior beings, and our repentance as an occasion of their joy." (p. 184) Of course, this attempt to rationalize by Common Sense philosophy "the mystery of our redemption" was in fact "mere conjecture" and was hardly relieved by Beattie's speculation that by "comets

. . . it is perhaps probable, that our solar system may be connected with other systems." (p. 183)

The high regard in which Beattie's *Essay on Truth* was held by contemporary churchmen is reflected in the fact that in 1774 Beilby Porteus (1731–1808), the rector of Lambeth, offered him ordination. Beattie did not accept, but he was pleased to note in his *Evidences* the encouragement that Porteus, by then bishop of Chester and later of London, had given him in writing that book. In a sermon "On the Christian Doctrine of Redemption," Porteus himself proposed pluralist views similar to Beattie's. Treating the delicate topic of the degree to which the redemption can be illuminated by human reason, Porteus responds to the pluralist objection to Christianity by rhetorically asking: "on what ground is it concluded, that the benefits of Christ's death extend no further than to ourselves?"[62] To justify his belief that Christ's crucifixion is efficacious elsewhere, Porteus draws on the Pauline epistles:

We are expressly told, that as "by him were all things created that are in heaven and that are in earth, visible and invisible; and by him all things consist: so by him also was God pleased (having made peace through the blood of his cross) to reconcile *all things unto himself, whether they be things in earth, or things in heaven:* that in the dispensation of the fulness of times, he might gather together in one *all things in Christ, both which are in heaven, and which are on earth, even in him.*"[63]

The heavy reliance Porteus places on the foregoing passage is reflected in his summarizing statement that "if the Redemption wrought by Christ extended to other worlds, perhaps many besides our own; if its virtues penetrate even into heaven itself; if it gathers together *all things* in Christ; who will then say, that the dignity of the agent was disproportioned to the magnitude of the work . . . ?" (p. 81) Porteus's position is doubly interesting; not only does the English bishop, by urging that Christ's crucifixion is efficacious for extraterrestrials, go beyond the Scottish layman's position that the redemption could act elsewhere only by "example," he even seems to state in the foregoing passage that if the Son's terrestrial incarnation and redemption are seen as extending to extraterrestrials, this adequately adjusts the proportions. Put starkly and in words Porteus probably would have rejected, he seems to say that the universe espoused by the pluralists removes the wonderment from God taking on and suffering in a corporeal form. On Porteus's behalf it may be noted that his wordings show a measure of modesty less easily discerned in Beattie's discussion; nonetheless, his claims are far more radical.

Edward King (1735?–1807) was a contemporary of Beattie and Porteus whose advocacy of a plurality of worlds may be dated from 1780, when he published anonymously his *Hymns to the Supreme Being. In Imitation of the Eastern Songs.* The first hymn begins

Thou, O Lord, hast made all things in Heaven and in Earth: and Thy tender care is over all.
2. Innumerable Worlds stood forth at Thy command; and by Thy word they are filled with glorious works.
3. Who can comprehend the boundless Universe? or number the Stars of Heaven?
4. Are they not the Habitations of Thy Power? filled with manifestations of Thy Wisdom, and Goodness?[64]

Another pluralist passage occurs in Hymn III, where he praises the sun: "Many worlds are nourished by it: and its glory is great." (p. 15) King's concern with pluralist ideas is also evident in his *Morsels of Criticism* (1788), a work of scriptural exegesis based partly on the premise that the Septuagint translation of the Old Testament contains foreshadowings of important doctrines. Among these King includes pluralism, which he presents in such a way as to have each star be the heaven for the resurrected inhabitants of its system of planets. After arguing that the solar rays are not themselves hot, but only produce heat in interaction with material bodies, King urges his readers to join him in viewing "our sun, and all the other fixed stars, merely as so many mansions, and habitations of residence; merely as so many *Islands* (as it were) of Bliss, placed in the vast ocean of space."[65] Not only does King turn the stars into material heavens, he also proposes a new theory of hell. As a reviewer of King's book remarked:

Ilive, a crazy journeyman printer, placed hell in this earth; and Mr. Swinden in the sun; and each wrote a book about it. As our author set out with finding heaven in a new place, so, in the last section of the Appendix, he inclines to fix not only our hell in the centre of the earth, but the hell of every other planet also in the centre of that planet.[66]

Although King's *Morsels of Criticism* went through a second edition and was discussed in a few later pluralist books, its main effect may have been that predicted by that same reviewer, who lamented "that, in all ages, good men, of the best intentions, have done more harm to the cause of Revelation by their speculations . . . than the whole body of sceptics who object without knowledge or reflection." (p.145)

Among British scientists of the latter half of the eighteenth century who endorsed pluralism, Wright, Ferguson, and Herschel have already been discussed. To them may now be added Roger Long (1680–1770), George Adams (1750–95), and Olinthus Gregory (1774–1841). In 1750, Long was chosen as the first Lowndean professor of astronomy and geometry at Cambridge, no doubt in part because of the excellence of the first volume (1742) of his *Astronomy*. A second volume followed in 1764, complete with a chapter on the plurality of worlds theme. Convinced that God would not have created the heavens simply for us, he advocates the pluralist position, dealing especially with the difficulty of the differing intensities with which solar rays fall on the inner and outer planets. To resolve this problem, Long proposes that "mercury and venus contain mines of saltpeter, which may cool their surfaces . . . ," while "in the globes of jupiter and saturn subterraneous fires . . . may compensate their distance from the sun. . . ."[67] Nonetheless, Long injects two reservations: First, he dismisses Huygens's claim that planetarians must be similar to men, and second, if not quite consistently, he endorses Raphael's advice to Adam in Milton's *Paradise Lost:*

> Dream not of other worlds; what creatures there
> Live in what state, condition, or degree.[68]

Although his years of experience may have dampened Long's enthusiasm for extraterrestrials, this did not happen to George Adams, who had half a dozen scientific books to his credit when in 1794 he published his five-volume *Lessons on Natural and Experimental Philosophy*. In that work he judges his largely traditional and teleological arguments for pluralism so convincing that from them "we may . . . safely conclude that *all* the planets of *every* system are inhabited."[69] Adams's penchant for confident proclamation in convoluted prose is evident when he asks whether it is conceivable that

. . . the Almighty, who has not left us with a drop of water unpeopled [!] . . . should leave such immense bodies destitute to inhabitants? It is surely much more rational to suppose them the possession of human beings . . . blest and provided with every object conducive to their happiness, and many of them in a greater state of purity than the inhabitants of our earth, . . . and placed in situations, furnishing them with scenes of joy, equal to all that poetry can paint, [!] or religion promise. . . . (pp. 242–3)

After stressing that the Christian should rejoice at the special redemptive benefaction bestowed by God on earth, he adds:

. . . since the inhabitants of . . . other planets . . . must equally be objects of the Divine favour with ourselves; and since the rational inhabitants of some few or more among so many myriads may have been found disobedient; is a man to blame for thinking that if they stand in need of restoration, they must be full as worthy of it as ourselves; and may for anything that we know, have been already redeemed, or may yet be to be redeemed . . . ?[70]

Whereas Long and Adams were established scientists when they discussed pluralism, Olinthus Gregory, an autodidact who eventually taught mathematics at the Royal Military College, was only nineteen when he presented that doctrine in his *Lessons Astronomical and Philosophical* (1793). Gregory's favorite argument derives from the observed abundance of terrestrial life: "seas, lakes and rivers teem with living creatures; mountains and vallies; trees and herbs; . . . even the blood and humours of animals themselves, all have their respective inhabitants; surely, then, the most numerous and large bodies of the universe, are furnished with beings adapted to their several situations."[71] This leads Gregory to conclude that the millions of suns are "attended by tens of millions of worlds . . . inhabited by millions of millions of rational creatures formed for endless felicity!" (p. 75)

Looked at broadly, British pluralism in the eighteenth century showed itself to be both extremely popular and exceedingly pliable. Barristers and bishops, philosophers and physicists, poets and preachers, as well as a shepherd and a secretary of state, published discussions of pluralism, albeit frequently for very different purposes. While Derham, Pope, Young, and other widely translated authors carried it to the Continent and Whiston to coffeehouses, Ferguson and "Tom Telescope" taught it to children. While at Cambridge Baker and Jenkin expressed reservations about it, Bentley, Newton, and Long endorsed it, and Plume, impressed by Huygens's *Cosmotheoros*, endowed an astronomy professorship for that university. While at Oxford Halley and Keill supported it, the Oxford-associated Hutchinsonians attacked it. While Watts put it in a hymn, Pope in a prayer, Ilive in an oration, and Cappe and Porteus in sermons, Chambers and Ferguson included it in encyclopedias. While Herschel sought selenites but settled for solarians, Pye prepared a Jupiterian's Genesis. While Wesley, founder of Methodism, was wary of it, Swedenborg placed pluralism in the scriptures of the Church of the New Jerusalem. While Young and others used it to embellish traditional Christian conceptions and Beattie, Hay, and Porteus to expand them, Blount, Bolingbroke, and Toland borrowed from it to advocate deism. And while Derham, Wright, Herschel, and others were extending it to the stars, pluralism was crossing the Atlantic to the British colonies, where that popular and pliable doctrine received a warm welcome.

2. Pluralism across the Atlantic: from "Poor Richard" to President Adams

By 1750, European science had barely begun to cross the Atlantic; at Harvard, the systematic teaching of physical science was then scarcely two decades old; at Yale, even less. The passage of pluralism was more rapid; Cotton Mather (1663–1728) had endorsed it in his *Wonderful Works of God Commemorated* (1690) and in his *Christian Philosopher* (1720). In the latter work, the prolific Puritan divine presents his religion as "a PHILOSOPHICAL RELIGION; And yet how Evangelical!"[72] Both these features, as well as his acceptance of pluralism, appear in a sort of prayer that concludes his discussion of the stars:

Great GOD, what a Variety of *Worlds* hast thou created! . . . How stupendous are the Displays of thy *Greatness,* and of thy *Glory,* in the Creatures, with which thou has replenished those Worlds! Who can tell what Angelical Inhabitants may there see and sing the *Praises* of the Lord! Who can tell what *Uses* those *marvellous Globes* may be designed for! Of these *unknown Worlds* I know thus much, *'Tis our Great GOD that has made them all.* (p. 19)

Mather, whose scientific attainments secured his election to the Royal Society, took most of his materials for his *Christian Philosopher* from the physicotheological writings of such authors as Cheyne, Derham, Grew, and Ray.[73] From Derham, for example, he draws the argument, supportive of pluralism, that "the *Whiteness* of the *Milky-Way* is not caused by the great Number of the *Fixed Stars* in that Place, but partly by their *Light,* and partly by the Reflections of their *Planets,* which blend their *Light,* and mix it." (p. 20) Cheyne is cited not only in support of other inhabited solar systems (p. 20) but also on behalf of comets being "the Habitation of *Animals* in a State of *Punishment.*" (p. 44) Mather even advocates lunar life, preferring on that issue the conclusion of Derham to that of Huygens: "For Mr. *Derham* has confuted *Hugenius* with his own *Glasses,* and has demonstrated, that there are great Collections of *Waters* in the *Moon,* and by consequence Rivers, and Vapours, and Air; and in a word, a considerable *Apparatus* for *Habitation.*" (p. 51) To this Mather adds, but without further elaboration, the remark: "But by what Creatures inhabited? A Difficulty this, that cannot be solved without *Revelation.*" (p. 51) Because many colonial readers of the astronomical sections of Mather's book were introduced thereby to the Copernican system as well as to the theories of Huygens and Newton, it has been assigned a significant role in the development of American science.[74] His advocacy of pluralism probably also helps explain the facts that by the 1740s that doctrine was taught at Harvard and in 1752 was the subject of a disputation held at Yale.[75]

That both the public and the profound were becoming acquainted with pluralism is illustrated by consideration of Benjamin Franklin (1706–90), who in 1750 was the leading American scientist, as well as the widely read author of the "Poor Richard" almanacs, which appeared between 1733 and 1758. In 1728, Franklin drew up his "Articles of Belief," a document abounding in deistic and pluralistic ideas. It begins:

I Believe there is one Supreme most perfect Being, Author and Father of the Gods themselves.

For I believe that Man is not the most perfect Being but One, rather that as there are many Degrees of Beings his Inferiors, so there are many Degrees of Beings superior to him.

Also, when I stretch my Imagination thro' and beyond our System of Planets, beyond the visible fix'd Stars themselves, into that Space that is every Way infinite, and conceive it fill'd with Suns like ours, each with a Chorus of Worlds for ever moving round him, then this little Ball on which we move, seems . . . to be almost Nothing, and my self less than nothing. . . .

When I think thus, I imagine it great Vanity in me to suppose, that the *Supremely Perfect,* does in the least regard such an inconsiderable Nothing as Man. More especially . . . I cannot conceive otherwise, than that He, *the Infinite Father,* expects or requires no Worship or Praise from us, but that he is even INFINITELY ABOVE IT.[76]

In a subsequent section, Franklin sets out his conception that

. . . the Infinite has created many Beings or Gods, vastly superior to Man, who can better conceive his Perfections than we, and return him a more rational and glorious Praise. . . .

It may be that these created Gods, are immortal, or it may be that after many Ages, they are changed, and Others supply their Places.

Howbeit, I conceive that each of these is exceeding wise, and good, and very powerful; and that Each has made for himself, one glorious Sun, attended with a beautiful and admirable System of Planets. (I, p. 103)

Bizarre though this belief in multiple gods, each with his own solar system, may be, Franklin scholars have traced its source to Isaac Newton's conversation (discussed earlier) with Conduitt that took place in March 1725 while Franklin was in London.[77]

Franklin's interest in pluralism is also evident in the famous almanacs that he began to issue seven years after composing his creed. For example, in his *Almanack* of 1749, Poor Richard (i.e., Benjamin Franklin) states:

It is the opinion of all the modern philosophers and mathematicians, that the planets are habitable worlds. If so, what sort of constitutions must those people have who live in the planet Mercury? where, says Sir Isaac Newton, the heat of the sun is seven times as great as it is with us; and would make our Water boil away. (III, p. 345)

He enriched his almanacs of 1753 and 1754 by reprinting large portions of James Burgh's *Hymn to the Creator of the World* (London, 1750), which proclaimed pluralism in both poetry and prose.[78] The thousands who read these almanacs thereby learned:

> And thence the glorious Sun pours forth his Beams,
> Thence copious spreads around his quick'ning Streams
> Each various Orb enjoys the golden Day,
> And Worlds of Life hang on his chearful Ray.[79]

The long prose portion of Burgh's book reprinted by Franklin consists of a presentation of the astronomy of the day, including a discussion of the likelihood of life on each of the planets, their satellites, and even comets.

Although mentioning that Mercury is so near the sun that waters on its surface will be kept constantly boiling, Burgh stresses: "it does not follow, that Mercury is therefore uninhabitable; since it can be no Difficulty for the Divine Power and Wisdom to accommodate the Inhabitants to the Place they are to inhabit. . . ." In discussing why the stars were created, Burgh explains that it was "without all Doubt, to enlighten innumerable Systems of Worlds. . . ."

In his almanac of 1757, Franklin discusses the predicted return of Halley's comet. His short essay describing the catastrophic results that would follow from its collision with the earth concludes with a consolation taken from pluralism and from Pope (who is slightly misquoted):

In the mean time, we must not presume too much on our own Importance. There are an infinite Number of Worlds under the Divine Government, and if this was annihilated it would scarce be miss'd in the Universe.

> God sees with equal Eye, as Lord of all
> A Hero perish, or a Sparrow fall.
> Atoms, or Systems, into Ruin hurl'd,
> And now a bubble burst, – and now a World!

Franklin was far from being the only almanac author advocating pluralism; Nathaniel Ames (1708–64), for example, in his 1728 almanac urged on religious grounds that an "innumerable number of heavenly Bodies are . . . stock'd with proper Inhabitants . . . ," and in his 1737 issue he discussed inhabitants in more detail. Unsure of life on the moon, he expressed his desire to travel there to see for himself.[80] Pluralism was also endorsed by Benjamin West (1730–1818), a mathematics professor at Brown University and the author of numerous almanacs, in some of which he used the pseudonym Isaac Bickerstaff. In his *Bickerstaff's Boston Almanack for . . . 1778* he wrote of Saturn:

> Strange and amazing must the difference be,
> 'Twixt this dull planet and bright Mercury;
> Yet reason says, nor can we doubt at all,
> Millions of beings dwell on either ball
> With constitutions fitted for that spot.
> Where Providence, all-wise, has fix'd their lot.[81]

Samuel Ellsworth, in his 1785 almanac, went so far as to describe the inhabitants of Mercury as "very sprightly, small in body, maintaining the upright posture of men, much given to talking, eloquent of speech, good lawyers and pettifoggers. . . ." That mythology motivated his assigning of such characteristics to his extraterrestrials is indicated by his Venusians being "much given to licentious love" and his Martians having "a warlike disposition."[82] Some of the best almanacs were prepared by Benjamin Banneker (1731–1806), a Negro tobacco planter who was introduced to astronomy by Ferguson's writings and then pursued more advanced knowledge of astronomy. That Banneker shared Ferguson's enthusiasm for pluralism may be seen from the first of his almanacs, that of 1792, where he states that Venus "is a planetary world, which with the four others . . . have fields, and seas and skies of their own, are furnished with all accommodations for animal subsistence, and are supposed to be the abodes of intellectual life. . . ."[83] His 1794 almanac contains these lines:

> View yon majestic concave of the sky!
> Contemplate well, those glorious orbs on high –
> There Constellations shine, and Comets blaze;
> Each glitt'ring world the Godhead's pow'r displays![84]

Philip Freneau (1752–1832), who is known as America's "poet of the Revolution," published an almanac, *The Monmouth Almanac, for the Year M,DCC,XCV,* for which he wrote two pluralist essays. In his "Of the Planetary System," he states that all stars are suns, each surrounded by planets populated by creatures "all comfortably provided for by the benevolence of the Creator."[85] Freneau's pluralism extends to the sun, which he suggests is "perhaps peopled with beings of nature infinitely superior to any of those on the neighboring planets." (p. 11) Religious sentiments underlie his presentation, which stresses God's ability to adjust extraterrestrials to the various conditions encountered elsewhere. This is especially evident in his second essay, "Philosophical Speculation," which is aimed at arguing that on the moon, nature is "animated and wound up to a degree of rationality, equal or perhaps superior to the reason of man." (p. 40) Concerning the problem of a lunar atmosphere, he admonishes his readers to remember that: "Endless are the modes of life. There may be millions of mediums by which it may be . . . supported. Air is the medium by the respiration of which, man exists. It would be folly, then, to insist that the inhabitants of the planets necessarily exist by the same medium also." (p. 41) Despite his earlier remark about the superiority of the selenites, the poet records his suspicion that the moon is

. . . in what we would call a *chaotic* state. I never direct my observations towards her with a good glass, but I feel a sort of horror thrilling through my limbs. What melancholy scenes of ruin and devastation do I there survey! What species of created being can possibly take delight in inhabiting those dismal abodes! (p. 41)

Freneau's two essays give the impression of a man informed about astronomy but insufficiently attentive to the evidence opposing his extreme pluralist position.[86]

Pluralist themes also appear in a number of Freneau's poems. Among these are his "On the Honourable Emanuel Swedenborg's Universal Theology" (1786), his "Reflections on the Constitution, or Frame of Nature" (1809), and his "On the Noctural View of the Planet *Jupiter,* and Several of His Satellites, through a Telescope" (1809). In the last of these, Freneau rhapsodizes about the revelations of that instrument:

> This tell-tale tube to me displays
> Vast rolling oceans, boundless seas,
> That on our ponderous planet, Jove,
> To other moons obedient move.

The telescope was not his only source of information:

> And Reason's eye, another race
> Of Nature's children there may trace.
>   The mountains and the spangled plain
> Were surely not designed in vain:
> Who knows but different men are there,
> Joys of their own, or griefs to share.

Moreover, in comparing his Jovians with earthlings, he asks whether the former may not be

> A race in rank that higher stand,
> In Nature's happier humour planned,
> And favoured more, and more caressed
> Than man, discordant and unblest?[87]

Looking broadly at Freneau's life, it seems probable that he encountered pluralism early in such orthodox authors as Derham, Hervey, Thomson, and Young and retained it in his later years as he moved closer to the deism of Thomas Paine, whose death he mourned, and to the Epicureanism of Lucretius, some of whose verses the aging Freneau translated.[88]

The accuracy of the information in American almanacs varied greatly, depending on the author's astronomical sophistication or his success in securing the services of an expert. To the latter end, a number of almanac publishers turned to the man whom Freneau labeled "that prince of astronomers," David Rittenhouse (1732–96).[89] By 1775, in which year the American Philosophical Society invited Rittenhouse to give an address, he had already, through his magnificent orreries and his Venus transit observations, established himself as having scientific and technical talents rivaled among Enlightenment Americans only by those of his fellow Philadelphian Franklin. Rittenhouse's *Oration,* purportedly a survey of the history of astronomy, is a remarkable document that Rittenhouse's most thorough biographer has described as his "public confession of faith."[90] Such it was, and specifically it was the "confession of faith" of a Christian and a pluralist.

In his *Oration,* after recounting the early history of astronomy, Rittenhouse remarks: "Astronomy, like the Christian religion, . . . has a much greater influence on our knowledge in general, and perhaps on our manners too, than is commonly imagined. Though but few men are its particular votaries, yet the light it affords is universally diffused amongst us. . . ."[91] Turning to Copernicus, he notes a number of implications of that astronomer's theory, including that the "planets . . . might be habitable worlds." (p. 9) He quotes Huygens's description of a region of Orion as appearing "as if there were an opening, through which one had a prospect into a much brighter region" and adds: "Here some have supposed old night to be entirely dispossessed, and that perpetual daylight shines among numberless worlds. . . ."[92] The foregoing statement, surely extraordinary, is no more noteworthy than his claim that "The doctrine of a plurality of worlds, is inseparable from the principles of Astronomy. . . ." (p. 19) On one level, his confidence in this claim, which pervades his *Oration,* rests on his belief in the great chain of being, concerning which he states:

. . . when we consider this great variety [of beings] on *our* globe, . . . we shall find sufficient reason to conclude, that the visible creation, consisting of revolving worlds and central suns . . . is but an inconsiderable part of the whole. Many other and very various orders of things unknown to, and inconceivable by us, may, and probably do exist, in the unlimited regions of space. (p. 26)

On another level, his pluralism was supported by his conviction that it could be reconciled with Christianity:

. . . I must confess that I think upon a proper examination the apparent inconsistency will vanish. Our religion teaches us what Philosophy could not have taught; and we ought to admire with reverence the great things which it has pleased divine Providence to perform . . . for man. . . . But neither Religion nor Philosophy forbids us to believe that infinite wisdom and power, prompted by infinite goodness, may throughout the vast extent of creation and duration, have frequently interposed in a manner quite incomprehensible to us, when it became necessary to the happiness of created beings of some other rank or degree. (p. 19)

Within this context, Rittenhouse ruminates on the possibility that the inhabitants of other globes may have avoided the faults of mankind:

Happy people! and perhaps more happy still, that all communication with us is denied. We have neither corrupted you with our vices nor injured you by violence. None of your sons and daughters . . . have been doomed to endless slavery by us in America, merely because *their* bodies may be disposed to reflect or absorb the rays of light, in a way different from *ours.* Even you, inhabitants of the moon, . . . are effectually secured, alike from the rapacious hand of the haughty Spaniard, and of the unfeeling British nabob. Even British thunder impelled by British thirst of gain, cannot reach you. . . . (pp. 19–20)

Repeatedly Rittenhouse stresses the vastness of the universe of the telescope, but he also urges that we consider "the wonderful discoveries of the microscope . . . lest . . . we should be tempted to conclude that man [is too insignificant to attain the] attention . . . of divine Providence." (p. 20)

Providence, patriotism, and pluralism: all three were blended in Rittenhouse's *Oration* and, it seems clear, in his deepest thoughts. In the present context, the point to be stressed is that this "prince of astronomers" viewed the universe within a thoroughly pluralist perspective, seeing comets as possibly "worlds yet in formation or once habitable worlds in ruin," sunspots as "like the cavities within the moon," and Mars as having a "dense atmosphere." That the astronomy of his day supplied little support for such judgments did not deter this enthusiast for pluralism, who was so firmly convinced that "The doctrine of a plurality of worlds, is inseparable from the principles of Astronomy" that he made it part of his "confession of faith."

Comets, long viewed as omens of the awful, presented a special threat to physicotheologians intent on specifying the final causes of each facet of nature. Newton had exacerbated the problem by presenting arguments in his *Principia* (bk. III, prop. 41) to the effect that a comet at perihelion could receive heat "about 2000 times greater than the heat of red-hot iron." This no doubt struck him as sufficient reason to seek other uses for them besides habitation. However, not all were satisfied with his suggestion that "from their exhalations and vapors condensed, the wastes of the planetary fluids spent upon vegetation and putrefaction, and converted into dry earth, may be continually supplied and made up. . . ." Although at Harvard John Winthrop (1714–78) had in 1759 expressed his doubts about the habitability of comets,[93] Hugh Williamson (1735–1819), a physician who had observed the Venus transits with Rittenhouse, set about salvaging comets as abodes of life. In his "Essay on the Use of Comets" (1770), he rejects both Newton's view of comets, citing as his reason that "since the Creation, this Globe has not sustained the loss of one ounce of water," and Whiston's "irreverent conjecture" that "fifty or a hundred worlds [comets] were created [as hells for] the inhabitants of this little globe."[94] His own position is that "Comets are doubtless inhabited," some possibly by "an order of being, greatly superior to this short-lived race of mortals. . . ." (p. 32) His first argument is teleological: "Are we to suppose that this little globe is the only animated part of the Creation, while the Comets, many of which are larger worlds, and run a nobler course, are an idle chaos, formed for the sole purpose of being burned and frozen in turns." (p. 32) To overcome the obvious physical problem pointed out by Newton, Williamson develops the theory that

. . . *all the heat which is caused by the Sun, depends on a tremulous motion excited by the rays of light, in the particles of the body which is heated. Hence . . . the heat of any body will not be according to its distance from the Sun, but according to the fitness of that body, to retain and propagate the several vibrations which are communicated to its particles by the rays of light.* (p. 35)

He urges that this view is supported by the observation that in the thin air on mountaintops, even in tropical regions, low temperatures prevail. Applying this to comets, he suggests that as they approach the sun, the particles of light drive much of the atmosphere of the comet behind it, leaving its remaining atmosphere at low density and hence less responsive to the heat-causing rays of the sun. On the other hand, at great distances from the sun, the large mass of the very extensive atmosphere of a comet may be sufficiently compressed to derive maximum heat from the rays reaching it (p. 36). Williamson's efforts on behalf of his "Cometarians," who, he speculates, may enjoy "antedeluvian" (sic) life spans, immediately won him election to the "Society of Science of Holland"[95] and, two hundred years later, secured a place for his paper in an anthology of primary sources in American science as an example the "fancies" of early scientists.[96]

Andrew Oliver (1731–99), a prominent Massachusetts jurist, had been introduced to astronomy by John Winthrop; in fact, he dedicated his *Essay on Comets* (1772) to his former Harvard teacher. This did not, however, deter him from centering his slim volume around the un-Winthropian claim that comets are inhabited bodies. Stimulated to compose his cometary treatise by Williamson's paper, the Salem jurist proceeds far beyond the Philadelphia physician by urging that all solar system objects – the sun, planets, and comets – are surrounded by atmospheres of air. Accepting Newton's view that air particles repel each other,[97] Oliver explains that as a comet approaches the sun, its air atmosphere is temporarily repelled to the side away from the sun, leaving the cometary nucleus with an atmosphere of sufficiently low density to offset the increased number of solar rays falling on its surface (pp. 30–2). On the other hand, as the comet departs the hot solar neighborhood for the cold and remote regions of its orbit, "[t]he cometic atmosphere being gradually re-condensed round its nucleus as before, would provide it with a suitable garment for winter quarters . . . ; agreeable to the ingenious hypothesis of Doctor *Williamson*." (p. 33) Oliver's analysis relies also on Williamson's theory that the heating effect of the solar rays in an area is conjointly the result of the number of rays reaching the region and the susceptibility of its atmosphere to being heated by those rays (pp. 60–8).

Oliver's explanation of cometary tails as due to a repulsive force between the atmospheric air particles of the sun and comet is, he admits (p. iv), not original with him; it had been put forth previously by Gowin Knight (1713–72), the English author of *An Attempt to Demonstrate That All the Phenomena of Nature May Be Explained by Two Simple Active Principles, Attraction and Repulsion* (1754). He also notes that Knight had suggested in this volume that the sun and stars may be sufficiently cool to accommodate life. In fact, he quotes Knight's statement concerning the sun and stars that according to his theory: "Their globes are no longer frightful Gulphs of Fire, but inhabitable Worlds: Those Philosophers who thought them too hot for the Habitation of Salamanders, and those sublime Genii, who thought them to be Hells, will now perhaps be in Pain, lest the inhabitants should freeze with Cold." (p. 18n) A thoroughgoing pluralist, Oliver extends his theory of comet atmospheres to the planets by speculating that the inner planets may remain tolerably cool by having atmospheres of low density, whereas the outer planets may have more dense atmospheres that retain more solar rays (p. 70).

Oliver's treatise is not unimpressive; it even includes experiments with an "artificial Comet, consisting of a small, gilt cork ball, with a tail of leaf-gold" (p. 46), and some ingenious arguments, based on Halley's work with a diving bell, to explain how cometarians may overcome the respiratory problems of an atmosphere of varying density (pp. 75–8). The merits of Oliver's essay were recognized; it went through a second edition as well as a French translation, and Jedidiah Morse (1761–1826), the "father of American geography," praised its contents.[98] Nonetheless, it suffers at a number of points from an underlying circularity in its arguments and excessive reliance on analogies. Both these characteristics can be illustrated from Oliver's argument for an airy atmosphere for comets. This argument begins from the assumption that all planets are inhabited, derives from it the conclusion that they are surrounded by air, and then uses the supposition that planets and comets are analogous to bestow air on the latter objects (p. 17).

Two other Harvard graduates, both prominent politicians, were involved with pluralism. James Bowdoin (1726–90), who eventually became governor of Massachusetts, was chosen in 1780 to be the first president of Boston's American Academy of Arts and Sciences. In fact, the first item in that society's *Memoirs* is Bowdoin's "Philosophical Discourse." No doubt with the thought of ending his address on a lofty and pious note, Bowdoin informs the assembled academicians:

. . . when we raise our view to the heavens, and behold the beauteous and astonishing scenes they present to us – unnumbered worlds revolving in the immeasurable expanse; systems beyond systems composing one boundless universe; and all of them if we may argue from analogy, peopled with an immense variety of inhabitants: When we contemplate these works of nature . . . , they force upon us the idea of a SUPREME MIND. . . .[99]

The first volume of the *Memoirs* contains two additional papers by Bowdoin, in the second of which he argues for the existence of a giant shell surrounding the entire universe, which shell would reflect back light lest it be lost and would also gravitationally hold stars in position. Moreover, he suggests that this shell could provide a residence, on both its interior and exterior surfaces, for vast numbers of beings.[100]

John Adams (1735–1826) graduated in 1755 from Harvard, where his interest in science frequently brought him into association with Winthrop. Judging from Adams's diary for 1756, the future U. S. president was at that time above all concerned to formulate a religious position somewhere between the Calvinism of his ancestry and the deism of his day. On a number of occasions in 1756 he pondered pluralism in this context. On April 24, for example, he wrote:

> Astronomers tell us . . . that not only all the Planets and Satellites in our Solar System, but all the unnumbered Worlds that revolve round the fixt Starrs are inhabited. . . . If this is the Case all Mankind are no more in comparison of the whole rational Creation of God, than a point to the Orbit of Saturn. Perhaps all these different Ranks of Rational Beings have in a greater or less Degree, committed moral Wickedness. If so, I ask a Calvinist, whether he will subscribe to this Alternitive [sic], "either God almighty must assume the respective shapes of all these different Species, and suffer the Penalties of their Crimes, in their Stead, or else all these Being[s] must be consigned to everlasting Perdition?"[101]

Probably Adams discussed pluralism with William Herschel while visiting the astronomer in 1786[102] and later with his vice-president, whose library contained the pluralist books of Fontenelle, Huygens, and Derham.[103] This was Thomas Jefferson (1743–1826), to whom Adams sent a letter in 1825, one passage of which reveals Adams's state of mind, one year before he died, on the question raised in his 1756 diary entry quoted earlier. Adams, in 1825, urged Jefferson not to hire European professors for the University of Virginia because "They all believe that great Principle which has produced this boundless universe, Newton's universe and Herschell's [sic] universe, came down to this little ball, to be spit upon by the Jews. And until this awful blasphemy is got rid of, there never will be any liberal science in the world."[104] Adams's harsh language suggests that Jefferson was not unsympathetic to this argument.

Another American who accepted the pluralist position was Ezra Stiles (1727–95), who served as Yale's president from 1775 to his death. Besides referring to the topic periodically in his *Literary Diary*, he touched on it in at least one of his sermons, where he was careful to include as part of the moral order not only spirits but also "those inhabitants of this earth & the planetary starry universe, all bodied beings, capable of *morals*."[105] John Bartram (1699–1777) was among the leading botanists of his age, but, as he wrote in 1762 to Alexander Garden, he believed that God's majesty was also revealed in "[o]rbs beyond orbs, without number, suns beyond suns, systems beyond systems, with their proper inhabitants of the great Jehovah's empire. . . ."[106] As such passages suggest, the idea of a plurality of worlds was so linked to natural theology that few doubted it. Herbert Leventhal, in his research on occultism and Renaissance science in the American Enlightenment, has uncovered unpublished writings by Manasseh Cutler (1742–1823), a Yale-educated Congregational clergyman, and by Nathaniel Dominy, wherein each expressed reservations about the pluralist position.[107] Although Samuel Johnson (1696–1772) endorsed the possibility of a plurality of worlds in a 1739 sermon and, according to the list he compiled of his readings, had a strong interest in pluralist literature, it is possible that after about 1750, as he grew ever more enthusiastic about the writings of the Hutchinsonians, he may have come to question the pluralist position.[108]

In a historical survey of pluralism published by an American magazine in 1792, it was claimed that the great majority of European thinkers, ancient and modern, had accepted the pluralist position.[109] Although the author mentioned no Americans in his exposition, a knowledgeable member of the new nation could note that many of its leading poets, preachers, natural philosophers, and politicians had also endorsed this doctrine. Religious sentiments more than scientific results lay behind this, but the same was largely true in the older civilization. Whatever pride an American Christian may have taken in his countrymen's early espousal of pluralist ideas must soon have been tempered when in the mid-1790s Tom Paine proclaimed that a thinking man could not accept both Christianity and pluralism. But this is a matter for a subsequent section.

3. Pluralism and the French Enlightenment: philosophes, savants, and abbés

In his *Lettres philosophiques* (1733), Voltaire contrasted French and British views of the universe:

A Frenchman who arrives in London finds things indeed changed in philosophy and in everything else. He had left the world a plenum, he finds it a vacuum. At Paris the universe is seen as composed of vortices of subtle matter; at London one sees nothing of that. . . .

You will note furthermore that the sun, which in France does not enter at all into [the tides], contributes around one fourth [of the effect]. . . . At Paris, the earth is represented as melon-shaped; at London, it is flattened at both poles.

Voltaire goes so far as to claim: "Even the essence of things is completely changed."[110] Perhaps so. But the universe of the Fontenelle-influenced French was no less full of extraterrestrials than that of the British, this being due in part to the prominence of pluralism in the writings of the most famous of the *philosophes,* Voltaire himself.

By 1717, when François-Marie Arouet (1694–1778) assumed the name Voltaire, he had already received seven years of education from the Jesuits, as well as an eleven-month prison term from French officials.

Although the successes of his drama *Oedipe* (1718) and of his epic *Henriade* (1723) might have inspired him to dream of becoming the new Sophocles or the French Homer, the young deist announced himself in 1722 as the "new Lucretius." (XV, p. 13) Back in the Bastille by 1726, he was freed on the condition that he accept exile. Thus, he came to spend two years with the English, whose liberal politics, Lockean philosophy, and Newtonian science he espoused with enthusiasm and soon urged upon his countrymen in his *Lettres philosophiques*. He may have left France a poet and returned a philosopher, as some have suggested, but his June 1, 1721, letter to Fontenelle suggests that the provenance of his pluralism was Gallic.[111] In fact, if this letter be believed, he had read Fontenelle's *Entretiens* lest he be lost among the ladies who took such delight in that book.[112] Although Voltaire's Newtonian conversion forced him to forsake Fontenelle's Cartesian vortices, his pluralism persisted; indeed, it appears in a substantial number of the more than 25,000 pages of his publications.

The influence of Fontenelle, in particular his idea that extraterrestrials may possess additional senses, can be detected in a draft of the thirteenth of his *Lettres philosophiques,* which contains the statement that "it is to be presumed that in other worlds there are other animals who enjoy 20 or 30 senses, and that other species, still more perfect, have senses to infinity."[113] Voltaire also availed himself of pluralism's literary and conceptual uses in his *Traité de métaphysique,* drafted in 1734 and dedicated to Madame du Châtelet, whose charms inspired it and whose prudence postponed its publication. That book, for example, includes a report made by an extraterrestrial after seeing the diverse races domiciled on our earth. Voltaire certainly intends the reader to share that visitor's skeptical reaction when told that all those races descended from a single pair (XLII, p. 124).

Voltaire's ambivalence in the 1730s to Fontenelle – a continuing admiration for that author's literary skills combined with an increasing hostility to the Cartesian content of his *Conversations* – is evident even from the opening sentence of Voltaire's *Éléments de la philosophie de Newton* (1738), where he chides his predecessor in scientific popularization by noting: "There is nothing here of a marquise, nor of an imaginary philosopher." (XLI, p. 23) Written to champion Newtonian philosophy, Voltaire's *Éléments* appeared in three forms in 1738, containing two rather different statements on the question of a plurality of worlds. The first printing, that from an Amsterdam publisher to whom Voltaire had given his manuscript, was made without Voltaire's approval or revisions and moreover contained materials added by another author. It was this version that John Hanna immediately translated into English. Both these editions contain the same endorsement of pluralism in which Huygens is

cited as the authority; in a discussion of comets, it is remarked that "if
. . . there are animate Creatures in the Comets, as *Mr. Huygens* has
proved there are in the Planets, they must of necessity retire into the
interior Cavities of these Comets, to secure themselves from the general
Conflagration of their exterior Surfaces."[114] Before the end of 1738,
Voltaire had published his authorized French edition (which does not
contain this passage) and also his *Eclaircissements nécessaires,* wherein
he sets forth his relationship to the two earlier editions. In this work,
Voltaire states that he does not wish to claim that living matter exists on
comets "as Monsieur Huygens has proved that there is in the planets.
. . ." Moreover, he adds the qualification that

. . . I do not see that Mr. Huygens has given more proofs of this pleasant and
intelligent conception than Cardinal Cusa, Kepler, Bruno, and all the rest, and
especially Mr. de Fontenelle, have given. It is one thing to render a likely opinion,
another thing to prove it. We can suppose that the planets, similar to our own,
are populated by animals, but we do not have on that point any other degree of
probability, speaking exactly, than a man will have who has fleas and who
concludes that all those whom he sees passing in the street also like him have
fleas. . . . (XLI, p. 42)

This passage suggests that Voltaire was skeptical of at least some of the
pluralists' proofs.

In his *Discours en vers sur l'homme* (1738), Voltaire uses an extraterrestrial commentator, a device also employed in a manuscript entitled *Le
voyage du baron de Gangan* that he sent in 1739 to Frederick the Great.
The latter piece, now lost, has been shown by Ira O. Wade to have been
almost certainly the first draft of Voltaire's famous *Micromégas* (1752).
Another source of information concerning Voltaire's views at this time of
pluralist speculations is his letter of August 10, 1741, to Maupertuis, a
fellow French Newtonian, who had recently returned from an expedition
to Lapland, where he had found observational evidence for the flattening
of the earth near its poles, as Newton had predicted. Maupertuis's result
so impressed Voltaire that he composed a poem to be placed at the base
of Maupertuis's portrait. In the letter transmitting these verses, Voltaire
contrasts Maupertuis's empirical approach with the speculations on the
height of Jupiterians that Voltaire had encountered in reading Christian
Wolff's *Elementa matheseos universae.*[115] Voltaire's witty letter states in
part:

I will not place such quatrains, my dear flattener of planets and of Cassinis, at
the base of the portrait of Christianus Volfius. For a long time I have viewed with
the stupor of a monad the height which that Germanic babbler assigns to the
inhabitants of Jupiter. He judges by the size of our eyes and by the distance of the
sun from the earth. . . . That man has brought back to Germany all the horrors
of scholasticism surcharged with *sufficient reasons, monads, indiscernibles,* and

all the scientific absurdities which Leibniz has in his vanity put on the world, and which the Germans study because they are Germans. (LXXII, pp. 471–2)

Voltaire expressed his dislike for the Leibnizians[116] most effectively in that genre, "les contes philosophiques," in which his talents were displayed, by common consensus, at their best. This is true not only of his most famous philosophic tales *Zadig, Micromégas,* and *Candide* but also of his *Memnon ou la sagesse humain,* which, like *Zadig,* dates from the late 1740s. In this story, Memnon's well-intentioned way of living leaves him destitute, humiliated, and blind in one eye. An extraterrestrial consoles him with the Leibnizian message that "in the ten thousand millions of worlds which are dispersed in space, all succeed each other in degree. Less of wisdom and pleasure exist in the second than in the first, and similarly up to the last, where the entire world is completely insane." When the distraught Memnon confesses his suspicion that the earth is precisely that lowest world, the extraterrestrial counters that such is not the case, but that nonetheless "everything should have its place" in the chain of being. Memnon then asks: "are then certain poets, certain philosophers in the wrong in saying that all is good?" The extraterrestrial assures him that for this view, "they have grand reason, considering the arrangement of the entire universe." The story ends as the dejected Memnon replies: "Ah! I will believe this only when I will no longer be one-eyed." (LIX, p. 39)

Voltaire's *Zadig, ou la destinée* (1747) deals with the identical problem, but with less pessimism than is found in the tales Voltaire told after Madame du Châtelet's death in 1749. Voltaire sets out the problem by having the virtuous but much afflicted Zadig leave Babylon to travel among the stars, where he reflects that the earth is only an "imperceptible point" in the universe and men are only "insects devouring each other on a tiny atom of mud." These images initially console him, but later when "responding to his heart, he thought that Astarté was perhaps dead because of him, the universe disappeared from his eyes, and he saw in all of nature nothing but the dying Astarté and the ill-fated Zadig." (LIX, pp. 100–1) The difficulties in deriving consolation from this pluralist vision are also suggested later in *Zadig,* when an emissary from God reveals essentially the same message. To this, Zadig replies with only a single word: "But. . . ." (LIX, p. 168)

Voltaire's *Micromégas* (1752) is filled with extraterrestrial elements – and with witty comments on prominent terrestrials. The story centers on Micromégas, an itinerant inhabitant of a satellite of Sirius, who stands 120,000 feet tall. After studying at his planet's Jesuit college and discovering on his own fifty theorems of Euclid,[117] he is exiled and begins to visit various planets, including Saturn and the earth. Saturnians he finds to be dwarfs, standing a mere 6,000 feet tall. Nonetheless, he becomes a friend of the secretary of the Academy of Saturn (Fontenelle is meant), "a man of good spirit, who in truth had invented nothing, but who rendered a very good account of the inventions of others, and who made passably good small verses and large calculations."[118] This figure laments that Saturnians have only seventy-two senses, a deprivation with which Micromégas sympathizes, realizing that his 1,000 senses would leave him ill-equipped on planets where even more sensationally endowed beings surely exist.

A comet carries the Sirian and the Saturnian to Jupiter and then to earth, Mars being too small for their comfort. They do, however, note the two Martian moons, necessary because of the distance of that planet from the sun.[119] The minuteness of earthlings tests the visitors' ingenuity, but eventually they establish communication with a group (the Maupertuis expedition to Lapland is meant) sailing in the Baltic. When a physicist among the sailors measures Micromégas by means of a sextant, the Sirian exclaims: "I see more than ever that it is necessary to judge nothing on the basis of its apparent size." (LIX, p. 200) Voltaire's intentions in his story run deeper than this lesson; indeed, he uses this framework to suggest that brilliant though men's physical inquiries may be, their philosophical sophistication has not kept pace. This point is stressed as representatives of various schools of philosophy expound their doctrines of the soul to Micromégas, to whom they in most cases appear defective. The conversation is interrupted by the speech of a theologian, who "said he knew the entire secret; that it was found in the Summa of St. Thomas; he looked at the two celestial inhabitants from tip to toe, and maintained that their persons, their worlds, their suns, their stars were all made uniquely for man." (LIX, p. 207) This provokes such convulsions of laughter from the two giants that the ship is temporarily lost. Despite Micromégas's anger "that these infinitely small beings had a pride almost infinitely large," he bestows upon them a book that he promises will give ultimate knowledge. It did teach a lesson: when opened in Paris, it was seen that all its pages were blank! (LIX, p. 207) The skeptical thrust of this "conte philosophique" is evident in the fact that its pyrrhonism begins where Protagoras's "Man is the measure of all things" leaves off.

The merits of Micromégas will continue to be debated; some will see it as the message of a despairing misanthrope, intent on ridiculing leading thinkers by a literary device long available; others will praise it as an ingenious integration of diverse elements and ideas, composed to temper the pride of earthlings. Be that as it may, it should not be seen, as it sometimes is, as primarily a pioneering effort in science fiction; it belongs rather to the genre "conte philosophique" and as such stands next to Voltaire's Candide as the most powerful of his philosophical tales.

The tragedy of the Lisbon earthquake provided Voltaire with yet another opportunity to assail the Leibnizian doctrine that "all is good." Early in the poem he composed on that event, Voltaire asks:

> "All is good," you say, "and all is necessary."
> What! The entire universe without the infernal abyss,
> Without engulfed Lisbon, would be less evil![120]

The poem provoked many responses, not least a long letter from Jean-Jacques Rousseau (1712–78), to whom Voltaire had sent his poem. As he recorded in his autobiographical *Confessions,* Rousseau, convinced by this poem that "Voltaire, while always appearing to believe in God, has never really believed in anything but the devil," "formed the senseless plan of bringing [Voltaire] to himself again, and proving to him that everything is good."[121] The many differences between the philosophies of Rousseau and Voltaire are sharply etched on the pages of this letter, which was published in 1759. Rousseau's acceptance of the Leibnizian claim that temporary evil is a necessary concomitant of long-range good was by no means complete; he was, after all, a champion of the view that most of man's misfortunes are of his own causing. Nonetheless, for a score of pages he attempts to combat Voltaire's pessimism. Strongly as they disagreed, however, they shared a conviction in the plurality of worlds; this point of agreement as well as one of Rousseau's major arguments appear in the following paragraph:

> On the good of the whole being preferable to that of the part, you make man say: *"As a thinking and sensing being, I ought to be as dear to my master as the planets, which probably sense nothing at all."* Without doubt this material universe ought not be more dear to its author than a single thinking and sensing being; but the System of this universe which produces, conserves, and perpetuates all thinking and sensing beings, ought to him be more dear than a single one of these beings. He then . . . can sacrifice something of the good for the conservation of the whole. I believe, I hope that I am worth more in the eyes of God than the earth of a planet, but if the planets are inhabited, as is probable, why should I be worth more in his eyes than the inhabitants of Saturn?[122]

Voltaire's immediate response to Rousseau's letter was only a brief note. However, as Rousseau suggested in his *Confessions,* a longer response may later have appeared in the form of Voltaire's severely satirical *Candide.*[123]

At least four of Voltaire's publications from the 1760s present pluralist themes. The earliest of these, his *Traité sur la tolérance* (1763), includes his "Prayer to God," in which deist and pluralist views are so intimately mixed that one can scarcely doubt the centrality of pluralism in his religious thought. His prayer begins: "It is no longer to men that I address myself; it is to you, God of all beings, of all the worlds. . . ." (XXXVIII, p. 291) Although "plurality of worlds" was not one of the entries in Voltaire's *Dictionnaire philosophique* (1764), the topic is touched on repeatedly, even in such an unlikely entry as "Chinese Cate-chism," in which Voltaire introduces pluralism into a critique of Confu-cian philosophy by suggesting that it is ridiculous to speak, as Confucians do, of "heaven and earth," because our earth "is infinitely less than a grain of sand in comparison with the millions of billions of universes before which we disappear. All that we are able to do is to join our weak voice to that of the innumerable beings who render homage to God in the Abyss of space." (XLIX, p. 249) In his entry "Chain of Created Beings," Voltaire reveals his disbelief in that doctrine: "when one regards it atten-tively, this great phantom evaporates. . . . Oh much admired Plato, I fear you have told nothing but fables. . . ." (LIII, pp. 157–60) In the miniature celestial voyage contained in "Dogmas," the traveler receives a message about divine justice:

BY ORDER OF THE ETERNAL, CREATOR, CONSERVER, REMUNERA-TOR, AVENGER, PARDONER, etc., etc., be it well known to all the inhabitants of the hundred thousand millions of billions of worlds that it has pleased us to form, that we will never judge any of these inhabitants on their empty ideas, but only on their actions, for such is our justice. (LIV, p. 106)

Finally, in "Providence," readers meet Sister Fessue, fresh from saving a sparrow by administering nineteen Ave Marias. Voltaire attacks the idea of a special providence by having a metaphysician inform her that God cannot be concerned about sparrows because he "has billions and bil-lions of other suns, planets, and comets to govern. . . ." Moreover, he tells her: "If the *Ave Marias* had made the sparrow . . . live an instant longer than it was to live, these *Ave Marias* would have violated all the laws set up from all eternity by the great Being; you would have dis-rupted the universe; you would have necessitated a new world, a new God, a new order of things." (LVII, p. 491)

Voltaire's *Le philosophe ignorant* (1766), which was his last major philosophical work and his most despairing, employs pluralist perspec-tives to suggest how little man can know with certainty. It opens with the questions: "Who are you? Whence do you come? What are you becom-ing? These are questions which ought be asked of all the beings of the universe, but to which none could provide us an answer." (XLIV, p. 1) Voltaire goes on to say that although he believes the planets are inhabited by intelligent beings, they cannot communicate with us. Nonetheless, their existence teaches us the wrongness of the view, ascribed to Abbé Pluche's *Spectacle de la nature*, that the universe was made for man. In fact, so far are we from being the lords of nature that if animals do not devour us before we die, they will afterward (XLIV, p. 2). In a section entitled "Feebleness of Mankind," Voltaire writes: "Is that which is impossible to my so feeble, so confined nature . . . also impossible in other globes, in other species of being? Are there superior intelligences

. . . who think and sense all that they wish? I know nothing; I know only my weakness and have no notion of the powers of others." (XLIV, pp. 14–15)

As frequently as Voltaire's extraterrestrials appear in his writings, one would expect to find many arguments justifying their existence. Such rarely occur, an apparent exception being his *Tout en Dieu* (1769), written as a critique of Malebranche's philosophy. The pluralist argument in *Tout en Dieu* takes off from the assertion that light from the stars is essentially identical with light from our sun and obeys the same laws of refraction. On this basis, Voltaire claims:

> Since this refraction is necessary to sight, it follows that there are in the planets beings who have the faculty of seeing. It is not probable that this good use of light should be lost for other globes. Since the instrument [light] exists, the usage of the instrument ought likewise exist. (XLVI, p. 246)

This argument is so obviously weak that one wonders whether or not Voltaire, who presented it in a work filled with irony, meant it to be taken seriously.[124]

By 1778, when the Sage of Ferney died, he may not have been regarded as a "new Lucretius," but his employment of pluralism in a dozen or more works had assured him a place with the Latin poet in the history of extraterrestrial life ideas. The surprising fact that this lifelong opponent of speculative systems never in his writings critically analyzed pluralist arguments admits of two explanations. Possibly Voltaire set aside whatever doubts he had about pluralism lest they detract from his literary, philosophical, and religious uses of it. Alternatively, perhaps some of his pluralist writings, such as *Micromégas,* should be seen as satirizing, along with much else, the pluralist claims of such authors as Fontenelle, Leibniz, and Wolff, as well as works in the cosmic voyage genre. Either explanation fits the characterization of Voltaire presented in a recent study of his life and thought:

> No one certainly was more contradictory than Voltaire. A staunch defender of truth, he was at the same time one of the world's consummate masters of the art of lying. He was . . . a religious sceptic who refused to doubt the existence of God; an experimentalist whose "passions dictated his conclusions." He was a powerful proponent of toleration for everyone's ideas except those who attacked his own; a lover of humanity who hounded to death those whom he loathed. . . . Courageous and cowardly, affectionate and suspicious, original and conforming, sincere and hypocritical, rational and impulsive, an Apollonian and a Dionysian – Voltaire was all this and more.[125]

On August 18, 1765, Denis Diderot (1713–84) jubilantly wrote his mistress to proclaim "Terre! Terre!" The occasion was the completion of the *Encyclopédie,* that vast ark that he and d'Alembert had launched in 1751. Left alone at the helm by d'Alembert's 1758 departure, Diderot nonetheless steered a course that avoided the Scylla of censorship and the Charybdis of financial ruin. Designed to delight by its eleven volumes of plates, to instruct by its eighteen volumes of text, the *Encyclopédie* was also intended – as Diderot candidly admitted – "to change the general way of thinking."[126] D'Alembert, although not attracted to Diderot's materialism and atheism, shared with him the goal of using the *Encyclopédie* to inculcate an empirical, skeptical, and antimetaphysical philosophy. This did not, however, preclude their acceptance of the pluralist position in the *Encyclopédie*.

Because of his long-standing admiration for the philosophy of Epicurus and Lucretius, Diderot must have written the article "Épicureisme" with special relish.[127] While praising it for anticipating various modern scientific views, including pluralism, he states: "There is nothing repugnant in the notion of a plurality of worlds. Worlds similar to ours, as well as ones that are different, can exist. It is necessary to consider them as great vortices supported against each other which compress smaller ones between them, and which together fill the infinity of space."[128] Diderot's interest in the pluralist position dates back at least to 1749, when he employed it in his *Lettre sur les aveugles,* a book that brought him both fame and a prison sentence. Although a deist in 1746, by the time he published his *Lettre,* Diderot had become an atheist, a transition influenced partly by his acceptance of the evolutionary naturalism advocated in Benoît de Maillet's *Telliamed* (1748).[129] Diderot's *Lettre* includes an attack on deism, formulated in one instance in pluralist terms. He introduces a blind man whose inability to see the spectacle of nature, so frequently cited in support of belief in God's existence, equips him to recognize the correctness of the Epicurean cosmology. After a discussion of defective forms in living beings, the blind man makes a statement that seems to foreshadow Darwinian theory:

> I conjecture, then, that in the beginning, when matter in a state of ferment brought this world into being, creatures like myself were of very common occurrence. But might not worlds too be in the same case? How many faulty and incomplete worlds have been dispersed and perhaps form again, and are dispersed at every instant in remote regions of space . . . where motion continues and will continue to combine masses of matter, until they have found some arrangement in which they may finally persevere? O philosophers, travel with me to the confines of this universe, beyond the point where I feel and you behold organized beings; cast your eyes over this new ocean, and search in its aimless and lawless agitations for vestiges of that intelligent Being whose wisdom fills you with such wonder and admiration here![130]

The harsh pluralism of this cosmos strikingly contrasts, as Diderot no doubt was aware, with that of theistic pluralists.

The most interesting comments in the *Encyclopédie* on the question of a plurality of worlds are those of Jean Le Rond d'Alembert (1717–83), whose prominence as a scientist led to his composing many of its scientific articles. Having endorsed the idea of a great chain of being in his article "Cosmologie,"[131] d'Alembert devotes part of "Étoile" to filling out that chain. After evidencing the idea that stars are suns, he urges that

. . . it is very natural to think that each star . . . has planets which make their revolutions around it . . . ; this is to say that it has opaque bodies which it illuminates, heats, and supports by its light: for why would God have placed all the luminous bodies at such great distances from each other, without there being around them some opaque bodies which receive their light and heat?[132]

He then refers the reader to the article "Pluralité des mondes," but in fact no article with this title appeared.[133] The topic is, however, taken up by d'Alembert in both "Monde" and "Planète." In "Monde," after remarking on the many meanings of that word, d'Alembert describes the books of Fontenelle and of Huygens. By means of the argument from analogy, Fontenelle had advocated pluralism, d'Alembert notes, while freeing himself from theological objections by claiming that the extraterrestrials are not "men." Huygens, on the other hand, urged that extraterrestrials "ought to have the same arts and area of knowledge as we have. . . ."[134] D'Alembert then remarks that although the pluralist position is "not improbable," it is beset by a number of difficulties:

1. One doubts whether some planets, among others the moon, have an atmosphere, and on the supposition that they have not, one does not see how living beings can respire and subsist there. 2. Some planets as Jupiter present changing figures and aspects on their surface . . . whereas it seems that an inhabited planet ought to be more tranquil. 3. Finally, comets are certainly planets . . . and it is difficult . . . to believe that comets may be inhabited. . . . What response ought then be given to those who ask whether the planets are inhabited? That one knows nothing.[135]

The cautious tone in "Monde" is less evident in "Planète," where d'Alembert recounts various astronomical observations (including some now known to be spurious) as a basis for believing that the planets have surfaces similar to that of the earth. For example, de la Hire's 1700 report of mountains on Venus is cited as evidence for other planets having mountains, and the changes reported on the surfaces of Venus, Mars, and Jupiter indicate that they and other planets possess atmospheres. He concludes:

Since Saturn, Jupiter and their satellites, Mars, Venus and Mercury are opaque bodies which receive their light from the sun, which are covered by mountains and surrounded by a changing atmosphere, it seems to follow that these *planets* have lakes, have seas . . . ; in a word, that they are bodies resembling . . . the earth. Consequently, according to many philosophers, nothing prevents us from believing that the *planets* are inhabited.[136]

Among such philosophers, d'Alembert lists Huygens, Fontenelle, and Wolff. He even quotes Wolff's "so singular" calculations of the height of

Jupiterians, describing them as "the aberrations which fall from the human spirit, when it opens itself to the madness of making systems. . . ."[137] The impression these articles leave is that d'Alembert accepted the pluralist position as probable, but was repelled by the extremes to which it had been carried by such authors as Christian Wolff. Moreover, he eschewed the Epicurean associations so dear to Diderot.

Among the most enthusiastic supporters of the *Encyclopédie*, and moreover the author of hundreds of its articles, was Paul Henri Thiry, Baron d'Holbach (1723–89), whose Paris salon served as a gathering point for many of the *philosophes*. During more than two decades as Diderot and others dined in the *coterie holbachique*, the conversation certainly centered on numerous occasions on religion and, one may conjecture, turned at times to the topic of extraterrestrials. D'Holbach, so hostile to religion that he relished referring to himself as a "personal enemy of God," directed from his estates an unrelenting attack on all forms of theism by encouraging others to write against religion and by composing numerous tracts himself. Not only did this affable atheist suggest the 1768 translation of Lucretius made by his children's tutor,[138] he also personally penned the "bible of atheism," his anonymous *La système de la nature* (1770), which presents the universe as a system of matter in constant motion, and mind as nothing more than a complex of atoms. Within this materialist context, d'Holbach endorses pluralism, urging that man is a particular product of the earth. In developing this point, he first proposes various ideas concerning the origin of the earth, for example, that it may be a "mass detached . . . from some other celestial body" or a "displaced comet," and then states: "Whatever may be the supposition adopted, planets, animals, men can only be regarded as productions inherent in and natural to our globe, in the position or in the circumstances in which it is actually found: these productions would be changed, if this globe by any revolution would happen to shift its situation."[139] D'Holbach illustrates his claim by stating: "Transport, in imagination, a man from our planet into *Saturn*, his lungs will presently be rent by an atmosphere too rarefied for his mode of being, his members will be frozen with the intensity of cold; . . . transport another into Mercury, the excess of heat will quickly destroy him." (p. 45) In stressing this idea, d'Holbach was not questioning the pluralist doctrine, but rather criticizing those "who are willing to conjecture that the other planets, like our own, are inhabited by beings like ourselves." (p. 45) So blatant was d'Holbach's overall message, so vigorous was his argumentation, that persons as different as Voltaire and Frederick the Great took it upon themselves to issue responses. Diderot, however, was delighted.

Only an exceedingly capacious definition of the term *scientist* would permit its application to Volatire, Diderot, or d'Holbach, although d'Alembert would surely qualify. That many scientists writing in French during the post-1750 period were pluralists can be illustrated from a discussion of Maupertuis, Euler, Bonnet, and Buffon. Pierre Louis Moreau de Maupertuis (1698–1759) possessed credentials so impressive that Frederick the Great named him in the 1740s to head the Berlin Academy of Sciences. It was with the prestige of that position that Maupertuis published his *Essai de cosmologie* (1750), citing the analogies between the earth and the other planets as a basis for inferring their habitation:

This planet [the earth] . . . can convince us that all the others, which appear to be of the same nature as it are not deserted globes suspended in the skies, but that they are inhabited. . . . Some authors have ventured concerning these inhabitants conjectures which can neither be proved nor falsified, but all is said, at least all that can be said with probability, when one has remarked that the vast bodies of the planets, having already many things in common with the earth, may also share with it in being inhabited. As to the nature of their inhabitation, it would be very rash to attempt to divine it. If such great varieties are observed already among those who populated the different climates of the earth, how can one conceive of those who live on planets so distant from our own? Their varieties probably exceed the scope of our imagination.[140]

That Maupertuis's pluralism dates from before his *Essai de cosmologie* is suggested by the fact that a manuscript of that book already existed in 1741, when Voltaire asked to read it,[141] and by a comment Maupertuis made in his *Lettre sur la comète* (1742). While discussing the possibility of a comet colliding with the earth, he asks, surely not seriously, "who would be the most surprised, ourselves or the inhabitants which the comet would throw upon our earth? What [odd] figure each would find in the others!"[142]

During Maupertuis's presidency of the Berlin Academy of Sciences, its chief luminary was Leonhard Euler (1707–83), the Swiss savant whose mathematical productivity was such that it is said that he ceased to calculate and to breathe at almost the same instant. But Euler was more than energetic: His brilliance established him as the most profound mathematician of his century. And he was more than a mathematician: His *Rettung der Göttlichen Offenbarung* (1747) shows the intensity of his Christian convictions, and his widely read *Lettres à un princesse d'Allemagne* (1768–72) testifies both to his insightfulness in natural philosophy and to his skill as an expositer. His sixtieth letter (dated September 19, 1760) in the latter work sets out his views on the question of life beyond the earth. After mentioning Fontenelle's pluralist book, he remarks: "The earth, with all its inhabitants, is sometimes denominated a world; and every planet, nay, every one of the satellites, has an equal right to the same appelation – it being highly probable that each of these bodies is inhabited. . . ."[143] Moreover, he notes that in addition to the twenty-nine worlds of our solar system, each of the vast number of stars has planets and satellites associated with it, so that "we have an almost infinite number of worlds, similar to our earth. . . ." Although this extravagant claim, for which Euler provides no evidence, suggests an overly speculative author, this impression is at least partly reversed when on the topic of the best of all possible worlds Euler states:

> In my opinion, a distinction must be carefully made, between the plans of a world which should contain corporeal substances only, and those of another world, which should contain beings intelligent and free. In the former case, the choice of the best would be involved in very little difficulty; but in the other, where beings intelligent and free constitute the principal part of the world, the determination of what is best is infinitely beyond our capacity; and even the wickedness of free agents may contribute to the perfection of the world in a manner which we are unable to comprehend. (p. 208)

After recommending that philosophers debating this Leibnizian topic be more attentive to this distinction between the material and the mental, he concludes his letter by remarking: "I am too sensible of my own incapacity to enter any deeper into this difficult question."

Like Euler, Charles Bonnet (1720–93) was a Swiss-born scientist with strong religious convictions. Discoveries in the life sciences, especially in entomology and botany, had already established Bonnet's reputation when in 1764 he published his *Contemplation de la nature,* a long, reverent discussion of the scale of being, chiefly in its terrestrial aspects. Its extraterrestrial extensions also enter; in fact, in his autobiography Bonnet reveals that his pluralism extended back to his student days, "when I was enraptured by reading Fontenelle's *Mondes,* and I did not restrict myself to a single reading. One may sufficiently judge the profound impression that this inimitable dialogue must have produced on a young man whose spirit longed for new ideas."[144] That his devotion to the pluralist doctrine had not diminished when more than two decades later he wrote his *Contemplation* is suggested by such passages as that in which he endorses pluralism on teleological grounds:

> Proud and ignorant mortal! Raise now your eyes to heaven and tell me: if some of those luminaries which hang in the starry vault were removed, would your nights become darker? Do not then say that the stars are made for me. . . . Insane! You were not at all the prime object of the liberality of the creator when he set Sirius in order and when he regulated the spheres.[145]

The speculative inclinations of the Genevan scientist emerge when in a short chapter specifically on pluralism he asks:

> Who knows . . . whether at the center of each of these worlds there is not another vortex, which has its sun, its planets, its satellites, its inhabitants? Who knows whether at the center of each of these small planets there is not yet another

proportional vortex? Who finally knows the point where this decreasing sequence breaks off? (p. 15)

He repeatedly links his pluralism to the great chain of being; for example, he states: "The different beings proper to each world can . . . be envisaged as so many particular systems tied to a principal system by diverse relations, and this system is itself enchained to other more extended systems, of which the combination composes the general system." (p. 18) His delight in diversity is demonstrated in the chapter "Varieties of Worlds," where he urges that just as no two terrestrial leaves are identical, so also we should not expect that any two planets or planetary vortices are the same. As he states:

The assortment of beings which is proper to our world will probably not coincide with that of any other world. Each globe has its particular economy, its laws, its productions.

There may be worlds so imperfect relative to our own that one would find there only [inanimate] beings.

Other worlds may on the contrary be so perfect that [in them] . . . the rocks are organized, the animals reason, the men are angels. (p. 25)

In fact according to Bonnet, numerous chains of being exist, one probably for each world, which chains all run from atom to angel, and unite to form a grand cosmic scale of being (p. 29).

In two later chapters on the "Celestial Hierarchies," Bonnet asserts that angels probably travel from world to world, examining the operations of each. He then asks whether the "INFINITELY GOOD BEING" will refuse us access to these worlds; his answer is "No; because you are called one day to take your place among the CELESTIAL HIERARCHIES, you will soar, as [the angels], from planet to planet; you will go from perfection to perfection, and each instant . . . will be marked by the acquisition of new knowledges." (pp. 85–6) Bonnet further elaborates this idea in his *La palingénésie philosophique, ou idées sur l'état passé et sur l'état futur des êtres vivans* (1769), where he advocates a form of the doctrine of transmigration of souls. He even states:

The same progression which we detect today among the different orders of organized beings will no doubt be observed in the future state of our globe. . . . Man, then, transported to another residence more matched to the eminence of his faculties will leave to the *monkey* or to the *elephant* that premier place which he occupies among the animals of our planet. In this universal restoration of animals, it will be possible to find among the *monkeys* or *elephants* Newtons and Leibnizes; among the *beavers,* Perraults and Vaubans.[146]

Arthur Lovejoy, in his *Great Chain of Being,* cited this passage to support his description of Bonnet's *Palingénésie* as "one of the most extraordinary speculative compounds in the history of either science or philosophy – an interweaving . . . of geology, embryology, psychology, eschatology, and metaphysics into a general view of the history, past and to come, of our planet and of the living beings thereon – a history which may be presumed to have its counterparts on other globes."[147]

Bonnet's *Contemplation* had such appeal to his contemporaries that within five years it had been translated into German by Johann Daniel Titius (1729–96), into Italian by Abbé Lazzaro Spallanzani (1729–99), and into English by John Wesley. Danish and Dutch renderings soon followed. Probably none of the other translators shared Wesley's scruples about pluralism that led him to excise from his translation many of Bonnet's pluralist passages, but not the last of those quoted earlier.[148] In an illuminating discussion of Bonnet's philosophy that stresses its quasi-evolutionary character, Robert R. Palmer describes Bonnet as "something between a sober scientist and a dreamer," as someone who "was prompted to cosmic dreaming by the intense fervor of his religion," and as a "Protestant [who] was most eccentric in his theology. . . ."[149] All this is true, but in the present context it should be clear that Bonnet's pluralism appears eccentric only because of the extreme to which he took it. Moreover, even his pluralism, as blended with his palingenesis doctrines, although not favorably received by many of his contemporaries, attracted a substantial following a century later in France when (as will be shown) A. Pezzani, C. Flammarion, L. Figuier, and others revived it. When they did so, however, they severed its connection with Christianity, which Bonnet had believed was among its most attractive features.

Bonnet's conception of biology differed greatly from that of Georges Louis Leclerc, Comte de Buffon (1707–88), whose forty-four-volume *Histoire naturelle* established him as the premier biologist of eighteenth-century France. So broad were Buffon's interests that even astronomy enters into the 1749 first volume of the great work; in particular, it contains his cometary collision theory of the origin of planetary systems.[150] That theory can be seen in the background to that section of Buffon's 1775 *Supplément* to his *Histoire naturelle* where he provides, within the context of a discussion of the temperatures of the planets, his most extensive contribution to the pluralist debate.[151] In that section, Buffon states:

> In all the places where the temperature is the same, there are found not only the same species of plants, the same species of insects, the same species of reptiles without them having been carried there, but also the same species of fish, the same species of quadrupeds, the same species of birds. . . . The same temperature everywhere nourishes, produces the same beings. . . .[152]

Jacques Roger, a leading expert on Buffon, has commented concerning this passage that "When Buffon says 'everywhere,' he does not wish to say 'everywhere on our planet,' but 'everywhere in the universe.' . . . Terrestrial animals and even man . . . ought to be found everywhere in the universe, everywhere at least where the same physico-chemical condi-

Table 3.1. *Beginning, end, and duration of the existence
of organic nature in each planet*[a]

Object	Beginning (dated from the formation of planets)	End (dated from the formation of planets)	Total duration (years)	Duration (years) (dated from the present)
5th satellite of Saturn	5,161	47,558	42,389	0
Moon	7,890	72,514	64,624	0
Mars	13,685	60,326	56,641	0
4th satellite of Saturn	18,399	76,525	58,126	1,693
4th satellite of Jupiter	23,730	98,696	74,966	23,864
Mercury	26,053	187,765	161,712	112,933
Earth	35,983	168,123	132,140	93,291
3rd satellite of Saturn	37,672	156,658	118,986	81,826
2nd satellite of Saturn	40,373	167,928	127,655	93,096
1st satellite of Saturn	42,021	174,784	132,763	99,952
Venus	44,067	228,540	184,473	153,708
Ring of Saturn	56,396	177,568	121,172	102,736
3rd satellite of Jupiter	59,483	247,401	187,918	172,569
Saturn	62,906	262,020	199,114	187,188
2nd satellite of Jupiter	64,496	271,098	206,602	196,266
1st satellite of Jupiter	74,724	311,973	237,249	237,141
Jupiter	115,623	483,121	367,498	

[a] Buffon's table of the times at which life began and ended on each of the planets and their satellites.[154]

tions reign, where the same temperature and the same suitable materials are found."[153]

Buffon elaborates his belief in the universality of terrestrial life forms in a table that serves as capstone to the long, highly mathematical discussion of planetary temperatures included in his *Supplément*. In working out the cooling rates and times for each solar system object, Buffon assumes that the planets and satellites were thrown from the sun in an incandescent state and that correspondingly most of their heat is internal, rather than being due to subsequent heating from the sun's rays. In addition, he recounts his experiments on the cooling rates of various metallic spheres heated to incandescence. From such information he calculates that the earth, for example, consolidated down to its core in 2,936 years. In 34,270½ years its temperature had declined to the point that it could be touched. At the end of a total of 74,832 years it had attained its present temperature, whereas after 168,123 years it will have cooled to ¹/₂₅ its present temperature, making it uninhabitable. Similar calculations are presented for the other planets and satellites, leading to Table 3.1, in which Buffon sets out dates for the beginning and end of life on each solar system object, these figures being based on the figure 74,832 years that he assigned as the age of the solar system. In his comments on this table, Buffon points out that from it we learn that "organized nature, such as we know it," has not yet appeared on giant Jupiter, but that when 40,791 years from now it arises, it will endure for 367,498 years. On the other hand, life has already ceased to exist on Mars, the moon, and Saturn's fifth (and nearly on its fourth) satellite. After noting that on every other solar system object listed, including the ring of Saturn, "living nature is actually in full existence," he adds that from this we can "infer the more than probable existence of these same beings in all the bodies which make up the systems of other suns. . . ."[155]

The concluding section of a recent book on Buffon cites Saint-Beuve's statement that "The genius of Buffon was as much that of a poet as a philosopher . . ." as a basis for asking whether he should be remembered primarily as stylist or as savant.[156] Buffon's own preference can be surmised from his own derisive characterization of Thomas Burnet's speculative book on earth history: "It is a well written romance . . . which one may read for amusement, but which should not be consulted for instruction."[157] Yet so extraordinary were Buffon's pluralist speculations in his *Supplément* that it is difficult to deny that this remark could with justice be turned back on its author.

Among the few eighteenth-century authors more prolific than Buffon was a Bruxelles-born Jesuit, François Xavier de Feller (1735–1802), who devoted one of the longest of his more than one hundred books to arguing against pluralism. Feller's *Observations philosophiques sur les systêmes de Newton, de Copernic, de la pluralité des mondes* (1771)[158] consists of four fictional dialogues, in the first of which a theologian responds to a philosopher's objections to belief in God's providential care by urging that God's ways are beyond man's knowing. This theme continues into the second dialogue, where Feller has the Christian skeptic Pierre Daniel Huet (1630–1721) debate Newton. Readers distressed at the anti-Newtonianism of this dialogue were no doubt dismayed by the anti-Copernican character of Feller's third dialogue, which pits Bellarmine against Galileo. These three dialogues set the stage for the fourth, in which Kircher combats Huygens and his pluralist claims. Some of Feller's antipluralist arguments in this dialogue are not unreasonable; for example, he justifiably notes the lack of evidence for planets orbiting other stars and rightly asserts that the lifeless moon weakens the case for inhabited planets. To the objection that God would not have created the universe only for man, Feller offers three replies: (1) the final cause of natural objects is difficult to determine, (2) the celestial bodies provide chronometric and navigational assistance, and (3) the heavens encourage piety in us now and may delight us in the afterlife. However, as he admits at the end his book, his primary objection is Christian: "The mystery of the Incarnation and in general the idea that Scripture and Faith give us of the creation of the world, of providence, etc. supposes only one world inhabited by rational beings." (pp. 175–6) To Huygens's reply that "writers very zealous for religion, as Malebranche, Pluche, Dulard, etc. have not thought as you" (p. 177), evidence is supplied that Dulard and Pluche had in fact opposed the doctrine, as had, among others, the author of a 1754 book entitled *Seule religion véritable*.[159] Feller's antipluralist position, which involved him in a controversy with Lalande,[160] appeared also in his *Catéchisme philosophique,* which went through many editions and translations.[161] Although a work of modest merit, Feller's *Observations* shows that the antipluralist position had at least some support in Enlightenment France. That it would be seriously mistaken, however, to equate Feller's view with that of the Catholic clergy in general is clear from consideration of a work, authored by another abbé, that appeared nearly simultaneously.

In 1769, after having circulated in manuscript for some years, a small work entitled *Traité de la infini créé* was printed, with Abbé Nicolas Malebranche (1638–1715) claimed as its author. This he almost certainly was not, the consensus of subsequent scholarship being that the volume was authored by Abbé Jean Terrasson (1670–1750),[162] who taught Greek and Latin philosophy at the Collége de France and who in 1732, shortly after the publication of his *Sethos,* was elected to the French Academy. That Terrasson's *Traité* contains extraordinary claims no one denies, but controversy about its interpretation persists. For example, Francisque Bouillier, over a century ago, described it as having been written by "a very sincere Cartesian and Malebranchist, completely convinced of the infinity of the universe, and making great efforts to demonstrate its reconcilability with Christian faith." Aram Vartanian, in 1953, specifically rejected Bouillier's view, claiming that Terrasson used "a mask of irony . . . to disseminate materialistic and pantheistic thinking under a rather diaphanous apologetic disguise."[163] In fact, the existence of a copy of the *Traité* manuscript dated 1692[164] suggests that the book may most appropriately be read as the production of a highly imaginative twenty-two-year-old, fresh from reading Fontenelle's 1686 *Entretiens,* captivated by his Cartesianism, and so enthralled by pluralism that one wonders at times whether the *Traité* might not be a lost work of Giordano Bruno.

The first two chapters urge the infinity of the universe in both its material and spiritual aspects. In his third chapter, the teleologically minded Terrasson claims that because man can derive no use from the matter beyond the earth, "it follows necessarily that there are other intelligent creatures who profit from the rest." (pp. 46–7) After conveying his total acceptance of Descartes's vortices, he remarks concerning Fontenelle that "all that [he] has advanced concerning the inhabitability of planets is for us a settled matter. . . . All that remains for us . . . is to supplement and to clarify." (pp. 47–8) What this entails begins to come clear as Terrasson states: "We say . . . that the inhabitants of the planets are men which is what [Fontenelle] wished not to say." (p. 49) With a boldness far beyond Fontenelle's, he then develops what he believes to be the theological consequences of Fontenelle's pluralism, commenting first that neither the Church nor Scripture advocates specific views on the pluralist issue. This he does: "It is asked . . . if the eternal Word [Christ] can unite himself hypostatically to a number of men; one responds without hestiation – yes. The men would all be men-God [hommes-Dieu], men in the plural, God in the singular, because these men-God would in effect be several in number as to human nature, but they would be only one in respect to the divine nature. . . ." (pp. 54–5) Turning to the question whether or not God would have become incarnate on planets free from sin, he argues the affirmative, suggesting that such sinless beings would be more deserving of this great honor than we are. In discussing how God became incarnate on our planet, he notes the anomaly that Christ had no human father. He explains that this was because Adam's sin was that he entered into intercourse before receiving permission from God, making copulation an impure act on this planet and necessitating that Christ be born of a virgin. In summary: "We infer from all this not only that the Word has incarnated himself on all the planets, but in those where sin has not at all entered, he is born as other men." (p. 67)

Terrasson was not unaware that serious objections could be marshaled against his speculations; in fact, he devotes the latter half of his third chapter to combating a number of these. Possibly the most fundamental of these involved the view that man should "add nothing, not only to religion, which would be a crime, but even to theology. . . ." To this he responds that philosophy can change and that his views belong not to theology but to philosophy as applied to theology (pp. 74–8). At another point he counters the claim that Scripture explicitly states that there is but one Lord by interpreting it as applying only to the divine part of Christ's nature. Admitting that Christ's terrestrial incarnation and redemption have sufficient merit for the entire universe, he nonetheless suggests that because Christ has a role both as savior and as teacher, his incarnation as teacher on sinless planets is fully appropriate (pp. 89–90). In concluding his third chapter, Terrasson suggests that once the reader has overcome his fear of novel views, he will find delight in the "admirable spectacle of not only the inhabitants of earth, who are only a small number of men, but also of men in infinite number, distributed in an infinity of planets, chanting the praises of the Lord. . . ." And he adds: "What an admirable spectacle is also presented by the advance of the infinite number of men-God, who in the last day of the planet present to the eternal Father this infinite number of bands of the elect." (pp. 92–3)

Terrasson's vision of life extending throughout a spatially infinite universe is developed still further in his final chapter, which stresses the infinity of time. Not only planets pass away in his ever-changing cosmos, but vortices as well, these being replaced as new systems form (pp. 100–7). All this takes time, he realizes; in fact, he boldly suggests that the six days of biblical creation may be interpreted as referring to "six years, six centuries, six millennia, six millions of years." (pp. 118–19) Terrasson's untiring imagination even supplies a theory of angels and demons; noting that Scripture does not discuss the origin of angels, he states: "I am strongly persuaded that the angels are only the resuscitated inhabitants of destroyed planets." (p. 126) Demons similarly are derived from the wicked, and both groups attempt to influence the inhabitants of planets. Although Terrasson was no deist, as is clear from his assertion that every planet has associated with it a particular providence (p. 98) and an individual man-God, he was nonetheless an enthusiast for the principle of plenitude: "We have for first principle . . . that God made all that he can make. . . . There is nothing purely possible, says our school, and all that can be has been, is, or one day will be." (p. 147) Terrasson concludes his *Traité* by conveying his hope that the Church will approve his ideas and by offering to forsake them, should they be found objectionable.

Terrasson's treatise received a very mixed reception. On the one hand, the *Journal encyclopédique* carried two very favorable reviews of it, in both of which Malebranche is accepted as the author. In fact, the first reviewer remarks: "it is impossible to fail to recognize the strong and sparkling imagination, the sublime and often poetic thoughts of that illustrious author."[165] The second review, even more enthusiastic in its endorsement, concludes with the assertion that readers will find in it "far reaching views, grand, new and truly sublime ideas; one admires also the art with which Father Malebranche reconciles different passages of holy scripture with his system, the chain of his principles, the fortunate application that he makes [of them], and the consequences he draws."[166] A probably more typical response came from H. F. d'Aguesseau (1668–1751), chancellor of France, who saw the work before he died and labeled it "an introduction of Spinozism" designed to ridicule "Descartes and Father Malebranche by the absurd consequences drawn from a part of their principles." Moreover, he approved of the theologians labeling the work "insidious, scandalous, reckless, impious, blasphemous, and every good philosopher will add those of chimerical in its ideas, frivolous and insolent in its purpose, false and absurd in its reasoning, insane and extravagant in the confidence with which it retails the dreams of an evil soul. . . ."[167] Overall, its influence was small, few subsequent pluralist authors referring either to it or to Hay's similar speculations. Perhaps the neglect of Terrasson's *Traité* was because those who read it variously dismissed it as a product of a no longer viable Cartesianism, as a malicious attack on Malebranche, or simply as the bizarre thoughts of a juvenile mind.

Another abbé who contributed to the pluralist debate was Étienne Bonnot de Condillac (1714–80), whose *La logique* (1780) contains some comments on the methodology of pluralist arguments. In discussing the reliability of reasoning by analogy, he asserts that the force of an analogical argument is a function of whether "it is founded on relations of resemblance, on relations to the end, or on relations of causes to effects, and of effects to causes."[168] Concerning the first of these forms, Condillac states that the inference "The earth is inhabited; therefore the planets are" is "the weakest of analogies, because it is only based on the relation of resemblance." Then he adds:

> But if it is remarked that the planets have diurnal and annual revolutions, and that in consequence their parts are successively illuminated and heated, does it not appear that these arrangements have been taken for the preservation of inhabitants? This analogy, which is founded on the relation of means to end, has then more force than the first. However, if this proves that the earth is not the only inhabited body, it does not prove that all the planets may be: for that which the Author of nature repeats in many parts of the universe for the same end, it is possible that he uses sometimes only as a result of the general system: it is consequently possible that a revolution may turn an inhabited planet into a desert. (p. 412)

In his discussion of analogy based on cause–effect, he refrains from discussing the question of extraterrestrial life. Although far from profound, Condillac's brief discussion of the logic of the pluralist argument from analogy was significant, because he seems to have been the first to have attempted an analysis of this crucial but complex issue. Mill, Olmsted, and others followed – with scarcely better success.

Whereas pluralism had been discussed by Feller and Terrasson from a theological point of view and by Condillac in a logical context, a fourth abbé, Jean Jacques Barthélemy (1716–95), treated it from the perspective of a classicist in his multivolume *Voyage de jeune Anacharsis en Grèce dans le milieu du quatriéme siècle avant l'ère vulgaire* (1788). Barthélemy's use of a fictional narration of the travels of a Scythian descendent of the philosopher Anacharsis as a device to detail Hellenic life and thought won him readers on both sides of the English Channel and election to the French Academy. From discussions with Callias and Euclid, young Anacharsis learns of the advocacy of extraterrestrials by Democritus, Leucippus, Petron, the Pythagoreans, Xenophanes, and others. In response, he exhibits amazement at the size of the universe and skepticism about their scantily supported pluralist claims.[169] Similar sentiments, one may suppose, were held by Barthélemy, whose chief concern was probably to remind his readers that the pluralist controversy, so common among their contemporaries, had originated more than two millennia earlier.

In the same year (1788) in which Barthélemy's *Voyage* was printed, the young poet Louis de Fontanes (1757–1821), who later became prominent as a politician, published the most famous of his poems. Entitled "Essai sur l'astronomie," its pluralism may have come from Pope, whose *Essay on Man* he had translated in 1783, but its proto-romantic pessimism was of Fontanes's forging. Both characteristics appear as he laments concerning other worlds:

> These changing states
> Yield as our own to the reign of the fates,
> And as our own have seen come to birth
> A thinking race, holding learning of worth; . . .

In fact, he opines that they, too, have had their "Pascal, Leibnitz, Buffons" and wonders whether or not on a nearby planet someone "Gives himself to transports as sweet as mine." He goes on to ask, were such a being to gaze on the earth,

> Could he surmise that in this doeful place
> Crawls an immortal being with tear-stained face?
> Unknown denizens of these distant domains,
> Do you sense our needs, our joys, our pains?
> Do you know our arts? To you has the Divine
> Given senses less impure, fates less confined?

Although disposed to believe that Plato's scale of being attains its fulfillment in other worlds, he nonetheless warns:

> Oh men, do not imitate earth's sad state!
> You would grieve for us, should you learn our fate;
> Your tears would dampen our dolorous displays.
> All ages in mourning, all in the same ways,
> Flow on without check, crushing from every side
> The thrones, the altars, the empires scattered wide;
> All without pause beset by irksome blows
> Pass on as I recount our suffering shows.
> You men, our equals, may you alas be!
> More wise, more united, more happy than we![170]

With Fontanes we see that the pluralist doctrine, long a staple of Enlightenment optimism, was showing its remarkable adaptability by becoming a resource of romantic melancholy.

When the pluralist debate in French writings from the 1750-to-1790 period is viewed in perspective, diversity emerges as its dominant characteristic. Diverse in background were the participants, and diverse were the views they developed. Nonetheless, some patterns can be perceived. The great majority of authors, be they atheists, deists, or clergy, Protestant or Catholic, philosophers, physicists, poets, or biologists, favored the doctrine, Feller being the sole clear-cut exception. Moreover, even among authors championing empiricist methodologies, teleology played a larger role than telescopes in their widespread acceptance of the great chain of being. Religion was repeatedly involved, although few correlations between an author's creed and his conceptions of extraterrestrials are evident. Many saw in pluralism an attractive medium for arguing views to which they were already committed. Voltaire, for example, availed himself of the doctrine to support deism and to attack Leibniz, Maupertuis, Wolff, and others whose views displeased him, whereas pluralism provided a basis for flights of fancy into palingenesis (Bonnet) or theologies involving multiple "men-God" (Terrasson). Despite all this, it is possible to detect among some of the deepest thinkers in this group – d'Alembert, Euler, Maupertuis, possibly Voltaire – an element of caution, a sense that the pluralist arguments were frail and elusive. This whisper, when combined with the tendency near century's end to question teleology and the great chain of being, suggests that confident proclamations of pluralism on such grounds were destined to encounter difficulties.

4. Advocacy of extraterrestrials elsewhere in Europe: from Klopstock's cosmic Christ to Jean Paul's "Speech of the Dead Christ"

The idea of a plurality of worlds, a basic element in the French Enlightenment, was scarcely less popular in the German *Aufklärung*. Kant, Lambert, Bode, and Schröter, as shown previously, were among its most prominent advocates, but their writings by no means adequately indicate the degree to which pluralism had permeated that cluster of countries in which German was the dominant language. Recent publications by Walter Schatzberg and Karl S. Guthke document that dozens of eighteenth-century German authors enthusiastically espoused ideas of extraterrestrial life.[171]

As Professor Schatzberg shows, by 1750, German translations of Derham's *Astro-Theology,* Fontenelle's *Entretiens sur la pluralité des mondes,* Huygens's *Cosmotheoros,* and Wilkins's *Discovery of a New World* had long been available. This helps explain the presence of pluralism in the four most popular German compendia of science published in the 1700–50 period.[172] In his 1729 compendium *Physica oder Natur-Wissenschaft,* Johann Jacob Scheuchzer (1672–1733) cites Huygens in support of pluralism, although this Swiss scientist goes beyond his Dutch predecessor by championing even lunar life. The endorsement of pluralism by Leibniz, on the other hand, was crucial in the compendia compiled by Christian Wolff (1709–54), the most widely read of these writers. With greater caution than Scheuchzer or Wolff, Johann Gottlob Krüger (1715–59) included pluralist ideas in his Wolffian *Naturlehre* (1740). The fourth of these authors, Johann Christoph Gottsched (1700–55), was certainly the most active in advocating pluralism. His *Erste Gründe der gesammten Weltweisheit* (1731), written to expound Wolffian natural philosophy, begins with a frontispiece featuring stars, each surrounded by planets. Beneath this appears the quatrain:

> At this stare mind and wit, the soul doth fade away,
> Amidst these wondrous worlds, amidst this grand display;
> Oh what in this is man? Nothing would he be named,
> Should he not see the grandeur in what God hath framed.[173]

In the book itself, he argues for extraterrestrials, urging with Fontenelle that they need not be similar to men.[174] The prolific Gottsched further contributed to the spread of pluralism by translating Fontenelle's *Entretiens* and Leibniz's *Théodicée,* as well as by employing pluralist themes in a number of his poems.[175] One example is his "Als der Verfasser sein funfzigsten Jahr zurücklegte" (1750), in which he gives thanks to God for the benefits that, despite his insignificance in the vast universe, he has enjoyed for fifty years. After mentioning the astronomical arrangements that produce our days, nights, and seasons, he notes that other planets possess comparable features:

> No less in worth the orbs on high
> Which still the sun doth glorify,
> As near and far they ever roam round.
> They warm themselves admidst her rays,
> Despite the changes in their ways;
> How could no creatures there be found?
> How could alone our planet be,
> Abode for beast and humanity?[176]

In short, "Where winter, spring, and summer are/There ensouled life will not be far." After citing the moons of Jupiter and Saturn as evidence for life on those planets, he affirms his belief in life on our moon. Later in the poem, he turns to comets:

> Although they be now cold, now warm,
> Embraced by Thine paternal arm,
> They can become by creatures covered,
> If they be spared Sol's scorching light
> Beneath that haze that stirs our fright.
> (lines 164–8)

Gottsched's efforts to increase the level of scientific learning deserve praise, but as Schatzberg notes and as the foregoing verses illustrate, he at times indiscriminately mixed solid science with speculative fancies.[177]

A tradition of employing pluralism in German poetry had already been established in the decades before 1750 by such authors as Brockes, Haller, Mylius, Lessing, Hagedorn, and Kleist. Most of the poetry of Barthold Heinrich Brockes (1680–1740) of Hamburg appeared in the nine volumes of his *Irdisches Vergnügen in Gott* (1721–48). Despite the "terrestrial" in their title, these volumes abound in astronomical allusions, many being pluralist. In his 1722 New Year's poem, for example, Brockes presents a discussion between a philosopher and a Christian, the latter asking: "Of more than one world what say you?/How do you prove that it is true?/The new heresy I cannot believe."[178] This "new heresy" raises such questions as

> Should Christ have died
> Solely for a single world
> Or how have the first Adams
> Fallen on all of them also?
> Have a thousand Eves also been deceived
> By a thousand snakes through a thousand apples?
>
> (I, p. 435)

To these and other questions, Brockes offers a number of replies, assuring the reader that the idea of our earth as a "grain of sand" (I, p. 432) in a vast universe filled with inhabited worlds supports religion and inspires praise for God. Other poems involve celestial voyages to planets where the inhabitants are more pious than earthlings even though they have but a single sense, or to Jupiter where a superrace dwells.[179]

Brockes's pluralism interestingly contrasts with that of the religiously more traditional, metaphysically more subtle, and scientifically more astute Albrecht von Haller (1708–77). Guthke's analysis of the pluralism of this famous Swiss scientist, philosopher, and poet portrays him as grounding his pluralism more on teleology than on the principle of plenitude on which Brockes primarily relied.[180] This characteristic as well as Haller's pessimism are apparent in the following lines from his "Ueber die Ursprung des Uebels":

> Perhaps this world of ours, which like a grain of sand
> Floats in the vast of heaven, is Evil's fatherland;
> While in the stars perhaps dwell spirits far more fair
> Vice reigning ever here, Virtue triumphant there.
> And yet this point, this world, whose worth appears so small
> Serves in its place to make complete the mighty All.[181]

In his "Unvollkommene Ode über die Ewigkeit" (1743), Haller extends pluralism to past and future worlds in a stanza that Kant quoted in his 1755 book:

Infinity! who measures thee?
To thee worlds are as days, and persons as moments.
Perhaps the thousandth sun passes over,
And thousands remain behind,
As a clock, given motion by a weight
So runs on a sun, moved by God's power:
Its drive runs down, and another starts,
You however remain, and number them not.[182]

When in March 1744 Abraham Kästner (1719–1800) published a
poem denying the possibility of cometarians, Christlob Mylius (1722–
52), fresh from scientific studies at Leipzig, rushed to their rescue with his
"Lehrgedicht von dem Bewohner der Cometen." To doubt the habitability of comets was to him as scandalous as questioning life on planets:

Why not remove life from the planetary sphere?
They make God's power and wisdom more clear
If populated they be. This also do the comets;
In fact, better, because they outnumber the planets.
Without design, a clever builder does not act.[183]

Not only teleology but also the variety exhibited by nature suggest to
Mylius the probability of forms of life comparable to our amphibians.
His cometary creatures "live in the air, but do not succumb to fire. . . ."
(p. 358) Mylius also pressed his pluralism in prose, for example, in
articles in *Der Naturforscher*, a short-lived weekly he edited in an effort
to popularize science. In one of its articles, he invokes the idea of the
great chain of being to support extraterrestrials, suggesting that the wide
array of terrestrial creatures "is without doubt only an infinitely small
link in the infinitely long chain of intelligent beings."[184] On the death
in 1754 of Mylius, who had discussed space travel in both poetry and
prose, his friend Kästner published an article praising him, as well as a
poem portraying Mylius's soul as traveling through the solar system,
beholding the distant side of the moon and the "eternal souls" of the
Martians.[185]

The demise of Mylius was also mourned by his cousin, Gotthold
Ephraim Lessing (1729–81), who prepared for publication the writings
of the short-lived poet-scientist. The free-thinking Mylius, who, Lessing's
biographers report, influenced Lessing's break with traditional religion,[186] probably encouraged his cousin to write the two pluralist poems
that Lessing published in Mylius's *Naturforscher*. Lessing's "Die Planetenbewohner" consists of this witty suggestion:

In charming fancies to take delight,
To populate the planets of the night,
Before on safe grounds you can infer
That on those planets wine is there:
That is called populating too soon.
Friend you should first divine
Whether in new worlds there is wine,
As on our earth we know there be:
Then, believe me, any child will see
That those worlds will have their drinkers.[187]

Lessing's "Die lehrende Astronomie" contains a more serious claim that suggests the influence of Haller. Contemplating the stars instills humility, he urges; moreover, in a more metaphysical vein, he exclaims:

There I see with astonished glance
A shining host of new worlds;
Confident that wealth will be
Not so valued there as virtue.
Perhaps there in Orion's regions
Will Truth, free of delusion, shine.

"Evil," a very wise man writes,
"Has among us won its kingdom."
Very well, but good is on the throne
In countless greater worlds.
With regret the former God created,
To complete the scale of things.[188]

Two of the more prominent German writers who published pluralist poetry in the 1740s were Friedrich von Hagedorn (1708–54) and Ewald von Kleist (1715–59). Hagedorn, whom Lessing in 1746 labeled "the greatest poet of our times,"[189] had a few years earlier published "Die Glückseligkeit," praising the new science, endorsing life on the planets, and even speculating that on them "Perhaps another Wolf [sic], another Newton teach."[190] Kleist's "Lob der Gottheit" appeared in the same decade, continuing the trend of praising God for his many inhabited worlds. At one point in that poem, Kleist asks:

Who bids millions of suns with majesty and splendour shine?
Who doth on their wondrous course to countless worlds their paths assign?
Who endows with life each circle? Who unites the wondrous band?
Thy lips' gentle breathings Lord! yea, thy most high and dread command.[191]

During the 1750s, no fewer than a dozen authors[192] published pluralist poetry in German, this trend being partly due to the appearance of German translations of Thomson's Seasons in 1745 and Young's Night Thoughts in 1752. A number of these authors were minor figures, but Friedrich Gottlieb Klopstock (1724–1803) ranks as a poet of classic stature. Only the first three of the twenty cantos of Klopstock's Der

Messias (1748–73) had appeared when Lessing labeled its author Germany's greatest genius and Bodmer proclaimed that a goldern age of German literature had begun.[193] Subsequent assessments of his poetry have varied, some preferring his odes to his epic, but his place in the pantheon of German literature has remained secure.[194] What gives Klopstock relevance in the present history is that throughout his poetry, pluralist themes occur with greater frequency than in the writings of any other major poet.

Frequently called the German Milton, Klopstock, in composing *Der Messias,* faced conceptual problems that had far less intensely troubled Milton when nearly a century earlier he had sought to find in Scripture the materials of epic poetry. Milton could assume acceptance of Christianity among his readers and faced few tensions with the pluralism of seventeenth-century astronomy. This was scarcely the case for Klopstock, writing when pluralism was highly popular and when, according to Albrecht Ritschl, there occurred "The Entire Disintegration of the Doctrines of Reconciliation and Justification by the German Theologians of the Illumination."[195] Klopstock's responses in his *Messias* to this pair of problems were bold and original. Pluralism he accepted fully and enthusiastically; astronomy abounds in his poem, where it is less the hills of Jerusalem than the infinite universe that forms the stage for the epic events he recounts.[196] What was still more daring was the degree to which he espoused a Christocentric cosmos. While others laid emphasis on Jesus as exemplary man, inspired teacher, and prophetic preacher, Klopstock centered his epic on Jesus as God-man, redeemer of sinful mankind, and (relying on the first chapter of Colossians) creator of the universe. Not nature and nature's God, not God the Father, but Revelation and Christ are central in *Der Messias.*[197] Moreover, Klopstock's Christocentrism made his pluralism all the more striking. A number of earlier German theologians (as noted before) had touched on the reconcilability of Christianity and pluralism, but none faced as fully as Klopstock the task of articulating the relations between a centerless physical universe and a spiritual universe centered on events occurring on a planet the size of a "sand grain."

Klopstock's *Der Messias* recounts in twenty cantos the suffering, death, and resurrection of Christ, presenting his actions within the context of a cosmos abounding with inhabited worlds. In the first canto, as Gabriel carries a message from Christ on the Mount of Olives to the Father, the angel passes by earths described as "Small, unremarkable, as under the foot of the wanderer/Lowly dust, inhabited by worms, rises and falls."[198] Our planet, although repeatedly portrayed in just such terms, is hailed from other worlds as

> . . . Queen amongst the earths,
> Focal point of creation, most intimate friend of heaven,
> Second home of the splendor of God, immortal witness
> Of those secret sublime deeds of the great Messiah!
>
> (I, 517–20)

Inhabitants of Venus, Jupiter, and Saturn also appear in this canto (I, 640–1). A crucial passage occurs in the fifth canto when the Father, descending to Christ on Mount Tabor and passing over "A thousand sun-miles, the space from sun to sun" (V, 147), comes to a particular planet "where persons [Menschen] were." These are described as

> Persons, as we in form, still fully innocent, not mortal persons,
> And their first father stood fully in masculine youth,
> Although centuries distant from his early years,
> Surrounded by his unaged children.
>
> (V, 154–7)

This unfallen Adam then reveals to his descendants the tragic events that occurred on earth and the fallen condition of earth's inhabitants:

> Far from us, on one of the earths, are persons such as we,
> Similar in form; deprived by themselves of innate innocence
> And of godly form – mortal in fact they are!
>
> (V, 205–7)

As this and other sections of the *Messias* suggest, and as Hans Wöhlert stated in his analysis of the epic, Klopstock reconciles pluralism with Christianity by assuming that "only the inhabitants of the earth have fallen into sin and they alone need salvation through a divine mediator."[199] Nonetheless, the poet stresses that Christ's actions on earth bring benefit throughout the cosmos: "The triumph of Christ extended to the stars of innocent persons/And of immortals."[200]

In writing his epic, Klopstock faced numerous problems of detail: where to place hell – at the edge of creation; where to place the patriarchs – in the sun; and how to portray the greatness of God, the latter problem being ameliorated by the infinitude of his cosmos. Such traditional themes as the chain of being appear, and the influence of Young is evident as well.[201] Delicate as the task he undertook was, bold as some of his approaches to it were, Klopstock's *Der Messias* attracted a wide readership. For example, the poet Christian Schubart (1739–91) wrote him in the mid-1770s to say:

I recited the "Messiah" in public in the concert-hall at Augsburg. I began with a few chosen listeners who were pleased with it. The company soon became much too large for my little room; then the magistrate fitted up a public hall for me, and the size of the audience soon rose to several hundred. All volumes of the "Messiah" . . . were soon bought up. . . . [H]igh and low, clergy and laity, Catholics and Lutherans came to the lecture with copies of the "Messiah" under their arms.[202]

Although some were surely attracted to Klopstock's epic by its sacred theme, others found delight in its stylistic richness, lofty tone, and conceptual originality. Its fame spread through much of Europe as various authors attempted translations.

The judgment of history has been that Klopstock's brilliance shone most brightly in his odes, a number of which develop pluralist themes. For example, in his "Die Genesung" (1757) he writes of a spirit having "Hailed the inhabitants of earths and suns/Greeted the countless denizens/Of distant comets."[203] The image of an "ocean of worlds" figures prominently in three early poems: "Das Landleben" (1759), "Die Frühlingsfeier" (1759), and "Die Welten" (1764). The second of these, his most famous ode, begins:

> Not into the ocean of all worlds
> Will I plunge! Nor hover
> Where the first created, the joyous choir of suns of light
> Worship, deeply worship! And fade into ecstacy!
>
> Only around the drop in the bucket,
> Only around the earth will hover and worship!
> Hallelujah! Hallelujah! The drop in the bucket
> Also flowed from the hand of the Almighty!
> When from the hand of the Almighty
> The greater earths gushed!
> The stream of light rushed, and became the seven planets.
> Then you, oh drop, escaped from the hand of the Almighty! (p. 89)

Slightly later, Klopstock seeks to relieve the tension between the minuteness of man and the vastness of space by the stanza:

> Who are the thousand thousands, who are the myriad,
> Who dwell and have dwelled on the drops? And who am I?
> Hallelujah to the Creator! More than the earths which flowed,
> More than the seven stars which streamed from the rays! (p. 89)

In his "Die Verwandelten" (1782) readers encountered a remarkable Saturnian system:

> Ring of Saturn, girdle of satellites,
> Remote and countless, they roll around
> The great star, lighting it and by it enlightened,
> Changing the heavens above! (p. 133)

Each of these satellites is portrayed as inhabited by creatures far more richly endowed than earthlings:

> The destiny of your denizens was a more joyous bliss
> Than we know: the bitter no doubt runs
> In their cup as in ours; yet easily dissolved
> It runs, and in drops.

> More delicate ears, a brighter eye, they see
> Streams and groves in nearby stars, hear
> One star loudly soar by, the echoes
> Sounding in a second. (pp. 133–4)

Passing over for the moment his "Psalm" (1789), we come to the last years of his life, when he wrote "Die unbekannten Seelen" and "Die höheren Stufen." The former poem contains a favorable reference to William Herschel's contention that life exists in the sun and stars (p. 171), whereas the latter, his last poem, presents a vision of the inhabitants of Jupiter:

> . . . I saw living beings
> Of very different form. Each form
> Frequently became another; it appears that in beauty
> It outdid itself each time it changed. (p. 179)

On March 14, 1803, Friedrich Klopstock died in Hamburg. Ambassadors of seven countries and a vast crowd attended his funeral, at which a chorus of over one hundred voices sang his "Psalm." That poem, composed as a meditation on the "Our Father," begins:

> Around suns circle moons
> Earths around suns
> Hosts of all suns travel
> Around a great sun:
> "Our Father, who art in heaven!"
>
> On all these worlds, illuminating and illuminated,
> Live spirits varying in vitality and bodily frames;
> But all contemplate and take delight in God.
> "Hallowed be Thy name!"[204]

So impressed was Johann Jakob Bodmer (1698–1783) by the first three cantos of Klopstock's *Der Messias* that this prominent Zurich professor invited the young poet to continue his composition while living in Bodmer's house. During the eight months (1750–1) Klopstock lived with Bodmer, they no doubt spoke of the admiration they shared for Milton's *Paradise Lost,* which Bodmer had translated, and for Young's *Night Thoughts,* and possibly of the plurality of worlds theme that Bodmer was including in his own epic poem, *Der Noah* (1752), treating the days of the scriptural deluge.[205] Astronomy enters that poem not only with the Whistonian comet-caused deluge but also through the telescope that Bodmer bestows on Sipha, Noah's scientifically inclined fellow patriarch. From that instrument and from Raphael's revelations to Noah comes information that stars are surrounded by inhabited planets, the sun itself is inhabited, and at least one planet has been spared the ravages of sin. Bodmer's solarians are described as

> Not of human form, and not of terrestrial dust;
> But with their own beauty adorned from the stuff of light,
> Worthy of inexhaustible skill, with finer limbs,
> In accord with their location, to endure the sun's heat. . . .[206]

Bodmer's determination to assign a knowledge of pluralist astronomy and, no less anachronistically, a powerful telescope to his patriarchs strikingly shows the centrality of pluralism in his thought.

Only a few months after Klopstock's departure, Bodmer opened his home to Christoph Martin Wieland (1733–1813), as yet in that period of "seraphic Christianity" that preceded his turn to a more sensual and skeptical approach to life and literature.[207] Four of Wieland's earliest works include pluralism, the first of these being *Die Natur der Dinge* (1752), which despite taking its title and format from Lucretius, urges an idealistic, largely Leibnizian philosophy. Nonetheless, this son of pietist parents shared the Roman materialist's belief in inhabited worlds. Endorsing the idea of a chain of being and even espousing a form of metempsychosis advocated by Georg Friedrich Meier (1718–77), who wrote a preface for the poem, Wieland expresses his enthusiasm for pluralism in such passages as this:

> Oh how amazed my mind, almost ceasing to think,
> When into these distant depths its sight does sink,
> Which creature cannot measure, where without thinking,
> Revolve citizened spheres beyond human counting.[208]

Wieland's *Zwölf moralische Briefe,* also published in 1752, supports pluralism; in fact, its concluding letter recounts a cosmic voyage to a distant planet. Shortly after joining Bodmer, the young poet composed a defense and explication of Bodmer's epic poem. Entitled *Abhandlung von den Schönheiten des epischen Gedichts Der Noah* (1753), it defends not only Bodmer's style but also his Old Testament pluralism: "there is nothing more believable than that the Patriarchs, so rational and free of all prejudices, had no objection to allowing the supposition that the heavenly bodies are separate worlds, especially since they would make certain of this from their contact with the angels. . . ."[209] In the same year, Wieland issued his *Briefe von Verstorbenen an hinterlassene Freunde,* at least three letters of which make use of a pluralistic perspective. For example, in the ninth and last letter, Theotima describes a planet whose inhabitants have not fallen into sin.[210] By 1754, Wieland had left Bodmer's home, and in 1758 he admitted to a friend: "There has been a time when I was charmed by [Edward] Young. That time has passed."[211] In 1759, Wieland left Zurich and by 1772 was on his way to Weimar, where his reputation reached new heights. Unlike Klopstock, however, Wieland seems to have made scant use of pluralism in the publications of his later years.

Not only poets but also philosophers and religious writers drew on pluralism to give cosmic scope to their writings. Among such authors, Gellert, Sturm, and especially Herder merit consideration. Christian Fürchtegott Gellert (1715–69), a philosophy professor at Leipzig, was known partly for his poetry, but above all for his ethical writings, in both of which he urged that "The heavens declare the glory of the Eternal." This was in fact the opening line of his most famous poem, "Die Ehre Gottes aus der Natur," which Ludwig van Beethoven (1770–1827) immortalized in music. The heavens of both Beethoven and Gellert were inhabited; the famous composer, for example, wrote: "when in the evening I look up at the sky and see moving eternally in their worlds the host of bodies of light which are called suns and earths, then my mind soars beyond these many millions of distant stars to the one source from which all things created stream and from which new creations will eternally continue to stream."[212] Gellert's pluralism is evident in his *Moralische Vorlesungen,* delivered to large groups of students in Leipzig during the last decade of his life. In one of his lectures, he asserts that earthlings "make up less than a thousandth part of the solar system's population," and given that stars are orbited by inhabited planets, the universe contains "an infinite crowd of creatures [whom] the Lord of all nature creates, knows, and conserves!" This leads to the pious exclamation: "great God, what myriads of nations praise your creating and sustaining hand. . . ."[213] Gellert's lectures were soon translated into English, a tribute also paid to *Betrachtungen über die Werke Gottes im Reiche der Natur* (1772), by Christoph Christian Sturm (1740–86), a prolific professor and preacher who had studied theology at Jena. In that book, Sturm asks whether or not it is conceivable that God created the "millions of worlds and of suns" merely to help man "measure time and ascertain the return of the seasons?" He answers:

Certainly not. We have every reason to believe that each of the fixed stars . . . are suns equally resplendent as that which beams on our horizon . . . ; have each worlds revolving round their centre. . . . We may also suppose that these spheres serve as abodes to different orders and species of living creatures, all rejoicing in the power and celebrating the magnificence of God.[214]

Sturm reveals that his purpose in presenting these "conjectures" is because they "fill the mind with awe and reverence, open it to a vast and boundless field of thought, do away the contracted and partial notions we may entertain of ourselves, and tend to soften and ameliorate our hearts. (p. 50)

In every study of German thought in the latter half of the eighteenth century, a central role is assigned to Johann Gottfried Herder (1744–1803) as critic of *Aufklärung* ideals, as the theoretician of the *Sturm und Drang,* and as precursor of many of the themes of the romantic movement. With great erudition and originality, Herder pioneered in an array of disciplines from comparative literature to world history. That Herder was moreover deeply interested in astronomy and the pluralist doctrines associated with it[215] is evident from his unpublished *Anfangsgründe der Sternkunde,* which consists of a student's notes from lectures the young Herder delivered in 1765 while teaching at the cathedral school in Riga. The source of the knowledge of astronomy manifested in this manuscript is certainly Immanuel Kant, with whom Herder had studied at Königsberg while preparing for the Lutheran ministry. Solid indications exist for believing that Herder at this time was among those few who had read Kant's *Allgemeine Naturgeschichte* (1775), the influence of which may be detected in Herder's remark concerning the moon that "its creatures can be of an entirely different species [Art] than we."[216] Associated with Herder's pluralism was a belief in palingenesis or metempsychosis. In a 1769 manuscript, he asks regarding the afterlife: "What does my soul do? it remains in the universe; . . . it begins equally to build itself again a body. where? how? at what time? of what form?"[217] To these questions Herder published answers in the 1780s, by which time he had moved to Weimar. In his *Ueber die Seelenwanderung* (1785), he states:

Perhaps also resting places, regions of preparation, other worlds are ordained for us, on which we, as on a golden celestial ladder, ever more easily, more actively, more happily climb to the source of all light, and always seek but never reach the center point of the pilgrimage, the lap of the Deity. . . . Where I meanwhile may be, and through whatever worlds I am led, I am and remain always in the hand of the Father, who brought me here and calls me further: always consequently in God's infinite lap.[218]

Herder rejected some versions of metempsychosis, such as the idea that human souls could assume animal forms or even other earthly human bodies; moreover, whereas Kant believed that the inhabitants of the planets more remote from the sun were of superior nature, Herder favored the idea that perfection was greater in the vicinity of the sun.[219]

Herder's magnum opus was his *Ideen zur Philosophie der Geschichte der Menschheit* (1784–91), designed to situate eighteenth-century European man not only within world history but also within the universe. He undertakes the latter task in its first book, urging that the earth must be "considered in the chorus of worlds in which it is placed." (XIII, p. 13) In regard to this cosmic perspective, he refers the reader to Kant's *Allgemeine Naturgeschichte,* Lambert's *Cosmologische Briefe,* and Bode's *Anleitung zur Kenntniss des gestirnten Himmels.* Turning to the solar system, he suggests that a journey to the other planets could provide us insights on the "formation of our earth" and on the "relationship of our terrestrial species to the organized beings on other worlds." (XIII, p. 17) Without such a journey, knowledge of the nature of the inhabitants of other planets is impossible, because we do not possess sound physical knowledge of them or of their formation and comparative structures. And he cautions:

What Kircher and Swedenborg dreamed on this matter, what Fontenelle jested, what Huygens, Lambert, and Kant, each in his own way, have supposed, are proofs that in this area we can know nothing, should know nothing. We may make our estimation ascending or descending; we may place perfect beings near to or far from the sun; yet it all remains a dream which is step by step destroyed by our lack of a progression in the variety of planets. . . . (XIII, pp. 17–18)

Despite these censures, Herder does not hesitate to set out – in the next paragraphs – his own contra-Kantian position, describing it as capable of encouraging us to believe that

. . . perhaps after we have attained the summit of the organization of our planets, it could be the destiny and the progress of our future to wander on more than one of the other stars, or that finally perhaps our destination may be to associate with all the perfected creatures of so many and so varied sister worlds.[220]

He also speculates that in this process, beings may all be converging on suns, adding in a footnote a reference to Bode's paper arguing that the sun may be inhabited (XIII, p. 20).

Herder's advocacy of a form of palingenesis continues later in his book. In this context, he notes that animals depend on plants for their food, colorfully illustrating this point by describing the elephant as "a grave of millions of plants; he is however a living, operating grave; he animalizes its parts into himself; the lower powers pass over into his form of life." (XIII, p. 178) He also endorses the idea of a great chain of being, finding support in it for metempsychosis:

Everything in nature is united: one state dies for another and prepares for it. If thus man culminates the chain of terrestrial organization as its highest and last link; so he thereby also begins as its lowest member the chain of a higher genus of creatures. . . . He can pass over into no organized being on the earth . . . ; there must be a step before him, which is close by and still as elevated above him as he is over the animals. . . . (XIII, pp. 194–5)

His confidence in and enthusiasm for this view are revealed in his claim that it "rests on all the laws of nature" and provides the key to man's place in history (XIII, p. 195).

Such passages are difficult to reconcile with Herder's previously cited criticisms of such authors as Swedenborg, Fontenelle, and Kant. They are all the more striking in a Lutheran clergyman who in his *Gott, einige Gespräche* (1787) excoriated those who on physicotheological grounds had specified God's reasons for assigning a ring to Saturn, a moon to the earth, and even a satellite to Venus. All this, he notes,

. . . had to be shamefully taken back when it was found that Venus had no moon and that with respect to the illumination of Saturn's inhabitants from its

ring of diamond as with our own moon, conditions were very different from
what they at first appeared. All these deceptions in which the name of God is
abused, escape the modest natural scientist who informs us not about particular
decisions from the chamber of the divine council, but rather investigates the
conditions of things themselves and notes the laws implanted in them. (XVI, p.
491)

In the last years of his life, by which time, according to H. B. Nisbet,
Herder's "religious beliefs no longer included any seriously transcendental or supernatural elements, but virtually coincided with his ideal of
'Humanität' . . . ,"[221] the aging scholar began editing (in fact, for the
most part, writing) a journal called *Adrastea*. Scientific essays, including
some on astronomy, abound in it, pluralism appearing as well. For example, in an 1802 essay entitled "Wissenschaften, Ereignisse und
Charaktere des vergangenen Jahrhunderts," he celebrates the advances
brought about by the telescopes of Herschel and Schröter. Bode and
Herder's friend Friedrich von Hahn are also praised for their solar theories. He speculates whether or not the moon was inhabited in the past, or
will be in the future: "On that, the reflecting telescope has instructed us!
How long will the future habitability of the moon be further prepared! Its
vegetation scarcely begins under a thin atmosphere. . . . the flowing
aether will one day also give life, and prosperity, and growth to the
moon."[222]

After Herder died in 1803, Bode quoted the final part of Herder's
Adrastea, remarking: "These are the last lines, with which the worthy
Herder in Weimar concluded the tenth part of his *Adrastea* and simultaneously his literary and terrestrial life." The lines, from a poem by H. W.
von Gerstenberg (1737–1803), are:

> Removed into new regions
> My inspired eye looks about – sees
> The reflection of higher divinity,
> His world, and this heaven, its canopy!
> My weak spirit humbled in dust
> Grasps not his miracle and grows silent.[223]

Herder's pluralism and associated palingenesis doctrines raise the intriguing question of the relationship of pluralism to the romantic movement. Professor Guthke has suggested that as German romanticism concentrated more on the individual than on the cosmos, on the inner rather
than outer, pluralism lost some of its appeal. He cites in this regard a
poem of Friedrich Schiller (1759–1803), entitled "An die Astronomen":

> Prate not to me so much of suns and of nebulous bodies;
> Think ye nature but great, in that she gives thee to count?
> Though your object may be the sublimest that space holds within it,
> Yet, my good friends, the sublime dwells not in regions of space.[224]

Such an interpretation would entail seeing Herder's pluralism as a remnant of *Aufklärung* rationalism, surviving even into his latest writings. This is not an implausible viewpoint, but in its history pluralism has shown a remarkable capacity to adapt to the most diverse movements of thought. Guthke has also cited on behalf of this position that passage from the writings of Johann Wolfgang Goethe (1749–1832) where he asks: "For what would serve all the pomp of suns and planets and moons, of stars and Milky Ways, of comets and nebulae, of worlds coming and ceasing to be, if finally a happy human does not unconsciously enjoy his existence?"[225] This passage raises the question whether or not Goethe accepted pluralism. The best answer that can be supplied at present is a cautiously affirmative response.[226]

As to the pluralism of Goethe's contemporary Freidrich Leopold Stolberg (1750–1819), there is no question. Moreover, Stolberg's "Die Fülle des Herzens" (1777) is discussed as being in the spirit of romanticism by Alexander Gode-von Aesch in his *Natural Science in German Romanticism*. In his essay, Stolberg warns that to the scientist "who is without feeling, the knowledge he possesses is like a dead treasure; to him who has that feeling his knowledge is a source of pure joys, of elevating emotions, of noble thoughts." Fullness of heart, Stolberg maintains, produces not only joy, but populated planets as well:

Without the warm participation of the heart the sciences are almost nothing. It is only through this participation that astronomy delights us in revealing to us the sparks of the heavens as thousands of suns, each one surrounded by terrestrial globes and each one of these inhabited by sentient and immortal beings.[227]

An intuition not unrelated to Stolberg's "fullness of heart" formed the basis of a 1792 paper entitled "Etwas aus der transcendenten Astronomie," by Ernst Gottfried Fischer (1754–1831), a science and mathematics teacher at a Berlin gymnasium. Written for Bode's *Astronomisches Jahrbuch,* this paper purports to provide a nonempirical, transcendental proof of pluralism. Toward this end, Fischer first calls into question a claim made by the Dutch physicist Willem Jacob s'Gravesande (1688–1742) that although the similarities of the other celestial bodies to the earth permit us to assert the possibility of their habitation, we cannot conclude for the probability of this because the purposes God could have had in creating them are infinite.[228] To this, Fischer replies: "The only possible purpose of habitability, *which I recognize,* is the existence of inhabitants. . . ." (p. 223) No other purposes are known, he insists, or even conceivable. That Fischer's position is more complex than this becomes clear when he urges that the observed similarities of the heavenly bodies to the earth are in fact very limited and are confined primarily to the planets. Such similarities, he claims, fail to explain why "almost every natural scientist is inclined to maintain that *every* heavenly body is habitable. . . ." (p. 228) He asserts that this conviction

. . . rests on an idea that lies developed or undeveloped in every human soul
. . . : that the purpose of all existence is organization, life, sensation, enjoyment, perfection of soul . . . and that lifeless nature exists solely for the sake of living nature . . . , that in all of nature the base is subordinate to the noble and the former is for the sake of the latter. . . . Who could with conviction contest [this] proposition on which our entire *knowledge of nature* . . . rests? For what are all natural scientific investigations but efforts to discover the purpose and interconnections of certain objects? (pp. 228–9)

In support of this claim Fischer does little more than to refer the reader to "the profound speculations which Kant carried out on the nature of our knowledge. . . ." (p. 229) So confident was Fischer of his *a priori* proof of pluralism that he asserts that it "could be worked out in a person's head before the discovery of telescopes. . . ." (p. 230) This does not, however, preclude his citing Schröter's observational claims for lunar life (p. 225).

In concluding this discussion of pluralist works written in German in the latter half of the eighteenth century, two sections from novels published in 1796 by Johann Paul Friedrich Richter (1763–1825), who used the pen name Jean Paul, effectively illustrate the diverse uses of the plurality of worlds theme.[229] Included in his *Leben des Quintus Fixlein* is a section in which two lovers and their young child await imminent death. To relieve the pathos of this situation, Jean Paul draws on belief in lunar life and also the doctrine of transmigration of souls. He has Eugenius and Rosamond exchange the following words as they watch their dying child:

"I feel so weary, and yet so well. Will it not be as if we left two dreams – the dream of life and the dream of death – if we enter the cloudless moon as the first shore beyond the hurricanes of life?"

"It will be still better," replied Rosamond, "for in the moon as thou has taught me, dwell the little children of this earth, and their parents remain with them till they themselves become as mild and tranquil as children." . . .

"Ay, from heaven to heaven – from world to world!" said Eugenius, ecstatically.[230]

This vision must have had a strong appeal to Jean Paul, who as a result of a transforming experience in 1790 had become obsessed with death and the question of an afterlife.[231]

Although widely read in the antireligious authors of the Enlightenment, Jean Paul by the early 1790s was seeking solace in religion. The depth of his earlier despair and his revulsion at the thought of a Godless universe were given powerful expression in another 1796 novel, his *Siebenkäs*. That novel contains a chapter entitled "Rede des todten

Christus," which became the most widely known and translated of all his writings. The narrator recounts a dream in which he awakes, finding himself in a graveyard where the bodiless dead wander among open coffins. These shadows gather in a church, where the last to leave his coffin, one wounded in the breast, is asked by them: "Christ! Is there no God?" Christ answers "There is none," and adds:

I have traversed the worlds, I have risen to suns, with the milky ways, I have passed athwart the great waste spaces of the sky; there is no God. And I descended to where the very shadow cast by Being dies out and ends, and I gazed into the gulf beyond, and cried, "Father, where art Thou?" But answer came there none. . . . And when I looked up to the boundless universe for the Divine eye, behold, it glared at me from a socket empty and bottomless. . . . Shriek on, then, discords, shatter the shadows with your shrieking din, for HE IS NOT![232]

This ghastly vision of a universe in which each world shakes forth "its glimmering souls on to the Ocean of death," in which "every soul in this vast corpse-trench . . . is utterly alone," and in which although there is no God, "the coils of the great serpent of eternity [are] all twined about those worlds" (pp. 263–5) was created by Jean Paul to show that "In all this wide universe there is none so utterly solitary and alone as the denier of God." (p. 260) Jean Paul's "Speech of the Dead Christ" haunted such subsequent authors as Carlyle, de Staël, Dostoevsky, Hugo, Nerval, and Vigny, who turned it to their own, sometimes quite different, literary purposes.[233]

The contrast between the *Fixlein* and *Siebenkäs* sections is striking and illuminating. The former exemplifies how pluralism could be used as a source of consolation, as a means of turning the universe into a divinely and benevolently ordered arena of life. The latter section, that of the speech of the dead Christ, suggests the possibility, just beneath the surface in the writings of some deists who embraced a richly populated universe while urging a distant and impersonal Deity, that the tragedy of man was not confined to our small earth but was replayed on millions of planets where without sense of ultimate purpose countless other intelligent creatures despairingly lived out their years. This was, in the words of Victor Hugo, echoing Jean Paul, the prospect that "The Universe is a monster and heaven is a dream. . . ."[234]

In Italy, interest in pluralism stimulated publication between 1711 and 1780 of four different translations of Fontenelle's *Entretiens*.[235] Derham's *Astro-Theology* does not seem to have been translated, but it nonetheless provoked Giovanni Cadonici (1705–86) to attack it in his lengthy antipluralist *Confutazione teologica-fisica del sistema di Guglielmo Derham inglese* (1760). Canon of the Cathedral of Cremona, an ecclesiastical historian, and a theologian of Augustinian inclination, Cadonici was less upset by Fontenelle's light-hearted *Entretiens* than by Derham's earnest attempt to bond pluralism to religion. Nonetheless, his position was totally antipluralist: Earthlings are the only rational physical beings in the universe.[236] Although at times attacking astronomical aspects of Derham's system, for example, its failure to provide observational evidence of planets circling stars (p. 5), Cadonici, who may not have been even a Copernican, formulates most of his arguments against Derham's pluralism in theological and philosophical terms. For example, he discusses whether or not men could be transported to other planets by devils (pp. 131ff), and he cites approvingly the arguments made by Thomas Aquinas five hundred years earlier for the existence of only one world (pp. 177–8). The doctrine of Christ's redemption is also presented as opposing pluralism: "What better reason do we have [for rejecting pluralism] . . . than that of [God's] ordaining the redemption of the human species . . . into which he descended, became incarnate, took flesh, lived, preached, died, and where he wanted his cross to replace the triumph of the devil and of death. . . ." (p. 202) Some of his theological arguments, however, are directed not so much against pluralism as against the theological conceptions of Derham, whose physicotheological enthusiasms had left him vulnerable in a number of ways. Moreover, Derham's determination to populate even the moon permitted Cadonici to cite Huygens against the Englishman's excessive claim (p. 269).

Derham's *Astro-Theology* was not the only pluralist publication of British origin criticized by Cadonici. In his most concentrated discussion of life on the planets (pp. 278–95), Cadonici holds up to ridicule the article "Planet" in Ephiam Chambers's *Cyclopaedia,* which employed Christian Wolff's method for calculating the average height of Jupiterians.[237] Such excessive claims made by pluralists scarcely justify Cadonici's retrogressive conservatism, but they help explain why a man of some learning felt justified in attacking a doctrine that he believed could entail the consequence that God the Son would "have become incarnate, lived, died, and resurrected in all the planets. . . ." (p. 340) No second edition of Cadonici's book was published, possibly because Italian Christians, like their contemporaries elsewhere in Europe, were learning to live with extraterrestrials. This possibility is suggested by the propluralist position advocated at almost the same time by another Italian priest who possessed an international reputation of the highest order.

This was Roger Joseph Boscovich (1711–87), the famous Jesuit scientist, philosopher, and poet, who was born in Dubrovnik to a father of Croatian and a mother of Italian ancestry. Although frequently traveling beyond Italy for extended periods, Boscovich spent most of his adult life in that country teaching at the Collegium Romanum and the University of Pavia and later playing a major part in the organization of the Brera

Observatory near Milan, where more than a century later Schiaparelli made his famous Mars observations.[238] Boscovich's first contribution to the pluralist debate was indirect and antipluralist in its implications; in 1753, he published *De lunae atmosphaera*, in which he analyzed whether or not the moon possesses an atmosphere. Led to his investigation by a 1748 paper by Euler, who had used observations of a solar eclipse to suggest that the moon possesses sufficient atmosphere to produce a refraction of 20 seconds of arc, Boscovich systematically analyzed possible and actual observations of objects (stars, planets, and the sun) occulted by the moon to conclude that Euler's result was wrong. In particular, he asserted that if the moon possesses any atmosphere, it is far less dense than Euler had supposed. Boscovich's book was in its third edition by 1776; moreover, already by 1764 the French astronomer A. P. D. Du Sejour (1734–1794) had published an analysis of solar eclipses supporting Boscovich in that Du Sejour set the upper limit of the density of the moon's atmosphere at 1,400 times that of the earth.[239]

That Boscovich was not in general opposed to pluralism is clear from his magnum opus, his *Philosophiae naturalis theoria* (1758). In his explanation of fire as a fermentation in which a sulfurous substance must be present, he suggests that "in the sun itself, & in the stars, . . . there may exist bodies altogether lacking in such a [sulfurous] substance; & these may grow & live without the slightest injury of any kind to their organic structure."[240] Remarkable as his claim for life in the sun and stars may be, Boscovich went even further in a suggestion based on his doctrine that matter ultimately consists not of hard, massy atoms but rather of point centers of force that at certain distances exert repulsive forces and at other distances attractive forces. Drawing on this hypothesis, Boscovich speculates about the interpenetrability of matter and even proposes that "there might be a large number of material & sensible universes existing in the same space, separated one from the other in such a way that one was perfectly independent of the other, & the one could never acquire any indication of the existence of the other." (p. 184) In elaborating this imaginative conception, he suggests multiple universes located in times different from our own eternity:

But, what if there are other kinds of things, either different from those about us, or even exactly similar to ours, which have, so to speak, another infinite space, which is distant from this our infinite space by no interval either finite or infinite, but is so foreign to it, situated, so to speak, elsewhere in such a way that it has no communication with this space of ours; & thus will induce no relation of distance. The same remark can be made with regard to a time situated outside the whole of our eternity. [p. 199]

In a work published only after his death, Boscovich entertained even the idea of "a sequence of similar universes," some of which were sand-grain-sized with respect to others.[241]

Even though the present state of historical research does not justify the conclusion that the plurality of worlds debate was extensive in Enlightenment Italy, the cases of Cadonici and Boscovich dramatically demonstrate its diversity.

That the idea of a plurality of worlds, so widespread in the West, reached Russia early in the eighteenth century seems not surprising; however, the way in which it did so and the commotion this caused constitute a remarkable story that has recently been reconstructed by Valentin Boss in his *Newton and Russia*. Boss shows that the revolutionary view of the universe that dawned on Europeans during the century and a half separating Copernicus's *De Revolutionibus* from Huygens's *Cosmotheoros* burst with startling suddenness on some Russians in the course of an evening's reading.

Intent to drag Russia into the modern world, Peter the Great (1672–1725) encouraged the translation of various European books into Russian. That this was needed is indicated by the fact that before 1717 there existed no published exposition in Russian of the Copernican system.[242] This situation was remedied not by one of the classic volumes of Copernicus, Galileo, Kepler, or Newton but rather by Huygens's *Cosmotheoros*, which was rendered into Russian by Jacob Daniel Bruce (1670–1735), a descendant of a Scottish family that had come to Russia in 1647. Friend and advisor to Peter the Great, the scientifically astute Bruce had toured Europe with him in 1697–8, introducing Peter to the technical learning of England and elsewhere. In the mid-1710s, Bruce completed his Huygens translation, which Peter the Great commanded Mikhail Petrovich Avramov to publish. However, on reading the book, Avramov found it work of "Satanic perfidy."[243] Decades later he described his dilemma in a letter to the more traditional Empress Elizabeth: "concealing his godless, frenzied, and atheistic heart, Bruce praised the book by the delirious author Kristofer Huiens, . . . pretending that it was very clever and wholesome for the educating of all the people, . . . and with such habitual and godless flattery deceived the sovereign." Avramov, however, was not duped; as he explained: ". . . I examined this book which was contrary to God in all ways, and with my heart quaking and my soul overawed, I fell before the mother of God with the sobbing of bitter tears, frightened to publish and frightened not to publish."[244] The daring decision he made was to subvert the tsar's order by printing only thirty copies of the book and making efforts to conceal these. Avramov's bold action was of little avail; in 1724, another publisher printed the book.[245] Russian readers encountered a second exposition of pluralism in the translation of Fontenelle's *Entretiens* prepared by Prince Antiokh Cantemir (1709–44). Although he had completed the translation in 1730, efforts made by the Holy Synod to suppress it delayed its publication until 1740, by which time Cantemir had forsaken Fontenelle's Cartesianism in favor of Newtonianism.[246] The Holy Synod also sought, with no better success, to suppress a translation of Voltaire's *Micromégas,* which between 1756 and 1793 was printed in Russian at least six times.[247]

Defense of the Copernican system and advocacy of pluralism in Russia did not come solely from foreign sources: Both were, for example, championed by Mikhail Vasilyevich Lomonosov (1711–65), the breadth of whose learning was such that Pushkin praised him as Russia's first university. The possible sources of Lomonosov's pluralism include Christian Wolff, with whom the young Lomonosov studied science for three years, and Fontenelle, whose *Entretiens* he read in French.[248] Lomonosov's pluralism was made public as early as his "Evening Reflections on Seeing the Aurora Borealis" (1743). In this famous poem he states:

> And Science tells us that each twinkling star
> That smiles above us is a peopled sphere,
> Or central sun, diffusing light afar;
> A link in nature's chain: – and there, even there
> The Godhead shines displayed – in love and light,
> Creating wisdom – all-directing might.

However, the poem concludes with the caution that

> Vain is the inquiry – all is darkness, doubt:
> This earth is one vast mystery to man.
> First find the secrets of this planet out,
> Then other planets, other systems scan!
> Nature is veiled from thee, presuming clod!
> And what canst thou conceive of Nature's God?[249]

But caution was not characteristic of Lomonosov, who in fact penned that poem while imprisoned for insulting some colleagues at the St. Petersburg Academy of Sciences.[250] Nor was it prudent for the deistic Lomonosov to publish in 1757 a poem satirizing Russian priests, who alone were allowed to wear beards. Lomonosov's "Hymn to a Beard" contains these lines:

True it be that all the planets
Resemble ours as earthlike objects.
Be on one of them a long-hair
Priest, or self-appointed prophet:
"By my beard, I swear to you,"
He said, "The earth is through and through
A lifeless planet; all is bare."
One who remonstrated: "Man lives there."
At the stake they burned him
To punish this free-thinker's sin.[251]

The irate Holy Synod reacted by asking the empress to issue a ukase, undoubtedly directed against Lomonosov, stipulating that "none henceforth shall dare to write or print anything either of the multitude of worlds, or of anything else in opposition to the sacred faith, or in disagreement with honorable morals, under penalty of most severe punishment."[252] Lomonosov replied in a paper discussing observations made by himself and others of the 1761 Venus transit, arguing on the one hand that his observations supported the idea of a Venusian atmosphere and hence inhabitants of the planet and, on the other hand, that texts by such church fathers as Basil the Great and John of Damascus show that Scripture does not preclude a plurality of worlds.[253] Such was the slow and painful parturition of pluralism in that country now most active in the search for extraterrestrials.

5. Conclusion: the century ends and new tensions arise

This section simultaneously treats two interrelated questions. First, what was the status of pluralism at the end of the eighteenth century in regard to science and in regard to religion? In particular, what major problems were passed to the nineteenth century? Second, how well do eighteenth-century developments in the debate square with the conception (mentioned at the beginning of this chapter) of the Enlightenment as "the hinge on which the European nations turned from Middle Ages to 'modern' times, marking the passage from a supernaturalistic-mythical-authoritative to a naturalistic-scientific-individualistic type of thinking?"[254] To illustrate the answers offered to these questions, this section includes a description of Tom Paine's dramatic entry into the debate in the 1790s.

Viewed in relation to the scientific community, pluralism passed through a remarkable transformation from 1600 to 1800. Many seventeenth-century scientists – Galileo, Descartes, Gassendi, Newton (in his published writings) – had viewed it as a speculation at the limits of legitimate science. During the eighteenth century, however, an international array of astronomers, including Lambert, Herschel, Schröter, Bode, Laplace, and Lalande, allied themselves with pluralism; in fact, for some it formed an integral part of their research programs. Moreover, while scientific journals published pluralist papers, astronomical texts and university courses regularly treated this topic. In short, pluralism made extraordinary progress with practitioners of the most ancient science.

However, if eighteenth-century pluralism is examined from the perspective of the present, it is clear that its conjectural component was large. Broad analogy more than detailed astronomy, physicotheology more than physics, and teleology more than telescopes had been used to erect a vast edifice on what nineteenth-century scientists gradually found to be a frail foundation. Although by the eighteenth century the medieval cosmos with its crystalline spheres, angelic planetary movers, and associated metaphysical and mythical elements had been discredited, it should not be forgotten that the cosmos championed by many Enlightenment figures was not free of comparable associations. Carl Becker did not discuss pluralism when in his *Heavenly City of the Eighteenth-Century Philosophers* he maintained that the Enlightenment "*philosophes* demolished the Heavenly City of St. Augustine only to rebuild it with more up-to-date materials,"[255] but he might have found in such authors as Wright, Kant, Lambert, Herschel, and Bode evidence to support a parallel claim for the heavens of Enlightenment astronomy. Crystalline spheres had disappeared, but Halley's inhabited subterranean spheres and Herschel's cool solar nucleus became available. Disputes about hierarchies of angels may have been abondoned, but a number of authors debated whether the superbeings of the solar system live on the outer planets or inner planets, or on the sun itself. Moreover, although some authors analyzed the question of other worlds in largely scientific terms, the great majority, explicitly or implicitly, invoked religious and/or metaphysical considerations. This suggests that whereas a shift from a "supernaturalistic-mythical" to a "naturalistic-scientific" mode of thought does describe the program of many Enlightenment pluralists, it is less descriptive of their practice.

Turning to the relations of religious authors to pluralism, we see a somewhat parallel situation. By 1790, most felt that the earlier tensions had largely abated. "An undevout astronomer is mad," Young had proclaimed, and many agreed. Physicotheologians embraced extraterrestrials as evidence of God's omnipotence and benevolence, while poets and preachers ransacked the new cosmos for ideas and images to embellish their writings. Numerous authors recommended to the religious that rather than fearing pluralism, they should see it as evidence of Divine power and generosity. Hopes to establish the "reasonableness of reli-

gion" seemed to have been fulfilled, even to the point of reconciling it to a
daring new doctrine.

Yet, as in the case of science, so in regard to religion: Serious difficulties lay just beneath the surface. The very success of the natural theological enterprise, whether practiced by Christians or deists, tended to em-
phasize "Nature's God" while downplaying the idea of an incarnated
redeemer. The importance of distinguishing theism in general from
Christian theism has recently been stressed by the French scholar Jean
Milet in his *God or Christ?*, where he describes a "religious bipolarity
peculiar to Christianity."[256] This bipolarity, Milet maintains, was especially evident when first-century Jews who worshiped a God "whose
name one did not even dare pronounce, whose face could not be seen
without dying . . ." were asked to believe that this same God "manifested himself in the form of a weak child in Bethlehem, . . . toiled on
the roads of Judaea and Galilee, . . . [and] – as the height of improbability – was to be recognized under the disfigured features of the man put
to death on Golgotha." Milet suggests that "a mental effort of such
magnitude was an unparalleled demand. . . ." (p. 15) The bipolarity
delineated by Milet highlights the fact that physicotheological proofs of
theism gave no support to the Christian conception of Christ as divine
redeemer. Structures of insects or solar systems may evidence God's existence, but they are mute as to a Messiah. Furthermore, pluralist physicotheology set off in even starker relief the radicalness of the Christians'
claim. Why would the God of all worlds select an insignificant planet for
his most remarkable actions? In short, whereas by the 1790s pluralism
had reached a rapprochement with theism, tensions with Christianity
had not yet been fully faced. This, at least, is the conclusion suggested by
the sensation that resulted when in the 1790s a vigorous attack on Christianity was launched on a pluralist basis.

Its author was Thomas Paine (1737–1809), who, according to Theodore Roosevelt, was a "filthy little atheist." Ralph C. Roper, in a 1944
paper, challenged this claim, citing many passages from Paine as a proof
that he had in fact campaigned against atheism. Moreover, by pointing
out parallels in the lives of Paine and Herschel, both of whom had been
influenced by James Ferguson, Roper showed the centrality of astronomy
in Paine's religious thought. Paine had, for example, lamented that astronomy and the other sciences were taught as accomplishments of man,
rather than "with reference to the *Being* who is the Author of them; for
all the principles of science are of divine origin."[257] The influence of
astronomy on Paine's thought is above all evident in Part I of Paine's *Age
of Reason*, written in 1793, just before he was imprisoned for ten months
by the French revolutionaries, whom he had supported in his *Rights of
Man*.[258] In his *Age of Reason*, written, according to Paine, to salvage

theism in the face of the radically irreligious attitude of many of the French revolutionaries (p. 235), Paine recounts his early education, noting that "The natural bent of my mind was to science. . . . As soon as I was able I purchased a pair of globes, and attended the philosophical lectures of Martin and Ferguson, and became afterwards acquainted with Dr. Bevis . . . an excellent astronomer." (pp. 273–4) Paine not only learned astronomy but also integrated it into his thought: "After I had made myself master of the use of globes and of the orrery, and conceived an idea of the infinity of space . . . , I began to . . . confront the internal evidence those things afford with the Christian system of faith."[259]

The result of that confrontation, when set out in Paine's *Age of Reason,* was one of the most violent attacks ever published, not on theism, but on Christianity. The *Age of Reason* is, after all, one of the classic works of deism. Majorie Nicolson, in a 1936 paper, searched out "what our generation misses as frequently as did the earlier – the real basis of Paine's 'deism' [and] the chief source of his theological beliefs."[260] As she urges, with much supporting evidence, "The first part of the *Age of Reason* . . . is the climactic and inevitable popularizing of . . . the controversy whether ours is not one of a plurality – even, some dared to think, of an infinity – of worlds. . . ."[261] Paine's prose supports her claim:

> Though it is not a direct article of the Christian system that this world that we inhabit is the whole of the habitable creation, yet it is so worked up therewith from what is called the Mosaic account of the Creation, the story of Eve and the apple, and the counterpart of that story – the death of the Son of God, that to believe otherwise, that is, to believe that God created a plurality of worlds at least as numerous as what we call stars, renders the Christian system of faith at once little and ridiculous and scatters it in the mind like feathers in the air. The two beliefs cannot be held together in the same mind; and he who thinks that he believes in both has thought but little of either. (pp. 276–7)

Paine recounts the results, as he saw them, of astronomical research during the previous three centuries. Astronomy had revealed the "probability . . . that each of those fixed stars is also a sun, round which another system of worlds or planets . . . performs its revolutions . . . and that no part of space lies at waste. . . ." (p. 281) This conception, he asserts, supports theism, but it alters, drastically, the Christian conception of God:

> From whence, then, could arise the solitary and strange conceit that the Almighty, who had millions of worlds equally dependent on his protection, should quit the care of all the rest and come to die in our world because, they say, one man and one woman had eaten an apple! And, on the other hand, are we to suppose that every world in the boundless creation had an Eve, an apple, a serpent, and a redeemer? In this case, the person who is irreverently called the

Son of God, and sometimes God himself, would have nothing else to do than to travel from world, in an endless succession of death, with scarcely a momentary interval of life. (p. 283)

In short, "every evidence the heavens afford to man either directly contradicts . . . the Christian system of faith . . . or renders it absurd." (p. 284) Paine presents his readers a stark choice: Either reject Christianity or reject pluralism. His own recommendation was clear: Reject Christianity. Elsewhere in Part I and in Part II of the *Age of Reason* Paine launches other attacks on Christianity, but central to the work is the section just discussed.

Paine's *Age of Reason* created a sensation. Within a few years, British and Irish readers had snatched up over 60,000 copies, despite government efforts to suppress the book. A French translation became available in 1794, and by 1795 Paine's *Age of Reason* was in its seventh American edition.[262] While some hung Paine in effigy, others composed replies – over fifty of these being listed in the *British Museum Catalogue*.[263] Some of these, such as Rev. Uzal Ogden's *Antidote to Deism* (Newark, 1795), dealt with Paine's pluralist objection to Christianity, but the breadth of Paine's attack generated a correspondingly wide array of replies.[264] The influence of Paine's pluralist argument even extended far into the nineteenth century, as will be shown in subsequent chapters.

Why did Paine's extraterrestrial objections to Christianity have such impact? One reason is suggested by Paine's remark that "I doubt not [that these objections] have occurred in some degree to almost every other person at one time or another. . . ." (p. 273) This supposition is supported by the fact that even before Paine, such authors as John Adams had been troubled by them. The evidence is even clearer in the case of Horace Walpole, fourth Lord of Orford (1717–98), who in his reminiscences revealed: "Fontenelle's Dialogues on the Plurality of Worlds, first rendered me an infidel. Christianity, and a plurality of worlds, are in my opinion, irreconcilable."[265] Whereas Walpole, Paine, and possibly Adams rejected Christianity because of pluralism, others questioned pluralism. They were a minority; pluralism apparently had won the day, and Christians were faced with the task of reconciling their religion with it.

PART II
From 1800 to 1860

4

The intensification of the plurality of worlds debate after 1800

1. Between the deism of Thomas Paine and the evangelism of Thomas Chalmers

As the nineteenth century began, many of the cherished doctrines of the Enlightenment were being called into question. Within the span of a few years centering on 1800, Wordsworth and Madame de Staël challenged, for example, the prevailing canons of literary excellence. Simultaneously, in the sciences, Young proposed a new theory of light, and Davy a new theory of heat, while Erasmus Darwin questioned the fixity of species. As Parson Malthus attacked Enlightenment optimism and its social philosophy, Pastor Schleiermacher and Yale's Preacher-President Dwight attempted to revive neglected or rejected features of Christian thought. And as Paine continued to press for radical republicanism, Maistre championed conservatism.[1] Yet it is a striking fact that diverse and daring as these heralds of the new century were, each of them was at one with the Enlightenment in endorsing extraterrestrial life.[2] Their readiness to accept pluralism does not imply that that doctrine was without difficulties. As noted in the previous chapter, eighteenth-century pluralists left two pressing problems for their nineteenth-century successors. The first, powerfully presented in Paine's *Age of Reason*, was whether or not Christianity and pluralism could be reconciled. The second, a less clearly recognized difficulty, was the paucity of astronomical evidence for extraterrestrial life. One or both of these problems were on the minds of nearly all the early nineteenth-century authors discussed in the present chapter, the first section of which examines the status of pluralism in the first and second decades of the nineteenth century. The second section focuses on the most influential pluralist book published in those two decades, Thomas Chalmers's *Astronomical Discourses* (1817), which many saw as alleviating the tensions with Christianity. Chalmers's book helped stimulate the wave of new pluralist publications discussed in section three. The fourth and final section examines early nineteenth-century efforts to salvage life on the moon, while suggesting, by means of a new interpretation of the so-called Moon Hoax, that the problems mentioned earlier were still far from resolution.

In surveying early nineteenth-century pluralism, it is convenient to commence with Britain and her scientists, some of whom displayed their commitment to that doctrine by accepting even Herschel's theory of life on the sun. For example, Robert Harrington, M.D., in 1796 published his *New System of Fire and Planetary Life: Shewing That the Sun and Planets Are Inhabited, and That They Enjoy the Same Temperment as Our Earth* (London), claiming that the two chief entities in nature are fire particles, which are mutually repulsive, and earth particles, which attract both air and fire particles. On the basis of this theory, he concludes that the sun and planets "all enjoy the identical same fire, or light, or heat; the same temperature, and, I make no doubt, the same men, animals, vegetables, and minerals; the same atmosphere and water; in short, every thing the same." And he adds: "What a vast idea! Reflect upon it *little man,* and humble thyself!" (p. 50) Harrington was on the fringe of British science, but Thomas Thomson, M.D. (1773–1852), was a leading Scottish chemist whose *System of Chemistry* contains in its 1807 edition an important exposition of Dalton's atomic theory. In that volume, Thomson also advocates Herschel's theory of the sun, without, however, mentioning solarians. In particular, Thomson states that Herschel's observations indicate that "the sun is a solid opaque globe, similar to the earth and other planets, and surrounded by an atmosphere of great density and extent [in which] float two regions of clouds. . . ."[3] Although in 1807 Thomas Young objected that the sun's great mass would make man-sized solarians weigh over two tons and that the solar clouds could not protect them from the sun's heat,[4] acceptance of Herschel's claims remained widespread.

Planets and satellites caused less severe problems for pluralists. For example, in *Of the Plurality of Worlds* (1813), which James Mitchell (1786?–1844) presented as a lecture at the London Mathematical Society, he argues for life on all the planets and their satellites. In doing this, he dismisses such problems as the nonterrestrial intensities with which solar rays reach the planets by urging the adaptability of their animals and the moderating effects of their atmospheres and soils.[5] Mitchell's stress on the prevalence of life on earth and his use of other forms of analogical argumentation embolden him to assert that to deny pluralism "would be to impeach the wisdom of our Maker." (p. 21) Whereas Mitchell dealt with the problem that Martians lack a moon by postulating its presence but assigning an unobservable size to it, Adam Walker

(1762–1841), a scientific author and lecturer who shared Mitchell's enthusiasm for extraterrestrials, suggested in his *System of Familiar Philosophy* that the Martian atmosphere may prolong sunlight.[6] After describing how Saturnians see the rings around their planet, Walker urges that their "eyes and constitution . . . are adapted [to making Saturn] as comfortable an abode as the worlds . . ." that receive more heat and light from the sun (II, pp. 284–5). Concerning stars, he states that they are suns and consequently "destined . . . to give light, heat, and vegetation, to various worlds . . . which are infinitely too remote to be perceived by us. . . . How much too vast for the human mind is this idea!" (II, p. 243)

Pluralism also attracted the attention of four famous English romantic poets. In 1798, William Wordsworth (1770–1850) included a cosmic voyage in his "Peter Bell, A Tale," in which his hero not only sees "the red-haired race of Mars" but also learns that

> The Towns in Saturn are decayed,
> And melancholy Spectres throng them; –
> The Pleiads that appear to kiss
> Each other in the vast abyss,
> With joy I sail among them.
>
> Swift Mercury resounds with mirth,
> Great Jove is full of stately bowers;
> But these, and all that they contain,
> What are they to that tiny grain,
> That little Earth of ours?[7]

Wordsworth's enthusiasm for extraterrestrials had not abated by 1835, when in his "To the Moon" he referred "To worlds unthought of till the searching mind/Of Science laid them open to mankind. . . ."[8]

Lord Byron (1788–1824) also employed the cosmic voyage theme. In his tragic peom *Cain: A Mystery* (1821), Lucifer gives the biblical Cain a tour of the celestial region, showing him "Myriads of starry worlds, of which our own/Is the dim and remote companion in Infinity of Life. . . ." (*Cain*, II, 2, 566–8) Byron's pluralism is even more evident in what has been called his "credo," which contains the assertion: "The Night is also a religious concern; and even more so, when I viewed the Moon and Stars through Herschell's [*sic*] telescope, and saw that they were worlds."[9] Herschel was not the only pluralist author known to Byron; Manfred Eimer, in a book-length monograph, has shown that Lucretius, Fontenelle, Pope, and Edward Young were also influential in leading Byron to embellish his poetry with cosmic imagery.[10] Moreover, a passage from a letter Byron wrote in 1813 suggests that pluralist astronomy led him to question the soul's immortality:

I am no Bigot of Infidelity and did not expect that, because I doubted the immortality of Man, I should be charged with denying the existence of a God. It was the comparative insignificance of ourselves and *our world*, when placed in competition with the mighty whole, of which it is an atom, that first led me to imagine that our pretensions to eternity might be over-rated.[11]

Pluralism is especially prominent in the writings of Percy Bysshe Shelley (1792–1822), whose active imagination was stirred by the astronomical ideas Adam Walker presented in lectures at Syon House Academy, where the young poet studied before entering Eton.[12] Shelley again heard Walker lecture at Eton, where he also encountered pluralism in the writings of Lucretius[13] and of Erasmus Darwin (1731–1802), a physician and poet who, like his famous grandson, proposed evolutionary ideas. That Erasmus Darwin's writings, especially his long poem *The Botanic Garden,* influenced Shelley has been demonstrated in detail.[14] That poem, in which Darwin espouses both evolutionary and pluralist ideas, contains such passages as

> "*Let there be light!*" proclaim'd the *Almighty* Lord,
> Astonish'd Chaos heard the potent word; –
> Through all his realms the kindling Ether runs,
> And the mass starts into a million suns;
> Earths round each sun with quick explosions burst,
> And secondary planets issue from the first;
> Bend, as they journey with projectile force,
> In bright ellipses their reluctant course,
> Orbs wheel in orbs, round centres centres roll,
> And form, self-balanced, one revolving Whole.[15]

This may be compared with a passage in Shelley's *Prometheus Unbound* (1820):

> Then, see those million worlds which burn and roll
> Around us; their inhabitants beheld
> My spher'ed light wane in wide heaven. . . .[16]

In the same poem, Darwin's influence is also seen in Shelley's use of Darwin's ideas that the moon may have been torn from the side of the earth and, although not inhabited now, may become so.[17]

Shelley's pluralism and a major conclusion he drew from it were first made public in notes to his *Queen Mab,* a poem privately printed in 1813 because of its shocking passages. One of these occurs in a note he added in explanation of his reference in the poem to "Innumerable systems." He states that the plurality of worlds makes it

. . . impossible to believe that the Spirit that pervades this infinite machine begat a son upon the body of a Jewish woman; or is angered at the consequences of that necessity, which is a synonym of itself. All that miserable tale of the Devil, and Eve, and an Intercessor, with the childish mummeries of the God of the Jews, is irreconcilable with the knowledge of the stars. The works of His fingers have borne witness against Him.[18]

This passage supports Ifor Evans's claim that "astronomical knowledge [led Shelley] to a rejection of the Christian faith."[19] This is also indicated by Shelley's "On the Devil, and Devils," written at most two years before his death in 1822.[20] In this satiric essay, which shows the influence of ideas of Herschel and Laplace, Shelley warns bishops "of the laxity among the orthodox . . . respecting a belief in the Devil,"[21] which he traces to the long-standing association of that doctrine with the discredited geocentric cosmology. Reminding his readers that modern astronomers have discovered that the earth "is a comparatively small globe, in a system consisting of a multitude of others . . . ," Shelley discusses in detail whether earthling or extraterrestrials, such as Jupiterians, are more deserving of direct visitation by the Devil! (pp. 397–8) Having thereby ridiculed the theologians who had attempted to answer Paine's anti-Christian pluralist polemic, Shelley cites Herschel's theory that the sun possesses a cool interior against such authors as Swinden, who had designated the sun as the location of hell (p. 404). The bitterness evident in this essay was sufficient that it was withheld even from the 1839 edition of Shelley's collected prose writings.

The fourth English romantic poet involved with pluralism is Samuel Taylor Coleridge (1772–1834), who during his student days at Cambridge composed an ode expressing his desire to visit "worlds whom elder Suns have vivified."[22] Nonetheless, on two later occasions, he expressed reservations about extraterrestrial life. One of these consists of a note in the margin of his copy of Nehemiah Grew's *Cosmologica sacra*. Where Grew had asserted that "the moon may be inhabited – but has . . . perhaps a different Furniture of Animals . . . ," Coleridge wrote in the margin: "But why, of necessity, any? Must all possible Planets be lousy? None exempt from the *Morbus pedicularis* of our verminous man-becrawled Earth?"[23] His second and more revealing remark occurs in his *Table Talk* for February 22, 1834. Apparently forgetting that he had accepted pluralism in his Cambridge poem, he asserts:

I never could feel any force in the arguments for a plurality of worlds, in the common acceptation of that term. A lady once asked me – "What then could be the intention in creating so many great bodies, so apparently useless to us?" I said – I did not know, except perhaps to make dirt cheap. The vulgar inference is *in alio genere*. What in the eye of an intellectual and omnipotent Being is the whole sidereal system to the soul of one man for whom Christ died?[24]

This passage suggests the conclusion that whereas Shelley and possibly Byron saw pluralism as reason for rejecting Christianity, Coleridge drew precisely the opposite inference.

A number of British religious writers took up the task of responding to Paine's *Age of Reason*. One of the most widely read replies was *The Gospel Its Own Witness*, published in 1799–1800 by the Baptist minis-

ter Andrew Fuller (1754–1815), who devoted one chapter of this antide-
ist book to Paine's pluralist arguments.[25] That chapter received substantial attention in later writings, one author even claiming that Chalmers
had only elaborated Fuller's position.[26] In fact, Fuller presents a number
of not necessarily mutually consistent claims. For example, he urges that
pluralism is at most probable, that it can be reconciled with both Christianity and scripture, that if pluralism is correct, then the Christian doctrine of redemption is "STRENGTHENED AND AGGRANDIZED" by
it, that possibly only men and angels have apostasized, and that even if
extraterrestrials sinned, Christ's incarnation and redemption on the earth
*"are competent to fill all and every part of God's dominions with ever-
lasting and increasing joy."* (p. 272) A noteworthy feature of Fuller's
arguments is that they assume the uniqueness of Christ's terrestrial incarnation and redemption.

Whereas Paine's *Age of Reason,* a work of a hundred pages, had
brought forth from Fuller a similarly sized book, Rev. Edward Nares
(1762–1841), never known for his conciseness,[27] responded to Paine's
dozen pages on pluralism by a volume of over four hundred pages. In
1801, Nares, who later became regius professor of modern history at
Oxford, published ΈΙΣ ΘΕΟΣ, ΈΙΣ ΜΕΣΙΤΗΣ; *or, An Attempt to
Shew How Far the Philosophical Notion of a Plurality of Worlds Is
Consistent, or Not So, with the Language of the Holy Scriptures* (London).[28] Nares's volume is free of the polemical approach of Paine; in fact,
he fairly describes his book as one "in which neither the author's own
opinions are advanced dogmatically, nor the opinions of others censured
uncharitably. . . ." (p. v) Nonetheless, it is clear that he is out to do
battle with the deists, at least to challenge their complaint "that man
should be so arrogant, as to think himself of importance enough in the
scale of being, to have a messenger sent from God, expressly to enlighten,
and even to redeem him." (p. 12) They have a point, Nares admits; man
is but "a speck in the Universe; and God is infinite and [moreover is the
maker] of myriads and myriads of such worlds as this, and perhaps of
myriads and myriads of different and distinct races of intelligent beings."
(pp. 13–14) But the deist goes too far in that he "can admire and acknowledge the hand of God in the wings of an ephemeron, though he
thinks our poor earth too paltry for the Son of God to visit!" (p. 14)

Nares reveals the purpose of his book as being to take the deists on
their own ground by accepting pluralism and to investigate whether or
not "the whole mediatorial scheme may not serve exceedingly to raise
and exalt our notions of God's greatness and magnificence. . . ." (p. 16)
Nares gives scant attention to the astronomical evidence for and against a
plurality of worlds; nonetheless, he accepts the doctrine as a "very well
founded conjecture." (p. 50) Although the earth may not be the unique

abode of life, Nares asserts that it is the only planet on which God became incarnate and performed redemptive actions: "the ONE GREAT manifestation . . . to accomplish the redemption of *all flesh*. . . ." The effects of Christ's redemptive efforts on this earth, however, are spread "in some way inscrutable to us, to every rational creature throughout the mightly firmament. . . ." (p. 18) Nares summarizes his position by rhetorically asking if God, "though necessarily one and unalterable," may not convey to extraterrestrials "such a knowledge of his ways and will, as their several wants and infirmities may need and require?" (pp. 19–21) Satan, he believes, may have worked his wiles in other worlds, although some planetary populations may have avoided sin (pp. 57–60). But as the first words of his title declare, there is only one God and one mediator or messiah: "Upon this Earth [Christ's] body was bruised, and his blood was shed; if there are other worlds in the universe, it is impossible for us to know how it may have pleased God to notify to them the sacrifice of the body and blood of Christ. . . ." (pp. 268–9) In support of this position, Nares claims that Henry More and Bishops Butler and Porteus held similar ideas. He frequently urges that pluralism, despite being conjectural, based on analogy, and not demonstrable from Scripture (pp. 77–9), provides a more sublime conception of God's operations in the universe than the nonpluralist view.[29]

Nares deserves special credit for two contributions. First, he repeatedly quotes from and carefully examines the views of earlier authors. Second, he provides the most detailed analysis ever attempted of the question of the relation of pluralism to Scripture. On this issue, Nares rejects the claim of Edward King that the pluralist position can be proved from Scripture (p. 48, but see p. 273), while urging (1) that at least nothing contradictory to the pluralist position can be found in Scripture (pp. 171, 344) and (2) that many passages acquire an enriched and more natural meaning if interpreted in terms of the plurality of worlds (pp. 193, 218, 248, 302–3). In this effort he studies the Hebrew and Greek words that have been translated into such English words as "heaven(s)" and "world(s)," attempting to show that they admit an expanded interpretation. On this basis he suggests that Nehemiah 9:6 "might *literally* be translated" as "Thou, even thou, art God alone; thou hast made the WORLDS, the UNIVERSE OF WORLDS; with ALL THEIR INHABITANTS; the EARTH, with all things that are therein; and thou fillest the whole with life; and THE INHABITANTS OF THE WORLDS worship thee." (pp. 177–8) Nares's erudite efforts were no match for Paine's pyrotechnics; while the *Age of Reason* ran through dozens of editions, Nares's tome achieved only one.

Prolific as Nares was, his writings scarcely rival in magnitude those of the Methodist theologian Adam Clarke (1762?–1832), who endorsed

pluralism in his multivolume commentary on the Bible.[30] Readers of
Clarke's *Commentary*, even before reaching Genesis 1:2, find an elaborate table of data on the planets and satellites, and by Genesis 1:16 are
informed that "Dr. Herschel's discoveries, by means of his immensely
magnifying telescopes, have, by the general consent of philosophers,
added a new *habitable world* to our system, which is the SUN."[31] After
elaborating on this, he comments concerning the moon: "There is
scarcely any doubt remaining in the philosophical world, that the moon
is a *habitable globe*." Moreover, analogy implies that "all the *planets* and
their *satellites* . . . are inhabited; for matter seems only to exist for the
sake of intelligent beings." The stars, he explains, are "*suns* . . . ; each
having an appropriate number of *planets* moving round it . . . ; consequently, there are innumerable worlds. . . ." He expands on this point
in his commentary on Deuteronomy 10:14, stressing the use of the plural
in the phrase "heaven of heavens" and adding: "Every *star* [is] a *sun*,
with its peculiar and attendant worlds. Thus there may be systems of
systems, in endless gradation, up to the throne of God!" In I Kings 8:27,
he finds reason to stress again the plural form of "heavens" and to
remark that "although the earth has been created nearly *six thousand*
years ago, suns, the centres of systems, may have been created at so
immense a distance that their light has not yet reached our earth. . . ."
Proceeding from the assumptions (1) that God typically uses seven days
for the creation of a solar system and (2) that our system was created in
4004 B.C., Clarke calculates that by 1819 God could have created "three
hundred and three thousand five hundred and seventy-five mundane systems. . . ." The inclusion of such statements as these in Clarke's *Commentary* and in his published sermons[32] no doubt disseminated the pluralist message to a wide audience. Scriptural commentaries and sermons
were, after all, still primary sources for the education of many, and
Clarke wrote with much erudition and with the authority of a man who
on three occasions had been chosen president of the Wesleyan body.
Nonetheless, one may suspect that in his pluralist statements, he would
have been wise to have heeded Wesley's caution to eighteenth-century
pluralists: "Be not so positive. . . ."[33]

As the eighteenth century ended in the United States, deism was making major inroads against Christianity. The most widely read of the
deistic works was Paine's *Age of Reason*, which by 1796 had gone
through seventeen American editions. According to one authority,

The Age of Reason was scattered the length and breadth of the land. Newspapers
advertised it, together with the counterblasts that conservatives wrote to over-
throw it. . . . The democratic clubs and deistic societies used it as a textbook.
College students swallowed it whole, to the great alarm of their preceptors; and

humble men in villages from New Hampshire to Georgia and beyond the Alleghenies discussed it by tavern candlelight.[34]

As this quotation suggests, deism was attractive to college students. At Princeton, a visitor in 1799 was scandalized that only a few students "made any pretensions to piety," and William Hill found "rudeness, ribaldry, and infidelity" among William and Mary students.[35] At Harvard, every student was given Bishop Watson's *Apology for the Bible*, lest they succumb to the new ideas. Lyman Beecher described Yale as "in a most ungodly state."[36]

This situation did not persist; in fact, what has been called a "Second Great Awakening" produced a "revitalized American Protestantism after 1800."[37] Evangelical preachers stormed against infidelity and carried congregations with them as they warned against the dangers of French thought, pointing to the excesses of the French Revolution. Such strategies had effect even among college students, especially at Yale, which Benjamin Silliman reported in 1803 to have become "a little temple" in which "prayer and praise seem to be the delight of the greater part of the students. . . ."[38] The person most responsible for this transformation at Yale was Rev. Timothy Dwight (1752–1817), who served as president of that institution from 1795 until his death.

Although active in strengthening the sciences at Yale, Dwight's chief concern – as befitted the grandson of Jonathan Edwards – was the religious life of the students. To save them from infidelity, to improve their morality, and to instruct them in Christianity, Dwight developed a series of one hundred seventy-three sermons, which he delivered in a four-year cycle, lest any undergraduate miss his message. Dwight's sermons, published posthumously in five volumes in 1818, carried the title *Theology Explained and Defended*.[39] Delivered with an evangelical enthusiasm but bolstered by many detailed arguments and a utilitarian stress on good action, Dwight's sermons reveal an author open to the sciences, attracted to natural theology, and convinced of pluralism. The latter two features figure prominently in his fifth sermon, as he discourses on the immutability of God, who created

. . . the countless multitude of Worlds, with all their various furniture. With his own hand he lighted up at once innumerable suns, and rolled around them innumerable worlds. All these . . . he stored, and adorned, with a rich and unceasing variety of beauty and magnificence; and with the most suitable means of virtue and happiness. Throughout his vast empire, he surrounded his throne with Intelligent creatures, to fill the immense and perfect scheme of being. . . . (I, pp. 78–9)

In the next sermon, turning to the omnipresence and omniscience of God, Dwight stresses the powers of a God who can comprehend a universe "inhabited by beings . . . emphatically surpassing number" (I, p.

93), yet in which he knows the thoughts of every mind. With the universe of the pluralists as backdrop, Dwight concludes the sermon by envisioning the last judgment and warning: "How different will be the appearance, which pride, ambition, and avarice, sloth, lust, and intemperance, will wear in the sight of God, in the sight of the assembled universe. . . ." (I, p. 105) The omnipotence of God, Dwight remarks in his seventh sermon, is "so easily demonstrated by Reason" that he need not attempt to prove, but only to impress it upon his audience (I, p. 107). This he does, drawing on the "Herschellian Telescope," which shows that "every star . . . is no other than a Sun, a world of light, surrounded by its own attendant planets, formed into a system similar to ours." (I, p. 109) More magnificent than all matter, however, is an individual mind, and it is for minds that these splendors were created. With his thirteenth sermon, that on the wisdom of God, Dwight concludes his discussion of the attributes of God, reminding his audience that astronomy has revealed a far more magnificent universe than that of the prophets.

Dwight's seventeenth sermon contains a full exposition of pluralism. In contrast to Paley, who pronounced astronomy a poor field for the exercise of the natural theologian, Dwight writes: "To the most uninstructed mind . . . , the firmament, with the innumerable and glorious bodies which it contains, has ever been far the most wonderful part of the visible creation. . . ." (I, p. 285) Superior intelligences are even more deeply moved, because they see the planets as "inhabited, in all probability, by endless multitudes of beings, rational and immortal." (I, p. 287) Concerning the moon, Dwight declares: "it is most rationally concluded, that Intelligent beings in great multitudes inhabit her lucid regions, beings probably far better and happier than ourselves." (I, p. 287) Such is Dwight's confidence in pluralism that he claims that the stars "are known, with absolute certainty, to be universally Suns, resembling our own. . . ." (I, p. 287) The teleological underpinning of his views becomes clear when after asking why God made the stars, he responds that they formed "to give light, and motion, and life, and comfort, to systems of worlds [which are] the residence of Intelligent beings, of incalculable numbers. . . ." (I, p. 288) Much of this was acceptable to deists, but Dwight departs from them by repeatedly stressing that the greatest of God's gifts to us is the mediative and redemptive action of Christ (I, p. 295). Moreover, Christ is the central subject of the second volume of his *Theology.* His forty-second sermon, that on the "Incarnation of Christ," leaves no doubt but that Dwight espouses a traditional view of Christ; as he states, "*Christ is,* notwithstanding the sneers of Unitarians, *God and Man.*" (II, p. 199) Dwight follows the Scriptures in urging Christ's direct action in creating and sustaining the universe:

Throughout immensity, [Christ] quickens into life, action, and enjoyment, the innumerable multitudes of Intelligent beings. The universe, which he made, he also governs. The worlds, of which it is composed, he rolls through the infinite expanse with an Almighty and unwearied hand. . . . From the vast store-house of his bounty he feeds, and clothes, the endless millions . . . and from the riches of his own unchangeable Mind informs the innumerable host of Intelligent creatures with ever-improving virtue, dignity, and glory. (I, p. 203)

Although such passages, penned in the post-Paine period, could not but give rise to the question whether or not Christ became incarnate on other planets, Dwight seems to have avoided this issue in his first two volumes.[40]

The third, fourth, and fifth volumes of Dwight's *Theology* focus on moral theology and grace. Late in his fifth and final volume, in discussing the end of the world and immortality, Dwight takes up the delicate problem raised by Paine. The context is an analysis of the precise meaning of Peter's prediction that at the second coming "the heavens shall pass away." (V, p. 504) Dwight urges that in such phrases "heavens" means but a portion of the universe. This interpretation leads him to declare: "Other Intelligent beings, therefore, and the worlds which they inhabit, may be concerned in this wonderful production, only in a mediate and remoter sense." (V, p. 506) Then comes the crucial passage:

This world was created, to become the scene of one great system of Dispensations toward the race of *Adam*. . . . It was intended, also, to be a theatre of a mysterious and wonderful scheme of providence. The first rebellion in the Divine Kingdom commenced in Heaven: the second existed here. The first was perpetrated by the highest, the second by the lowest, order of the Intelligent creatures. These two are with high probability the only instances, in which the Ruler of all things has been disobeyed by his rational subjects. (V, p. 508)

This passage indicates that Dwight's position was that Christ's incarnation and atonement are unique to this planet, and moreover that only earthlings need redemption. As he states: "in this world there exists a singular and astonishing system of Providence; a system of mediation between God and his revolted creatures. . . . This system, never found elsewhere, is accomplished here. . . ." (V, p. 509) Whatever may have been the theological soundness of this position that Dwight deemed the only one that could be "rationally argued," it restored to our planet the primacy it lost in the Copernican revolution. This and other dramatic features of his position, not least the contrast between earth's fallen population and the coming to it of the incarnate God, could be and were used by Dwight and later with even more effect by Chalmers to stir the minds and hearts of even the highly educated. Dwight's sermons at Yale were very influential; in some years, as many as one-third of the Yale students entered the ministry.[41] No doubt, as these men ascended their own pulpits, many preached a theology that incorporated the pluralist position. An especially significant example of Dwight's influence is his former theology student Denison Olmsted (1791–1859), who returned to Yale to teach and who during the 1850s made an important contribution (discussed later) to the pluralist debate. Furthermore, Dwight's influence extended far beyond Yale; his *Theology Explained and Defended* went through at least twelve American editions and continued to be reprinted in Britain as late as 1924.

Enthusiasm for the idea of a plurality of worlds, widespread in Britain and the United States in the decades before 1817, was also common on the Continent. Among its champions were some of the most prominent French figures of the period. For example, Pierre Samuel Du Pont de Nemours (1739–1817), statesman and political economist, espoused the doctrine in his *Philosophie de l'univers* (1796). In fact, his system includes not only traditional extraterrestrials but also superbeings possessing extra senses who move among the planets and influence their inhabitants. Transmigration of souls is also advocated in his book, which shows the continuing influence of the idea of a great chain of being. Bizarre though his book may seem to some, Du Pont wrote with deep conviction, stating in his final paragraph that his doctrine "is the fruit of numerous meditations during thirty-five years. Such are the thoughts which . . . have guided my public and private conduct since the age of eighteen. Such is *my Religion.*"[42]

Even more distinguished politically than Du Pont was Jacques Necker (1732–1804), the Swiss finance minister of Louis XVI. After fleeing revolutionary France in 1790 for his estate near Geneva, Necker lived out his last years under the care of his brilliant daughter, Anne-Louise-Gemaine Necker, famous in the literary world as Madame de Staël (1766–1817). In 1800, Necker published his *Cours de la morale religieuse,* which reveals an author devoted to natural theology, the great chain of being, and the doctrine of a plurality of worlds. The first discourse, "Of the Existence of a Supreme Being," stresses God's power and goodness in maintaining a universe composed of "innumerable suns which serve to illuminate planetary spheres, to direct their rotation, and . . . to illuminate everywhere animate beings, sensitive beings, and perhaps more wise, more grateful than we are."[43] Necker goes on to urge, citing Lambert as his source, the "probable existence of five hundred million terrestrial masses traversing in an elliptical orbit the vast extent of our solar system. . . ." (p. 49) These habitable comets "show us the relation of this immense space to the multiplication of beings, with this end so superb and so generous."[44] That Necker's pluralism was passed on to Madame de Staël is shown by her *Corinne* (1807). In that novel, Corinne has

Oswald read from his father's manuscript reflections on death. These contain the message that

. . . death will be for you only a change in habitation; and that which you leave may be the least of all. Oh innumerable worlds, which to our eyes fill the infinity of space! unknown communities of creatures of God! communities of his children, scattered in the firmament and arrayed under his vaults! let our praises join to yours; we do not know your situation; we are ignorant of your first, second, and last portions of the generosities of the supreme being; but in speaking of death, of life, of times past and to come, we attain, we touch on the interests of all intelligent and sentient beings. . . . Families of people, families of nations, assemblies of worlds, you speak with us; Glory to the master of them all, to the king of nature, to the God of the universe![45]

Necker's book suggests that physicotheology held no less fascination for the French than for their British brethren; this point is illustrated even more effectively from the writings of Jacques Henri Bernardin de Saint-Pierre (1737–1814), whose *Études de la nature* (1784) is a work squarely in the natural theology tradition. The teleological discussions of the wonders of nature characteristic of his *Études* were extended in his posthumous *Harmonies de la nature,* which after its 1815 publication in both French and English attained widespread popularity.[46] The ninth and final book of *Harmonies,* that devoted to astronomy, contains an enthusiastic endorsement, based on analogy and teleology, of life not only on all the planets but also on the moon, sun, and comets.[47] Regarding the planets, Bernardin de Saint-Pierre asserts that they ought to be inhabited because "Nature has made nothing in vain, and what would be the use of desert globes? There must be vegetable products in them, because there is heat; there must be eyes, because there is light; and there must be intelligent beings, because intelligence is displayed in their formation." (III, p. 256) This former director of the Jardin des plantes was not scientifically uninformed; he draws heavily on Herschel's writings, for example, in support of life on the sun (III, pp. 238–55). But his approach frequently leads him beyond the boundaries of science, as in his suggestion that the sun "should be the receptacle of the earth's inhabitants in a future stage of existence. . . ." (III, p. 234) The extravagant character of some of his techniques for salvaging the habitability of planets is illustrated by his bestowal on Uranus of "an immense atmosphere" (III, p. 307) and "an animal of the rein-deer kind, feeding on moss and combining in itself the advantages of the fleece of sheep, the milk of the cow, the strength of the horse, and the lightness of the stag." (III, p. 310) Although his Uranians are so remote from the sun that they are unable to see its other planets, he suggests that "they may see those of a neighboring system. . . ." (III, p. 308) This proto-romantic naturalist shows himself so attached to pluralism that he assigns souls to even the celestial bodies themselves (III, pp. 315–18) and so taken with teleology that he attributes parabolic shapes to lunar mountains that they might more effectively reflect light (III, p. 340).

Although containing no discussion of the relations of pluralism to Christianity, Bernardin de Saint-Pierre's *Harmonies* was written to convey the message that God's ways are revealed in the grandeurs of nature. The same sentiments animated Louis Cousin-Despréaux (1743–1818) in his *Leçons de la nature* (1802), the romantic and religious temper of which is suggested in its subtitle by the phrase "présentées à l'esprit et au coeur." Proclaiming that "all these world are populated by an infinite multitude and infinite variety of sentient, intelligent beings who make the name of Omnipotence resound in all the spheres," Cousin-Despréaux even asserts that "comets, without doubt, are not vast solitudes."[48]

Poets also availed themselves of the uses of pluralism, though few as fully as Paul Gudin de la Brenellerie (1738–1812). In 1801, Gudin published a long didactic poem, *L'astronomie,* and in 1810 expanded it for a new edition, to which he appended a discourse on pluralism. That doctrine Gudin describes as having "become so much the fashion that there is at present no person who, were he to arrive at the moon or Saturn, would feel less at home than in arriving at China or Mexico."[49] Gudin, whose sources were Fontenelle, Huygens, Buffon, Herschel, Lalande, and probably Voltaire, with whom he had been friends, separates himself from this sentiment by stating that he believes "all the globes are populated, even suns and comets, but . . . by beings very different from us; some [are] far above us, others much below our weak intelligence." (p. 193) In developing this position, he asserts that the earth possesses two atmospheres, the air and also the oceans in which may swim beings "intelligent and susceptible to instruction." (p. 195) With such liberal logic it is not surprising that Gudin takes the moon's lack of an atmosphere and of seas not as evidence against lunarians but as proof that they "have then no need to respire or to drink." (p. 196) Gudin's planetary populations range from Venusian "troglodytes," dwelling in caves so as to escape the sun's intense rays, to his Jupiterians, living low in their planet's atmospheres so as to avoid the terrible tides produced by its five moons. The problems involved in telescopic observation of the planetary satellites and the asteroids lead Gudin to avoid describing their inhabitants, but not to doubt their existence. Aware of the difficulties facing inhabitants of comets, he nonetheless endorses their existence, using them as an example of the diversity for which nature strives in creating intelligent beings (p. 210).

Probably the most prominent French literary figure of the early nineteenth century was Vicomte François-René de Chateaubriand (1768–1848). That astronomical imagery abounds in his writings was shown by Edmond Grégoire, who also portrayed him as a pluralist, in fact, as "preoccupied" by that doctrine.[50] As evidence of Chateaubriand's plural-ism, Grégoire cites the following passage from his *Les martyrs* (1809), in which passage Christ is described as traveling through the universe: "From globe to globe, from sun to sun, his majestic steps had traversed all those spheres which the divine intelligences inhabit, and perhaps [peut-être] men unknown to men."[51] Chateaubriand's use of the word "perhaps" indicates that he held life on other worlds as possible. That he viewed it as no more than this is suggested by his speculation in his *Le génie du Christianisme* (1802) that God created other worlds as future habitations for "the race of Adam," but because man sinned, those worlds "have remained only sparkling solitudes."[52]

No doubts, however, exist concerning the acceptance of pluralism by Comte Joseph de Maistre (1754–1821). Sometimes called the "French Burke," Maistre had fled his native Savoy after its attack by French forces, eventually coming to Saint Petersburg as representative of the Sardinian monarchy. There he composed his antirevolutionary writings, most notably his *Soirées de Saint-Pétersbourg* (1821). Although critical of science in that work, Maistre espouses pluralism, censuring "certain theologians" who reject it

. . . for fear that it disturbs the doctrine of the redemption; that is to say that, following them, we are to believe that man voyaging in space on his sorrowful planet . . . is the only intelligent being of the system, and that the other planets are only globes, *without life and without beauty,* which the Creator has flung into space to amuse himself as a juggler throws balls. No, never has a more paltry thought been presented to the human spirit.[53]

In no less strong language, he warns: "Do not miserably belittle the infinite Being by setting ridiculous limits to his power and to his love. What is more certain that this proposition: *everything has been made* by and for *intelligence*? What else can a planetary system be than a system of intelligences, and each planet than the abode of these families?" (II, p. 319) Concerning the problem of the redemption, he writes:

If the inhabitants of the other planets are not like us guilty of sin, they have no need of the same remedy, and if, on the contrary, the same remedy is necessary for them, are the theologians of whom I speak then to fear that the power of the sacrifice which has saved us is unable to extend to the moon? The insight of Origen is much more penetrating and comprehensive when he writes: "The altar was at Jerusalem, but the blood of the victim bathed the universe." (II, pp. 319–20)

Maistre's confidently phrased discussion, based on other passages from Origen, continues beyond this point, but the cited passages are sufficient to show that this Savoyard sage, so outraged by many of the doctrines of his time, was committed to the pluralist position.

Elsewhere in Europe during the first two decades of the nineteenth century, enthusiasm for pluralism remained high. In Germany, Bode and Schröter (as shown previously) championed the doctrine, and in Italy the young Giacomo Leopardi (as shown subsequently) was adopting it. Amidst immense political upheavals following the French Revolution, and despite the cultural revolution associated with the romantic movement, the idea of extraterrestrial life was seldom called into question. Accepted by intellectuals as diverse as the antireligious Shelley and the ultramontane Maistre, by Yale's evangelical Dwight and by Oxford's scholarly Nares, by Bernardin de Saint-Pierre in France and by Bode in Germany, in short, by an array of authors who may have agreed on nothing terrestrial, the idea of extraterrestrials was no less popular at the beginning of the new century than it had been at the end of the old. Deep disagreements continued concerning how pluralism was to be interpreted and integrated into prevailing systems of thought. No single publication from the pre-1817 period won widespread adherents, but this situation was soon to change as the result of a remarkable series of events initiated at noon on November 23, 1815, in Glasgow.

2. "All the world is acquainted with Dr. Chalmers' splendid Astronomical Discourses"

The day was Thursday, November 23, 1815; the place was Tron Church, Glasgow; the circumstance was the practice of the Glasgow ministers of offering Thursday noon sermons for the busy burgers of their city; the preacher was Thomas Chalmers. The topic treated in the series of seven sermons Chalmers began that day was the relation of Christianity to extraterrestrial life. Their effect, when published, was so extraordinary that four decades later Rev. Edward Hitchcock asserted: "All the world is acquainted with Dr. Chalmers' splendid Astronomical Discourses."[54]

Thomas Chalmers was born March 17, 1780, in Fifeshire, Scotland.[55] An early decision for the ministry brought him to St. Andrews University in his eleventh year, where his enthusiasm for ministerial work came to be equaled, if not surpassed, by his love for mathematics, which subject became the focus of his studies during his eight years there. Being only nineteen when his studies ended, he was permitted to be licensed to preach by the Presbytry of St. Andrews only because he was judged to be "a lad o' pregnant pairts."[56] His licensing took place in 1799, but at least four years were to pass before preaching became a central concern in his life. His passion for science and mathematics, so intense that until around 1815 he was known as "Mr. Chalmers, the mathematician,"[57] led him to spend two years at Edinburgh University studying those subjects as well as Scottish Common Sense philosophy. Chalmers found this philosophy, especially as presented in James Beattie's *Essay on Truth,* quite attractive. In particular, it provided him a way around the materialism of d'Holbach's *System of Nature,* a work that had strongly moved him during his days at St. Andrews. As the nineteenth century began, Chalmers supported the synthesis of Common Sense philosophy and religious moderatism popular in Edinburgh.[58] The most serious temptation to his ministry at this time was toward seeking a chair in mathematics. In fact, his call to the rural congregation at Kilmany in Fife, which led to his ordination in 1803, was chiefly attractive to him because it allowed him to accept an assistantship in mathematics at St. Andrews. Although he lost the latter position after a year, he kept in contact with students in the sciences and made unsuccessful attempts to secure a chair in mathematics at Edinburgh University and the chair in natural philosophy at St. Andrews.

About halfway through his dozen years at Kilmany, Chalmers underwent a deep and lasting religious reorientation. Invited by David Brewster to write for the *Edinburgh Encyclopaedia,* Chalmers not only undertook such articles as "Trigonometry" but also that on "Christianity." The latter assignment, coming at the time of the deaths of some persons close to Chalmers and a serious illness he himself suffered, led him to investigate in new depth the Christian religion. Moved both by Pascal's *Pensées* and by the Frenchman's forsaking mathematics for religion, Chalmers shifted from his earlier moderatism and rationalism toward an evangelical religious position centered on Christ and the atonement as well as man's sinfulness and dependence on grace. Whereas formerly he had distrusted the emotionalism of the evangelicals of his church, he now saw their biblical Christianity with its stress on the individual human soul as close to his own convictions. His sermons, which from 1811 on displayed greater emotional appeal, drew a more intense response from and expanded the size of his congregation. Word of Chalmers's new dynamism spread to Glasgow, where the Tron pulpit had become vacant, leading to Chalmers being proposed to fill the opening. Dr. Robert Balfour, who first visited Kilmany in 1814, commented on Chalmers's candidacy in words that while too harsh on the pre-1811 preacher were echoed by many in describing the post-1811 Chalmers:

. . . his talents are of the first order, and now distinguished grace adorns them. He has long been known as a celebrated philosopher and scorner of the peculiar doctrines of Christianity; now, from conviction and with a warm heart, he preaches the faith which once he destroyed. I . . . am astonished at a man of such superior powers so modest and humble. He is indeed converted and like a little child.[59]

Although some in Glasgow feared the evangelical enthusiasm of this young preacher, and at least one person charged that "Much learning and religion has made Chalmers mad,"[60] his supporters prevailed, albeit by a slight margin. In July 1815, he left his Kilmany parish and became minister to Tron Church.

Four months later, as Chalmers began the series of sermons so important in the pluralist debate, he did so not only as a preacher of intense evangelical fervor but also as person commanding scientific credibility. This he needed, because in those sermons he boldly blended Enlightenment ideas of extraterrestrial astronomy with evangelical sentiments distasteful to the cultured despisers of religion. His sermons, published in 1817 as *A Series of Discourses on the Christian Revelation Viewed in Connection with the Modern Astronomy*, produced a remarkable response extending even beyond Britain. Samuel Warren provided an eyewitness account of their initial reception:

One or two of these "Discourses" . . . were heard by the writer of this paper, then a boy. He had to wait nearly four hours before he could gain admission as one of a crowd in which he was nearly crushed to death. It was with no little effort that the great preacher could find his way to the pulpit. As soon as his fervid eloquence began to stream from it, the intense enthusiasm of the auditory became almost irrestrainable; and in that enthusiasm the writer, young as he was, fully participated. He has never since witnessed anything equal to the scene.[61]

William Hanna, who shortly after Chalmers's death memorialized him in a four-volume biography, provided a similar account of the "congregated thousands" of Glasgow merchants who on those days left their desks to hear Chalmers describe "heaven and its high economy."[62]

The success of these sermons when published was even more noteworthy. Eschewing the suggestion that subscriptions be sought to underwrite their publication, Chalmers decided that they should take their chances with the general public. Published on January 28, 1817, Chalmers's *Astronomical Discourses* sold 6,000 copies within the first ten weeks, and by year's end 20,000 copies in nine editions had been purchased. Writing in 1851, Hanna commented: "Never previously, nor ever since, has any volume of sermons met with such immediate and general acceptance. . . . It was, besides, the first volume of Sermons which fairly broke the lines which had separated too long the literary from the religious public."[63] The critic William Hazlitt recorded that the sermons "ran like wild-fire through the country,"[64] and George Canning, already a superb orator and soon to become prime minister, read and praised the volume.[65] The enthusiasm evoked by Chalmers's *Discourses* at their first appearance scarcely declined in subsequent decades; as late as 1851, by which time Chalmers's publications came to over twenty volumes, Hanna stated that Chalmers's *Astronomical Discourses* "to this day . . . command a larger sale than any other portion of Dr. Chalmers' writings."[66] Published in America already in 1817, they continued to be republished there as late as 1860, while British editions appeared in the 1870s. Moreover, in 1841, a German translation was published.[67]

Chalmers's eloquence has defied both explanation and imitation, even among those who themselves experienced it. As Hugh Watt related, those "young men, and they were not few, who tried to ape his manner and apply his method became the bores of their generation."[68] Given this, it seems likely that quotations, no matter how extensive, or descriptions, no matter how detailed, will fail to conjure up among present-day readers the intensity of feeling experienced by thousands of Chalmers's nineteenth-century readers. Nonetheless, let us turn to his book, with the suggestion that Chalmers's genius consisted in seeing that the doctrine of a plurality of worlds, long a staple of rationalism, could be used to evoke an intense evangelical response.

In the preface to his *Astronomical Discourses,* Chalmers notes that the "astronomical objection against the truth of the Gospel" is seldom countered by those who have written against infidelity, even though this objection has caused "serious perplexity and alarm."[69] The form that this objection takes in the mind of the infidel has two parts: (1) the "assertion . . . that Christianity is a religion which professes to be designed for the single benefit of our world . . . ," and (2) the "inference . . . that God cannot be the author of this religion, for he would not lavish on so insignificant a field such peculiar and such distinguishing attentions as are ascribed to him in the Old and New Testament." (pp. 3–4) To the first contention, Chalmers replies that this assertion is only that of the infidel, not of the Christian. In response to the second, he evokes a vision of God as so powerful and so generous that his ministrations and solicitude are almost without limit.

The first sermon begins with a quotation from the Eighth Psalm: "When I consider thy heavens, the work of thy fingers, the moon and the stars, which thou hast ordained; What is man, that thou are mindful of him? and the son of man, that thou visitest him?" (p. 2) Commenting that it is "truly a most Christian exercise, to extract a sentiment of piety from the works and appearances of nature" (p. 13), he describes the heavens, not as seen by the ancient psalmists but as beheld by modern astronomers with their far-reaching telescopes. Such instruments have revealed planets that dwarf the earth and that, like the earth, possess seasons and satellites. Stars in startling number have also been sighted and shown to be suns that may surpass our luminary in brilliance and rival it in richness of planetary retinue. Moreover, these planets "must be the mansions of life and of intelligence." (p. 24) Nebulae of immense proportion also glow within the heavens of the moderns. Even were the earth and all the celestial orbs visible to us to disappear, yet "there are other worlds, which roll afar; the light of other suns shines upon them;

and the sky which mantles them, is garnished with other stars." (p. 30)
Still more powerful imagery is evoked when he writes: "The universe at
large would suffer as little, in its splendour and variety, by the destruction of our planet, as the verdure and sublime magnitude of a forest
would suffer by the fall of a single leaf." (p. 30) Turning to the question
of the psalmist – "What is man, that thou art mindful of him?" – he
suggests that the Christian, unlike the infidel, conceives of a God whose
generosity extends from the smallest insect crawling on this planet to the
inhabitant of the most distant orb in the universe. With this view of God,
the Christian need not fear to face the question why God would "send his
eternal son, to die for the puny occupiers of so insignificant a province in
the mighty field of his creation." (p. 32) Chalmers's facility in handling
celestial distances as well as planetary detail bolsters his claim that religion has nothing to "fear from the ingenuity of her most accomplished
adversaries." (p. 33)

Humility is the main virtue encouraged in his second sermon, entitled
"On the Modesty of True Science." Herein he praises Newton as an
empiricist whose modest conception of the powers of intellect would
have led him to avoid the inconsistency of the infidel who while speculating on the botany of distant planets, denies the possibility of Christianity
being communicated to those locales. Newton, contrasted with Voltaire,
is portrayed as approaching both nature and Scripture with a childlike
modesty.

In the third discourse, "On the Extent of the Divine Condescension,"
Chalmers attempts to resolve the earlier question how God can care for
this earth, which modern astronomy has revealed to be but a minute part
of a vast cosmos. To this he suggests that the microscope, invented nearly
simultaneously with the telescope, provides much of the answer: "The
one [the telescope] led me to see a system in every star. The other leads
me to see a world in every atom." (p. 66) After asserting that the microscope shows God's solicitude extending even to the smallest of animals,
Chalmers urges the acceptance of a God who "while magnitude does not
overpower him, minuteness cannot escape him, and variety cannot bewilder him. . . ." (p. 61) Similarly, Chalmers asks how the man who
knows that God oversees his every thought can doubt that God's beneficence extends also to minds on other orbs. Concerning Christ and the
redemption, he asks his auditors to contemplate the truth that God, who
created this vast universe, "came to this humblest of its provinces, in the
disguise of a servant, and took upon him the form of our degraded
species, and let himself down to sorrows, and to sufferings, and to death,
for us." (p. 71) At this point the question whether or not Christ came
as redeemer to other planets is faced by Chalmers only in the negative
sense:

. . . for any thing we can know by reason, the plan of redemption may have its influences and its bearings on those creatures of God who people other regions, and occupy other fields in the immensity of his dominions; that to argue, there-fore, on this plan being instituted for the single benefit of . . . the species to which we belong, is a mere presumption of the Infidel himself. . . . (p. 73)

Entitled "On the Knowledge of Man's Moral History in the Distant Places of Creation," his fourth sermon suggests that just as the effects of Christ's redemptive action have not diminished through millennia of time on this earth, so also may they reach to remote planets. Learning what Christ has done for this puny planet, how could minds elsewhere doubt God's generosity? The darkness of our minds and the ambiguity of Scrip-ture prevent us, Chalmers notes, from knowing the redemptive plan in its fullness, but some passages from Scripture are quoted that he suggests can be reinterpreted in terms of the enlarged cosmos of modern astron-omy.

An example of this is the passage from Luke (15:7) that heads the fifth discourse: "I say unto you, that likewise joy shall be in heaven over one sinner that repenteth, more than over ninety and nine just persons who need no repentance." (p. 95) Taking this as his text, he sketches its significance on a cosmic scale. Poetic imagery helps him evoke an image of the entire universe rejoicing at the repentance of one earthling, yet without asserting that such acclamation occurs. Introducing angels, ex-traterrestrials with scriptural warrant, he portrays them as ringing "throughout all their mansions the hosannas of joy, over every one indi-vidual of [earth's] repentant population." (p. 100) He suggests other ways in which the tension postulated by the infidel can be relieved: "For anything I know, the every planet that rolls in the immensity around me, may be a land of righteousness." (p. 107) This leads to the awesome thought that the universe may be "one secure and rejoicing family [wherein our] alienated world is the only strayed, or only captive mem-ber. . . ." (p. 109) Such thoughts should do more than silence the infi-del; they should call us to repentance.

Satan enters the sixth discourse, and with him the "Contest for an Ascendency over Man, amongst the Higher Orders of Intelligence," as this sermon is entitled. Chalmers presents the earth as "the actual theatre of a keen and ambitious contest amongst the upper orders of creation" (p. 117) and speculates that the earth "has become the theatre of such a competition, as may have all the desires and all the energies of a divided universe embarked upon it. It involves in it other objects than the single recovery of our species. It decides higher question." (p. 118) Although he admits that "I know not if our rebellious world be the only strong-hold which Satan is possessed of, or if it be but the single post of an extended warfare, that is now going on between the powers of light and of darkness . . ." (p. 118), he later uses this idea to explain "why on the salvation of our solitary species so much attention appears to have been concentrated. . . ." (p. 124) After reading passages such as these, one suspects that the burgers of Glasgow returned to their work with an elevated sense of the significance of their every action. Donne's "No man is an island . . ." pales in comparison with Chalmers's imagery in loftiness and grandeur of scope.[70]

Chalmers's seventh and final discourse stresses that a sensitivity to majestic sights and luxurious imagery is different from a readiness to accept Jesus as personal Lord. Such sensitivity is commended, but this skillful preacher, who could not have been unaware of the impact of his eloquence, presents his sermons as above all a call to a humble life of Christian commitment. It is the Christian's will, not the fineness of his sensitivities, that brings redemption. This serves as another reminder that Chalmers spoke and wrote not as an astronomer, not as a philosopher or theologian, not even as a poet, but as an evangelical preacher striving to reach the hearts and souls of those who came to him. We cannot know whether he succeeded in this, but it is clear from the reception of his book that many left it convinced pluralists, certain that Christianity not only could be reconciled with that doctrine but also could thereby attain new grandeur.

When these sermons are considered as a whole, it is clear that for all their eloquence, they are scanty in information, vague on details, and inconclusive in argumentation. One searches in vain among them for information on the precise nature or exact locales of extraterrestrial life. Moving metaphors, an earnest tone, richness of imagery, and skillful integration of the parts are among their virtues. Persons fearful of the new discoveries of science, bewildered by the universe that Herschel and others were revealing, disturbed by Paine's polemics, or unsatisfied with the low level of feeling evoked by many preachers, all had reason to become caught up in Chalmers's response to the question of the psalmist.

An important aspect of Chalmers's *Discourses* has been stressed by David Cairns, whose claim is that if the primary goal of the natural theologian is taken to be to prove independent of revelation the existence of God and to demonstrate some of his attributes, then Chalmers's book should not be seen as a work of natural theology, as is frequently done. Instead, Cairns suggests, Chalmers was undertaking two tasks that can be seen as meritorious even by Christians who reject the traditional natural theological approach. In particular, Chalmers was responding to serious objections made by the infidel: "he is seeking, not so much to prove theism, as to remove difficulties in the way of belief."[71] Second, Chalmers was trying to work out the implications for Christian belief of a "scientific" doctrine; in doing this, he did not avoid information provided by revealed religion, but gave it central consideration.[72] Cairns's analysis seems fully reconcilable with the feature of Chalmers's book stressed in the present study, that is, its character as a work of evangelical religion. In these terms, Chalmers's sermons may be seen as a brilliant attempt to tranform a doctrine of rationalistic natural theology into a heart-moving and soul-saving vision of a bountifully generous God so deeply concerned about each of us as persons that he sent his Son for our salvation. Put differently, Chalmers's *Discourses* was less the last phase of his early rationalism than the first great success in the campaign for evangelical Christianity that characterized the remaining years of his life.

Brought into prominence first by his *Astronomical Discourses,* Chalmers subsequently became ever more widely known. Chosen for the chair in moral philosophy at St. Andrews in 1823, five years later he ascended to the professorship in theology at Edinburgh. In 1829, Francis Henry Egerton, the eighth Earl of Bridgewater, died, leaving 8,000 pounds for a series of lectures on natural theology. Eight prominent figures, in most cases scientists, were chosen, each to write one treatise. Not only was Chalmers selected, his name was placed at the head of the list. Another person chosen was William Whewell, later so important in the pluralist debate. In the post-Humean and post-Huttonian days of the 1830s, severe difficulties faced authors attempting to expound natural theology, but only two authors, Chalmers and Whewell, showed themselves sensitive to these difficulties.[73]

Chalmers's Bridgewater Treatise appeared in 1833 as *On the Wisdom, Power, and Goodness of God.*[74] Significantly, it concluded with a section "On the Defects and the Uses of Natural Theology." Here, again, is found evidence for seeing Chalmers as the man chiefly renowned as "the restorer of Evangelicalism in the Church of Scotland."[75] In that section, he stresses that "Natural Theology is overrated by those who would represent it as the foundation of the edifice. It is not that, but rather the taper by which we must grope our way to the edifice." (p. 270) Natural theology has no formal primacy, but only historical primacy, in the life of a Christian; it "creates an appetite which it cannot quell; and he who is urged thereby, seeks for a rest and a satisfaction which he can only obtain in the fullness of the gospel." (p. 271) It gives us the "probability" of God's existence, but it proves itself helpless "when it undertakes the question between God and man. . . ." (p. 268) Natural theology is limited, having within it "enough of manifestation to awaken the fears of guilt, but not enough to appease them. . . . Natural theology may see as much as shall draw forth the anxious interrogation, 'what shall I do to be saved?' The answer to this comes from a higher theology." (pp. 268–9) Although Chalmers did not discuss pluralism in his 1833 book, the passages cited show that his commitment to an evangelical approach remained central in his thought.

The event for which Chalmers is most widely known occurred in 1843, when, as the result of a long-developing dispute, he and 470 other ministers withdrew from the Church of Scotland at great personal sacrifice. Chalmers was chosen as the first head of the thereby formed Free Church of Scotland. He lived only four more years, by which time he was widely recognized as the central figure in Scottish religious history in the first half of the nineteenth century. His rise to that position began with his 1815 sermons, which also mark an epoch in the extraterrestrial life debate.

3. Responses to Chalmers, especially the one world of Alexander Maxwell and the many, many worlds of Thomas Dick

Responses to Chalmers's *Astronomical Discourses* came in many forms; at least seven reviews and four books resulted from its publication. Moreover, the topic to which Chalmers had devoted a matter of months became a continuing and central concern in the career of his fellow Scotsman Thomas Dick.

Most of the seven reviews of Chalmers's book address five questions: (1) Should the pluralist objection of Chalmers's infidel be taken seriously? (2) Is pluralism itself acceptable? (3) How should it be related to Christianity? (4) Is Chalmers's style deserving of praise? (5) What overall assessment of his book is justified? The seven authors, all of whom wrote anonymously, by no means always agreed in their responses. For example, in regard to the first question, whereas the reviewer for the *British Review* comments that he cannot recall ever having encountered this objection to Christianity,[76] and the reviewer for the *Christian Observer* suggests that it never "obtained a currency sufficient to render it deserving of so elaborate a refutation,"[77] a very different judgment is presented in the *Evangelical Magazine:* "Perhaps there are few minds accustomed to reflection and inquiry, to which the plausible objection to the truth of Christianity, combated by Dr. Chalmers, may not, at some period, have suggested itself."[78] Intermediate positions appear in *Blackwood's Edinburgh Magazine* and the *Eclectic Review*. In the former, the author, probably the Baptist preacher Robert Hall (1764–1831), notes that some have claimed that Chalmers "has combatted a phantom," but he responds:

The objections which he combats are not so much the clear, distinct, and decided averments of infidelity, as they are the confused, glimmering, and disturbing fears and apprehensions of noble souls bewildered among the boundless magnificence of the universe. Perhaps there is no mind of any strength, no soul of any nobility, that has not often been beset by some of those majestic terrors. . . .[79]

Another prominent Baptist divine, John Foster (1770–1843), writing for the *Eclectic Review,* quotes with approval Chalmers's own remark that the objection "does not occupy a very pre-eminent place in any of our Treatises of Infidelity, but is often met with in conversation; and we have known it to be the cause of serious perplexity and alarm in minds anxious for the solid establishment of their religious faith."[80]

Concerning the second question, most of the authors treat the pluralist position as accepted fact, although no one else went as far as Foster's extraordinary statement that

. . . we think that, excepting to minds repugnant to magnificent ideas, the probability that the other orbs of our system are inhabited worlds, must appear so great, that a direct revelation from heaven declaring the fact, would make but very little difference in our assurance of it.[81]

The more cautious reviewer in the *Christian Observer* comments that a reader of Chalmers's book can scarcely leave it without a feeling that the pluralist position "rests upon almost as firm a foundation as the most rigid truths of geometry." But he adds: "what after all is the *proof* of this fact?" (p. 608) The author in the *British Review* not only concurs with the pluralist doctrine but goes so far as to state: "In all that Dr. Chalmers has said upon this subject, we most cordially and unreservedly agree." (p. 26) The two reviews in *Blackwood's* express no disagreement with Chalmers's conclusions; in fact, they scarcely mention his ideas, concentrating instead on his style.[82]

Because only one of the reviewers (and he weakly) questions the pluralist doctrine, it is not surprising that none of them expresses any hesitancy at Chalmers's attempts to associate it with Christianity. Indeed, Foster's view is that Chalmers was not so much attempting to refute the infidel as to "co-extend the truths and feelings of revealed religion, with the demonstrations and speculations of astronomy. . . ." (p. 206) Foster does, however, express reservations concerning some of the ways in which Chalmers interweaves pluralism and Christianity, especially in his fourth, fifth, and sixth discourses. He questions the legitimacy of Chalmers's suggestion that Christ's redemptive actions and the other religious developments of our planet are known on other planets. In a statement that seems to suggest the view that Christ became incarnate on other planets, Foster urges that extraterrestrials have religious events of comparable significance to any on this planet, making it unlikely that events of our planet would be noteworthy on other orbs (p. 469). Foster also suggests that the pluralist doctrine helps explain the problem of evil on this planet by making it possible to claim that the amount of sin and misery on this planet is atypical of the rest of God's creation (pp. 470–1). The *Monthly*

Review article also faults Chalmers for carrying his speculations too far, but unlike the Foster review, it contains no alternate speculations. In this most negative of the reviews, the author does not reject the reconcilability of pluralism and Christianity, but he claims that Chalmers's efforts are both misplaced and unnecessary.[83]

Chalmers's style is assessed very differently in the seven reviews, the range of judgments extending from that expressed in *Blackwood's* to that in the *Monthly Review*. John Wilson, writing in the former periodical, describes Chalmers's book as possessing "an air of philosophical grandeur and truth" and as "written with an enthusiasm, and an eloquence, to which we scarcely know where to find any parallel."[84] Wilson's views are shared in the later *Blackwood's* article, where it is even suggested that Chalmers's *Discourses* reads like an "inspired book." In the *Monthly Review,* on the other hand, his book is described as "impotent in point of argument, and vicious in point of style." (p. 73) The *Evangelical Magazine* reviewer admits that he does not know "how to do justice to the feelings which have been excited in our hearts in reading this extraordinary production." (p. 269) Even stronger praises appear in the *British Review:* "perhaps, in sparkling vigour of expression, opulence and control of diction, and a profound feeling of his subject . . . , scarcely any writer, ancient or modern, can stand a comparison with the Author of these discourses."[85] A middle position is taken in the *Eclectic Review,* where Foster states: "No readers can be more sensible to its glow and richness of colouring . . . ; but there is no denying that it is guilty of a rhetorical march, a sonorous pomp, a 'shewy sameness;' a want, therefore, of simplicity and flexibility. . . ." (p. 476) In the *Christian Observer,* Chalmers is described as having a style "which will always be applauded by the world at large, while it furnishes the critic with considerable matter for animadversion and complaint." (p. 606)

In overall judgment of the work, four reviews are very favorable; the *Eclectic Review* and *Christian Observer* are moderately so; and the *Monthly Review* is decidedly negative. Wilson recommends the volume "in the strongest manner" and predicts that it will have a "great and salutary" effect, whereas Hall asserts that the book is "not to be praised as a mere work of human genius. . . ." Readers of the *Evangelical Magazine* are promised that "high as their expectations of this book may be, they will not rise from its perusal disappointed." The article in the *British Review* concludes by the statement that "when we class it among the finest productions of modern genius, we shall do it but feeble justice, unless, at the same time, we observe that it has an eminent worth and dignity, to which modern genius seldom rises or aspires." In the *Eclectic Review,* Foster praises Chalmers's treatment as far superior to that of Fuller, with which he compares it at some length. The reviewer for the

Christian Observer expresses the wish that Chalmers had been less ready to assume a plurality of worlds, but overall his judgment is favorable. Finally, the *Monthly Review* article faults Chalmers for triumphalism, poor style, weak arguments, and much else. But the author attests to his influence:

> When the reader has taken time to peruse this farrago, he will perhaps no longer wonder that the preacher, in his late visit to our metropolis, excited a burst of admiration, and occupied as much of the eager buz . . . as if a new missionary of the Gospel had actually arrived from the moon, to bring us news of other worlds and to strike infidelity dumb. (p. 72)

Chalmers once remarked that "It has been singularly the fate of my publications to be torn to pieces in the journals, but at the same time to be extensively bought and read."[86] This characterization may be elsewhere applicable, but it scarcely describes the reception given his *Astronomical Discourses* in these seven reviews.

Three small treatises and a large book were written in response to Chalmers's *Astronomical Discourses,* all opposing it. Of the treatises, only that published anonymously in 1818 by Henry Fergus (1765?– 1837) deals significantly with Chalmers's pluralist ideas.[87] Although Fergus faults a number of Chalmers's particular claims about comets, stars, and volcanoes, he expresses no reservations about Chalmers's acceptance of extraterrestrial life (p. 22). Clearly a champion of traditional natural theology, Fergus seems disturbed at the introduction of such Christian notions as the atonement into this area. Moreover, he asserts that Chalmers's infidel objection is "little known, and . . . no where stated with clearness and precision." (p. 24) In his concluding discussion, Fergus makes some negative remarks on the style of the *Discourses,* describing them as "not unlike a comet, set off with the GARNITURE of a long flaming tail, and a glaring body, but having only a small and ill-defined nucleus." (p. 41) He does hold out hope that in the future Chalmers will become what Scotland in his view so needed, a preacher both erudite and popular, but no such attribution is made in this tactful treatise written by a rationalist in response to the work of an evangelical.

The most elaborate response to Chalmers's *Discourses* was written by Alexander Maxwell, about whom almost nothing is known beyond a few details uncovered by Augustus De Morgan over a century ago. From a correspondent De Morgan learned that Maxwell was "a law-bookseller and publisher (probably his own publisher) in Bell Yard. He had peculiar notions, which he was fond of discussing with his customers. He was a bit of a Swedenborgian."[88] Entitled *Plurality of Worlds: or Letters, Notes & Memoranda, Philosophical and Critical; Occasioned by "A Series of Discourses on the Christian Revelation, Viewed in Connection with the Modern Astronomy." By Thomas Chalmers, D.D.,*[89] Maxwell's book consists of twelve letters that reveal a chatty, good-natured, and highly opinionated author who delights in embellishing his book with long quotations placed in numerous footnotes.[90] Unable to share the "universal applause given to Dr. Chalmers," and fearing that Chalmers's "plan of reasoning, or rather high wrought declamation, . . . is calculated . . . to strengthen the arguments of infidels" (p. 8), Maxwell attacks Chalmers and also the whole system of modern astronomy, which "has been a strong fort and rampart with the Infidel for many generations. . . ." (p. 9) Maxwell cites among his qualifications some study of astronomy (p. 1), but above all "courage enough to push against wind and tide. . . ." (p. 13)

In tracing the history of pluralist ideas from antiquity on, Maxwell comes to William Herschel, whose writings, he claims, show "wildness of speculation, and the pregnant associations, of a disordered mind. . . ." (p. 25) But others, he notes, have not accepted pluralism; in particular, Wesley and the Hutchinsonians have "expressed some very just doubts upon the subject." (p. 24) As this comment suggests, and as subsequent statements put beyond doubt, Maxwell was not a pluralist, nor a Swedenborgian (as De Morgan's correspondent suggested), but a latter-day Hutchinsonian who, as De Morgan correctly put it, "does not admit modern astronomy."[91] Maxwell's first attack on astronomy resurrects the Hutchinsonian claim that astronomers are unable to measure accurate parallaxes (solar, planetary, and stellar). Maxwell's skepticism on this issue was not without basis; as he notes, an error of only one second of arc can throw off a determination of the distance of the sun by millions of miles (p. 31). Moreover, Maxwell notes that Herschel had measured the diameter of Pallas as 80 miles, whereas Schröter "makes it no less than 2099 miles." (p. 37)[92] Disputes among astronomers whether or not a stellar parallax had been determined (it had not) are also cited by Maxwell, as well as disputes (exaggerated by him) concerning solar parallax.

When in his next letter Maxwell challenges theoretical astronomy, his critique goes beyond good sense. Bringing charges of infidelity and failure to follow Bacon against a number of authors, including Chalmers (pp. 61–9), Maxwell follows the Hutchinsonians in calling into question even Newton and his system (pp. 71–118). He laments Newton's Arianism as well as his attachment to such "occult" ideas as gravitation. Moreover, Maxwell maintains that speculators of all sorts, from Locke and Laplace to Hume and Hutton, as well as mathematical studies in general, lead to infidelity. As he states, "it would appear that many eminent mathematicians and skilful [sic] astronomers are devoid of sensibility, even upon the first elements of our holy religion." (p. 160) Such charges would hardly dispose intelligent readers to take Maxwell's views seriously when in his ninth letter he provides his most detailed critique of pluralism. This begins with a discussion of a number of authors who had tried to reconcile it with Christianity, including Beattie, Porteus, Fuller, King, Nares, Swedenborg, and of course Chalmers, all of whom he finds unacceptable. So uncritical is Maxwell that he claims that between Chalmers's and Swedenborg's ideas, the "points of coincidence are remarkable." (p. 196) Commenting to his readers that "you are not to suppose . . . that no men of science and literary reputation have sanctioned the opposite sentiment" (p. 197), Maxwell presents the antipluralist views of Thomas Baker as well as of such Hutchinsonians as Catcott and Parkhurst.

The final three sections of Maxwell's book deal with Scripture, philosophy, and Chalmers's style. On Scripture, Maxwell shows himself to be a fundamentalist, distressed at the geological theories of his day. Concerning philosophy and all thought not based on Scripture, Maxwell reveals the retrogressive nature of his position by claiming that Adam was "not only the first, but the greatest of philosophers." (p. 237) Turning to Chalmers's style, Maxwell mentions that he has heard it described "by men of taste, knowledge, and piety, with mingled emotions of pity and astonishment." More specifically, he reports:

One had received "a shock of holy electricity;" another had been "sublimated in the crucible of this spiritual chemist;" another had been "carried to the third heaven;" and a fourth had been so attracted, that "he could have sat for ever under the droppings of the skies." (p. 252)

Such was not Maxwell's reaction. Arguing that Chalmers's book lacks simplicity, variety, and harmony, he predicts that it will soon fade from popularity.

At the end of the second (and last) edition[93] of Maxwell's book, he excerpts praises of his first edition from three reviews it received. An examination of these reviews, however, suggests that its reception was far from uniformly favorable. Although the review in *Gentleman's Magazine* was supportive, that in *Evangelical Magazine* shows a clear preference for Chalmers, while that in *Monthly Review,* which had earlier been so critical of Chalmers, finds little to praise in Maxwell except his dislike of Chalmers![94]

What is to be made of Maxwell's book? A pluralist of that day would have been justified in labeling it superficial in its arguments and retrogressive in its approach. And a sophisticated antipluralist would no doubt have deemed it an embarrassment. If the widespread acceptance of extraterrestrial life was to be effectively challenged, it would scarcely be accomplished by resurrecting the never very popular arguments of the Hutchinsonians.

Although Chalmers's entry into the pluralist debate was extremely influential, it involved only one of his books and engaged him for but a few months of his career. In these latter respects, he stands in striking contrast to Rev. Thomas Dick (1774–1857), who devoted years to writing numerous books in which pluralism appears as a prominent theme. Born in Hilltown, Dundee, Dick was raised in the Secession Church of Scotland. Like Chalmers, he was powerfully drawn to both science and religion. After studies at Edinburgh University, he was licensed to preach, but pursued instead a career as a teacher, first in the Secession School in Methven, then at Stewart's Free Trade School in Perth. In 1827, at age fifty-three, Dick forsook teaching for full-time writing. This he did from near Dundee, where a large house, complete with observatory, was constructed according to his design. The occasion for this change in career was the success of Dick's first book, his *Christian Philosopher* (1823). Dick later authored at least nine major books, all requiring numerous reprintings and at least three of which were translated into Welsh and one into Chinese. His warmest reception came from the United States, where one writer stated that "No author of the nineteenth century has a higher claim upon the respect and gratitude of the world than the Christian philosopher, Dr. Thomas Dick," and Union College, in Schenectady, New York, bestowed upon him an honorary doctorate.[95] What makes Dick significant in this history is that nearly all his books were laced with ideas of extraterrestrial life.

Dick's subject in his *Christian Philosopher* is suggested by its subtitle: *The Connection of Science and Philosophy with Religion*.[96] Pluralist passages abound in it. For example, in praising the wisdom of the Deity for creating the sun of just such a size and at just such a distance as to "refresh and cheer us, and to enliven our soil . . ." (p. 30), Dick adds that this does not preclude the atmospheres and physical constitutions of the other planets from being equally beneficial to their inhabitants, some of whom may have no need of sleep. He manifests no less confidence in his pluralist vision when he moves beyond the solar system to assert that "every star . . . is the centre of a system of planetary worlds, where the agency of God may be endlessly varied, and perhaps, more strikingly displayed than even in the system to which we belong." (p. 31) Dick does not hesitate to suggest, following and even going beyond Herschel, that God has placed within the sun "a number of worlds . . . and peopled them with intelligent beings. . . ." (p. 81) Turning to the moon, he expresses hesitancy about Herschel's "observation" of lunar volcanoes, suggesting instead that "It would be a far more pleasing idea, and perhaps as nearly corresponding to fact, to suppose that these phenomena are owing to some occasional splendid illuminations, produced by the lunar inhabitants, during their long nights." (p. 84) Moreover, he predicts that *"direct proofs"* of the moon's habitability will be forthcoming, supplementing this by two appendixes in which he expresses doubts that the observations of Schröter and Gruithuisen have provided such proofs (pp. 150–2). Although populating all the planets, Dick hesitates concerning comets: That they were not created in vain we know, but their specific purposes we do not know. The facts of astronomy, with which Dick was well acquainted, are not the only source of his confidence in extraterrestrial life; he assures his readers: "the language of scripture is not only consistent with the doctrine of a plurality of worlds, but evidently supposes their existence to all the extent to which modern science can carry us." (p. 91) In an appendix, he even claims: "A plurality of worlds is more than once asserted in Scripture, and in numerous passages is evidently taken for granted." (p. 153) In support of this, he cites such passages as that from the Eighth Psalm, which had been Chalmers's text. Near the end of his book, Dick turns to the planetoids to discuss the hypothesis that they are fragments of a larger body. The explosion from which they resulted, he suggests, "would seem to indicate, that a moral revolution has taken place among the intelligent beings who had originally been placed in those regions. . . ." (p. 143)

Between his *Christian Philosopher* (1823) and his first primarily astronomical book, his *Celestial Scenery* (1837), Dick published five large books:

1. *The Philosophy of Religion; or, An Illustration of the Moral Laws of the Universe* (1826)
2. *The Philosophy of a Future State* (1828)
3. *On the Improvement of Society by the Diffusion of Knowledge* (1833)
4. *On the Mental Illumination and Moral Improvement of Mankind* (1836)
5. *An Essay on the Sin and Evils of Covetousness* (1836)

Diverse as the subjects of these books were, Dick devised ways to incorporate pluralist pronouncements into at least four of them. It takes him only until the second paragraph of his *Philosophy of Religion* to remind his readers: "We have the strongest reason to believe, that the distant regions of the material world are also replenished with intellectual beings . . . in which there may be a *graduation* upwards, in the scale of intellect above that of man, as diversified as that which we perceive in the descending scale from man downwards to the . . . animalcula." After setting out his ideas on morality, he urges: "The grand principles of morality . . . are not to be viewed as confined merely to the inhabitants of our globe, but extend to all intelligent beings . . . throughout the vast universe. . . ."[97] He goes on to state that *"there is but one religion throughout the universe."* (p. 68) Pluralism plays an even larger part in his *Philosophy of a Future State,* which he dedicated to Chalmers. For example, Dick calculates that because we can see 80 million stars and because each star has at least thirty satellites encircling it, 2,400,000,000 inhabited worlds must exist within the *visible* universe.[98] Our immortal souls, he suggests, will devote much of eternity to studying the scenery and history of these other worlds. At one point, he speculates that celestial messengers will convey to us information on these worlds (p. 73); at another, he draws on the "highly probable, if not certain" idea of some astronomers that a massive body is located at the center of the universe to suggest that this body may be "THE THRONE OF GOD." (p. 103) "Here," he adds, "deputations from all the principal provinces of creation, may occasionally assemble, and inhabitants of different worlds mingle with each other, and learn the grand outlines of those physical operations and moral transactions, which have taken place in their respective spheres." (p. 103) Dick's readiness to reduce the spiritual to the material extends even to the suggestion that angels are probably material beings (p. 93).

A book entitled *On the Improvement of Society by the Diffusion of Knowledge* that abounds in hints on how to keep women's dresses from catching fire and methods for curing smoking chimneys would scarcely seem a favorable format for a discussion of celestial beings. Yet this Dick manages. He stresses that the doctrine of a plurality of worlds will give us an expanded conception of the Divine[99] and again uses this doctrine to illustrate one of his central themes: that science and religion are in harmony (pp. 130–2).

In 1836, this teacher-turned-writer set out his philosophy of education in his *On the Mental Illumination and Moral Improvement of Mankind.* While ranging over topics from the instruction of infants to the establishment of mechanics' institutions, Dick explains how astronomy should be taught, stressing that pluralist ideas instill humility and reverence.[100] He recommends pluralism to the religion teacher and preacher as well (pp. 154, 212). Dick's *Essay on Covetousness* (1836) may contain no references to pluralism, but this doctrine was given prominence when in 1837 he published the first of his three major astronomical treatises.

This was his *Celestial Scenery; or, The Wonders of the Planetary System Displayed; Illustrating the Perfection of the Deity and a Plurality of Worlds.*[101] Having promised in his preface to present an "overlooked department of astronomical science . . . *the scenery of the heavens as exhibited from the surfaces of the different planets and their satellites* . . . ," Dick devotes his eighth chapter to discussing the heavens as seen from Mars, the planetoids, Jupiter, and elsewhere. This approach, employed more than a century earlier by Huygens, is hardly innovative, but Dick does break new ground in assigning specific numbers for the populations of the planets and planetoids and even for the edges of the rings of

Table 4.1. *Thomas Dick's population table for the solar system*

	Population
Mercury	8,960,000,000
Venus	53,500,000,000
Mars	15,500,000,000
Vesta	64,000,000
Juno	1,786,000,000
Ceres	2,319,962,400
Pallas	4,000,000,000
Jupiter	6,967,520,000,000
Saturn	5,488,000,000,000
Saturn's outer ring Inner ring Edges of the rings	} 8,141,963,826,080
Uranus	1,077,568,800,000
The Moon	4,200,000,000
Jupiter's satellites	26,673,000,000
Saturn's satellites	55,417,824,000
Uranus's satellites	47,500,992,000
Amount	21,891,974,404,480

Saturn! Basing his calculation on the population density of England being 280 persons per square mile, and neglecting the possibility of oceans elsewhere, Dick constructs a table (p. 135) in which he assigns every planet and planetoid in the solar system, except Vesta, a higher population than the earth. Dick's omission of a population figure for the sun does not indicate that he doubted solarians; in fact, after citing Herschel on their behalf, he warns: "it would be presumptuous in man to affirm that the creator *has not placed* innumerable orders of sentient and intelligent beings . . . throughout the expansive regions of the sun." (p. 107) Moreover, he all but carries out the calculation by noting that the surface area of the sun is thirty-one times the combined surface areas of all the other solar system objects (p. 136).

Having discussed pluralist ideas in nearly every section of his book, Dick devotes his last chapter to five arguments for a plurality of worlds:

I. *[T]here are bodies in the planetary system of such MAGNITUDES as to afford ample scope for the abodes of myriads of inhabitants.* (p. 161)

II. *There is a GENERAL SIMILARITY among all the bodies of the Planetary System, which tends to prove that they are intended to subserve the same ultimate designs in the arrangements of the Creator.* (p. 163)

III. *In the bodies which constitute the solar system, there are SPECIAL AR-RANGEMENTS which indicate their ADAPTION to the enjoyments of sensitive and intelligent beings; and which prove that this was the ultimate design of their creation.* (p. 166)

IV. The scenery of the heavens, as viewed from the surfaces of the larger planets and their satellites, forms a presumptive proof that both the planets and their moons are inhabited by *intellectual* beings. (pp. 171–2)
V. [I]n the world we inhabit, *every part of nature is destined to the support of animated beings*. (p. 172)

Dick was sufficiently informed astronomically to know that serious objections could be brought against these arguments. For example, he knew that some astronomers objected to populating particular celestial bodies because no evidence of atmospheres on them could be found. This objection Dick dispatches by proposing that their atmospheres are "invisible" and "purer" than ours, and from this he concludes: "the moral and physical condition of their inhabitants is probably superior to that enjoyed upon earth." (p. 168) What supplied Dick with such conclusions and the confidence with which he asserts them was not telescopes but teleology. Convinced that we can know final causes even in the remote regions of space, Dick warns his readers:

Without taking pluralism into account, we can form no *consistent* views of the character of Omnipotence. Both his wisdom and his goodness might be called into question, and an idea of the Supreme Ruler presented altogether different from what is exhibited by the inspired writers in the records of Revelation. (p. 176)

Dick's determination to prove and promulgate pluralism was no less evident in his 1840 *Sidereal Heavens and Other Subjects Connected with Astronomy, As Illustrative of the Character of the Deity and of an Infinity of Worlds,*[102] in which he describes pluralism "not merely as conjectural or highly probable, but as susceptible of moral demonstration." (p. iv) Moreover, his seventeenth chapter is entitled "A Plurality of Worlds Proved from Divine Revelation." Although devoted to the sidereal system, as his *Celestial Scenery* had centered on the solar system, this book contains a discussion of comets, which shows that Dick's propensity to make excessive pluralist claims had increased in the seventeen years since he wrote his *Christian Philosopher*. Having refrained in his 1823 book from declaring comets habitable, he now suggests that "some of the comets . . . may be peopled with intelligences of a higher order than the race of man." (p. 168) This is followed by long quotations of some of the most bizarre of Lambert's speculations on cometary inhabitants. He draws this together in a remark about comets that have solid nuclei: "If this position be admitted, then we ought to view the approach of a comet, not as . . . a harbinger of evil, but as a splendid world . . . conveying millions of happy beings to survey a new region of the Divine empire. . . ." (p. 174) Various approaches employed in his *Celestial Scenery* appear also in this book; although he provides no table of the populations of individual stellar systems, he does suggest

60,573,000,000,000,000,000,000 as the likely population of the visible universe (p. 148). He also supplements his earlier five arguments for pluralism by three new arguments and an entire chapter of scriptural "proofs." His three new arguments are:

I. *That the doctrine of a plurality of worlds is more worthy of the perfections of the Infinite Creator, and gives us a more glorious and magnificent idea of his character and operations than to suppose his benevolent regards confined to the globe on which we dwell.* (p. 118)

II. *[W]herever one perfection of Deity is exerted, there also ALL his attributes are in operation,* and must be displayed, in a greater or less degree, to certain orders of intelligences. (p. 122)

III. *There is an absurdity involved in the contrary supposition* – namely, that the distant regions of creation are devoid of inhabitants. (p. 123)

Actually, Dick finds not one but many absurdities in the antipluralist position, not the least of these being that its acceptance "would virtually deprive the Creator of the attribute of *wisdom*. . . ." (p. 125)

Another feature of Dick's *Sidereal Heavens* is a chapter "On the Physical and Moral State of the Beings That May Inhabit Other Worlds." Herein Dick rejects the view recently put forward by a reviewer who proposed "that in some worlds the inhabitants may be as large as mountains, and in others, as small as emmets." (p. 135) Mountainous size, he reasons, would make for overcrowding and be "injurious to the exercise of intellectual faculties." (p. 135) Turning to the question of "whether we may ever enjoy an intimate correspondence with beings belonging to other worlds . . . ," he suggests that although this is not now possible, man "is destined to a future and eternal state of existence, where the range of his faculties and his connexions with other beings will be indefinitely expanded." (p. 139) Even though evidences abound that Dick had read with care the writings of such astronomers as the Herschels, Lambert, and Laplace,[103] it is obvious that the ideas in such quotations as those just cited originated less from astronomy than from Dick's own somewhat peculiar religious mentality. This is especially clear, as is the utopian character of much of his pluralist thought, when he writes: "It is probable that the greater part of the inhabitants of all worlds are . . . in that state of moral rectitude in which they were created. . . ." (p. 141)

In his section on novae, Dick discusses the views of Samuel Vince (1749–1821), the Plumian professor of astronomy at Cambridge, who in his *Complete System of Astronomy* had urged that the disappearance of a star may be "the destruction of that system at the time appointed by the Deity for the probation of its inhabitants. . . ." (p. 47) A similar view is attributed to Dr. Mason Good, who warned: "What has befallen other systems will assuredly befall our own." (p. 47) From these views Dick

dissents, suggesting that although God may create worlds, he never annihilates them (pp. 47–8).

Before his death in 1857, Dick published a third astronomical treatise,
his *Practical Astronomer* (1845). Pluralist passages rarely occur in it,
probably because of its instrumental orientation. Another possible reason is that Dick may have been becoming sensitive to criticisms of his
practice of blending religion and astronomy.[104] Such criticisms had merit;
the "Christian philosopher" of Dundee may have injured both Christianity and natural philosophy by his overconfident pronouncements about
God's design and his bizarre speculations about extraterrestrials. This
should not, however, obscure the fact of Dick's far-reaching influence.
Persons as prominent as Ralph Waldo Emerson, William Lloyd Garrison,
and Harriet Beecher Stowe came to Dundee to meet him, and Emily
Brontë seems to have borrowed from his *Sidereal Heavens* to enrich her
poetry.[105] Furthermore, E. E. Barnard and J. A. Brashear testified to the
role of Dick's writings in leading them to the astronomical careers in
which they so distinguished themselves, and Dr. David Livingstone revealed in 1857 that it was Dick's *Philosophy of a Future State* that
inspired him to missionary work.[106] Important as these cases may be,
they scarcely exhaust the examples of Dick's influence, as subsequent
discussions will show.

4. Saving the selenites, including evidence that R. A. Locke's "moon hoax" was not a hoax

Although evidence against lunar life was already substantial when the
nineteenth century began, not a few authors endeavored to save the
"selenites," as the supposed inhabitants of our satellite were called. As
noted previously, both Dwight and Dick had endorsed them, the latter
even prophesying that "*direct proofs*" might soon be secured. Selected
for discussion in this section are such prominent astronomers as Gruithuisen, Gauss, Littrow, Bessel, and Mädler, who discussed the question
of lunar life, as well as a New York journalist, R. A. Locke, who in 1835
created a sensation by publishing a report that rational beings had been
sighted on the moon. Although Locke's publication, widely known now
as the "moon hoax," may seem to belong to the history of science fiction,
the point of view advocated in the present study is that it has a special
place in the pluralist debate and in fact *was not a hoax*.

Pluralism lost two of its most prominent advocates with the death of
Schröter in 1816 and the death of Bode in 1826. This double loss might
have been more severe had there not emerged in the decade separating
their deaths an astronomer scarcely less active as an observer than Schrö-

ter or as a publicist than Bode. This was Franz von Paula Gruithuisen (1774–1852), who although trained in medicine at Landshut University was above all interested in astronomy. So prolific was Gruithuisen that in a list compiled in 1882 of the most prolific astronomers up to that time, he ranked seventh, with 177 papers, and so active was he as an editor that this former physician edited three astronomically oriented journals.[107] Moreover, in enthusiasm for pluralism he surpassed both Bode and Schröter, the latter of whom he may have taken as his hero. Such is suggested by the fact that in 1821–3 he published a long paper entitled "Selenognostische Fragmente" in which he makes repeated reference to Schröter's *Selenotopographische Fragmente,* including his "observations" of lunar industries. In fact, Gruithuisen quotes with approval his predecessor's claim that a part of the Maris Imbrium region of the moon is "just as fruitful as the Campanian plain."[108] Much of this paper is devoted to saving Schröter's selenites, on behalf of whom Gruithuisen argues for lunar lakes and an attenuated atmosphere.

Although Gruithuisen published a number of subsequent discussions of lunar life, he would no doubt rank as most important a lengthy paper issued in 1824 under the title "Discovery of Many Distinct Traces of Lunar Inhabitants, Especially of One of Their Colossal Buildings."[109] In the first part of this paper, Gruithuisen presents his evidence, based mainly on tints he observed on the lunar surface, for various lunar climates and for corresponding forms of vegetation, claiming that lunar flora extend 55 degrees to the south and 65 degrees to the north (p. 142). The second part presents his evidence, based on "paths" he saw on the moon, for lunar animals, which, he claims, roam "from 50° northern latitude up to 37° or possibly 47° southern latitude." (p. 153) The third part, although commencing with the caution that one must not expect the impossible from telescopes, contains Gruithuisen's observations of various geometrically shaped features on the moon that he labels roads, walls, fortifications, and cities. One of these observations, he reports, so excited him that he exclaimed: "Oh Schröter, here is what you have always sought in vain." (p. 163) Another observation, of a star-shaped structure that he labels a temple, sets him speculating about the religion of his lunarians (pp. 168–9). Gruithuisen's paper, which concludes with urgings that others extend his observations, soon attracted attention in France, Britain, and elsewhere.[110] That this publication, bizarre though it may be, contributed to his career is suggested by the fact that two years after its publication, he was chosen professor of astronomy at Munich University. Nonetheless, its long-range effect was to discredit his observational work and to make him an object of ridicule among astronomers. Writing in 1868, T. W. Webb described Gruithuisen as having "assuredly thought, and published, an uncommon amount of nonsense." Webb, however, added: "this man made good use of a keen eye and sharp instrument, and saw much, and if he had spared us his inferences, would have been accepted as an observer of no little weight."[111]

Gruithuisen's passion for pluralism was by no means confined to the moon or to the 1820s; during the 1833–5 period, for example, he published a series of papers arguing for life on Mercury, Venus, and comets.[112] Of all Gruithuisen's planetary fantasies, the one that attracted the most attention occurs in his discussion of Venusians.[113] His speculation concerns the "ashen light," a faint illumination in the dark portion of Venus reported in 1759 by J. T. Mayer and in 1806 by K. L. Harding. The best method of explaining these illuminations, suggests Gruithuisen, is to attribute them to

. . . general festivals of fire given by the Venusians, which are so much more easily arranged, because on Venus the tree growth must be far more luxuriant than in the virgin forest of Brazil. . . . [S]uch festivals may be celebrated either to correspond to changes in government or to religious periods. The period from Mayer's to Harding's observation is 76 Venus years or 47 earth years. If the period is religious we cannot comprehend a cause for this number of years. If however it corresponds to the time when another Alexander or Napoleon comes to supreme power on Venus, then it becomes somewhat easier to understand. If we take the ordinary life span of a Venusian to be 130 Venus years . . . , then the reign of an absolute monarch may easily last 76 years.[114]

Should a detailed study of Gruithuisen's career ever be undertaken, it seems probable that he will be found to have been a man of vast energy, extensive learning, excellent eyesight and instrumentation, and little sense. His undoing was an overly active imagination that he could not or would not control. In all this, he may be seen as a successor to Schröter.

Gruithuisen's weaknesses did not go unrecognized by his scientific contemporaries. Among these was Carl Friedrich Gauss (1777–1855), the director of the Göttingen Observatory and the foremost mathematician of the nineteenth century. Gauss is known to have complained of the "mad chatter" of his Munich colleague and on learning of a dispute between Gruithuisen and Schelling commented: "The two opponents seem to me to be completely worthy of each other."[115] Gauss's disdain for Gruithuisen was shared by the Bremen astronomer Wilhelm Olbers (1758–1840), who in his correspondence with Gauss labels the Munich astronomer "peculiar" ("wunderlich") and laments the largeness of his "power of imagination."[116] Johann Joseph von Littrow (1781–1840), director of the Vienna Observatory, was also not impressed; in his widely read *Die Wunder des Himmels,* Littrow declared that the claims of Schröter and Gruithuisen that they had observed lunar cities were "entirely unproven."[117] Gauss, Littrow, and Olbers were no doubt correct in their assessments of Gruithuisen, but evidence exists that they were themselves pluralists and even proponents of lunar life.

Gauss and Littrow receive mention in some present-day discussions of extraterrestrial life because it is believed that each made a proposal concerning sending signals to the moon or Mars. The most widespread version of this story attributes to Gauss the suggestion that in Siberia there be erected a giant figure in the shape of the "windmill" diagram used in Euclid's demonstration of the Pythagorean theorem. Because mathematics on the moon would be identical with its earthly counterpart, the selenites, on seeing this figure, would recognize our orb as inhabited and respond accordingly. Littrow, it is reported, proposed the alternative of a huge circular or square canal cut in the Sahara. Onto the waters of the canal kerosene would be poured and set ablaze, making an even more striking display of terrestrial intelligence.[118] The history of this proposal, especially the Gauss portion of it, can be traced through two dozen or more pluralist writings reaching back to the first half of the nineteenth century. When this is done, however, it turns out that the story exists in almost as many forms as its retellings. Furthermore, these versions share one characteristic: Never is reference supplied to where in the writings of Gauss and Littrow the proposals appear![119]

Let us examine some early versions of the story. In his *Wunder des Himmels,* Littrow approvingly presents the proposal, attributed to "one of our most distinguished geometers," that a geometrical figure, "for example, the well known so-called square of the hypotenuse, be laid out on a large scale, say on a particular broad plain of the earth. . . ."[120] Sometime before 1847, François Arago (1786–1853), in his lectures at the Paris Observatory, discussed this proposal, attributing it, like Littrow, to a "German geometer." However, in the Arago version, the locale is specified as Siberia and the signaling method as mirrors.[121] In 1853, the English encountered it in Patrick Scott's *Love in the Moon,* where the author is "a learned man," the locale "the great African dessert," the signal formed by a "great plantation of trees," and the figure that of the "47th Proposition of Euclid."[122] Two years later, readers of the *Christian Remembrancer* were warned against this "strange and monstrous conception [which is], we believe, due to the originality of the gallant and eccentric Col. Thompson [who] proposed to carve the [Pythagorean] figures . . . upon Salisbury Plain. . . ."[123] In 1878 the American astronomer Asaph Hall and in 1892 the English astronomer J. Norman Lockyer both attributed the plan for the geometrical figure to a "German astronomer" who had recommended "fire signals" sent from Siberia.[124] In the interim, an author writing in *Chambers's Journal* had mentioned "the proposal of a Russian savant to communicate with the moon by cutting a large figure of the forty-seventh proposition of Euclid on the plains of Siberia, which, he said, any fool would understand."[125] Simon Newcomb, in 1902, ascribed the idea of a Siberian figure to the astronomer "Zach." Newcomb's "triangle" had sides "several hundred miles in extent," this increase in size no doubt being necessary because the signal is specified as being sent to Mars rather than the moon.[126]

The strangest and most interesting report may also be the earliest, dating from the October 1826 issue of the *Edinburgh New Philosophical Journal.* An anonymous author revealed:

> Olbers considers it as very probable that the moon is inhabited by rational creatures, and that its surface is more or less covered with vegetation. . . . Gruithuisen maintains that he has discovered . . . great artificial works in the moon, erected by Lunarians. . . . Gruithuisen, in conversation with the great astronomer Gauss, after describing the regular figures he had discovered in the moon, spoke of the possibility of a correspondence with the inhabitants of the moon. He brought, he says, to Gauss's recollection, the idea he had communicated many years ago to Zimmerman. Gauss answered, that the plan of erecting a geometrical figure on the plains of Siberia corresponded with his opinion, because, according to his view a correspondence with the inhabitants of the moon could only be begun by means of such mathematical contemplations and ideas, which we and they have in common.[127]

What is to be made of this report? The editor of the *Annals of Philosophy,* where it was promptly reprinted, suggested that Gruithuisen and his supporters should be labeled "downright lunatics," and he added: "As to the alleged conversation between MM. Gruithuisen and Gauss, the latter must, we conclude, have intended to laugh in his sleeve at the strange speculations of the former, whilst he seemed to enter into his wild, extravagant views."[128]

Although my efforts to trace the claims made in this report to the published papers of Olbers and Gauss have not succeeded, materials have been uncovered showing that the report faithfully represents the views of Olbers and is at least partially correct concerning Gauss. In a letter sent to Gruithuisen on September 18, 1825, Olbers states: ". . . I hold it to be very probable that the moon is inhabited by living, even rational creatures, and that something not wholly dissimilar to our vegetation occurs on the moon." Olbers goes on to encourage Gruithuisen in his research and, without specifically endorsing his lunar cities, praises him for his lunar observations: "You . . . have already made known many very noteworthy results. . . ."[129] That Olbers's pluralism extended throughout the universe is suggested by the fact that in the 1823 paper in which he put forth the famous paradox named after him, Olbers quotes approvingly a passage from Kant's *Allgemeine Naturgeschichte und Theorie des Himmels* in which the philosopher argued for an infinite universe on the basis of the principle of plenitude. Moreover, Olbers adds that it is "most highly probable" that "all of infinite space is filled with suns and their retinues of planets and comets."[130]

Olbers could, however, be critical of pluralists; in a number of his letters to Gauss, Olbers remarks negatively on Gruithuisen. His letter of June 22, 1824, for example, contains the comment:

Have you seen the representation of Gruithuisen's alleged lunar city and his avenues of trees and his roads in the moon? The power of imagination of the man is large; but nonetheless what he maintains is a city, even though it has no similarity to such, is surely noteworthy if otherwise his drawing, which I have no reason to doubt, is correct.[131]

On June 12, 1826, Olbers wrote Gauss to lament that Prince Metternich was becoming interested in Gruithuisen's observations, and in his February 25, 1827, letter Olbers complains: "That peculiar Gruithuisen is said now to have compromised you and also me through his indiscretion. At least we are both named with him in an English journal article, in which a person expresses himself sharply on [Gruithuisen's] fantasies."[132] Not only Olbers was a pluralist; so also Littrow, as is clear from his *Wunder des Himmels,* [133] even though this work does not seem to be the source, if such exists, of the proposal of kerosene-filled canals as a signal to the selenites. Gauss was also a pluralist; in fact, one form of the idea of sending a signal to lunarians can be traced to him. In 1818, while involved in a geodetic survey of the region near Hamburg, Gauss invented the heliotrope, a device in which a mirror reflects sunlight over long distances. Gauss calculated that even a small mirror would reflect enough sunlight to be seen at great distances. In a March 25, 1822, letter to Olbers, Gauss suggests: "With 100 separate mirrors, each of 16 square feet, used conjointly, one would be able to send good heliotrope-light to the moon."[134] Furthermore, Gauss is quoted as saying: "This would be a discovery even greater than that of America, if we could get in touch with our neighbors on the moon."[135]

This information concerning Gauss, Gruithuisen, and Olbers suggests that the 1826 *Edinburgh New Philosophical Journal* report is fundamentally correct and that it probably originated directly or indirectly from Gruithuisen himself. Olbers's opinion was known to him from their correspondence, and Gruithuisen may have learned Gauss's view during his visit to Gauss in August and September 1825.[136] The fact that Gauss had speculated on sending signals to the moon by means of a giant heliotrope makes it not improbable that in conversation with Gruithuisen he made the additional comments attributed to him in the report. On the other hand, nothing in the report precludes the conjecture of the editor of the *Annals of Philosophy* that Gauss may have been laughing in his sleeve as he discussed this project with Gruithuisen.

In the last year of his life, Gauss once again became involved with pluralism. William Whewell sent copies of his 1853 antipluralist book to Gauss and to Alexander von Humboldt. On March 4, 1854, Humboldt wrote Gauss that Whewell had claimed in his book that only the earth can be inhabited, because "all intelligent beings are by their nature sinful and the redemption (crucifixion) can not be repeated on the many millions of nebulae observed by Rosse."[137] Gauss responded on May 5, 1854, urging that the idea that life is limited to the earth could not be maintained even by a person "who strongly believes in the literal truth of the Christian dogmas. . . ." Gauss also faults Whewell, and with him Bessel, whom Whewell had cited in support of his antipluralist position, for their rejection of lunar life: "it would . . . be very precipitous to deny without elaborate argumentation all inhabitants to the moon. Nature has more means than a poor man may divine."[138] Gauss also discussed Whewell's book with his close friend and colleague Rudolf Wagner (1805–64), professor of comparative anatomy and physiology at Göttingen. Wagner's records of their conversations not only contain further evidence of Gauss's attachment to pluralism but also show that Gauss had adopted the doctrine that after death our souls take on new material forms on other cosmic bodies, including even the sun.[139] That Gauss held such an extreme form of pluralism is also evidenced in the biography of him written immediately after his death by Baron Wolfgang Sartorius von Waltershausen. This intimate friend revealed that Gauss

. . . held order and conscious life on the sun and planets to be very probable and occasionally called attention to the action of gravity on the surface of heavenly bodies as bearing preeminently on this question. Considering the universal nature of matter, there could exist on the sun with its 28-fold greater gravity only very tiny creatures . . . whereas our bodies would be crushed. . . .[140]

The evidence is, in short, very strong that Gauss, Gruithuisen, Littrow, and Olbers all accepted life on the moon and elsewhere in the universe, even while differing markedly in readiness to make their convictions public.

Given the acceptance of selenites by such eminent German astronomers, one suspects that no small amount of courage may have been required of Friedrich Wilhelm Bessel (1784–1846), director of the observatory at the University of Königsberg, when in an 1834 lecture on the physical properties of the heavenly bodies, he launched an attack on the selenites.[141] Although Bessel's entry into astronomy had been overseen by Olbers, and despite the fact that he had been trained as an observer by Schröter, Bessel had become convinced by the sharpness with which stars are occulted by the moon that the moon lacks any significant atmosphere.[142] Pointing out that evidence against a lunar atmosphere had long been available, Bessel asks:

Why, notwithstanding all reasonable evidences, have some wished to assert the existence of the moon's atmosphere? It is really not a matter of indifference, because with it simultaneously fall many beautiful dreams of the habitability of the moon and the conditions for persons there; I say persons for regardless of all protestations of feeling hearts who have wished to find sympathy on the moon, they imagine the inhabitants of the moon to be as similar to earthlings as one egg is to another. The moon has no air; thus also no water, because without the pressure of air, water at least in the liquid state would evaporate; thus also no fire, for without air nothing can burn. (p. 81)

He goes on to argue that not even infusorial animals could live on the moon and to ask: "What concept do we have of a life which is not terrestrial life?" To speculate about lunar life without recognizing this is to renounce the assistance of the astronomer. Furthermore, the former assistant to Schröter condemns those who seek "traces of industry of lunar inhabitants." (p. 82) The remainder of Bessel's lecture is a survey of the sun and planets aimed at determining how similar they are to the earth. His conclusion is that

The moon is decisively different from the earth in the primary point of an atmosphere; the sun is of an entirely different nature; for Mercury and Venus we have found no basis for assuming a similarity; Mars appears to possess an atmosphere and summer and winter, even snow and ice; the new planets are small bodies without properties for us; Jupiter and Saturn are very dissimilar to the earth in the matter of which they consist. . . . (p. 92)

Having advocated a conception of the solar system radically different from that espoused by pluralists, Bessel warns against the pluralists' propensity to exceed the limits of astronomy by populating all the planets: "the fruitless attempt to transgress these limits [which activity] blackens with fantasies a science which is so rich in attainable facts . . . can only have come from such persons who did not know how to discover the way to [attainable] facts." (p. 93)

The illusion of lunar life was also disappearing among those German astronomers most intent on making high-quality lunar maps. The most outstanding selenographers of this period were Wilhelm Lohrmann (1796–1840), who published his map in 1838, and J. H. Mädler (1794–1874), who, working in conjunction with Wilhelm Beer (1797–1850), published during the 1830s a map of and book on the moon. Lohrmann adopted a very different approach from Gruithuisen's; in describing the Sinus Aestuum region of the moon, Lohrmann states:

In this region . . . Mr. Gruithuisen believes he has seen a city, a fortress, and other artificial works. He hopes soon to recognize the lunar inhabitants themselves, if they parade en masse through their forest glades, and he tells much in his selenographical writings of hot springs, animals and plants. But these famed discoveries and the elaborate hypotheses based on them have no place in a straightforward book on lunar topography.[143]

Beer and Mädler showed a similar professionalism, presenting the moon as "no copy of the Earth."[144] The later nineteenth-century selenographer Edmund Neison correctly described their approach, although he overestimated their influence, when he characterized the situation that prevailed it midcentury as being that "it was generally regarded as demonstrated that the moon was to all intents an airless, waterless, lifeless, unchangeable desert. . . ."[145]

The efforts made by Bessel, Lohrmann, Mädler, and Beer to exorcise the selenites were not without effect, at least among professional astronomers. By a curious coincidence, just a year after Bessel's important 1834 lecture, a series of events began to unfold in New York City that in a very different way led some to rethink the question of lunar life.

The August 25, 1835, issue of the New York *Sun,* a penny daily less than two years old, carried the initial installment of perhaps the most remarkable report ever printed. So fascinated were New Yorkers by its revelations that they purchased over nineteen thousand copies of the August 26 *Sun,* giving it the largest circulation of any paper on this planet. The news creating this sensation was nothing less than that living beings had been sighted on our satellite by Sir John Herschel, the already famous son of Sir William Herschel. So began what has come to be known as the "great moon hoax," the events of which would perhaps not justify another recounting were this not necessary as a preliminary to the position advocated in the present study, which is that compelling evidence exists to show that the moon hoax ascribed to R. A. Locke was not in fact a hoax at all and that this persistent misinterpretation of it has obscured its place in the pluralist debate.[146]

The story begins with the Tuesday, August 25, 1835, issue of the *Sun,* the front page of which carried the headline:

GREAT ASTRONOMICAL DISCOVERIES
Lately Made
By Sir John Herschel, LL. D., F. R. S., &c
At the Cape of Good Hope

Then follows the first portion of a paper purportedly reprinted from the *Edinburgh Journal of Science.* It is true that in 1835 John Herschel was at the cape making astronomical observations; however, sophisticated readers may have been puzzled by their lack of familiarity with that journal, which, as is now known, did not exist. After some rhetoric about Herschel's pious feelings as he began his observations, feelings "nearly akin to those with which a departed spirit may be supposed to discover the unknown realities of a future state," the article previews the results to be announced:

. . . by means of a telescope, of vast dimensions and an entirely new principle, the younger Herschel . . . has already made the most extraordinary discoveries in every planet of our solar system; has discovered planets in other solar systems; has obtained a distinct view of objects in the moon . . . ; has affirmatively settled the question whether this satellite be inhabited, and by what order of beings; has firmly established a new theory of cometary phenomena; and has solved or corrected nearly every leading problem of mathematical astronomy. (p. 51)

As a preliminary to the discussion of these astounding discoveries, the first installment describes in detail the telescope used by Herschel in his observations. Because this instrument included an objective mirror twenty-four feet in diameter, permitting a magnification of 42,000, Herschel believed that with it he would be able to "study even the entomology of the moon. . . ." (p. 64)

With the issue of August 26, readers begin to learn of Herschel's remarkable results. At first, only items of geological and botanical interest are reported, but then, as Dr. Andrew Grant, Herschel's amanuensis, announces: "our magnifiers blest our panting hopes with specimens of conscious existence." (p. 78) They find first a creature similar to the terrestrial bison, which, however, is seen to possess

. . . one widely distinctive feature, which we afterwards found common to nearly every lunar quadruped we have discovered; namely, a remarkable fleshy appendage over the eyes, crossing the whole breadth of the forehead and united to the ears. We could most distinctly perceive this hairy veil, which was shaped like the upper front outline of the cap known to the ladies as Mary Queen of Scot's cap, lifted and lowered by means of the ears. It immediately occurred to the acute mind of Dr. Herschel, that this was a providential contrivance to protect the eyes of the animal from the great extremes of light and darkness to which all the inhabitants of our side of the moon are periodically subjected. (p. 79)

Their amazement increases as they behold a bearded goatlike animal equipped with a "single horn." They also espy some remarkable birds and are on the track of lunar fish as the installment ends with a sinking moon.

Despite containing some fantastic observations of lunar seas and volcanoes, the August 27 installment may have disappointed readers anxious for the observation of intelligent beings. The closest Herschel comes to this is a "biped beaver [that] carries its young in its arms like a human being. . . . Its huts are constructed better and higher than those of many tribes of human savages, and from the appearance of smoke in nearly all of them, there is no doubt of its being acquainted with the use of fire." (p. 87) The hundreds who waited in the streets for the August 28 issue had their hopes fulfilled. Herschel and his fellow observers see clusters of creatures who

. . . averaged four feet in height, were covered, except on the face, with short and glossy copper-colored hair, and had wings composed of a thin membrane, without hair, lying snugly upon their backs, from the top of the shoulders to the calves of the legs. The face, which was of a yellowish flesh-color, was a slight improvement upon that of the large orang-outang, being more open and intelligent in its expression, and having a much greater expansion of forehead. . . . In general symmetry of body and limbs they were infinitely superior to the orang-outang; so much so, that, but for their long wings, Lieut. Drummond said they would look as well on a parade ground as some of the old cockney militia! (pp. 95–6)

Doubts about their intelligence are dispelled when these creatures are seen

. . . evidently engaged in conversation; their gesticulation, more particularly the varied action of their hands and arms, appeared impassioned and emphatic. We hence inferred that they were rational beings, and, although not perhaps of so high an order as others which we discovered the next month on the shores of the Bay of Rainbows, that they were capable of producing works of art and contriv-ance. (p. 96)

The author, thereby reassured of their rationality, adds: "We scientifi-cally denominated them the Vespertilio-homo, or man-bat; and they are doubtless innocent and happy creatures, notwithstanding some of their amusements would but ill comport with our terrestrial notions of deco-rum." (p. 98)

The August 29 installment reports the sighting of various oceans and especially a "magnificent. . . temple – a fane of devotion, or of science, which, when consecrated to the Creator, *is* devotion of the loftiest or-der. . . ." (p. 105) The final installment, that of August 31, tells of the observation of yet higher beings and of the "universal state of amity among all classes of lunar creatures. . . ." (p. 109) Saturn is also briefly studied, but Herschel's calculations are omitted "as being too mathemat-ical for popular comprehension." (p. 113) As the article ends, they see the highest species of lunar bat-men: "In stature, they did not excel those last described, but they were of infinitely greater personal beauty, and ap-peared, in our eyes, scarcely less lovely than the general representation of angels by the more imaginative school of painters." (pp. 115–16)

This amazing report reached not only the readers of the *Sun;* its August 29 number noted that reprints of the series in pamphlet form were avail-able. Of these, 60,000 copies were sold, and soon lithographs of the lunarians were being offered for purchase.[147] The *Mercantile Advertiser* began reprinting the series, noting that "It appears to carry intrinsic evidence of being an authentic document." The *Daily Advertiser* ex-pressed its enthusiasm by stating: "No article, we believe, has appeared for years, that will command so general a perusal and publication. Sir John has added a stock of knowledge to the present age that will immor-talize his name, and place it high on the page of science." The *Albany Daily Advertiser* called it a "Stupendous Discovery" and told of having read the story "with unspeakable emotions of pleasure and astonish-ment. . . ." The *New York Times* pronounced the discoveries "proba-ble and possible," and the *New Yorker* described them as creating "a new era in astronomy and science generally."[148] According to a contemporary report, "Some of the grave religious journals made the great discovery a subject of pointed homilies . . . ," and according to another report, an American clergyman warned his congregation that he might have to solicit them for funds for Bibles for the inhabitants of the moon.[149] It has even been claimed that "the philanthropists of England had frequent and crowded meetings at Exeter Hall, and appointed committees to inquire . . . in regard to the condition of the people of the moon, for purposes of relieving their wants, . . . and, above all, abolishing slavery if it should be found to exist among the lunar inhabitants."[150] A person who had been in New Haven in 1835 described the situation there: "Yale was alive with staunch supporters. The literati – students and professors, doctors in divinity and law – and all the rest of the reading community, looked daily for the arrival of the New York mail with unexampled avidity and implicit faith."[151] Timothy Dwight was dead by then, but Yale professors Loomis and Olmsted went to New York to examine the deleted mathematical sections, but were sent on a wild-goose chase.[152] Edgar Allan Poe later reported:

Not one person in ten discredited it, and (strangest point of all!) the doubters were chiefly those who doubted without being able to say why – the ignorant, those uninformed in astronomy, people who *would not* believe because the thing was so novel, so entirely "out of the usual way." A grave professor of mathematics in a Virginia college told me seriously that he had *no doubt* of the truth of the whole affair![153]

Finally the bubble burst. A reporter for the *Journal of Commerce* was sent to the *Sun* to secure a copy so that his journal might republish it. He met with a *Sun* reporter named Locke, who told him: "Don't print it right away. I wrote it myself." The *Journal of Commerce* then denounced the articles as a hoax, and the *New York Herald* specified Locke as its perpetrator.[154]

Richard Adams Locke (1800–71), an American-born collateral descendant of the philosopher John Locke, studied at the University of Cambridge. Before returning to the United States, he worked in England as a writer and editor for various liberal publications. On coming to New York, he wrote for the *Courier and Enquirer* before joining the *Sun* in the summer of 1835. In later years he worked for other papers and in the New York customs house.[155] Locke's interest in scientific matters is attested to by a number of biographical sources, but especially by the impressive detail included in his *Sun* articles. He also wrote with skill; Edgar Allan Poe, who had written a fictional moon voyage shortly before the articles appeared, praised Locke's ability to write with "true imagination" as well as with "concision, luminousness, completeness – each quality in its proper place."[156]

Although Locke possessed the literary skills Poe attributed to him, compelling if long-neglected evidence points to the conclusion that what he wrote was not a hoax; he had instead employed his skills at the delicate task of writing satire. Locke himself told friends: "If the story be either received as a veritable account, or rejected as a hoax, it is quite evident that it is an abortive satire; and, in either case, I am the best self-hoaxed man in the whole community."[157] If his writing be approached as satire, the question immediately arises as to what or whom he was satirizing. The answer is: advocates of a plurality of worlds, and above all Thomas Dick. William Griggs, in his 1852 account of the incident, made this clear, even if later authors failed to see the point. As Griggs wrote,

. . . we have the assurance of the author, in a letter published some years since, in the *New World,* that it was written expressly to satirize the unwarranted and extravagant anticipations upon this subject, that had been first excited by a prurient coterie of German astronomers, and thence aggravated almost to the point of lunacy itself, both in this country and in England, by the religo-scientific rhapsodies of Dr. Dick. At that time the astronomical works of this author enjoyed a degree of popularity, in both countries, almost unexampled in the history of scientific literature. The scale of the editions republished in this country was unbounded until nearly the whole of his successive volumes found a place in every private and public library in the land.[158]

Griggs, moreover, described Dick in terms that Locke would have endorsed:

. . . it would be difficult to name a writer who, with sincere piety, much information, and the best intentions, has done greater injury, at once, to the cause of rational religion and inductive science, by the fanatical, fanciful, and illegitimate manner in which he has attempted to force each into the service of the other, instead of leaving both to the natural freedom and harmony of their respective spheres.[159]

That Locke was satirizing Dick is also shown from Griggs's account of how Locke came to write his articles. In the summer of 1835, Locke was reading the 1826 issue of the *Edinburgh New Philosophical Journal,* where according to Griggs he found an article by Dick recounting the idea attributed to Gauss and Gruithuisen of erecting an immense geometrical signal to the selenites. The absurdity of these ideas led Locke, in Griggs's words, to view them as "fair subjects of sedate and elaborate satire." Locke then turned to the works of Dick, finding in them a rich field for satire.[160] That Locke's articles were satirical was apparently recognized at the Académie des sciences, where the astronomer Arago read them to members amidst "repeated interruptions from uncontrollable and uproarious laughter. . . ."[161] Herschel, when shown the articles by a visitor to the cape, laughed at them, although he later lamented the inconvenience caused by the large number of English, French, German, and Italian inquiries he received from readers who took them seriously.[162] Lady Herschel may have represented her husband's views when

she wrote his Aunt Caroline about the articles, concluding with the reflections:

. . . the whole description is so well *clenched* with minute details of workmanship and names of individuals boldy referred to, that the New Yorkists were not
to be blamed for actually believing it as they did for forty eight hours – It is only a
great pity that it *is not true* but if grandsons stride on as grandfathers *have* done,
as wonderful things may yet be accomplished.[163]

The ultimate irony is that Locke's satire failed. He had underestimated
the gullibility of a generation raised on the pluralist writings of Dick,
Dwight, Chalmers, and their followers. As one author, writing in 1852
and referring especially to Dick, put it: "The soil had been thoroughly
ploughed, harrowed and manured in the mental fields of our wiser people, and the seed of farmer Locke bore fruit a hundred fold."[164] It was not
that Locke lacked the skills of a satirist; it was rather that pluralist
preachings and pronouncements had so permeated the thought of his
contemporaries that they first failed to see the articles as satire, and failed
again as they branded them a "hoax." One may wonder whether readers
of the French, German, Italian, and Spanish translations of Locke's work
possessed the wit Americans lacked. But one need not wonder how
Thomas Dick received it, for he discussed Locke's "hoax" in his *Celestial
Scenery*. It provided him with an occasion for a lecturette directed at
people whose credulity resulted from insufficient knowledge of astronomy, as well as a sermonette directed at the unnamed "young man" who
wrote it. Dick urged him to remember "that all such attempts to deceive
are violations of the laws of the Creator, who is the 'God of Truth' . . .
and therefore, they who wilfully and deliberately contrive such imposi-
tions ought to be ranked in the class of liars and deceivers."[165] One may,
however, trust that the God of Truth is also a God of Satirical Truth,
who may even have a place for persons like Locke whose satires, brilliant
though they may be, fail at first to find an appropriate audience.

What significance should be seen in Locke's satire? Besides inspiring at
least two additional satires,[166] it shed light on both the problems for
pluralism presented at the opening of this chapter: the frailty of its scientific foundations and its tensions with religion. By exposing the credulity
of readers of pluralist books and the overstated claims of their authors, it
no doubt instilled a needed note of caution in the debate. In regard to
religion, Locke taught a lesson that Paine and Shelley, on the one side,
and Dwight, Chalmers, and Dick, on the other, all needed to learn: that
pitfalls await those who rashly pronounce on the ways of God. These
contributions entitle Locke to a permanent if peculiar place in the history
of the extraterrestrial life debate, and possibly also in the history of
satire.

5

The decades before Whewell

1. Pluralism in Britain: Would nature "tip a hogshead to fill a wineglass"?

Richard Locke's article, dicussed in the previous chapter, provided a warning to pluralists that they had overextended themselves; in 1853, William Whewell's *Of the Plurality of Worlds: An Essay,* discussed in the next chapter, dealt them a far more severe blow. The present chapter sets the stage for consideration of Whewell's book by surveying the extraterrestrial life debate in the decades before 1853, during which period pluralism continued to spread on a number of fronts, producing many remarkable developments, as some of the leading intellectuals of the time adapted it to their teachings or their teachings to it. The range of materials that resulted stretches from serious religious discourses and somber philosophical discussions to speculative poetic diversions and satirical artistic drawings. Analysis of these developments continues to illustrate pluralism's extraordinary appeal as well as its exceptional flexibility.

Of numerous British authors who entered the pluralist debate in the decades before 1853, none spoke with more authority and influence than John Herschel (1702–1871), the sole offspring of William Herschel and heir to his astronomical instruments. That the father's genius had passed to the son was evident by 1813, when John graduated from Cambridge as Senior Wrangler and First Smith's Prizeman and also won election to the Royal Society.[1] By 1833, when the son set off for the Cape of Good Hope to extend his father's surveys to the southern heavens, his scientific achievements had already brought him knighthood, as well as medals from the Royal Society and nearly its presidency. In 1833 he also published his *Treatise on Astronomy,* expanding it in 1849 as his *Outlines of Astronomy,* which long remained the leading astronomy text in English and was translated even into Arabic, Chinese, and Russian.

Readers must have turned to Herschel's *Treatise,* or later his *Outlines,* with great interest in whether or not the elder Herschel's pluralist views had passed to his son. They perhaps realized that, for all his achievements, the autodidact father had to some extent remained on the fringes of astronomy, whereas his son, having had the finest scientific education available, was at its very center. Such readers, knowing that John had published in 1830 a book expounding an empiricist methodology of science, would not have been surprised to find in the opening paragraphs of his *Treatise* and *Outlines* an exhortation to set aside prejudice so as to be ready to accept astronomical conclusions "supported by careful observation and logical argument. . . ."[2] A critical reader might, however, have been taken aback on finding among the immediately presented examples of such carefully supported conclusions the following two: (1) "the planets . . . are . . . spacious, elaborate and habitable worlds; several of them vastly greater and far more curiously furnished than the earth . . .," and (2) "the stars . . . are . . . suns of various and transcendent glory – effulgent centers of life and light to myriads of unseen worlds. . . ."[3] Later in both books, he reveals the chief source of his pluralist conviction:

Now, for what purpose are we to suppose such magnificent bodies scattered through the abyss of space? Surely not to illuminate our nights . . . nor to sparkle as a pageant void of meaning and reality, and bewilder us among vain conjectures. Useful, it is true, they are to man as points of exact and permanent reference; but he must have studied astronomy to little purpose, who can suppose man to be the only object of his Creator's care, or who does not see in the vast and wonderful apparatus around us provision for other races of animated beings. The planets, as we have seen, derive their light from the sun; but that cannot be the case with the stars. These doubtless, then, are themselves suns, and may, perhaps, each in its sphere, be the presiding center round which other planets, or bodies of which we can form no conception from any analogy offered by our own system, may be circulating.[4]

This passage reveals that John Herschel's belief in extraterrestrial life rested largely on metaphysical and religious assumptions concerning the plans and purposes of the Creator, a point he failed to see even after he had read Whewell's 1853 book that demonstrated both the paucity of pluralism's scientific evidence and the metaphysical character of many of its doctrines.[5]

By 1833, William Herschel's claims concerning the sun and moon and their inhabitants had become increasingly problematic, as his son no doubt realized. Nonetheless, in both his *Treatise* and *Outlines* he endorses his father's doctrine that the sun has a large solid nucleus, which becomes visible through the "openings" (sunspots) in its exterior layer or "luminous ocean."[6] He also champions a layer of clouds separating this "luminous ocean" from the solid interior, finding evidence for it in the appearances at the edges of sunspots. Although admitting the extraordinarily high temperature of the sun's exterior and also that "the most intensely ignited solids appear only as black spots on the disk of the sun when held between it and the eye," he adds:

. . . it follows, that the body of the sun, however dark it may appear when seen through its spots, *may*, nevertheless, be in a state of most intense ignition. It does not, however, follow of necessity that it *must* be so. The contrary is at least physically possible. A *perfectly reflective* canopy would effectually defend it from the radiation of the luminous regions above its atmosphere, and no heat would be conducted downwards through a gaseous medium increasing rapidly in density. That the penumbral clouds are highly reflective, the fact of their visibility in such a situation can leave no doubt.[7]

That the source of this passage lay not only in filial fondness for his father's ideas, some of which John did not accept, but also in his own solicitude for solarians is suggested by John Herschel's theory of Nasmyth's "willow leaves," to be discussed shortly. John Herschel does not directly discuss solar inhabitants in his *Treatise* and *Outlines,* being content with having supplied the previously indicated provisions for their existence, of which Brewster and others availed themselves later in the century.

The moon, John admits, "has no clouds, nor any other indications of an atmosphere,"[8] but he nonetheless claims that in the past, water may have been present. He notes, for example, that "there are regions perfectly level, and apparently of a decided alluvial character."[9] Nor does it exclude for him the possibility of water remaining there. In fact, after discussing the periodic alternation between bright sunlight and near total darkness experienced by each region of the moon, he notes:

The consequence must be absolute aridity below the vertical sun, constant accretion of hoar frost in the opposite region, and, perhaps, a narrow zone of running water at the borders of the enlightened hemisphere. It is possible, then, that evaporation on the one hand, and condensation on the other, may to a certain extent preserve an equilibrium of temperature, and mitigate the extreme severity of both climates.[10]

This speculation was soon seized on by Patrick Scott, who used it to legitimize the lunarians in his dreamy poem *Love in the Moon.*[11] Herschel's summary statement of the possibility of lunar life and its detectability is:

Telescopes . . . must yet be greatly improved, before we could expect to see signs of inhabitants, as manifested by edifices or by changes on the surface of the soil. It should, however, be observed, that, owing to the small density of the materials of the moon, and the comparatively feeble gravitation of bodies on her surface, muscular force would there go six times as far in overcoming the weight of materials as on the earth. Owing to the want of air, however, it seems impossible that any form of life analogous to those on earth can subsist there. No appearance indicating vegetation, or the slightest variation of surface which can fairly be ascribed to change of season, can any where be discerned.[12]

Despite this, he ends this chapter with a paragraph that begins "If there be inhabitants of the moon . . . ," and then describes the earth as seen by them.[13]

Late editions of *Outlines* show Herschel espousing what Willy Ley described as "probably the wildest astronomical hypothesis ever advanced."[14] In 1856, Peter Andreas Hansen (1795–1874), a prominent mathematical astronomer, published a paper in the *Royal Astronomical Society Memoirs* in which he accounted for certain discrepancies between lunar observation and theory by hypothesizing that the center of gravity of the moon is about thirty miles more distant from the earth than the moon's center of figure. This claim implied that any water and air on the moon would be drawn to its remote side, where the gravitational field would be strongest. Although by 1868 Simon Newcomb had shown that Hansen's hypothesis was "without logical foundation,"[15] this left time for Herschel to embrace and embellish it in the 1858 fifth edition of his *Outlines,* describing it as "not improbably what takes place in the moon" and using it to argue for the possibility of "animal or vegetable life" on the moon's far side.[16] Moreover, as will be shown in Chapter 8, this claim for life on the moon's far side became widespread in pluralist literature of the 1860s, at which time Herschel again championed it.

In discussing the planets, Herschel notes that the intensity of solar radiation on Mercury is 7 times greater than that for the earth, which receives 330 times more than Uranus. He also states that planetary densities vary greatly, that of Saturn being so low that "it must consist of materials not much heavier than cork." Such information led others to deny the habitability of the planets; to Herschel, it suggests

. . . what immense diversity must we not admit in the conditions of that great problem, the maintenance of animal and intellectual existence and happiness, which seems . . . to form an unceasing and worthy object for the exercise of the Benevolence and Wisdom which presides over all![17]

Turning to particular planets, he notes that both Venus and Mercury seem to show uniform brightness, which he suggests is most naturally explained by assuming that both are surrounded by atmospheres "much loaded with clouds." This conjecture (now known to be wrong for Mercury) might seem scientifically based, were Herschel not to have added the pluralist comment that these atmospheres "may serve to mitigate the otherwise intense glare of their sunshine."[18] The pattern apparent here is seen repeatedly in his writings: He goes to great lengths to show the possibility of life on the various solar system objects and to remove or alleviate the most serious objections to their habitability. In doing this he tends not only to neglect theories unsupportive of life but also to pass over observations not conducive to this view. For example, Herschel fails to note that Mercury is so small and distant and its viewing conditions so poor that we could not see distinct surface features on Mercury unless they were extremely large. In Mars, the most favorable case for pluralists, Herschel sees "with perfect distinctness, the outlines of what may be continents and seas."[19] He notes also the "greenish" color of the seas and that the polar caps are "conjectured with a great deal of probability to be snow."[20] In treating Saturn, Herschel states:

The rings of Saturn must present a magnificent spectacle from those regions of the planet which lie above their enlightened sides, as vast arches spanning the sky from horizon to horizon, and holding an invariable situation among the stars. On the other hand, in the region beneath the dark side, a solar eclipse of fifteen years in duration, under their shadow, must afford (to our ideas) an inhospitable asylum to animated beings, ill compensated by the faint light of the satellites. But we shall do wrong to judge of the fitness or unfitness of their condition from what we see around us, when, perhaps, the very combinations which convey to our minds only images of horror, may be in reality theatres of the most striking and glorious displays of beneficent contrivance.[21]

Herschel's discussion of the asteroids is so brief in his *Treatise* that one finds no information on their diameters. Nonetheless, he tells readers that Pallas "is said to have somewhat of a nebulous or hazy appearance, indicative of an extensive and vaporous atmosphere. . . ."[22] He adds that because of low gravitational forces, "On such planets giants might exist; and those enormous animals, which on earth require the buoyant power of water to counteract their weight, might there be denizens of the land. But of such speculations there is no end."[23]

In the sidereal portion of Herschel's *Treatise,* pluralist ideas naturally enter only rarely. However, traces of Herschel's beliefs may be seen in his discussion of double stars, which presented problems for pluralists because the gravitational forces between them create conditions unfavorable to stable planetary orbits. This Herschel admits, but he also proposes that their planets are "closely nestled under the protecting wing of their immediate superior. . . ." Unless this is the case, "the sweep of their other sun in its perihelion passage . . . might carry them off; or whirl them into orbits utterly incompatible with the conditions necessary for the existence of their inhabitants. . . ."[24] This, Herschel states, is "a strangely wide and novel field for speculative excursions, and one which it is not easy to avoid luxuriating in." Nor does he hesitate to do so, as he considers binaries of different colors; imagine, he suggests, "what variety of illumination *two suns* – a red and a green, or a yellow and a blue one must afford a planet circulating about either; and what charming contrasts and 'grateful vicissitudes,' – a red and a green day, for instance. . . ."[25]

Herschel's *Treatise on Astronomy* and his *Outlines of Astronomy* contain no special section on or systematic argument for a plurality of worlds. This is partly because Herschel realized that this speculative topic ill-suited the type of presentation he wished to make, but another reason may be that Herschel viewed the doctrine as scarcely in need of proof. Herschel did, however, have his speculative moments; in fact, in a lecture delivered in late 1861 and subsequently twice published, he put forth a pluralist speculation that went beyond even those of his father. Around 1860, James Nasmyth, a respected astronomer with one of the best telescopes of the period, reported that he had observed the surface of the sun to be covered with numerous objects shaped like willow leaves. These were intensely luminous objects of immense size and in constant motion. In his 1861 lecture, Herschel not only accepts this observation, which by the mid-1860s had been shown to be erroneous, but goes beyond it to argue for the solidity of the willow leaves and to state that they are *"evidently the immediate sources of the solar light and heat. . . ."* Then he adds the remarkable claim that "we cannot refuse to regard them as *organisms* of some peculiar and amazing kind; and though it would be too daring to speak of such organization as partaking of the nature of life, yet we do know that vital action is competent to develop both heat, light, and electricity."[26] Two considerations help explain how the premier British astronomer of that period could make such a fantastic assertion. The first, which is supported by the materials presented earlier, is that the younger Herschel had inherited not only his father's instruments and abilities but also his pluralist metaphysics. This John once summarized by stating: "If matter exists in the universe for the purpose of life, Nature would seem to tip a hogshead to fill a wineglass, when it makes life possible only on a little planet."[27] The second factor is the openness of Herschel's astronomical contemporaries to pluralist claims.

An excellent example of this is Admiral W. H. Smyth (1788–1865), whose publication in 1844 of his *Cycle of Celestial Objects* (London) won him a gold medal from the Royal Astronomical Society. In that book, Smyth, without directly advocating William Herschel's theory of life on the sun, responded to Thomas Young's objection that solarians could not overcome the sun's gravitation by stating:

. . . the mysterious WORD which formed the Laplander and the Negro, the condor and the whale, the mosquito and the elephant, for the several portions of one and a small globe, is surely not to be limited to the fashioning of creatures of our constitution or conception. The inhabitants of every world will be formed of the material suited to that world, and also for that world; and it matters little whether they are six inches high, as in Lilliput, or as tall as [Voltaire's] inhabitants of Sirius . . . ; whether they crawl like beetles, or leap fifty yards high. (I, p. 92)

Like Smyth, Sir William Rowan Hamilton (1805–65), Ireland's Astronomer Royal, ascribed to pluralism. In fact, in an 1842 publication, he speculated that the ten days between Christ's ascension and Pentecost were taken up in a cosmic procession through other worlds: "May not [Christ's] transit from the cloud to the throne have been but one continued passage, in long triumphal pomp, through powers and principalities made subject? May not the only begotten Son have then again been brought forth into the world, not by a new nativity, but as it were by proclamation and investiture, while the Universe beheld its God, and all the angels worshipped Him?"[28]

Comets the Herschels left lifeless; however, some of their contemporaries in science were made of stouter stuff. The pluralism discussed by Mrs. Jane Marcet (1769–1858) in her widely read *Conversations on Natural Philosophy* (1819) extends not only to all the planets but to comets as well. She nonetheless qualifies her claim for cometary life by admitting that if comets are inhabited, "it must be by a species of beings very different, not only from the inhabitants of [the earth], but from those of any of the other planets, as [comets] must experience the greatest vicissitudes of heat and cold. . . ."[29] David Milne (1805–90) expresses no such scruples in his *Essay on Comets* (1828); in fact, within three pages he thrice repeats his claim that cometarians may "possess a constitution not very dissimilar to that of the human species."[30] The possibility of cometary life does not in Milne's analysis extend to all comets, only to those "whose advanced state of maturity renders the sun's influence incapable of materially affecting the surface of the nucleus. . . ." (pp. 139–40) Nor does he require all his cometarians to be terrestrial in form: "Comets may be the residence of beings widely different from those which fall within the narrow sphere of human observation. What though these beings, from the peculiarities of their situation, be endowed with neither lungs, nor eyes, nor the feelings which afford the sensations of heat and cold, like our bodily organs?" (p. 144) He is quick to caution against conceiving such creatures as disadvantaged; indeed, they may enjoy extraordinary sights; for example, they may be "brought so close to the planet Saturn, that they can examine the wonderful phenomena of his rings even with the naked eye. . . ." (p. 144) Milne, alas, does not explain how his eyeless cometarians can see such splendors, nor is it clear how his not very profound *Essay,* according to its title page, "Gained the First of Dr. Fellowes's Prizes Proposed to Those Who Had Attended the University of Edinburgh in the Last Twelve Years."

Milne's enthusiasm for life on comets should be judged within the context of his time. It is in this regard relevant that cometarians were also endorsed by Sir Humphry Davy (1778–1829), the brilliant electrochemist and president of the Royal Society.[31] This was done in his posthumous *Consolations in Travel, or The Last Days of a Philosopher,* which consists of a set of dreamy dialogues in one of which an invisible "Genius" takes Davy on a tour of the cosmos. The lesson learned is: "The universe is every where full of life, but the modes of this life are infinitely diversified. . . ."[32] Davy beholds inhabited comets and is told that their inhabitants "so grand, so glorious . . . once belonged to the earth. . . ." (pp.

55–6) As this suggests, transmigration of souls is also taught; in proportion to the degree with which a soul has loved knowledge, "it rises to a higher planetary world." (pp. 57–8) Saturnians are seen swimming in that planet's atmosphere: "they had systems of locomotion similar to those of the morse or sea-horse, but I saw with great surprise that they moved from place to place by six extremely thin membranes which they used as wings." (p. 47) Repelled by the sight of the Saturnians, Davy is informed that they are intellectually superior to earthlings and "possess many modes of perception of which we are wholly ignorant. . . ." (p. 48)

Not the least remarkable feature of Davy's book is that he meant it to be taken seriously; in a letter to Lady Davy, he described it as containing "certain truths that cannot be recovered if they are lost, and which I am convinced will be extremely useful. . . ."[33] Moreover, he characterized his *Consolations* as containing "the essence of philosophical visions."[34] Actually his celestial vision was at least in part the record of some of his dreams, which he may have viewed as comparable to those of Swedenborg.[35] The public took considerable interest in Davy's *Consolations*. By 1869 it had gone through seven English as well as a number of American editions. Moreover, in that year Flammarion published a French translation, which was in its ninth edition by 1883. Nonetheless, the judgment of history must be that Davy, a scientist of whom Berzelius said that had he written more systematically, "he would have advanced chemistry by a whole century,"[36] added little of significance to the extraterrestrial life debate, or to his own reputation, by the pluralist fantasies in his *Last Days of a Philosopher*.

Mary Somerville (1780–1872), whose knowledge of technical astronomy was exceptionally deep, discussed the planets in her *Connexion of the Physical Sciences* (1834), concluding that "the planets, though kindred with the earth in motion and structure, are totally unfit for the habitation of such a being as man."[37] Charles Lyell (1797–1875), in his 1837 presidential address to the London Geological Society, also expressed reservation about pluralism:

Astronomy had been unable to establish the plurality of habitable worlds throughout space, however favourite a subject of conjecture and speculation; but geology, although it cannot prove the other planets are peopled with appropriate races of living beings, has demonstrated the truth of conclusions scarcely less wonderful, the existence on our own planet of many habitable surfaces, or worlds as they have been called, each distinct in time, and peopled with its peculiar races to terrestrial beings.[38]

Charles Darwin (1809–82), on the other hand, accepted pluralism, and in fact its cosmic perspective may have supported his attempt to explain by naturalistic means the origin of various terrestrial forms of life. This conjecture is suggested by a passage in his 1844 essay on the species question: "It is derogatory that the Creator of countless Universes should have made by individual acts of His will the myriads of creeping parasites and worms, which since the earliest dawn of life have swarmed over the land. . . ."[39]

Similar sentiments probably influenced Robert Chambers (1802–71), an Edinburgh publisher and naturalist, who in 1844 created a sensation by his anonymous *Vestiges of the Natural History of Creation,* which by 1853 was in its tenth English edition and which produced a flood of rebuttals to its evolutionary and deistic doctrines.[40] *Vestiges* begins with two astronomical chapters in which Chambers points out that the stars are suns and form "astral" systems[41] and argues at length for a modified form of Laplace's nebular hypothesis. Except for a section in which Chambers urges that the moon is not inhabited but may become so (pp. 36–9), these two introductory chapters do not discuss pluralism. Nonetheless, his advocacy of the nebular hypothesis and of the associated idea that each star is surrounded by planets formed from the same nebula from which it came sets the stage for the pluralist discussion found in a later chapter. This is especially true because in Chambers's version of the nebular hypothesis, all the planets have approximately the same surface temperature as the earth (p. 31).

By his eleventh chapter, Chambers has set out his evolutionary ideas. He then urges that his doctrine applies "not merely for the origination of organic being upon this little planet, third of a series which is but one of hundreds of thousands of series, the whole of which form but one portion of an apparently infinite globe-peopled space, where all seems analogous." (pp. 160–1) Chambers presents pluralism as an accepted doctrine that supports his evolutionary ideas by being a natural conclusion from them. In a passage strikingly similar to that cited from Darwin, Chambers states:

. . . every one of these numberless globes is either a theatre of organic being, or in the way of becoming so. . . . Is it conceivable, as a fitting mode of exercise for creative intelligence, that it should be constantly moving from one sphere to another, to form and plant the various species which may be required in each situation at particular times? (p. 161)

Convinced that all planets contain the same chemical elements and are subject to the same physical laws, Chambers boldly infers:

Where there is light there will be eyes, and these, in other spheres, will be the same in all respects as the eyes of tellurian animals, with only such differences as may be necessary to accord with minor peculiarities of condition and of situation. It is but a small stretch of the argument to suppose that, one conspicuous organ of a large portion of our animal kingdom being thus universal, a parity in all the other organs – species for species, class for class, kingdom for kingdom –

is highly likely, and that thus the inhabitants of all the other globes of space bear not only a general, but a particular resemblance to those of our own. (pp. 163–4)

This also fits well with Chambers's deism, in which God is seen "to have all done by the employment of the smallest possible amount of means." (p. 164)

Chambers's book had two important effects on the history of the nebular hypothesis: It made that hypothesis far better known, and yet by having presented it in association with what many saw as antireligious evolutionary views, it led many to question that hypothesis. In later editions of *Vestiges,* Chambers backed down substantially from his advocacy of the nebular hypothesis.[42] This he must have done reluctantly, because that hypothesis not only provided the framework for his theory but also had been the idea that in the 1830s had spawned in Chambers's mind the book itself.[43] That Chambers did not significantly alter his pluralist passages as he revised *Vestiges* suggests that his critics did not object to that doctrine. It is possible, however, that some readers may have been led to question not only the nebular hypothesis but pluralism as well when they encountered these doctrines tied to the radical theses that made *Vestiges* so controversial.[44]

Among mid-nineteenth-century British biologists, few were more hostile to evolutionary doctrines than Richard Owen (1804–72), the prominent comparative anatomist who conceptualized animate nature in terms of archetypes or ideal forms. This might lead one to expect Owen to be opposed to pluralism, but in his *On the Nature of Limbs* (1849) he not only endorses pluralism but also suggests that his theory of archetypes can be used to draw inferences about extraterrestrials. The idea is that because the "conceivable modifications of the vertebrate archetype are very far from being exhausted by any of the forms that now inhabit the earth . . . ,"[45] one may suppose that some of the remaining forms attain realization on other planets. Few finer examples of the pliability and attractiveness of pluralism can be found than that both Chambers and Owen, so opposed on many matters, accepted that doctrine and strove to integrate it into their systems.

Acceptance of pluralism is one issue, diffusion another. Among the scientific writings mentioned in this section, only those of Chambers, Davy, Herschel, and Marcet secured wide readership. Even more successful were John Pringle Nichol (1804–59) and Dionysius Lardner (1792–1859), both of whom studied for the ministry, but found lecturing on and writing about science more to their tastes. This they did with singular success, one author naming Nichol the "*prose* laureate of the stars" and praising him for having "done more than any man living to uncase science from its mummy confinements, and to make it walk abroad as a free and living thing."[46]

Nichol, the professor of astronomy at Glasgow, extolled the wonders of the celestial realm both in his native Scotland and in the United States, where he lectured in the winter of 1848–9. Americans were already acquainted with his ideas from his *Views of the Architecture of the Heavens* (1838), which recent research has shown "was the single book most responsible for bringing the nebular hypothesis to the attention of the American reading public."[47] In that book, Nichol champions the nebular hypothesis and uses it to argue that stars are orbited by planets.[48] These planets he views as in most cases populated: "there too . . . in splendid hieroglyphics the truth is inscribed, that the grandest forms of present Being are only GERMS swelling and bursting with a life to come!" (p. 198) Even the Orion nebula, which Nichol up to 1846 believed to be a shining fluid, is presented as having "laid up in its dark bosom – the germs, the producing powers of that LIFE, which in coming ages will bud and blossom, and effloresce, into manifold and growing forms, until it becomes fit harbourage and nourishment to every varying degree of intelligence, and every shade of moral sensibility and greatness!" (pp. 123–4) Pluralist passages also occur in Nichol's *Phenomena and Order of the Solar System* (1838). For example, he describes the moon as shining

. . . with the most various *colours*. In the *Mare Serenitatis* . . . , large districts shine with a most beautiful green, intermingled with grey portions, and several other tints. These varieties . . . [suggest] the different aspects of light reflected from a desert of Zahara, a golden Savannah in the new world, or a field of luxuriant and freshening Fern.[49]

Nichol also asks if all planets are inhabited, and noting that regions of the earth are now or have been bereft of life, he answers negatively (p. 196). However, one suspects that Nichol, who envisioned the universe as evolving, would hold that they will attain life.

In 1840, *Athenaeum* lauded Dionysius Lardner's *Steam Engine* as the "most popular mechanical treatise ever published," and of Lardner himself it stated that "he was, we think, decidedly the most popular, and the most deservedly popular, of all the popular writers of his day."[50] Despite the obituarial past used in the *Athenaeum,* Lardner had not died; he had only left England, taking with him another man's wife. This did not end his career as an expositor of science, which by 1840 had involved him in authoring over a dozen volumes and editing "Dr. Lardner's Cabinet Cyclopaedia" in 133 volumes.[51] In 1841, he began a lecture tour in the United States, drawing audiences as large as two thousand, giving substance to his publisher's claim that "Probably no public lecturer ever continued for the same length of time to collect around him so numerous audiences."[52] By 1844, when Lardner concluded his lecture tour, citizens of cites as remote as Vicksburg and St. Louis had heard him extol the benefits of astronomy, physics, chemistry, and steam engineering. Lardner himself reaped some of these benefits: It is reported that he earned $200,000 from his lectures, not including the profits from their 1846 publication as his *Popular Lectures on Science and Art*. Leaving America for France, he continued to publish works, including his twelve-volume *Museum of Science and Art* (London, 1854–6).

Obviously Lardner knew how to please an audience. One source of his success is that both his *Popular Lectures* and his *Museum* begin with sections on extraterrestrial life. Entitled "The Plurality of Worlds," the first chapter of his *Popular Lectures* asks: "Are those shining orbs which so richly decorate the firmament peopled with creatures endowed like ourselves with reason to discover, with sense to love, and with imagination to expand toward their limitless perfection the attributes of Him of 'whose fingers the heavens are the work?' " (I. p. 51) Admitting that the telescope could not directly answer this question, Lardner nonetheless draws upon analogy for an affirmative answer. Finding the earth adapted to its inhabitants, he urges that astronomy reveals that the planets have "provisions in all respects similar"; they are, in fact, "habitations similarly built, ventilated, warmed, illuminated, and furnished. . . ." (I, p. 53) He reconciles Saturn receiving 400 times less light and heat per unit surface area than the earth by giving Saturn an appropriate heat-retaining atmosphere and Saturnians eyes of large aperture. Seeing clouds on Mercury, Venus, and Mars, and (possibly) on Jupiter and Saturn, he concludes: "whenever the existence of clouds is made manifest, *there* WATER must exist; *there* EVAPORATION must go on; *there* ELECTRICITY . . . must reign; *there* RAINS must fall; *there* HAIL and SNOW must descend." (I, p. 60) Excessive as are some of the pluralist statements in his *Popular Lectures,* none surpasses that in the introductory section of his *Museum of Science and Art:* "The numerous analogies which we have indicated give the highest degree of probability, not to say moral certainty, to the conclusion that the three planets, Mars, Venus, and Mercury . . . are like the earth appropriated [by God] to races very closely resembling, if not absolutely identical with, those by which the Earth is peopled." (I, p. 23) Lardner's pluralism was not without limitations; in his *Museum,* he rejects inhabitants for the moon, sun, satellites, comets, and planetoids because these bodies fail to fit the analogy with the earth (I, pp. 63–4). The *Athenaeum*, in the review cited previously, commented that Lardner "floated into popularity on the very crest of the tide of the diffusion-of-knowledge" movement. Lardner's breadth of learning and vigorous style further contributed to his popularity. So also, one suspects, did his decision to begin his *Popular Lectures* and *Museum* with the exciting question of life elsewhere in the universe.

Discussions of pluralism appeared not only in scientific publications but also in literary, philosophic, and religious writings published in Britain in the period between Chalmers and Whewell. Among the religious writers was Samuel Noble (1779–1853), who, on encountering Paine's *Age of Reason* at age sixteen, felt "nothing less than the striking of a dagger into my vitals. The agonising thoughts that took possession of my mind . . . are indescribable."[53] His Christian convictions having been restored by reading Swedenborg, Noble joined the New Jerusalem church, becoming in time one of its ministers. His frequently reprinted *Appeal in Behalf of the . . . New [Jerusalem] Church* (1826) long remained an important exposition of its dogmas, and his *Astronomical Doctrine of a Plurality of Worlds* (1828) sought to reconcile Christianity with that "almost overwhelming discovery," that "absolutely certain" doctrine of a plurality of worlds.[54] After citing Paine's position that no such reconciliation is possible, and after asserting that the Hutchinsonian system derived all its followers from this problem, Noble examines the responses of Chalmers and Nares. He finds the former response "completely nugatory" (p. 26), but the latter more satisfactory, because he believed it had been influenced by the writings of Swedenborg (p. 33). These writings he interprets as asserting that Jehovah took the form of humanity and came to our earth because it possessed the most corrupt beings in the universe. His actions here brought salvation not only to earthlings but to extraterrestrials as well.[55] Given its message and the context in which it appeared, Noble's *Plurality of Worlds* may be seen as a second "appeal" on behalf of Swedenborgianism.

The need to find a place for pluralism in religious systems also affected Sharon Turner (1768–1847), a pioneer in the study of Anglo-Saxon history. In his *Sacred History of the World* (1832), which went through at least eight editions, he endorses pluralism, although expressing puzzlement at the "peculiar seclusion" of the planets of our system from all other worlds. He includes the speculation that "our Earth may be a nursery of the immaterial principle; that it is here brought into its first state of being in animal forms . . . in order that it may be elsewhere used in some advanced or ulterior condition, and in other modes of material existence."[56] This claim, which Turner does not develop, suggests that transmigrational ideas, not uncommon on the Continent, and sometimes associated with pluralism, were entering England.

Sharon Turner was among the wide array of authors cited in support of extraterrestrials when in 1834 Alexander Copland, a Scottish "Advocate" and a son of Professor Copland who taught pluralism *"ex cathedra,"* published *The Existence of Other Worlds.*[57] Originality is not a noteworthy feature of Copland's book, nor did he intend it to be; his stated goal was simply to survey religious and scientific literature on his topic (p. 6). His book is embellished with quotations from such pluralists as Beattie, Chalmers, Fontenelle, the Herschels, King, Nares, Porteus, and Sturm. Although admitting that he has "no doubts" about that doctrine, Copland includes discussions of such opponents of pluralism as Maxwell and the physician Edward Walsh (1756–1832), who in an 1830 essay had argued that planets in most cases lack atmospheres and thermal conditions suitable for life.[58] Copland's critical comments are largely reserved for antipluralists; for example, he reminds Walsh that God could create extraterrestrials adaptable even to the high temperatures probable on Mercury (p. 121). Here, as elsewhere, Copland shows himself unwilling to separate the religious and scientific aspects of the question. Furthermore, Copland's excessive credulity is shown by his acceptance of life on the sun (pp. 87–8) and on the moon. He even predicts that lunar life will soon be verified "by actual sight of the natives or their works. . . ." (p. 160) On Copland's behalf it can be stated that he avoids the wild speculations into which some pluralists fell. Ironically, this may be why his book attained only a single edition.

In sharp contrast to Copland and his book stand Isaac Taylor (1787–1865) and his *Physical Theory of Another Life* (1836). Whereas Copland was an obscure Scotsman, Taylor was a prolific English historical and religious writer, known for his *Natural History of Enthusiasm*, translations of Greek authors, and publications on the patristic period. Whereas Copland's *Other Worlds* was derivative, informational, and published in only one edition, Taylor's *Physical Theory* was original, speculative, and repeatedly reprinted for thirty years. Taylor does not deny the conjectural character of his book; indeed, he openly states that its goal is to provide "conjectures" about the "future life."[59] Within this capacious context, he devotes most of its latter half to conjectures about extraterrestrials. The method used is analogy, which, despite a chapter-long discussion of its powers and limitations, he employs with almost unrestrained freedom. The analogies, for example, between our own and other celestial bodies assure him not only that they are inhabited but also "that rational and moral agents, in whatever world found, and whatever diversity of form may distinguish them, would be such as that we should soon feel ourselves at home in their society. . . ." (p. 213) In material conditions we no doubt differ from them, but "it is not possible to doubt that all, in all worlds, who are capable of reason, . . . are tending . . . toward the same eternal truths. . . ." (p. 215) Indeed, we shall eventually "be brought to compare histories, and so to receive severally the benefit derivable from their common experience." (p. 215) This will probably entail our communicating with an advanced race, which will greatly enlighten us (p. 221). The means for this communication are left unspecified, but this vision is developed into a physical theory of the universe, presented in three conjectures.

Taylor's first conjecture, based partly on Herschel's solar theory, is that our sun, and similarly every star, "is the home of the higher and ultimate spiritual corporeity, and the center of assembly for those who have already passed their preliminary era upon the lower ranges of creation." (pp. 222–3) Taylor supports this conjecture by noting that God acts in such a way as to produce a plurality of forms. If in the vegetable kingdom we see "a free exuberance, a copiousness, a versatility, and an unchecked love of embellishment and beauty" (pp. 229–30), should we expect less variety in the celestial realm? When Taylor speculates that a central body may exist in our sidereal system that "may be the home of a still higher order of life" (p. 234), one suspects the influence of Thomas Dick. However, neither Dick nor most other pluralists, except Chalmers (p. 208), receive mention in his book. Within the context of this first conjecture, Taylor even suggests that the planets may be hollow and inhabited (p. 234).

Taylor's second conjecture is that "within the field occupied by the visible and ponderable universe . . . there is . . . another species of life – corporeal indeed, and various in its order; but . . . not to be seen, not to be heard, not to be felt by man." (pp. 237–8) He justifies this by reference to the ether, which was then becoming more accepted in physical science, by urging that no one can deny that God could create such an invisible universe (p. 241), and by suggesting that our inability to perceive these beings may be explained by the failure of our five senses, of which many more are conceivable, to register their material embodiments (pp. 245–51). This leads to a theory of the action of the mind (pp. 251–69), in which he maintains that because these invisible beings are material, they may be drawn by gravitation to congregate around stars and planets (pp. 269–70).

Taylor's third conjecture is that "the visible universe . . . is to fill one period only in the great history of the moral system, and that it is destined . . . to disappear . . . , giving place to new elements and to new and higher expressions of omnipotence and intelligence." (pp. 285–6) This idea of a temporal succession of universes is elaborated partly in terms of Boscovich's theory that matter consists of point centers of force, which theory, by reducing matter to motion, permits him to suggest that God's mind, as the source of that motion, may destroy the universe simply by ceasing to will the continuance of that motion (pp. 290–3).

Taylor's *Physical Theory* is so speculative that when he states in the preface (p. xi) to its 1858 edition that in the more than two decades since its first publication he has found little, if anything, in it needing change, one suspects that this is due less to the correctness of his conjectures than to their being so fanciful that factual information can have no bearing on them. The excessively speculative character of his book derived from his use of Boscovich and the elder Herschel, his acceptance of a form of the principle of plenitude (p. 318), and above all his reliance on analogical argumentation. He shows little awareness that whereas analogical analysis may productively suggest new ideas, more rigorous methods are needed to prove them. For this failure, Taylor should not be severely faulted; many pluralists were oblivious to this distinction, and philosophers had at that time rarely attempted to formulate the logic of analogical reasoning.

When the much neglected problem of the logic of analogy was finally treated by John Stuart Mill (1806–73) in his *System of Logic* (1843), it was probably no accident that he chose as his main illustration the question of life on the moon and planets. The first point Mill makes is that "a fact *m,* known to be true of A, is more likely to be true of B if B agrees with A in some of its properties . . . than if no resemblance at all could be traced between B and any other thing known to possess the attribute *m.*"[60] Every such resemblance, provided that it is not known to be irrelevant to *m,* increases our confidence that *m* is likely to be found in B. As an illustration, Mill considers the question of life on the moon, noting that persons seeing many resemblances between the earth and the moon have tended to infer that the moon is inhabited. Mill stresses, however, that every dissimilarity found between A and B, the earth and the moon in his example, "furnishes a counter-probability on the other side." (p. 557) What is determinative in such a case is whether the resemblances or the dissimilarities preponderate. Mill provides no analysis of how this weighing is to be accomplished, but he does state that because the moon lacks some features such as water that on the earth are "indispensable conditions of animal life, we conclude that if the phenomenon [life] does exist in the moon . . . it must be as an effect of causes totally different from those on which it [life] depends here. . . ." (pp. 557–8) This ironically leads him to conclude that "all the resemblances which exist become presumptions against, not in favor of, her being inhabited."

A more important point emerges from his discussion of the planets, which he admits have stronger resemblances to the earth than does the moon. However, in this case, he stresses concerning the planets that "when we consider how immeasurably multitudinous are those of their properties which we are entirely ignorant of, compared with the few which we know, we can attach but trifling weight to any considerations of resemblances in which known elements bear so inconsiderable a proportion to the unknown." (p. 558) This statement in effect specifies that in Mill's time, the question of life on the planets would not admit of rigorous analysis, because the region of "unascertained properties" was still far too large. Mill, moreover, seems to have realized the lack of rigor involved in most reasoning by analogy, for near the end of his analysis of analogy he comments that its highest scientific value is "in suggesting experiments or observations that may lead to positive conclusions." (pp. 559–60) Put in more modern terms, Mill's statement consists of the claim that the method of analogy is primarily a method of discovery rather than of proof. To this he adds: "When the agents and their effects are out of the reach of further observation and experiment, as in the speculations already alluded to respecting the moon and planets, such slight probabilities are no more than an interesting theme for the pleasant exercise of imagination. . . ." (p. 560) All in all, Mill's analysis of analogy was a telling indictment of the reliance of pluralists on analogical reasoning, at least as a method of verification. However, despite the high esteem in which Mill's *System of Logic* was held, the research for this volume has not uncovered a single case in which later pluralists discussed Mill's view.

The possibilities in pluralism for poetry were taken up by two young British poets, Philip James Bailey (1816–1902) and Alfred Tennyson (1809–92), both of whom achieved prominence in the 1830s. In 1839, Bailey, attracted by Goethe's *Faust*, but distressed at its message, published *Festus*, a long poem set against the background of the Christian pluralists' universe. Early in the poem, Christ tells an angel assigned to earthlings:

> Think not I lived and died for thine alone,
> And that no other sphere hath hailed me Christ.
> My life is ever suffering for love.
> In judging and redeeming worlds is spent
> Mine everlasting being.[61]

Bailey's *Festus*, which went through dozens of editions in Britain and the United States, was both his first and last successful production. His later poems won few praises from the literary and religious public.

Among all Victorian poets, none was more interested in astronomy, none more intent to assay its significance than Tennyson, who as a child suggested to his brother, who was beset with shyness: "Fred, think of Herschel's great star-patches, and you will soon get over all that."[62] No doubt Tennyson's early astronomical interests intensified while he studied at Trinity College, Cambridge, where his tutor was Whewell, whose Chalmersian sermons of early 1827 may still have echoed in University Church when late in that year Tennyson came to Cambridge. One suspects as well that the young Tennyson read Whewell's 1833 Bridgewater Treatise with its pluralist passages and approval of the nebular hypothesis.[63] Be that as it may, the poet's pluralism can be traced back to his "Timbuctoo," which won the Chancellor's Medal at Cambridge in 1829, as well as to his "Armageddon," the poem on which it was based, and to *The Devil and the Lady,* which Tennyson wrote at age fourteen. "Timbuctoo," for example, contains references to the "Moon's white cities" and to the

> . . . harmony of planet-girded suns
> And moon-encircled planets, wheel in wheel,
> Arch'd the wan sapphire. Nay – the hum of men.
> Or other things talking in unknown tongues,
> And notes of busy life in distant worlds
> Beat like a far wave on my anxious ear.[64]

In his "The Two Voices" (1833), Tennyson reflects on man's place in the universe of pluralist astronomy:

> I said: 'When first the world began,
> Young Nature thro' five cycles ran,
> And in the sixth she moulded man.
>
> 'She gave him mind, the lordliest
> Proportion, and, above the rest,
> Dominion in the head and breast.'
>
> Thereto the silent voice replied:
> 'Self-blinded are you by your pride;
> Look up thro' night; the world is wide.
>
> 'This truth within my mind rehearse
> That in a boundless universe
> Is boundless better, boundless worse.
>
> 'Think you this mould of hopes and fears
> Could find no statelier than his peers
> In yonder hundred million spheres.'[65]

Having used pluralism to counsel a cosmic stoicism in this poem, Tennyson turned it to another purpose when in 1852, as the new poet laureate, he wrote "Ode on the Death of the Duke of Wellington." Near its end, he invokes the vastness of geological time and of astronomical space to suggest the spiritual:

> And Victor he must ever be.
> For tho' the Giant Ages heave the hill
> And break the shore, and evermore
> Make and break, and work their will,
> Tho' world on world in myriad myriads roll
> Round us, each with different powers,
> And other forms of life than ours,
> What know we greater than the soul?[66]

As astronomy advanced, pluralism took on new forms. These are reflected in Tennyson's later poetry, which is discussed in a subsequent section.

The popularity of pluralism in Britain in the third, fourth, and fifth decades of the nineteenth century is linked with many factors. A still flourishing, if increasingly frail, physicotheology supported it, while with Chalmers, evangelical religion assimilated it. Paine's polemics had received an array of responses, and the poetic resources of pluralism were being exploited by Shelley, Wordsworth, Tennyson, and others. Astronomers as prominent as Herschel and Smyth had endorsed it, and popularizers as successful as Dick, Lardner, and Nichol had appropriated and propagated it. Cautions and qualifications were not unknown, but as yet there had been no sustained attack on it. However, such an attack was in preparation as the period ended, coming from William Whewell, who had been among its most articulate spokesmen.

2. Extraterrestrials and Americans: given modern astronomy, "Who can be a Calvinist or who an Atheist"?

Other than Locke's satire, the pluralist publications most widely read in America in the 1817-to-1853 period were the books of Chalmers and Dick. Derivative[67] though American pluralism was, a remarkable diversity characterized it, as can be illustrated by consideration of Ormsby MacKnight Mitchel, Ralph Waldo Emerson, Thomas Lake Harris, Ellen G. White, and Joseph Smith.

It is a remarkable feature of the development of American astronomy that by 1845, Cincinnati, Ohio, possessed the second largest refracting telescope in the world. This was the achievement of Ormsby MacKnight Mitchel (1809–62), a West Point graduate, who combined a gift for oratory with a passion for astronomy to persuade Cincinnati laborers and lawyers, paperhangers and plumbers to fund the purchase and erection of that telescope. Much of Mitchel's success must be attributed to his skill in challenging the democratic sentiments of the Cincinnati citizenry, but another factor was no doubt the effectiveness with which he proclaimed pluralism. One evidence for this is that his *Orbs of Heaven*, which was based, as he stated, on the "lectures . . . to which the observatory owes its origins," concludes with a long pluralist passage.[68] Moreover, it is known that Mitchel drew "Great Applause" when in one of his later lectures he asserted:

Around us and above us rise Sun and System, Cluster and Universe. And I doubt not that in every region of this vast Empire of God, hymns of praise and anthems of glory are rising and reverberating from Sun to Sun and from System to System – heard by Omnipotence alone across immensity and through eternity![69]

Enthusiastic as Americans were for a pious pluralism, the spectre of Paine's objection hovered over the period, leading the statesman and orator Daniel Webster (1782–1852) to doubt Christianity.[70] A more striking example of the influence of that objection may be uncovered from the career of Ralph Waldo Emerson (1803–82), who, after studies at Harvard Divinity School, served as minister to Boston's Second Unitarian Church until he resigned his pastorate in September 1832, being unable to reconcile his altering beliefs with administering the Lord's Supper. By the late 1830s he had established a new career as writer and lecturer; in fact, his "Nature" (1836) and "Harvard Divinity School Address" (1838) are classics of American literature. The former was the first full statement of American Transcendentalism, of which Emerson became the leader, while the latter advocated such a total reorientation of religious thought that it shocked even the liberal Harvard theologians and made Emerson unwelcome at Harvard for over two decades.

All this is well known; what is less widely recognized is the role that astronomy played in Emerson's thought. This may be seen in his "Nature," the most seminal of his writings, which begins: "Our age is retrospective. It builds the sepulchres of the fathers. It writes biographies, histories, and criticism. The foregoing generation beheld God and nature face to face; we, through their eyes. Why should not we also enjoy an original relation to the universe?"[71] Emerson suggests how to do this; begin with solitude:

But if a man would be alone, let him look at the stars. . . . One might think the atmosphere was made transparent with this design, to give man, in the heavenly bodies, the perpetual presence of the sublime. . . . If the stars should appear one night in a thousand years, how would men believe and adore; and preserve for many generations the remembrance of the city of God which had been shown! [p. 5]

Such themes recur in other essays by Emerson and explain why he visited Dick's observatory in Scotland and Amici's observatory in Italy.[72] They also suggest why in 1832 Emerson would write "I hope the time will come when there will be a telescope in every street,"[73] and why Emerson scholars have described him as "obsessed with astronomy throughout his life."[74] Astronomy was more than a rich source of metaphor for Emerson; it also exerted a powerful force, not only on his own thought but also, according to him, on the development of thought generally. He stated: "The most important effect of modern astronomy has been the tapping our theological conceit, and upsetting Calvinism."[76]

The most striking evidence of the influence of astronomy on his thought comes from a sermon he first delivered on May 27, 1832, a few days before revealing his difficulties with the Lord's Supper to his congregation.[77] This sermon entitled "Astronomy," and first published only in 1938, shows that Emerson's astronomy was that of the pluralists.[78] Emerson begins by noting that both nature and Scripture are viewed as coming from God. Because science extends our understanding of nature, it illuminates Scripture. Astronomy is especially useful because of its many associations with religion. He suggests that "the song of the morning stars . . . , the lights of the skies, are to a simple heart the real occasions of devout feeling more than vestries and sermon hearings. . . ." (p. 171) From earliest times to the present, the study of astronomy has acted "to correct and exalt our views of God, and humble our views of ourselves." (p. 172) Astronomy has corrected man's tendency to view God in anthropomorphic ways by demonstrating "that whatever beings inhabit Saturn, Jupiter, Herschel [Uranus], and Mercury . . . must have an organization wholly different from man." (p. 173) Man could not breathe on the moon or walk on Jupiter, nor could his blood circulate on Uranus; yet God has populated these bodies with beings possessing "perhaps far more excellent endowments than he has granted to mankind." (p. 173) Moreover, astronomy "has had an irresistible effect in modifying and enlarging the doctrines of theology." (p. 170) Recounting Copernican astronomy, he states that it brought about "an equal revolution in religious opinion." In particular, whereas the Ptolemaic system could be reconciled with the Christian notion of atonement, "I regard it as the irresistible effect of the Copernican astronomy to have made the theological *scheme of Redemption* absolutely incredible." (pp. 174–5) This, he urges, was its effect on Newton, who became a Unitarian. It is true, he states, quoting Edward Young, that "The undevout astronomer is mad," but the "gross creeds" of Christianity "so revolted . . . the profound astronomers of France" that they felt forced to forsake them.[79] Moreover, seeing no alternative to the Christian conception of God, many rejected theism. But this, he explains, goes too far, for astronomical "investigations of the last two hundred years have brought to light the most wonderful proofs of design. . . ." (p. 175) The point is put most sharply in his journal for May 23, 1832, when he writes that given the teachings of astronomy, "Who can be a Calvinist or who an Atheist[?]"[80]

Emerson then suggests how modern astronomy affects the doctrines of the New Testament. It produces, he claims, "not contradiction but correction . . . , not denial but purification." God remains, but "no mystic sacrifice, no atoning blood." (p. 177) If Christ as savior be banished, what remains except the God of natural theology? He answers: "if we could carry the New Testament to the inhabitants of other worlds we might need to leave Jewish Christianity, and Roman Christianity . . . , but the moral law, justice and mercy would be at home in every climate and world where life is. . . ." (p. 177) This will make "moral distinctions still more important," but it "will not teach any expiation by Jesus; it will not teach any mysterious relations to him." Jesus "only saves us, by inducing us to save ourselves. . . ." (p. 177–8)

Such was Emerson's sermon. It leaves one puzzled as to why the Harvard theologians, if they knew of this sermon, which Emerson repeated on a number of occasions, were so shocked when in his Divinity School address he asserted: "Historical Christianity has fallen into error. . . . It has dwelt, it dwells, with noxious exaggeration about the *person* of Jesus. The soul knows no persons. It invites every man to expand to the full circle of the universe. . . ."[81] Emerson's "Astronomy" may, in short, reveal how one of the most spiritually inclined men of his time came to forsake pulpit for lecture platform and to espouse a position similar to that propounded by Paine.

The influence of pluralism on Emerson may help explain its presence in the writings of the most important of Emerson's fellow Transcendentalists, Henry David Thoreau (1817–62), whose *Walden* (1854) contains such passages as: "We might try our lives by a thousand simple tests; as, for instance, that the same sun that ripens my beans illumines at once a system of earths like ours. If I had remembered this it would have prevented some mistakes. This was not the light in which I hoed them."[82] Elsewhere in the same book, Thoreau reflects: "This whole earth which we inhabit is but a point in a space. How far apart, think you, dwell the two most distant inhabitants of yonder star . . . ? Why should I feel lonely? Is not our planet in the Milky Way?"[83]

The sources of Emerson's pluralism were no doubt numerous; for example, William Herschel is mentioned in a number of his writings. Another source was Emanuel Swedenborg, whose ideas Emerson initially encountered in the 1820s and assessed in his 1850 essay "Swedenborg; or, The Mystic." Emerson's evaluation is mixed. On the one hand, he states that Swedenborg, "who appeared to his contemporaries a visionary, and elixir of moonbeams, no doubt led the most real life of any man then in the world . . . ." On the other hand, Emerson urges: "For the anomalous pretension of Revelations of the other world, – only his probity and genius can entitle it to any serious regard. His revelations destroy their credit by running into detail."[84]

Emerson's interest in Swedenborg was not an isolated phenomenon in America; by 1850, fifty-four Swedenborgian societies had been established, with thirty-two ordained ministers serving them.[85] No doubt these societies spread the pluralist message. Among those who around 1850 became involved with Swedenborgian ideas, none carried pluralism further than Rev. Thomas Lake Harris (1823–1906). A Baptist-turned-Universalist-turned-spiritualist, Harris, in 1850, while in a series of trances, dictated a book-length pluralist poem called *An Epic of the Starry Heaven*. Its approach is indicated in these lines:

A new-born language trembled on my tongue,
   Whose tones accorded with the singing stars;
A company of spirits, blithe and young
   From Jupiter, and Mercury, and Mars,
Drew near and said to me, "Three days, dear friend,
   Thou art our guest; come, wing thy blessed flight
   Through the unvailing ocean of sweet light."[86]

Strange as the pluralist themes in this and other poems by Harris may be, his most remarkable pluralist production is his 1858 *Arcana of Christianity*,[87] which, as the title suggests, was written in the spirit of Swedenborg. Five times longer than the Swedish seer's *Earths in the Universe*, although written as a commentary only on the first chapter of Genesis, Harris's *Arcana* presents a detailed discussion of life on the sun as well as on various planets, including some otherwise unknown. Whereas Isaac Taylor labeled his system conjectural, Harris presents his as a revelation from the Lord, whose angels show Harris various celestial regions. Fantastic features abound; the inhabitants of the star (!) Cassiopeia, for example, "feed chiefly upon the aromas of exquisite flowers" (p. 58), while Mercury, he finds, is "a Phalansterian world of Christian Platonic Philosophers. . . . Had Charles Fourier lived [there] he would have seen the good, the useful, the beautiful and the true, in all his speculations, far transcended. . . ." (p. 69) The highest civilization he encounters is on Titania, a planet 200,000 miles in diameter, orbiting a star in the Little Bear. His Titanians are clothed in light, do not marry until over a century old, and also live on aromas (pp. 275–7). The earth, he learns, is the single planet on which evil is now present (p. 1) and the only one on which Christ became incarnate (p. 376). He also includes details on the sex life of his planetarians, as well as interviews with Cromwell, Saints Peter and Paul, and Swedenborg, who is known on Saturn as the "Planet-walker" (p. 9) and who apologizes to Harris for having given wrong information in some of his visions (p. 285).

Such writings and his personal charisma won Harris a substantial following in America and Britain, perhaps as many as two thousand persons at one point associating themselves with his "Brotherhood of the New Life," which he ruled as "Father." The Swedenborgians, however, distanced themselves from him, and charges of sexual scandal and of financial abuses gradually diminished his influence. Nonetheless, in his *Varieties of Religious Experience* (1902), William James referred to Harris as "America's best-known mystic."[88] James's judgment in this instance seems excessive, possibly because of his own interest in the Swedenborgians. In any case, Harris's Brotherhood of the New Life scarcely survived him, and the judgment of one expert is that Harris, from the time of his *Arcana,* "went from excess to excess even to the hour of his death."[89]

Although Harris failed to establish a new religion capacious enough to include extraterrestrials, two other Americans, Ellen G. White and Joseph Smith, succeeded. In fact, the denominations that these two founded have survived into the twentieth century and presently have millions of members.

Ellen G. Harmon (1827–1915), the "prophetess" of the Seventh-day Adventist church, was born in Gorham, Maine, the daughter of a hatter.[90] The intense religious concerns of her youth heightened as she became involved in the Millerite movement, centered on the prediction of William Miller that Christ's second coming would occur in 1843 or 1844. Following the failure of this prediction, the Millerite movement splintered, with different groups adopting various interpretations of the so-called Great Disappointment. The most favored theory was that on October 22, 1844, a cosmic event had occurred – Christ had entered into the most holy region of the heavens in preparation for his second coming, which would soon begin. A version of this view was adopted by Harmon, who in late 1844 began to experience visions.

Commanded in one of these visions to spread the gospel, she traveled through New England, coming in the process to marry a Millerite minister, James White (1821–81). She also became acquainted with Joseph Bates (1792–1872), a former sea captain and Millerite, whose attainments included a knowledge of astronomy. In fact, in 1846 Bates published a booklet entitled *The Opening Heavens* that contains references to the writings of Dick, Ferguson, and other astronomical writers. The theme of Bates's booklet is that the Orion nebula is the opening through which Christ will return for his second coming. That Bates was also a proponent of pluralism is shown by his citing, with approval, Ferguson's claim that in the Orion nebula "there seems to be a perpetual uninterrupted day among numberless worlds. . . ."[91] Bates, who through contacts with the Seventh Day Baptists had become convinced that the Lord's day should be observed as Saturday rather than Sunday, was initially skeptical of Ellen White's visions. However, his views changed dramatically when in November 1846 he was present as the young prophetess, despite professing ignorance of astronomy, experienced a vision with astronomical content. An eyewitness described the event:

Sister White was in very feeble health, and while prayers were offered in her behalf, the Spirit of God rested upon us. We soon noticed that she was insensible to earthly things. This was her first view of the planetary world. After counting aloud the moons of Jupiter, and soon after those of Saturn, she gave a beautiful description of the rings of the latter. She then said, "The inhabitants are a tall, majestic people, so unlike the inhabitants of earth. Sin has never entered here." It was evident from Brother Bates' smiling face that his past doubts in regard to the source of her visions were fast leaving him.[92]

Bates's impression of her description of the "opening heavens," also seen in that vision, is recorded as follows:

While she was talking and still in vision, he arose to his feet, and exclaimed, "O how I wish Lord John Rosse was here tonight!" Elder White inquired, "Who is Lord John Rosse?" "Oh," said Elder Bates, "he is the great English Astronomer. I wish he was here to hear that woman talk astronomy, and to hear that description of the 'opening heavens.' It is ahead of anything I ever read on the subject."[93]

Captain Bates's confidence in Mrs. White's powers grew still greater when in a subsequent vision she confirmed his views of the Sabbath.

The vision of 1846 was not Mrs. White's only vision concerning extraterrestrials; in 1849 she revealed the following:

The Lord has given me a view of other worlds. Wings were given me, and an angel attended me from the city to a place that was bright and glorious. The grass of the place was living green, and the birds there warbled a sweet song. The inhabitants of the place were of all sizes; they were noble, majestic, and lovely. . . . I asked one of them why they were so much more lovely than those on the earth. The reply was, "We have lived in strict obedience to the commandments of God, and have not fallen by disobedience, like those on the earth." . . . Then I was taken to a world which had seven moons. . . . I could not bear the thought of coming back to this dark world again. Then the angel said, "You must go back, and if you are faithful, you with the 144,000 shall have the privilege of visiting all the worlds. . . ."[94]

Mrs. White was not the only prophetess of the Seventh-day Adventist church, the name adopted in the early 1860s, by which time the group associated with Ellen and James White and Joseph Bates had grown to over three thousand;[95] she was also a prolific writer of articles and books in which she developed the doctrines of the Adventists. One of these is that fallen angels attempted to spread evil throughout the universe, but succeeded only on the earth. Moreover, Christ became incarnate solely on our planet and will return, bringing eternal life and happiness to the faithful. Her discussions of this doctrine frequently set it within a cosmic, pluralist context. For example, in *The Story of Patriarchs and Prophets,* she states: "When Satan was thrust out of heaven, he determined to make the earth his kingdom." With Adam's fall, Satan precipitously concluded that he had succeeded, but God "gave his own dear Son . . . [who] undertook to redeem man. . . . The great controversy begun in heaven was to be decided in the very world . . . that Satan had claimed as his." This remarkable event is known in other worlds:

It was the marvel of all the universe that Christ should humble himself to save fallen man. That he who has passed from star to star, from world to world, superintending all . . . [took] upon himself human nature, was a mystery which the sinless intelligences of other worlds desired to understand. When Christ came to our world . . . , all were intensely interested in following him as he traversed . . . the bloodstained path from the manger to Calvary. . . . And as Christ in his expiring agony upon the cross cried out, "It is finished!" a shout of triumph rung through every world. . . .[96]

In discussing God's testing of the terrestrial Abraham, she presents it as significant and instructive for the "sinless intelligence of heaven and of other worlds."[97] By 1889, some 72,000 copies of her *Great Controversy between Christ and Satan* were in print, detailing in its concluding paragraph the delights destined for the redeemed of our planet:

All the treasures of the universe will be open to the study of God's redeemed. Unfettered by mortality, they wing their tireless flight to worlds afar, – worlds that thrilled with sorrow at the spectacle of human woe, and rang with songs of gladness at the tidings of a ransomed soul. . . . With undimmed vision they gaze upon the glory of creation, – suns and stars and systems, all in their appointed order circling the throne of Deity.[98]

Such powerful images expressed with such eloquence are reminiscent of Klopstock and Chalmers. The attractiveness of her writing as well as her energetic missionary efforts in both Europe and Australia were major factors in the growth of the Seventh-day Adventist church.

Questions about the role of Ellen White and the degree of trust to be placed in her visions and writings are currently under consideration by Seventh-day Adventist theologians, precipitated partly by evidence that some of her writings contain sections nearly identical with sections in earlier publications by non-Adventist authors.[99] However this issue may be resolved, two points are clear: She showed remarkable gifts as the spiritual leader of a denomination that now includes millions of believers. Moreover, she provided her church with what few congregations can claim: a theology incorporating extraterrestrial beings. Perhaps this was prophetic; or perhaps this was a product of the pervasiveness of pluralism in the period in which she wrote.

The story of Joseph Smith, the "First Prophet, Seer, and Revelator" of the Church of Jesus Christ of Latter-day Saints, is filled with high drama. Born in 1805 to a farm family in Sharon, Vermont, Smith, in 1830, despite possessing only a rudimentary education, published the *Book of Mormon*. This he claimed he had translated from the "Egyptian, Chaldic, Assyrian, and Arabic" characters on a set of gold plates that Moroni, an angelic being from the New World, had commanded him to dig up from a site near Palmyra, New York, and subsequently to return to the angel. Presented as a supplement to the traditional Christian Scriptures, the *Book of Mormon* narrates the ancient activities in the New World of peoples who had originally migrated from Palestine to the New World in two migrations. Within months of its publication, the charismatic Smith had drawn around him a group of disciples. Growing in size and separateness and repeatedly encountering hostility, this band of believers moved westward, first to Kirtland, Ohio, then to Jackson County, Missouri, and finally in 1839 to Commerce, Illinois, where the Latter-day Saints' community of Nauvoo was established. Again troubles developed; in 1844, Joseph Smith was murdered – martyred, his followers felt – by an angry mob. After Smith's death, Brigham Young (1801–77) became head of one branch of the Saints, leading them in the late 1840s from Nauvoo to the Great Salt Lake area. Controversies too numerous to narrate, ranging from disputes as to the existence of the gold plates to the now abandoned practice of polygamy, surrounded the Latter-day Saints for decades. Despite these, Smith, aided by Young, Oliver Cowdery, Orson and Parley Pratt, Sidney Rigdon, and other early disciples, established a church that has grown in numbers and vitality over the century and a half since its inception.[100]

The Book of Mormon, for all the newness of its narrative contents, does not differ substantially in doctrinal matters from the traditional Christian Scriptures. Nor does it include the idea of a plurality of worlds. Most of the features that have made Mormon theology distinctive were set out in writings published later, including *The Doctrine and Covenants,* which was first printed in 1835 but was later supplemented by new sections, and *The Pearl of Great Price,* first published in 1851, but compiled almost entirely from materials previously printed in magazines.[101] Both *The Doctrine and Covenants* and *The Pearl of Great Price* contain many statements in support of a plurality of worlds. In fact, B. H. Roberts, in what is described as the official history of the church, states: "The Prophet Joseph Smith taught that these worlds and world-systems . . . are or will be inhabited by sentient beings. This is assumed in all his revelations. It is everywhere taken for granted."[102] *The Doctrine and Covenants* records Smith's conviction that many important points "had been taken from the Bible, or lost before it was compiled . . . ," and in particular that "if God rewarded every one according to [his] deeds . . . , the term heaven, as intended for the Saints' eternal home, must include more kingdoms than one." (Section 76) Smith and Sidney Rigdon, after study of and prayer on the matter, received a revelation clarifying this point. They learned that the Son sits at the right hand of God and "That by him, and through him, and of him, the worlds were created, and the inhabitants thereof are begotten sons and daughters unto God." (76:24; also 93:10) Later in the revelation they were told that after death, souls move either to the outer darkness or to one of the three realms of glory: the celestial, terrestrial, or telestial, depending on the lives they led. In a revelation later in 1832, Smith learned: "And there are many kingdoms; for there is no space in which there is no kingdom; and there is no kingdom in which there is no space, either a greater or a lesser kingdom." (88:37) Both the context of this passage and later commentaries point to this statement as indicating a plurality of celestial worlds.[103] On April 2, 1843, Smith received revelations on such topics as time, angels, and other worlds:

4. In answer to the question – Is not the reckoning of God's time, angel's time, prophet's time and man's time, according to the planet on which they reside?
5. I answer, Yes. But there are no angels who minister to this earth but those who do belong or have belonged to it.
6. The angels do not reside on a planet like this earth;
7. But they reside in the presence of God, on a globe like a sea of glass and fire, where all things for their glory are manifest, past, present, and future, and are continually before the Lord. (130:4–7)

The Pearl of Great Price consists primarily of the "Book of Moses" and the "Book of Abraham." The former begins with a chapter entitled "Visions of Moses," which purports to recount materials revealed to Smith in June 1830. This section was, however, first published in January 1843 in the Latter-day Saint newspaper *Times and Seasons*.[104] Among the visions and revelations attributed to Moses were some relating to a plurality of worlds:

29. And he [Moses] beheld many lands; and each land was called earth, and there were inhabitants on the face thereof.
33. And worlds without number have I [God] created; and I also created them for mine own purpose; and by the Son I created them, which is mine Only Begotten.
35. But only an account of this earth, and the inhabitants thereof, give I unto you. For behold, there are many worlds that have passed away by the word of my power. And there are many that now stand, and innumerable are they unto man. . . .
38. And as one earth shall pass away, and the heavens thereof, even so shall another come; and there is no end to my works. . . .

The "Book of Moses" in a later section contains statements on the vast number of other worlds and on mankind as the most wicked race in the universe:

And were it possible that man could number the particles of the earth, yea, millions of earths like this, it would not be a beginning to the number of thy creations. . . . (7:30) Wherefore, I can stretch forth mine hands and hold all the creations which I have made; and . . . among all the workmanship of mine hands there has not been so great wickedness as among thy brethren. (7:36)

The Latter-day Saint scripture richest in astronomical lore is the "Book of Abraham," a short work of five chapters, first published in March

1842 and reprinted in *The Pearl of Great Price*.[105] It purports to be a translation of some papyri to which Smith gained access in 1833.[106] Its third chapter describes the throne of God and its relation to a star called Kolob in the revelation:

2. And I saw the stars, that they were very great, and that one of them was nearest unto the throne of God; and there were many great ones which were near unto it;
3. And the Lord said unto me: These are the governing ones; and the name of the great one is Kolob, because it is near unto me, for I am the Lord thy God: I have set this one to govern all those which belong to the same order as that upon which thou standest.

Pluralism was not only propounded in scriptures; it was also advocated in statements by some of Smith's most important disciples. For example, Parley P. Pratt, in his *Key to the Science of Theology* (1855), stated:

Gods, angels and men, are all of the same species, one race, one great family widely diffused among the planetary systems, as colonies, kingdoms, nations, etc.
The great distinguishing difference between one portion of this race and another, consists in the varied grades of intelligence and purity, and also in the variety of spheres occupied by each, in a series of progressive being. . . .[107]

Also, Brigham Young asserted in one of his discourses that "He [God] presides over the world on worlds that illuminate this little planet, and millions on millions of worlds that we cannot see; and yet he looks upon the minutest object of his creation. . . ."[108] Pluralist themes even appear in traditional hymns; the following is a hymn still sung by Latter-day Saints. It is by William Wine Phelps, one of Smith's earliest disciples, who is praised for writing verses "most characteristic of Latter-day Saint thought and aspiration."

> If you could hie to Kolob,
>   In th' twinkling of an eye,
> And then continue onward,
>   With that same speed to fly—
>
> D'ye think that you could ever,
>   Through all eternity,
> Find out the generation
>   Where Gods began to be?
>
> Or see the grand beginning,
>   Where the space did not extend?
> Or view the last creation,
>   Where Gods and matter end?
>
> Methinks the Spirit whispers,
>   "No man has found 'pure space,' "
> Nor seen the outside curtains
>   Where nothing has a place.

The works of God continue,
   And worlds and lives abound;
Improvement and progression
   Have one eternal round.

There is no end to matter,
   There is no end to space,
There is no end to "spirit,"
   There is no end to race.[109]

Before terminating this discussion of pluralism in Joseph Smith's writings, mention should be made of two relatively recent claims concerning this topic. In a number of discussions of extraterrestrial life written within the Latter-day Saint perspective, a tendency can be detected to view Smith as having transcended his time in advocating life elsewhere. If it should be found that that was the case, belief in his prophetic gifts would be enhanced.[110] On the other hand, Fawn Brodie, in her 1945 biography of Smith, claimed that he took certain metaphysical notions, and especially his pluralist views, from Thomas Dick's *Philosophy of a Future State,* which was quoted in an 1836 Latter-day Saint newspaper. She alleged, as did Klaus J. Hansen in a recent reassertion of this claim, that Smith's Kolob, for example, derived from Dick's idea that a larger central body exists in the universe and governs the motion of other stars.[111] Problems with Brodie's claim, which claim would weaken the view of Smith as receiving special revelations, have been pointed out by two Latter-day Saint scholars, Edward T. Jones and E. Robert Paul.[112]

Based on my analysis of the relevant primary and secondary documents, I have drawn the following conclusions: First, a number of ideas espoused by Dick were also endorsed by Smith, but overall their positions differ substantially. Second, numerous sources from which Smith could have drawn some, if not all, his pluralist doctrines were in print during the time when he wrote. In particular, solid evidence indicates that Smith had ready access to Dick's book and to Paine's *Age of Reason;* moreover, Smith could have had access to the writings of Chalmers, Dwight, Ferguson, Fuller, or others who advocated pluralism, although the thorough researches of Professor Paul have uncovered no direct evidence of this. Third, because both the view of Smith as prophet of pluralism and Brodie's view of him as quasi plagiarist of Dick have derived support from the belief that Smith's pluralism and associated doctrines were uncommon in and "quite advanced" for the period around 1840, whereas the opposite is nearer the truth, both views must be seen as incorporating an erroneous assertion in their premises. To cite one example: in putting forth his giant central body thesis, Dick did not present it as original; rather, he labeled it, albeit overly enthusiastically, as a theory "now considered by astronomers, as highly probable, if not certain. . . ."[113] In short, if Smith is seen within the context of his time, a period in which pluralism was popular if not pervasive, the force of these claims for him either as prophet of pluralism or as plagiarist of Dick is largely dissipated.

Ralph Waldo Emerson, Ellen G. White, and Joseph Smith, whose writings millions have read and revered, were three of the most remarkable figures of nineteenth-century America. Their successes were partly due to the skill they shared for selecting the deepest and most difficult issues of their era and boldly addressing them in ways that stimulated the minds and hearts of their disciples. Realizing that extraterrestrial life was emerging as such an issue, each advanced claims concerning it. Significances and systems that astronomers could not or would not see, they daringly detected. For example, while White and Smith revealed that earthlings are precisely the most corrupt inhabitants of the cosmos, Emerson opined that selenites, Jovians, and Uranians possess "far more excellent endowments than . . . mankind." Whereas White located Christ in Orion and Smith placed Kolob "nearest unto the Throne of God," the Transcendentalist Emerson assigned the "city of God" to the heavens as a whole. This suggests, and subsequent materials document, that speculative religious and philosophical writers were finding that the question of other worlds was one that they could not easily leave aside.

3. Conceptions on the Continent: "Who dwells on yonder golden stars?"

The question of a plurality of worlds, so widely discussed among British and American authors in the 1817-to-1853 period, was also exciting interest among Continental intellectuals. Among these were scientists as distinguished as Arago and Oersted, literary figures as widely acclaimed as Balzac, Heine, and Leopardi, and philosophers as influential as Comte, Fourier, and Hegel.

In France, the premier astronomer was François Arago (1786–1853), who from 1830 served both as perpetual secretary of the Académie des sciences and as director of the Paris Observatory. With the eloquence expected in the former position and the expertise requisite for the latter, Arago for decades delighted the Parisian public with his astronomical lectures. These appeared posthumously as his *Astronomie populaire*, a work soon translated into English and German.[114] Pluralist proclivities were not absent from Arago. For example, in the section of his *Astronomie populaire* headed "Is the Sun Inhabited?" he answers:

. . . I know nothing. But if one asked me whether the sun can be inhabited by beings organized in a manner analogous to those which populate our globe, I would not hesitate to make an affirmative response. The existence in the sun of a central dark nucleus enveloped in an opaque atmosphere, far from the luminous atmosphere, offers nothing in opposition to such a conception. (II, p. 181)

Arago then describes William Herschel's model of the sun, mentioning as an aside Dr. Elliot, whose already tarnished reputation was further darkened by Arago reporting that he killed Miss Boydell. Arago dryly adds: "The conceptions of a madman are today almost generally adopted." (II, p. 182) In a later chapter entitled "Of the Inhabitability of Comets," which objects he had earlier presented as varying greatly in temperature and density of atmosphere during their orbital motions, Arago reviews experiments in which men managed to breathe under very high and very low pressure. He then remarks:

I do not pretend to draw from these considerations the conclusion that comets are populated by beings of our species. I have presented them here only to render, as Lambert puts it, their *habitability* less problematic. I do observe, in addition, that all the celestial bodies have raised the same question and the same doubts. If the solution has presented some difficulties, it is only that our views of organized beings are in fact very limited; it is that we conceive only with difficulty of animals which differ from those of which we have studied the form, movements and nutrition. (II, p. 482)

Arago's nearly book-length discussion of the moon in his third volume leaves the question of its habitability unanswered. He mentions Gruithuisen's observations of lunar "fortifications," but notes that subsequent observations by Beer and Mädler, as well as by Lohrmann, have revealed only features comparable to those sighted elsewhere on the moon (III, p. 427). It is surprising that having allowed life on the sun and comets, Arago should hesitate at the moon. The final section of Arago's *Astronomie populaire* consists largely of a discussion of the celestial appearances as viewed from various bodies in our solar system. In this section, Arago remarks:

Some very pious persons have imagined that to examine what will be the astronomy of an observer situated on diverse planets was to put onself into a culpable disregard of holy Scripture. I do not share this view. In effect, in transporting an observer to different planets, and even to the center of the sun, we are not saying that he resembles the inhabitants of our globe. Besides, some very wise theologians, for example, Dr. Chalmers, have proved that nothing in the holy books forbids the supposition that the planets are inhabited. (IV, pp. 759–60)

Turning first to the solar system as it would be seen by solarians, Arago offers further religious assurances to his readers by noting that Cardinal Cusa in the fifteenth century had accepted solar inhabitants and that the Jesuit Hervas y Pandura in the eighteenth century had written a long cosmic voyage involving solarians (IV, p. 760). Arago's discussion of the heavens as seen from different planets no doubt contributed to the appeal of his presentation; nonetheless, it had the disadvantage that readers could easily be misled by it into believing that he was endorsing their inhabitation.[116]

The enthusiasm evoked by Arago's lectures and his *Astronomie populaire* led Camille Flammarion to praise him as the "Founder of Popular Astronomy."[117] Should a history of the popularization of astronomy be prepared, Arago would certainly deserve an important place, although not the primacy Flammarion proposed. Such a study would reveal that one of the techniques most frequently employed by popularizers of astronomy has been to embellish their writings with pluralist passages. Fontenelle, Lalande, Bode, and others had discovered this device before the director of the Paris Observatory, but he showed a special flair, illustrated in the quotations cited, for such phrasings as there is "nothing in opposition to" and "nothing . . . forbids." Such linguistic legerdemain tended to lead his readers to believe that he supported solarians, whereas his statements were more cautious than that.

Interesting and attractive as Arago's lectures were, they did not carry all his auditors into the pluralist camp. In 1847, François Édouard Plisson, a physician and professor at the Athénée royal in Paris, published his *Les mondes,* which raised important questions about the pluralist position.[118] In Plisson's "Préamble," he dissociates himself from Fontenelle and Huygens; the former he describes as "having sought more to amuse his reader than to instruct him," whereas the latter went "beyond the limits of reasonable conjectures." (p. 2) Fontenelle and Huygens having written "romances" (pp. 318, 320), Plisson prefers to base his volume on writings "which can provide us with the best established principles of physics, mechanics, and physiology." (p. 4) The last mentioned science is almost certainly the source of the refreshingly skeptical approach taken by this physician toward pluralism. No evidence indicates that religious matters, which are never mentioned, contributed to his caution. A modestly positivistic orientation, evident in his repeated attacks on the use of final causality, is present, but this seems to have its source more in good sense than in religious scruple.

Plisson does not totally reject pluralism; rather, he examines each of the bodies in our solar system in regard to whether or not it possesses an atmosphere, soil, temperature, surface features, and so forth, capable of supporting living beings, which, although different from terrestrial beings, nonetheless must in some sense conform to the laws of biology. Amidst repeated reminders of our limited knowledge of the planets, Plisson presents forceful evidence that at least some solar system objects cannot be inhabited. For example, he writes concerning the moon: "Disinherited of all fluid and of an atmosphere of air, the moon . . . has neither rain, nor storm. It is a solid, arid, deserted, silent mass, without the slightest vestige of vegetation, and where it is evident that no animal can find subsistence." (p. 227) A tendency to irony, found frequently in his book, is exemplified in this statement:

If however one wishes, at all hazards, that [the moon] have inhabitants, we will freely consent, provided that one makes them beings deprived of all sensibility, of all sentiment, of all movement, in a word, that one reduces them to the condition of inorganic bodies, of inert substances, of rocks, of stones, of metals, which from our point of view, are the only Selenites possible. (pp. 227–8)

The idea of sending signals to the selenites he labels "bizarre, a veritable scientific folly" (p. 75), and he also rejects for an array of reasons inhabited comets (e.g., p. 83). Solarians fare scarcely better than selenites; after discussing the story of Dr. Elliot, Plisson echoes Arago by commenting: "Oh well, almost all the astronomers of our day, and the most eminent among them, freely adopt the opinions that not long ago were viewed as being able to spring only from the mind of a madman." (p. 44)

Scientific analysis, however, rather than derision, characterizes the book, which is not entirely opposed to pluralist ideas. For example, Plisson finds evidence that the "fourth satellite of Jupiter" (Callisto) possesses some conditions favorable to life (p. 167). All of the then known planets, satellites, and asteroids are examined insofar as their conditions are known; moreover, he speculates on the factors that may determine the level of civilization on any bodies that may be inhabited (pp. 297–305). His final discussion summarizes his overall conclusion and exemplifies his open, cautiously skeptical approach. After noting that the widespread presence of living forms on the earth suggests that many celestial bodies, especially the planets, can be inhabited, he warns that we should "not forget that this idea . . . is in fact only a conjecture, and . . . rests in the main only on relations of analogy, and not on direct, unquestionable, decisive, and convincing proofs." To readers wondering if such a conclusion justifies the efforts that went into his long treatise, he responds: "our end was not to prove the plurality of worlds, but rather only to render account, according to the generally admitted principles of science, of the special astronomical conditions in which such inhabitants would have to live." (pp. 306–7) Plisson's volume is now almost unknown, even though a German translation appeared in 1851. The reasons for its failure to attract a wider audience probably include the pervasiveness of pluralism, the insensitivity to the need for a physiological analysis of the issues, and Plisson's avoidance of the bold claims characteristic of many popular writers.

Arago's astronomical lectures were far from being the sole source of pluralism for Parisians; another was interest in Swedenborg. Among French intellectuals drawn to the writings of the Swedish seer was Honoré de Balzac (1799–1850), whose involvement can be traced in his largely autobiographical novel *Louis Lambert*.[119] The most Swedenborgian of Balzac's novels is his *Seraphita,* of the mid-1830s, one of the most curious and most poetic of his works. Seraphita, who appears as a lovely young woman to Wilfred, but as a handsome man to the female Minna, is a human metamorphosing into an angel who both embodies and preaches Swedenborgian doctrines throughout the novel. The third chapter, which is largely an exposition of Swedenborg's ideas, contains a number of pluralist passages, including the following:

According to the Seer, the inhabitants of Jupiter do not affect the sciences, which they call Shades; those of Mercury object to the expression of ideas by words, which they think too material, and they have a language of the eye; those of Saturn are persistently tormented by evil spirits; those of the Moon are as small as children of six years old, their voice proceeds from the stomach, and they creep about; those of Venus are of gigantic stature, but very stupid, and live by robbery; part of that planet, however, is inhabited by beings of great gentleness, who live loving to do good.[120]

Shortly thereafter, the prediction is made "that some day, perhaps, scientific men will drink of these luminous waters." The success of Balzac's *Seraphita* may partly be explained by the statement contained in it that Swedenborgian "followers now number more than seven hundred thousand souls, partly in the United States of America . . . , and partly in England, where there are seven thousand Swedenborgians in the city of Manchester alone." (p. 56) These figures seem excessive but, if correct, are related to the fact that persons frequently combined Swedenborgianism with their own faiths, as was done by Balzac himself.[121]

Another French advocate of pluralism was the utopian social reformer Charles Fourier (1772–1837). From his first major work, *Theorie des quatres mouvements,* of 1808, to his posthumous "Cosmogonie" and "Analogie et cosmogonie," of 1848, Fourier was a veritable fountain of cosmological speculations that have been described by modern scholars in terms ranging from "mad" to "a mask" for his radical social critique.[122] That pluralism was central to Fourier's system is evident even from his 1808 book on the "four movements" (i.e., social, animal, organic, and material). In setting out the particular principles governing the theory of each movement, Fourier formulates them in such a way that they apply not only on the earth but also in the extraterrestrial realm. For example, he specifies concerning the social movement: "Its theory must explain the laws following which God orders the structure and the succession of different social mechanisms on all the inhabited celestial bodies."[123] Fourier states that before his book, only the laws of the material movement had been discovered, that having been done by Newton, whose work Fourier viewed himself as continuing.[124] Not only are the planets inhabited in Fourier's system; they themselves are living bisexual beings. In his 1808 work, a planet is seen as "a being which has two souls and two sexes and which procreates like an animal or vegetable by the joinging of two generative substances."[125] Stars also copulate, the seed being nourished in the Milky Way and transported through space by comets.[126]

Planets in Fourier's system emit aromas, the qualities of which influence neighboring celestial bodies. The aromas from the earth are of special importance. Because mankind had failed to establish phalanxes, the social units of 810 persons of differing talents prescribed by Fourier as the cure for the ills of society, the earth could not emit the proper aromas, which upsets the sun, which "has now been for six thousand years in a state of slow fever, of consumption. . . ."[127] Should phalanxes be established, Fourier promises that harmony will come to society, the oceans will turn into a sort of lemonade, our dead moon will be replaced by five satellites, and the sun will be restored, with further beneficial effects extending to the biuniverses, triuniverses, and so forth, of his system.[128] Mixed with all this was an elaborate doctrine of metempsychosis, according to which human souls alternate lives between our planet and its aromal atmosphere. On the death of our planet, the souls associated with its great soul would pass with that soul to form a new planet or even universe. Because of man's failure to establish phalanxes, the earth had been placed in celestial quarantine.[129]

Dispute continues concerning how seriously Fourier intended his cosmology, cosmogony, and therapeutics of the universe to be taken. However this may be resolved, at least some of Fourier's followers found this portion of his philosophy unattractive, if not an embarrassment. For example, when in 1844 Parke Godwin (1816–1904), an American journalist and Fourierist, published his *Popular View of the Doctrines of Charles Fourier,* he positioned Fourier's cosmological speculations at the end of his exposition.

They were treated in an even less friendly fashion by Jean-Ignace-Isidore Gérard (1803–47), a noted French illustrator who used the name J. J. Grandville. Some of Grandville's most memorable illustrations appear in a curious 1844 volume entitled *Un autre monde,* which came about in the following way. Grandville had prepared about one hundred miscellaneous and delightful drawings that were later loosely tied together by means of a story composed for the illustrations by Taxile Delord, a minor French writer.[130] Among Grandville's illustrations are a number that satirize the utopian dreams and cosmic fantasies of Fourier. For example, Figure 5.1 represents a solar eclipse, or, in Fourier's views, a conjugal embrace of the sun and moon. Figure 5.2 represents the north polar region turned to lemonade, and our worn-out moon replaced by the five fresh moons promised by Fourier. Figure 5.3, which may be less directly aimed at Fourier, portrays a cosmic juggler of worlds. These

*Figure 5.1. Grandville's satirical portrayal of Charles Fourier's idea
that a solar eclipse is a conjugal embrace of the sun and moon.*

fantastic drawings show that Grandville's artistic and satiric imagination
was a match for the wild social and cosmogonic imagination of Fourier.
It is an interesting fact that although authors of present-day pluralist
books have rarely, if ever, referred to Fourier's strange ideas, a number of
them have included some of Grandville's drawings, without, however,
mentioning his satirical intent.[131]

The uncritical acceptance of extraterrestrials by Fourier and others

Figure 5.2. Grandville's satirization of Fourier's claim that if univer-sal harmony were attained, the North Sea would turn to lemonade and our exhausted satellite would be replaced by five fresh moons.

suggests that a rigorously empirical analysis of the issues was needed. One might expect to find this in Auguste Comte (1798–1857), the founder of positivism, who in the second volume of his *Cours de philoso-phie positive* (1830–42) set out to provide a systematic analysis on em-piricist grounds of the methodology of astronomy, the only science that Comte believed had as yet transcended the theological and metaphysical stages of development.[132] Despite its advanced state, Comte prescribes

Figure 5.3. Grandville's juggler of worlds: possibly satirizing Fourier as an advocate of a plurality of worlds.

further purification for astronomy. For example, he warns astronomers against excessive speculation concerning the celestial bodies: "we will never by any means be able to study their chemical composition or their mineralogical structure. . . ." (p. 2) The imprudence of this pronouncement, coming but a few decades before the development of spectroscopy, has often been noted. So has Comte's lack of insight in rejecting sidereal astronomy: "the countless stars scattered in the sky . . . have but little interest to us for astronomy except to serve as markers in our observations. . . ."[133] What has not been noted is that Comte's strictures are not extended to the most highly speculative area of astronomy, the question of extraterrestrials. Yet one finds in the same discussion Comte's assertion that "If, what is highly probable, the planets provided with an atmosphere, as Mercury, Venus, Jupiter, etc. are in fact inhabited, we can regard those inhabitants as being in some fashion our fellow citizens, since from what is a sort of common fatherland, there ought of necessity result a certain community of thought and even of interests, whereas the inhabitants of other solar systems will be complete strangers." (p. 9)

Comte's critique of astronomy was carried even further in his *Système de politique positive,* the first volume of which appeared in 1851, five years after the discovery of the planet Neptune. In that volume Comte criticizes astronomers for their interest in remote regions of the solar system, lamenting the "mad infatuation . . . which some years ago possessed, not only the public, but even the whole group of Western astronomers, on the subject of the alleged discovery [of Neptune], which, even if real, would not be of interest to anyone except the inhabitants of Uranus."[134] One cannot be struck by the inconsistency of Comte in banishing the stars and Neptune from the domain of astronomy, while embracing, without supporting arguments, the existence of extraterrestrials. The explanation of the latter anomaly may lie in his use of pluralism in criticizing theology. Comte excoriates in this context those who see astronomy as allied with religion, "as if the famous verse 'The Heavens declare the glory of God' had preserved its meaning. It is however certain that all true science is in radical and necessary opposition to all theology. . . ." Moreover, he adds that for those familiar with the true philosophy of astronomy, "the heavens declare no other glory than that of Hipparchus, of Kepler, of Newton, and of all those who have cooperated in the establishment of laws."[135] In particular, what has shown the unacceptability of theology is the realization that the earth, rather than being the center of the universe, is only a secondary body circling the sun "of which the inhabitants have entirely as much reason to claim a monopoly of the solar system which is itself almost imperceptible in the universe."[136] If it is correct that Comte's enthusiasm for extraterrestrials arose from a belief that their existence invalidated any theology in which man has a primacy, then it appears legitimate to suggest that Comte's own astronomy may not have transcended even the antitheological stage.

In Italy, pluralism attracted figures as different as the statesman Count Camillo Cavour (1810–61) and the lyric poet Giacomo Leopardi (1798–

1837). Cavour's friend Michelangelo Castelli recalled that while he accompanied the count on long walks in the Alps, "the beauty and grandeur of nature turned his conversation toward religion, toward discussing the plurality of worlds and various resulting hypotheses."[137] Leopardi's involvement in pluralism went far deeper. Although Leopardi's lyric gifts made him one of the most prominent poets of his century, few are aware of his intense interest in astronomy. One of his longest writings is a history of astronomy written when he was but fourteen or fifteen, yet showing scholarly skills that would do credit to a much older man.[138] One feature of this book is a richly referenced section on the history of the "most famous and most unsolvable question," that of a plurality of worlds.[139] Pluralist themes also appear in a number of Leopardi's mature writings, for example, in one of his *Operette morali* entitled "Dialogo della terra e della luna."[140] Early in this conversation, the earth asks the moon whether she is inhabited, as "thousands of philosophers, ancient and modern ones from Orpheus to De la Lande" have asserted. The moon answering affirmatively, the earth asks about the men inhabiting her surface, provoking the moon to respond: "What men?" The earth, portrayed as "not . . . very bright," is surprised to learn that the fortresses and roads reported by terrestrial astronomers do not exist on the moon. Leopardi's pessimistic view of life emerges as the earth learns that vice, moral evil, and unhappiness are far more frequent than goodness and happiness on the moon. "Likewise down here," the earth replies; "in this we're quite the same." Moreover, the moon relays the melancholy message that unhappiness and evil dominate on all the planets and comets and probably on the sun and stars. At this point the conversation concludes, the earth fearing that it may rob earthlings of their sleep, "the only comfort they have." Leopardi's playful, albeit deeply pessimistic, dialogue demonstrates once again the fact that authors found pluralism sufficiently pliable that they could project upon it their personal philosophies, in Leopardi's case an intense aversion to optimism.

In Denmark, pluralism received the support of Hans Christian Oersted (1777–1851), the most distinguished Danish scientist of the first half of the nineteenth century. Attracted throughout his life to both philosophy and physics, Oersted in his last years collected his philosophical essays and addresses into a volume that was soon translated into English and published in 1852 as *The Soul in Nature*.[141] The idea of a plurality of worlds appears prominently in three essays in that collection. In his 1837 "Christianity and Astronomy," a dialogue demonstrating the superiority of the Copernican system, Oersted introduces pluralism to invoke humility: It teaches us "that we are nothing compared *with* God, but that we are something *through* God." (p. 446) In his 1844 address "The Comprehension of Nature by Thought and Imagination," Oersted draws on the doctrine to urge that scientific truths "contain rich material for the imagination." (p. 43) Although promising to "prove" pluralism (p. 52), Oersted centers his argument on the claim that "it belongs . . . to the nature of things, that reason should develope itself into self-consciousness . . . in every member of the system, although in different degrees." (p. 53) While urging a uniformity in mental and spiritual characteristics among extraterrestrials, Oersted also champions a cosmos rich in diversity:

If we regard the whole of existence as a living revelation of Reason in time and space, we can conceive that the most varied degrees of developement may be found distributed through all time, and that some bodies are still spheres of vapour, others have reached fluidity; while others have gained a solid nucleus, and so onwards to the highest point of development; and then backwards again even to those bodies which are on the verge of final destruction. (p. 54)

The third and latest of Oersted's essays, that entitled "All Existence a Dominion of Reason," is designed to demonstrate that extraterrestrials must possess not only the same science as earthlings but also fundamentally similar laws of aesthetics and morality, even while they exhibit "the greatest variety in the forms of existence." (p. 92) Although they may have eyes more or less sensitive than ours and be capable of seeing solar rays of wavelengths invisible to us, their optical laws must be identical (pp. 99–102). So also for their laws of heat, mechanics, electricity and magnetism, and chemistry. This is because "God reveals himself to these beings through the surrounding universe, and rouses their slumbering reason by that Reason which reigns through the sensible world. . . ." (p. 109) Ultimately Oersted's analysis is directed less at extraterrestrials than at urging that "Eternal Reason," assisted by the Divine, is gradually dawning everywhere in the universe. This is a particularly curious claim in an essay that begins with the assurance that metaphysics will be avoided (p. 92). In the concluding section of this essay, Oersted suggests that as science advances, we may eventually be able to attain knowledge of life at other locations in the universe, and that some extraterrestrials may already have knowledge of life on the earth (p. 132).

Because of the eminence and eloquence of Oersted, his pluralist ideas had considerable influence. At Oxford, they were enthusiastically received by Rev. Baden Powell and probably were the source of the pluralist position advocated by Ludwig Colding (1815–88), Oersted's most prominent pupil.[142]

As shown previously, most German astronomers of the first half of the nineteenth century approved of pluralism. While Bode, Gruithuisen, Littrow, Schröter, and other astronomers endorsed the doctrine, a number of German philosophers, especially those of idealist inclination, resisted or attacked it. The source of their opposition was not the science of the period, nor did it result from their religious views; rather, it arose in most cases from the primacy they sought to bestow upon man in the universe.[143]

An excellent example of the aversion among German philosophers to extraterrestrials is G. W. F. Hegel (1770–1831), who in 1817 presented his philosophical system in his *Enzyklopädie der philosophischen Wissenschaften*. This was expanded for an 1827 edition and an 1830 edition, and in 1847 one of Hegel's disciples, C. L. Michelet, brought out an edition with much supplementary material. In this work Hegel disparages the importance of the stars:

> Matter, in filling space, erupts into an infinite plurality of masses. . . . This eruption of light is as little worthy of wonderment as an eruption on the skin or a swarm of flies. . . . The multitude of stars . . . means nothing to Reason; this is externality, the void, the negative infinitude to which Reason knows itself to be superior.[144]

Allied to Hegel's view of the stellar region is his assertion that among the celestial bodies, "the planetary nature is the most perfect. . . . Consequently it is only on the planet that life can appear." (p. 81) Moreover, he adds:

> The sun is the *servant* of the planet; and sun, moon, comets, and stars generally are mere conditions of the earth. . . . If there is a question of pride of place, we must give the place of honour to the earth we live on. Quantitatively regarded, we may indeed let the earth sink beneath our notice, seeing it as "a drop in the ocean of the infinite"; magnitude, however, is a very external determination. We have now come therefore to stand on the earth as our home, and not only our physical home but the home of the Spirit too. (pp. 103–4)

Hegel later repeats this point in an even more straightforward statement: "The earth is the most excellent of all the planets. . . ." (p. 279) Lest any doubt remain concerning his position he adds: "It has been rumoured round the town that I have compared the stars to a rash on an organism . . . or to an ant-heap. . . . In fact, I do rate what is concrete higher than what is abstract, and an animality that develops into no more than a slime, higher than the starry host." (p. 297) In support of such statements, Hegel provided no substantive scientific arguments; rather, his claims stem from his emphasis on Humanity throughout his writings. Although Hegel's rejection of the sidereal realm is strikingly similar to that of the positivist Comte, who also elevated Mankind to primacy, their positions differ in that the French philosopher based his claims on methodological arguments and did not incorporate an antipluralist component into his system.

Hegel's disciple Carl Ludwig Michelet (1801–93) echoed his master's views in an 1841 volume. In response to Fontenelle's argument that were the heavenly bodies not inhabited, they would serve no purpose, Michelet states: "How many islands are . . . scattered in the oceans of the earth which have produced nothing. What do they serve? we may ask. Are the stars not likewise scattered bare rocks of light?"[145] The earth, Michelet claims, "has priority in honor over the sun, and also if not the material it is the spiritual center of the system." (p. 228) He also states: "The earth is . . . the most beautiful feature not only of the planetary system, but of the entire starry heavens as well." Except on it, "no trace of a spiritual nature is to be found." (p. 230) Although his reasons for holding this position were largely philosophical, he urges that "our religious traditions" also attest to the primacy of the earth (p. 231).

In 1830, one of the most brilliant of Hegel's former students anonymously attacked the Christian doctrine of personal immortality. This controversial volume launched the writing career and ended the teaching career of Ludwig Feuerbach (1804–72). A major portion of his book consists of a philosophical critique of the pluralist position, against which Feuerbach made such strongly worded statements as "it is absolutely certain that, in all of creation, there exists but one animated and ensouled point, and . . . this point is the earth which is the soul and purpose of the great cosmos."[146] Feuerbach presents a number of antipluralist arguments; for example, in opposition to the claim that the planets must be inhabited, for otherwise they would be wasted, he asserts: "superfluous, useless, purposeless existence strikes me everywhere, even here on earth. . . ." (p. 60) Another argument begins from the assertion that extraterrestrials must be identical with man or of lower or higher nature (p. 66). The first possibility is ruled out because such beings would be "totally superfluous," whereas the second is rejected for the reason that lower forms, if identical with the lower forms of terrestrial life, would likewise be superfluous: "A repetition of lowliness would be totally purposeless and irrational." (p. 67) The possibility of higher beings is rejected because their existence would make terrestrial life meaningless; in this context, Feuerbach states that

. . . humanity itself is the ultimate of all individual beings, the highest of all individuals. Thus, the highest life is life in religion, science, art, in the world-historical totality of humanity. . . . Reason, will, freedom, science, art, and religion are the only real guardian angels of humanity, are the only actually higher and more perfect beings. Infinite, everlasting life exists in these alone, but not on Saturn or Uranus or anywhere else. (p. 71)

Allied to this argument is his conviction that the chain of being is already complete on this earth (p. 72). This view, which would have astounded earlier proponents of the chain of being doctrine, seems acceptable to Feuerbach because of the lofty position he assigns not to men as individuals but to mankind as a collectivity. Believing that only the latter can attain immortality, he shows himself intent on refuting the idea of a personal immortality on some other planet (pp. 57, 82). Whatever one may think of Feuerbach's overall position, it contains one fully acceptable feature: his recognition that the conceptions of extraterrestrials common among his contemporaries were primarily projections of man's view of himself. This claim Feuerbach repeated in his most famous book, *Das Wesen des Christentums* (1841): "In fact, we people the other planets, not that we may place there different beings from ourselves, but *more* beings of our own or of a similar nature."[147]

German philosophers of the Hegelian school were not the only opponents of pluralism. Friedrich Schelling (1775–1854) opposed it,[148] as did his disciple Henrich Steffens (1773–1845), whose criticisms are contained in an 1839 book in which Steffens, who had extensive training in the sciences, states:

. . . we may here express the view that present day astronomy is fast approaching the time in which our planetary system will be recognized as the most organized point in the universe, and that then also the time will not be far away in which similarly our earth will be recognized not as the apparent, but as the internal, spiritually considered central point of the planetary system, as man is in the total organism. . . . The hallowed place, on which the Lord appeared, will be recognized as the absolute middle point of the universe.[149]

For revealing to man the true nonphysical structure of the universe, Steffens praises Schelling as "the spiritual Kepler of our day." (p. 206)

One of Hegel's philosophical rivals was Arthur Schopenhauer (1788–1860), who like Leopardi was attracted to pessimism and on occasion presented it in an astronomical context. In his *Parerga and Paralipomena,* he laments:

If we picture to ourselves . . . the sum total of misery, pain, and suffering of every kind on which the sun shines in its course, we shall admit that it would have been much better if it had been just as impossible for the sun to produce the phenomenon of life on earth as on the moon, and the surface of the earth, like that of the moon, had still been in a crystalline state.[150]

Elsewhere in that book he accepts the existence of extraterrestrials, but this scarcely mitigates his pessimism, for he asserts that within the scale of beings present in the universe, "Man occupies the highest point . . . ; and his existence also has a beginning, and in its course there are many grievous sorrows and few joys sparingly meted out; and then . . . it has an end after which it seems as though it had never been."[151] This passage brings to mind the remark of James Branch Cabell: "The optimist proclaims that we live in the best of all possible worlds; and the pessimist fears this is true."[152]

Among German theologians who in this period discussed the question of other worlds, probably none treated the topic more fully than Johann Heinrich Kurtz (1809–90), a Lutheran theologian who taught at the German-dominated University of Dorpat in Russia. In 1842, Kurtz published *Die Astronomie und die Bibel,* which he enlarged in later editions.[153] Although the first and longer portion of his book is devoted to the relations of the Old Testament to astronomy, his deepest concern is that treated in the second part: the reconcilability of Christianity with the post-Copernican cosmos. Well read in the astronomical researches of the Herschels, Bessel, Mädler, and Struve on dark stars, double stars, globular clusters, nebulae, and novae, this scholarly author argues that the sidereal realm as seen by such astronomers refutes "the view so fondly advanced by *Fontenelle,* that all the fixed stars were suns like our sun, with solid bodies similar to it, and like it encircled by planets, moons, and comets. . . ." (p. 382) Admitting that other stars may be orbited by planets and thus form systems comparable to our own, Kurtz urges that "not a *single fact* can be brought forward *in proof of it.*" (p. 384) Moreover, he claims that our solar system "is comparatively quite near the central spot" of the universe and possesses an "independent and peculiar character; in fact, it is "perhaps the *only one* of the kind in the universe. . . ." (p. 414) He later adds: "From the aids of science it has become probable, that the like of our planetary system is nowhere to be found in all the known universe. . . ." (p. 482) Kurtz does not, however, present the stars as purposeless; they function as abodes of angels (pp. 439–56), who, Kurtz had earlier claimed, are physical beings (pp. 191–207). This permits him to co-opt the rhetoric of some of the pluralists, for example, Gotthilf Henrich von Schubert (1780–1860), to whose writings Kurtz repeatedly refers.[154] It nonetheless leaves him with the problem of the planets in our solar system. He argues against these being the abodes of angels or of the spirits of the terrestrial dead, preferring the view that they were designed for man's use – had he not sinned (pp. 456–61).

Kurtz's conceptions, which he put forth with commendable caution, are clearly aimed at alleviating the astronomical difficulties the deists had urged against "the fundamental and leading doctrine of all Christendom . . . the doctrine of the *Incarnation of God in Christ.* . . ." (p. 471) Turning directly to this topic, he examines the views of Chalmers, finding them "incomplete and unsatisfactory." (p. 478) He also rejects the position, attributed to Christian Hermann Weisse (1801–66), of "an incarnation of Deity upon *all* worlds, corresponding to the incarnation upon the earth . . ." (p. 491), partly because this idea seems to imply that, irrespective of sin, the incarnation was necessary to bridge the distance between the human and divine (p. 492). This discussion leads him to consider the paradox of man's "felix culpa," as Augustine had called it. Kurtz maintains that it was indeed man's sin that brought God to this planet and that destined man for a position at the right hand of God, above even the angels (pp. 498–505). Convinced that only men and angels are mentioned in Scripture, and contending that fallen angels cannot be redeemed, Kurtz summarizes his position by stating "that an incarnation of God can have occurred upon the earth only . . . ; and that the inhabitants of other worlds either do not *require* a redemption . . . since they *have not* been the subjects of a fall, or that they are *incapable* of redemption if they be fallen beings." (p. 507) The book concludes with a scripturally based argument that after the second coming, man will be elevated and the universe renovated, with man and Christ living together on the earth (p. 525).

Overall, Kurtz attempted to formulate a structure for the universe in harmony with Christianity as he conceived it. Although he showed an interest in, knowledge of, and respect for the conclusions of astronomers, his ideas go far beyond their writings. This points to the conclusion that pluralists had no monopoly on speculation. That Kurtz succeeded in supplying an attractive cosmos is reflected in the fact that his book went through a number of editions and was translated into English and Russian.

No doubt many German poets of the period pondered pluralism, but few, if any, did so with the intensity of Heinrich Heine (1797–1856), whose 1827 poem "Fragen" asks

> . . . what is the meaning of man?
> Whence does he come? Whither does he go?
> Who dwells on yonder golden stars?[155]

The depth of feeling behind Heine's final question is suggested by his account in his *Confessions* of a conversation he had at age twenty-two with Hegel. As they stood gazing at the nocturnal heavens, Hegel shocked the young poet by responding to his rhapsodic ruminations on the sidereal realm by the remark: "The stars! Hm! Hm! the stars are only a brilliant eruption on the firmament."[156] More revealing still of the penetration of pluralism into the poet's religious life and even his dreams is the following fascinating fragment in which the poet reflects on his childhood:

I grew entirely confused by all the information learned from astronomy, which subject even the smallest child was not spared in that period of enlightenment. I could not get over the wonder of it, that all these thousands of millions of stars were great and beautiful globes, like our own, and that one simple God ruled over all these gleaming myriads of worlds. Once in a dream, I remember, I saw

God, in the farthest distance of the high heavens. . . . He was scattering handfuls of seeds, which as they fell from heaven opened out . . . and grew to tremendous size, until they finally became bright, flourishing, inhabited worlds. . . . I have never been able to forget this face; I often saw this cheerful old man in my dreams again, scattering the seeds of worlds out of His tiny window. . . . I could only see the falling seeds, always expanding to vast shining globes: but the great hens, which were possibly lying in wait somewhere with wide-open beaks, to be fed with these worldspheres, those I could never see.[157]

This much, at least, may be made of this enigmatic account: Pluralism had not only spread widely in the West, it had penetrated deeply into the consciousness of sensitive and imaginative souls.

4. Conclusion: the half-century surveyed

Remarkable in many ways were the developments regarding the idea of a plurality of worlds in the first half of the nineteenth century. Remarkable above all is the extent to which this idea was discussed. From Capetown to Copenhagen, from Dorpat to Dundee, from Saint Petersburg to Salt Lake City, terrestrials talked of extraterrestrials. Their conclusions appeared in books and pamphlets, in penny newspapers and ponderous journals, in sermons and scriptural commentaries, in poems and plays, and even in a hymn and on a tombstone. Oxford dons and observatory directors, sea captains and heads of state, radical reformers and ultramontane conservatives, scientists and sages, the orthodox as well as the heterodox – all had their say.

Remarkable as well is the fact that although nearly all accepted, indeed many proclaimed, pluralism as a revelation of astronomy rich in meaning, the significances seen in it were nearly as numerous as the authors who assayed it. While Chalmers incorporated it into the evangelical tradition and Dwight drew on it to inspire students to the ministry, Shelley used it to attack Christian traditions, and Emerson found in it reason to renounce his pastorate. While Gruithuisen, under its spell, saw fortresses on the moon and John Herschel giant organisms on the sun, Mrs. White enjoyed visions of Jupiterians, and Heine dreamt of the Divine scattering world-seeds before cosmic hens. While Locke and Grandville saw in some pluralist writings the makings of satire, such poets as Wordsworth and Tennyson drew images and ideas from it to enrich their lines and possibly their lives. While Milne populated comets and Kurtz associated angels with the stars, while Harris placed life on Titania, Smith on Kolob, and Taylor in the interiors of planets, Herschel, with help from Hansen, sent the selenites to the far side of the moon, and Dick took a cosmic census, extending even to the edges of Saturn's rings. Within the empiricist camp, Comte dismissed much of modern astronomy but saw no reason to doubt pluralism, whereas Mill exposed methodological problems in the pluralists' reliance on analogical argumentation. Idealists were no closer to agreement; whereas Feuerbach and Hegel denied extraterrestrials, Oersted and Owen accepted them. While such pessimists as Leopardi and Schopenhauer lamented that life — filled with suffering and beset with evils — might exist throughout the universe, Fourier optimistically promised that his socialist system would bring universal harmony.

Not least remarkable is the contrast between the confidence with which many accepted pluralism and the paucity of evidence available to support it. Some did express doubts, but Coleridge, the Hegelians, Kurtz, Maxwell, and Walsh raised only weak or scarcely relevant objections to it. Bessel and Plisson made more promising forays, but few followed. By midcentury, although selenites and solarians were less than secure, planetarians — suitably adjusted to the heat of Mercury or the chill of Neptune by the God of physicotheology — seemed to have little cause for fear. However, their days were numbered, even if their demise (at least in our solar system) remained decades distant. As the second half of the century began, a plot against the planetarians was in preparation, launched from a very unlikely location by a most improbable person. In precisely the same rooms at Trinity College, Cambridge, where the pluralists Wilkens and Bentley had lived, a later master of Trinity who had twice defended the fortress of pluralism began to believe that its battlements were built of balsa and its treasures tarnished. With altered allegiance, this "lionlike man," as Tennyson had called his tutor,[158] donned new armor and, with visor down, rode forth in 1853 against the extraterrestrial establishment. At first dubbed a "sidereal Quixote," he was soon revealed as William Whewell.

6

William Whewell: pluralism questioned

1. Whewell's pluralist period: "No one can resist the temptation . . ."

One of the most intense phases of the extraterrestrial life debate began in late 1853 when William Whewell (Figure 6.1) anonymously published *Of the Plurality of Worlds: An Essay.* Whewell shocked his contemporaries by claiming that many of the pluralists' arguments were scientifically defective and religiously dangerous. The controversy he created is treated in the next chapter. the present chapter being devoted to recounting how he came to write his remarkable book and to analyzing its arguments.

The perplexing question of why Whewell, for many years a proponent of pluralism, attacked it in 1853 has been debated since the appearance of his book. The problem is intriguing because Whewell never explicitly revealed what led to this transformation. Moreover, despite the growing literature on Whewell and two lengthy biographies of him,[1] no one has yet provided a widely accepted account of the major changes of mind Whewell experienced during his life. Among the theories of Whewell's conversion to the antipluralist position, one of the oldest is that put forward by Isaac Todhunter, Whewell's most thorough biographer. Yet few, if any, scholars favor his claim that Whewell's position did not essentially change.[2] Equally open to criticism is Otto Zöckler's suggestion that German philosophical speculation influenced Whewell's change of mind.[3] On the one hand, Kant can scarcely be considered a source; although Whewell held his writings in high esteem, Kant was propluralist. On the other hand, Hegel's antipluralist views were never mentioned by Whewell, and they probably would have had little appeal to Whewell, who in 1849 complained that Hegel's thought "on every subject . . . is equally fanciful and shallow. . . ."[4] Moreover, it is clear that no other pre-1853 antipluralist significantly influenced Whewell, his book contains not a single reference to Kurtz, Maxwell, or Plisson, and only passing references to Bessel.

Figure 6.1. Portrait of William Whewell made in 1836 by G. F. Joseph. (Courtesy of the Master and Fellows of Trinity College, Cambridge)

Two recent theories deserve more serious consideration. Frederic B. Burnham has linked Whewell's change of mind to the emphasis in his 1853 *Essay* on the nomological, as opposed to the teleological, form of the design argument.[5] It is true that in his 1853 book Whewell stressed that God's design in nature should be understood not primarily in terms of the purposefulness of specific adaptations (the teleological approach) but rather in regard to overall laws and patterns (the nomological approach). It is also correct that in his book Whewell used the nomological position to argue for the possibility of uninhabited celestial bodies, suggesting that they could be seen as side effects of processes that proved fruitful in at least one case (earth). Burnham's claim suffers from at least one problem; as will shortly be shown, Whewell in 1833 accepted both pluralism and a nomological form of the design argument. John Hedley

Brooke, in a detailed study of the Whewell debate, advanced the thesis that Whewell's change of mind resulted primarily from his determination to dispose of the position developed in Chambers's *Vestiges of Creation*. In particular, Brooke asked: "Is it too far-fetched to suppose that Whewell wrote one anonymous book in reply to another, or at least partly so, sacking life on other planets to make room for the final causes that went with special creation on this one?"[6] The interpretation presently to be developed is that while the theories of both Burnham and Brooke have merit, a careful examination of Whewell's thought in the period before 1853 shows that his doubts about pluralism arose from his coming to believe that serious difficulties stood in the way of reconciling it with Christianity.

William Whewell (1794–1866) was born in Lancaster, England, the son of a master carpenter. He studied at Cambridge University, winning a number of prizes, and in 1817 was elected a fellow of Trinity College, Cambridge. By that time he had studied Kant's *Critique of Pure Reason*, which influenced Whewell in important ways, for example, in leading him away from the empiricism dominant in Britain at that time. That Whewell's own philosophical position was a form of idealism is clear, although controversy continues as to whether he had stronger affinities with Kant or with Plato.[7]

By about 1820 or earlier, Whewell had become acquainted with the question of a plurality of worlds. Todhunter recorded that in an "old note-book" that had belonged to Whewell he found evidence that Whewell had read "Fontenelle's celebrated work at an early period, but adds no opinion respecting it."[8] Around 1822 Whewell wrote his sister: "I am glad that you have been reading more of Chalmers's sermons; I have read only one or two of the astronomical ones, but I think they are all that you describe them."[9] In 1827, the year after his ordination in the Church of England, Whewell delivered in University Church in Cambridge a series of four sermons, the third of which contains passages that show Whewell's enthusiasm for the pluralist message of the Scottish divine. Chalmersian echoes are evident, for example, in Whewell's statement that

. . . the earth . . . is one among a multitude of worlds . . . with resemblances and subordinations among them suggesting . . . that [they may be inhabited by] crowds of sentient . . . beings – these are not the reveries of idle dreamers or busy contrivers; – but, for the main part, truths collected by wise and patient men on evidence indisputable, from unwearied observation and thought; and for the rest, founded upon analogies which most will allow to reach at least to some degree of probability.[10]

Lest this awesome spectacle promote "despondency," Whewell presents what was in effect Chalmers's microscope argument: "Does not even science herself compel us to expand our notions of God's vivifying and cherishing care, when she . . . shews us the myriads of animals that live in the blade of grass or in the drop of water?"[11] No direct mention is made of Chalmers, but Whewell's indebtedness to him is suggested by this remark: "Hastily and imperfectly we pass over this portion of our subject – a matter already touched by other hands than ours."[12] Whewell was not a person to elaborate another preacher's theme without adding a thought of his own, in this case one of particular significance, because twenty-five years later Whewell was to reverse this idea, using it as his "atom of space–atom of time" or geological argument against the pluralists. In 1827, however, Whewell's goal was to overcome his auditors' aversion to geological theories about extinct animal forms purportedly present on the earth in its remote past. His argument is that just as we allow astronomers the existence of almost unimaginable animal forms on spatially remote planets, so also we should permit geologists to suppose that strange beasts roamed the earth during temporally far distant epochs.[13] This sermon shows that Whewell found pluralism, especially as presented by Chalmers, an attractive topic with which to enrich sermons. Whewell had other contacts with Chalmers, whom he attempted to visit in 1834 and did visit in 1843.[14]

In 1828, Whewell's scientific gifts won him the professorship of mineralogy at Cambridge. Shortly thereafter, he was invited to write one of the Bridgewater Treatises. This appeared in 1833 as *Astronomy and General Physics Considered with Reference to Natural Theology*, which Todhunter described as "perhaps . . . the most popular" of the eight treatises.[15] In writing it, Whewell faced four problems, all of which were relevant to his later antipluralist book. First, Hume's penetrating critique of the design argument in his *Dialogues Concerning Natural Religion* (1779) challenged the validity of the argument itself. Whewell, the most philosophically sensitive of the eight authors, felt this problem with special severity. A second problem was that even if the design argument and natural theology were valid, the tension between natural and revealed religion remained. Third, in writing a tract in natural theology centered on astronomy, Whewell encountered the problem that astronomy, as Paley had suggested, "is not the best medium through which to prove the agency of an intelligent Creator [because we] are destitute of the means of examining the constitution of the heavenly bodies . . . ,"[16] and also the problem that astronomy as a branch of physical science seemed barred, as Bacon had urged, from the use of final causality, that is, explanations in terms of the purpose of the entity to be explained. The fourth problem was that such authors as Buffon, Kant, and Laplace had introduced evolutionary modes of explanation into astronomy, in particular the nebular hypothesis, which many saw as conflicting with the design argument.

In his *Astronomy and General Physics,* Whewell grapples with each of these problems. He deals with the first largely by limiting the validity of natural theological claims to believers. For example, he states in his dedication that his book is written to lead "friends of religion" to look favorably on "the progress of the physical sciences, by showing how admirably every advance in our knowledge of the universe harmonizes with the belief of a most wise and good God."[17] In a section almost certainly directed at Hume, Whewell admits that we cannot in general infer causes from effects, and especially we cannot infer a Divine artisan from an analogy of nature with the works of human artisans. Whewell deals with this problem by stating: "It is not . . . among remote conclusions, but among original principles, that we must place the truth, that such arrangements, manifestations, and proceedings as we behold about us imply a Being endowed with consciousness, design, and will, from whom they proceed." (p. 258) Thus, Whewell's position is that the design argument cannot be expected to convince an atheist, but may aid the believer.[18]

Whewell's sensitivity to the tensions between revealed and natural theology becomes clear at many points, such as when he allows that the latter, unlike the former, "is utterly insufficient for . . . the purpose of reforming men's lives, or purifying and elevating their characters, or preparing them for a more exalted state of being." (p. 9) Moreover, he later avers that only revealed religion shows us God as "lawgiver and judge of our actions; the proper object of our prayer and adoration; the source from which we may hope for moral strength here . . . and the elevation of our nature in another state of existence." (p. 274) In this treatise on natural theology, he ironically urges that science above all leads us to see God "as incomprehensible; his attributes as unfathomable!" (p. 278)

Concerning the third problem, Whewell admits that design arguments "drawn from cosmical considerations, labour under some disadvantages when compared with the arguments founded on those provisions and adaptations . . . of organized creatures." (p. 119) Following Bacon, Whewell furthermore states: "Final Causes are to be excluded *from physical inquiry.*" (p. 264) Despite these difficulties, Whewell urges that the design argument can be retained for physical science by focusing not on specific adaptations but on general laws. In answer to Paley's objection, he suggests that "in considering the universe, according to the view we have taken, as a collection of *laws,* astronomy . . . possesses some advantages." (p. 120) Compared with biological science, astronomy has greater precision of statement and superior clarity concerning causes. Moreover, the contemplation of these laws leads men to the idea of a lawgiver, for "men conceive these laws to be evidence of thought and mind. . . ." (p. 224) Whewell cites John Herschel and even Lucretius to show how readily men move from the perception of law, order, and regularity in the universe to a belief in "a calm and untroubled intellect presiding over the course of events." (p. 225) In another passage, Whewell states: "with regard to the physical world, we can at least go so far as this; — we can perceive that events are brought about, not by insulated interpositions of divine power exerted in each particular case, but by the establishment of general laws." (p. 267) This statement is especially significant because Whewell was later to encounter it, probably to his distress, as the lead quotation in the book that did the most damage to the cause so fully financed by the Earl of Bridgewater. This book was Charles Darwin's *The Origin of Species*.

Whewell's emphasis on laws, as opposed to specific adaptations, is also central in his treatment of the problem of final causality as applied to physics and astronomy. Laplace, Whewell notes, had stated: "Let us run over the history of the progress of the human mind and its errors: we shall perpetually see final causes pushed away to the bounds of its knowledge."[19] Whewell's response is: "We have shown, we trust, that the notion of design and end is transferred by the researches of science, not from the domain of our knowledge to that of our ignorance, but merely from the region of facts to that of laws." (pp. 261–2) Whewell's position is that in investigating physical and astronomical phenomena, only physical, not final, causes are to be sought. When physical causes have been found and expressed in laws, teleological considerations may enter; specifically, we may be able to assign the final causes of the laws themselves. Thus, already in 1833 Whewell was invoking a nomological form of the design argument, using a teleological approach only after the laws had been brought to light.

These considerations shed light on Whewell's treatment of his fourth problem, the nebular hypothesis. Some readers must have been surprised at his openness to this hypothesis, which many viewed as hostile to religion. Whewell does stress the conjectural character of the nebular hypothesis (or hypotheses, for there were a number of such), and he notes that God must have been the First Cause not only of the nebulous matter but also of the *laws* that produced the solar system (pp. 143–50). Thomas Chalmers, in his Bridgewater Treatise, also commented on the nebular hypothesis, but unfavorably.[20]

Whewell did not view all aspects of the science-religion area as beset with problems. For example, he expresses no reservations in his *Astronomy and General Physics* about extraterrestrial life. Discussions of this topic are not numerous in his book, but those that occur are sufficiently positive that some reviewers of his anonymous 1853 *Essay* cited them against him. In a chapter entitled "Of the Vastness of the Universe" in his 1833 treatise, Whewell notes that six planets besides the earth orbit the sun, which planets are, "so far as we can judge, perfectly analogous in their nature." Because of this, "No one can resist the temptation to conjecture, that these globes, some of them much larger than our own, . . . are, like ours, occupied with organization, life, intelligence." (p. 206) Turning to the stars, he suggests that

> . . . they may . . . have planets revolving round them; and these may, like our planet, be the seats of vegetable and animal and rational life: — we may thus have in the universe worlds, no one knows how many, no one can guess how varied: — but however many, however varied, they are still but so many provinces in the same empire, subject to common rules, governed by a common power. (p. 207)

He adds that the "beneficence and love of the Creator" are shown by the observation that, so far as we can tell, the celestial arrangments serve "to support the existence, to develope the faculties, to promote the well-being of these countless species of creatures. . . ." (p. 208) At no point in his treatment does Whewell present evidence for extraterrestrial life, nor does he discuss in any detail the problems some felt it posed for Christianity. He does repeat Chalmers's microscope argument, but this was aimed only at preserving the idea of a personal God, not necessarily one involved in an incarnation and atonement (pp. 212–22).

In summary, Whewell's book is noteworthy in the present context for its acceptance of extraterrestrial life, its development of natural theology in relation to physics and astronomy, its cautiousness in specifying limits for natural theology, and its openness to the nebular hypothesis. It was favorably received by many persons, including Davies Gilbert, John Herschel, Robert Malthus, and H. J. Rose, and it went through at least seven editions, the last appearing in 1864 with a new preface.[21] Among the few who criticized it was David Brewster, whose attack in the *Edinburgh Review* shows that already in the 1830s Brewster had developed an antipathy to Whewell's ideas.[22] This was in the background of their clash in the 1850s over extraterrestrial life. Although he did not publicly criticize Whewell, Charles Darwin is now known to have written in one of his 1838 notebooks: "Mayo (Philosophy of Living) quotes Whewell as profound because he says length of day adapted to duration of sleep in man!! whole universe so adapted!!! & not man to Planets. — instance of arrogance!!"[23] This comment serves as an effective reminder of the precariousness of the program taken up in the Bridgewater Treatises.

The period of the 1830s was the most productive portion of Whewell's career. In 1837, he not only won a gold medal from the Royal Society for his memoirs on tidal theory but also published his three-volume *History of the Inductive Sciences*. This was followed in 1840 by the two large volumes of his *Philosophy of the Inductive Sciences*. These books show the idealist inclinations of Whewell, who in the third edition of his *Philosophy* stated that "The chapters on the Ideas of Space and Time in [it are] almost literal translations of chapters in [Kant's] *Kritik der reinen Verknuft*."[24] Whewell's idealism set him off from most of his contemporaries; it placed him, for example, in opposition to the empiricism of John Herschel, to whom Whewell dedicated his *History*. Even at Cambridge, most preferred the empiricist orientation of John Stuart Mill's *System of Logic* (1843) to Whewell's approach. Nonetheless, the breadth of Whewell's learning made him one of the wonders of the Victorian period, John Herschel remarking: "A more wonderful variety and amount of knowledge in almost every department of human inquiry was perhaps never in the same interval of time accumulated by any man. . . ."[25] As readily as his contemporaries recognized the range of Whewell's learning, few adequately appreciated its depth and originality. Such recognition has, however, come in recent decades, most present-day historians and philosophers of science find greater richness in Whewell's methodological writings than in the empiricist publications of Herschel and Mill. In 1838, Whewell became professor of moral philosophy at Cambridge, planning to devote his subsequent writings less to the sciences than to ethical issues. In 1841, he married, receiving on that day word of the resignation of the master of Trinity College. Whewell was rapidly selected as his successor, placing him in the leadership position in the most scientifically distinguished of the Cambridge colleges. From that position, Whewell became, as Harvey Carlisle put it, "the most prominent feature of the University until the day of his death."[26]

Although Whewell's *History of the Inductive Sciences* and *Philosophy of the Inductive Sciences* contain no discussion of pluralism, they do include endorsements of final causality, at least as applied to organic nature. In his *History,* for example, Whewell urges that final causality has served as "the instrument of some of the most important discoveries which have been made."[27] Whewell contrasts the teleological approach with the doctrine of unity of plan or theory of analogues, as put forth by Etienne Geoffroy Saint-Hilaire. According to the latter position, the structures and functions of animals are to be interpreted by means of analogies among different animal groups; we are to see in animals variations on an original type. Geoffroy Saint-Hilaire, in advocating this position, rejected final causes, as did many other scientists of the day, albeit for other reasons. In his *History,* Whewell argues that the physiologist cannot help but refer to final causes, supporting his position by noting that Kant had stated that this must be so, because the assumption of an end is involved in the very definition of an organized being.[28] Developing this position more fully in *Philosophy of the Inductive Sciences,* Whewell states: "This idea of Final Cause is not *deduced* from the phenomena by reasoning, but is *assumed* as the only condition under which we can reason on such subjects at all."[29] As an example of a case comprehensible in terms of final causes but inexplicable on a "unity of plan" approach, Whewell cites Richard Owen's study of the elaborate and apparently unique structures by which kangaroos provide milk to their young before their mouths are large enough to grasp their mother's nipples.[30] Whewell, however, limits application of final causality to organic matter:

> That there is purpose in many other parts of the creation, we find abundant reason to believe, from the arrangements and laws which prevail around us. But this persuasion is not to be allowed to direct our reasonings with regard to inorganic matter, of which conception the relation of means and end forms no essential part.[31]

Whewell's statements on final causality in these two works seem in no way to contradict, although they significantly supplement, the position he had espoused in his Bridgewater Treatise. His views of final causality were related to his rejection of the transmutation of species. Whewell, who served as president of the Geological Society of London in 1838, did not deny extinction of species, nor that breeding and external factors could modify species to the extent of producing new varieties,[32] but along with the majority of scientists of his day, he did reject transmutation of species, citing as evidence that "If we allow such a *transmutation* of species, we abandon that belief in the adaptation of the structure of every creature to its destined mode of being, which . . . has constantly and irresistibly impressed itself on the minds of the best naturalists, as the true view of the order of the world."[33] Whewell saw only two ways of explaining the succession of animal forms revealed by geology:

> . . . either we must suppose that the organized species of one geological epoch were transmuted into those of another by some long-continued agency of natural causes; or else, we must believe in many successive acts of creation and extinction of species, out of the common course of nature; acts which, therefore, we may properly call miraculous.[34]

Whewell rejects the first of these alternatives, partly because the evolutionary mechanisms known in his day were extremely problematic. This leaves him with the latter alternative, but he is cautious to point out that theological accounts of origins must not be mixed with geological accounts. Biology and geology must be allowed to proceed as far as they can in investigations of the origins of things. But he was convinced that although they could go far in this direction, ultimate origins were beyond their ken. Whewell did not significantly change his views on the transmutation of species, even after Darwin's *Origin of Species,* but his views on final causality did undergo modification in the 1840s. This will be discussed shortly.

A book of particular significance for Whewell's thought during the 1840s was Robert Chambers's *Vestiges of the Natural History of Creation* (1844). Upset at its errors and even more distressed because of the evolutionary doctrines contained in *Vestiges,* Whewell nonetheless realized that it was sufficiently timely and attractive that it would win a wide readership. Adam Sedgwick, the Cambridge geology professor, was so upset by it that he wrote Macvey Napier, the editor of the *Edinburgh Review:* "If the book be true, the labours of sober induction are in vain; religion is a lie; human law is a mass of folly, and a base injustice; morality is moonshine; our labours for the black people of Africa were works of madmen; and man and woman are only better beasts!"[35] Sedgwick did so "loathe and detest the Vestiges" that he sent off a very critical review to the *Edinburgh Review,* and still not satisfied, he subsequently published an even longer refutation of *Vestiges.*[36]

A number of Whewell's friends urged him to respond to *Vestiges.* Even the *Edinburgh Review,* which had published Brewster's critical reviews of his books, asked Whewell to review *Vestiges.*[37] This request he declined, choosing to respond in a rather unusual form. His 1845 *Indications of the Creator*[38] consists of a preface followed by selections from his *Astronomy and General Physics, History of the Inductive Sciences,* and *Philosophy of the Inductive Sciences.* He selected this approach for two reasons. First, he felt that for an attack on *Vestiges* to succeed, it would be necessary not only to draw on but in fact to create in readers a sense of scientific method.[39] Second, by using materials published before *Vestiges,* and by not mentioning that book in his first edition, he believed that he could avoid the appearance of a direct attack that might give *Vestiges* all the more notoriety. In responding to that book, Whewell urges that science cannot reveal origins and that the author's attempts to do so led him into excessive speculations. Among the many examples cited by Whewell is the nebular hypothesis, which Chambers in his 1845 *Explanations* had put forth as among "ascertained truths." Whewell asserts that this hypothesis remains "a mere conjecture" and in fact has lost ground in astronomical thought (pp. 25–7). An aspect of *Vestiges* that especially upset Whewell was its claim that God acts by general laws, this being a position that Whewell himself had espoused. Whewell deals with this by urging that those who reject *Vestiges* do so not because they necessarily reject this view but rather because "they do not believe that the wrtier in question is able to tell them what those laws are. . . ." (p. 12) Their positions were farther apart than even this would suggest, because Chambers's belief was that God acts on matter only according to those laws, whereas Whewell's conviction was that God at times directly intervenes. This is evident in passages in his *Indications* and is even more clear in a letter to Rev. F. Myers in which Whewell contrasts his book with *Vestiges:*

I have attempted to show that, dim as the light is which science throws upon creation, it gives us reason to believe that the placing of man upon the earth (including his creation) was a supernatural event, an exception to the laws of nature. The Vestiges has, for one of its main doctrines, that this even was a natural event, the result of a law by which man grew out of monkey.[40]

It is unnecessary to summarize Whewell's remaining arguments against *Vestiges,* for most of these were discussed earlier as part of the analysis of the three books from which Whewell took his selections for his *Indications.*

During the 1840s, Whewell's views concerning the argument from design underwent an important expansion. Whereas before the 1840s he favored both a teleological approach such as that employed by Paley and a nomological approach as presented in his 1833 Bridgewater Treatise, Whewell also came to recognize the merits of what one scholar has labeled the "idealist design argument."[41] The proximate source of Whewell's openness to this point of view was his friend Richard Owen, whose theory of archetypes derived support from his detailed biological and paleontological research. Owen urged that the vertebrates, for example, could best be understood as modifications of an ideal, an archetype that could be seen in each vertebrate species. These archetypes had never existed on the earth; they were rather plans or patterns in the "Divine Mind." In this context, Owen focused attention on homologies. It was well known that various animals possess skeletal similarities that could not be explained in functional terms. Why should the bones of a bird's wing and the forelimbs of a quadruped be similar when the purposes of wings and forelimbs are so different? Owen argued that these homologous structures were due to variations of the overall vertebral archetype and developed this point in painstaking detail in such books as *Anatomy of Fishes* (1846), *On the Archetype and Homologies of the Vertebrate Skeleton* (1848), and *On the Nature of Limbs* (1849). This was a form of idealism, a sort of Platonism, applied to biology, which included among its advantages the ability to explain various structures such as male teats, which seemed inexplicable from a purely functional approach. Owen's position had fundamental similarities to Geoffroy Saint-Hilaire's unity of plan theory, but with at least one difference crucially important to Whewell: Owen, unlike the French naturalist, retained final causality. As Whewell put it, Owen possessed "the teleological turn of the Inductive Mind."[42] The importance for the pluralist debate of Whewell's acceptance of Owen's archetypal approach was that when he came to write his antipluralist *Essay* of 1853, it enabled Whewell to argue that planets could be uninhabited without violating design, because they could be seen as products of an overall pattern or design that in at least one instance (earth) proved purposeful. Whewell's employment of Owen's archetypal analysis against pluralism was in one sense ironic; as noted previously, Owen was himself a pluralist, having suggested that some forms of archetypes not embodied on earth might exist on other worlds.

In none of Whewell's many writings from the fourteen years following his *Astronomy and General Physics* (1833) did he return to the topic of a plurality of worlds, nor is it touched on in any of his letters from that period that have been published. However, a long paragraph in his 1848 "Second Memoir on the Fundamental Antithesis of Philosophy" contains interesting hints as to his thought at that time. Having designed that paper to illustrate that "the Progress of Science consists in a perpetual reduction of Facts to Ideas,"[43] Whewell cites as an example the developments in mechanics by which empirical information concerning levers was eventually subsumed under the law of the lever. The case of present relevance is the nebular hypothesis, which Whewell describes as still under discussion. For full acceptance, it would have to succeed in "taking into itself, as necessary parts of the whole Idea, many Facts which have already been observed. . . ." (p. 615) Among these, Whewell mentions the forms and changes of nebulae. Moreover, it would have to account for "many facts which, so far as we can at present judge, are utterly at variance with the Idea. . . ." In the latter group of facts, Whewell includes "the existence of vegetable and animal life upon [the planets'] surfaces." (p. 615) Whewell then warns, in what may be seen as an implicit reference to the view set forth in *Vestiges,* that just because an "Idea . . . is a large and striking Idea," we must not assume its correctness, especially if "another Idea, no less large and striking," is available to be compared with the facts. What Whewell had in mind is revealed in his statement that

. . . if we take into our account . . . not only vegetable and animal life, but also human life, this other idea appears likely to take into it a far larger portion of the known Facts, than the Idea of the Nebular Hypothesis. The other Idea . . . is the Idea of Man as the principal Object in the Creation; to whose sustenance and development the other parts of the Universe are subservient as means to an end; . . . we may say that, taking into account the Facts of man's intellectual and moral condition, and his history, as well as the mere Facts of the material world, the difficulties and apparent incongruities are far less when we attempt to idealize the Facts by reference to this Idea, of Man as the End of creation, than according to the other Idea, of the World as a result of Nebular Condensation, without any conceivable End or Purpose. (p. 615)

Todhunter's claim that this passage anticipates the central idea of Whewell's 1853 antipluralist *Essay*[44] has merit, but needs qualification. The passage does show Whewell recognizing a tension between traditional Christian cosmogony and the nebular cosmogony he had accepted in

1833. This passage also suggests that Whewell's reading of *Vestiges,* which began with the nebular hypothesis and ended with a radical evolutionary theory, led him to question the former hypothesis. Where Todhunter's claim falls short is in its failure to note that in 1848 Whewell was classifying extraterrestrial animal life as a fact and moreover as being in conflict with the nebular hypothesis. Whewell does not specify the source of this conflict, which is surprising, because Chambers had adopted both pluralism and the nebular hypothesis. Whewell's thought must have undergone further development by 1853, when in his *Essay* he described even extraterrestrial animal life as problematic and moreover advocated one form of the nebular hypothesis.

With this, we have followed Whewell from his arrival at Cambridge as a carpenter's son to the later 1840s, when he adorned the university as master of Trinity. By then broadly learned in both scientific and humane disciplines, he was also a doctor of divinity deeply concerned about theological issues. Known as a pluralist, he apparently persisted in that persuasion, although he had forged a set of positions and perspectives that by 1853 he would marshal against pluralism with a skill that surprised and a determination that distressed his contemporaries. The transformation of Whewell from pluralist to antipluralist, if the analysis that follows is correct, culminated in 1850.

2. Whewell's dialogue "Astronomy and Religion": a way around a "desolate" and "dark" thought

In the summer of 1850, Whewell set off for Kreuznach in Rhenish Prussia, hoping that its baths would aid his wife, whose health was failing and who in fact died in 1855. While there, he wrote an uncompleted and never published thirty-five-page manuscript entitled "Astronomy and Religion" that is devoted to the question of a plurality of worlds and its implications for religion. Although it seems to be the chief transition piece between his early pluralism and his 1853 book, it has never previously received a published analysis. The manuscript consists of three dialogues between two interlocutors labeled only A and B. Whereas both present ideas that were on Whewell's mind, B's dominance suggests that he should be identified as Whewell's spokesman.

The first dialogue opens with the passage from the Psalms that had formed the text for Chalmers's *Astronomical Discourses:* "When I consider the heavens, the work of thy hands, the moon and the stars, which thou has ordained; what is man that thou art mindful of him, and the son of man, that thou regardest him?"[45] Speaker A in the dialogue comments that this text disturbs him, leading B to ask him to explain why. A responds that modern astronomy, unlike ancient astronomy, has made difficulties for this passage. They then examine various astronomical theories to see which have created tensions with the psalmist's views. The Copernican theory caused a shock, but was ultimately accommodated to Christianity. Laplace's nebular hypothesis of planetary formation is then mentioned, A commenting that the Copernican and Laplacian doctrines disturb him by placing man on one small planet among a number in the solar system, some of which are better provided with moons, and some or all of which may be inhabited. Does this not diminish man's dignity and foster a loss of confidence in God's providential care? A states that he finds this thought "oppressive," "desolate," and "dark," to which B responds that others have experienced the same feeling. But A adds that other astronomical discoveries increase this distress. We find that the earth occupies but a minute portion of space, to which B responds that these vast spaces between stars prevent perturbations. The discovery of Uranus and Neptune makes our position seem still less significant in the universe. Moreover, stellar parallax measurements, if correct, place stars at incredible distances and support the view, based on analogy, that the stars have their own planets. Must not the inhabitants of the planets of distant stars also be concerned whether or not they are being overlooked in God's administration? A asks: "Is it not difficult for you in thought to bring God down from such a heaven to attend to the concerns of the earth and to the destiny and actions and thoughts of you and me? And yet if we have not a God who does this, how blank a desert this world is!" (I, p. 15) The problem is further aggravated by Herschel's discoveries that stars are more numerous and seem more distant than had been thought, that in fact many of the stars we see may, given the finite speed of light, have ceased to shine long ago. The speculation that nebulae are forming planetary systems over immense periods of time suggests that we are located not only in an immensity of space but also in an immensity of time, making human history appear insignificant.

Late in the dialogue, B asks A if he has now disclosed the whole length and breadth of his concern. A responds that he has, but not its depth. The statement of this final difficulty forms the climactic and key section in the dialogue. A states that what has gone before, he finds difficult to reconcile with his belief that

. . . God has interposed in the history of mankind in a special and personal manner; . . . that one, having a special relation to God, came from God to men in the form of a man; that he showed himself by manifest signs to be thus sent; and being [both] divine and human underwent death and rose again [and thereby provided men with] the means of their preservation from the greatest possible evil to their moral and spiritual being. . . . (I, p. 21)

But this raises a severe problem: If this belief be true,

. . . what are we to suppose concerning the other worlds which science discloses to us? Is there a like scheme of salvation provided for all of them? Our view of the saviour of man will not allow us to suppose that there can be more than one saviour. And the saviour coming as a man to men is so essential a part of the scheme . . . that to endeavour to transfer it to other worlds and to imagine there something analogous as existing, is more repugnant to our feeling than to imagine those other worlds not to be provided with any divine scheme of salvation at all. The one and only Saviour came as a man to men and the humanity which he thus assumed is, as we must conceive, essential to his being. It seems to me that it would be an offense against God and his scheme of salvation as it has been revealed to us to imagine a like scheme to be carried into effect for the inhabitants of Jupiter or of one of the planets which revolve round Sirius. . . . (I, p. 22)

It is noted that God did not take on the form of an angel when the angels fell and that it is difficult to see that man has special merits that justify his selection. A further difficulty is that "we cannot conceive how the scheme of salvation which was carried into effect upon the earth can be the means of any advantage to the inhabitants of other worlds. . . ." (I, p. 23) B responds that it would be a "very precarious and fanciful notion to assume such an effect," and agrees that it is difficult "to conceive how God, the father of all the worlds, should provide such a scheme for one only among them all, and that so far as we can find one of the smallest." (I, pp. 23–4) B, no doubt speaking for Whewell, states: "I know that such thoughts may become very painful and disturbing. I have myself dwelt among such trains of thought, and I will gladly give you any reflexions which have occurred to me and which offer to bring any comfort and consolation. . . ." (I, p. 24) With B noting that this will take some time, the dialogue ends. The important conclusion that this summary of Dialogue I suggests is that by 1850 Whewell was doubting the reconcilability of pluralism with Christianity.

The second dialogue begins with a discussion of the relation of God to material creation, with some views of Pascal and Newton being noted. A comments that the notion of God as a lawgiver who acts throughout all space by means of mechanical, chemical, and physiological laws seems as unsatisfactory to him as does pantheism. Rather, what men seek is a personal God, a "Lord," as Newton had put it. "We need a God who watches over the careers of nations and of individuals, who warns and judges, rewards and punishes. A Deity who is merely the complex combination of all the laws of nature . . . does not come up to the measure of such a need." (II, p. 3) B asks whether or not A can accept the notion of a God who works not only by mechanical, chemical, and physiological laws but also by psychological laws that could, for example, lead to vices generating their own punishments. This would seem to imply a notion of providence "as universal as gravitation"; moreover, this notion would not be "disturbed or weakened by the necessity of extending it to other worlds. . . ." (II, p. 3) But, again, A finds this unsatisfactory. He sees on the earth the influence "of a Supreme Being whose Intelligence has a sympathy with our Intelligence, of a Master and Judge whom we are to obey as servants and subjects, not of a Moving Principle by which we are actuated . . . whether we know it or not." (II, p. 4) Such questions as these, A says, "harass and torment me." The idea of a plurality of worlds "disturbs my faith in a Providence" and leads him to be "thrown off the road altogether." (II, p. 5) B sympathizes with these feelings, noting that it is difficult to accept a notion of providence applied only to the earth. The second dialogue breaks off shortly thereafter when B asks A whether he has examined the views of Chalmers.

The brief and apparently uncompleted third dialogue consists of an attempt to combat one of the chief arguments of the pluralists. In particular, Whewell (B becomes dominant in this dialogue) presents his geological argument. In setting the stage for it, Whewell notes that geology reveals that man appeared on the earth only after immense periods of time during which the earth was without intelligent life. But this raises a problem, because it seems to imply colossal wastage and hence a violation of design. To circumvent this problem, Whewell stresses that in nature waste is in fact common. Thousands of seeds fall to the ground, with but a few fructifying; a fish produces millions of embryos, with only a tiny minority reaching adulthood. Consequently, vast expenditures of time and of generative materials are reconcilable with God's creative design. The relevance of this analysis is that Whewell turns it against the pluralist argument that the planets would be wasted and design violated were they without intelligent life. Wastage, even extraordinary apparent wastage, is shown by geology to be fully in harmony with nature's ways and thus with God's design. Moreover, if we see that God has placed man on earth during only a tiny portion of the earth's history, is it not possible that intelligent life may exist in only a proportionately minute portion of space? The manuscript breaks off with Whewell citing evidence against life on the moon and some of the planets (III, p. 5).

Whewell's third and last dialogue, incomplete though it is, possesses special significance because it shows Whewell advocating as a resolution of the tensions he saw between pluralism and Christianity nothing less than the rejection of pluralism. Whewell understood his geological argument not as proving but as permitting this by providing an answer to one of the pluralists' most powerful arguments. It seems a legitimate inference that in writing this manuscript, Whewell realized the dangers involved in any categorical statement that pluralism could not be reconciled with Christianity. Such an assertion would have been theologically presumptuous and, moreover, would have tempted critics to dismiss his arguments as motivated by particular religious concerns. The task he consequently faced was to formulate a case against pluralism that featured scientific and philosophical arguments, even though his antipluralism had its *origin* in religious concerns. This task Whewell took up, armed with his geological argument and his nomological and archetypal forms of the design argument. His delight must have been great and his surprise not small as he began to see numerous ways in which the pluralist position could be called into question.

This theory of Whewell's motivation for challenging pluralism gains additional support from an exchange of letters in October 1853 between Whewell and Sir James Stephen, to whom Whewell sent proof sheets of his book. Stephen wrote Whewell that certain sentences in his book almost seem to imply that if pluralism is accepted, "the orthodox opinions as to the Christian system must be wrong." Stephen warns: "Such an admission ought hardly to have so much as the shadow of a support from you, unless [you can show] that the doctrines of the Incarnation and the Trinity cannot be reconciled at all . . ." with pluralism.[46] Stephen must have been shocked when Whewell responded: "*the difficulties* belonging to that relation [i.e., between Christianity and pluralism] are the starting point of my whole essay and of Chalmers's speculations, which are my text. I will soften the expressions which you notice, but the topic cannot be excluded; for it is in fact *the* topic of my essay."[47]

Within the context of the claim that Whewell's antipluralist position originated in this way, it is illuminating to reconsider the theories put forward by Drs. Burnham and Brooke. The former is certainly correct in stressing the importance for Whewell's antipluralism of his adoption of the nomological form of the design argument. This agrees with the materials already discussed and is further documented in the account of Whewell's *Essay* that follows. To be more specific, Whewell's nomological approach helped him accept, although it by no means required, the antipluralist position. Brooke is also on solid ground in pointing to the powerful effect of Chambers's book on Whewell's thought; that book seems to have set starkly before him the contrast between the deist and Christian conceptions of the cosmos. Moreover, Brooke is no doubt right in emphasizing Whewell's concern that the Christian notion of God's providential guidance was, at least implicitly, called into question by Chambers. Nonetheless, Whewell's previously unanalyzed dialogue suggests that Whewell's greatest concern was that pluralism appeared to undermine the most remarkable of God's providential actions: the incarnation and redemption. That Whewell was slow to come to this conclusion is understandable in light of the large number of individuals who either felt that this was not a major difficulty or accepted Chalmers's resolution of it. On the other hand, the impact of the problem on Walpole, Paine, Shelley, Emerson, and others shows that this difficulty could be seen as nothing less than crucial. And that, I have suggested, is precisely how Whewell came to see it.

3. Whewell in "combat against all the rational inhabitants of other spheres"

According to a reviewer for the London Daily News,

We scarcely expected that in the middle of the nineteenth century, a serious attempt would have been made to restore the exploded idea of man's supremacy over all other creatures in the universe; and still less that such an attempt would have been made by one whose mind was stored with scientific truths. Nevertheless a champion has actually appeared, who boldly dares to combat against all the rational inhabitants of other spheres; and though as yet he wears his vizor down, his dominant bearing, and the peculiar dexterity and power with which he wields his arms, indicate that this knight-errant of nursery notions can be no other than the Master of Trinity College, Cambridge.[48]

Such was the surprise that greeted Of the Plurality of Worlds: An Essay when in late 1853 it rolled from the presses; such was the astonishment as word spread that the challenger was William Whewell. In the major controversy that resulted, the book ran through at least five English and two American editions,[49] generated over seventy published responses, many of them of book length, and was soon followed by Whewell's early 1854 anonymous Dialogue on the Plurality of Worlds, written to answer some of his critics.[50] Whewell wrote his book during the summer of 1853 while at Kreuznach,[51] and in the fall of 1853 he sent the proof sheets to Sir James Stephen, whose comments led Whewell to delete substantial portions of the latter half of the book. Fortunately, a copy of these sheets was preserved.[52] Because this book contains highly original arguments, some of which may still be of relevance, and because it serves effectively as a focus for a discussion of pluralism in the middle of the nineteenth century, it justifies a detailed analysis.

In the preface to his first edition, Whewell states: "It will be a curious, but not a very wonderful event, if it should now be deemed as blameable to doubt the existence of inhabitants of the Planets and Stars as, three centuries ago, it was held heretical to teach that doctrine." (p. iii) Moreover, in a November 1853 letter to J. D. Forbes, Whewell expressed his fear that the book might be seen as containing "heresies."[53] Both predictions proved correct. Whewell identifies one reason for such reactions when later in his preface he points out that although Scripture says nothing of pluralism and although Christianity for most of its history had no need for the doctrine, nonetheless "at the present day . . . many persons have so mingled this assumption with their religious belief, that they regard it as an essential part of Natural Religion." (pp. iii–iv) However, some persons, Whewell suggests, find it in conflict with their religious convictions. Expressing the hopes that his book may reduce this conflict and that scientists may find it of interest, Whewell brings his preface to a conclusion.

Whewell's first chapter, "Astronomical Discoveries," begins as Chalmers's book and Whewell's unpublished dialogue had earlier, with a quotation from the Eighth Psalm: "When I consider the heavens, the work of thy fingers, the moon and the stars, which thou has ordained; What is man, that thou art mindful of him? and the son of man, that thou visitest him?" Whewell notes that the psalmist was not questioning whether or not God watches over us, but rather asking "*What* is man?" that he should be so blessed. The confidence of the psalmist, writing beneath the simpler skies of Old Testament times, is less easily attained under the heavens of modern astronomy. The chief results of modern astronomy, especially as they bear on the question of life elsewhere, are then sketched: Planets are analogous to the earth and hence are probably inhabited, stars are analogous to the sun and therefore are encircled by planets, nebulae are made up of suns, the Milky Way is a nebula, and so forth.

In Chapter II, "The Astronomical Objection to Religion," Whewell states this objection as follows:

. . . if this world be merely one of innumerable worlds . . . occupied by intelligent creatures, [then] to hold that this world has been the scene of God's care and kindness, and still more, of his special interpositions, communications, and personal dealings with its individual inhabitants . . . is, the objector is conceived to maintain, extravagant and incredible. (p. 19)

Whewell treats this claim "not as an objection, urged by an opponent of religion, but rather as a difficulty, felt by a friend of religion." (p. 20) In a Pascalian passage, Whewell asks, given modern astronomy, "How shall the earth, and men, its inhabitants, thus repeatedly annihilated, as it were by the growing magnitude of the known Universe, continue to be anything in the regard of Him who embraces all?" (p. 21) The possibility of a Pascalian influence on Whewell's book is supported by Whewell's comment concerning it in a September 7, 1853, letter to Stephen: "I have some speculations of the nature of those of Pascal, though I have not referred to him in them."[54]

Chapter III, "The Answer from the Microscope," recapitulates and critiques Chalmers's argument that whereas the revelations of the telescope may lead persons to doubt that God's care can extend to man in a cosmos so vast, the discoveries made by the microscope give assurance that the solicitude of the Creator reaches even to the smallest animals on earth (p. 25). Although referring to the "remarkable work of Chalmers,"

Whewell suggests that this argument responds to a difficulty rarely experienced. The religious perplexity arising from pluralism stems from belief not in life elsewhere but rather in intelligent life. And in this form of the difficulty, discoveries made by the microscope could help only if they were discoveries of intelligent beings.

Having earlier sketched the religious questions raised by extraterrestrial life, Whewell turns in his fourth chapter, "Further Statement of the Difficulty," to the deeper issues that emerge if we assume that distant planets are inhabited by "creatures analogous to man; – intellectual creatures, living . . . under a moral law, responsible for transgression, the subjects of a Providential Government." (p. 30) This leads to a discussion of what the intellectual history of the inhabitants of another planet would be. Whewell asserts: "Wherever pure intellect is, we are compelled to conceive that, when employed upon the same objects, its results and conclusions are the same." (pp. 31–2) From this it follows that intelligent extraterrestrials, if such exist, must have "the intellectual history of the human species. . . . They must have had their Pythagoras, their Plato, their Kepler, their Galileo, their Newton. . . ." (pp. 36–7) Whewell suggests that the views of extraterrestrials propounded by such authors as Fontenelle show that we tend to conceive of intelligent beings elsewhere as essentially identical with man in regard to their nature and history. In support of this he states that "the great astronomer Bessel had reason to say that those who imagined inhabitants in the Moon and Planets, supposed them, in spite of all their protestations, as like to men as one egg to another."[55] Whewell intended this argument as a reduction to absurdity. To postulate that extraterrestrials have a nature and intellectual history identical with ours is, he claims, "an act of invention and imagination which may be as coherent as a fairy tale, but which, without further proof, must be as purely imaginary and arbitrary." (p. 37)

Whewell avers that man has advanced not only intellectually, but morally as well. This he presents as evidence that God has been our "Lawgiver" and has supplied "Divine Government." Moreover, God has sent a "special Messenger" who not only has revealed God's law in a perfected form but also has established means for man's reconcilation with God. Whewell adds:

The arrival of this especial message . . . forms the great event in the history of the earth. . . . It was attended with the sufferings and cruel death of the Divine Messenger thus sent; was preceded by prophetic announcements of his coming; and the history of the world, for the two thousand years that have since elapsed, has been in a great measure occupied with the consequences of that advent. Such a proceeding shows, of course, that God has an especial care for the race of man. (p. 44)

Whewell's inference from this startled his contemporaries:

The earth . . . can not, in the eyes of any one who accepts this Christian faith, be regarded as being on a level with any other domiciles. It is the Stage of the great Drama of God's Mercy and Man's Salvation. . . . This being the character which has thus been conferred upon it, how can we assent to the assertions of Astronomers, when they tell us that it is only one among millions of similar habitations . . . ? (pp. 44–5)

Whewell is in effect agreeing with Thomas Paine as to the alternatives, although disagreeing as to which should be selected. The preceding passages shed much light on Whewell's thought; they are, for example, impossible to reconcile with John Herschel's comments that Whewell's *Essay* "can hardly be regarded as expressing his deliberate opinion, and should rather be considered in the light of a *jeu d'esprit,* or possibly, as has been suggested, as a lighter composition, on the principle of '*audi alteram partem,*' undertaken to divert his thoughts in time of deep distress."[56]

Whewell's next six chapters present his scientific arguments against pluralism. In the first of these chapters, "Geology," he reviews the results of that science relevant to the age of the earth and the extinction of species so as to set the stage for his next chapter, "The Argument from Geology." Here Whewell suggests that although man's late appearance on earth might be taken as evidence that a monstrous wastage of God's energies occurred until then, we should see this rather as an indication of how God works. His point is that apparent wastage of time is reconcilable with our image of the Creator. If this be so, if human history is but an "atom of time," why should we hesitate to believe that intelligent life occupies no more than an "atom of space," that is, the earth. Such is the respectability of geology, Whewell notes, that her analogy should offset the analogy of the astronomers.[57] This argument was original with Whewell, although it had roots in his 1827 sermons. Depending on how the argument is interpreted, it is either weak or strong. If interpreted as an argument that the earth is the only planet with intelligent life, the argument is obviously weak. However, if interpreted as a sort of reduction to absurdity of the pluralist natural theologian's argument that if God did not populate distant planets, his efforts would have been wasted in creating them, then the argument attains greater force. Whewell meant the argument in the second sense, whereas many of his opponents, taking it in the first sense, criticized him for it.

Chapters VII through X of Whewell's *Essay* move from "The Nebulae" (VII) to "The Fixed Stars" (VIII), "The Planets" (IX), and "The Theory of the Solar System" (X). Whewell's position concerning nebulae stood in sharp contrast to that of his contemporaries, most of whom believed that all nebulae could be resolved into individual stars. Astronomers were not without reasons for these views. During the 1840s, reports came from Lord Rosse, with his giant reflecting telescope, and from Harvard, where the largest refractor of the day had been erected, that many previously unresolved nebulae, including even Orion, had been seen to be composed of individual stars.[58] This result carried significance for the pluralists' position because it tended to support the existence of island universes comparable to the Milky Way and hence additional domains for extraterrestrial life. Whewell's position was that "we appear to have good reason to believe that nebulae are vast masses of incoherent or gaseous matter, of immense tenuity, diffused in forms more or less irregular, but all of them destitute of any regular system of solid moving bodies." (p. 138) Among his arguments for this claim, the most important derived from John Herschel's report that within the Magellanic Clouds one could see nebulae and clusters immediately adjacent to stars of comparable magnitude. As Whewell puts it, "There are such things as nebulae side by side with stars, and with clusters of stars. Nebulous matter resolvable occurs close to nebulous matter irresolvable. [Consequently these] are different kinds of things in themselves, not merely different to us." (p. 120) In other words, if in looking at the Magellanic Clouds one sees a nebulous patch adjacent to a star of comparable magnitude, it makes no sense to assume that the nebula is composed of thousands of stars awaiting resolution. Nebulae and stars, in this case at least, must be comparable in size, although distinct in character (i.e., nebulae were not just clusters of stars). Whewell also suggests that when nebulae are resolved, they are not resolved into stars but into "dots," which according to the "very bold" assumption of astronomers are stars, whereas they may be lumps of matter not comparable to stars. Nebulae consequently may not be island universes at all.

This argument seemed forced to many of Whewell's contemporaries. However, Whewell's theory of the nebulae gained support in the 1860s when William Huggins used the newly devised spectroscope to prove that many nebulae, including Orion, could not be resolved into stars because they are giant gas clouds. Moreover, as the island universe theory fell into almost complete disfavor in the late nineteenth century, Whewell's Magellanic Cloud argument was cited by astronomers as evidence against it. Writing in 1885, Agnes Clerke described Whewell's argument as "unanswerable,"[59] and three years later R. A. Proctor pointed to both the dramatic character and prophetic quality of Whewell's argument against all nebulae being resolvable when he urged that the plurality of worlds debate of the 1850s would have had great value even if "nothing else had been evolved by it than the first clear and definite suggestion (made by Whewell) that the theory of our galaxy and of external galaxies, advanced by William Herschel, supported by Humboldt, and Arago, and a host of others . . . cannot possibly be correct. . . ."[60] This is but one example of how Whewell's *Essay* influenced pure astronomy.

Whewell's discussion of the stars centers on whether or not satisfactory evidence is available for the widespread belief that stars are comparable in nature to our sun. Considerations parallel to those raised for nebulae suggest that it is a "very bold assumption" that the objects in "star" clusters are actually stars. In assessing this argument, it is important to keep in mind that the distances of only a few stars and no nebulae or clusters were known in Whewell's day, the first successful stellar parallax determination having been made in 1838. Thus, it was possible to argue that clusters were either distant groupings of huge stars or nearby pieces of a single sun, possibly in the process of formation. Concerning double stars, Whewell notes that for both the Alpha Centauri and 61 Cygni systems, approximate distances and periods had been determined, thus permitting estimates of their masses. When such were made, the combined mass of the former star and its companion was found to be less than that of our sun, whereas the mass of the 61 Cygni pair was set at less than a third that of our sun.[61] Such evidence emboldened Whewell to assert that "so far as we yet know, the Sun is the largest sun among the Stars. . . ." (p. 160) Moreover, Whewell suggests, the gravitational field around a star in a double star system would be ill-suited to provide a stable condition for planets. Single stars, Whewell asserts, may have planets revolving around them, but we have no evidence of this; indeed, for all we are able to see of them, many such stars may be greatly extended bodies of small density, perhaps condensing into objects comparable to our sun. Many stars are known to be variable and thereby not strictly analogous to our sun. Some stars have disappeared altogether and others appeared, making it seem that the sun may be unusual in having reached "a permanent condition." The variable star Algol is given special attention, because its periodicity had been explained by its having a planet that regularly eclipses it. But Whewell notes that for this to occur, Algol's companion must be beyond planetary size.[62] In general, Whewell argues that the analogy between the sun and stars is less strong than many had supposed, and he cites Alexander von Humboldt, who in his *Cosmos* had asked: "is the assumption of satellites to the fixed stars so absolutely necessary? If we were to begin from the outer planets, Jupiter, &c., analogy might seem to require that all planets have satellites. But yet this is not true for Mars, Venus, Mercury."[63] Whewell concludes by urging readers to refrain from speculating about the inhabitants of planets circling stars until evidence is found that such planets exist.

Turning to the solar system, Whewell begins with Neptune, which, because it is so much farther from the sun than we are, seems ill-provided with the heat and light necessary for life. But he confesses it might be possible to devise inhabitants for it, if it were found to be the rule that the objects of the solar system have inhabitants. This leads him to proceed to the moon as the body most favorably situated both for life and for telescopic observation. Drawing especially on Bessel's researches, he shows that the moon lacks not only life but also water and an atmosphere. From this he concludes that it is unnecessary to postulate life on the planets. Jupiter's density is shown to be approximately that of water and only about one-fourth that of the earth. Its great mass makes the gravitational force on its surface over twice as great as on our planet. Probably covered with clouds, Jupiter may in fact be "a mere sphere of water," and if it is inhabited at all, it is by "cartilaginous and glutinous masses," "boneless, watery, pulpy creatures" floating in a fluid (pp. 179–83). Saturn, which receives only about one-nineteenth as much heat and light per unit surface area as the earth and which is less dense than Jupiter, is found to be an even less likely abode for intelligent creatures. Uranus and Neptune are dispatched with similar ease. Mars is a harder case; although it receives less heat and light than the earth and is smaller in size, it does resemble the earth, having (as some then thought) seas, continents, and polar ice caps. Whewell suggests that it may be comparable to the earth long ago and hence may be populated by dinosaurs and such. Whewell does not seriously consider the asteroids as abodes of life, but Venus and Mercury receive mention, although astronomers had little to report concerning them. Their closeness to the sun makes them unlikely locales for life. If Venus is tenanted at all, it is probably by "microscopic creatures, with siliceous coverings" protecting them from heat (p. 192).

In his survey of the universe from the nebulae to Mercury, Whewell's arguments are sometimes excessive; nonetheless, it is clear that the pluralists, by their use of weak analogies and their readiness to speculate on flimsy evidences, had provided him with a field day. Having rejected their metaphysical and theological arguments for pluralism, Whewell was able to see that many of their astronomical claims were extremely weak. Whewell was, after all, correct in believing the solar system bereft of higher forms of life beyond the earth, and he was justified in doubting that stars are in every case encircled by habitable planets. Nor did he err in questioning that all nebulae are resolvable into stars, as Huggins soon showed. Moreover, whatever the merits of the theological position that motivated his attack, he no doubt made some pluralist astronomers see the frailty of their conjectures. This was an important contribution to astronomy.

Whewell's tenth chapter is an attempt to support the conclusions of his previous chapter by integrating them into a theory of the solar system. This theory, although it had affinities with the nebular hypothesis, was largely Whewell's creation. Having previously suggested that the planets between the sun and Mars are solid, whereas Jupiter and the planets beyond are fluid, Whewell speculates that water and vapor either were driven out from the center to form the giant planets or were retained there as the solar system formed. Having argued that the region beyond the earth's orbit is extremely cold, whereas the inner region is excessively hot, Whewell infers that the earth's orbit is "the Temperate Zone of the Solar System." (p. 196) Moreover, the planet-forming powers of the solar system are said to be most vigorous in the region near the earth, growing feebler farther out. On the one hand we have the nebulous zodiacal light inside the earth's orbit; on the other hand are the plane-toids and watery spheres beyond Mars. He summarizes his view by stating:

> The Earth is really the domestic hearth of this Solar System; adjusted between the hot and fiery haze on one side, the cold and watery vapour on the other. This region only is fit to be a domestic hearth, a seat of habitation; and in this region is placed the largest solid globe of our system; and on this globe, by a series of creative operations, entirely different from any of those which separated the solid from the vaporous, the cold from the hot, the moist from the dry, have been established, in succession, plants, and animals, and man. . . . the Earth alone . . . has become a World. (p. 203)

Describing his views as new and difficult, he pleads for their careful consideration, urging that they will explain numerous phenomena. He also suggests that his theory coincides at some points with the nebular hypothesis:

> The Nebular Hypothesis . . . is that part of our Hypothesis, which relates to the condensation of luminous nebular matter; while *we* consider, further, the causes which, scorching the inner planets, and driving the vapours to the outer orbs, would make the region of the earth the only habitable part of the system. (p. 208)

This chapter is, in short, a very speculative attempt to modify the nebular hypothesis so as to make it reconcilable with his ideas of the planets.

At the end of his tenth chapter Whewell suggests that because plural-ism has been entertained not for physical reasons but in spite of them, it is important to examine the theological and philosophical reasons that have given it currency. The final three chapters of Whewell's book are devoted to this, with the eleventh chapter, "The Argument from Design," focused on showing how his antipluralist position can be reconciled with natural theology. Whewell begins by lauding the teleological design argu-ment, but noting problems that have arisen for it. While exulting in the adaptation of the wing of a bird and the arm of a man to their functions, it fails to explain why those appendages, so different in function, should have strikingly similar skeletons. It is no more successful in explaining the functions of such bodily parts as nipples on the male. These problems Whewell proposes to remove by the consideration (which he owed largely to Owen) that God acts according to overall patterns, modified to fit individual circumstances; thus, the homologies of the arm and wing indicate that God designed them on the same plan. This view does not destroy the argument from design. As he puts it,

> If the general law supplies the elements, still a special adaptation is needed to make the elements serve such a purpose; and what is this adaptation, but design? The radius and ulna, the carpal and metacarpal bones, are all in the general type of the vertebrate skeleton. But does this fact make it less wonderful, that man's arm and hand and fingers should be constructed so that he can make and use the spade, the plough, . . . the lute, the telescope . . . ? (p. 214)

Moreover, this use of overall patterns, of general laws, may evidence "some other feature of the operation of the Creative Mind." We should avoid placing too much confidence in our teleological explanations of nature's adaptations, but when these explanations fail, we should see how readily these features may be explained as the result of God working by means of general laws and patterns. The relevance of this is that the planets and stars may be explained not by God having created them for living beings but as resulting from a general plan of creation of which the most noteworthy result is our inhabited planet. To the argument that if the other planets are not inhabited, then God created them in vain, Whewell is now able to respond that God, because he works in general patterns, frequently appears to have worked in vain. All around us, after all, we see "embryos which are never vivified, germs which are not developed. . . ." And he adds: "Of the vegetable seeds which are produced, what an infinitely small portion ever grow into plants! Of animal ova, how exceedingly few become animals, in proportion to those that do not; and that are wasted, if this be waste!" (p. 222) On this analogy, it is possible to view the earth as the only "fertile seed of creation," as the "one *fertile* flower [of] the Solar System. . . . One such fertile result as the Earth, with all its hosts of plants and animals, and especially with Man . . . is a worthy and sufficient produce . . . of all the Universal Scheme." (p. 224) If a person be troubled by the thought that this would leave great portions of the universe without intelligent life, let him reflect that that was the situation on earth for most of its history (his argument from geology).

How are we to reconcile the splendors of the heavens with the thought that man may be the sole intelligent being in the universe? On one level, his colorfully worded answer is that

> The planets and the stars are the lumps which have flown from the potter's wheel of the Great Worker; — the shred-coils which . . . sprang from His mighty lathe: — the sparks which darted from His awful anvil when the solar system lay

incandescent thereon; – the curls of vapour which rose from the great cauldron of creation when the elements were separated. If even these superfluous portions are marked with universal traces of regularity and order, this shows that universal rules are his implements, and that Order is the first and universal Law of the heavenly work. (p. 243)

On another level, he urges that the heavens are nothing compared with man:

The majesty of God does not reside in planets and stars . . . which are . . . only stone and vapour, materials and means. . . . the material world *must* be put in an inferior place, compared with the world of mind. If there be a World of Mind, *that* . . . must have been better worth creating . . . than thousands and millions of stars and planets, even if they were occupied by a myriad times as many species of brute animals as have lived upon the earth since its vivification. (p. 244)

After citing poets to the same point, Whewell again stresses that mankind has such potencies that our existence can serve to justify the universe. Not only mankind but even "one soul created never to die . . . outweighs the whole unintelligent creation." These considerations, Whewell urges, show that natural theology has little to lose by forsaking pluralism.

Whewell's Chapter XII, "The Unity of the World," was something of a patchwork, reflecting its manner of composition. On Sir James Stephen's recommendation, Whewell dropped one chapter from his original version and substantially reduced four others because they were deemed too metaphysical for his readers.[64] The remnants from these chapters were combined to form the published chapter. This was a wise choice, because much of the excised material, although interesting in itself and for what it shows about Whewell's thought, had only marginal bearing on the topic under discussion. As noted previously, Whewell was very attracted to the nomological and archetypal forms of the design argument. In this chapter he returns to that approach, admitting its Platonic character, but stressing that Richard Owen had adopted it. By studying the laws of nature and the archetypes, we can, Whewell asserts, come to knowledge of the "Divine Mind." The significance of this is that if man is such a creature that he can bring his intellect into a measure of harmony with the Divine Mind, then we have a reason "why, even if the earth alone be the habitation of intelligent beings, still the great work of creation is not wasted."[65] Because God acts on all parts of the universe by means of laws, we can say that the "remotest planet is not devoid of life for God lives there." Whewell adds that, given this, it is doubtful that "the dignity of the Moon would be greatly augmented if her surface were ascertained to be abundantly peopled with lizards. . . ."[66]

God has given us not only physical laws but also moral laws, which we can come to know and to obey. Because man has a capacity for moral goodness, he is sufficient justification for the whole universe: "One school of moral discipline, one theatre of moral action, one arena of moral contests for the highest prizes, is a sufficient center for innumerable hosts of stars and planets, globes of fire and earth, water and air, whether or not tenanted by corals and madrepores, fishes and creeping things." (p. 256) Not only is man given the opportunity for such moral elevation, his trial is so significant that it can lead to immortal life with God. In noting that God has aided man in remarkable ways, Whewell again refers to Christian themes:

The Interposition of God, in the history of man . . . is an event entirely out of the range of those natural courses of events which belong to our subject; and to such an Interposition, therefore, we must refer with great reserve. . . . But this, it would seem, we may say: – that such a Divine Interposition . . . is far more suitable to the Idea of a God of Infinite Goodness, Purity, and Greatness, than any supposed multiplication of a population . . . not provided with such means of moral and spiritual progress.[67]

In concluding this chapter, Whewell expresses his respect for those who espouse pluralism for religious reasons. But he warns: "we cannot think ourselves authorized to assert cosmological doctrines, selected arbitrarily by ourselves, on the ground of their exalting our sentiments of admiration and reverence for the Deity, *when the weight of all the evidence which we can attain respecting the constitution of the universe is against them.*"[68]

Whewell's final chapter, entitled "The Future," raises the question what we can predict for the future of man, doing this partly to answer an objection that could be brought against his geological argument; that is, a person might urge that although human history is at present only a minute portion of earth history, man may continue for many millennia. Whewell suggests that science can tell us that the great event in the past history of the earth was the appearance of man, but it has no basis for a conjecture that a being as superior to man as man is to the brute will eventually appear on the earth. Furthermore, Whewell suggests that no one contemplating animal creation could have predicted the appearance of man. Whewell does admit that man may make immense technological progress, but such would seem to have little correlation with man's moral advancement: "Men might be able to dart from place to place, and even from planet to planet, and from star to star, on wings, such as we ascribe to angels in our imagination: they might be able to make the elements obey them at a beck: yet they might not be better, or even wiser, than they are." (p. 274) The most productive method for man's advancement, Whewell urges, is the formation of societies, especially international societies. In this way, both man's intellectual state and moral condition could be greatly improved.

4. Whewell's first critic, his earliest ally, and "the most curious of all [his] unpublished pieces"

Having composed the manuscript for his *Essay* in secrecy, in the fall of 1853 Whewell apparently came to feel the need for the comments of an astute critic. Consequently, he had the proof sheets of his volume sent to his friend Sir James Stephen (1789–1859), former under secretary for the colonies, who in 1849 had become regius professor of modern history at Cambridge. The fascinating correspondence that resulted consists of twenty-seven letters from Stephen and seventeen responses from Whewell.[69]

As early as Stephen's letter of September 20, by which time he had seen only about three of Whewell's chapters, it becomes clear that Stephen possessed acute sensitivity to the difficulty for Christianity presented by pluralism. He suggests that Chalmers's astronomical objection applies "almost exclusively" to revealed as opposed to natural theology. The problem arises when a man whose "mind has been stretched almost to agony with the effort to conceive of the Creator of all this infinity of worlds" is told to believe that this God "appeared on this speck of His own boundless universe, not 'despising the Virgin's womb': passing thirty years in an obscure village as a Carpenter; and that He was then 'dead and buried and rose again the third day, and ascended into Heaven'. . . ."[70] To most persons, these "Trinitarian doctines will at first seem irreconcilably at jar with the Newtonian philosophy, and with the Telescopic and Microscopic discoveries." They will say, Stephen suggests, that the "illusion" that man is at the center of creation permitted the Israelites to believe in God's special care for them. Had this illusion not persisted, Christ and his doctrines "could never have gained a footing in the world." Stephen says that he has overcome this feeling, but not because of Chalmers: "If I had been left to Chalmers, I should have been incurably a convert to the opinions which he controverts. . . . I deliberately think that his book is far better calculated to nourish infidelity than to propagate belief in the Evangelical, or at least in the Athanasian doctrine." Stephen concludes by urging Whewell to make every effort to show his readers how to overcome this difficulty "not merely by assailing the assumption of a Plurality of inhabited worlds, but by shewing how, even on that assumption, it may be invalidated." Stephen stresses the latter need because "the world must and will always lean" to pluralism, which "can never be disproved, and must ever hang like a heavy weight on that Christian system. . . ."

Stephen's letter of September 26, 1853, is rather remarkable; although he wrote it after receiving only part of Whewell's "Geology" and none of his "Argument from Geology" chapter, Stephen nonetheless succeeded in divining, for the most part, Whewell's argument from geology. Whewell congratulated his friend on this "divination."[71] Actually, what Stephen had done was to realize that by analogy with the geological evidence that man appeared only very late in the history of the earth, one might infer that other planets would be bereft of life, at least at present. Stephen warns Whewell that if this is his conclusion, he should realize that such a thought "comes to the mind as a sudden shock [and] appears to derogate from the infinite Majesty and Glory of the creator. It seems to represent him who 'is Love' as existing in an Isolation [and to] represent the Universal Fountain of Life as either not salient at all, or as flowing in a stinted, an occasional, and a much interrupted stream." If Whewell is to make this inference, he should provide his readers with some means of overcoming the distress that they will feel. Whewell responded that the thought of a universe inhabited only by man produced no such feelings in him; in fact, quite the reverse.[72] By October 6, Stephen had received Whewell's "Argument from Geology" chapter, confirming his fears as to Whewell's position. In his letter of that date, Stephen states that if intelligent life appeared in the universe only some few thousands of years ago, then "I am in the presence of a God of whom in reference to that anterior eternity it seems impossible to say that he was Love; for Love, when not objective, is a nonentity."[73] Stephen suggests ways of avoiding this harsh conclusion. Perhaps intelligent beings without bodies or with totally decomposable bodies came before man; perhaps geology is inapplicable to other planets, and so forth. In any case, if geology claims intelligent life exists only on the earth, "we shall scarcely listen to Geology. . . ."

In his October 8 letter, Stephen cautions that because the "Plenum" versus "Vacuum" of worlds question cannot at present be decided, Whewell should appear less the advocate and more the impartial judge.[74] Similarly, his October 10 letter, along with much else, recommends "some softening of the tone."[75] In his October 12 letter, written in response to Whewell's attempts to banish habitable planets from around stars, Stephen praises the power of Whewell's style as, among other things, "inducing a feeling (an awful and almost oppressive feeling) that the case may be even so."[76] Stephen's minimal scientific background prevented him from supplying detailed criticism of Whewell's physical chapters, but in his October 13, 1853, letter he suggests: "you should somewhere indicate (briefly and dogmatically of course) what are the rules of logic applicable to this debate, for nothing is more evident to me than that in this particular subject those rules are almost universally unheeded; – I suppose because they are almost universally unknown."[77] To this suggestion Whewell responded that to lay down such rules would only "add a new class of subjects of dispute. . . ."[78]

Stephen's October 15 letter contains an especially moving passage:

Can it really be that this world is the best product of omnipotence, guided by omniscience and animated by Love? – that the Deity has called into existence one race of rational beings only, and that one race corrupt from the very dawn of its appearance? – that of this solitary family "many are called, but few are chosen"? – for the vast majority of them, as far as we can judge, it had been infinitely better that they had never been born?[79]

If the universe is not to be seen as a "desolate Cosmos," we need to be able to envision a better world than our own, a "world sinless, wise, holy, and happy." Whewell's response includes the point that Christians, before pluralism gained currency, did not seem to need this doctrine.[80]

In his letter of October 18, Stephen asks what our theology would be like if the earth were surrounded by an atmosphere making the heavenly bodies invisible to us. Lacking a view of the heavens to "'declare the glory of God,'" our theology would be impoverished and our love of God "unstable and irreverent." At the conclusion of the letter, he urges:

On behalf of my clients, the inhabitants of the extraterrestrial universe, I am pleading like some of my old Bar associates – "Gentlemen of the Jury, before you give credit to that evidence, think on the misery that your verdict must inflict on the Prisoner's family." If the learned logician had said, not "be incredulous lest you should give pain," but "be cautious where your error might lead to such painful consequences," he would not have been so far wrong. It is in this latter sense that I oppose your terrible artillery (for I acknowledge myself to be alarmed by it). . . .[81]

By October 24, Stephen was receiving sheets from Whewell's "Argument from Design" chapter. He characterizes its contents as "strangely ingenious and eloquent and beautiful – but . . . hardly less unwelcome, and depressing."[82] He urges that we cannot ascribe to God "Power, Wisdom and Benevolence in the highest degree" if we also assert that his "*only* moral, intellectual and spiritual workmanship has been a wreck from the beginning and is tending to a yet more fatal wreck. . . ." A person confronted with the prospect of unpeopled planets "may be tempted to regret that this planet could not be unpeopled also. . . ." Stephen's next letter contains the previously cited passage urging Whewell to avoid stating that Christianity necessarily entails rejection of pluralism. In his letter of October 31, Stephen comments that Whewell's chapter on "The Argument from Law" is liable to aggravate the Englishman, "the mortal foe of Metaphysics, and especially of German Metaphysics."[83] In his November 10 letter, Stephen concludes that pluralism

. . . aims formidable blows at the foundation of our faith in Christianity. The opposite doctrine aims blows scarcely less formidable at the foundations of our faith in natural religion. . . . If one or the other of the two must be abandoned, it is impossible not to see that [men will tend] . . . to disbelieve the Evangelists, rather than to disbelieve the Natural Theologians.[84]

Whewell, in responding, expresses surprise at "reflections so grave."[85] With Stephen's November 15 letter wishing the book a "prosperous voyage," the correspondence ends.

As pleased as Whewell must have been by Stephen's care, comprehension, and candor, he cannot but have been shocked by the degree to which his book startled Stephen and provoked repeated pleas for caution. The vision of a universe with life only on the earth may have delighted Whewell, but it dismayed Stephen. Perhaps Whewell had hoped to find his first convert in his colleague; in fact, Stephen's letters collectively constituted a massive warning that the book would receive no warm welcome. The isolation of Whewell's position is indicated by his remark in his October 9, 1853, letter to Stephen that he does not know of anyone who "has been the advocate of one world against many worlds . . . in recent times."[86] Believing this, and with Stephen's strictures echoing in his mind, Whewell must have felt anxious and alone as the book's publication date drew near. Then, in Stephen's last letter, came word of an ally: "Here is a little book by a former fellow of your College. . . . If you have not seen it, the correspondence of his views and yours . . . will seem curious to you."[87] Curious as are the parallels between Whewell's book and the small volume sent by Stephen, they are no more surprising than those between Whewell's career and that of the volume's author, Rev. Thomas Rawson Birks (1810–83), who had studied at Trinity College, winning, as had Whewell before him, seconds in both the Wrangler and the Smith Prize competitions. After a time as a fellow of Trinity, Birks held various positions in the Church of England before returning in 1872 to Cambridge as professor of moral philosophy, a chair earlier held by Whewell. Having praised some of Birks's earlier writings, Whewell was no doubt doubly pleased as he read Birks's book, finding even in its existence confirmation that "men's minds are ready for a fuller discussion of the subject. . . ."[88]

Birks wrote his *Modern Astronomy* (1850) to show the harmony between astronomy and religion. The influence of Chalmers is obvious in it, as when Birks borrows Chalmers's microscope argument to deal with the distress some feel at their insignificance in a universe of millions of stars and thousands of nebulae.[89] Nonetheless, when treating pluralism in the final third of his book, Birks departs from the position of the Scottish divine. After noting that belief in pluralism is widespread, although extraterrestrials are never mentioned in Scripture, Birks turns to his central concern, which is how belief in Christ's redemption can be reconciled with the apparent unimportance of our planet. Two unsatisfactory responses, according to Birks, have been made to this difficulty. It is claimed that "ours is the only world where sin has entered," but because we know of only two races, angels and men, both fallen, this suggestion violates "the plainest lessons of moral probability" (pp. 53–4) The other possibility is that Christ's coming to us is but one in a "series of revelations." But this must be rejected because the incarnation bears "the plainest impress of eternity. . . . Christ . . . is the Son of God and the Son of man, in two distinct natures and one person, forever." (pp. 54–5) Birks's solution is to call pluralism into question. It is only conjecture that stars are surrounded by planets, that these planets have inhabitants. Such a belief rests not on science but on inferences we draw from the "moral presumption . . . that God has made nothing in vain," which could as well be fulfilled by associating angels with planets. Birks then develops the "argument from geology," as Whewell was to call it. He suggests that the late appearance of man on earth suggests that other worlds may not yet, or ever, be populated. The cosmic situation may be comparable to Christ at the wedding feast; in this context, he imagines Christ saying:

My hour to people these worlds of light with myriad worshippers is not yet come. Your planet, little though it is . . . , is the Bethlehem where I now choose to reveal the mystery of my love to sinners, the guilty and the despised Nazareth of the wide universe from which streams of light and heavenly wisdom shall go forth to gladden the countless worlds I have made.[90]

Birks is suggesting that "when the work of redemption is complete, a celestial emigration may begin from our little planet. . . . It may be, that as fresh planets are prepared . . . to receive a race of inhabitants, unborn patriarchs may be sent forth, like Noah, to people its desolate heritage. . . ." (p. 63) This Birks clearly labels a conjecture. In discussing his geological argument, Birks notes that the geological writer Hugh Miller had earlier "made use of a similar line of reasoning. . . ." (p. 61) This is correct, although Miller's employment of it left him closer to the pluralists than to Birks or Whewell.[91] Although Birks's book exhibits a fine style, a competent knowledge of astronomy, and substantial originality, it attracted little attention.

In the years after his *Essay* of 1853, Whewell wrote various responses to his critics, most notably his *Dialogue on the Plurality of Worlds,* which is treated in the next chapter. While preparing his biography of Whewell, Todhunter discovered the manuscript of one additional extraterrestrial endeavor that Whewell undertook sometime between 1853 and his death in 1865. What Todhunter found and described as "the most curious of all the unpublished pieces"[92] was a long but incomplete lunar life fantasy written by Whewell himself! It narrates the longings of a Cambridge man to travel to the moon, the only celestial object besides Venus and Mars that he thinks may be inhabited. His hopes for making the voyage are frustrated, but he meets Mono, who had made the trip from the moon to the earth and who tells him of life on our satellite.[93]

This was the last of the extraterrestrial efforts of that remarkable man who, according to one wit, had tried to prove that "through all infinity, there was nothing as great as the master of Trinity."[94]

5. Conclusions concerning Whewell's Essay, "the cleverest of all the author's numerous writings"

Among the more important conclusions to emerge from this study of Whewell and his antipluralist *Essay* is, first of all, that he faced as his primary opponents not the materialists, who from antiquity had advocated pluralism, but rather the physicotheologians, who in the century and a half before 1853 had incorporated pluralism into natural theological writings. In attacking their writings, Whewell proposed fundamental reorientations in physicotheology, such as greater stress on laws and overall patterns, as well as increased restraint in claims about God's specific modes of operation.

Second, Whewell's book and the controversy it produced show, as Dr. Brooke has stressed,[95] that natural theology was already in disarray before Darwin's *Origin of Species* (1859) created a larger, although in some ways parallel, controversy. Moreover, Whewell's *Essay* contains clues as to why Whewell, who was scientifically well prepared to appreciate evolutionary theories and whose philosophical writings helped Darwin in developing his doctrines,[96] reacted very cautiously to Darwin's book.

Third, evidence in this chapter shows how misleading it is to characterize Whewell as a "semi-deist."[97] In particular, Whewell's "Astronomy and Religion," his *Essay*, and his letters to Stephen all indicate that his Christian convictions were the origin of his antipluralist position. This is not to deny that having attained this "starting point," as he called it, he was aided by the ideas of Owen and stimulated by the troubling vision in *Vestiges*. Whewell's transformation was not sudden. As he stated:

. . . the views which I have at length committed to paper have long been in my mind. The conviction which they involved grew gradually deeper, through the effect of various trains of speculation; and I may also say, that when I proceeded to write the Essay, the arguments appeared to me to assume, by being fully unfolded, greater strength than I had expected.[98]

Fourth, although the origin of Whewell's *Essay* was religious, its arguments were primarily philosophical and scientific. It would be a serious mistake to believe that Whewell felt he could convince readers of his antipluralist message on the basis of the tensions he perceived between pluralism and Christianity. Such an assumption would scarcely explain why over half of his book consists of scientific materials or why he devoted so much space to a methodological critique of the pluralists' arguments.

Fifth, the magnitude of the challenge Whewell faced deserves particular note. For many decades, pluralism had been taught in classrooms, preached from pulpits, and incorporated into an array of books ranging from astronomy texts to religious tomes. Moreover, his book was nearly without precedent; it was the first sustained critique of pluralism prepared by a man of broad learning and established reputation. In addition, Whewell faced the methodologically complex task of attacking an unfalsifiable hypothesis.

Sixth, difficult as his task was from the foregoing points of view, from another he was presented with excellent opportunities. The pluralists in many cases had proceeded on the basis of scanty empirical information, overly extended analogies, excessively bold claims about the ways of the Divine, and insufficiently sophisticated methodological assumptions. Of all these opportunities he availed himself. The book that he produced cannot be considered his most important, although it retains great interest as the fullest record of Whewell's views during the 1850s on leading issues of the day. Moreover, Todhunter stated that "many persons" came to consider Whewell's *Essay* "the cleverest of all the author's numerous writings."[99]

7

The Whewell debate: pluralism defended

1. Sir David Brewster: "why is he so savage?"

William Whewell's *Of the Plurality of Worlds: An Essay* generated an intense debate that by 1859 included over fifty articles and reviews (listed in the appendix to this chapter) and twenty books, as well as miscellaneous other materials, such as sections of books, letters, and so forth. Many figures of prominence participated, the controversy continuing even after 1859, when Darwin's *Origin of Species* touched off a larger but not unrelated furor. The present reexamination of the Whewell debate, while drawing on previous analyses of it, especially those of Isaac Todhunter, J. H. Brooke, and F. B. Burnham, surveys a larger number of responses than were known to earlier authors, including possibly even Whewell himself.[1]

Among Whewell's opponents in the debate was Sir David Brewster (1781–1868); in fact, the central feature of the controversy is frequently seen as the clash between these two aging titans. Although now usually remembered as the last and staunchest opponent of the wave theory of light, Brewster was a man of many achievements. Awarded the Copley, Royal, and Rumford medals of the Royal Society of London and the Keith Prize of the Edinburgh Royal Society, chosen as one of eight foreign associates of the French Institute, Brewster published over three hundred technical papers and books. Recent research has shown that his opposition to the wave theory of light did not blind him to many of its merits and was based on arguments that received a measure of respect, if not support, from his contemporaries.[2] The most interesting question about Brewster's two published reactions to Whewell's book was raised by Whewell himself. Writing to Roderick Murchison, Whewell asked concerning Brewster's criticisms: "why is he so savage?"[3] The answer to this question emerges from an examination of Brewster's background.

In 1794, Brewster entered Edinburgh University to study for the ministry. The breadth of his interests manifested itself as he learned natural philosophy from Professor Robison and Scottish Common Sense philosophy from Dugald Stewart. Licensed in 1804 to preach in the Church of Scotland, Brewster chose not to be ordained, although his deep concern for religious matters continued even to his deathbed, from which he praised "the grand old orthodox truths."[4] Brewster's entry into the extra-terrestrial life debate can be dated from 1811, when he edited an edition of and contributed twelve supplementary chapters to James Ferguson's *Astronomy,* noted earlier for its advocacy of pluralism. A few years later, while editor of the *Edinburgh Encyclopaedia,* Brewster asked Chalmers to write the article on Christianity that so transformed Chalmers's life. Their friendship deepened during the events in the early 1840s that led to the founding of the Free Church, with Chalmers at its head. As Brewster's daughter noted, her father "has taken part in every step of the long conflict," coming to be called "THE suffering elder of the Free Church."[5] His close association with Chalmers suggests that in 1854 Brewster may have felt that Chalmers's mantle as chief defender of pluralism had passed to his shoulders.

Among the many Brewster-Whewell clashes before 1854, the most significant for the present study occurred in 1834, when Brewster reviewed Whewell's *Astronomy and General Physics,* faulting it for many reasons, including Whewell's stress on the changing form and " 'imperfect and scanty character of natural religion.' "[6] Moreover, Brewster criticized Whewell for advocating in that book the wave theory of light and its associated ether, urging the dangers of linking natural theology with possibly changing theories (p. 429). This concern and Brewster's empiricist orientation go far, as E. W. Morse has shown,[7] toward explaining Brewster's opposition to the wave theory of light. Overall, Whewell's openness to mathematical and hypothetical methods in science distanced him from Brewster, whose experiments and instruments had revealed many of nature's secrets and whose Common Sense empiricism had provided him a way around Hume's objections, a task that in a very different way Kantian thought had accomplished for Whewell. Brewster also feared that Whewell's acceptance of the adaptation of plants to their environments as evidence of design came close to setting limits on God's powers. These would be shown to greater advantage by the assumption that plants may possess an "elastic energy" allowing them to adjust to many environments, including "to a residence on every planet in the system." (pp. 435–6) Whereas Whewell had devoted but a few paragraphs in his 1833 book to pluralism, Brewster discusses it at length in his review, stressing that it illustrates God's omnipotence, and faulting Whewell for suggesting that the universe may be finite and that the sun may be the largest body in the universe (pp. 442–4). The most significant feature of Brewster's review is the irony of his advocating an extreme empiricism while championing pluralism, a doctrine far more speculative than the wave theory of light. This inconsistency is also evident in Brewster's later pluralist writings.

Brewster's battle with Whewell continued into the 1840s as the Scottish scientist wrote lengthy, negative reviews of Whewell's *History of the Inductive Sciences* and his *Philosophy of the Inductive Sciences*. The latter book Brewster labeled the "Philosophy of Scholastic Metaphysics rather than of physical sciences."[8] Brewster also continued to press the pluralist cause. For example, his 1847 review of books by Nichol and Smyth contains numerous pluralist passages,[9] and in his 1850 presidential address to the British Association for the Advancement of Science, he urged that if men had more knowledge of and faith in science, "they would see in every planet around them, and in every star above them, the home of immortal natures . . . of souls that are saved, and of souls that are lost."[10]

In January 1854, Brewster, by then in his seventies, was invited by the *North British Review* to review Whewell's anonymous *Essay*. Believing that it supported pluralism, Brewster accepted, and by February 4, 1854, he was well into it, "groaning at every line [and saying that it is] 'quite disgusting,' and displays great ignorance. . . ." His curiosity as to who had written it was dispelled when a few days later he learned that its author was "Whewell!!!"[11] Setting aside the life of Newton he was then composing, Brewster fired off the promised review and followed it with a book entitled *More Worlds than One: The Creed of the Philosopher and the Hope of the Christian*.[12] The review appeared in May 1854; the book soon followed, for by June 1854 reviews of it had begun to appear. The intensity of Brewster's reaction to Whewell's ideas is evident in the rhetoric running through the review. For example, he describes the denial of life on other words as a notion "which could be harboured only by an ill-educated and ill-regulated mind, – a mind without faith and without hope . . . ," and as indicative of "a mind dead to feeling and shorn of reason." The very publication of such a book so incensed him that he states: "did we believe in the proximity of the millennial age, we would rank it among the lying wonders which are to characterize the latter times." (31, p. 10) Whewell's geological argument he portrays as involving an "inconceivable absurdity which no sane mind can cherish, but one panting for notoriety . . . ," as "too ridiculous even for a writer of romance," and in general as "utterly inept and illogical." (31, pp. 21–2) Whewell's chapter on nebulae is characterized as containing not "the trace of an argument," as being an attempt to "concuss" the reader by "the grapeshot of assertion, banter, and ridicule." (31, pp. 24–5) Labeling the author's opinions as "degrading to astronomy and subversive to the grandest truths" (31, p. 31), Brewster even compares him to the author of *Vestiges*.

The tone of Brewster's book is no less immoderate; in fact, a number of the foregoing passages reappear, with new passages equally extreme being added. Brewster describes some of Whewell's views in his "Fixed Stars" chapter as ascribable "only to some morbid condition of mental powers, which feeds upon paradox, and delights in doing violence to sentiments deeply cherished, and to opinions universally believed." (p. 227) Whewell's atom of space–atom of time argument is branded the "most shallow piece of sophistry which we have ever encountered in modern dialectics." (p. 202) The depth of Brewster's anger is also revealed from records of exchanges within his household. A visitor, seeing some passages in the review, commented that they seemed "calculated *to hurt his* [Whewell's] *feelings*." To this Brewster replied: "Hurt *his* feelings! why it is he that has hurt *my feelings*.[13] Furthermore, Brewster's daughter reported that Brewster's book as published had been purged of all passages that she and a Miss Forbes found to be too severe.[14]

Brewster's arguments for extraterrestrials are as extreme as his rhetoric. His daughter recorded his "ardour"[15] on finding the scriptural passage that states: "For thus saith the Lord that created the heavens, God himself that formed the earth, and made it; he hath established it, created it not in vain, he formed it to be inhabited." (Isaiah 45:12) Taking this to imply that planets without life would have been created in vain, Brewster asserts that scriptural prophets knew of the existence of other inhabited worlds (pp. 9–14). As one of his ways around Whewell's geological arguments, Brewster proposes that evidence of pre-Adamite races may eventually be detected. He suggests that Jupiter not only is inhabited but also may contain "a type of reason of which the intellect of Newton is the lowest degree." (p. 73) Advocating life on both the moon and sun, he populates the latter body with "the highest orders of intelligence." (p. 102) Despite the evidence to the contrary, he claims *that every planet and satellite in the Solar system must have an atmosphere.*" (p. 108) In a chapter on stars, he asks: "Can we doubt, then, that every *single* star . . . is the center of a planetary system like our own . . . ?" For multiple star systems, he finds the evidence for planetary systems still stronger. He argues that the conclusion is inevitable "that all nebulae are clusters of stars" (p. 176), and he rejects the claim that any stars are changing. In short, Brewster's confidence was excessive, his tone severe if not savage, his views extreme. These features of Brewster's book were admitted by a number of reviewers, including many who preferred his position to that of Whewell. Personal elements, his persisting antipathy to Whewell's thought, and his association with Chalmers were major causes of these characteristics, but more systematic reasons can also be cited.

One such reason was that although Brewster stated that pluralism rests on analogical argumentation (31, p. 23), he possessed an inadequate awareness of its limitations. For example, he failed to understand that any two objects, as C. S. Peirce later stated, "are alike in numberless respects. . . ."[16] This failure led Brewster to rely excessively on arguments that, although empirically rich, carried little force. Moreover, Brewster fell into another fallacy in the analogical area when he adopted William Herschel's argument that the partial breakdown of the analogy between the earth and the sun and moon is actually a proof that they are inhabited, because this shows that God favors a diversity in the types of regions he populates (pp. 94–100). When an argument allows one to infer habitability from either affirmation or denial of its primary premise, it is easy to understand a pluralist's enthusiasm for it. Whewell's greater sensitivity to the degree to which the plurality of worlds question was underdetermined by the facts made him more aware than Brewster of the extent to which metaphysical, methodological, and religious factors permeated the positions adopted.

Brewster's reaction was also influenced by the teleological character of his natural theology. That God would create a celestial body of significant size without bestowing life on it was to him both inconceivable and an impious idea. Repeatedly in his review and book, Brewster argues: "In peopling such worlds with life and intelligence, we *assign the cause of their existence;* and when the mind is once alive to this great truth, it cannot fail to realize the grand combination of *infinity of life with infinity of matter.*" (p. 183) Typical of this tendency is his remark that "the size or bulk of Jupiter is about 1300 times greater than that of the Earth, and this alone is a proof that it must have been made for some *grand* and *useful* purpose [i.e.] . . . of being the seat of animal and intellectual life." (pp. 65–7) Another factor was that Brewster's eschatology was linked to pluralism by his claim that the Christian should see the celestial systems "as the hallowed spots in which his immortal existence is to run." (p. 258) Moreover, as J. H. Brooke has shown, Brewster's evangelical conception of man as completely dependent on God made the pluralists' stress on the insignificance of man more attractive to him than the anti-pluralist position of Whewell, who viewed man as lofty enough to think in harmony with the "Divine Mind."[17]

Brewster did not share Whewell's conviction that pluralism created tensions for Christianity. Brewster could, after all, fall back on Chalmers's views, and to some extent he did this. But his preferred way around the problem was linked to the accepted belief that Christ's redemptive actions extend to past and future terrestrial populations as well as to those of remote regions of the earth. Given this, why, he asks, may we not assume that the "force" of the atonement does "not vary with any function of distance. . . . Emanating from the middle planet of the system, because, perhaps, it most required it, why may it not have extended to them all – to the planetary races in the past . . . and to the planetary races in the future . . . ?" (pp. 149–50) In his *Dialogue,* Whewell responded that this claim lacks scriptural warrant and assigns the earth a primacy in the universe that ill-accords with the pluralist system.[18] In his review, Brewster urges that Whewell's chapter on the tensions between Christianity and pluralism "is a mere display of ingenuity, obliterating metaphysically the brightness of our perceptions, and coming over our minds like an Eastern fog. . . ." (31, p. 16) Whewell's trenchant reply was: "I am sorry that he has allowed me to obliterate metaphysically the brightness of his perceptions."[19]

By outliving Whewell, Brewster had the last word in their debate. In an obituary of his adversary, Brewster notes that some have suggested that Whewell wrote his *Essay* "to occuppy his mind" while his wife was severely ill, but proposes the alternative idea that it was to alleviate doubts his wife had about the reconcilability of Christianity and pluralism.[20] Brewster's remark, which is without support in Whewell's writings, is doubly suggestive. First, it shows Brewster's continuing inability to believe that Whewell took his own book seriously. Second, it may contain an unconscious hint about the circumstances in which Brewster composed his review and book. According to Brewster's daughter, the death in 1850 of his wife precipitated in him a religious crisis lasting for some years, which included "strong crying and tears" and "doubt about the inspiration of God's Word."[21] These facts suggest that Whewell's book caused such a shock to Brewster because he perceived it as calling natural theology into question precisely when his assurances of revealed theology were in a weakened state.

Brewster's *More Worlds than One* secured a wide readership, larger in fact than that for Whewell's *Essay.* It continued to be republished into the 1890s and probably sold at least fourteen thousand copies.[22] One reason for the greater success of Brewster's book may be drawn from the statement with which Brewster began his book: "There is no subject within the whole range of knowledge so universally interesting as that of a Plurality of Worlds." (p. 1)

2. Rev. Baden Powell's attempt "to hold the balance"

In 1855, Rev. Baden Powell (1796–1860, the Savilian professor of geometry at Oxford, as well as the author of various studies on the history and methodology of science, published his *Essays on the Spirit of the Inductive Philosophy, The Unity of Worlds, and The Philosophy of Creation*

(London). The middle part, that on *The Unity of Worlds,* consists of Powell's attempt "to hold the balance" (p. iv) between Brewster and Whewell in the pluralist debate. Although Powell's approach in it appears to be more that of the judge than that of the advocate, it is wise in reading his works to keep in mind Owen Chadwick's statement that in Victorian religious disputes, Powell

. . . delivered his assaults impartially upon the enemies of Hampden, the Tractarian doctrines, the critics of science, the Mosaic cosmogonists. He delivered these assaults with so sedate a tone, so impassive an urbanity, so quiet a confidence, that for years readers failed to perceive that behind this neutral mask was the *most radical mind of contemporary divinity.*[23]

Another striking feature of Powell's thought is that although among Victorian writers on the methodology of science Powell is rarely ranked with Herschel, Mill, or Whewell, he alone among them endorsed Darwin's theory of evolution. In 1860, Powell praised ". . . Darwin's masterly volume on *The Origin of Species . . .* which substantiates on undeniable grounds the very principles so long denounced by the first naturalists, – *the origination of new species by natural causes. . . .*" And he predicted that Darwin's book would "soon bring about an entire revolution of opinion in favour of the grand principles of the self-evolving powers of nature."[24] This endorsement appears in Powell's contribution to *Essays and Reviews,* a collection of essays by prominent Anglicans. Although published when religious disputes were common, this book caused a crisis that, according to one authority, surpassed them all.[25] In the resulting controversy, which centered on the Higher Criticism, Powell's endorsement of Darwin and denial of miracles as evidence of Christianity might have played an even larger part had he not died in June 1860. Powell's position in *Essays and Reviews* should have surprised no one; he had been espousing evolutionary doctrines for many years. While Brewster, Whewell, and others were castigating the author of *Vestiges,* Powell was writing him that as early as 1838, he had himself entertained "a strong leaning toward the very views you have since so much more ably expounded."[26] Moreover, in his 1855 *Essays,* Powell commented that among all the charges leveled against the author of *Vestiges,* "nothing can be more utterly and palpably unjustifiable than the charge of an *irreligious* tendency against a work *in which almost every page is replete with expressions of the most devout homage to the Divine power, wisdom, and goodness.*" (p. 453) As Milton Millhauser has stated, "Powell must rank as the most influential and courageous of the few outspoken friends of *Vestiges* [and as] its one considerable clerical advocate."[27] Despite Powell's importance, he has received scant attention from historians.[28]

The first of the three parts of Powell's 1855 *Essays,* that devoted to the methodology of science, is fundamental for the other two parts, in which he discusses the plurality of worlds and evolutionary questions. Throughout the first part of his *Essays,* he attempts to steer a course between the idealism of Whewell and the empiricism of Mill. In his discussion of induction, he accepts the Whewellian thesis that "every induction is seen essentially to involve *a certain amount of hypothesis*" (p. 6), but he rejects his Cambridge contemporary's espousal of *a priori* necessary truths in science. For Powell, "THE SOUL OF INDUCTION IS ANALOGY" (pp. 18–19), which in turn rests on our confidence in the uniformity of phenomena, our belief that examined phenomena are similar to those unexamined. Powell's stress on the "Unity of Nature" separated him dramatically from Whewell. For example, in contrast to Whewell's contention that different methodologies are appropriate to different areas of science, Powell accepts the legitimacy of attempts to reduce biology to physics and chemistry. Whewell's catastrophism in geology is faulted by Powell as violating the principle that explanations of past events should be uniform with explanations of present events. Concerning man, Powell presses for a distinction, crucial to his acceptance of evolution, between man's animal nature and his moral and spiritual natures. Admitting that man as animal is "very little superior to brutes," Powell adds that nonetheless man's *"moral and spiritual nature* [refers] . . . wholly to a DIFFERENT ORDER OF THINGS. . . ." (p. 76)

Whereas Whewell had favored the teleological approach of Cuvier, Powell praises Geoffroy Saint-Hilaire's principle of the "unity of composition." Rejecting the use of final causes in science and in natural theology, Powell claims that their employment in the latter area has led to much criticism of it. Natural theology, he urges, can retain the design argument while rejecting final causes by focusing on nature's *"symmetry* and *arrangement,* but not the *end for which* the adjustment is made." These features, seen in the laws governing the heavenly bodies, show design because they are *"results of mind."* Moreover, he adds: "A mere numerical relation invariably preserved, . . . or a systematic arrangement of useless parts or abortive organs on a regular plan, are just as forcible indications of intelligence, as any results of immediate practical utility." (p. 135) As this indicates, Powell favored the nomological design argument. He was espousing a natural theology far different from Paley's: "Paley expressly held that the mechanism of the heavens was a branch of science the least susceptible of this kind of application; according to the principle here advocated, it forms the highest and most satisfactory." (p. 142)

Powell admits that his conception of natural theology "leads us only to a very limited conception of the Divine perfection;" to have a fuller knowledge of God, we must have recourse to *"other* and *more spiritual sources."* (pp. 162–3) What is lost in scope is gained back in certainty. Given what has been said of Powell's position up to this point, it is surprising that the philosophical views he most praises in this portion of his book are those in Oersted's *Soul in Nature*. One does not expect to find chapters aimed at criticizing the Kant-influenced Whewell ending in endorsements of the Kant-influenced Oersted. Perhaps Powell's enthusiasm for Oersted indicates a Platonic tendency in the Savilian professor of geometry, but whatever the precise nature of his position in natural theology, he saw it as capable of assimilating Darwinian doctrines. Yet few of his contemporaries gave Powell's philosophy serious consideration, possibly because they thought his liberal theological stance made his entire system suspect.

In attempting "to hold the balance between the two disputants" on the question of other inhabited worlds, Powell directly mentions Whewell only as an earlier advocate of pluralism. Nonetheless, he was well aware of Whewell's authorship: "it would be absurd to pretend ignorance of his real eminence." (p.vi) Powell appears less passionately concerned about the plurality of worlds question than Brewster or Whewell; in fact, he comments that the question "is in itself of a very secondary and unimportant character" as compared with "the broad principle involved in *any or all cosmical speculations."* (p. 298) Urging that the plurality of worlds issue is first of all a scientific question and one of probabilities, Powell suggests that therefore "the argument must be based on an extension of *inductive analogies"* (p. 183), but he never analyzes what conditions must be satisfied for two entities to be called analogous. Despite Whewell's strictures, Powell does not hesitate to state that all stars are suns with planetary systems around them. In support of this thesis, he offers little more than a "pourquoi non" argument and the nebular hypothesis. Although he does not champion life on the sun and moon, his criteria for the "extension of inductive analogies" are capacious enough that he refuses to rule this out and even provides some mechanisms by which life might be possible on these bodies. The conditions on the planets differ, he urges, only in degree, not in kind, from those on earth. This seems overly generous, given that Mercury receives more than six times as much heat per unit surface area as the earth and Neptune nine hundred times less. Powell's evolutionary view leads him to suggest that planets now lacking life may in the future attain it or may have had it in the past. In summary, he states:

Looking at the subject solely as a question of plausible philosophical conjecture, and guided as we should be by the pure light of inductive analogy, all astronomical presumption, taking the truths of geology into account, seems to be in favour of progressive order, advancing from the inorganic to the organic, and from the insensible to the intellectual and moral in all parts of the material world alike, though not necessarily in all at the same time or with the same rapidity; in some worlds one stage being reached, while in others only a comparatively small advance may have been made. (p. 231)

In this section, Powell appears far closer to Brewster than to Whewell; moreover, this passage suggests that "the pure light of inductive analogy" was sufficiently bright for Powell that he could see just about what he wished. It was far less astronomy and geology than Powell's confidence in the "progressive order" and "uniformity of nature" in the universe that made Powell a pluralist.

When Powell turns to the theological aspects of his topic, he describes Whewell's approach as *"retrograde"* and his "masquerade" as a *"mediaeval* costume." He adds that Whewell's "speculations evince, if not a literal and physical, yet a moral *Ptolemaism:* they seem conceived in the spirit of the dark ages. . . ." (pp. 239–40) Noting that Oersted had criticized some writers for rejecting pluralism on narrow religious grounds, Powell cites a long passage from him, the "general conclusion" of which he endorses. Typical of Oersted's argument is this statement: "If we regard the whole of existence as a living revelation of Reason in time and space, we can conceive that the most varied degrees of development may be found distributed through all time. . . ." (p. 251) Returning to the question of final causes, Powell praises Owen for separating arguments based on final causality (these Powell finding objectionable) from those grounded on inductive analogy. Concerning Jupiter, Owen had suggested that its inhabitants may have eyes analogous to ours but modified to suit the differing conditions on that planet; as Powell states, "such creatures, he says, '*may* exist to profit by such sources of light; and *must* exist IF the only conceivable purpose of these beneficent arrangements is to be fulfilled.' " (p. 254) Powell finds that Owen's "reference to the archetype is simply one of the highest forms of *conjecture* from *analogy,"* and as such, fully legitimate. The inference that because there are forms of the eye not developed on earth, they have developed elsewhere "is simply and strictly of an inductive character." (p. 255) Able to accept Owen's archetypes, Powell backs off only from the portion of Owen's argument after the "IF."

Brewster's reliance on final causality draws criticism from Powell: "even if disposed to admit the *truth* of the conclusion, I should still have much doubt as to the *mode* of arriving at it." (pp. 259–60) Powell finds Whewell's limitations on the use of final cause arguments more acceptable, although he claims that Whewell employs arguments that rest on final causality, for example, in explaining the late appearance of man on the earth. Returning to the principle of the uniformity of nature, Powell applies it on a cosmic scale, relying once again on Oersted. After noting in this chapter, in which he had repeatedly stressed the need to keep theological and physical questions separate, that Oersted had contended that there is "a community of *moral* laws throughout the inhabited universe" (p. 268), Powell expresses his assent, "warranted as I think by the soundest inductive principles," and says that this allows persons to recognize "THE TRUE UNITY OF WORLDS." (p. 269) As this suggests, Powell's preference for pluralism was linked to his belief that it is more spiritually elevating (pp. 266–7).

The problems that pluralism, in the eyes of Whewell and others, presented for Christianity caused Powell little distress. He suggests that Whewell's as well as Wesley's concern in this regard "seems . . . to refer too much to those narrow humanised ideas of the Divinity and His dealings with man derived so commonly from too literal an interpretation of the anthropomorphisms of the Hebrew Scriptures. . . ." (pp. 287–8) Moreover, he avers that such difficulties "arise wholly from the inconsistency of *attempting to reason at all* on subjects which the writers themselves . . . pronounce *to be above all reason.* . . ." And he adds: "If it be an inscrutable mystery *wholly beyond human comprehension* that God should send His Son to redeem this world, it cannot be a *more* inscrutable mystery . . . that He should send His Son to redeem ten thousand other worlds." (p. 291) The concluding section of Powell's *The Unity of Worlds* consists of a broad discussion of the science-religion relation. As such, it is not of concern in the present study, nor is the final portion of his book, that entitled *The Philosophy of Creation.*

Powell's capacious conception of the use of inductive analogy and his stress on the unity of nature led him to tip the balance toward Brewster. Although an Anglican divine, an experienced scientist, and a philosopher of ability, Powell felt little attraction to Whewell's claims in these areas. Ironically, Powell's endorsement of evolutionary ideas in the final portion of *Essays* may have led many readers to view the earlier sections with hesitation. Whatever the cause, his discussion of pluralism attracted less attention than those of Brewster and Whewell.

3. *Responses from astronomers and mathematicians: "a plurality of opponents" of Whewell's "singular" book*

Among all Whewell's contemporaries, the one he no doubt would have most welcomed as an ally of his *Essay* was Sir John Herschel, and yet he must have sensed the unlikelihood of winning over his friend for many decades. Nonetheless, on January 3, 1854, Whewell sent Herschel his *Essay*, describing it as the work of a "friend" whose ideas, although "so much at variance with opinions which you have countenanced," deserve at least not to be "suppressed."[29] In words that scarcely reveal the cosmic holocaust attempted in the *Essay,* Whewell suggests: "Perhaps you would not take it much to heart if the inhabitants of Jupiter, or of the systems revolving about double stars which you have so carefully provided for, should be eliminated out of the universe." Herschel's letter in response is remarkable. Concerning the *Essay,* which Herschel notes "people very oddly persist in attributing to yourself," the great astronomer comments: ". . . I should not have thought there was so much to be said on the non-plurality side of the question."[30] And he adds that "though somewhere I have myself stated that taken in a lump Saturn might be regarded as made of cork – it *never did* occur to me to draw the conclusion that *argues* the *surface* of Saturn must be of extreme tenuity. . . . But to dispossess Jupiter of his solidity and make him a huge Arctic Ocean I confess never once occurred to my thoughts. Yet so it would seem it must be." Herschel, who was not inexperienced in extricating extraterrestrials from difficult situations, was scarcely daunted by that admission. His speculative gifts were at once set to work populating the depths of that sea with "What fishes! there may be. What Crystal Palaces they may construct. . . . What water organs in the nature of Sirenes they may construct and to what a perfection they may have brought the science of hydropathy. . . ."[31] Herschel may not have been entirely serious in such speculations, but he was sincere in lamenting:

> So *this* then is the best of all possible worlds – the *ne plus ultra* between which and the 7th heaven there is nothing intermediate. Oh dear! Oh dear! 'Tis a sad cutting down. Look only at the Russians and Turks – Look at the revelations of the Blue Book and the Police Courts. I can't give in my adhesion to the doctrine that *between* this and the angelic there are not some dozen or two grades of intellectual and moral creatures.[32]

Warning Whewell of the falsifiability of his position, Herschel exclaims: "Dissentions and Protesting. – The whole theory is destroyed if there can be two cases produced in which the process has gone on to its completion in the production of that ne plus ultra – An Earth! inhabited by Men!! for if two why not 2000." Herschel's conclusion contains words of praise: "The book is full of striking things. The geological argument is very pointedly put. – The Magellanic Clouds are very availably brought into action – Time and Space are duly and properly scorned and reduced to their true value." So, in private, Herschel and Whewell contested over the cosmos as the Turks and Russians battled over the Crimea. Possibly from friendship, more probably from dismay, Herschel wrote no review of Whewell's *Essay.* However, he did suggest in his obituary notice on Whewell that he had written his book as a *"jeu d'esprit."*

Whewell must have opened with great interest the March 1854 letter he received from William Parsons, Lord Rosse (1800–67), the leading expert on the resolvability of nebulae and the discoverer of spiral nebu-

lae, drawings of which adorned the frontispiece of Whewell's *Essay*.
Rosse, who was president of the Royal Society, wrote: "There seems to
be strong evidence in support of your opinion that the nebulae are not
immensely distant in proportion to the Fixed stars, and in my last R. S.
[Royal Society] Address which I enclose I have taken that view."[33]
Rosse's presidential address, although not published, served as a source
of support for Whewell; he cited its conclusions in both the preface to the
second edition of his *Essay* and in his *Dialogue*.[34] A letter from another
officer of the Royal Society soon followed. Serving then as treasurer and
foreign secretary, Edward Sabine (1788–1883) would in 1861 begin a
ten-year term as its president. His concern in his letter of March 4 was to
tell Whewell that he was reading "with much interest" a copy of Whewell's *Essay* loaned him by George Airy, the Astronomer Royal, and that
he agreed with Whewell's theory that solar and lunar tides exist in our
atmosphere.[35] It would be interesting to know more about the reactions
of Rosse, Sabine, and Airy, but no further information appears to be
available. Nonetheless, the Rosse and Sabine letters show that Whewell's
volume contained ideas judged significant by important scientists of the
period.

In response to Whewell's *Essay* and *Dialogue,* the astronomer William
Stephen Jacob (1813–62) published in 1855 a short book entitled *A Few
More Words on the Plurality of Worlds*. Claiming that most replies to
Whewell's book paid insufficient attention to astronomy,[36] Jacob sets out
to remedy this, pointing out a number of errors in fact and calculation
committed by Whewell, whom Jacob does not identify by name. Much of
Jacob's argument in this area concerns cases in which he interprets observations or data differently than Whewell. For example, it is possible to
estimate the mass of a star if its parallax (hence its distance) is known and
if it is a component of a double star system for which the motions of the
component stars are known. In the case of the Alpha Centauri system, it
had been calculated that the mass of the main star was between one-half
and four times the mass of the sun. Whewell opted for the former, Jacob
for the latter figure as equally probable (p. 17). Similarly for the question
of a lunar atmosphere: Whewell urged that the evidence pointed to absence of an atmosphere, whereas Jacob contends that we can say only
that the moon has little or no atmosphere (pp. 36–7). The distinguishing
feature of Jacob's volume, however, was his stress on a probabilistic
approach. Important as this approach may be, Jacob made serious errors
in applying it.

Consider Jacob's example of an urn filled with a thousand balls of
"colour unknown." A person withdraws one ball, finding it to be black.
Jacob comments: "This, then, is the case of those who hold that *one*
planet being inhabited makes it likely that *all* are inhabited . . . ; *they*

are as likely to be right as not." (p. 35) He adds that our situation on earth is equivalent to that of a being, roaming the universe, who happens upon a star and finds it encircled by planets. This being, Jacob asserts, can draw the conclusion that the odds are essentially even that all stars are surrounded by planets, an inference analogous to that in the case of the urn (p. 36). But Jacob's analysis is pervaded by errors, some more obvious than others. Surely it is evident that we cannot validly infer after withdrawing a black ball from the urn containing 1,000 balls of "colour unknown" that there is a 50:50 chance that all the remaining balls will be black. It is legitimate to claim before any balls are withdrawn that the appropriate probability to assign to the hypothesis "the next ball removed will be black" is one-half, provided that we know that all the balls fall into two classes, say black and white. But this statement must not be construed as a statement about the world, only about our ignorance. Being a statement about hypotheses rather than the world, it is of interest only if no means are available for overcoming our ignorance. In the plurality of worlds dispute, this had long ceased to be the case. Jacob makes a further mistake when he implies that the being who, finding one star encircled by planets, infers that others are also, is making an inference equivalent to that which we make when, finding the sun surrounded by planets, we infer that other stars are also orbited by planets. The mistake is that in the latter instance, the inference is influenced by the presence of the observer; it could not be drawn were the observer not present. Put differently, the inference that there are planets (or intelligent life) in other solar systems can be drawn only in systems in which planets (or life) exist. In this regard, our situation is not analogous to that of the being who *by chance* finds a system with planets.[37]

Perhaps Jacob's errors arose from the major role that natural theology played in his thought. Believing with Brewster that the planets would be wasted if uninhabited, he states that "men cannot believe it probable than an intelligent Creator would make an unlimited number of *mimic* worlds . . . , leaving all but one unproductive." (p. 37) Given this statement, it is surprising that he declares seven pages later that the "theological aspect of the question I have not as yet touched upon. . . . (p. 44) Jacob's concluding discussion of the theological issues is of interest only in that it makes clear that he did not comprehend Whewell's difficulty about reconciling Christianity and pluralism. In fact, he states: "I agree with [Whewell] that our religious belief is in no way affected by either view" (p. 44) of the plurality of worlds question. Whewell had not said this, nor would he.

Whereas Jacob rejected Whewell's position, James Breen (1826–66), an astronomer at the Cambridge Observatory, attempted to steer a course between Brewster and the essayist. In *The Planetary Worlds*

(1854), which may have been occasioned by the debate[38] and which is rich in information about the telescopic appearances of the objects of the solar system, Breen devotes one section to the clash between Whewell and his Edinburgh adversary. He concludes that "there is no positive evidence on either side," urging pluralists to keep in mind that "the saying of D'Alembert is still applicable, '*Qu'on n'en sait rien*,'" and their opponents to ponder "the old saying, slightly altered . . .: 'Man *supposes*, and God disposes.'" (pp. 259–60)

Three American astronomers contributed to the Whewell debate, the youngest being Arthur Searle (1837–1920), whose 1855 review of the books of Brewster and Whewell was written while he was an undergraduate at Harvard and decades distant from the Phillips professorship he later attained there.[39] Writing for *Harvard Magazine*, Searle describes Whewell's book as having been read by many recipients of that periodical and Brewster's response as "showing great ability, but still greater violence." (22, p. 167) Searle notes the very speculative nature of the issue debated and attempts, without reaching a conclusion, to assess whether the burden of proof falls chiefly on the pluralists or their opponents. The pluralists are the dominant party, he suggests, but they should be expected to justify their claims. Brewster is faulted for his solar inhabitants and Whewell praised for his geological argument, which Searle, unlike many reviewers, realized was designed to refute a pluralist claim, not to prove the impossibility of life elsewhere. Overall, Searle's review was noncommittal or slightly Whewellian.

The second American astronomer was Maria Mitchell (1818–89), who in 1857 toured Europe, meeting with a number of scientists, including Whewell and Mary Somerville. Mitchell had attained fame as well as a gold medal from the king of Denmark for her discovery in 1847 of a comet. Her meeting with Whewell went poorly, she perceiving him as arrogant. This impression is conveyed in her journal, for example, by her remark: "They say in Cambridge that Dr. Whewell's book, 'Plurality of Worlds,' reasons to this end: The planets were created for this world; this world for man; man for England; England for Cambridge; and Cambridge for Dr. Whewell!"[40] Her reading of Whewell's book led her to conclude: "There is nothing from which to reason. The planets may or may not be inhabited."[41] She also met with Mrs. Somerville, describing her as speaking "with disapprobation of Dr. Whewell's attempt to prove that our planet was the only one inhabited by reasoning beings; she believed that a higher order of beings than ourselves might people them."[42]

Among the most important reviews of the books of Brewster and Whewell was that written by Denison Olmsted (1791–1859), who had studied theology with Yale's pluralist president, Timothy Dwight, and who taught astronomy at that institution for a number of decades.[43] Early in his 1854 *New Englander* review, Olmsted presents six *"principles of reasoning"* as applicable to the dispute. The most interesting of these are his fourth principle, *"the argument from analogy is apt to be delusive, and is often abused,"* and his sixth, *"men readily believe any doctrine which is supposed to be favorable to their religious faith, or to their settled opinions on any other subject, and as readily reject what is subversive to such opinions."* (29, pp. 574, 576) Brewster is cited for violating the sixth principle; by labeling his position "the hope of the Christian," he revealed "a state of mind unfavorable to sober argumentation." A section on the observational evidences against pluralism is summarized in Olmsted's statement that "The telescope . . . has added nothing to the amount of evidence in favor of the doctrine that the planets are inhabited; it has, in fact, greatly diminished that amount, since the points of dissimilarity to the Earth . . . have increased faster than the points of resemblance." (29, p. 578) Such admissions lead one to expect Olmsted to announce himself a Whewellite, but the rest of the review reveals that he had remained a disciple of Dwight.

Turning to evidence for pluralism, Olmsted avers that when we study nature, we see a uniformity of plan; this leads us to "predicate a uniformity of purpose." Consequently we need only determine the purpose of the planet earth to be able to infer "the grand purpose for which Nature designed the others." This Olmsted had already done in an 1849 essay in the *New Englander* in which he "endeavored to prove that the *World* was made for Man*."* (29, p. 580) Animals, in a passive way, profit from light and heat, but man actively uses such forces in his steam engines and telescopes. Man alone can benefit from such productions of the earth as gems and coal; indeed, "the *animal* kingdom . . . was made for man." From his uniformity of plan and purpose principles and his determination of the purpose of the earth, Olmsted concludes that the other planets "are also the abodes of life and intelligence." (29, p. 583) Apparently Olmsted's sixth principle, while barring theological inferences, was not so restrictive as to banish natural theological or metaphysical argumentation. Nor, it seems, did he view his fourth principle as preventing him from arguing that, in analogy with the human body and society, "the sun and planets . . . compose a system [in which] every part has important functions to perform. . . . But if the planets are barren wastes . . . , they present to us a totally different view of the economy of Providence from that which is everywhere set before us in the world." Olmsted's approach leads him to remark that the fact that Jupiter's moons "husband [sunlight] and turn it upon the planet, implies the presence of eyes to behold it. We could hardly be more certain of the presence of water when we see . . . a ship under sail." Moreover, he states, this "same course of reasoning" justifies the conclusion that *"the stars are centres of other solar systems which, like ours, are filled with sentient beings."* (29, pp. 583–5)

Reviewing the empirical evidence against planets being inhabited, Olmsted urges that although the physical properties of materials throughout the universe must remain constant, the Creator could establish "very great modifications" in the "relations which subsist between them and the animal creations" present on other planets. Among the difficulties that could be overcome by such modifications, Olmsted includes even "the want . . . of air and water." (29, p. 586) Stating that up to this point he has considered only "purely scientific" evidence, Olmsted turns to the "moral and religious" aspects of his topic. Reviewing the ideas of the "infidel Thomas Paine" and, with more enthusiasm, those of Fuller and Chalmers, Olmsted characterizes Whewell's position as "obscure and difficult of comprehension" and in any case as "not conclusive." (29, p. 592) Whewell's geology argument he dismisses with the remark that it shows only that other planets probably acquired inhabitants after an extended period of formation. Turning briefly to Brewster, he admits his agreement with Brewster's use of a teleological approach. In his final section, Olmsted makes explicit his preference for pluralism: "we are inclined to believe in the affirmative, although we do so with a full conviction that there is much to be said on the other side." (29, p. 602) The restrained tenor of that statement is less easily seen in his final sentence: "To infer that rational beings cannot exist without water, or vegetables, or air, is as erroneous as to infer that a clock cannot go without a pendulum." Moderate in tone, ingenious in some of its arguments, Olmsted's review was nonetheless that of an enthusiast who slipped into some of the traps he had warned against in his six principles.

Augustus De Morgan (1806–71), a witty English mathematician who in 1854 was secretary of the Royal Astronomical Society, wrote Whewell two letters in that year about his *Essay*. In his letter of January 24, De Morgan suggests that Whewell should have called his *"singular"* book: "on the singularity of the world."[44] And he adds: "I have always held that when the phrase 'there is a good deal to be said on both sides' applies, it means that we do not know much about the matter. Your book is a converse instance: to wit, that when we do not know much about the matter, there is always a good deal to be said on both sides." Writing four months later to thank Whewell for having sent a copy of his *Dialogue,* De Morgan comments that while "[you] deny a plurality of worlds, [you must] admit a plurality of opponents."[45] Among these opponents was De Morgan, who admits: "I cannot divest myself of the idea that [the planets] have uses independently of us − and these uses are inhabitants." Perhaps the most interesting aspect of this comment, which was grounded in natural theology, was that it was made by a man who had been barred from a Cambridge fellowship and twice resigned his professorship at University College, London, because of the liberality of his religious views.

Nearly a decade later, De Morgan wrote Whewell again about his books, the occasion being that De Morgan was writing on paradoxes, which he defines in the letter as "a thing strange to general opinion."[46] As examples, De Morgan mentions "circlesquarers," proponents of "anti-gravitation," and "Wilkins, who in his time conjectured the moon *might* be a world, and yourself, who in our time have conjectured it may *not*." He also asks Whewell for the date of his *Dialogue:* "The omission is awful: for after all, there may be more worlds, and how is ours to be the best of all possible worlds, as Leibniz said, if books go without dates?" Whewell, possibly displeased that he would appear in such company, may not have responded. Whatever the case, when De Morgan's *Budget of Paradoxes* appeared, Whewell was given a place, although as Todhunter noted, De Morgan misdated his *Dialogue*.[47]

The volley that Baden Powell launched at the master of Trinity was one of two that came from the Oxford direction in 1855. The other was fired by H. J. S. Smith (1826–83), a mathematician who later succeeded Powell in the Savilian professorship. Smith, who Todhunter claimed had written "by far the ablest of the unfavourable reviews,"[48] wrote with the same tact and apparent detachment as Powell. These characteristics, as well as Smith's motivation for writing, are suggested by his comment concerning Whewell's book: "A work so brilliant and so suggestive deserved a more elaborate reply than that which Sir David Brewster has given to it." (53, p. 110) In discussing Whewell's *Dialogue,* Smith states that many of the early interlocutors respond on the level of a "simpleton." Todhunter's response to this charge is correct: Most of the arguments made by those interlocutors were originally advanced by men of "eminent distinction."[49] Commenting on both Brewster's and Whewell's religious positions, Smith writes: "Our knowledge of the plan of Nature is more partial, and the results to which it leads are more ambiguous, than either of them seem willing to concede." (53, p. 143) Although criticizing Brewster for his reliance on final causes, Smith rhetorically asks "whether that crowning evidence of goodness and power can be wanting everywhere but here?" (53, p. 144) Smith's repeated attempts to salvage the teleological design argument (53, pp. 145–50) show his Brewsterian biases.

In discussing Whewell's astronomy, Smith asserts that Lord Rosse's discoveries have convinced most astronomers that all nebulae are in principle resolvable into stars. But as Whewell had already revealed in his *Dialogue,* and as Smith does not mention, Lord Rosse was himself in doubt about this generalization.[50] In general, Smith describes Whewell's astronomy as that of a *"minimi-fidian"* who errs by assuming that life can exist only in situations comparable to those on the earth. Smith's discussion of the moon begins with this admission: "The Moon we are compelled to surrender at discretion; and we own ourselves, on this point, unable to withstand the desolating rhetoric of the Essay." (53, p. 136) However, Smith soon qualifies this seemingly unqualifiable statement; noting that Peter Hansen had claimed that the moon's center of figure does not coincide with its center of mass, he suggests that life may exist on the moon's remote side. One of Whewell's responses to this idea, which Herschel and Powell also advanced, was to ask if an inhabitant of the moon's far side who wandered to our side would not be "woefully perplexed (especially if he were a philosopher jealous of *waste* in the creation) to see this great luminary placed exactly in that single point of the universe in which it could possibly be of use to his race?"[51] No single passage in Smith's essay more effectively suggests his overall reaction to Whewell than when he states that

We cannot imagine a more painful spectacle of human presumption than that which would be afforded by a man who would sit down to arrange, 'in a satisfactory way,' a scheme for the extension of Divine mercy to some distant planet, and who, when he found 'great difficulty in conceiving' such an extension . . . , instead of desisting from his vain attempt, should . . . infer that no such scheme can exist, because he fails to discover a *modus operandi* for it. (53, p. 117)

The *Christian Examiner,* published in Boston and of Unitarian origins, carried a review of the books of Brewster and Whewell in its September 1854 issue. The author was probably Thomas Hill (1818–91),[52] a Harvard-educated mathematician and Unitarian minister who in 1862 became president of Harvard. More rhetoric than reason is found in this review, which begins by describing pluralism as "giving rise to many beautiful passages of oratory, many strains of noble verse. It has exalted our conceptions of the Almighty Power, and thrilled our souls with a wider sense of brotherhood with the sons of God." In contrast, the discourse of Whewell is portrayed as "prolix and almost tedious" and as laying "merciless hands upon this fair fabric of speculation." (7, p. 209) This Unitarian reviewer finds no problems for Christianity in the pluralist position, despite the "more specious than real . . . objections [of] the anonymous writer." (7, p. 212) Whereas many had found Brewster's book excessive in its style, Hill presents it as "a book in which brevity, clearness and accuracy of statement, and cheerful confidence of tone, form a pleasing and refreshing contrast to the tedious obscurity of many passages in the anonymous essay." (7, p. 218) Brewster is faulted for excessive reliance on Scripture and for his theory of the asteroids, concerning which Hill remarks in his final paragraph that we "can see no reason why they may not be as inhabitable as the earth." Whewell at times had argued that if the pluralists would populate the planets, why not asteroids as well – a sort of reduction to absurdity. But clearly this argument would have carried little force with Hill.

Among the astronomers and mathematicians discussed in this section, all except Searle were closer to Brewster than to Whewell. This, as we shall see, is typical of the scientists, who, whatever their specialities, for the most part found Whewell's position unacceptable if not incomprehensible.

4. Responses from geologists: "Geology versus Astronomy"

Perhaps it was the prominence Whewell gave his geological argument, possibly it was the fact that in earlier decades geologists had repeatedly been forced to think about the relations between science and religion, but whatever the reason, many geologists commented on Whewell's book. Among these was Rev. Adam Sedgwick (1785–1873), the Woodwardian professor of geology at Cambridge, who on February 27, 1854, wrote John Herschel:

> What wonderful health [Whewell] has! And indeed he ought to be strong, to destroy a plurality of worlds, as he is trying to do. Have you seen the big pestle and mortar by which he has pounded 500,000 worlds into comet-tail-dust, and the big snuffers by which he has put out the *lights* of all *livers* above and below the earth? I was much amused by it, but not convinced.[53]

Whewell's efforts to win over Sedgwick included a June 8, 1854, letter in which he responds to one of Sedgwick's objections and recommends that he read the printed but unpublished portion of the *Essay*.[54] Part of the reason that Sedwick read Whewell's book was its broad scope; this feature is also illustrated by Charles Lyell's comments on it in his journals; these refer primarily to Whewell's views on the nature of terrestrial species.[55]

As early as December 1853, Whewell sent his *Essay* to the Scottish geologist Sir Roderick Murchison (1792–1871), asking for his reactions but requesting that Murchison not reveal his authorship.[56] On January 15, 1854, Murchison responded, stating that he was pleased by Whewell's use of geology and adding; "As to the nebulae, we men of the earth are too earthy to have any opinions worth having; but I may say that I greatly admire your ingenious and well reasoned views."[57] Murchison also comments: "What some of the over-stiff laced & matter of fact biblicists – men void of brains and imagination – may say of your effort I know not, but I like the book for nothing more than for its boldness & sincerity." After agreeing with Whewell that a "*progressive design*" may be seen in earth history, Murchison concludes by asking: "Why did you not call the book 'The Unity of the World'?" Whewell responded in his *Dialogue* by suggesting that had he done so, persons would probably have assumed that he was writing to suggest "a connexion of all the parts, rather than exclusion of a plurality of similar parts."[58] On May 30, 1854, Whewell sent a note to Murchison that ends with these interesting remarks:

> I do not think I have been so unsuccessful in making converts even among those bigoted people the astronomers. As to the nebulae all the best astronomers are with me. Herschel, Lord Rosse, Airy, Challis. For the rest I can wait.
>
> I should be curious to know if Brewster makes converts to his fancies. But why is he so savage?[59]

These exchanges suggest that Murchison's sympathies may have been more with Whewell than with Brewster.

This was less true for two other Scottish geologists, J. D. Forbes and Hugh Miller. James David Forbes (1809–68), professor of natural philosophy at Edinburgh, had corresponded with Whewell for many years when on November 4, 1853, Whewell wrote to tell him that he was sending a copy of his *Essay,* which would appear anonymously because of the "heresies" contained in it.[60] On November 7, 1853, Forbes responded: "I shall be interested in your Anti-Chalmersian speculations. I can well believe that your logic will tear down rather rudely his elaborate oratory. I know that the arguments he uses are not of a nature to bear exact scrutiny."[61] Forbes wrote again on December 26, mentioning that he had already read five chapter "with great and *increasing* interest" and commenting concerning Whewell's hopes for anonymity: "It would not be easy to conceal the authorship – but I find it is already perfectly understood here, so that I presume that you have been at no pains to conceal it."[62] In his long letter of February 16, 1854, Forbes begins to comment on the *Essay,* which, as he tells Whewell, he had been asked to review for *Fraser's Magazine.* As this letter reveals, and the review shows, Forbes had reservations about Whewell's astronomical arguments but accepted much of the remainder of his book. He especially praises Whewell's treatment of the design argument, but laments his advocacy of the nebular hypothesis, which Forbes describes as at "the utmost limits of vague conjecture or wild hypothesis" and which he "fears may be taken advantage of by some favourers of the cosmogony of the 'Vestiges.' "[63] Responding on February 19, Whewell states concerning the nebular hypothesis: "you may pull it to pieces . . . with my perfect goodwill. I have never, I think, argued upon it, except as an hypothesis. . . . I shall be surprised if any one thinks I give support to the man of the *Vestiges.* It seemed to me that my book might have some value as a strong case exactly opposed to his."[64]

The March 1854 issue of *Fraser's* carried Forbes's review, which was subsequently reprinted in the *Eclectic Magazine* and, for American readers, in *Littell's Living Age*. Early in his review Forbes claims that "the preponderance of belief in all ages has been in favor of the Plurality of Worlds. . . . The history of this opinion would be a curious one. . . ." (23, p. 51) Whewell did not dispute Forbes's last point, but in his *Dialogue* he urged that the pluralist doctrine "has always been regarded as extravagant, and has been maintained by only a few men, who have become by-words on that account."[65] Forbes even argues for an "involuntary prepossession of mankind at large" toward this doctrine, citing as "strong proof" the case of Dr. Elliot. To urge as evidence for this "prepossession" the case of a man judged insane in a court of law, partly because of this belief, does not speak well of Forbes's analytical abilities.[66] Forbes suffers another logic lapse when, after admitting that the absence of air and water on the moon tells against its habitability, he claims that evidence of air and water on the planets has "at least equal force" in arguments for their habitability (23, p. 53). This conclusion conflates necessary with sufficient conditions, as Whewell pointed out in his *Dialogue*.[67] Forbes attempts to save Venus and Mars for the pluralists, but Jupiter he surrenders with a flourish:

> Alas! for the imagined seat of higher intelligence; alas! for the glories of the most majestic planet of our heavens, the stern will of the ruthless destroyer has dissipated with no sparing hand the threads on which we hung the net-work of our imagery. No unsentimental housemaid ever made with relentless broom a cleaner sweep of a geometrical cobweb! (23, pp. 53–4)

As he had earlier in correspondence, he criticizes Whewell's use of the nebular hypothesis: "the wildest imagining which ever emerged from the mind of a mathematician."[68] Among the commentators considered up to this point, Forbes is unique in that while rejecting some of Whewell's astronomical ideas, he gives "hearty assent" to the "remaining arguments." These include Whewell's geological argument and his discussion of design, which, according to Forbes, frees "natural theology from some of the shackles with which it has commonly been trammelled." (23, p. 59) He expresses sympathy with Whewell's Christian objection, but apparently prefers the resolution of it offered by Chalmers, whose book Forbes found to be enhanced by a second reading (23, p. 59). Although Todhunter classified this review as unfavorable,[69] Whewell was no doubt pleased that Forbes both agreed with and praised substantial parts of his *Essay*.

Hugh Miller (1802–56), as noted previously, anticipated Whewell's geological argument in an 1846 article in the *Witness*. In the September 20, 1854, issue of that evangelical periodical, Miller returned to the debate with an article entitled "Geology versus Astronomy," which in

1855 appeared as the first of the four parts of a book also entitled *Geology versus Astronomy*.[70] Presenting in that book the Whewell debate as a conflict arising from tensions between the maturing science of geology, represented by Whewell, and the established science of astronomy, represented by Brewster (pp. 4–5), Miller soon makes clear his allegiance to Brewster. The basis for Miller's position is his belief that astronomy supports pluralism, whereas the vast time periods discovered by geology, rather than refuting the pluralists' arguments, show only that other planets gradually attain higher life forms. Although we cannot in the case of any specific planet indicate the level of life reached on it, we can say that ultimately it will become inhabited. As he puts it, we can at least infer that "other planets *may* have been ripening, nay, in all likelihood, *have* been ripening as certainly as our own; and the period of rational inhabitancy may have arrived in not a few of them." (p. 13) Miller faults Whewell for being overly severe in denying life on the planets (pp. 14–21), and he states that "Almost the only deductions of Sir David Brewster in which we are not disposed to acquiesce are some of his geological ones."

In the final section of his book, Miller turns to the relations between pluralism and Christianity, bringing up the Brewsterian idea that the first inhabitants of some planets may come from the earth. He even speculates that "Our globe may be the great nursery, not of the solar system only, but of the whole material universe." (p. 31) Miller also considers the possibility that such planets as Neptune may serve as places of punishment for resurrected earthlings. Taking a second tack, Miller suggests that intelligent beings on the earth or elsewhere need the mediation of an incarnated God in order to feel the love due the Creator. Miller does not, however, propose multiple incarnations; rather, he suggests that Christ of Calvary may in time provide that mediation for the inhabitants of other planets, although he leaves the mechanism of that mediation vague. He states that "though only one planet and one race may have furnished the point of union between the Divine and the created nature, the effects of that junction may extend to *all* created nature. . . . If it was necessary that the point of junction be somewhere, why not here?" (p. 33) What one finds in Miller is, in short, a desire to preserve both natural and revealed theology.

After Miller's death in 1856, two of his works were edited for publication by Rev. William Samuel Symonds (1818–87), a Cambridge graduate who ministered to a small congregation in Pendock in Worcestershire and wrote on natural history. Symonds's contribution to the debate, a long article in the *Edinburgh New Philosophical Journal,* was reprinted as a book entitled *Geology As It Affects a Plurality of Worlds*.[71] Symonds sets the stage for his discussion of the books by Whewell (not identified) and Brewster by characterizing Whewell's *Essay* as either "an enormous error or a great *truth*; there is no *via media* about it." (16, p. 268) Opting for the former alternative, Symonds presents a long list of Whewell's most controversial claims, following it with the remark: "No man in his senses can believe in astronomy and agree with the Author of the Essay without also believing [that all celestial bodies] are . . . fulfilling no purpose. . . ." (16, p. 274) Turning to Brewster, Symonds notes that he has been "vividly struck" by the "*enormous* difference of opinion" not only between the essayist and Brewster but also among astronomers themselves, even on such issues as whether or not the moon has an atmosphere (16, p. 276). Such ambiguous results in astronomy give point to Symonds's chief complaint against Brewster, that he neglected geology, which (for Symonds) sheds the most light on the issues. Symonds, we learn, accepts a great age for the earth, but sees its significance very differently than Whewell. In particular, Symonds asserts that the productions of each of the great geological periods fulfill specific purposes. Minerals from Plutonic rocks, for example, nourish our crops. The geologist, moreover, being aware of the progressive order in the fossil record, should conclude, according to Symonds, that a similar progression occurs on other planets.

In discussing Whewell's design chapter, with its stress on the primacy of mind, Symonds asserts that this aspect of the *Essay* should, at least among geologists, act to promote pluralism. The reason he cites is that because nearly all astronomers agree that the planets are of "superior construction . . . and possibly of earlier creation" than the earth, "it is more than probable that they are [already] tenanted by INTELLECTUAL beings. . . ." (16, p. 225) Symonds also charges that "The principle that pervades the whole of the Essay has been rightly characterized as 'the glorification of man.' "[72] Symonds sums up by stating: "Putting together the facts of geology, with the general beliefs of astronomers, we have no faith in the arguments of the Essay, or that there is the slightest ground for believing this earth anything more than the insignificant portion of the universe it has generally been believed to be." (16, pp. 236–7) For all of Symonds's stress on geology, as well as that of Miller, to whom Symonds frequently refers, it is odd that neither saw that their geology was, insofar as it was relevant, largely natural theology. This in the long run weakened their positions, not least because their teleologically oriented natural theology, although flexible enough to incorporate many geological discoveries from earlier in the century, could not survive the greater challenge soon to come from Darwin's *Origin of Species*.

The same spectre, it can be seen in retrospect, hung over Rev. Edward Hitchcock (1793–1864), the American geologist of the period most concerned to reconcile his science with religion. Hitchcock's *Religion of*

Geology (1851) has, in fact, been described as "the culminating document of the pre-Darwinian controversy between natural science and theology in America. It brought Paley up to date."[73] By 1854, Hitchcock had served as a Congregationalist pastor, as first chairman of the Association of American Geologists and Naturalists, and as president of Amherst College. All this made him a natural choice to write the "Introductory Notice" to the first American edition of Whewell's *Essay*.[74] Although praising parts of the *Essay*, Hitchcock wrote his introduction chiefly as a refutation of Whewell's central conclusion. After stating that "All the world is acquainted with Dr. Chalmers' splendid Astronomical Discourses," he asserts that the *Essay's* antipluralist thesis "gives us a very narrow idea of the plans and purposes of Jehovah, and one not sustained in our opinion by the analogies of science." Like Miller and Symonds, Hitchcock claims that geological evidence of stages prefatory to man's appearance on earth allows the inference that when other "worlds have passed through these prefatory changes, rational and immortal beings may be placed upon them." Moreover, seeing the wonderful adaptations of creatures on earth to their environments, "can we doubt that rational and intelligent beings may be adapted to physical conditions in other worlds widely diverse from those on this globe?" Although approving Whewell's treatment of geology and the transmutation of species problem, Hitchcock claims that his "favorite notions narrow our conceptions of the Divine plans and purposes." Prefaced by such comments, Whewell's *Essay* made its appearance in America.

From Berlin, the scientist and explorer Alexander von Humboldt (1769–1859) wrote Whewell on February 21, 1854, to thank him for having sent his *Essay* and to praise him for having attacked "a problem that has often received barbarous treatment." Humboldt adds:

It has become almost dangerous to treat it for fear of encroaching upon an entirely poetic imagination or of mingling too directly with the high principles of religion. . . . All of [your book] that relates to the solar system and to the nebulae is replete with imposing and often very new ideas; the geological survey, the theory of the solar system, the 11th, and especially the 12th chapter on the Unity of the World, appear to me to be full of attractions.[75]

That courtesy more than candor characterized Humboldt's letter is revealed by the comment he made but a few days later in a letter to Gauss:

How eternally marvelous the English are! Although thoroughly sensible in his History of the Inductive Science, Professor Whewell . . . in a peculiar writing "On the Plurality of Worlds" maintained that it is necessary on the basis of Christianity that no celestial body except the earth can be inhabited by intelligent beings because all intelligent beings are by nature sinful and the redemption (crucifixion) could not be repeated on so many million Rossean nebulae.[76]

As the comment suggests, geologists were no more inclined than astronomers to agree with Whewell. Nonetheless, the tendency that Whewell shared with such geologists as Miller and Symonds to view planets as changing entities became widely accepted in the remaining decades of the century.

5. Responses of other scientists: "mercurial in Mercury, saturnine in Saturn, and anything but jovial in Jupiter"

Interest in Whewell's book was widespread in the scientific community, extending not only to the astronomers, mathematicians, and geologists already discussed but also to biologists, chemists, physicists, and physicians, as well as to persons who may be identified as "science writers."

Among biologists, Charles Darwin (1809–82) read Whewell's *Essay* shortly after it appeared; in fact, one authority on Darwin has suggested that the unnamed upholder of the creationist position whom Darwin attacked in his *Origin of Species* was Whewell, whose views Darwin knew from his *Essay* and his earlier *History of the Inductive Sciences*.[77] Moreover, good evidence exists that Thomas Henry Huxley (1825–95) not only read the *Essay* but also reviewed it. In 1854, he was employed editing all and writing most of the scientific reviews published in the *Westminster Review*.[78] The April 1854 issue of that periodical carried a joint review of Lardner's *Museum of Science and Art* and Whewell's *Essay*. The former work receives scant attention, although Lardner's endorsement of pluralism is noted to provide a contrast to the antipluralist position of Whewell, whose authorship is suggested by the comment that the *Essay* "seems to us to indicate a mind far more deeply imbued with the principle of the 'Philosophy of the Inductive Sciences' than with those of the 'System of Logic'. . . ." (48, p. 314) Unfavorable to Whewell, despite his "cleverness and ingenuity," the reviewer accuses him of "anthropomorphism" and of "excessive carelessness" in quoting Owen.[79] Whewell's geological argument is rejected, the author urging a far greater antiquity for man than Whewell: "we live in hope yet to see a bit of palaezoic pottery. . . ." On a more general level, the author avers that pluralist "speculation and argument" are "unprofitable."

Three months later, the *Westminster Review* carried a review of Whewell's *Dialogue* and of Brewster's book. That the same author, presumably Huxley, wrote it is indicated by the fact that much of the review is given to answering Whewell's response in his *Dialogue* to the charge that he had misquoted Owen (49, pp. 243–5). Nonetheless, the reviewer praises Whewell for the "dignified good temper" exhibited in his *Dialogue* and for his argument that when the scientific evidence for and against pluralism is carefully analyzed, its force is seen to be " 'so small that the belief of all thoughtful persons on this subject will be determined by moral, metaphysical, and theological considerations.' "[80] Brewster is briefly but vigorously criticized, especially for his tendency to claim "that 'Heaven' is identical with 'the Heavens.' " But the reviewer's dominant theme is that the question of a plurality of worlds, being concerned with "the most hyper-hypothetical of speculations" is "unfitted for discussion." (49, p. 246) His concluding sentence reiterates this point: "surely, our scientific Alexanders are not yet justified in crying for other worlds to conquer."

That Huxley authored these reviews is supported by the similarity between the position adopted in them and the position advocated by Huxley in his *Science and Christian Tradition* (1894):

> Looking at the matter from the most rigidly scientific point of view, the assumption that, amidst the myriads of worlds scattered through endless space, there can be no intelligence, as much greater than man's as his is greater than a blackbeetle's; no being endowed with powers of influencing the course of nature as much greater than his, as his is greater than a snail's, seems to me not merely baseless, but impertinent.[81]

Moreover, Huxley urges: "Until human life is longer and the duties of the present press less heavily, I do not think wise men will occupy themselves with Jovian, or Martian, natural history. . . ." (p. 40)

Robert James Mann (1817–86), a physician-turned-science-writer, reviewed Whewell's *Essay* along with the volumes of Brewster, Jacob, and Powell for the *Edinburgh Review*.[82] Beginning with a sketch of the pluralist debate, Mann falls into an array of errors. For example, he misdates Copernicus's discovery by nearly a century, misrepresents the method by which he made it, and wrongly identifies him as a pluralist (17, pp. 436–7). Whewell is described as arguing from "preconceived ideas" and "from the obscure to the obvious." (17, pp. 441–2) His response to Whewell's geological argument consists of a recapitulation of Brewster's pre-Adamite speculation and of his prediction that man will survive far into the future. He treats Whewell's theory of the nebulae by rehashing the arguments of Brewster and of Jacob, while dismissing the essayist's analysis of the planets as "a mass of gratuitous and unsupported assumption." (17, p. 456) Attempts are made to preserve warmth for the most distant planets by postulating internal heat (following Jacob), whereas vision for their inhabitants is saved by modifications of the eye (following Brewster). No efforts are made to secure habitability for the sun, but Mann marshals most of the ideas then available to salvage lunar life. Mann summarizes Jacob's probability arguments and, by reasoning even more fallacious than that of Jacob, concludes that Whewell's position is *"one thousand times more likely to be wrong than to be right."* He also errs by urging that because life on earth occurs where air is present, planets having an atmosphere must be inhabited. Failing to realize that this mistakes a necessary condition for a sufficient condition, Mann states that pluralism is entitled to rank henceforth among the dogmas of science. . . ." (17, p. 465) The source of Mann's confidence is indicated by his claim that it is inconceivable that God's creative powers would not result in extraterrestrial life. Nonetheless, Mann, in his concluding paragraphs, faults Whewell for mixing science and religion. Overall, Mann's review is less an assessment of the books of Brewster, Jacob, Powell, and Whewell than an attack on the last author by means of arguments found in the first three.

Whewell, perhaps feeling that by responding to this largely derivative article he could combat an array of adversaries, published a reply in the *Saturday Review*.[83] Early in this article, identified as "by the author of the Essay," Whewell asks: "why does not the Reviewer consider the question, whether there *are* religious grounds for believing the unique position of man?" To this Whewell responds: "Not because he excludes religious arguments; for his main argument *for* plurality of worlds is the ordinary theological one, that otherwise the planets and stars are wasted. . . ." (42, p. 9) The problems in Jacob's probability arguments are exposed, as is Mann's distortion of Rosse's position concerning nebulae. Mann's provisions for selenites are shown to be unscientific, Bessel's researches being cited. Overall, Whewell makes a good case for his claim that the reviewer was guilty of "extreme abuses."

Whewell was not without allies in Edinburgh, as he learned in a letter of October 24, 1854, from George Wilson (1818–59), who in 1855 became the first director of the Scottish Industrial Museum, as well as regius professor of technology at Edinburgh University. Religiously concerned, he viewed himself as a Baptist, although he was active in the Congregational church. Wilson wrote to thank Whewell, whose identity he seems not to have known,[84] for a copy of the new edition of his *Essay,* to praise its arguments, and to describe its reception in Edinburgh. Wilson states that to many persons in Scotland, the *Essay*

. . . has been most welcome for its reverential, earnest, and thoughtful style, and although few have been prepared to accept the special conclusions still fewer among the thoughtful and educated have read with pleasure the accusations of irreligiousness brought against it.

Sir David Brewster's bitter book is popular among those, whom he and others have induced to believe that orthodoxy is committed to his side of the question; which can scarcely be the case seeing that the author of the Vestiges and he, are on this point at one. . . .

Why had Whewell sent Wilson his *Essay?* The probable answer is that he somehow learned of Wilson's 1852 book entitled *Electricity and the Electric Telegraph: Together with the Chemistry of the Stars; An Argu-*

ment Touching the Stars and Their Inhabitability (London). The volume consists of two small books, joined for reasons that are by no means clear. Only the fifty-page *Chemistry of the Stars* deals with pluralism. It is a curious text, employing extravagant but engaging language to discuss whether or not extraterrestrials, if such exist, resemble the denizens of earth (pp. 3–4). Wilson attempts to show that far greater diversity exists in the universe than is usually assumed and thereby to cast doubts on the pluralist position, especially as presented in *Vestiges*. In his first approach to his task, Wilson sets out to determine the "common sense" answer by summoning a jury of ordinary folk, before whom he parades the planets, stars, and nebulae, noting the different sizes, densities, and forms within each class of objects. The jurors render judgments in the jargon of their professions; a sailor, for example, finds as much diversity in the celestial parade as in a collection consisting of "a man of war, . . . an African slaver with its doleful passengers and demon-crew, . . . light Tahitian schooners . . ." and other vessels of all varieties (pp. 13–14). Wilson's point is to suggest the diversity among celestial bodies and to justify the unanimous verdict of this quaint crew: *"There are celestial bodies, and bodies terrestrial: but the glory of the celestial is one, and the glory of the terrestrial is another. There is one glory of the sun, and another glory of the moon, and another glory of the stars: star differeth from star in glory."* [1 Corinthians 15:41] (p. 20)

Wilson then turns to "the inner court of the priests," that is, to the astronomer, chemist, and biologist of his day. The "Astronomer," on examining the planets, finds that the telescope does not favor "the idea that a telluric or terrestrial character is common to the members of the solar system." (p. 25) The "Chemist," working from such sources as spectroscopic data and analyses of the compositions of meteorites, reaches the same conclusion. Whereas the author of *Vestiges* had claimed a homogeneity in the universe from the supposed fact that meteorites *"contain the ordinary materials of the earth,"*[85] Wilson's chemist, noting that only about a third of the sixty-odd terrestrial elements are present in meteorites, finds them to be further evidence of the diversity in the celestial realm. Theological arguments are also used to support Wilson's conviction of heterogeneity in the heavens; we are not, he urges, "to suppose that the Infinite One has exhausted the counsels of his wisdom in arranging the chemistry of our globe, and could only therefore repeat that endlessly throughout space. . . ." (pp. 38–9) The conclusions of his "Biologist" are similar; terrestrial life is not to be expected on the planets, for "we should be literally mercurial in Mercury, saturnine in Saturn, and anything but jovial in Jupiter. . . ." (p. 42) Wilson draws this together by urging, in opposition to the author of *Vestiges*, that the great diversity among the heavenly bodies makes it "seem impossible to hold the belief that the stars are all but so many Earths." (p. 43) In summary, Wilson opposed the idea that extraterrestrials similar to man are likely. Whether or not he was open to the possibility of beings of a very different sort on other worlds is unclear because of his metaphorical manner and playful approach.[86] Whatever Wilson's views on the latter issue were, Whewell, in the preface to the third edition of his *Essay,* praised Wilson's book as a "lively tract" containing "ingenious reflections." William Miller, an antipluralist of the 1880s, went even further, describing it as "the most important support to the position taken by Whewell."[87] Probably because of the increased attention being given to the question of other worlds as a result of Whewell's *Essay,* a second edition and a third revised edition of Wilson's volume were published before the end of the 1850s.

Sir Benjamin Collins Brodie (1783–1862), sergeant-surgeon to the queen, and from 1858 to 1861 president of the Royal Society, wrote to thank Whewell for having sent a copy of the *Essay.*[88] Correctly divining Whewell's authorship from its contents, Brodie praises the "extensive and various knowledge combined with great sagacity" shown in the book, but expresses the reservation that the author "has not shown that the mental principle, & that one and indivisible essence which each of us feels himself to be, *may* not exist in combination with other forms of matter besides that with which it is associated on this earth." Whewell also received a letter from Sir Henry Holland (1788–1873), who besides serving as physician to Queen Victoria and as president of the Royal Institution, attained some renown as a writer. A number of the objections raised in Holland's letter were taken sufficiently seriously by Whewell that he responded to them in his *Dialogue.* Although Holland's letter has been lost, Whewell's practice of marking in his personal copy of the *Dialogue* the initials of the person who raised each objection makes it possible to reconstruct to some extent what Holland wrote.[89] Moreover, other sources reveal Holland's overall reaction to the *Essay:* "it is the most brilliant of all Dr. Whewell's writings, but the theory is false."[90]

Both the *American Journal of Science and Art* and the London-based *Mechanics' Magazine* carried anonymous reviews in which the books of Brewster and Whewell were assessed together. In the former journal, Whewell's argument is praised as being "conducted with consummate skill and great power, usually with fairness, although sometimes sophistical when direct reasoning was insufficient, and in all parts with ennobling thoughts of man's religion and destiny." (1, p. 450) His evidence for the planets being uninhabited the reviewer finds "very nearly conclusive," but Whewell's ideas concerning the stars are judged of "far less, we should say, very little, probability." Brewster's approach is characterized as resting "much on the ground that the making of a world is waste labor unless the surface is afterward stocked with inhabitants. . . ." The author notes with surprise Brewster's advocacy of life on the sun, moon, and Neptune. Readers of the *Mechanics' Magazine* may have been surprised to find a four-part review of the books of Brewster and Whewell. Their perplexity could scarcely be diminished by the extravagant style of the reviewer, who admits to digressions "as wide as the revolutions of the solar system." (25, p. 513) Presenting the debate as having arisen between "the representative of academic culture in England" and "the pet of Modern Athens," the reviewer describes it as having spread to the platforms of "our Mechanics' and Literary Institutes." (25, pp. 441–2) The "dull" and "heavy" style of Whewell is contrasted with the no less ultimately unsatisfying "pyrotechnic display" of Brewster, for whose book a wider circulation is predicted because it is "so pleasant to find, by irrefragable proofs, a great name given over to infidelity. . . ." (25, p. 443) Using convoluted arguments, the reviewer claims that both Brewster and Whewell have forsaken the path of true inductive philosophy and have advocated positions on an issue the reviewer deems unresolvable. Todhunter's comment on this reviewer still stands: "it is difficult to say what is his main purpose."[91] But the reviewer's harsh conclusion concerning both Brewster and Whewell is clear enough: "they have obstructed the path of truth, impeded the advance of science, and even done damage to the interests of true religion. . . ." (25, p. 513)

Montagu Lyon Phillips, who taught science at Manchester Academy, entered the debate in 1855 with his *Worlds beyond the Earth* (London). The final words of his preface encapsulate not only his position but also one of the main strands of his argument: "There are more Worlds than Man's World. WHY NOT?" (p. v) As this indicates, Phillips is a pluralist; indeed, he favors life on all the planets, on the moon, and possibly on the sun, as well as on planetoids and meteorites. His pluralism rests partly on his belief that heavenly bodies without inhabitants are wasted. Phillips also champions a form of the nebular hypothesis that he claims enables him to deduce the size, mass, and density of each planet. A special feature of his book, which reveals the level of his scientific sophistication, is his conclusion that all planets are hollow. He even supplies a table giving the size of the central cavern of each planet (pp. 130–1). Phillips's pluralism makes him hostile to Whewell's ideas, but his enthusiasm for Brewster does not extend to his rejection of the nebular hypothesis. Phillips's book, which shows the extremes to which even a person with a measure of scientific knowledge would go to save the pluralist position, received few reviews, and no later edition appeared.

The year 1855 also saw the publication of *Scientific Certainties of Planetary Life: or, Neptune's Light As Great As Ours* (London), by Thomas Collins Simon, who, like Phillips, never published a scientific paper.[92] In his preface, Simon promises to solve the plurality of worlds question by "science alone" (p. x), but it soon becomes clear that despite having read the Herschels, Humboldt, Lardner, and Nichol, he had a very odd conception of science. For example, after discussing for a hundred pages whether or not the stars are suns, he turns to the planets and announces that although they move in different orbits, they are all in every way similar to the earth (p. 121) and even have "the same vegetable, animal, and intellectual life." (p. 127) The culminating portion of Simon's book strikingly demonstrates its author's aberrant views; in it he claims that Neptune, thirty times more distant from the sun than the earth, receives the same amount of light. The source of Simon's mistake is his belief that light from a central source decreases as the square of the distance only if the medium hinders its transmission. In fact, the inverse square law for light is geometrical, rather than physical, as a reviewer pointed out, thereby demolishing his main argument (28, p. 73).

The preface of Simon's book contains the fascinating claim that Whewell's attempt to refute pluralism "is, perhaps, the first of the kind that has ever been publicly made; for that of Dr. Cullen, the Roman-catholic Archbishop, [who wrote] something to the same purpose a few years ago, seems to have been suppressed, either by the authority of his Church, as excess of zeal, or in consequence of his own discernment of its unsuitableness to the present state of science in this country." (p. xii) Despite a number of attempts, I have been unable to confirm this claim concerning Cullen, who eventually became cardinal archbishop of Dublin.[93]

The books of Brewster and Whewell were reviewed in France by the physicist and meteorologist Jacques Babinet (1794–1872). His review begins with a history of the pluralist debate, described as culminating in the "immense sensation" created by Whewell and Brewster, the former being faulted for taking a position contrary to the metaphysical principle that "nature does nothing in vain." (41, p. 372) Brewster is treated far more favorably, although Babinet rejects life on the moon and sun. He is critical of both authors, but especially Whewell, for having written works that are "essentiellement théologiques." Whewell is described as making the sermons of Chalmers his starting point and as taking the words of the psalmist too literally. On a more technical level, Babinet states in opposition to Whewell that Sirius, according to John Herschel, gives 146 times more light than the sun. In response, Whewell noted that in 1833 Herschel specified the luminosity of Sirius as at least double that of the sun, whereas in 1849 he raised it to sixty-three times brighter. Finding that figure to have more than doubled in Babinet, who cites no reference, Whewell stressed the looseness in such estimates.[94] Although rejecting theology, Babinet accepts a metaphysical approach, claiming that astronomy and metaphysics combine to make the pluralist position quite certain (41, p. 383).

Viewed as a whole, the scientists discussed in this section, like those analyzed earlier, were mainly pluralists, the exceptions being Wilson, the reviewer for the *American Journal of Science and Art,* and possibly Huxley.

6. Religious responses: "a Bethlehem in Venus, a Gethsemane in Jupiter, a Calvary in Saturn"?

As must already be evident, few writings in the Whewell debate can be clearly categorized as either scientific or religious. In fact, a central issue in the debate was whether or not such a demarcation was possible or desirable. Nonetheless, one can specify which publications appeared in religious journals or were written by individuals, such as clergymen, who focused on religious issues. Collected for consideration in this section are items of this type. In examining them it is interesting to ask whether a correlation can be found between an author's creedal convictions and his position in the debate.

Two High Church Anglican journals,[95] the *Theologian and Ecclesiastic* and the *Christian Remembrancer,* carried unsigned reviews of Whewell's book. The review in the former journal was favorable, Whewell being praised for providing "a considerable confirmation of our belief that this earth is the only habitable . . . World. . . ." (47, p. 278) The reviewer cites Whewell's discussion of the planets as a "sound and brilliant ratiocination" and his design chapter as a "masterpiece of close and logical reasoning." That High Church Anglicans were divided on the pluralism problem is suggested by the very different review in the *Christian Remembrancer.* Although the books by Whewell, Brewster, Fontenelle, Huygens, and Wilkins are listed in the heading of the review, the critic concentrates on his contemporaries. He describes the *Essay* as written with "very clumsy and disorderly notions of logic, and a marvellous inelegance" of style (10, p. 60). Whewell, the reviewer avers, advocates an overly exalted notion of man: "if there be such a religion as Anthropolatry, we conceive our essayist to be its Hierophant." (10, p. 70) Although viewing Brewster's position as correct, the reviewer remarks: "a highly discursive strain of querulous remonstrance feebly advocates the established conviction . . . not seldom by mere bombast." (10, p. 53) In support of pluralism, the author cites the insignificant size of the earth, the need to explain why the celestial bodies were created, and the pious feelings produced by it. Although not sympathetic to Whewell's Christian objection to pluralism, the reviewer praises his discussion of design.

A High Churchman reading these two reviews might have found them similar only in offering few solid arguments.

Low Church Anglicans had access to more judicious, though no less opposed, assessments in one of their organs, the *Christian Observer*, where anonymous reviews of the Whewell and Brewster books were published. The reviewer of Whewell's volume rejects his position as based on "reasonings which do not seem very conclusive. . . ." (8, p. 422) Although writing with moderation and showing evidence of a sincere effort to understand, to recount, and to refute Whewell's views, he fails to convey the meaning and force of Whewell's Christianity objection. For example, he incorrectly describes the *Essay* as "an attempt to exalt the character of the human race by shewing that it is not numerically an unimportant element of the rational creation." (8, p. 424) The actual difficulty was better stated in a passage, which the reviewer quotes, from James Anthony Froude's 1849 autobiographical novel *Nemesis of Faith,* which led to Froude's forced resignation of his Oxford fellowship, Wrote Froude in reference to the earth:

> This miserable ball, not a sand-grain in the huge universe of suns, and yet to which such a strangely mysterious destiny was said to have been attached. I had said to myself, Can it be that God, Almighty God, He, the Creator Himself, went down and took the form of one of those miserable insects crawling on its surface, and died Himself to save their souls? I have asked the question. Did ever any man ask it honestly, and answer *Yes*? (8, p. 419)

Seven months later the *Christian Observer* carried a review of Brewster's book, obviously by a different reviewer. In fact, whereas the reviewer of Whewell's volume showed Brewsterian tendencies, the reviewer of Brewster's book was a Whewellite. After stating that Chalmers's writings on pluralism "have done more, perhaps, than all previous writings, to spread the impression of its certain truth," the reviewer suggests that the doctrine has led to "a tendency to distort the facts of science." (9, p. 36) Moreover, Brewster's belief that the Bible supports pluralism is wrong: "Our conviction . . . is exactly the reverse; the only natural conclusion to be drawn from . . . Scripture, and, above all, from . . . the Incarnation, is that Man really is the central object in the moral government of the Creator. . . ." (9, p. 42) This reviewer also accuses Brewster of writing with "asperity of tone," of arguing with "the zealotry of a partizan," and of falling into scientific "inaccuracies." (9, pp. 58–9) In particular, he criticizes Brewster for his pre-Adamites, for failing to distinguish among the various forms of the nebular hypothesis, and for accepting life on the sun. Despite questioning some of Whewell's statements concerning stars, he supports his treatment of nebulae by citing Rosse, J. Herschel, and other authorities. Although siding with Whewell's view that multiple star systems are unsuited for habitable

planets and faulting pluralists for uncritical reliance on analogical reasoning and on God's omnipotence, he admits the possibility of extraterrestrial life. His concluding recommendation is that because neither science nor Scripture permits us to decide the pluralist question, we should
suspend judgment for a time and pursue "the real fields of scientific
inquiry. . . ." (9, p. 61) Whewell, pleased by this well-argued essay,
called attention to it in the preface to the fourth edition of his *Essay*.

That the Anglican community was deeply divided in the Whewell debate is also suggested by a survey of the views expressed by a number of
other Anglican authors. For example, Rev. Dr. George Croly (1780–
1860) pressed the pluralist position in letters to London newspapers.
That Croly's pluralism was grounded in natural theology is shown by the
summarizing sentence in his long letter to the *Standard:* "If the planets
and stars are not fitted for such habitancy, what is their conceivable
use?"[96] On the other hand, Rev. Dr. Charles Musgrave wrote Whewell of
the "vast gratification" he derived from the *Essay*.[97] Musgrave's support
led a member of his congregation, Frederick William Cronhelm, to publish in 1858 a short book, *Thoughts on the Controversy As to a Plurality
of Worlds* (London), dedicated to Musgrave and siding with Whewell.
Cronhelm begins by tracing the history of the pluralist position, noting
the widespread acceptance of such sentiments as those of the mathematician Peter Barlow (1776–1862), who, according to Cronhelm, wrote
that "reason and imagination point out to us millions of suns and millions of worlds, each peopled by intelligent inhabitants. . . ." (p. 8)
Then came Whewell: "Inexpressible was the astonishment, and dire the
commotion . . . when one of the most eminent living philosophers
arose, and boldly called this popular creed in question." (p. 10) A David
arose to do battle with the "giant of Cambridge [but] the arrows in his
quiver were but railing and contumely." (p. 10) The main reason for
Cronhelm's opposition to pluralism was that if inhabitants of other planets exist, they must in many cases have fallen into sin, necessitating "a
Bethlehem in Venus, a Gethsemane in Jupiter, a Calvary in Saturn." (p.
17) This being absurd, men and angels must be the only intelligent beings. Lest the planets be without use, Cronhelm proposes that they function as abodes of angels and resurrected humans (pp. 23–4).

Rev. Josiah Crampton (1809–83), a graduate of Trinity College, Dublin, and rector of Killesher in Ireland, published in 1857 a book comparable to Cronhelm's in size but different in purpose. Like a number of his
earlier and later writings, Crampton's *Testimony of the Heavens*[98] argued that astronomy supports religion. To this end he invokes pluralist
themes, urging, for example, that God's design and power are shown by
the reflection that "all the vast number of those glittering spheres which
we behold are not desert solitudes, but peopled worlds. . . ." (p. 19) He

recounts how deeply he was moved, while using one of Lord Rosse's telescopes, by the thought that "every brilliant star [is] attended by a . . . train of satellite worlds. . . ." (p. 23) And each nebula he saw as a firmament, a universe! Although not mentioning Whewell, to whom he sent his book, Crampton no doubt had him in mind in discussing how to reconcile pluralism with Christianity. Crampton's response is basically Chalmersian, although in concluding his book he adds that Christ's ascension into heaven and promise to prepare a place there for his disciples further prove that the "material heavens [are] places of habitation." (p. 30) Crampton's book shows what Whewell no doubt knew: Many preachers would not readily relinquish Chalmers's cosmos. [99]

That poets were no more prone than preachers to surrender pluralism was the implicit message Whewell received when John Peat, a Cambridge graduate and "Incumbent of the Weald, Sevenoaks, Kent," sent Whewell his *Thoughts, in Verse, on a Plurality of Worlds* (London, 1856). In this work, Peat asks:

> Stupendous Universe! gigantic plan!
> Wast thou conceiv'd alone for earth-born man? (p. 7)

And he answers:

> Then surely there are worlds, as well as Earth!
> All matter is prolific, since its birth!
> In all its courses, – all its graceful curves,
> Wherever Matter is, it LIFE subserves. (p. 9)

Not only does the principle of plenitude indicate pluralism, but also

> Impartial science! thou canst teach the soul
> That Earth is but a part of one great Whole. . . . (p. 11)

Peat's poem attained a second edition, but this did not justify the prediction of Brewster, to whom it was dedicated, that it " 'could not fail to become popular.' "[100]

Whewell would have been disappointed to learn the verdict passed on his volume by his former student Alfred Lord Tennyson. In 1854, Tennyson, as his son recorded, "carefully studied" Whewell's *Essay* and concluded: "It is to me anything but a satisfactory book. It is inconceivable that the whole Universe was merely created for us who live in this third-rate planet of a third-rate sun."[101] Whewell fared better with another former student, William Carus, a canon of Winchester and in 1854 vicar of Romsey. Writing Whewell in 1854, Carus describes the *Essay* as "extremely interesting" and admits being "carried away" by its argument. "I feel myself . . . still your pupil," he adds, noting that he had spoken of the volume to a "Bishop and a host of Clergy."[102] Whewell also received a letter from the Oxford-educated James Carden, at one time curate of

Oddington. Carden conveys "deep agreement" with Whewell's book, but follows it with fifteen pages of nonsensical scriptural commentary.[103]

The Anglican vicar of Polesworth in central England, Rev. Robert Knight, published anonymously in 1855 his *The Plurality of Worlds: The Positive Argument from Scripture, with Answers to Some Late Objections from Analogy* (London).[104] Writing in reaction to Whewell, Knight compares the *Essay* to the volumes of Paine and Maxwell, even hinting that Whewell had plagiarized the latter author. Determined to reconcile pluralism with the Bible, he urges that although the Scriptures tell us that the "Incarnation is unique," they also reveal that its "influences . . . are universal." (p. 16) Knight's first argument is that because angels function as messengers and because one or a few angels could accomplish that task for terrestrials, we must either assume other inhabited worlds or ascribe "indolent inactivity [to] myriads of heavenly messengers." (pp. 30–8) Knight next argues that passages in the New Testament, especially in the Pauline epistles, refer to a plurality of worlds (pp. 39–48). The pluralist doctrine brings a sense of proportion to Christian thought; for example, it helps answer those who ask why so few persons seem to follow Christ (pp. 48–57). Turning to objections to pluralism, Knight, in a short chapter, sweeps away all the astronomical problems, relying especially on the argument that whereas it is possible that astronomers "may discover the actual existence of inhabitants of other worlds – they can never disprove it." (pp. 91–2) The weakness in this type of argument is illustrated by Knight's parallel response to the possibility that astronomers might show that there is no evidence whatsoever of "sentient existence" on the sun and moon. To this he replies: "Are there no vehicles adapted to be receptables . . . of spiritual existence, but such bodies as are palpable to human sight?" (p. 88) Knight's treatment of Whewell's geological argument is scarcely more satisfactory, being based on pre-Adamites and on the supposition that the history of the earth before man may have served as preparation for him. Whewell's reinterpretations of the design argument Knight partly neglects and partly rejects, in the course of trying to prove that God never acts in vain (pp. 110–28). After recapitulating his arguments, Knight urges that pluralism helps us to avoid setting limits to our charity and to see that we are involved, perhaps in a special way, in the religious history of the entire universe. Overall, Knight's book exhibits moderation of tone, but a minimal understanding of science, an uncritical theological sense, and a propensity to misuse Scripture.

In concluding this discussion of Anglican writings, mention may be made of Rev. Ebenezer Cobham Brewer (1810–97) of Trinity Hall, Cambridge. In the final part of his *Theology in Science*,[105] Brewer sets out the issues in the form of a debate that he leaves unresolved. This more or less accurately reflects the situation among Anglicans; while most favored pluralism, although not necessarily Brewster's formulation of it, no consensus had emerged.

A sample of English Roman Catholic thought can be found in the *Rambler,* a Roman Catholic journal edited by Richard Simpson (1820–76), an Oxford-educated former vicar who had converted in 1845. Simpson probably wrote at least the first two of the three *Rambler* reviews.[106] The first review, that of Brewster's book, begins with the information that Catholic authorities had made no *de fide* declaration for or against pluralism. Although St. Augustine and St. Philastrius of Brixen included this doctrine in lists of heresies, St. Clement of Rome, Clement of Alexandria, St. Irenaeus, Origen, and St. Jerome all affirmed it (38, p. 130). Theologically there is nothing against it, but it may conflict with the philosophy supporting that theology, even though admittedly that philosophy may be false. The reviewer accepts extraterrestrial "seats of animal life," but rejects pluralism as "the hope of the Christian," as Brewster had called it. In fact, he states that Brewster's principles would incline him "to abandon Christianity." Brewster's treatment of the afterlife and of the possibility of Christ becoming incarnate on other planets provokes him to a charge of "Gnosticism." He labels Brewster's teleologically based claim that wherever matter exists, life must also exist, as "the very essence of materialism." In general, Brewster's position contains "evidences of a stupidity and narrowness of mind" produced by too little attention to metaphysics. This reviewer returned to the debate two months later with a brief review of Whewell's *Essay,* revealing that he "had formed a high estimate of it from misrepresentations of Sir David. . . ." (39, p. 60) On reading the *Essay,* he found it "really remarkable," but he warns against what he describes as the Kantian philosophy of its author.

In August 1855, the *Rambler* published notices of the books by Phillips and Powell. Negative toward both, the reviewer asks concerning Powell: "What will [Oxford] do to her Professor, who writes a book inconsistent not with the thirty-nine Articles, but with Christianity altogether? Probably nothing at all." (40, p. 151) Describing Powell as the "rev. but infidel gentlemen," the reviewer criticizes the first part of Powell's *Essays* for reductionist and positivist tendencies, while faulting the second part for the claim that the Old Testament account of the creation should be read as mere poetry. The third part of Powell's book being found even more objectionable, he concludes that the book "is a melancholy but instructive evidence of the necessary tendencies of the Protestant principle." Overall, the *Rambler* reviews opposed pluralism and contained more name-calling than careful analysis.

Powell's book was also reviewed in the *London Quarterly Review,* a journal of Methodist orientation. The reviewer, accepting Powell as a believing Christian but viewing his book "with the greatest apprehension" (24, p. 220), concentrates on the first and third essays, describing them as philosophically and theologically dangerous. In regard to the pluralist second essay, he observes that Powell's "perilous predeliction for the discovery of analogies" is but one of the factors that made it "scarcely possible" for him to "hold the balance" between Whewell and Brewster "with a steady hand." Influenced possibly by the excesses he perceived in Powell's book, this reviewer opts for the antipluralist position of Wesley, whom he quotes.

The *British Quarterly Review,* a Baptist and Congregationalist journal, published in 1854 a long review that carried the titles of the books of both Whewell and Brewster in its heading; yet, as the reviewer reveals in his final paragraph, he saw Brewster's book only after his review was being printed (6, p. 85). The reviewer is very negative to Whewell, accusing him of sophistry, of "perverted ingenuity," and of making the solar system "a clumsy, slaternly contrivance" and the universe "a sublime failure." He asserts that the " 'religious' argument against plurality is founded upon the most irreligious view which can be taken of the divine operations." The arguments brought against Whewell are frequently based on biased analogies; for example, in arguing for life on some of the planets, the reviewer compares them to houses seen at a distance. The reviewer gives up the habitability of the moon only amidst a host of cautions and qualifications, but he confidently declares, in opposition to Whewell's geological argument, that when nature seemed to be "idling away her time in growing huge *calamites* and *stigmariae,"* she was in fact "constructing coal-cellars for future men. . . ." (6, p. 55) He concludes by proclaiming Brewster's book to be "as emphatic a protest against the essayist's views as it would be possible to pronounce."

Two members of the Catholic Apostolic or "Irvingite" church published books in the debate. This denomination was founded shortly after 1832 when Edward Irving (1792–1834) was expelled from the Presbyterian church. Rev. William Tarbet sent Whewell a copy of his anonymous *Astronomy and Geology As Taught in the Holy Scriptures.*[107] In the accompanying letter, Tarbet praises Whewell's *Essay,* adding that it had led him to write his own book. Whatever delight Whewell derived from Tarbet's letter was no doubt dispelled when, on turning to Tarbet's booklet, Whewell found him to be an extreme scriptural literalist who rejected modern geology and accepted creation in six days (p. 13). Tarbet's tone and approach are reflected in this statement: "It is by faith, then, and not by science that we can understand aright *how* God made the world. He has revealed it to us. And *He* ought to know." (p. 17) Henry Drummond (1786–1860), who served in Parliament, was a founding member of the Irvingites. In 1855, he published *On the Future Destinies of the Celestial Bodies* (London), which is primarily a work of scriptural exegesis and eschatological speculation. In it, he argues that resurrected men will be placed at various locales in the universe, with "the higher locality, the Throne of the Eternal, the dwelling-place of God [being] this globe, to the dust of which the Son of God united Himself." (p. 63) He also discusses pluralism, seeming to accept it, but not because of science (pp. 20, 45).

Many pluralist Christians were no doubt displeased when in 1855 Edward Higginson (1807–80), a Unitarian divine, published his *Astro-Theology; or, The Religion of Astronomy* (London). Describing pluralism as "the grandest thought . . . in the whole range of Natural Science and Religion" (p. 28), Higginson rejects Whewell's Christianity-based objection as "the most perverse and gratuitous of religious difficulties." (p. 47) Brewster is also criticized; his "special pleadings of a most unphilosophical kind" and his claim that the psalmist received pluralism by divine inspiration have done "the most serious damage to the credibility of the Scriptures." (pp. 13–17) Finding that comets and planetoids are exceptional bodies and that the sun and moon serve other functions, Higginson does not populate them. However, he deems it "impossible" to doubt that nearly all planets are inhabited (p. 29) and is scarcely less sure that almost every star is "encompassed by a system of worlds." (p. 33) Higginson then attacks Christianity. Because the man of science accepts pluralism, he should see that "the orthodox Atonement is proved to be a paralogism in astronomy, as palpably as the Athanasian Trinity is an absurdity in arithmetic." (p. 51) He devotes his final section to urging that the pluralist creed elevates and refines Christian thought, and using Brewster's eschatological speculations, he argues against the traditional Christian belief in the resurrection of the body (pp. 73–96).

The Unitarian *National Review* published in 1855 a review of the books by Brewster, Higginson, Powell, Simon, and Whewell. Although describing Powell's *Essays* as "in advance of them all" (28, p. 72), the reviewer says little of it or of the books by Higginson or Simon, the latter's central scientific error being exposed in a footnote (28, p. 73). The pluralist author focuses on Whewell, who is criticized above all for his claim that an analogy is destroyed when it is not exact. In particular, Whewell is charged with believing that a difference in degree nullifies an analogy, whereas only an absolute difference will suffice (28, p. 74). The author's argument would have been stronger had he not asserted that the fact that Neptune receives only one nine-hundredth of the light and heat reaching the earth constitutes a difference only in degree. The reviewer refers to the improbability of Whewell's position and dispatches his religious objection by citing Brewster's response. He also speculates that the beings on planets nearer the sun have intellects superior to those of beings on more remote planets (28, pp. 88–90).[108]

In 1858, a rather curious book, *The Stars and the Angels,* appeared anonymously in London.[109] Its author claims (1) that very few planets support vegetable and animal life and (2) that among these, "not one in a million is inhabited by a race of intelligent and moral beings . . . ," and (3) fewer still by "fallen intelligences." (p. 35) On one level, he bases these claims on his conviction that Christ came as redeemer only to the earth (p. 122). Nonetheless, on behalf of the first two claims, he marshals cogent scientific arguments, some no doubt derived from Whewell's *Essay.* Although adopting a Whewellite solar system, the author champions a cosmos spatially and temporally more extended than that of the essayist. For example, he speculates that "there are millions of extinct suns . . . , each with dead planetary systems still revolving around them" (p. 89), and having identified nebulae as other universes, he suggests that there may be "millions of extinct nebulae" as well as "materials for millions of millions more of nebulae . . . , which have not yet begun to be kindled." (pp. 87–8) Despite the rarity of inhabited planets in his cosmos, its magnitude makes their number significant, leaving him with the problem of how to deal with their inhabitants. In this regard, he urges that extraterrestrials, being made in God's image, are similar in form to man and in fact are members of the "human species." (pp. 156, 188) Moreover, angels belong to the same species and possess the form that men will eventually attain (pp. 156–7). Resurrected men will consequently be able to travel from planet to planet, perhaps populating them: "may we not suppose that many of the stars are, as it were, the cradle of angelic hosts, from which, it may be, thousands of other stars are peopled . . . ?" (p. 195) This book's blend of scientific sophistication, moderateness of tone, and religious speculation helped it attain possibly two English editions and certainly two American editions.

From Scotland came contributions by James Morison (1816–93) and George Gilfillan (1813–78), both of whom had been licensed to preach by the Presbyterian church, Gilfillan serving a congregation in Dundee until his death. Morison, suspended in 1841 for non-Calvinist views of the atonement, founded the Evangelical Union or "Morisonians" and edited the *Evangelical Repository,* for the first issue of which he reviewed Whewell's and Brewster's books.[110] Although describing Whewell's *Essay* as a "seasonable recall from certain extravagant conceptions" (18, p. 24), Morison assails it, arguing, for example, that "observations of Sir William Herschel, Arago, and others" indicate that the sun is habitable. The moon, he maintains, may also be inhabited, but by beings who are "no analogues" of earthlings. Labeling Whewell's view of nebulae as "nebulous" and his theory of stars as "far from shining by its own light," he concludes by citing Whewell's Bridgewater Treatise against him. Morison's review of Brewster's book shows scarcely more restraint; he describes it as an "onslaught" written by a man who "seems to be in heat" and attacks Brewster's claims that evil is " 'a necessary part of the general scheme of the universe' " and that " 'all the attributes of God, except omnipotence, are inferences.' " Morison's reviews probably derived from his attachment to the Scottish pluralist position, as formulated by Chalmers and by Morison's father-in-law, Thomas Dick.

By 1854, when George Gilfillan reviewed Whewell's *Essay* for the *Eclectic Review,* he had already begun to establish his reputation as a skillful writer and respected critic. His *Gallery of Literary Portraits* (1845) was the earliest major success in a literary career that included a hundred or so books and pamphlets. In his review, Gilfillan recounts the intense emotions and dreamy delights he experienced when at age eleven he encountered Fontenelle's book. These raptures recurred a few years later when he read Chalmers's *Astronomical Discourses,* which "more certainly than any volume of sermons ever published . . . electrified the reading world." (15, pp. 515–16) As Gilfillan's biographers state, his reading of Chalmers "gave his thought . . . spiritual direction. . . . Now he saw that one might have literary power, might be eloquent and command the ear of the age, might even be scientific, and none the less be Christian."[111] Chalmers, it seemed to Gilfillan, had provided a "bright bridge, for the first time uniting Earth and Heaven." Such might seem the making of an ardent pluralist, but, writes Gilfillan, "after the momentary madness of admiration had passed away, it was found that the bridge was only a rainbow, beautiful, evanescent, unreal. . . ." (15, p. 516) In fact, as he notes, he had in his 1845 *Gallery* already criticized Chalmers's book, claiming that its ideas had been " 'conveyed' . . . bodily from Fuller's 'Gospel Its Own Witness,' " even though not "worth the trouble . . . being made up, on the whole, of assumptions and truisms." (15, p. 519) Moreover, in his *Christian Bearings of Astronomy* (1848), Gilfillan raised further objections.[112] Coming to Whewell's *Essay* with this background, he found it, as he states in his review, "a remarkable production" characterized by "manly energy, clear precision, and philosophic calm" and "exquisite . . . reasoning." After a thorough summary of the *Essay,* he concludes his review by praising it as "a blow in the face of Natureworship," a "demolition of the development theory," and for showing that the value of a man exceeds that of the entire material universe. Gilfillan was not, however, totally opposed to pluralism; as he wrote Whewell, he had in his *Gallery* commented: "May not this Universe be only beginning to be peopled, and the Earth be the first spot selected for the great colonization?"[113]

American Presbyterians also entered the debate. In 1855, the *Presbyte-*

rian Quarterly Review carried a discussion of the Brewster and Whewell books and a book by Lardner. Its anonymous author's command of astronomy is shown by its contents, and an editors' note prefacing it identifies him as an "officer of the Church." The author, who marshals much astronomical information on behalf of Whewell, criticizes Brewster's extravagant claims for life on the sun and moon. Proceeding from the assumption that conditions known to be incompatible with terrestrial life preclude life elsewhere (35, p. 587), he argues that, at least at present, no planets are inhabited (35, p. 598). He concludes by stating: "the general tendency of telescopic discoveries during the *last* three quarters of a century, has been decidedly adverse to the theory, that the planets and stars are the abodes of sentient and rational existence." From this it might seem that American Presbyterians had attained a consensus – had the editors not prefaced the review by the remark that despite the "valuable facts" in it, they were "unconvinced by the argument." And they add: "whatever may be the truth in regard to suns and moons, the planets and those bodies occupying analogous positions in other solar systems, are inhabited by rational creatures." (35, p. 572)

Few inferences about the preferences of American Lutherans or Methodists can be drawn from a notice of Brewster's book in the *Evangelical Review* (20, p. 438), an Evangelical Lutheran journal, or a notice of Whewell's book in the *Methodist Quarterly Review* (26, pp. 617–18), each being but a paragraph long. However, this was space enough to make clear that the pluralist position was favored. Americans ecumenical enough to have read the longer reviews of the Whewell and Brewster books in the Baptist-oriented *Christian Review,* the *Methodist Quarterly Review,* and the *New York Quarterly* may have been struck by the consensus among them, but this is easily explained – all were written by the Congregationalist minister Joshua Leavitt (1794–1873).[114] In the most religiously oriented of his reviews, that in *New York Quarterly,* Leavitt accepts Whewell's Christian objection to pluralism and praises "the profound sagacity and the great learning" in his *Essay* (30, p. 467). Leavitt, far more ready than Whewell to specify God's plan for the universe, argues that man is the culmination of the Creator's efforts and proceeds to deal with why God created the rest of the cosmos, asserting (erroneously) that the other planets are necessary for the solar system's stability and (more erroneously) that the stars "keep the planets in their places." From this, he infers that "the stars are for man." (30, pp. 448–50) Designating as "drivel" some of Brewster's conjectures, he faults Brewster for claiming scriptural evidence for his position, but succumbs to the same temptation himself (30, pp. 457–8). In his contribution to the *Christian Review,* Leavitt lists many cases in which Brewster and Whewell, although "standard authorities" (11, p. 205), disagree on astronomical detail. Again he attempts to justify uninhabited celestial bodies, showing in this that despite denying Brewster's pluralism, he was fully Brewsterian in seeking teleological explanations. Leavitt's best review is that in the *Methodist Quarterly Review,* where he provides a primitive if promising criticism of the pluralists' use of analogical argumentation (27, pp. 360–5). Nonetheless, readers must have been puzzled as to how he could simultaneously assert that "It is impossible to make any theory of a plurality of worlds square with the scheme of redemption" (27, p. 378) and that, were pluralism proved, "it would not destroy a single argument in favour of Christianity. . . ." (27, p. 359) It is also unclear how, after having strived mightily in his first two reviews to provide a teleological explanation of many features of the universe, Leavitt could attack pluralists by claiming "we have no competency whatever to determine, *a priori,* what is wise or fitting for the Creator to do in the creation of the world." (27, p. 377)

David Nevins Lord (1729–1880), a Yale graduate who edited the interdenominational *Theological and Literary Journal,* certainly wrote the longest and probably wrote all three reviews that appeared in his journal.[115] The earliest is a short but otherwise unrestrained denunciation of Whewell's book (44, pp. 165–7). The attack was taken up three months later in a review, identified as by Lord, of the books of Brewster and Whewell. The intensity of Lord's aversion to Whewell's volume is illustrated by his assertion that the "postulate . . . from which [Whewell] starts is not that of a Christian believer; it is not that of an ordinary deist. . . . It is the postulate of one who . . . thinks the deity is essentially like himself, immeasurably imperfect. . . ." (45, p. 281) This leads Whewell, according to Lord, to put forward "the meanest, the falsest, and the most contemptible form of theism that has ever been devised." (45, p. 287) Lord concludes by calling Whewell's views "cold, barren, senseless, and atheistic." Although describing Brewster's book as "able and pleasing" (45, p. 288), he criticizes its author for admitting as a hypothesis the great age of the earth. The review scarcely mentions astronomy, which was probably felt to be unnecessary, Lord being convinced that both the Old and New Testaments proclaim the pluralist position. The principle of plenitude is implicitly invoked when Lord asserts that God's nature necessitates that he "should raise his empire to an extent so immense as to display . . . his infinite greatness. . . ." (45, p. 299) Lord's wild charges and misrepresentations suggest how deeply he felt concerning the issues. The third review, that of Powell's *Essays,* if not by Lord, is in line with his fundamentalism. It is aimed primarily at Powell's evolutionary views, these being labeled "the revolting result [of] rejecting the word of God, and following what he regards as the lights of science." (46, p. 613) Powell's pluralism is favorably noted, but he is faulted for not supporting it by scriptural proofs and teleological arguments.

Two large books discussing pluralism within a religious perspective appeared in America. In 1855, William Williams anonymously published his two-volume *The Universe No Desert; The Earth No Monopoly; Preceded by a Scientific Exposition of the Unity of Plan in Creation* (Boston).[116] The first volume, devoted to an exposition of the "Unity of Plan in Creation," consists primarily of a survey of biology and geology as presented in natural theology books. Williams's references in it suggest that he read widely, but his lapses into such notions as numerological analogies between plants and planets (v. I, pp. 86–7) reveal an author who understood far less than he read. Except for the twenty pages of quotations from Chalmers that conclude the first volume, pluralism is postponed to the second volume, in which Williams seeks to justify his final words: "THE UNIVERSE IS NO DESERT, AND THE EARTH IS NO MONOPOLY." (v. II, p. 239) His second volume, written to respond to the "ruthless onslaught . . . of [Whewell's] subversive work" (v. II, pp. 3–4), begins by discussing the solar and stellar regions. Then he recounts many stock pluralist arguments, especially that in which opponents are charged with advocating a Divinity of limited powers or wasteful tendencies. Williams espouses life on the moon (v. II, pp. 46–54, 126–31) and sun (v. II, p. 70); in regard to the planets, he endeavors to show both that terrestrial conditions prevail on them and that if they did not, God could create suitably adjusted inhabitants for them. His survey of the history of the pluralist debate contains such claims as "Galileo received many suggestive hints from Giordano Bruno. . . ." (v. II, p. 162) The fact that Whewell is listed as a pluralist indicates that Williams was unaware of the identity of the essayist, who with W. S. Plumer[117] is the only antipluralist mentioned. Williams's final arguments begin with the claim that men possess "inner senses" that when developed give access to otherwise unobtainable knowledge. He then enthusiastically presents twenty-seven pages of testimony from a person with such developed inner senses: Emanuel Swedenborg.

In 1859, Rev. Charles Louis Hequembourg published his *Plan of the Creation; or, Other Worlds, and Who Inhabit Them* (Boston). His starting point is the problem of reconciling pluralism with Christ's incarnation and atonement, concerning which he comments: "No difficulty . . . was ever made a source of sceptical opposition of half such formidable magnitude." (pp. 36–7) Moreover, despite Chalmers, Foster, and Fuller, it has remained without "decisive solution." Fundamental to Hequembourg's resolution is his distinction between "whether the universe was *designed* to be inhabited and whether it *is* so at present." (p. 39) Hequembourg's thesis is that Christians must reject pluralism as applying to the present time when *"the universe is in its comparative infancy"* (p. 58), but that God's plan is that eventually the other planets will be populated by resurrected men of the earth (p. 388). In short, man "is to emerge in successive generations . . . to overspread the universe. . . ." (p. 384) Hequembourg develops his thesis in great detail, relating it to many scriptural passages. Although Whewell is not mentioned, Hequembourg refers to his *Essay* (p. 60) and may have drawn arguments from it. Despite the efforts that went into the large volumes of Williams and of Hequembourg, neither attained a second edition.

The final publication in this group appeared anonymously in France in 1858 under the title *Rêveries et vérités, ou de quelques questions astronomiques envisagées sous le rapport réligieux, en résponse à l'ouvrage du Docteur William Whewell sur la pluralité des mondes* (Paris). Its author was J. J. Larit, a French Protestant, who sent Whewell a copy accompanied by a long letter.[118] In the first part of his book, the "Rêveries" section, Larit, who admits never having studied science (p. 1), theorizes that planets continually cool and undergo compaction; as their density increases, they approach the sun in a gradual spiral motion. Because of this motion, the earth will eventually burn up, but the outer planets will solidify and come sufficiently close to the sun to be habitable (pp. 51–6). A similar process occurs in other solar systems, giving each a single habitable planet at any given time (p. 48). The author's chief interest, however, is in the "Vérités" section (pp. 85–316) of his book, where he uses pluralist imagery and many scriptural quotations to stress God's love for creatures on all worlds. He also discusses Whewell's Christian objection, faulting him above all for having too limited a conception of God's love (pp. 137–40). This and his feeling that Christian humility cannot easily be reconciled with Whewell's position seem to have been the sources of Larit's book.

The feature of the religious component of the Whewell debate that most strikingly sets it off from other science-religion controversies is the diversity of the responses, even among those coming from a single denomination. Of persons identified as Anglicans, six favored and seven opposed Whewell, with one (Brewer) remaining neutral. Congregationalists, Irvingites, Methodists, and Presbyterians were also in each case essentially equally divided. Overall, fourteen authors preferred Whewell, whereas eighteen opposed his position. One cause of this diversity was the ambiguous testimony of both science and Scripture on the issue of other worlds. Moreover, many authors (consciously or not) probably perceived the decision as involving such traditionally perplexing problems as whether they should give primacy to teleological or nomological design arguments, to revealed or natural theology, and even to Christ or God.[119] The lack of a decisive commitment in the Christian *community*, however, contrasts dramatically with the stances taken by *individuals,* nearly all of whom boldly defended one side or the other.

7. Pluralism and the public: other responses to Whewell "who has so electrified us all"

A number of items in the Whewell debate do not fit the categories used up to this point; some came from the general public and/or appeared in publications serving it. This does not imply their insignificance; in fact, some of the most widely read reviews appeared in journals of this type.

The *Athenaeum,* which was such a journal, published brief reviews of the books of Brewster, Phillips, and Powell. Although neutral in tone, the review of Brewster's volume concludes with a comment that suggests a serious problem in his approach. Noting his claim that the planets are now or will eventually be inhabited, the reviewer remarks: "Against an assertion so elastic it is not easy to make head." (2, p. 710) Present-day methodologists, sensitized by Karl Popper to the dangers of unfalsifiable claims, would agree. Stylistic similarities suggest that the later joint review of the Phillips and Powell books came from the same pen. Rating Powell's pluralist essay the "most important" of the essays in his "earnest and graceful volume," the reviewer appears positively disposed toward it (3, p. 639). This may not have been his reaction to Phillips; one suspects irony in his comment that Phillips "gives fearless expression to every thought," especially when he then quotes Phillips's claim that the absence of an atmosphere is not a serious obstacle to a planet's habitability.

In the August 19, 1854, issue of *Notes and Queries,* the *Essay* is attributed to the person who in the first Bridgewater Treatise endorsed pluralism. The article is propluralist, Bruno, Brahe, and Kepler being cited as earlier advocates of that doctrine (32, p. 140). The issue of December 9, 1854, includes a response to the query from "W. W." of Malta, who found the earlier hint as to the authorship of the *Essay* insufficient. He is informed that the British Museum has cataloged the *Essay* under Whewell (33, p. 466).

A potentially very effective way of presenting the issues was used in the *British Controversialist,* which during 1855 carried a debate between six disputants, all employing pseudonyms, with three arguing the positive or pluralist position and three favoring Whewell. The debate begins with the affirmative side, represented by "Philalethes,"[120] who maintains that the abundance and diversity of terrestrial life point to the probability of life elsewhere. Moreover, the nebular hypothesis and God's generosity and power, as well as the "fact that some [planets] possess an atmosphere and volcanic mountains," support the same conclusion (5, pp. 16–20).

"H. D. L.," arguing the negative, shows rather more skill. Contrasting the views of that "mighty veteran" Brewster with those of Whewell "who has so electrified us all," he draws on Whewell's geology and the Christian arguments, while disparaging such Brewsterian excesses as his pre-Adamites, his stress on size as an indication of habitability, and his passion for populating the sun and moon. The problem of the purpose of the planets is met by adopting Brewster's idea that resurrected men may inhabit them.

"Threlkeld" appears next for the affirmative, using natural theology to claim that "the orbs of heaven . . . are wasted if . . . tenantless." (5, p. 51) He admits some exceptions: Because other purposes may be assigned to the sun and moon, they may lack life. He also cites Scripture in support of the pluralists, who alone can make sense of such biblical passages as "The heavens shall declare thy wonders, O Lord." Lastly, he suggests that the antipluralist position would diminish the delight and devotion in their lives. "S. S.," next for the negative, points out the inconsistency of "Threlkeld" in using analogical arguments for planetary life but abandoning them for the sun and moon. He also urges that appearances and analogies give ambiguous testimony, and Scripture none at all, on the question of other worlds (5, pp. 91–4).

"L'Ouvrier" of Birmingham then comes forth for the affirmative, recommending an empiricist approach: "We like facts, there is something so tangible, so real, and life-like about them. . . ." (5, p. 143) But many of the facts he cites range from dubious to absurd. For example, the first "fact" mentioned after the foregoing statement is that we receive 85 percent as much heat from the stars as from the sun. He also asserts that science shows "the high probability that those portions of the universe which are most definitely observable, possess land to walk on, air to breathe, water to drink, a fire for warmth." He responds to the Christianity-based objection by repeating the suggestion of "Threlkeld" that only earthlings have sinned. "Vincat Veritas" urges, against "Philalethes," that extraterrestrials, being formed in God's image, must resemble men and that consequently terrestrial criteria as to conditions of life must govern the debate. He also attacks "Philalethes" for relying on the nebular hypothesis and uses chemical studies of meteorites against the pluralists (5, pp. 174–5).

The concluding statements of "Philalethes" and of "H. D. L." can be treated together. The former alleges that his opponents bias the question by claiming that what is at issue is whether or not men exist on other planets (5, p. 207). This seems reasonable until he asserts that this error leads them to believe that planetary "density and temperature . . . size and surface" are relevant. In response, "H. D. L." asserts that the antipluralists' insistence that extraterrestrials must be assumed to resemble men is necessary because "it is the only way in which an argument can be conducted at all." (5, p. 249) The reason is that if we assume that God creates beings adapted to any environment, then we are "incapacitated from reasoning about them in any way." Both debaters misconstrue Whewell's geological argument as a positive argument, and a parallel problem plagues them in regard to teleology: Despite their differing conclusions, they assume that God's purpose for each feature of nature is determinable. A more modest view on the teleological issue might have kept "Philalethes" from claiming that were the planets uninhabited, God would have *"failed to execute his purpose,"* and his opponent from countering that the purpose of the planets is to provide man a lovely sky and final residence. Such problems suggest that the high potential of the approach used in the *British Controversialist* was far from fully realized.

Blackwood's Edinburgh Magazine* for fall of 1854 contains a long, anonymous review of the books by Brewster and Whewell as well as Lardner's *Museum of Science and Art.* Its author was Samuel Warren (1807–77),[121] whose career included study of both medicine and law, election to the Royal Society and to Parliament, and publication of a number of legal manuals and novels. Warren praises Whewell's *Essay* as free of "dogmatism and arrogance" and as "replete with subtle thought, bold speculation, and knowledge of almost every kind, used with extraordinary force and dexterity. . . ." (4, p. 400) In contrast, Brewster's book is a "slight performance . . . disfigured . . . by an overweening confidence." He also suggests how Brewster felt when he saw

. . . Dr. Whewell go forth on his exterminating expedition through Infinitude! It was like a father gazing on the ruthless slaughter of his offspring. Planet after planet, . . . star after star, . . . nebula after nebula, all disappeared before this sidereal Quixote! . . . This could be borne no longer; so thus Sir David pours forth the grief and indignation of the Soul Astronomic. . . . (4, p. 380)

Quoting frequently and paraphrasing effectively, Warren reconstructs Whewell's overall argument, giving special attention to his geological argument, which he suggests "was the origin of the . . . Essay" (4, p. 300) and which he correctly describes as designed "not to prove an opinion, but *to remove an objection.*" (4, p. 387) Warren cites some of Brewster's excessive claims and, drawing on Whewell's *Dialogue,* responds to objections raised by the Scottish scientist. Warren's endorsement of Whewell's position, which he also communicated in letters they exchanged,[122] greatly pleased Whewell, who praised Warren's review in the preface to the third edition of his *Essay.*

"Within the last few months a remarkable controversy has arisen . . ." (12, p. 246); so begins an anonymous 1854 review of Brewster and Whewell in *Dublin University Magazine.* The reviewer favors Whewell, who "has conducted his argument, in the main, with discretion and logical precision. . . ." (12, p. 256) Concerning Brewster, however, than whom "no living philosopher . . . commands more universal respect," the reviewer finds himself "refusing to admit that his treatise is a satisfactory or even a judicious answer to the Essay itself." He illustrates the "over-eagerness with which speculations . . . are grasped" by Brewster by citing his advocacy of pre-Adamites and of other ideas that "impugn the received doctrines" of geology. Whewell is criticized for his treatment of stars and nebulae, but praised for his "striking, indeed noble passage" dealing with apparent wastage in the universe, for his lofty view of man, and for his "most dignified, philosophic, and moderate spirit."

Five years later, the same journal published an article by Rev. James Wills (1790–1868), an Irish Anglican known for his poetry, periodical essays, and historical writings. Whereas the reviewer of 1854 favored Whewell, Wills attacks him on the basis of traditional natural theology. However, the time had passed when one could credibly claim that God "cannot be at a loss to people the most elaborate of his planets with suitable intelligences, suitably framed." (13, p. 337) Nor could informed readers take seriously Wills's explanation of the belt of asteroids as a "planetary wreck . . . an impressive memorial . . . telling an awful history of rebellion, revolt, and Divine justice, to the eyes that are privileged to read it." (13, p. 338) Wills did learn from Whewell: The latter's geological argument led Wills to admit that we need not suppose "each individual planet, or any particular planet, to be the present habitation of rational beings." (13, p. 341)

Preserved in Whewell's papers are three anonymous reviews that Todhunter discussed, but without being able to identify the authors who wrote them or the journals that published them. Two of these have already been discussed and their sources and authors specified.[123] The third, which is headed "The Plurality of Worlds" and which begins with the words "The third edition of the 'Plurality of Worlds' was advertised in our last number . . . ," describes the Essay as showing "good writing, ingenious thinking, and very extensive scientific knowledge" and as dealing with a problem raised in "the most brilliant volume by the greatest of Scottish clergymen," that is, Thomas Chalmers (54, p. 1). The reviewer concludes by recommending the books of both Brewster and Whewell "as two of the most remarkable pieces of controversy which the present age has produced." Curiously, however, little is said between these two statements about either book, the author preferring to present his own speculation that after the resurrection we may inhabit the other planets, that our planet "may be the great nursery, not of the solar system only, but of the whole material universe." Despite being somewhat open to pluralism, the author maintains that Christ's incarnation must be unique to the earth, although its influence may extend to other worlds.

Overall, the review is slightly pro-Whewell but largely irrelevant to the essayist's arguments.

Among the letters Whewell received in regard to his *Essay,* the most remarkable came from seventy-four-year-old G. W. Featherstonhaugh (1780–1866).[124] After praising the *Essay,* he reveals an 1845 experience that he had never previously communicated even to his "most intimate friends." While attending a table-rapping seance, Featherstonhaugh received astronomical information from Newton and the elder Herschel, including "the real object and intention of all the planetary bodies." This, alas, he did not share with Whewell, although he assured him of its "cutting off every possibility that any one of them was a habitation for mortal beings." Whewell's response, if any, was not preserved. Whewell no doubt answered the letter of Francis Egerton, first earl of Ellesmere (1800–57), whose uncle had endowed the Bridgewater Treatises.[125] Egerton expresses agreement with Whewell's position and thanks him for relieving his "mind of a load not of willing scepticism but of reluctant doubt." The long and thoughtful letter of the distinguished historian Henry Hallam (1777–1859) is summed up in this statement: "Upon the whole, it is an original and very remarkable book, and will probably mark an epoch in those speculations. I cannot deny that it leaves considerable difficulties and will, I dare say, be unfavourably viewed by the majority. Chalmers's theories were well received, though they were very arbitrary."[126] Ellen Henslow wrote in appreciation of the book, stating that the arguments gave "strength to my own previous leanings to your side of the question."[127] Gilbert Rorison, a Scottish minister, had similar leanings; in fact, he mentions in his letter that he had expressed anti-pluralist sentiments in sermons written before Whewell's book.[128] Whewell also received letters from Stephen Edmond Spring Rice and Arthur Henry Dyke Troyte, both of whom had studied at Oxford. Neither of their letters contains anything more important than general praise for Whewell's book. Joseph Toomer, on the other hand, expresses disappointment at seeing the lofty pluralist position under attack. The last of this group is an anonymous letter in which the writer admits that Whewell's arguments have converted him, and he attacks the ideas, said to be held by some, that the *Essay* is derogatory to God's powers and "ministers to overweening conceit in the created. . . ."[129]

Four reviews, three of Whewell's American edition and one of Brewster's book, appeared in American periodicals published for the general public. Although none exceeds a page in length, all contain vigorously expressed views. The earliest, a review in *Putnam's Monthly* of the *Essay,* is a description of Whewell's ideas and of the objections in Hitchcock's introduction. Although leaning toward the latter, the author's main suggestion is that "we should leave the stars to settle their own business. . . ." (36, p. 678) The cynical tone of this comment is also evident in the review of Brewster's book that appeared in *Putnam's* two months later. Although favoring the pluralist position and praising Brewster for writing "with much plausibility and earnestness," the author admits that "no positive proofs [and] no scientific grounds" support pluralism. All in all, he states, these books present us "with the spectacle of a most elaborate and spirited controversy between two of the most eminent men of science in Great Britain, in respect to a subject on which there are not facts to argue." (37, p. 222)[130] Between the two *Putnam's* reviews, *Peterson's Magazine* reviewed Whewell's *Essay*, which is praised for "the force of its reasoning and the felicities of its style." (34, p. 65) Although mentioning Hitchcock's hesitancy, the reviewer places "great store on the book." In 1855, the *Southern Quarterly Review* noted a new American edition of Whewell's book. The reviewer prefers it to the book by Brewster, who "writes more like a man repelling personal abuse, than a philosopher refuting a scientific theory." (43, p. 264)

That the debate reached not only the periodicals but also the parlors of the Victorian period can be illustrated by Anthony Trollope's novel *Barchester Towers* (1857). A conversation in it is a miniature of the debate, which shows that Trollope knew of Whewell's "race of salamanders in Venus" and his "pulpy" Jupiterians. One participant describes talk of such creatures as "almost wicked," and another asks "why shouldn't the fish [in Jupiter] be as wide awake as men and women here?" The last words go to Charlotte Stanthrope, whose response to this question is: "That would be saying very little for them." And she adds: "I am for Dr. Whewell myself; for I do not think that men and women are worth being repeated in such countless worlds."[131]

8. Conclusion: "a most elaborate and spirited controversy"

When the Whewell debate is looked at broadly, a number of significant generalizations emerge. The most striking is that the debate was very extensive, more extensive than is suggested by Todhunter's earlier analysis.[132] Whereas he referred to nine books published in response to Whewell's *Essay*, twenty are treated in this chapter. Moreover, the number of periodical publications has risen from twenty-two for Todhunter to over fifty. If letters be counted, the views of nearly one hundred participants in the debate have been documented. Although future research will no doubt locate more primary sources, the number known at present is sufficient to justify some quantified conclusions:

1. Of the twenty books published in the Whewell debate, 70 percent supported pluralism.

2. Of the articles and reviews, about two-thirds favored pluralism.
3. Of publications by Anglicans, about 71 percent opposed Whewell. If Unitarians are excluded, a similar percentage applies to items by non-Anglican Protestants.
4. Among the scientists discussed in the first five sections of this chapter, 83 percent favored pluralism.
5. Among the religious writings discussed in the sixth section of this chapter, 56 percent opposed Whewell, showing that religious writers were signficantly less inclined than scientists to adopt the pluralist position.
6. British and American authors were about equally likely to support pluralism.
7. Although nearly all the periodical publications appeared anonymously, the authors of about half have now been identified.
8. Of the published items known to Todhunter, 72 percent opposed Whewell. This indicates that his sample was not unrepresentative, even though it consisted largely of items sent to or collected by Whewell.

As these data show, Whewell made relatively few converts. However, as Herschel noted, he may have prevented "a doctrine from crystallizing into a dogma."[133] A number of considerations help explain why Whewell's book created, in the words of one of its reviewers, "a most elaborate and spirited controversy." (37, p. 122) First, the boldness of Whewell's position and the brilliance with which he advanced it attracted a wide audience. Even those denying his conclusions frequently admitted the ingenuity of his arguments.

Second, the degree to which the pluralism of the astronomers had become vulnerable provided Whewell with an opportunity for his attack. Dubious observations, overly extended analogies, questionable theoretical claims, and problematic methodological assumptions permeated much of the pluralist literature prepared by astronomers in the decades preceding the *Essay*. Yet pluralism had persisted, championed both by such fringe figures as Dick and Gruthuisen and by such prominent professionals as Bode, Arago, and John Herschel. Nonetheless, its foundations were so frail that they could not withstand the close scrutiny Whewell provided.

Third, the time was also right for an attack on the pluralism of the natural theologians, who had been no less persistent than astronomers in pressing pluralism to its limits. As an aid to understanding their reaction to Whewell, recall how distressed natural theologians were by Darwin's *Origin of Species* (1859), which they perceived as claiming that widespread wastage occurs in nature, that teleological approaches are no longer acceptable, and that God's design can be deciphered only with difficulty. What is significant in the present context is that six years earlier Whewell had advanced these claims in his *Essay*. Moreover, in doing so, he seemed to forsake many lofty goals of the physicotheologians: to display God's omnipotence and generosity, to present conceptions of the Divine reconcilable with science, and to write of God without involving sectarian or scriptural disputes.

Fourth, one of the most religiously controversial features of Whewell's book was his claim in regard to Christianity. Its radical character emerges most clearly if it is compared with Paine's contentions (1) that we must reject either pluralism or Christianity and (2) because astronomy has established pluralism, Christianity must be false. What distressed many pluralists was that they saw Whewell as endorsing Paine's first claim, which for them entailed the second. At the least, Whewell seemed to undermine such authors as Chalmers, who in opposing Paine denied his determination of the alternatives.

A fifth factor contributing to the size and severity of the debate was the lack of agreement concerning the precise status of and proper support for pluralism. Not the least controversial claim made by Whewell was that the plurality of worlds question would be determined less on scientific grounds than "by moral, metaphysical, and theological considerations."[134] Although it would be excessive to suggest that, in general, scientists accepted pluralism primarily for theological reasons while religious writers sought its support in science, it is probable that in some cases this occurred, and it is certain that most pluralists believed that both science and religion supported pluralism. The contrast between the Whewell debate and the Darwin debate in this regard is striking: Whereas the latter dispute is usually interpreted as a conflict between science and religion, the Whewell controversy consisted of a battle within religion and a battle within science. Put differently, it was a sort of night fight in which the enemy could not readily be recognized. As Whewell must have realized after sending copies of his *Essay* to friends, it was distressingly difficult to predict who would support and who reject his conclusions. Allies in many previous battles, such as Forbes, Owen, and Sedgwick, opposed Whewell's position on pluralism.

These were the factors that fired this curious debate in which men of deep religious convictions were challenged not by unbelievers but by men of equally sincere religious beliefs to debate what some saw as a question of astronomy. If this chapter and the excellent analysis published by Dr. Brooke have a prime significance for historians of nineteenth-century thought, it is that the Whewell debate deserves their careful consideration because of all that it reveals concerning both astronomy and science-religion interactions during the middle years of the nineteenth century.

Appendix: bibliography of articles in the Whewell plurality of worlds controversy published between 1853 and 1859

The items in this bibliography are chiefly reviews of Whewell's book or of books written in response to it. Where the author of the article is known, his first initial and last name are given. If his name is followed by an asterisk, his authorship is known from a source other than the article itself. Where the article is a review, the last name of the author of each book reviewed is listed in parentheses. Otherwise the title of the article is given. An asterisk following Whewell's name indicates that in the review Whewell is not identified as the author of his *Essay*. References in this chapter to these articles are given in the form (*x*, p. *y*), where *x* is the article number and *y* is the page number.

1 *American Journal of Science and Art* (Brewster, Whewell*), 18 (November 1854), 450.
2 *Athenaeum* (Brewster), no. 1389 (June 10, 1854), 709–10.
3 *Athenaeum* (Powell and Phillips), no. 1440 (June 2, 1855), 639.
4 *Blackwood's Magazine* (S. Warren*) (Whewell, Brewster, Lardner), 76 (September–October 1854), 288–300, 370–403.
5 *British Controversialist,* "Is the Notion of a Plurality of Inhabited Worlds Consonant with Science and Revelation?" 6 (1855), 16–25, 50–4, 91–4, 141–5, 173–7, 207–10, 249–52.
6 *British Quarterly Review* (Whewell*, Brewster), 20 (1854), 45–85.
7 *Christian Examiner* (T. Hill?*) (Whewell*, Brewster), 57 (September 1854), 208–20.
8 *Christian Observer* (Whewell*), 54 (June 1854), 407–25.
9 *Christian Observer* (Brewster), 55 (January 1855), 35–62.
10 *Christian Remembrancer* (Whewell*, Brewster, Fontenelle, Huygens, Wilkins), 29 (1855), 50–82.
11 *Christian Review* (J. Leavitt*) (Whewell*, Brewster), 20 (April 1855), 202–19.
12 *Dublin University Magazine* (Whewell*, Brewster), 44 (August 1854), 246–56.
13 *Dublin University Magazine* (J. Wills), "Other Worlds," 53 (1859), 330–46.
14 *Eclectic Magazine* (R. Mann*) (Whewell*, Brewster), 37 (1854), 25–45; reprinted from *Edinburgh Review*.
15 *Eclectic Review* (G. Gilfillan*) (Whewell), 7 (May 1854), 513–31.
16 *Edinburgh New Philosophical Journal* (W. Symonds), "Astronomical Contradictions and Geological Inferences Respecting a Plurality of Worlds," 2nd ser., 2 (1855), 267–83; 3 (1856), 39–58, 212–38.
17 *Edinburgh Review* (R. Mann*), (Whewell*, Brewster, Powell, Jacob), 102 (October 1855), 435–70.
18 *Evangelical Repository* (J. Morison*) (Whewell), 1 (September 1854), 22–8.
19 *Evangelical Repository* (J. Morison*) (Brewster), 1 (1854), 60–5.
20 *Evangelical Review* (Brewster), 6 (January 1855), 438.
21 *Fraser's Magazine* (J. Forbes*) (Whewell*), 49 (March 1854), 1–12.
22 *Harvard Magazine* (A. Searle*) (Whewell, Brewster), 1 (April 1855), 166–72.
23 *Littell's Living Age* (J. Forbes*) (Whewell*), 41 (1854), 51–60; reprinted from *Fraser's*.
24 *London Quarterly Review* (Powell), 8 (1857), 219–38.
25 *Mechanics' Magazine* (Whewell, Brewster), 61 (1854), 440–4, 461–6, 486–91, 508–13.
26 *Methodist Quarterly Review* (Whewell), 36 (October 1854), 617–18.
27 *Methodist Quarterly Review* (J. Leavitt*) (Whewell*, Brewster), 37 (July 1855), 356–79.
28 *National Review* (R. Hutton?*) (Powell, Whewell, Brewster, Simon, Higginson), 1 (July 1855), 72–91.

29 *New Englander* (D. Olmsted*) (Whewell, Brewster), 12 (November 1854), 570–603.
30 *New York Quarterly* (J. Leavitt*) (Whewell, Brewster), 3 (October 1854), 435–67.
31 *North British Review* (D. Brewster*) (Whewell*), 21 (May 1854), 1–44.
32 *Notes and Queries*, "Plurality of Worlds," 9 (August 19, 1854), 140.
33 *Notes and Queries*, " 'Plurality of Worlds': Its Author," 9 (December 9, 1854), 465–6.
34 *Peterson's Magazine* (Whewell*), 26 (July 1854), 65.
35 *Presbyterian Quarterly Review* (Brewster, Lardner, Whewell), 3 (March 1855), 572–601.
36 *Putnam's Monthly* (Whewell*), 3 (June 1854), 678.
37 *Putnam's Monthly* (Whewell*, Brewster), 4 (August 1854), 222–3.
38 *Rambler* (R. Simpson*) (Brewster), N.S., 2 (August 1854), 129–37.
39 *Rambler* (R. Simpson*) (Whewell), N.S., 2 (October 1854), 360–1.
40 *Rambler* (Phillips, Powell), N.S., 4 (1855), 151–3.
41 *Revue des deux mondes* (J. Babinet) (Whewell, Brewster), 1 (January 15, 1855), 365–85.
42 *Saturday Review* (W. Whewell*), "The Plurality of Worlds," 1 (November 3, 1855), 8–10.
43 *Southern Quarterly Review* (Whewell), 28 (1855), 263–4.
44 *Theological and Literary Journal* (Whewell*), 7 (July 1854), 165–7.
45 *Theological and Literary Journal* (D. Lord) (Whewell*, Brewster), 7 (October 1854), 276–303.
46 *Theological and Literary Journal* (Powell), 8 (April 1856), 593–615.
47 *Theologian and Ecclesiastic* (Whewell*), 16 (1854), 271–8.
48 *Westminster Review* (T. Huxley?*) (Lardner, Whewell*), 61 (April 1854), 313–15.
49 *Westminster Review* (T. Huxley?*) (Brewster, Whewell*), 62 (July 1854), 242–6.
50 *Witness* (H. Miller) (Brewster, Whewell) (September 20, 1854); republished in his *Essays, Historical and Biographical, Political and Social, Literary and Scientific* (Edinburgh, 1862), pp. 364–79; also in Miller's *Geology versus Astronomy* (Glasgow, [1855]), pp. 1–14.
51 *Witness?* (H. Miller*) (Brewster); date unknown; copy preserved in Whewell review volume at Trinity College, Cambridge (Adv. C. 16. 35); also published in Miller's *Geology versus Astronomy* (Glasgow, [1855]), pp. 14–22.
52 Dr. G. Croly, a letter to the *Standard* (1854 or 1855); preserved in the Whewell review volume at Trinity College, Cambridge (Adv. C. 16. 35).
53 H. J. S. Smith, "The Plurality of Worlds," *Oxford Essays of 1855* (London, n.d.), 105–55.
54 Review of Whewell's *Essay*, beginning "The third edition of the 'Plurality of Worlds' was advertised in our last number . . . ;" a review running six columns, preserved, without source given, in Whewell review volume at Trinity College, Cambridge (Adv. C. 16. 35).

PART III

From 1860 to 1900

8

New approaches to an ancient question

1. Developments from the 1860s, especially the "new astronomy"

Writing in 1867 in what may have been the last review of Whewell's *Of the Plurality of Worlds: An Essay*, Rev. Theodor Appel (1823–1907), an American professor of astronomy and mathematics, described it as "like a bomb-shell thrown into an army resting on its victorious march."[1] Although Appel's comment is basically correct, it should not obscure the fact that some publications in the years after Whewell's *Essay* show no awareness even of its existence, let alone its arguments.[2] Moreover, a significant number of the pluralist articles from the 1860s were centered on issues dominant in the 1850s.[3] The slowness with which ideas change is dramatically indicated by the fact that life on the sun was championed in the late 1850s and in the 1860s by a number of American, Belgian, British, French, and German authors. Some supporters of solarians came from the fringes of science,[4] but in other cases, prominent scientists advocated this idea. For example, the English chemist Dr. Thomas Lamb Phipson (1833–1908) asserted in 1867 that the sun "must indeed be a region of eternal life and perfect happiness."[5] The next year the English astrophysicist J. Norman Lockyer (1836–1920), in his *Elements of Astronomy,* presented solar life as a possibility.[6] Also in the 1860s, Mungo Ponton (1802–80), a pioneer of photography and a founder of the Bank of Scotland, attacked Whewell's "notorious book" and championed both William Herschel's theory of a cool, habitable core for the sun and John Herschel's view that Nasmyth's "willow leaves" consist of giant organisms.[7] In an 1859 address to the Belgian Academy, Jean Baptiste Joseph Liagre (1815–91), an astronomer and mathematician, concluded his discussion of the sun by stating that it ought no longer be seen "as a devouring furnace and destroyer, but as the most imposing of the planetary globes; [as a] majestic abode where the perfection of organized beings ought to be . . . in harmony with the magnificence of the habitation."[8] Most energetic on behalf of solarians was Fernand Coyteux

(1800–?), who in 1866 published a massive book arguing for life on the sun, thereby creating a controversy in a learned society in Poitiers, France.[9]

New approaches, however, did arise, the most important being associated with what came to be called the "new astronomy." Central in this development was Sir William Huggins (1824–1910), an English astronomer, who in an 1897 paper surveyed his more than three decades of discoveries as one of its leading practitioners.[10] There Huggins recounts his dissatisfaction during the 1850s with the "routine character of ordinary astronomical work" and his hope of finding new methods of research. This wish was fulfilled when he learned of the publications dating from 1859 of Bunsen and Kirchhoff, who showed that a prism or diffraction grating added to a telescope transformed it into a new instrument, the spectroscope, which in turn made the astronomer an astrophysicist and astrochemist. Kirchhoff's 1861 spectroscopic determination of the chemical composition of the sun dramatically demonstrated the potential in this new research technique, the availability of which gave Huggins a feeling "like the coming upon a spring of water in a dry and thirsty land." (p. 911) Collaborating for a time with the chemist William Allen Miller (1817–70), Huggins found that "nearly every observation revealed a new fact, and almost every night's work was red-lettered by some discovery." (p. 913) The story of the remarkable results attained by Huggins and such other pioneers of astronomical spectroscopy as Ångstrom, the Drapers, Janssen, Lockyer, Schellen, and Secchi has frequently been recounted[11] and need not be repeated here. What is noteworthy is that many of these discoveries were seen as significant for the pluralist debate; in fact, to a greater extent than has previously been noted, the pioneers of astronomical spectroscopy saw themselves as involved in that debate.

Huggins's first major spectroscopic paper, written in conjunction with Miller, appeared in 1864 as "On the Spectra of Some of the Fixed Stars."[12] It is a report on the spectroscopic study of nearly fifty stars, with special attention being given to a few of the brightest (p. 48). They present their detection of stellar spectral lines coinciding exactly with those of many terrestrial elements as "some proof that a similar unity of operation extends through the universe. . . ." And they add: "The differences which exist between the stars are of the *lower order,* of differences of *particular adaptation,* or special modification, and not differences of the *higher order* of distinct *plans of structure.*" (p. 60) From the structural similarities of these stars to our sun they infer that they "fulfill an analogous purpose, and are, like our sun, surrounded by planets. . . ." And the presence of terrestrial elements in those stars suggests that such chemicals are probably "present in the planets genetically connected with them. . . ." Moreover, they assert:

It is remarkable that the elements most widely diffused through the host of stars are some of those most closely connected with the constitution of the living organisms of our globe, including hydrogen, sodium, magnesium, and iron. . . . On the whole we believe that the foregoing spectrum observations on the stars contribute something towards an experimental basis on which a conclusion, hitherto but a pure speculation, may rest – viz. that at least the brighter stars are, like our sun, upholding and energising centres of systems of worlds adapted to be the abode of living beings. (p. 60)

That this claim went beyond observational evidence was admitted by Huggins in a footnote he appended to this passage in 1909. Although stating in that footnote that Miller had persisted in maintaining that position, Huggins asserted that by 1866 he had freed himself "from the dogmatic fetters of my early theological education," which had motivated that passage.[13]

Welcome as his first 1864 paper was to pluralists, Huggins's next discovery distressed them by jeopardizing the idea of island universes. Most astronomers had by the 1850s concluded that all nebulae are in principle resolvable into individual stars and consequently are structures comparable to the Milky Way. The reports in the late 1840s from Lord Rosse in Ireland and W. C. Bond at Harvard that Orion had been resolved seemed to confirm this belief, Whewell and Herbert Spencer being among the few dissenters. In 1864, Huggins published the first in a series of papers that in effect showed that their dissent was justified. This paper recounts spectroscopic observations of six planetary nebulae, as well as the ring nebula in Lyra and the nebula in Vulpecula, for all of which he found bright line spectra, indicating that rather than being island universes, all were composed of glowing gases (pp. 108–16). An 1865 paper on Orion revealed that it gives a bright line spectrum, but it also showed Huggins's hesitancy to reject the Rosse-Bond resolution. Suggesting in 1865 that Orion consists of glowing gaseous points (p. 119), Huggins by 1866 was drawing the more natural conclusion (pp. 131–2) and by 1868 was reporting that his spectroscopic study of seventy of the brightest nebulae had shown that "about one-third give a spectrum of bright lines." (p. 137) These researches of Huggins were a major reason why the island universe theory was rejected by the end of the century, only to be successfully resurrected around 1920. Another argument used against it was Whewell's analysis of the Magellanic Clouds. Huggins's determination of the gaseous nature of some nebulae did not by itself destroy the island universe theory; he had, after all, found continuous spectra for many nebulae, including Andromeda (p. 500). This left open the question of island universes, although it certainly did not prove their existence. This is clear from the fact that in 1889 Huggins urged that Andromeda was not a star cluster, but rather a single star with an adjacent planetary system (p. 173).

Although the spectroscope cannot provide information on the chemical composition of such nonglowing objects as planets, it does enable astronomers to determine if a planet or moon possesses an atmosphere and, if so, its chemical composition. In his 1864 study of the moon and planets, Huggins states that the moon yields only a solar spectrum, confirming its lack of a sensible atmosphere (p. 360). A distinctive modification in the solar spectrum is reported for Jupiter and probably for Mars and Saturn (pp. 365–9). In an 1867 paper, Huggins describes his efforts, labeled partly successful, to find a characteristic spectrum for Mars (pp. 369–72). This paper (as discussed later) caused controversy in the 1890s when W. W. Campbell, despite superior instrumentation, was unable to duplicate Huggins's results.

Huggins's papers, which usually appeared in the Royal Society's *Philosophical Transactions,* reached only a limited audience. Yet as early as 1865 more accessible journals carried accounts of his discoveries, presented in the context of the plurality of worlds question. For example, Robert Hunt (1807–87) explained them to readers of *Popular Science Review,* quoting the pluralist passage already cited from the 1864 Huggins-Miller paper. Hunt concludes that these researches "appear . . . to have proved all the planetary and stellar worlds are made after the type of our own Earth."[14] William Carter was no less enthusiastically pluralist in his account of the Huggins-Miller results in his 1865 *Journal of Science* article. Entitled "On the Plurality of Worlds," it argues for life on Venus, Mars, Jupiter, and Saturn, even though the spectroscope had provided little direct information about any of them, except possibly Mars.[15]

The linkage of spectroscopic astronomy with pluralism, so strikingly evident in Huggins's writings, is found also in the publications of a number of the other pioneers of the new astronomy. Lockyer's openness to the possibility of life on the sun has already been noted, and (as shown subsequently) Rev. Angelo Secchi, the most prominent Italian astrophysicist, endorsed pluralism in many of his writings. Heinrich Schellen, a German who authored one of the earliest books on spectrum analysis, devoted a major part of it to astronomical researches. After noting that the spectroscope shows "a great difference in the constitution of individual stars," he adds: "we must assume that even these individual peculiarities are in necessary accordance with the special object of the star's existence, and its adaptation to the animal life of the planetary worlds by which it is surrounded."[16] The propensity to present spectroscopy as supporting extraterrestrials is even more evident in the leading French astronomical spectroscopist of the period, Jules Janssen (1824–1907). By 1869 he had reported the discovery of water in the Martian spectrum and had asserted:

To the close analogies which already unite the planets of our system, a new and important character has now been added. All the planets form, accordingly, but one family; they revolve around the same central body giving them heat and light. They have each a year, seasons, an atmosphere. . . . Finally, water, which plays so important a part in all organized beings, is also an element common to the planets. These are powerful reasons to think that life is no exclusive privilege of our little earth. . . .[17]

Moreover, in an 1896 address before the Académie des sciences, he claimed not only that spectroscopy had shown that the bodies in the universe form a "single family" but also that the question of extraterrestrial life

. . . has been worked out by an aggregate of facts, analogies, and rigorous deductions which leave no room for doubt. . . . It is infinitely probable that hydrogen, oxygen, nitrogen, carbon, and especially water, which are the indispensable constituents of vegetable and animal life, fill a like office not only in the planets of our system, but throughout the universe.[18]

Janssen even went from his belief in the material unity of the universe to suggest a "mental and moral unity, and . . . that, as there is only one physics and one chemistry in the universe, there are also only one logic, one geometry, and one moral, and that the beautiful, the good, and the true are identical and of the universal order everywhere. . . ." (p. 814) Janssen's confidence in his 1897 claims is all the more noteworthy because he made them three years after W. W. Campbell had published solid evidence, subsequently confirmed, that Janssen's report of water vapor in the Martian atmosphere was a spurious result.

The two leading American experts on astronomical spectroscopy during the 1860s were John William Draper (1811–82) and his son, Henry Draper (1837–82), both of whom participated in the pluralist debate. The elder Draper, in 1840, made the first daguerreotype of the moon and emerged over the next two decades as an important predecessor of Bunsen and Kirchhoff. Moreover, combining photography and spectroscopy, he succeeded in 1843 in photographing the solar spectrum. His son, Henry Draper, in 1872 made the first photograph of the spectrum of a star and in 1877 published the first report of spectroscopic detection of oxygen in the sun.[19]

John Draper participated in the extraterrestrial life debate primarily as a historian rather than as a scientist. Pluralist passages occur in his *History of the Intellectual Development of Europe* (1863) and in his *History of the Conflict between Religion and Science* (1874). The latter book touches on the debate only briefly, in a section in which Bruno is presented as a martyr, against whom the "special charge [was] that he had taught the plurality of worlds. . . ."[20] In the earlier work, the importance of pluralism is repeatedly stressed. For example, Draper states that the geocentric theory has been found to be "utterly erroneous and unworthy" and has been replaced by a new theory that entails this important lesson:

Man, when he looks upon the countless multitude of stars — when he reflects that all he sees is only a little portion of those which exist, yet that each is a light and life-giving sun to multitudes of opaque, and therefore, invisible worlds . . . may form an estimate of the scale on which the world is constructed, and learn therefrom his own unspeakable insignificance.[21]

Draper presents the emergence of the pluralist doctrine as among the most important developments of modern thought and adds a statement later endorsed by Frederick Engels: "The multipicity of worlds in infinite space leads to the conception of a succession of worlds in infinite time."[22] The physical insignificance of man was, for Draper, but one of the lessons to be learned; another concerns man's "intellectual principle."

What is it that has descended into the infinite abysses of space, examined the countless worlds that they contain, and compared and contrasted them together? . . . That which is competent to do all this, so far from being degraded, rises before us with an air of surpassing grandeur and inappreciable worth. It is the soul of man. (pp. 296–7)

The younger Draper entered the pluralist debate in 1866 with a published lecture entitled "Are There Other Inhabited Worlds?"[23] It is a sober discussion of this question in which Draper draws heavily on astronomical spectroscopy to urge that conditions for life exist on Mars, but not on the excessively hot interior planets. Jupiter and the planets beyond receive scant treatment, but the moon, which Draper was then studying photographically, is discussed in some detail. Lecturing in the city of the "moon hoax," which "left behind an unfortunate scepticism" (p. 50), Draper rejects life on the moon's near side, although leaving open by means of Hansen's hypothesis the question of life on its remote side. His overall conclusion exemplifies the cautiously propluralist tone of his presentation: "there may be worlds that have never passed the state in which the earth was in early geological times, while on others conspiring circumstances may have allowed life to develop even beyond our standard. . . ." (p. 54)

By 1882, in which year both Drapers died, other American practitioners of the new astronomy had appeared, including Samuel Pierpont Langley (1834–1906), whose invention of the bolometer contributed to astrophysics and astrochemistry. In his *New Astronomy* of 1884, Langley introduced the public to "Urania's . . . younger sister," embellishing his presentation with frequent comments on the question of extraterrestrial life. For example, his discussion of solar physics opens with John Herschel's giant glowing solar organisms, because "nothing else can so forcibly illustrate the field of wonder and wild conjecture solar physics presented even a few years ago. . . ."[24] Literary allusions abound in this book, as well as passages from the literary pluralists Fontenelle and Voltaire. Langley, in fact, had a special interest in Fontenelle, whom he praised in 1877 for writing "The First 'Popular Scientific Treatise.' "[25] The source of Langley's most striking statements was the spectroscopic revelation of that "memorable fact . . . perhaps the most momentous that our science has brought us . . . that, after reaching across the immeasurable distances, we find that the stars are like *us*, – like in their ultimate elements to those found in our sun, our own earth, our own bodies." (p. 233) This was the main message of his book and also an idea gaining prominence in many pluralist presentations.

If the phrase "new astronomy" be defined as equivalent to astrophysics and astrochemistry, then it encompasses more than spectrum analysis. Of the many advances in physics during the latter half of the nineteenth century, a number could be applied to astronomy and in some cases to the extraterrestrial life question. The Irish physicist G. Johnstone Stoney (1826–1911) contributed significantly to spectroscopy, but the most important advance in astrophysics associated with his name was introduction of the newly developed kinetic theory of gases as a method of analyzing planetary atmospheres. This was of such significance that a recent author has claimed that by 1870, it had "placed the odds heavily in Whewell's favor."[26] The story is more complicated than this, but the eventual influence of Stoney's approach is not to be denied.

Stoney's first major paper in this area was his 1869 "On the Physical Constitution of the Sun and Stars."[27] Much more important for the pluralist debate was the series of papers that Stoney presented to the Royal Dublin Society beginning in 1870. These were not printed at that time, their main conclusions being published only in 1898.[28] Stoney, however, stated in 1898 that his addresses from the 1870s were "known to many," even though "only imperfect printed accounts have appeared. . . ." (p. 307) In these papers, Stoney introduced the concept now called "escape velocity" and used it to draw conclusions about the compositions of the atmospheres of planets and satellites. Knowing from kinetic theory that the average velocities of molecules in a gas decrease as molecular weights increase, and deriving the velocities, which depend on the gravitational mass of the celestial body, necessary for a molecule of a gas to escape from that body, Stoney was able to draw inferences about which gases would eventually escape from each planet and satellite of known mass. For example, Stoney's first paper of 1870 attributes the moon's lack of an atmosphere to that relatively small body's inability to hold (overcome the escape velocity of) all gases (pp. 305–6). Shortly thereafter, Stoney explained the absence of hydrogen from the earth's atmosphere as due to the high average velocities of molecules of this light element. Also, at about the same time, he derived the conclusion that "it is probable that no water can remain on Mars" (p. 306), Mars having substantially less mass than the earth. This leads to the suggestion that the Martian polar caps consist of carbon dioxide. Stoney's 1898 paper draws these results together, showing that Mercury will lose water vapor and probably both nitrogen and oxygen from it surface (p. 318), that Venus can retain an atmosphere similar in composition to that of the earth (p. 320), that Jupiter can retain all known gases, and that the other large planets beyond it can probably retain all gases except hydrogen (pp. 322–5). Most moons and all asteroids will, like our moon, be bereft of an atmosphere.

Stoney's conclusions were challenged to some extent in later decades of the nineteenth century (p. 305) and when published in 1898 created a controversy (subsequently discussed), but they are now accepted doctrine. Although his ideas were somewhat hypothetical, it is surprising that they entered so slowly into the pluralist debate, at least if Stoney's statement is accurate that they were, from the 1870s, "known to many." Opponents of pluralism, even after his 1898 paper, rarely cited him on behalf of their position. In the context of Stoney's work it is noteworthy that J. J. Waterston (1811–83) in an 1845 paper presenting the fundamentals of the kinetic theory of gases anticipated Stoney in the application of that theory to planetary atmospheres. However, Waterston's paper was rejected by the Royal Society and was first published only in the 1890s.[29]

Another area in which astrophysics advanced was photometry, to which Friedrich Zöllner (1834–82) of the University of Leipzig made important contributions. Among these was an 1874 paper reporting on the reflectivity of Mercury's surface, which is shown to be nearly identical with that of the moon. As Zöllner states: *"Mercury is a body, the surface condition of which corresponds very nearly to that of the moon, [and] which consequently, like the moon, probably possesses no significant atmosphere."*[30] Disappointing as this may have been to pluralists, they could take consolation in the fact that the reflectivities reported for the other planets (with the partial exception of Mars) were substantially higher (p. 641).

Whereas spectrum analysis was generally seen as supportive of the pluralist position, and the kinetic theory of gases as applied by Stoney was largely antipluralist, the theories of energy conservation and dissipation developed during the 1840s and 1850s were for the most part neutral or only indirectly relevant. Mechanical difficulties in the nebular hypothesis, especially the problem that the sun has less angular momentum than that theory predicts, placed it in jeopardy and thereby weakened the pluralists' claim for extrasolar planetary systems.[31] Important as were the advances in physics in the 1860s, the dominant topic of scientific discussion in that decade was the theory of evolution by natural selection, as set out in 1859 by Darwin in his *Origin of Species*. Although containing no discussion of pluralism, that book influenced the extraterrestrial life debate in a number of ways. Chief among these was that it made a teleological approach to nature, precisely the approach taken in most pre-1859 pluralist writings, ever more dubious. Pluralists in many cases persisted in using that approach, but it became less frequent and less overt and/or was confined to authors at the fringes of science. On the other hand, Darwinian theory gave support to a naturalistic and evolutionary approach to astronomy that some pluralists adopted. For example, R. A. Proctor increasingly tended to ask not whether a given planet is now inhabited but whether it had been or would be in its long history. It is an interesting question whether or not Darwinian theory provided scientists a basis on which to speculate on the bodily forms of extraterrestrials. Because of its stress on chance variations and the survival value of modifications, its initial impact was to negate such systems as that of Owen, which clearly had greater predictive value than Darwinian theory, to which even today many deny predictive powers. Although in subsequent decades one frequently finds authors simultaneously espousing pluralistic and evolutionary doctrines, or opposing both, this alignment was far from universal. The most striking exception is A. R. Wallace, codiscoverer of the theory of natural selection, who in 1903 came forth as a latter-day Whewellite.[32] As for the new astronomy, so also for the "new biology": Its relations to the pluralist debate were many and varied and frequently were seen differently by various persons. Important as these relations were, they were rarely decisive.

2. Richard Proctor: Anglo-American popularizer of astronomy and pluralist of evolving views

The new astronomy reset the stage for the extraterrestrial life debate during the 1860s and thereafter. Moreover, a new cast of characters appeared in the 1860s, headed by Richard Proctor and Camille Flammarion, two exceptionally prolific authors whose careers were intimately involved with the pluralist debate. When Richard Anthony Proctor (1837–88) died, the London *Times* asserted that he had "probably done more than any other man during the present century to promote an interest among the ordinary reading public in scientific subjects."[33] The American astronomer C. A. Young agreed: "As an expounder and popularizer of science he stands, I think, unrivaled in English literature."[34] This claim, coming from both sides of the Atlantic, certainly seems justified, at least if its scope is limited to the popularization of astronomy in the English-speaking world. In the twenty-five years between Proctor's first publication and his sudden death in 1888, he authored fifty-seven books, mainly on astronomy, with most of these consisting of essays previously published in periodicals. Yet so prolific was Proctor that these republished essays represented, as he once stated, less than one-fourth of his total output.[35] This estimate is no doubt correct, for he published at least five hundred essays, not counting his eighty-three technical papers in the *Royal Astronomical Society Monthly Notices*. Ranking seventh in the list of the most prolific astronomical authors publishing before 1882 and surpassing all but Flammarion and Secchi in rate of publication,[36] Proctor moreover gave lecture tours in Britain, the United States, Canada, Australia, and New Zealand. After marrying an American widow and taking up residence in the United States around 1881, he also edited, first from St. Joseph, Missouri, then from Orange Lake, Florida, the London-based scientific periodical *Knowledge,* which he had founded in 1881. So productive was Proctor that he wrote under one or more pseudonyms, possibly to confuse critics who accused him of writing too much.[37] Not only was Proctor prolific; he also was the most widely read participant in the pluralist debate in Britain and America during the 1870-to-1890 period. The following survey of Proctor's writings is designed to answer three questions: (1) What led to his almost unprecedented choice of a career as an author of astronomical writings? (2) How did he attain such remarkable success in it? (3) What position(s) did he adopt in the extraterrestrial life debate? The answer to the last question is especially fascinating, for his views shifted in a surprising direction.

Born in London in 1837 of well-to-do parents, Proctor graduated from St. John's College, Cambridge, in 1860 as twenty-third wrangler, his modest ranking being ascribed by some contemporaries to his having lost his mother and acquired a wife during his college years. Having ample means, he prepared for the bar, devoting some of his leisure to astronomy to ready himself for the instruction of his firstborn. He had up to that time, as he said, "read absolutely nothing in astronomy as a science. . . ."[38] Inspired by reading Nichol's *Architecture of the Heavens* and O. M. Mitchel's *Popular Astronomy,*[39] he set himself to composing a popular essay on "Colours of the Double Stars." Easy reading, he found, required hard writing; six weeks of work produced a nine-page piece, eventually published in *Cornhill Magazine* in 1863. Two personal tragedies, the loss of his first child in 1863 and of his fortune in 1866, led to his career as a science writer. To relieve the distress due to his son's death, Proctor on his physician's advice took up the project of writing *Saturn and Its System,* a scholarly treatise that on its publication at his own expense won him the respect of the astronomical community. Such was his financial situation that its failure to sell[40] would have been no serious problem had not a bank in which he was the second largest shareholder failed in 1866, leaving him with a staggering debt of 13,000 pounds. Faced with feeding a family of five, this author whose first book had been a financial failure had, as he wrote, "not for literary capital creative powers in a subject in which all men take interest, but simply a power of generalization in a subject regarded as abstruse, with a possible chance of learning how, by special care of exposition, to interest a small section of the general public in the popular study of that subject."[41] A few days before the bank failure, the editor of *Popular Science Review* invited Proctor to write two astronomical articles, which request he had already set about declining. But his desperate financial situation forced him to this opportunity, and in fact from the day of the bank's failure "onward for five years, I did not take one day's holiday from the work which I found essential for my family's maintenance."[42]

The work of writing he found so difficult that he confessed that he "would willingly have turned to stone-breaking on the roads, or any other honest but unscientific labour, if a modest competence in any such direction had been offered me."[43] According to Agnes Clerke, "The struggle was severe. Article after article was sent back to him. . . . Book after book, atlas after atlas, was refused by the publishers."[44] His efforts eventually bore fruit; in 1873, an American author noted: "Ten years ago, the name of Richard Anthony Proctor was absolutely unknown; five years later, it was familiar in scientific circles in London; and to-day it is familiar as household words to every educated man in England, and to many thousands in this country."[45]

One reason for Proctor's success was his scientific expertise; for example, his impact on stellar astronomy led one contemporary to credit him with "the final demolition of that theory of the Stellar Universe which so long held place in text-books as Sir William Herschel's. . . ."[46] Elected to the Royal Astronomical Society in 1866 and subsequently selected as its honorary secretary, he possessed the credibility requisite to win the confidence of both professionals and the public. However, technical sophistication is at most a necessary condition for successful popularizing. Motivation for publication and a suitable style and subject matter are also essential. Motivation Proctor had in abundance; financial need kept in constant motion the pen first employed to express leisure interests and to relieve personal grief. A successful style he acquired gradually; by 1870 he had followed the "dismal failure" of his *Saturn* by six more books, most faring better, and an array of articles. One of these books, his first major success, became in time probably his most widely read work, continuing in print until 1909. The tactic Proctor employed in this, his *Other Worlds than Ours* (1870), was to weave pluralist themes into astronomical presentations, an approach other authors had used before, but none, except Flammarion, with greater success.

In 1888, Proctor explained how he had come to compose his *Other Worlds*. It was *not* from personal interest in the pluralist debate; in fact, he wrote: "I had been rather wearied by the over warm discussions of Whewell and Brewster. . . ." It was rather that he saw it "as a convenient subject with which to associate scientific researches which I could in no other way bring before the notice of the general reading public."[47] In the opening sentence of an essay of 1878, and repeatedly in the essay itself, Proctor made the same point: "The interest with which astronomy is studied by many who care little or nothing for other sciences is due chiefly to the thoughts which the celestial bodies suggest respecting life in other worlds than ours."[48] Having discovered the public's appreciation of pluralist publications, Proctor included discussions of extraterrestrial life ideas in a dozen later books.

Proctor's *Other Worlds than Ours,* his fullest treatment of the pluralist question, was explicitly written against the background of the Brewster-Whewell controversy, although he reminds his readers that "We stand in a position much more favourable for the formation of just views. . . ."[49] That Proctor's preference at this time was for a basically Brewsterian and teleological approach is evident from such statements as: "we see proofs on all sides, that besides the world on which we live, other worlds exist as well cared for and as nobly planned." (p. 2) This does not deter him from declaring that it is "all but certain . . . that no part of the moon's globe is inhabited . . ." (p. 7) or from urging that arguments for life on the sun are "too *bizarre* [for] consideration." (p. 20) The spectroscope is credited with showing that the stars are composed of the same elements as the earth, which leads Proctor to infer that planets, whether of our own sun or another star, are similarly constituted chemically (p. 45). His methodology in discussing life on planets is to assume that "Until it has been demonstrated that no form of life can exist upon a planet, the presumption must be that the planet is inhabited." (p. 57) However, the intensity of the solar radiation reaching Mercury makes him doubt its habitability. Venus fares better: "On the whole, the evidence we have points very strongly to Venus as the abode of living creatures not unlike the inhabitants of earth." (p. 81) Mars, his most favorable case, "exhibits in the clearest manner the traces of adaptation to the wants of living beings such as we are acquainted with. Processes are at work out yonder in space which appear utterly useless . . . unless . . . they subserve the wants of organised beings." (pp. 84–5) In support of his view of Mars, he cites Huggins's report of water vapor in the Martian atmosphere, as well as the observations of "ice caps," "oceans," "seas," and "inlets." These features are also represented in his included map of Mars, first published in 1867. This map provoked charges of chauvinism, because the better locations were named after British observers; however, few faulted him for going far beyond observation in applying such terrestrial terms as "oceans" and "inlets" to features scarcely discernible on the surface of Mars.

As Proctor proceeds to the giant outer planets, he moves farther from Brewster and nearer to Whewell by beginning his Jupiter chapter with the rejection of Brewster's claim that such a large planet, if uninhabited, would be a colossal waste (pp. 110–12). Yet this chapter ends with a harsh indictment of Whewell: "Surely no astronomer worthy the name can regard this grand orb as the cinder-centered globe of watery matter so contemptuously dealt with by one who, be it remembered thankfully, was not an astronomer." (p. 145) Between these sections, Proctor presents a new theory of the Jupiterian system, the origins of which he revealed in a later publication, where he stated that in writing *Other Worlds than Ours,* he

. . . set out with the idea of maintaining . . . that all the eight known planets . . . are inhabited worlds. But even as I wrote that work I found my views changing. So soon as I began to reason out the conditions of life in Jupiter and Saturn, so soon I began to apply the new knowledge which would, I thought, establish the theory that life may exist in these worlds, I found the ground crumbling beneath my feet. The new evidence . . . was found to oppose fatally . . . the theory I had hoped to establish.[50]

The theory that Proctor arrived at and presented in his 1870 book is that Jupiter is not "at present a fit abode for living creatures," but that it is "in a sense a sun . . . a source of heat" serving its four satellites on which "life – even such forms of life as we are familiar with – may still exist." (p. 141) Not content with making Jupiter a quasi sun, he urges that Jupiter "*must* be intended to be one day the abode of noble races."[51] A similar theory of Saturn is presented, even though in his *Saturn and Its System* (1865) he had argued for its habitability.[52] Concerning Uranus and Neptune, Proctor wavers: Perhaps they are also quasi suns to their satellites, or possibly they are inhabited by creatures suited to their "arctic" conditions.

The astronomy of the sidereal and nebular realms is treated in some detail in the final fourth of his book, despite its tenuous connection with his central question. These sections are noteworthy because in them Proctor accepts a number of Whewell's arguments while rejecting his anti-pluralist conclusions. For example, he agrees with Whewell's argument that John Herschel's observations of the Magellanic Clouds make it doubtful that nebulae are extragalactic (p. 290). Huggins's spectroscopic determination that at least some nebulae are gaseous further supported Whewell, leading Proctor to conclude that all visible nebulae are in our galaxy (pp. 277–82). The spectroscope had, however, gone against Whewell's claim that all we know is that stars are "dots of light"; it had revealed that their spectra are comparable to that of our sun and hence that stars are "real suns." (p. 240) Although agreeing with Whewell that visible nebulae are members of our galaxy, Proctor does not let this admission significantly influence his pluralist position. For example, he claims that nebulae may contain "other classes of worlds peopled with their own peculiar forms of life. . . ." (p. 278) In rejecting nebulae as galaxies, Proctor was probably influenced by his having "no doubt whatever that galaxies, resembling our own, exist" (p. 306) at distances beyond the reach of telescopes. Such statements, which preserve pluralism while severing its relationship to observational astronomy, become easier to understand if one accepts Proctor's later view that "the question of life in other worlds . . . is not a scientific question at all, belonging rather to the domain of philosophy than of science."[53] That teleological ideas influenced Proctor in his 1870 book is repeatedly evident, especially when he infers from the presence of spectral lines indicating metals in stars that "the orbs circling around those distant suns are not meant merely to be the abode of life, but that intelligent creatures, capable of applying these metals to useful purposes, must exist in those worlds."[54]

In his final chapter, Proctor avoids the much disputed question of the relation of pluralism to revealed religion (pp. 296–7), commenting instead on the religious-astronomical ideas in a popular booklet of the day entitled *The Stars and the Earth*.[55] Proctor's restraint on religious matters probably spared his book the attacks launched against some pluralist presentations. Had he shown a comparable reticence in making pluralist claims, many of which went far beyond the abundance of astronomical information presented in his book, he would have increased its quality, although, one suspects, diminished its audience. His *Other Worlds* went through at least twenty-nine printings between 1870 and 1909, making it one of the most widely read books in the entire extraterrestrial life debate, even though, as will soon become clear, it had ceased by the mid-1870s to represent Proctor's position.

In 1872, Proctor published another pluralist book. Entitled *The Orbs around Us*, it contains, according to Proctor, "matter which, if space had permitted, I should have included in 'Other Worlds than Ours'. . . ."[56] Actually, this book, made up mainly of essays published before 1870, reveals almost nothing about the new directions his thought would soon take. The only significant new material is his criticism of William Thomson's theory (discussed later) that terrestrial life may have originated from life-bearing meteorites (pp. 217ff). Thus, it seems probable that he published *The Orbs around Us* not so much to advance the pluralist debate as again to bring before the public various astronomical materials embellished by pluralist themes.

In 1873, Proctor, by then the author of fourteen books, published three more, two of which, *The Expanse of Heaven* and *The Borderland of Science,* contain chapters relevant to the pluralist debate. The former work restated the conclusions of his 1870 volume in a style that one reviewer accurately described as reminiscent of Nichol. *The Borderland of Science* includes four Brewsterian essays, one of which recounts a celestial voyage to Mimas, a Saturnian satellite. The last of the four essays, an argument for "Life on Mars," is followed by "A Whewellite Essay on the Planet Mars," in which Proctor asks what arguments a latter-day Whewellite might marshal against Martian life. He suggests that because of Mars's greater distance from the sun, it receives far less heat than the earth, and because of its smaller mass, it has less atmosphere and retains less heat than the earth. Proctor concludes that "Neither animal nor vegetable forms of life known to us could exist on Mars" and that if living beings somehow exist there, they "must differ so remarkably from what is known on earth, that to reasoning beings on Mars the idea of life on our earth must appear wild and fanciful. . . ."[57] Atmospheric problems are thoroughly treated; at one point, Proctor notes that the thinness of the Martian atmosphere must depress the boiling point of water by about seventy degrees, producing the distressing result that "A cup of good tea is an impossibility in Mars, and equally out of the question is a well-boiled potato. It does not make matters more pleasant that the teaplant and the potato are impossible. . . ." (p. 149n) This essay, published anonymously in *Cornhill Magazine* in July 1873, must have shocked readers of *Borderland,* but he did not develop its Whewellite approach further in that volume.

This he did in two 1875 books. Readers of the first three essays in Proctor's *Our Place among Infinities* and of the first essay in his *Science Byways* could scarcely doubt that his views had undergone a major transformation since 1870. In the essay in the latter volume, he traces this process that began in the late 1860s when he jettisoned Jupiter and Saturn as inhabited planets. After that, "gradually . . . both the Brewsterian and Whewellite theories of life in other worlds gave place in my mind to a theory in one sense intermediate to them, in another sense opposed to both. . . ."[58] The Whewellite aspect of his new position is exhibited in *Byways* by his exclamation: "Millions of uninhabited orbs for each orb which sustains life!" (p. 35) But Brewsterian features remain; he accepts it "at least as probable that *every member of every order – planet, sun, galaxy, and so onward to higher and higher orders endlessly – has been, is now, or will hereafter be, life-supporting 'after its*

kind.' " (pp. 34–5) Underlying these statements and reconciling them is
Proctor's new theory of planetary evolution:

Each planet, according to its dimensions, has a certain length of planetary life,
the youth and age of which include the following eras: – a sunlike state; a state
like that of Jupiter or Saturn, when much heat but little light is evolved; a
condition like that of our earth; and lastly, the stage through which our moon is
passing, which may be regarded as planetary decrepitude.[59]

His emphasis on time, in particular on the vast age of the universe and on
planets as evolving entities, is the crucial factor in Proctor's new position.
Passages in these essays suggest the sources of his new perspective. Whewell's stress on the vast period during which the earth was uninhabited
by higher forms surely influenced Proctor. Evolutionary theories of the
solar system, whether by nebular condensation or by accretion of small
masses, both of which Proctor accepted, were cited in his arguments
(*Byways,* p. 17). Darwinian passages, but without mention of Darwin,
likewise occur (*Byways,* p. 11). It is interesting that the evolutionary
doctrines of Darwin, opposed by Whewell, pushed Proctor in a Whewellite direction. The doctrine of the degradation of energy, as well as concern about the depletion of Britain's coal resources, drove him to see the
earth as tending toward the decrepitude seen on the moon.

On a philosophical level, he criticizes the teleological approaches taken
by Chalmers, Dick, Brewster, John Herschel, and other pluralists, whom
he describes as overestimating the degree to which God's design for the
universe can be known (*Infinities,* pp. 45–8). This leads him to stress
such apparent evidences of waste as that "less than the 230 millionth"
part of the sun's heat and light falls on the planets (*Byways,* p. 22). Also
on the philosophical level, Proctor had come to realize that the pluralist
position rested in part on astronomical data, but no less on planetary
analogies with the earth, which he felt had been misconceived by persons
insufficiently attentive to the age of the earth (*Byways,* pp. 5–6). All solar
system life, except on earth, seems surrendered in these essays. Mars once
had life, although "the development of higher forms of life may have
been less complete than on our earth. . . ." Jupiter, on the other hand,
as yet lifeless, is seen as evolving a habitat suitable for life, which would
include "creatures far higher in the scale of being than any that have
inhabited, or may inhabit, the earth. . . ." (*Byways,* pp. 27–8) Even the
sun, he says, will eventually attain life (*Infinities,* p. 68).

In a sense, Proctor's 1875 essays may be read as a vindication of
Whewell, who is praised for his philosophic, calm, and dispassionate
force of reasoning" and for having broken through "all these old-fash-
ioned methods" of argument used by pluralists (*Infinities,* pp. 50–1).
Proctor even admits that the "balance of evidence" favors Whewell's
positon rather than Brewster's (*Infinities,* p. 52). Yet, in a larger sense,

Proctor emerges more as a latter-day Brewsterian than as a Whewellite; in the last paragraph of his third essay in *Our Place among Infinities,* he states:

Have we then been led to the Whewellite theory that our earth is the sole abode of life? Far from it. For not only have we adopted a method of reasoning which teaches us to regard every planet in existence, every moon, every sun, every orb in fact in space, as having *its period* as the abode of life, but the very argument from probability which leads us to regard any given sun as not the centre of a scheme in which at this moment there is life, forces upon us the conclusion that among the millions on millions, nay, millions of millions of suns which people space, millions have orbs circling round them which are at this present time the abode of living creatures. (pp. 69–70)

Had Whewell been alive, he might have been puzzled as to how to respond to an opponent who praised nearly all his arguments and yet by falling back on the "infinities" of space and time retained the extraterrestrial life hypothesis.

Throughout the remainder of his life, Proctor persisted in the evolutionary pluralist position he first presented in the 1870s.[60] Religious considerations, which had powerfully influenced Whewell in his initial questioning of pluralism, do not seem to have been involved in Proctor's change of mind, although the pluralist position may have influenced his religious beliefs. According to one obituary notice, Proctor, when "prostrate with grief at the loss of a darling boy, . . . sought refuge in the Catholic Church, and some letters he wrote to me at that time are charged with the zeal of the proselyte But the illusion held him not long."[61] The period of his association with Catholicism must have lasted about a decade, for according to another obituary notice, Proctor "for a time was a Roman Catholic, but in 1875 he severed his connection with that faith on the ground . . . that church theologians had told him that some of his theories and scientific views were not in conformity with loyalty to the church. He was so convinced of the truth of his ideas, that he left the church."[62] It would be interesting to know more of this matter, but his writings rarely contain commentary on his doctrinal convictions or on the reconcilability of Christianity and pluralism.[63]

The book by Proctor from the late 1870s most significant for the pluralist debate is his *Myths and Marvels of Astronomy* (1877), which contains three relevant essays. Among these is "Swedenborg's Vision of Other Worlds," which provoked a book-length response (discussed subsequently) from a Swedenborgian theologian, Augustus Clissold. Suggesting that pluralism is a doctrine especially suitable to the "visionary," Proctor cautions that the spectroscopic methods and improved telescopes of the late nineteenth century forbid many of the speculations indulged in during Swedenborg's century. Noting that despite improved methods, only a "slight change" in views has occurred, he suggests a reason for this:

If men no longer imagine inhabitants of one planet because it is too hot, or of another because it is too cold, of one body because it is too deeply immersed in vaporous masses, or of another because it has neither atmosphere or water, we have only to speculate about the unseen worlds which circle round those other suns, the stars; or . . . we can look backward to the time when planets now cold and dead were warm with life, or forward to the distant future when planets now glowing with fiery heat shall have cooled down to a habitable condition.[64]

In another essay, "Other Worlds and Other Universes," he develops the point that except for professional astronomers, people have essentially no interest in astronomical matters, unless they shed light on the question of extraterrestrial life (pp. 134–5). Clearly, Proctor could speak from experience on this point. The lunar hoax is the subject of another chapter in which one can detect a growing cynicism on Proctor's part toward the public's passion for pluralist pieces. Proctor, taunted at times for writing overly popularized presentations of astronomy, could not but have been anxious to separate himself from the fantasies of Swedenborg and the sensationalism of the moon hoax.

During the 1880s, pluralist features continued to be present in Proctor's books; in fact, some were given pluralist titles, even when, as in his *Other Suns than Ours* (1887), they contain few pluralist passages. His *Universe of Suns* (1884) begins with a pluralist poem, apparently of his own composition, and includes an essay entitled "Life in Mars," which turns out to be only an abbreviated version of his 1873 essay with the same title. His *Poetry of Astronomy* (1881) contains more pluralist material than any of his other books from the 1880s. In an essay "Is the Moon Dead?" from that volume, he answers affirmatively, but then explores whether or not it formerly had life. In that essay he employs a distinction, too frequently neglected in pluralist writings, between habitability and actual habitation,[65] confining his analysis solely to the question whether or not the moon was habitable at an earlier time. Yet his pluralist tendencies emerge in that essay when, under the stimulus of *The Unseen Universe,* by Stewart and Tait, he suggests:

May there not be a higher order of universe than ours, to which ours bears some such relation as the ether of space bears to the matter of our universe? and may there not, above that higher order, be higher and higher orders of universe, absolutely without limit? And, in like manner, may not the ether . . . be the material substance of a universe next below ours, while below that are lower and lower orders of universe absolutely without limit? (pp. 180–1)

As an aside, he presents evidence that his French counterpart, Camille Flammarion, had plagiarized his writings.[66] The same volume contains an essay on what life would be like on the newly discovered moons of

Mars. His later volumes also contain presentations of his general position in the pluralist debate; his *Illusions of the Senses and Other Essays* (1886) includes, for example, a slightly shortened version of his lead essay from *Science Byways,* whereas *Mysteries of Time and Space* (1883) has an apparently new presentation of his views. Another of his activities in this period was the retranslation and serial publication in his journal *Knowledge* of Fontenelle's *Entretiens*.

In early September 1888, Proctor set off from his Florida home for New York City, whence he planned to proceed to England for yet another lecture tour. Arriving in New York, he became gravely ill, dying shortly thereafter of what was diagnosed as yellow fever. His fifty-seventh book, intended as his magnum opus, was then nearly finished. This was his immense *Old and New Astronomy,* completed by A. C. Ranyard and published in 1893. Ironically, this book, which contains almost no pluralist passages, probably secured publication and readership in large part because of the international reputation he had established as a popularizer of astronomy who rewarded readers of his writings by embellishing many of them with pluralist themes. He had learned this technique in writing his first major success, *Other Worlds than Ours,* and he repeatedly returned to it in subsequent publications.

The impact of his publications was immense; thousands were introduced to astronomy by them. His influence was also felt within astronomy proper, for example, in the dominance for many decades of his view of Saturn as a "hot, distended, gaseous globe."[67] His role in the pluralist debate was also extremely large; more than anyone before him, he stressed the need to take an evolutionary approach, to see the planets as changing, possibly developing entities. This was prophetic of the future form of the debate, as was his tendency to retain pluralism even as the other planets of the solar system were declared lifeless. It is ironic that he was led to adopt this approach in large part because he took Whewell's antipluralist arguments seriously.

Because of the nature of Proctor's final illness, he was buried in an unmarked grave. However, an enthusiast for his writings, George W. Child, erected a monument to him and secured reburial in Brooklyn. Moreover, Proctor, whose views had so changed that in 1888 he asserted that "we must regard it as at the very least highly probable that on Mars . . . few of the higher forms of life were (or have been) developed . . . ,"[68] was made the focus in 1896 of efforts both to establish an international Proctor Memorial Association and to erect in California a telescope in his honor. The proposed telescope of 100-foot aperture was to be so powerful that earthlings, as the *New York Times* headline proclaimed, "WILL SEE MEN ON MARS."[69]

3. Camille Flammarion: a "French Proctor"?

In 1894, *McClure's* carried an article on the author "who has done more toward popularizing the study of astronomical science than any of his contemporaries. . . ."[70] The person so warmly praised was not Proctor, but his French counterpart, Camille Flammarion (1842–1925). Writing in the same year, Simon Newcomb stated that Flammarion at first "wrote so much like a French Proctor that, could a man have a legal copyright on his own personality, the Englishman might have brought suit on the ground of infringement."[71] Newcomb was no doubt not alone in noting similarities between these two authors. Both were remarkably productive; in a list of the most prolific astronomers writing up to 1881, Proctor ranked seventh, Flammarion fifth; in rate of publication, Proctor was third, the Frenchman first.[72] However, whereas in 1881 Proctor's death was less than a decade distant, Flammarion was at the end of only the second of the six decades during which he published profusely. By his death in 1925, he had authored over seventy books, many attaining translations, a rarity for Proctor's writings. Like Proctor, Flammarion had dozens of technical papers to his credit. And, like Proctor, Flammarion first caught the public eye by a pluralist book and was similarly successful in retaining it by writings frequently rich in pluralist passages. These and other parallels will be noted and nuanced in what follows – and a major difference between them will be suggested.

Born in Montigny-le-Roi in France, Flammarion at age ten began four years of study in a seminary. Already drawn to astronomy, his interest greatly increased when an opera glass revealed "mountains in the moon, as on the earth! And seas! And countries! Perchance also inhabitants!"[73] At fourteen, he and his family moved to Paris, where he worked for a time as an engraver; by his fifteenth year, he was, as he stated, "above all, taken up with cosmographical questions, and wrote a big book on the origin of the world. . . ."[74] This manuscript of over five hundred pages, entitled *Cosmogonie universelle,* came to the attention of Leverrier, director of the Paris Observatory, who arranged for its sixteen-year-old author to work at the observatory as an apprentice astronomer.

In 1862, Flammarion published his first book, *La pluralité des mondes habités,* on the title page of which the twenty-year-old author presents himself as "Ancien calculateur à l'observatoire impérial de Paris, professeur d'astronomie, membre de plusiers sociétés savantes, etc." This booklet of fifty-four pages, according to its author, "at once made my reputation."[75] Moreover, it placed him at the center of the pluralist debate, a position he retained for over six decades. By 1864, he had expanded it to 570 pages for its second edition. By 1865, it had attained at least twenty-four reviews[76] and had been supplemented by a second, even longer pluralist volume. By 1870, when Proctor's *Other Worlds than Ours* first appeared, Flammarion's *La pluralité* was already in its fifteenth edition, and his supplemental *Les mondes imaginaires et les mondes réels* was in its ninth. Both went through numerous later editions, being republished as late as the 1920s. *La pluralité* was translated into at least six and possibly as many as fifteen languages.[77]

A number of factors influenced Flammarion in composing his plurality of worlds book. As he recounts in his autobiography, he found Leverrier's emphasis on mathematical and positional astronomy too confining for his tastes, which were better suited to the earlier directorship of Arago. Physical astronomy, with its attention to the nature of the planets and stars, became his passion. This, along with his interest in the extraterrestrial life debate, led him to write a manuscript entitled *Voyage extatique aux région lunaires: correspondance d'un philosophe adolescent,* which was never published. He also read such pluralist authors as Fontenelle, Cyrano de Bergerac, Huygens, Voltaire, Lalande, Brewster, Herschel, and Jean Reynaud.[78] Also during his four years at the observatory, he lost the Catholic faith of his youth, citing its irreconcilability with post-Copernican astronomy.[79] The void thus created was filled by the religious doctrines espoused by Reynaud in his *Terre et ciel* (1854). Jean Reynaud (1806–63), trained at the École polytechnique and a teacher at the Écoles des mines, contributor to various encyclopedias, and French undersecretary of state in 1838, set out in this volume a religious system that he believed reconcilable with Christianity, a judgment rejected in 1857 by a council of bishops. Reynaud advocated the transmigration of souls, urging that after death we pass from planet to planet, progressively improving at each stage. This doctrine of indefinite perfectibility with its associated idea of the lowly nature of terrestrial life appealed strongly to Flammarion, who while discussing nineteenth-century mystical pluralists in a later book commented: "Of all the works written on this subject during the period . . . , the most important is without a doubt that of our master and friend, Jean Reynaud."[80]

Flammarion thus began his first pluralist book; as he stated: "I consecrated the year 1861 to this composition, enflamed with a fiery ardor as one has at age nineteen, not doubting for an instant that I would demonstrate to myself that my conviction in extraterrestrial life was well founded."[81] The approach taken in it emerged in Flammarion's mind as "for me the apotheosis of astronomy and its supreme end." In fact, it became "in some sense the program of all my literary and scientific life."[82] The first edition of *La pluralité* was published by Mallet-Bachelier; the second expanded version was published by Didier, who accepted not only Flammarion's book but also his brother Ernest, who after Didier's death eventually became head of the famous firm that now bears the family name. The history of that firm, which published over fifty of the astronomer's books, is consequently intimately involved with the extraterrestrial life debate.[83]

In its 1862 form, Flammarion's *La pluralité des mondes habités* is divided into historical, astronomical, and physiological sections, with a preface promising a truly scientific approach. His historical section is aimed at showing that "the heroes of thought and of philosophy have ranged themselves under the banner which we are going to defend."[84] By lumping pluralism with teachings on transmigration of souls, Flammarion finds the pluralist tradition beginning "contemporaneously to the establishment of man on earth" (p. 8) and espoused by the Indians, Chinese, Arabs, Egyptians, and Greeks. Forty-seven figures from the seventeenth and eighteenth centuries are packed into a single paragraph, showing the breadth if not the depth of his reading (p. 14). Quotations from such authors as Kant, Laplace, and John Herschel argue the inconceivability of the opposing position.

In his astronomical section, after setting out the basic data of the solar system, Flammarion asserts: "the earth has no marked pre-eminence in the solar system of such a sort for it to be the only inhabited world. . . ." (p. 33) Poorly provided with moons, small in size and mass, our planet compares unfavorably with the giant planets and the sun. The problem that the other planets are poorly positioned to receive solar heat and light is dismissed, mainly by urging that atmospheric conditions are crucial in this regard (p. 25). This proponent of a scientific approach warns his readers that no evidence exists to prove that the atmospheres and fluids on other planets are "of a chemical composition analogous" to those on earth; in fact, his opinion is that they are "essentially different." (p. 26n) This claim shows Flammarion's readiness to set aside science, which can, of course, say nothing of materials postulated to be different in nature from those that it studies.

The physiological section suffers to some extent from the same problem, for in it he urges readers not to assume that known laws of physiology govern life on other planets. Those persisting in this practice are accused of "hurling a gross insult in the shining face 'of the infinite Power who fashions the worlds.' " (p. 43) However, one feature of terrestrial life, the "inexhaustible fecundity of nature," is applied to the planets, where he supposes that nature acts analogously in producing an array of animal and plant forms (p. 37). He uses passages from Pascal to develop his theme of the miserableness of man's position on a minute planet that "is far from being the world most favorably established for the maintenance of existence. Differences of ages, positions, masses, . . . biological conditions, etc. place a great number of other worlds at a degree of habitability superior to that of the earth." (p. 50) In this context, he also refers to Reynaud's *Terre et ciel* (pp. 48–9), although his transmigrational teachings are only touched on in this edition. Whewell's book is never mentioned; Plisson's presentation is cited, but his cautious approach is not followed. Flammarion finds his own exposition so convincing that he claims that pluralism is probably innate to our minds (pp. 7, 45). His final paragraph concludes with the suggestion that his readers join in declaring to God that "we were insane to believe that there was nothing beyond the earth, and that our poor abode alone possessed the privilege of reflecting your grandeur and your power!"

What explains the success of the book? Historical and scientific detail, as well as its author's confident approach, gave it credibility, while existentialist echoes of Pascal and its pious and poetic tone must have attracted readers repelled by the materialism, positivism, and pessimism of the period. Although transmigrational doctrines lay beneath its surface, the book's overall message of the humbleness of man and the grandeur of God could have offended few. Perhaps some turned to it to see the erudition attained by an author so young. Its lack of well-developed scientific arguments left it open to attack by such scientists as Abbé Moigno, but such deficiencies had not deterred readers of Fontenelle's or Chalmers's volumes, to which it was in a number of ways comparable.

Flammarion recounts in his autobiography the praises bestowed on his book by some prominent persons of the period. He presented a copy to Reynaud, "which he received sympathetically, read without delay, and adopted as his own."[85] Allan Kardec (1804–69), the leading French figure in spiritualism, a subject that later became a consuming interest of Flammarion, praised it in the *Revue spirite,* suggesting that the author's youth "is for us clear proof that his spirit is not at its beginning, or that without his knowledge he has been assisted by another spirit."[86] J. A. Pezzani, a proponent of metempsychosis, not only wrote a book (discussed later) combining that doctrine with pluralism but also considered Flammarion "the master of a school."[87] Even Napoleon III took an interest in the book, discussing it with the empress and others, and indirectly inviting Flammarion to discuss it with him.[88] One person not pleased was Leverrier, who dismissed him from the observatory.[89] Flammarion then secured a position at the Bureau des longitudes, where he remained until 1866. During this period he succeeded Reynaud as scientific editor of *Magasin pittoresque* and also wrote regularly for *Cosmos.*

Relatively few of the thousands of readers of *La pluralité* knew it in its fifty-four-page first edition; they encountered it in the format (over 500 pages) of its forty or more subsequent editions. The majority of this expansion came with the second edition; Flammarion's views, unlike Proctor's, underwent few substantial changes in later years. An examination of the thirty-third edition of around 1885 shows the form of presentation he settled on and retained. The fourfold increase in the historical section of that edition is supplemented by forty-five pages of passages from various pluralist authors. Both the astronomical and physiological discussions are increased, as well as being supplemented by a section on the sidereal realm, scarcely mentioned in the first edition. Technical notes, placed with the pluralist quotations at the end of the volume, supplement those sections. The radical character of Flammarion's pluralism is indicated in his astronomical section by his advocacy, albeit not without qualifications, of life on the sun and moon (pp. 81–5). In addition, a section of more than one hundred pages entitled "L'humanité dans l'univers" appears in this work, as well as a long discussion of the relationship between Christianity and pluralism.

Flammarion's "L'humanité dans l'univers" chapter begins with a discussion of the planetarians presented by such writers as Huygens, Wolff, Swedenborg, Kant, Locke, and Fourier. He accuses these authors of "anthropomorphism," their planetarians being only remodeled men. The differing conditions on the planets and the tendency of nature to diversity lead him to stress that our conceptions of extraterrestrials must be relative. Having secured planetarians partly by denying that terrestrial laws govern elsewhere, he would seem to be barred from speculating on the forms of such creatures, but speculate he does. Stressing the lowliness of the earth in the hierarchy of worlds and urging the divine origin and hence absolute character of ideas of goodness, beauty, and truth, he asserts that extraterrestrials, as children of God, form a celestial family. Man is a "citizen of the sky"; humanity in a broad sense is everywhere in the universe. This idea is tied to the transmigration of souls: "the earths which hover in space have been considered by us . . . as the future regions of our immortality. There is a celestial home of many dwellings, and there . . . we recognize those places which we will one day inhabit." (p. 320) Moreover, the planets are "studios of human work, schools where the expanding soul progressively learns and develops, assimilating gradually the knowledge to which its aspirations tend, approaching thus evermore the end of its destiny." (p. 328) He appears in a sense as a double pluralist: "Plurality of worlds; plurality of existences: these are two terms which complement and illuminate each other." (p. 324) Flammarion's linking of pluralism with metempsychosis, although not original with him, helps explain why these doctrines are frequently found together, especially in France, in writings from the final decades of the nineteenth century.

Although making no efforts to reconcile transmigration with Christianity, he devotes forty pages to the problems involved in relating that religion to pluralism, claiming that these were at the core of the Bruno and Galileo conflicts (p. 340). According to him, four solutions to these problems had been proposed: (1) God simultaneously became incarnate and died on all planets where sin had occurred. (2) God became incarnate on various planets at different times. (3) God came only to the earth, because only there did sin arise. (4) Christ's earthly actions brought redemption to all the planets. The third is described as Chalmers's solution, the fourth as Brewsters's, with Flammarion favoring the latter. Whewell's book is discussed, seriously misquoted, and rejected as based on "specious arguments" and "sophisms" and as being a "deep entrenchment made on the venerable ramparts of the sacred citadel!"[90] In a review of this edition, Charles Augustin Saint-Beuve, while praising Flammarion's style and faulting his reliance on final causality, describes his book as "no longer a popular treatise on astronomy [but rather] a book of transcendent philosophy and quasi-theology. Bernardin de Saint-Pierre is gone beyond in his dreams of celestial harmony; Jean Reynaud has become the prophet and the John the Baptist of the system of which Flammarion is the mystical evangelist. . . ."[91]

In 1865, Flammarion brought out both his *Les mondes imaginaires et les mondes réels* and his *Merveilles célestes*. The former book, longer even than his revised *La pluralité*, complemented it by being a history of extraterrestrial life speculations and of works in the cosmic voyage genre. More distinguished by the breadth of its coverage than the depth of its analyses, it rivaled his 1862 book in frequency of republication. His *Merveilles célestes*, an elementary presentation of astronomy with a chapter devoted to extraterrestrials, sold over sixty thousand copies[92] and reached English readers in a translation by the wife of Norman Lockyer. In 1866, using the pseudonym "Hermes," he published a second volume on psychic phenomena, and in 1867 he issued the first of the nine volumes of his *Études et lectures sur l'astronomie* (1867-80). Included in it were many of the public talks that established Flammarion as, like Proctor, a widely appreciated speaker. Also in 1867, he chanced upon Humphry Davy's *Consolations in Travel*, finding in its author "a singular identity of convictions on certain points of the philosophy of science and even of astronomy. . . ."[93] So impressed was he that he translated the book and arranged for its publication. Consequently, in 1868 Davy's book entered the pluralist debate in France, where it went through at least nine editions. During the late 1860s, Flammarion began a study of the atmosphere of the earth, a project involving colorful balloon ascensions and culminating in his *L'atmosphere* (1872). When James Glaisher edited that book for English translation, he felt it necessary to reduce its size by half and to remove the "rhapsodies" so characteristic of Flammarion's writings.[94]

In addition to his atmospheric writings, Flammarion published during the early 1870s a number of books with pluralist features. For example, in his *Histoire du ciel,* he traces the history of pluralism and of metempsychosis back to the Druids of early Gaul! Over half of his *Récits de l'infini* consists of a long dialogue that was later frequently republished and translated separately as *Lumen.* Not a novel, for it has no plot, not a scientific treatise, for it is obviously fictional, this quasi-religious fantasy consists of a conversation in which "Lumen" expounds both pluralism and transmigration in the course of recounting to "Quarens" his celestial travels after he had left his body (died) on earth. Having acquired the ability to move faster than light, Lumen witnesses the past of earthly life as well as his previous existences on other planets, the strange inhabitants of which he describes. These features make this book even more fantastic than the lunar voyages of Jules Verne, who was himself influenced by Flammarion.[95]

In the mid-1870s, Flammarion was permitted to use one of the telescopes of the Paris Observatory, which he employed to compile a respected catalog of double stars. In 1877, he published his large *Les terres du ciel,* which by 1881 was in its tenth edition. As he states, this book marks a return to the pluralist thesis, which, he asserts, "can now be greatly developed and absolutely confirmed. Such is the end of this book."[96] The form this development takes is a discussion of the habitability of each planet of the solar system. The radical character of his position is shown by his final sentence, in which he proclaims "this henceforth imperishable truth: LIFE develops without end in space and in time; it is universal and eternal; it fills INFINITY with its harmonies, and it will reign for ever and ever, during endless ETERNITY" (p. 769) Of course, the habitability of all the planets was neither "absolutely confirmed" nor made an "imperishable truth" by Flammarion's book. Proctor, far less intent on pushing a quasi-metaphysical claim increasingly contradicted by astronomical observation, realized this, but his particular criticism of Flammarion's book was that portions had been plagiarized from his own writings.[97] English reviewers, such as W. H. M. Christie, soon to become Astronomer Royal of England, were also distressed at Flammarion's tendencies "to mix fact and fancy," to indulge in "hasty inferences from doubtful observations," to use "sensational numbers," and to grasp at "terrestrial analogies." As he states: "Whilst there is so much doubt as to these points, it seems idle to discuss the conditions of life and the character of the inhabitants of these planets."[98] Flammarion, however, knew what would win an audience, that pluralist speculations would help sell even such a large and costly book as his *Les terres du ciel.* This must not be misunderstood; it was not greed but rather a quasi-religious vision of astronomy that inspired his rhapsodic writing.

That a more sober way of popularizing astronomy was possible is shown by Proctor's writings.

Flammarion's most successful work was his gigantic *Astronomie populaire,* which according to the astronomer J. E. Gore, who translated it into English, "sold . . . no fewer than one hundred thousand copies . . . in a few years – a sale probably unequalled among scientific books."[99] More recently it has been claimed that it did "more than any other book ever written [to] spread interest in astronomy."[100] Pluralist passages abound in it, an entire chapter being devoted to arguing for lunar life, to which end Flammarion even cites Schröter's long discredited observation of industrial smoke. Those who argue that the absence of a lunar atmosphere precludes life are accused of using "the reasoning of a fish."[101] Lunar life is also advocated by linguistic legerdemain; referring to supposed changes on the moon's surface, he uses the following double negative: "we cannot affirm that . . . there are not some changes which can be due to the vegetable kingdom or even the animal kingdom, or – who knows? – to some living formations which are neither vegetable nor animal." (p. 188) He describes Mars as "an earth almost similar to ours [on which exist] water, air, heat, light, winds, clouds, showers, brooks, fountains, valleys, mountains. . . . This is certainly a place little different from that which we inhabit." (p. 487) One sentence from J. E. Gore's annotated translation of *Astronomie populaire* epitomizes Flammarion's bold and buoyant style and the problems it presented its translator:

> We have already seen twenty-five stars blazing out in the sky with a spasmodic gleam, and relapsing to an extinction bordering on death [the number of *well-authenticated* cases of 'temporary stars' is much less than twenty-five. — J. E. G.]; already bright stars observed by our fathers have disappeared from the maps of the sky [that any *bright stars* have really disappeared is very doubtful. — J. E. G.]; a very great number of red stars have entered on their period of extinction [that red stars are really cooling down has been disputed. — J. E. G.].[102]

Such passages suggest why a reviewer of Flammarion's next major book, *Les étoiles et curiosités de ciel* (1882), described him as "enthusiastic and imaginative to a fault."[103] Written as a supplement to his *Astronomie populaire,* its eight hundred pages also support the reviewer's reference to Flammarion as the "most prolific of all writers on astronomy of the present day." The success of these books was creating a high level of interest in astronomy in France, on which Flammarion was quick to capitalize by founding in 1882 the journal *Astronomie: Revue d'astronomie populaire,* the first volume of which was enriched by various pluralist writings, including a series of articles by Flammarion on Martian life. Also in 1882, an admirer of Flammarion's writings offered him a chateau and estate in Juvisy near Paris. The chateau he turned into a lavish observatory and museum, the words "Ad Veritatem per Scientiam" being inscribed in gold on the gate.[104] Despite publishing seven astronomical books in the 1885–7 period, he found time to create and serve as first president of the Société astronomique de France, which published a journal and which by 1894 had over 600 members. One significance of this society is suggested by the recent remark that its "activities . . . created a reservoir of scientists from which emerged most of the outstanding French astronomers of this century."[105] Moreover, Flammarion societies devoted to astronomy sprang up in the French provinces, elsewhere in Europe, and in South America.[106] The paradox in all this is that although in his excessive pluralism he violated his motto "Ad Veritatem per Scientiam," Flammarion nonetheless created great interest in and institutions supportive of astronomy. In the latter regard, he was a "French Proctor," but in the former feature of his career, his radical pluralism, he differed significantly from his English contemporary.

In 1892, Flammarion published the first of the two volumes of his *La planète Mars et ses conditions d'habitabilité,* which like a number of his other writings is discussed in subsequent sections. In the final paragraph of this book, aimed at proving that life exists on Mars, he expresses his delight at having written a book on "the first world explored in the heavens." He felt this particularly strongly, he states, because of being a person "whose scientific and literary career had its beginnings . . . precisely in the defense of the doctrine of a plurality of worlds, and who consecrated [his] entire life to showing that the end of astronomy transcends celestial mechanics, and extends even to knowledge of the *conditions of life,* present, past or future, in the immense Universe. . . ."[107] In reading this statement it is well to remember that when Flammarion made it, his public career was less than half over, and dozens of his books remained to be written. Thirty years later he was still active, being awarded in 1922 a commandership of the Legion of Honor. That his pluralist convictions continued even to the time of his death in 1925 is suggested by the fact that the December 12, 1923, *New York Times* carried an article captioned "Flammarion Predicts Talking with Mars." The method by which this was to be accomplished: telepathic waves.[108]

4. The continuing quest for lunar life and some surprising side effects

Persons broadly acquainted with the progress of astronomy from 1860 to 1910 might expect that by then all controversy about lunar life had ceased. Yet, as the present section shows, claims for lunar life lingered in astronomy even beyond 1910 and moreover played a major role in the renaissance in selenography that began in the late 1860s and possibly had a role in the founding of one of the world's greatest observatories.

An unfortunate effect of the publication during the 1830s of the highly respected studies of the moon made by Beer and Mädler was that detailed observation of our satellite largely ceased for over two decades. The selenographer Neison, writing in 1876, suggested that this was because it was widely believed that Beer and Mädler had "finally solved . . . the great questions" in regard to the moon; in particular, they had "demonstrated that the moon was to all intents an airless, waterless, lifeless, unchangeable desert. . . ."[109] The spectroscope did nothing to alter that image of the moon; as early as 1865, Huggins and Miller reported: "the spectrum analysis of the light reflected from the moon is wholly negative as to the existence of any considerable lunar atmosphere."[110] This conclusion was supported a few years later by G. J. Stoney, who used the kinetic theory of gases to argue that the less massive celestial bodies lack a sufficiently strong gravitational field to retain at least the lighter gases.[111]

Among the few selenographers active in the 1850s and early 1860s was J. F. Julius Schmidt (1825–84), director of the Athens Observatory, who on October 16, 1866, made a "discovery" so exciting that, according to Neison, "Nearly every astronomer was led to study the moon, and for months all the principal telescopes of Europe were turned upon our satellite. Moreover, many of our present amateurs were then for the first time led to purchase telescopes, and take up the study of astronomy."[112] Another result, noted by Patrick Moore, was that "Systematic lunar observations began once more, and has continued ever since."[113] Schmidt's "discovery," which is now recognized as having been both extremely influential and almost certainly spurious, concerned the lunar crater Linné. Schmidt claimed that Linné, drawn earlier by Beer and Mädler as well as by Lohrmann and by Schmidt himself, had disappeared, with only a whitish patch being visible in its place.[114] By 1880, the Linné literature had grown very large, making it the most discussed lunar feature of the period.[115] Announcements of other supposed changes soon followed; on June 5, 1868, Schmidt wrote William Radcliff Birt (1804–81), secretary of the British Association Lunar Committee, to report that a small crater in the Alpetragius region of the moon had also disappeared.[116] Moreover, in the early 1870s, Birt published materials that indicated that the region inside the crater Plato grew darker as the sun's rays approached the perpendicular.[117] Changes in the region of the crater Messier were also reported by various observers,[118] and late in the 1870s Hermann J. Klein (1844–1914), director of the Cologne Observatory, announced his "observation" in May of 1877 of a new crater in the Hyginus N region.[119]

The reactions of astronomers to these supposed changes differed dramatically, some denying their reality and others seeing in them evidences of a lunar atmosphere or even of lunar organisms. Flammarion, in his *Astronomie populaire* (1880), notes that Klein had reported greenish tints in the lunar Sea of Serenity, attributing them to a "vegetable carpet."[120] Moreover, referring to the supposed changing tints in Plato, Flammarion states: "It is highly probable that this periodic change of tint . . . is due to a modification of a vegetable nature caused by the temperature. . . ." And he adds: "we have facts of observation which are difficult, not to say impossible, to explain, if one admits only a mineral soil, and which, on the contrary, are easily explained by admitting a bed of vegetation. . . ." (pp. 197–8) This was a strategy avoided by Birt, who in 1871 urged that the "existence of a lunar atmosphere or the existence of a lunar flora are not worth agitating in 'our' columns, which may be better filled with the publication of facts. . . ."[121] It is noteworthy, however, that under his classification "facts," Birt would have included the changes reported on the moon.

These and related issues were treated in three books on the moon published in England during the 1870s. In 1874, James Nasmyth (1808–90) and James Carpenter (1840–99) published *The Moon: Considered as a Planet, a World, and a Satellite*. Despite the word "World" in their title, they reject lunar life, describing the lunar landscape as "a realization of a fearful dream of desolation and lifelessness – not a dream of death . . . but a vision of a world upon which the light of life has never dawned."[122] In arguing for this conclusion, they deny that water or an atmosphere is present on the moon (p. 56) and that any actual changes have been observed on it (p. 174). They also adopt the principle that they will not begin "by assuming forms of life capable of existence under conditions widely and essentially different from those pertaining to our planet [for were this done] there would be no need for discussing our subject further: we could revel in conjectures, without a thought as to their extravagance." (pp. 175–6)

The lunar landscape presented in 1876 by Edmund Neison (1851–1938) in his massive *The Moon* and in journals of the period is somewhat more cheery. His book, for example, includes an extended discussion supportive of the existence of a thin lunar atmosphere.[123] Moreover, referring to the supposed changing tints in Plato, he invokes organic processes: "it does not appear how it can justly be questioned that the lunar surface in favourable positions may yet retain a sufficiency of moisture to support vegetation of various kinds. . . ." (p. 129) In that book and in an 1877 paper, he endorses the actuality of the changes reported in the Linné, Messier, and other regions, stating that although "the general opinion of astronomers appears to be against any such physical changes having occurred . . . , scarcely any astronomer known to have devoted time to the study of selenography doubts that many processes of actual lunar change are in progress. . . ."[124]

The third lunar treatise is that of R. A. Proctor, who published *The Moon: Her Motions, Aspect, Scenery, and Physical Condition* in 1873 and supplemented it by various papers. What is interesting in the present context is not that Proctor in his book takes a primarily negative view of lunar life but that he denies that "any *variation* of colour" has been witnessed on the moon and expresses serious doubts about the supposed changes in Linné, suggesting an optical origin of both effects.[125] His supporting arguments were more fully developed in a number of essays, including one from the late 1870s in *Belgravia,* in which he also criticizes the Hyginus case. In regard to Plato, he explains the effect as due to failure of the eye to compensate for changing contrasts, whereas in the other instances he uses detailed arguments to show the dubious character of the observations on which the claims rest. In doing this he does not deny the honesty of the selenographers who reported and accepted these changes, but suggests that "they are strongly prejudiced. Their labours, as they well know, have *now* very little interest unless signs of change should be detected in the moon."[126] Both Birt and Neison were aware of Proctor's analysis, but rejected it.[127] However, Proctor has been proved correct, although the Linné and Plato dispute persisted into the twentieth century and even today his achievements in this regard do not seem to be adequately appreciated.[128] Neison's arguments for a thin lunar atmosphere were one of the frail supports relied on by Rev. Timothy Harley in concluding, albeit cautiously and partly on religious grounds, for lunar life in his *Moon Lore* (1885). Rich in references, his section on "Moon Inhabitation" leaves the reader wishing that Harley had been more critical in discussing and better at balancing the testimonies of the many sources he cited.[129]

By the early 1880s, Birt and Schmidt had died, and Neison had departed for Durban, South Africa, to direct the Natal Observatory. Progress in selenography nonetheless continued, aided by improved photographic methods and by the increased interest in the moon that had developed in the previous two decades. That much of this interest had arisen from the spurious detection of changes on the moon is but one of the paradoxes in this history. The debate over lunar life seems to have led to a second, even more strikingly benefit for astronomy – the Lick Observatory. At the time of its opening in 1888, E. S. Holden, its first director, discussed the factors that influenced the uneducated millionaire James Lick (1796–1876) to donate $700,000 for an observatory containing the largest refracting telescope constructed up to that time. Desiring to leave a memorial of himself, Lick originally planned to build, according to Holden, "a marble pyramid larger than CHEOPS on the shores of San

Francisco Bay," but was dissuaded from this by the fear that it would be destroyed in a bombardment. Lick then hit on the idea of an observatory; as Holden recorded, "The instruments were to be so large that new and striking discoveries were to follow inevitably, and, if possible, living beings on the surface of the moon were to be described, as a beginning."[130]

Hopes for lunar life were sustained during these decades not only by selenographers but also by Peter Andreas Hansen, whom Simon Newcomb praised as "the greatest master of celestial mechanics since Laplace."[131] As noted earlier, Hansen published a paper in 1856 arguing that a certain feature of the moon's motion previously unaccounted for could be explained by assuming that the moon's center of figure is located about thirty-three miles farther from us than its center of mass. Such an asymmetric distribution of the moon's mass would cause any atmosphere or fluids on the moon to retreat to its remote side. Hansen made explicit the pluralist implications of his hypothesis: "one can no longer conclude that the [remote] hemisphere may not be endowed with an atmosphere, and that it has no vegetation and living beings."[132] Hansen's hypothesis, which (as discussed previously) was seized on during the 1850s by such pluralists as Herschel, Liagre, Mann, Powell, and Smith to save the selenites, continued to attract attention during the 1860s and beyond. Let us examine briefly the form this development took.[133]

In 1860, Hervé Faye pointed out a probable consequence of Hansen's theory: After the atmosphere of the remote side has been heated by prolonged periods of sunlight, it should spread to the point that its effects should become visible to us.[134] In 1862, Cornhill Magazine carried an essay, almost certainly by John Herschel, in which support was found for Hansen's hypothesis in an 1860 study by the Russian astronomer H. Gussew, who from an examination of stereoscopic photographs of the moon had concluded that the moon's figure is asymmetric, in particular that the moon is shaped somewhat like an egg, with its narrow end inclined toward the earth. Gussew, according to Herschel, estimated the moon's center of figure to be about fifty-nine miles nearer to us than its center of mass, or about twice Hansen's value, which broad agreement Herschel saw as supportive of Hansen's bold claim. Herschel was pleased: "Either result, but especially M. Gussew's, . . . would be quite compatible with the existence of [air and water], and of a habitable hemisphere on the [moon's] opposite side. . . ."[135] Hansen's hypothesis and its pluralist implications were the subjects of an entire chapter of William Leitch's God's Glory in the Heavens (1862); moreover, in the mid-1860s Henry Draper endorsed it, and Jules Verne used it in From the Earth to the Moon and Around the Moon.[136] In 1869, John Watson, a fellow of the Royal Astronomical Society, gave two lectures favoring lunar life. Although the reports on his lectures make no mention of Hansen's hypothesis, it no doubt influenced Watson, who argued that the lunar surface shows evidence of sea beds, and hence water, and that lunar volcanoes prove the presence, at least at earlier times, of an atmosphere. This water and atmosphere could not disappear, he urged; hence, they must have moved to the moon's remote side. On this claim, Watson "staked his reputation as a chemist and physicist. . . ."[137] R. Kalley Miller, mathematics professor at the Royal Naval College, Greenwich, adopted Hansen's hypothesis, stating in his *Romance of Astronomy* (1873) the "unquestionable conclusion that on [the moon's] other side . . . the atmosphere must attain a very considerable density, such as we have every reason to suppose would render it perfectly well fitted for the support of animal life."[138] Even before his book appeared, Miller regretted his rash remark; as he states in his preface, while his pages were in press, he "learned that Professor Adams and others have thrown grave doubts upon the accuracy of [Hansen's] calculations. . . ." Because the collected papers of John Couch Adams contain no refutation of Hansen's hypothesis, one may infer that his role was to call attention to the criticism of it published by Newcomb.

In 1868, Hansen's hypothesis suffered a mortal blow when Simon Newcomb showed that Hansen's claim was "without logical foundation," in particular that "the supposed discordance between theory and observation would not follow from Hansen's hypothesis, and, therefore, even if it exists, cannot be attributed to that hypothesis."[139] Newcomb's analysis was soon endorsed by the French astronomer Charles Delaunay, but drew a sharply worded response from Hansen.[140] Proctor on five or more occasions between 1865 and 1873 criticized the Hansen and Gussew theories. In 1870, for example, he endorsed Newcomb's analysis and moreover repeatedly noted flaws in Gussew's presentation.[141] Such attacks discredited Hansen's hypothesis, which Willy Ley, whose writings are rich in accounts of bizarre theories, described as "probably the wildest astronomical hypothesis ever advanced."[142] His description seems justified, but what is remarkable is the enthusiasm shown for it by such scientifically informed individuals as Draper, Herschel, Leitch, Liagre, Miller, Powell, Smith, and Watson. This is difficult to understand, unless one assumes that the selenites were the sirens that drew them to it.

At the beginning of the twentieth century, the question of lunar life was brought before the public by a prominent Harvard astronomer, William H. Pickering (1858–1938), who after training at Massachusetts Institute of Technology and a period of teaching there, joined the staff of the Harvard Observatory, which was directed by his older brother, E. C.

Pickering. By 1900, the younger Pickering had played a part in establishing the Harvard Observatory in Arequipa, Peru, and the Lowell Observatory in Arizona, as well as discovering the ninth satellite of Saturn. In 1902, one year before the appearance of his book *The Moon,* which included a photographic atlas of our satellite, Pickering published a pair of lunar papers in *Century* magazine. The first is entitled "Is the Moon a Dead Planet?" and begins with a discussion of the craters Linné and Plato, Pickering asserting that the former provides evidence of continuing lunar activity.[143] Taking this to indicate a thin lunar atmosphere, he urges that white areas on lunar mountain peaks and crater rims are best explained as snow or hoarfrost (pp. 91–5). This implies water vapor, which he views as nourishing low forms of lunar vegetation. As evidence of the latter, he notes the darkening under perpendicular solar rays of regions in Plato and elsewhere, an effect that he denies can be optically explained (pp. 95–8). Thus, Pickering proposes "a new selenography . . . which consists, not in a mere mapping of cold dead rocks and isolated craters, but in a study of the daily alterations that take place in small selected regions, where we find real, living change. . . ." (p. 99) He had thus become, like Birt before him, a victim of the "Plato illusion,"[144] but with this difference: He had added the sensational claim that the darkening is due to lunar vegetation. This was by no means the first time he had proceeded far beyond his data. When in 1892 he reported lakes and snow on Mars, his brother, the respected Harvard Observatory director, wrote him that the

. . . telegram to the N. Y. Herald has given you a colossal newspaper reputation. A flood of cuttings have appeared, forty nine coming this morning. In my own case I should have restricted myself more distinctly to the facts in this and other cases. You would have rendered yourself less liable to criticism if you had stated that your interpretations were probable instead of implying that they were certain.[145]

William Pickering's tendencies to sensationalism are scarcely less evident in his "The Canals in the Moon" in a later issue in 1902 of *Century.* In it, he presents drawings of numerous canals on the moon, especially in the Eratosthenes region, claiming this "discovery" to be important in numerous ways, such as "in exemplifying the tenacity with which life will exist throughout the universe in situations that seem . . . most unfavorable. . . ."[146] Moreover, being easily "observable" and displaying features not seen previously in the Martian canals, the lunar canals, Pickering suggests, should illuminate Martian studies. He attributes these canals to vegetation, the thin atmosphere making higher life forms impossible. Pickering's claims were picked up and embellished by popular science writers such as Waldemar Kaempffert, who published a paper praising Pickering's vegetation theory as the "most satisfactory that has yet been advanced."[147]

Pickering's pluralist enthusiasms persisted throughout his life. In 1912, for example, while using two Harvard telescopes to observe the moon from Mandeville, Jamaica, William Pickering wrote his brother: "Whatever reputation . . . I lost when I published my former observations, will be nothing to the destruction produced when these get into print, especially the drawings. I have seen everything practically except the selenites themselves running round with spades to turn off the water into other canals."[148] Between 1919 and 1924, Pickering published six papers in *Popular Astronomy* concerning the crater Eratosthenes, which papers are among the most remarkable of his four hundred fifty publications. After claiming evidence of lunar vegetation in the earlier papers, he brings the series to a sensational conclusion by attributing changes in the positions of certain dark spots to migrations of swarms of lunar insects. The reason he gives for this paper being the final installment in the series is that "Harvard has decided to dismantle the Mandeville station. . . ."[149] Pickering, on reaching the age of sixty-five, retired from Harvard as an assistant professor, but he continued to publish on lunar life even as late as 1937,[150] the year before he died. As was noted in one of the obituaries on him, with his death "there passed away . . . one of the few remaining astronomers of the school of Flammarion, Schiaparelli, Lowell."[151] A partial explanation of Pickering's persistent claims for lunar life may lie in the pliability of the pluralist position, which, it seems, can almost always be reformulated so as to survive contrary evidence. Two contemporary illustrations come to mind. In the 1960s, even such a well-informed astronomer as Professor Carl Sagan speculated on lunar life by reducing it to microorganisms and placing it far below the moon's surface.[152] As Patrick Moore remarked on Sagan's proposal, "there is no evidence whatsoever in favour of anything of the kind. . . ."[153] And in 1976, George H. Leonard, as a "result of studying thousands of NASA photographs," published a book in which he claimed to have detected a government coverup of the "fact" that such items as "bridges," "vehicles," "super rigs," "plumbing objects," and an "obelisk" are shown in these photographs.[154]

5. The signal question: sending messages to the moon or Mars

Speculations about life on the moon or Mars encouraged discussions of the possibility of communicating with their inhabitants. Such discussions had occurred before the 1860s, but they intensified at that time and continued in every subsequent decade until it was realized that for such communication to occur, man would have to send not only signals but also inhabitants. The story may be taken up in 1866, when Victor Meunier (1814–1903), a French writer, argued for lunar life on the basis of observations made by an Italian astronomer, M. Pompolio de Cuppis, who claimed to have witnessed the bending of light from certain stars occulted by the moon. He explained this observation, so contrary to reports of other astronomers, by suggesting that the moon has a thin atmosphere that does not extend above the lunar mountain ranges that in most cases begin the occultation. Meunier not only agreed and noted that Arago had proposed a similar idea but also asked whether earthlings and lunarians "will remain for long without communicating. . . ."[155]

Meunier left the method of communication unspecified, but Charles Cros (1846–88) created something of a sensation when in 1869 he proposed a method for sending signals to either Mars or Venus. Cros, a prodigy who became a professor at age eighteen and later attained prominence in literary circles and as a possible inventor of the phonograph,[156] first revealed his ideas as a guest speaker in a series of lectures by Flammarion.[157] On July 5, 1869, members of the Académie des sciences were informed that Cros had submitted a memoir on this subject.[158] Readers of the August 1869 issue of Cosmos, a journal edited by Meunier, encountered an exposition by Cros of his ideas, which later in 1869 he published as a brochure entitled Études sur les moyens de communication avec les planètes.[159] In it, Cros suggests that rays from one or more electric lights could be focused by parabolic mirrors so as to be visible to inhabitants of Mars or Venus, if such exist and possess telescopic means. He also presents a method using periodic flashes of conveying messages to them and even ways of designating colors and plane figures. His presentation, which contains evidence of deeper research, shows high enthusiasm; he exclaims that if our signals receive a response, "It will be a moment of joy and pride. The eternal isolation of the spheres is vanquished." (p. 471) He also suggests that flashes reported by Messier, Schröter, and Harding on such planets as Venus may be signals to us (p. 475). Cros was apparently not a believer in lunar life, the moon not being mentioned in his memoir. By 1870, his ideas had spread to Italy and to England,[160] and Flammarion kept them before the public by quoting from Cros in his Excursions dans le ciel. But none of Cros's contemporaries seem to have been sufficiently enthused or adequately endowed to attempt his project.

The signal question again attracted attention when Flammarion announced in 1891 that a Frenchwoman who had been a devotee of his writings had died, bequeathing 100,000 francs as a prize to be named after her deceased son, Pierre Guzman. As Flammarion revealed, the conditions of her bequest were:

A prize of 100,000 francs is bequeathed to the Institute of France (Science Section) for the person of whatever nation who will find the means within the next ten years of communicating with a star (planet or otherwise) and of receiving a response.

The testatrix especially designates the planet Mars, on which the attention and the investigations of all scientists are already directed. If the Institute of France does not accept the legacy, it will pass to the Institute of Milan, and in the case of a new refusal, to that of New York.[161]

The Académie des sciences did accept responsibility for the prize, influenced apparently by the donor's stipulation that interest on the award could be used to fund astronomical research.[162] Flammarion extensively publicized the prize, stating in an 1892 paper that the idea "is not at all absurd, and it is, perhaps, less bold than that of the telephone, or the phonograph, or the photophone, or the kinetograph."[163] Mars, he suggests in the same paper, presents the best opportunities because "its intelligent races . . . are far superior to us." (p. 110) Moreover, he proposes "inter-astral magnetism" and telegraphy as modes of communication (pp. 112–14).

By no means all French astronomers agreed with Flammarion's views of extraterrestrial communication; in fact, Amédée Guillemin (1826–93) published a critique of them. He dismisses the idea of communicating with selenites – on the grounds that none exist. Mars offers scarcely better prospects, for, as he points out, when we are closest to Mars, we appear "lost in the sun's rays," whereas when "better situated at quadrature [we are] also at a much greater distance."[164] Moreover, the quantity of light needed exceeds what could be produced. Thus, he concludes that "the problem of interplanetary communication is still far from solution; and I believe I shall never be contradicted by real astronomers." (p. 363)

In England, an even more lively controversy developed, involving some leading intellectuals of the period. It began on August 6, 1892, when the statistician and meteorologist Francis Galton (1822–1911) published a letter in the London *Times*, suggesting that a combination of mirrors would reflect sufficient sunlight to be detected by Martian telescopes. Richard Holt Hutton (1826–97), writing anonymously in the August 13 *Spectator*, labeled this idea "in the highest degree extravagant. . . ."[165] His reasons were (1) that Martians may not exist, (2) that if they do exist, they may not have developed as far as we or may possess different faculties, and (3) that it seems impossible to formulate a method based on flashes by which anything more than arithmetic trivialities could be communicated. Hutton published a second essay in the next issue of the *Spectator*, asking whether communication with Mars, if established, "would be advantageous or injurious" to earthlings.[166] The position he develops is that such contact would lead (even though it should not) to a decrease in man's confidence and a lowering of his sense of moral responsibility, because he would take it as yet another sign of his insignificance. As Hutton states, we shall be ready to discuss planetary beings only "when we shall have mastered the tendency to regard our own insignificance as an excuse for treating our wills as impotent because they are not omnipotent, and our reason as all but imbecile because it is not omniscient."

In the week between the publication of Hutton's two essays, the *Pall Mall Gazette* carried a recommendation from a Mr. Haweis, who stated: "I infer from the astronomers that a signal on our earth about six miles in size of the nature of a bright light could be seen by the inhabitants of Mars, who by all accounts seem to be making the most systematic and herculean efforts to communicate with us by flashing triangular signals of presumably electric light."[167] Haweis suggested that a better way than Galton's reflected solar rays would be for the lights of London to be dimmed systematically as a signal. J. Norman Lockyer, astrophysicist and editor of *Nature,* discussed the Galton and Haweis proposals in the September 8, 1892, issue of his journal, favoring a form of the latter's method.

One reason for the increased interest in Martian signaling was that during the 1890s, a spate of reports appeared of bright spots on Mars. In 1895, for example, *Popular Astronomy* carried a note that revealed that recently "articles were noticed in such papers as the *New York Herald,* giving account of the so-called 'Signals from Mars,' the canals on the surface of the planet, and also the latest astronomical speculation, that some of the features of its surface spell out the name of The Almighty in Hebrew letters." Not surprisingly, they declare it "a burning shame that such nonsense finds place in our best and greatest daily papers."[168]

Galton let the question of a method of communicating with Mars drop until 1896, when health problems forced him, as he stated, "to spend a somewhat dreamy vacation" at the hot baths of Wildbad. From this came a manuscript, the conclusions of which he summarized in an essay in the *Fortnightly Review.*[169] Galton's paper is devoted to developing a language suitable for extraterrestrial communication, this being presented in the context of a hypothetical situation in which earthlings receive signals composed of dots, dashes, and lines from a "mad millionaire in Mars." The code is cracked by a "clever little girl" who urges that the signals are in the base eight, because "the Mars folk are nothing more than highly developed ants, who counted up to 8 by their 6 limbs and 2 antennae, as our forefathers counted up to 10 on their fingers." (p. 661) Despite such playful passages, Galton was serious about his conclusion that "an effective inter-stellar language admits of being established. . . ."(p. 664) In his unpublished manuscript, he even speculates about the senses and reproductive patterns of Martians.[170]

Although unable to communicate with Mars, earthlings showed skill in disseminating information about Mars on their own planet. For example, an article in *Figaro illustrée* on the sighting of a geometrical figure on Mars was discussed in the *Kaluga Herald,* where the Russian teacher and eventual rocket pioneer Konstantin Tsiolkovskii (1857–1935) saw it and was inspired to publish in 1896 a paper on extraterrestrial communication. Noting that we are able to see the moons of Mars even though they are reported to be only about ten kilometers wide, Tsiolkovskii suggests that a similar area on earth, if equipped with rotatable mirrors, would permit us to signal Mars.[171]

Near the turn of the century, the idea of signaling Mars again attracted attention in France. In 1899, A. Mercier published a booklet entitled *Communication avec Mars* (Orléans) in which he not only argues for Martian life on the basis of canal observations and other reasons but also suggests the practicality of establishing communication with that planet. Noting that luminous projections had been reported on Mars, he speculates that these may be signals sent in recognition of the extra illumination resulting from the Universal Exposition in Paris in 1889 (p. 13). Mercier also discusses various signaling methods, his preference being for an arrangement in which solar rays would be reflected at sunset from the sunlit side of a mountain to a mirror atop the mountain, thence to a mirror on the dark side of the mountain and finally to Mars (p. 16). The advantage of this method is that the solar rays would attain increased visibility by being seen against a dark background. The enthusiastic author also proposes the formation of a society for Martian signaling and requests subscriptions for the establishment of a fund of 50,000 francs (p. 5). Subsequently, he organized two conferences, the first on May 2, 1900, in Paris, the second on May 16, 1901, in Orléans, to discuss "The Project of the Study on the Practical Means for the Execution of Luminous Signals from the Earth to Mars," as he subtitled the booklet he published in 1902 to report on the conferences and on the responses to his first booklet.[172] From the later booklet one learns that Flammarion supported his efforts and that over twenty-five periodicals published reviews of his first book (p. 27). Mercier's second booklet includes discussions of a projection recently observed on Mars as well as of various methods of signaling. Although Mercier reported receiving a number of subscriptions for his project, these were apparently insufficient for its execution.

The climate created by the canal controversy, Flammarion's writings on the Guzman prize, the publications of Galton, Mercier, and others, and not least the appearance in 1898 of H. G. Wells's *War of the Worlds* tended to be supportive of reports that could be interpreted as signals from Mars. For example, when A. E. Douglass of the Lowell Observatory telegraphed that he had seen a "projection" on Mars, a number of newspapers took it to be a Martian signal. Douglass and Lowell issued denials, while scientific journals attempted to clarify the situation, probably with limited success.[173] Douglass even received a letter from a Denver attorney who asked: "Suppose the people of Mars have built a monument 10 miles square and a hundred miles high, covered exteriorly with polished marble (which is the fact). Would not the monument reflect a shaft of light? If you saw a shaft of light what was its color, and did the light scintillate?"[174]

An even more remarkable report of Martian or possibly Venusian signals came at this time from Colorado Springs, Colorado, where Nikola Tesla (1856–1943), a famous electrical inventor, had set up high-voltage equipment to experiment with wireless signals. In the February 9, 1901, issue of *Collier's*, Tesla published his "Talking with the Planets," which is full of oracular pronouncements, including the claim that he had detected an extraterrestrial signal.[175] He also predicts that interplanetary communication will "become the dominating idea of the century that has just begun," and he assures his readers that he has devised new methods whereby "with an expenditure not exceeding two thousand horsepower, signals can be transmitted to a planet such as Mars with as much exactness and certitude as we now send messages by wire from New York to Philadelphia." (p. 4) The *Colorado Springs Gazette* expressed its delight at Tesla's discovery:

If there be people in Mars, they certainly showed most excellent taste in choosing Colorado Springs as the particular point . . . with which to open communication. In fact, we may feel assured that if the mystical one-two-three which Tesla says may have been impulsed from Mars, should be translated . . . , it would read, "How is the weather in Colorado Springs?" . . .[176]

Not all were so enthusiastic. Edward S. Holden, former director of the Lick Observatory, commented: "It is a rule of sound philosophizing to examine all probable causes for an unexplained phenomenon before invoking improbable ones. Every experimenter will say that it is 'almost' certain that Mr. Tesla has made an error. . . ."[177] The editors of *Current Literature* noted that Tesla's report "is generally regarded as preposterous . . . ," and they added with approval a quotation from the *St. Louis Mirror*: "Nikola Tesla is the Ignatius Donnelly of physics."[178] Nonetheless, *Current Literature* reprinted Tesla's article! Tesla's ideas, as well as the whole idea of signaling Mars, were also sharply criticized in 1901 by Sir Robert Ball of Cambridge University. In his paper he surveys the signaling methods that had been proposed and finds all to be impractical. In reference to Tesla, he states: "no electrical signalling to Mars appears to me to be possible, for the simple reason that the apparatus would have to be sixteen million times as efficient as that which would suffice to do for wireless telegraphy far more than even its most ardent champions have yet, so far as I know, ventured to claim for it."[179] Tesla nonetheless persisted in his claims, and around 1920, the Nobel-prize-winning electrical genius Guglielmo Marconi (1874–1937) announced that he had detected extraterrestrial signals.[180]

The highest level of interest, however, was not in electrical signals but in visual signals. In 1909, stimulated by the Mars opposition of that year and by a proposal of W. H. Pickering that at a cost of $10 million a cluster of mirrors could be constructed capable of signaling Mars, an extensive debate broke out in the United States, especially in *Scientific American*. It began in the May 8, 1909, issue with an anonymous article commenting negatively on Pickering's proposal and on the less expensive idea of R. W. Wood of Johns Hopkins that an immense strip of black cloth be periodically unrolled on a desert in the southwestern United States. Being unconvinced that Martians exist, the author could not take either proposal seriously.[181] Other proposals appear in the next issue, including the money-saving suggestion that 5,000 four-inch mirrors per mile would do as well as Pickering's 5,000 ten-foot mirrors. The author also mentions the plan of David Todd of Amherst to ascend in a balloon with the most sensitive wireless telegraph receiver available so as to pick up Martian signals.[182] Two issues later, George Fleming states that optical principles dictate that four-inch mirrors will not do, that the larger mirrors of Pickering are necessary.[183] Wilfred Griffin then suggests that a "huge battery of powerful electric searchlights" be "winked" at Mars.[184] Todd's proposal is faulted in the same issue, the author asking how the Martian signals can be distinguished from those of the "about 2,000 wireless stations scattered over the earth."[185] The issue of three weeks later contains the comments of W. C. Peckham of Adelphi College, who urges the impracticality of Wood's black cloth proposal on the grounds that when Mars is in opposition, the side of the earth toward Mars is entirely dark. He quotes with approval Professor Moulton's remark that "The newspaper talk of communication between the earth and Mars by any imaginable means is utter foolishness."[186] The *Scientific American* signal series was brought to a close by Pickering himself, who reaffirms his method, assuring everyone that it "is merely a question of the most elementary mathematics."[187]

Although Flammarion approved of Pickering's proposal,[188] E. L. Larkin of Mount Lowe Observatory attacked it using very elementary mathematics. Larkin states that if we wish to produce a signal one-tenth of a second wide when seen from Mars when at quadrature, then "the reflector must be 52 miles wide. Human skill is now being taxed to the extreme in making a mirror 100 inches in diameter, in Pasadena, for the Mount Wilson Observatory."[189] On this basis he describes the situation as "hopeless." Larkin may have been too hard on Pickering's plan, which was probably more practical than the method suggested in *Collier's* by W. R. Brooks, director of the observatory at Hobart College. Brooks's idea was to set up "a great area of electric lights" to be flashed on and off, "not necessarily after the Morse code for it were idle to suppose that the Martians are familiar with this."[190] However, signaling by the Morse code was an acceptable procedure for "T. C. M." (probably Thomas Corwin Mendenhall), who injected some humor into the debate by a satirical article in *Science* that advocates the solution: "*A hole through the earth.*" He adds: "For our immediate purposes of wigwagging to Mars such a hole must necessarily be several miles in diameter. Although some minor difficulties in the way of the execution of this plan remain to be overcome, many of the details are already settled . . ."[191] An element of satire is probably also present in an editorial in the *Independent*, where, without mentioning Flammarion, his proposal for telepathic communication with the Martians is discussed. After mentioning that "Mediums are cheaper than mirrors," the editors add:

Professor Flournoy, of Geneva, has published a volume of the revelations of [a] lady . . . giving full details of Martian life and language. Professor Hyslop in this country has given Mrs. Smead's account of the same people. . . . The two descriptions do not agree at all in architecture, costume and language, but there is nothing in that to discourage a psychic reader. The two mediums may have been seeing different parts of the planet.[192]

The astronomer E. E. Barnard seems to have been no more enthusiastic about proposals for Martian signaling than Mendenhall. Sometime during this period, he published a story in which by means of paper letters 100 miles long placed in an African desert, earthlings finally send Mars the message: "Why do you send us signals?" At last the answer comes back: "We do not speak to you at all, we are signaling Saturn."[193]

In considering these proposals for signaling the moon or Mars, one is struck by how many of them originated with or were taken seriously by scientists as prominent as Flammarion, Galton, Lockyer, W. H. Pickering, and Tesla. Some of these ideas showed a high level of imagination, but a low level of responsibility. The proposals also illustrate the pervasiveness around the turn of the century of the "Mars mania." The story of that development is the subject of the concluding sections of this book.

6. *The message of the meteorites: "From World to World/The Seeds were whirled"?*

For centuries pluralists have dreamed of discovering direct evidence of extraterrestrial life. At times during the nineteenth century such proof seemed to appear in a most dramatic manner: falling from the skies in meteorites. But the message of the meteorites has been ambiguous, creating a controversy continuing to the present. Even in recent decades, the detection of organic materials in meteorites has provoked debates about whether such materials came with the objects or became present as terrestrial contaminants after their fall. Also at issue has been whether such materials, if extraterrestrial, were products of living organisms or, alternatively, were formed by complex chemical processes not involving living beings. Of the great number of meteorites known, only a few dozen are carbonaceous chondrites, the type most revelant to this topic. The present consensus is that no indisputable evidence of extraterrestrial life has yet been secured from meteorites. A recent authoritative text states, for example, that "Chondrites, and especially carbonaceous chondrites, are now recognized to be the relatively well-preserved samples of the non-volatile material of the nebula from which the Sun and planets formed."[194]

Shortly after 1800, scientists realized that meteorites originate beyond the earth's atmosphere, a discovery that so increased interest in them that over five thousand publications on meteorites appeared during the nineteenth century.[195] Enthusiasm was especially high after the occurrence on November 12, 1833, of the most brilliant meteor display in recorded history. One American witness exclaimed "The world is on fire," and another reported that "never did rain fall much thicker than the meteors. . . ."[196] Denison Olmsted of Yale observed that the display centered on a single point in the constellation Leo and speculated that the meteor shower was caused by the earth passing through a swarm of meteors orbiting the sun, noting on behalf of this conjecture that such displays regularly occur on November 12, although never previously with such brilliance. Olmsted's theory, and later modifications of it, rivaled the theory presented by Laplace and accepted by a number of authorities that meteorites come from volcanic eruptions in the moon.

The 1830s are also noteworthy for a chemical analysis of various meteorites published in 1834 by J. J. Berzelius (1799–1848). Included among these was a carbonaceous chondrite that fell in Alais, France, in 1806. Noting that the chemists Thenard and Vauquelin had earlier found carbon in this object, Berzelius investigated it with two questions in mind: "Does this carboniferous earth contain possibly humus or a trace of other organic compounds? Does this give possibly a sign of the presence of organic forms in other worlds?"[197] His cautious conclusion is that "The presence of carboniferous material in the meteoritic earth has analogy with the humus contained in terrestrial earths, but it is possibly added in a different way, has other properties, and seems not to justify the surmise that it has an analogous determination to that of carboniferous materials in terrestial earth." (p. 123) In 1836, the *Philosophical Magazine* carried a summary of Berzelius's memoir containing the statement on the Alais meteorite that "The carbonized substance that this earth contains, in a state of mixture, would not authorize the conclusion that in its original habitat, this earthy substance was of organic nature."[198] Berzelius's position may have been linked to his belief, expressed in his 1834 paper, that meteorites come from volcanic eruptions on the moon. He even adds: "most meteoric stones are so similar to one another in their composition that one can consider them as coming from the same [lunar] mountain. . . ." (p. 5) He does, however, consider the idea that they may be fragments from a planetary collision in the asteroid belt (p. 7).

In the 1850s, Friedrich Wöhler (1800–82) published chemical analyses of the Kaba (1857) and Cape (1838) meteorites. In both papers, he reports the detection of carbonaceous materials of "organic origin," the article on the Cape meteorite making the claim that "there is no doubt that this meteoric mass . . . contains a carbonaceous substance which can have no other than an organic origin."[199] On May 14, 1864, the largest carbonaceous chondrite known until 1950 fell at Orgueil, France, stimulating a controversy that continued for over a century. In the analysis of it published in 1864 by Stanislas Cloëz (1817–83), it is stated that it consists of materials "analogous to those of the organic part of several varieties of peats and of lignites." His specific results were 63.54 percent carbon, 5.98 percent hydrogen, and 30.57 percent oxygen. He adds that this seems "to indicate the existence of organic substances in the celestial bodies."[200] In 1868, Marcellin Berthelot (1827–1907) reanalyzed the Orgueil meteorite, reporting materials similar to hydrocarbons of the formula C_nH_{2n+2} that he describes as "comparable with the oils of petroleum."[201] Thus, by 1870, although no living material or fossil remains had been detected in meteorites, evidence was available that certain meteorites contain organic materials. On the assumptions that this material was not an earthly contaminant and (as is now known to be incorrect) that such materials must have come from living organisms, this could be seen as evidence of extraterrestrial life.

Meteors were much on the minds of the scientists who attended the 1871 meeting of the British Association for the Advancement of Science, because Sir William Thomson (1824–1907), later Lord Kelvin, brought his presidential address to a sensational conclusion by simultaneously discussing spontaneous generation, evolutionary theory, meteorites, and the plurality of worlds. Raising the question of the origin of terrestrial life, Thomson states that both "philosophical uniformitarianism" and Pasteur's experiments rule out spontaneous generation.[202] Given this, Thomson asks: "How, then, did life originate on earth?" The answer offered is that

. . . because we all confidently believe that there are at present, and have been from time immemorial, many worlds of life besides our own, we must regard it as probable in the highest degree that there are countless seed-bearing meteoric stones moving about through space. . . . The hypothesis that life originated on this earth through moss-grown fragments from the ruins of another world may seem wild and visionary; all I maintain is that it is not unscientific. (pp. 269–70)

Thomson, who had been introduced by the previous B.A.A.S. president T. H. Huxley, must have delighted that defender of Darwin by stating that "all creatures now living on earth have proceeded by evolution from [lower forms]." Huxley, however, was no doubt disturbed by Thomson's additional comment in regard to Darwinian theory: "I have always felt that this hypothesis does not contain the true theory of evolution. . . ." (p. 270) Citing John Herschel's claim that Darwin's theory does "not sufficiently take into account a continuously guiding and controlling intelligence," Thomson concludes by recommending a return to Paley's design argument (p. 270).

Reactions to Thomson's theory were mixed. One objection was put into poetry in *Punch:*

> But say, whence in all those meteors life began,
> From whose collision came the germs of man?
> Still hangs the veil across the searcher's track,
> We have but thrust the myst'ry one stage back.[203]

Another wit wrote: "From World to World/The Seeds were whirled/ Whence sprung the British Ass. [Association]."[204] Three days after the address, Joseph Dalton Hooker wrote to Darwin, commenting concerning Thomson's presentation:

What a belly-full it is, and how Scotchy! . . . The notion of introducing life by Meteors is astounding and very unphilosophical. . . . Does he suppose that God's breathing on Meteors or their progenitors is more philosophical than breathing on the face of the earth? I thought too that Meteors arrived on the earth in a state of incandescence. . . . For my part I would as soon believe in the Phoenix as in the Meteoric import of life.[205]

The somberness seen in public scientific discourses in Victorian Britain was also absent when on August 11, 1871, Huxley privately communicated to Hooker his comparison of Thomson's theory with a game of the day: "What do you think of Thomson's 'creation by cockshy' – God Almighty sitting like an idle boy at the seaside and shying aerolites (with germs), mostly missing, but sometimes hitting a planet!"[206] By September 1871, Proctor was pointing to the irony of the situation: "Regarded not so many years ago as probably the vehicles of the Almighty's wrath, comets are made by this new hypothesis to appear as the parents of universal life."[207] Not impressed, Proctor adds: "I can scarcely bring myself to believe, indeed, that the eminent professor was serious in urging his hypothesis of seed-bearing meteors." (p. 218) Thomson, however, was very serious about his theory and continued to press it in later years. At the 1877 B.A.A.S. meeting, his theory was criticized in the presidential address of the Glasgow embryologist Allen Thomson (1809–89),[208] but advocated in a paper by Sir William Thomson, who responded to the objection to his earlier paper that organisms on meteoric material would burn up during descent. The lively discussion generated by that paper was described by Walter Flight (1841–85), a specialist on meteorites, who noted that at one stage in the discussion,

. . . some one . . . introduced the Colorado Beetle, and this was held to be irresistibly funny; then someone else got up and said he was an Irishman, which was judged to be even funnier still. At length another speaker arose to breathe the hope that when Papa Colorado Beetle dropped down on a meteorite he would leave Mamma Colorado Beetle behind, which was felt to be far and away the funniest thing of all. . . . [A]lthough a gallant effort appears at this junction to have been made to win back their confidence by assuring them that meteorites really do not contain organic matter of any kind, the Section was not to be comforted. . . .[209]

The intensity of Thomson's attachment to his theory is also indicated by his comment in an 1882 letter to the Duke of Argyll: "I returned to the subject again and again in the British Association Meeting [in 1881] at York, and obtained the appointment of a Committee to investigate meteoric dust, chiefly with the view to ascertaining whether any of it contains either traces or actual specimens of life. . . ."[210] Finding that his views on the ultimate origin of life were frequently misunderstood, Thomson attempted to clarify them in an 1886 letter: "the 'star germ theory' which I put forward as a possibility does not in the slightest degree involve or suggest the origination of life without creative power, and is not in any degree antagonistic to . . . Christian belief."[211] Late in his long life, however, he may have come to question his theory; in 1903, he held that none of the other planets of the solar system was habitable, although he believed that inhabited planets might exist in other systems.[212]

William Thomson was not the only or the earliest advocate of the idea that terrestrial life is of extraterrestrial origin. In 1821, Count de Montlivault published his *Conjectures sur la réunion de la Lune à la Terre,* in which he suggested that life came to earth on meteorites from lunar volcanoes.[213] In 1865, Hermann E. Richter, a German physician and defender of Darwin, advocated the meteoric origin of life on earth. Richter, who cites Flammarion's *La pluralité* on behalf of extraterrestrial life, enthusiastically claims that his thesis "supplies the capstone to Darwin's daring edifice."[214] Moreover, those who refused to take Thomson's position seriously must have been surprised to learn that among its advocates was the premier German physicist of the period, Hermann von Helmholtz (1821–94), who revealed in 1875 that he had arrived at the same hypothesis slightly before Thomson and had presented it in a public lecture in the spring of 1871 at Cologne and Heidelberg.[215] Contained in the lecture is this statement:

Who can say whether the comets and meteors . . . may not scatter germs of life wherever a new world has reached the stage in which it is a suitable dwelling place for organic beings? We might, perhaps, consider such life to be allied to ours, at least in germ, however different the form it might assume in adapting itself to its new dwelling place.[216]

Helmholtz had been provoked to defend this thesis by an 1872 attack on Thomson's presentation by Johann Zöllner, who claimed the theory to be faulty for two reasons: first, because it transforms the straightforward question why life developed on earth into the question why life arose on another planet; second, because it fails to explain how organisms in meteorites can survive the extreme temperatures generated as the meteorite traverses the earth's atmosphere.[217] In response, Helmholtz admits the hypothetical character of his theory, but asserts that interior portions of the meteorite may remain cool and that the organisms could be blown free of the meteorite in the upper atmosphere before incandescence commences.[218] Other German scientists also espoused the hypothesis, including the botanist and bacteriologist Ferdinand Cohn (1828–98) and the physiologist William Preyer (1841–97). Walter Flight of England in 1877 showed himself open at least to considering it seriously, and in 1891 the French botanist P. E. L. Van Tieghem (1839–1914) endorsed it in his *Traité de botanique*.[219] Most scientists, however, were skeptical if not hostile to this theory. In the opening years of the twentieth century, Svante Arrhenius (1859–1927), a Nobel-Prize-winning Swedish scientist, put forth a new theory of the extraterrestrial origin of life on earth, but without involving meteorites. Probably because his theory, like those of his predecessors, was low in falsifiability, although high in explanatory power, it attracted only limited attention.[220]

In the interim, the question of a meteoritic origin of terrestrial life continued to be debated, especially after 1880, when the lawyer Dr. Otto Hahn, who had studied geology at Tübingen, created a short-lived sensation by his *Die Meteorite (Chondrite) und ihre Organismen*. Dr. D. F. Weinland, a German zoologist, publicized Hahn's book, stating in 1881 that in Hahn's illustrations "we can actually see with our own eyes the remains of living beings from another celestial body."[221] What Hahn had done was to cut thin slices of noncarbonaceous chondrites, grind them down to semitransparency, and examine them microscopically, finding what he believed to be minute fossils comparable to terrestrial corals, crinoids, and sponges. In his book, Hahn debates whether his fossils were of extraterrestrial origin or were thrown temporarily into the earth's orbit as a result of some body colliding with the earth.[222] Francis

Birgham, writing in *Popular Science Monthly* in 1881, endorsed Weinland's claim and stressed "the transcendent importance of this new and great discovery."[223] A later pluralist even reported that Hahn had set his samples before Charles Darwin, causing him to leap from his chair and exclaim: "Almighty God! What wonderful discovery! Now reaches the life down."[224] Hahn's "discovery" soon received decisive refutation. Carl Vogt (1817–95), for example, published a paper stating that the figures reported by Hahn were not similar to terrestrial fossils, that they were in fact crystalline formations, and that Stanislas Meunier (1843–1925), a French expert on meteorites, had produced comparable forms in his laboratory.[225] A dozen or more articles appeared on Hahn's "fossils," the *American Journal of Science* remarking that Hahn's "conclusions are so fanciful and so obviously without foundation in fact, that it would be unnecessary to refer to them except that a recent article [presumably Birgham's] has been published in this country in which the discoveries are described in the spirit of the author. . . ."[226] *Popular Science Monthly,* which had earlier printed Birgham's excessive claims, published a paper noting that Professor J. Lawrence Smith had disputed Hahn's position and that Professor Hawes of the Smithsonian had described Hahn as an observer whose "imagination has run wild with him."[227]

Harold C. Urey, in a 1966 review of reports of biological materials supposedly found in meteorites, remarked concerning Berzelius, Wöhler, Cloëz, and Berthelot that "during the 19th century chemists investigating these meteorites reported finding material which is found on earth and confidently described as decomposition products of biological material."[228] However, many of the scientists, for example, Berzelius and Cloëz, wrote with commendable caution, which was at times abandoned by extreme pluralists. For instance, Dr. Louis Figuier went from reports of material similar to peat in the Orgueil meteorite to state: "Peat is simply the product of the gradual decomposition of vegetables. . . . There are vegetables, then, in the planet from which came the Orguiel aerolite. By consequence, there are vegetables in the planets which neighbor our own."[229] No doubt the scandal of the Hahn-Weinland incident contributed to the fact that no further reports of extraterrestrial remains in meteorites appeared during the rest of the nineteenth century. When such reports began again in the 1960s, Hahn was cited as a warning that "the decision whether a certain form is of biological or inorganic origin is . . . quite subjective."[230]

9

Religious and scientific discussions

1. *French religious writings: is man a "citizen of the sky"?*

Rarely have ideas of extraterrestrial life been free from interactions with religion; these were common before 1860 and, as demonstrated in this chapter, continued in subsequent decades. In discussing post-1860 interactions, it is convenient to commence with France, focusing first on some pluralist systems separate from the Christian tradition and then on reactions of French Catholics to the idea of other worlds. Brief mention is made at the end of this section of some Spanish and Italian writings.

Although for a few hours in 1870 Louis Auguste Blanqui (1805–81) was head of the French government, by 1871 he had been thrown into prison. Incarceration was not new to Blanqui, who spent nearly half his life behind bars, but the cosmological and pluralist speculations he drafted in that dungeon and published in 1872 as *L'éternité par les astres: hypothèse astronomique* reveal his neophyte status in those areas. Assuming life on apparently every planet in a spatially and temporally infinite universe, he asserts:

> Our Earth . . . is the *repetition* of a *primordial* combination that reproduces itself identically and that exists simultaneously in billions of identical copies. Each examplar is born, lives, and dies in its turn. It is being born, it is dying by the billions in each second that passes. On each of them succeed all the material things, all the organized beings, in the same order, in the same place, at the same moment as they succeed on the other earths, their copies.[1]

As illustration, Blanqui adds: "That which I write in this moment in a dungeon of the fort of Taureau, I have written and I will write during eternity, on a table, with a pen, in such apparel, in entirely similar circumstances." (p. 73) In support of this view, he cites spectroscopic evidence that the celestial bodies are composed of the same elements as the earth. The number of such elements being finite, whereas space and time are infinite, duplication in detail must constantly occur. Yet it should not be supposed that the histories on all the planets in his universe are identi-

cal: He urges that although on billions of planets Napoleon lost at Wa-
terloo, on billions of others he achieved victory (p. 57). What led this
anarchist, atheist, and materialist to compose his book? Part of the explanation may lie in his confinement, which no doubt limited what he could
write. He probably also sought, as Epicurus and Lucretius had earlier, to
shore up his social philosophy with a supportive cosmology that would
provide solace in an impersonal and fortuitous universe in which defeat
was both a matter of chance and of minute significance. Blanqui describes his doctrine as "a simple deduction from spectral analysis and
from the cosmogony of Laplace." (p. 75) It was scarcely that, but it was
in violation of the second law of thermodynamics, Blanqui attributing
the formation of nebulae and thereby the renewal of stars to the perpetual collisions of burned-out stars (p. 31). Scientists and Christians, albeit
for different reasons, viewed Blanqui's fantasies with disdain, but the
case was not so clear for metempsychosis, which attained renewed popularity in France in the post-1860 period.

Flammarion was not the first figure in the history of theories concerning the transmigration of souls[2] to tie that doctrine to pluralism; Bonnet
and others had done this in the eighteenth century, and Flammarion
himself had learned of this linkage from Reynaud's Terre et ciel (1854).
Nonetheless, Flammarion's advocacy of it in a number of his widely read
writings was especially influential, as was recognized by André Pezzani
(1818–77), who viewed him as the "chef d'école" – the master of a
school.[3] In 1865, Pezzani sought to swell the size of that school by
publishing La pluralité des existences de l'âme conforme à la doctrine de
la pluralité des mondes, on the title page of which he identifies himself as
an "Avocat à la cour d'appel de Lyon et Lauréat de l'Institut." An enthusiast for metempsychosis since 1838,[4] he devotes his long book to showing the reconcilability of that doctrine with pluralism and to arguing that
"from a prodigious antiquity, man has resolutely believed in future destinies, and that the most persistent form of these destinies has been the
plurality of existences, the doctrine of transmigrations. . . ." (p. i) He
traces the history of transmigrational teachings among the Hindus, the
Druids, the ancient mystery cults, the Pythagoreans, the Jews of the
Zohar, the early Christians, and in such moderns as Bonnet, Dupont de
Nemours, Fourier, Reynaud, and Flammarion. Convinced that Flammarion had "irrevocably resolved" (p. 383) the pluralist debate, Pezzani
focuses on arguing for metempsychosis. He summarizes his claims as
follows:

1.   Hell, absolute and eternal, is an error, since it is contradictory simulta-
     neously to the nature of God and of man.
2.   Without a belief in previous lives [we cannot explain] the coming of the new
     soul in the evil world of the earth, nor the sometimes irremediable infirmities

of the body, nor the evils which afflict it, nor the disproportionate distribution of riches, nor the inequality of intelligences and of morality. . . .
3. Pre-existence . . . entails logically the plurality of successive existences in the future for all the souls that have not arrived at the end, still have . . . imperfections to efface. (pp. 484–5)

Pezzani's efforts met with some success; by 1872, his book was in its sixth edition, and moreover Spanish and Swedish translations were made. He also published *La nature et destination des astres,* devoted to proving that heavenly bodies serve as habitations for souls in their progression toward perfection.[5]

Among the earliest enthusiasts for Flammarion's blending of pluralist astronomy with metempsychosis was the foremost French literary figure of the nineteenth century, Victor Hugo (1802–85), who after reading Flammarion's *La pluralité* sent him a letter stating: "The matters which you treat are the perpetual obsession of my thought, and exile has only augmented this meditation in me, by placing me between the two infinities, the Ocean and the Sky. . . . Your studies are my studies."[6] Hugo's lifelong interest in astronomy was documented in a 1933 book by Edmond Grégoire, who illustrated Hugo's pluralist convictions by passages from his poetry; among these are:

> Every globe revolving round a star
> Is home to a humanity near yet far.
>
> *(Dieu,* 1, 1)
>
> On the globes gliding through the realms of space,
> Races such as ours are everywhere in place.[7]
>
> *(Dernière gerbe,* 3)

Although Flammarion repeatedly cited Hugo in support of his astronomical doctrines,[8] the source of Hugo's interest in astronomy, according to Grégoire, was above all Chateaubriand, whose writings the young Hugo had especially admired.[9] Moreover, Hugo's commitment to spiritualism can be traced to the 1850s, if not earlier.[10]

More widely read than Pezzani's books and far more astronomical in approach was *Le lendemain de la mort, ou la vie future selon la science,* published in 1871 by Louis Figuier (1819–94), a prolific French science writer and former professor with doctorates in physical science and in medicine. The death of his young son led Figuier around 1870 to ask "of the exact sciences what positive evidence they could render on . . . the new life that must open to us beyond the grave."[11] The results of his reflections are presented in *Le lendemain,* which, although purporting to be scientific, is primarily a work of spiritualism or, according to its author, "Spiritualism demonstrated by science." Like Flammarion, disturbed at the materialist philosophy of his day – "It is not petroleum which set fire to the monuments of Paris: it is materialism". (p. 5) –

Figuier recommends as remedy a blend of pluralist astronomy and trans-migrational religion. Opening his book with a brief exposition of metem-psychosis, he turns to the solar system, finding all planets inhabited (pp. 28–84). Figuier suggests that after death planetary inhabitants, if sufficiently perfected that they need not endure another cycle of life on their planet, ascend to the ether around their planet, where they acquire a new body, composed of ether and endowed with superhuman attributes. Spared the need for nourishment, free of fatigue, and united with their families, they remain there until further metamorphoses take them to the sun, their final home, where, freed from material bodies, they serve as the source of solar emanations that spread life throughout the planetary system (pp. 84–191). All this is embellished with much scientific detail – to which, of course, it was at most tenuously tied. Animals have souls, he urges, which in accord with the eternal law of progress in the universe, become souls of men (pp. 192–202). In fact, in an effort to explain the differing aptitudes humans possess, he suggests that the soul of the singer was once in a nightingale, the soul of an architect in a beaver (p. 247). In support of metempsychosis, he cites many of the arguments presented earlier by Pezzani, who, along with Reynaud, is mentioned. He responds to various objections, including the claim that his system *"is borrowed from Fontenelle's 'The Plurality of Worlds.'"* (p. 305) To this he replies by tracing the history of pluralism and arguing that his system of solar superhumans goes beyond all earlier authors (pp. 305–15). Ethical urg-ings and a sidereal section conclude his book.

Despite Figuier's claim that his system was based on science, he must have recognized its speculative character, and at least once in his book he admits that "metaphysical and moral" ideas underlie it (p. 315). Because of the criticisms of Catholicism contained in his volume, as well as its unorthodox doctrines, Figuier's book was placed on the Index, but this did not deter it from going through at least eleven French editions nor from being translated by 1872 into English, which translation was repub-lished into the 1890s. That Figuier held his doctrines until his death is indicated by the fact that during the last year of his life, he revealed to a visitor his theory that

. . . certain comets, notably those that return into our solar system, are agglom-erations of superhuman beings which have just finished a voyage in the profound depths of the sky and end their trip by returning into the sun. . . . According to this hypothesis, these comets are pleasure trains made up of the inhabitants of ethereal spaces.[12]

The position presented by Pezanni and by Figuier was also espoused and expounded by Victor Girard, who in 1876 published his *Nouvelles études sur la pluralité des mondes habités et sur les existences de l'âme* (Paris). Girard, who drew his astronomy from Flammarion's writings, was accurately described by Flammarion as having had in that book "no other object than to pour forth ever more fully the consoling ideas inspired by the spiritualist philosophy of astronomy."[13] Twelve years later, Girard published *La transmigration des âmes et l'évolution indéfinie de la vie au sein de l'univers* (Paris). Judging by its title, it is in the same tradition. The efforts of Reynaud, Flammarion, Pezzani, Figuier, and Girard to unite pluralist and transmigrational teachings led a number of orthodox Christians who favored pluralism to attempt to sever this linkage. To those attempts and to the larger question of the reaction of French Catholics to pluralism, we now turn.

A singularly dramatic moment occurred in 1863 when, just a year after Flammarion published his *La pluralité des mondes habités*, Père Joseph Félix (1810–91), a prominent Parisian preacher, came before thousands of Catholics in Notre Dame Cathedral to proclaim "to all scientists who would make [pluralism] a peremptory reason against Christianity" that

You wish . . . to discover inhabitants of the moon; you wish to find in the stars and suns brothers in intelligence and in liberty; and as certain spirits who pretend to the intuitive vision of all the worlds put it, you wish to greet across space astronomical societies and civilizations. So be it. If you have no other reason to break with us, nothing prevents our extending our hand to you. . . . Put into the sidereal world as many populations as you please, under such degree of material and moral temperature as you wish to imagine; Catholic dogma has here a tolerance that will astonish you and ought to satisfy you. . . .[14]

Flammarion was delighted by this statement, but Abbé Joseph Émile Filachou (1812–90?) must have been deeply disturbed, because two years earlier in his *De la pluralité des mondes* (Paris) he had asserted:

The opinion which assumes men or creatures of the same species as men in the stars is absolutely rejectable, not only because it continues to lack positive proof and rests only on the idea of the possibility of the thing in general, but still more because it entails with it consequences incompatible with all that one admits as most rational in metaphysics, aesthetics and physics. (p. 99)

Pluralism, Filachou asserts, is also contrary to Christianity, in which context he develops three points: "the importance presupposed [in Scripture] of the role of man on earth, the supreme dignity attributed to the Divine founder of the Christian Church, and finally the grandeur attributed to the Church itself." (p. 100) The striking contrast between the positions of Félix and Filachou is characteristic of the pluralist debate in late-nineteenth-century France. Never distant from the center of controversy was Flammarion, whose pluralism some Catholics sought to castigate – and others to baptize.

The latter strategy was adopted by Monseigneur de Montignez when in 1865–6 he published a series of nine essays under the title "Théorie chrétienne sur la pluralité des mondes."[15] This series, which forms a treatise longer and more learned than Filachou's book, begins with the statement that the fourth edition of Flammarion's *La pluralité* "treats with a superior expertise and a rare magnificence of style one of the most interesting questions that has ever piqued human curiosity." (9, p. 385) Quoting Flammarion frequently and praising him as being "endowed with an elevated spirit, profoundly moral, and of a heart almost Christian . . ." (9, p. 385), Montignez nonetheless faults him for in effect trying to found a new religion and for his transmigrational teachings, which "will have no chance either in the eyes of science or at the tribunal of reason." (9, p. 387) The overall position Montignez develops is that pluralism, rather than being in tension with Christianity, is supportive of it. Because our earth is of insignificant size and contains "probably the most disgraced" creatures in the cosmos, it served as the ideal locale for that "annihilation of the divinity" which is the incarnation.[16] As Christ chose "Bethlehem . . . the least among the cities of Judah" for his birthplace, so also he selected the earth as the location for the founding of his Church and his redemptive actions.[17] In summary, he states: "the relative smallness of the earth acts only to strengthen our belief in the mystery of the redemption . . . ; the more you represent the earth as a useless point, the more you make man a stunted, weak, pitiful, disgraced being, the more you justify the preference of which he is the object. . . ." (9, pp. 403–4)

In the third essay in the series, Montignez, after arguing against Flammarion's transmigrational teachings (10, pp. 102–8), comments that man should be seen as a member of a cosmic family united in the divine fatherhood rather than as a "citizen of the sky," as Flammarion had phrased it (10, pp. 108–12). So convinced was Montignez of the existence of extraterrestrials that he asserts: "Without the plurality of worlds, each chapter of the prophets and almost every verse of the psalms is an enigma full of obscurity and of mystery. . . ." (10, p. 142) Drawing mainly on New Testament passages, Montignez in his fourth essay develops the thesis that although Christ came only to the earth, he is nonetheless Lord of the universe, and moreover "the blood which flowed on Calvary has gushed out on the universality of creation . . . ; has bathed not only our world, but all the worlds which roll in space. . . ." (10, p. 272) Montignez maintains that Christ's incarnation and redemptive actions were designed not solely, or even primarily, to free earthlings from original sin, but rather for a cosmic purpose, even though that purpose did not include bringing extraterrestrials to redemption from sin, whether their own or that of Adam in which they had no part. His claim that although beings on other worlds would undergo a test, they would not fail it as man did, is used to overcome the problem of the "small number of the elect," which problem "disappears if the human race is considered as only a minimal and infinitely limited fraction among the universal number of intelligent creatures. . . ." (10, pp. 274–5)

In the final five of his nine essays, Montignez returns to his theme that Scripture contains many passages that reveal their meaning only when seen within the pluralist perspective (10, 297–307). He also articulates and develops four principles to be used in interpreting Scripture. The first is the solidarity or fellowship of earthlings with extraterrestrials (10, pp. 307–8, 374–85). The second is that the elect of the earth will finally attain a true royalty (10, pp. 308–13, 369–70). This he supports by citing, for example, the Eighth Psalm, which announces that "God has elevated man almost to the equal of the angel; that he has set him to the government of his works; that he has subdued everything under his feet." (10, p. 369) The beings to be governed by the elect of the earth are the inhabitants of other planets, this privilege coming to mankind as a result of Christ's extraordinary action of taking on human flesh. His seventh essay is devoted to his third principle, which is that many scriptural passages support pluralism. The fourth principle is: "The Holy Books say clearly and with many repetitions that the elect, in eternity, will command entire nations and numerous peoples." (10, p. 372) In this regard he interprets the Second Psalm as

. . . less a scene of the earth . . . than a complete drama of which the acts are unfolded in the skies. It becomes more and more apparent . . . that there is up there, for the celestial people, a harsh test to undergo. The submission of the heart to the supremacy of Jesus Christ and to that of man deified; the elevation to the first rank of the most small and the last of creation; the descent of the grand and robust under the sceptre of a ruler of small origin. . . . The nations of the sky will tremble, but they will submit. (11, p. 89)

The final essay also centers on his speculation that mankind, the one sheep among the hundred that was lost and required the special care of the Lord (11, pp. 176–7), will eventually provide the princes of the other planets, in light of which claim he reinterprets the Forty-ninth and Sixtieth Psalms.

The series of essays by Montignez is a bold and engaging attempt to develop and recast the increasingly popular doctrines of Flammarion within a Christian context. He was aided in his undertaking by the many ambiguities in Scripture and by the not fewer flexibilities in pluralism. The reaction to his treatise seems to have been disproportionately small compared with the effort and erudition put into it. This neglect may have been due to the fact that Montignez was a little known author who presented his highly controversial position in a specialized journal published in a small town (Besançon) far from the center of French culture.

Among other French clerics who supported pluralism in the 1860s and

1870s were Auguste Gratry, Louis Lescoeur, François Moigno, and Léger-Marie Pioger. Gratry (1805–72), a priest of the Oratory who was elected to the French Academy, stated in his *Les sources* that

. . . if [the idea of a plurality of worlds] does not enter into your astronomy, nor poetry, nor philosophy, nor religion, *nor hopes, nor conjectures concerning eternal life . . .* , and if, in the face of these grandiose characters and these fundamental traits of the visible work of God, you behold without seeing, without comprehending, without suspecting the possibility of intelligence, then, oh! then, I pity you![18]

Lescoeur, also a priest of the Oratory, in his *La vie future* (1872) not only included a chapter on "M. Flammarion et sa théologie astronomique" but also cited a long passage from Maistre on behalf of pluralism. Lescoeur states that "the hypothesis of a plurality of worlds, which one today strives to twist noisily in the guise of serious arguments against Christian theology, has never been condemned by it."[19] Abbé Moigno (1804–84), a prolific scientific author, editor, and translator, expressed his deep concern about the relations of science and religion in his multivolume *Les splendeurs de la foi* (1877–9). In his second volume, Moigno cites both Félix and Gratry in support of pluralism and adds in regard to Flammarion's pluralist objection to Christianity that he (Moigno) received approval from "the Commission of the Roman Index to declare formally to [Flammarion] that the creation and the redemption are by no means an obstacle to the existence of other worlds, of other suns, of other planets, etc., etc."[20] Abbé Pioger also accepted pluralism and advocated its reconcilability with Christianity. Although attacking Figuier's transmigrational theories in his *La vie après le mort; ou, la vie future selon le Christianisme* (1872), he recommends pluralism in his *Le dogme chrétien et la pluralité des mondes habités.*[21] Moreover, in his five-volume *Les splendeurs de l'astronomie ou il y a d'autres mondes que le notre* (1883–4), he repeatedly advocates the pluralist perspective. For example, the first volume, *Le soleil,* begins with a long pluralist introduction in which he quotes statements from Félix, Bishop Denis de Frayssinous, Abbé (later Bishop) Emile Bougaud, and Père Jacques Monsabré in support of the acceptability to Christians of the pluralist position.[22] The intensity of Pioger's attachment to pluralism is indicated by his openness to life on both the sun and the moon[23] and by his statement that the doctrine of a plurality of worlds "transfigures the universe [and] is the most beautiful and grandiose expression of the divine work."[24]

Despite the impressive array of Catholic clerics who had aligned themselves with pluralism, Jules Boiteux, in 1876, published a massive treatise entitled *Lettres à un matérialiste sur la pluralité des mondes habités et les questions qui s'y rattachent,* which was intended as an attack on the radical pluralism of Flammarion. Boiteux, a director of foundries at various locations in France,[25] wrote his book, as he states, "in the spirit of an orthodox Catholic."[26] It consists of sixty-two letters to a materialist named "Camille," who, Boiteux cautions, should not be identified as Camille Flammarion, even though that astronomer's writings serve as the source of most of Camille's views. Boiteux uses a variety of arguments to develop the position that the possibility of intelligent life on other planets has been seriously overestimated. The first of his book's three sections is largely astronomical, treating such topics as the thermal problems caused for life on planets by their differing distances from the sun and the unlikelihood that multiple star systems can support inhabited planets. The solid arguments in this section set the stage for the biological, geological, and meteorological considerations dominant in his book's second section, where he argues for the frailty of life on the earth and for the consequent improbability of intelligent life on even less favorably situated celestial bodies. Materialist biology and to some extent Darwinism are criticized, but Boiteux emerges more as the well-informed critic than as a reactionary. Some of his most powerful antipluralist arguments appear in this section, where his sensitivity to biological considerations is especially impressive.

The third section of his book is devoted to specific cases and to the relationships of the materials already presented to other areas, especially religion. Although much in Boiteux would justify detailed analysis, his fifty-ninth letter, subtitled "Christian Doctrine Is Not Irreconcilable with the Idea of a Plurality of Worlds," seems particularly significant. He begins it by stating that first plants, then animals, and finally the "august presence of man" appeared on our planet (p. 528). Turning to the problems pluralism was said to present for Christianity, especially that of multiple incarnations, Boiteux urges that the first two stages – plants and animals – need not necessarily be followed on other planets by the coming of intelligent beings, his earlier arguments having indicated that "superior or intelligent life will occur only in exceptional cases. . . ." (p. 529) But assume, he suggests, a limited number of "sidereal humanities"; if we do this, "is it probable that they are all equal among themselves in regard to morality?" (p. 530) God would, he remarks, probably make most extraterrestrials better than men, reducing thereby the need for repeated redemptions. Moreover, noting that the coming of Christ was designed both "to teach us [Christ's] perfect law and to purify us by his sacrifice" (p. 531), Boiteux suggests that the former purpose, the only one appropriate for unfallen extraterrestrials, could be accomplished elsewhere by angels. After mentioning as another possibility that Christ's redemptive actions on earth might have redemptive force beyond the earth (p. 532), Boiteux expresses his readiness to accept, if authorities should sanction it, even multiple incarnations. This analysis leads him to the conclusion that "the doctrine, full of unknowns, of the plurality of worlds furnishes no positive argument against Christian dogma." (p. 535)

Boiteux's appeal to Camille at the very end of his book to "return to the Christian faith and embrace it without reserve . . ." (p. 570) certainly went unheeded. However, had Flammarion read this book, he should have been impressed by the powerful and varied scientific arguments marshaled in it against his radically pluralist stance. Moderation of tone and breadth of coverage are among its other strengths. Yet the fate of most antipluralist writings, even those containing excellent arguments, awaited Boiteux's book: While Flammarion saw his books roll through dozens of editions, Boiteux followed Plisson, Whewell, and Filachou into obscurity. A second edition did appear after fifteen years, and a third in 1898, but then no more was heard in the pluralist debate from this erudite engineer.

Like Boiteux, Abbé Jean Boudon was anxious to combat the antireligious writings of Flammarion and other authors. In 1875, he published *Adam à son origine, roi et unique médiateur de tout l'univers planétaire. Question délicate touchant à la pluralité des mondes habités* (Bar-le-duc). Boudon's book, written primarily in a religious vein, advances the theses that "The fault of the first man [Adam] decided the fate of all the universe" (p. 209) and that after Christ's second coming, "man will be recognized as having been created to be the sole King and Mediator of all the visible universe." (p. 212) In short, Boudon was intent on arguing for the primacy of humanity in the universe. Although his book went through three editions, the latter two being more than double the size of the first edition, it drew little if any attention from later French contributors to the pluralist debate.

In 1894, Pierre Courbet published an essay in *Cosmos,* comparable in intent to Boiteux's fifty-ninth letter.[27] After admitting that neither science nor Scripture provides definite evidence of extraterrestrials, he attempts to show that their existence would at least not be "in contradiction with the formal teachings of Christianity. . . ." (p. 210) Sinless extraterrestrials, he suggests, would need no redeemer, nor would those so sinful as to be beyond redemption. Moreover, the test for some extraterrestrials may have been individually set, as it was for angels, rather than being communal, as it was for earthly inheritors of Adam's guilt (p. 273). Proposing that the problem of why Christ came to earth is analogous to the traditional dilemma of his choice of Bethlehem for his birth, Courbet suggests that Christ may have come to us because "the human race is perhaps . . . the most guilty of all [and had] the greatest need to profit directly from the redemption." (p. 273) Convinced of the unacceptability of the idea of multiple incarnations, Courbet proposes ways in which

Christ's terrestrial atonement may act elsewhere. He suggests that extra-terrestrials would not need to know of these actions to derive their benefits, any more than the infant must understand baptism (p. 274). If a physical connection is needed, this could take the form of a communication transmitted through the ether (p. 275). Courbet's paper, which was probably written for apologetic purposes, shows the continuing concern among Catholics to combat the pluralist objection to Christianity.

The same apologetic intent is evident in the large volume published in the same year by the priest and theologian Théophile Ortolan (b. 1861), whose *Astronomie et théologie* won the Prix Hugues awarded for apologetic writings by the Institut Catholique in Paris.[28] Early in his book, Ortolan urges that Christian theology was formulated by the fathers of the Church at a time when astronomy was little studied. Hence, its fundamental tenets were not significantly tied to astronomy, nor did they come to be in the writings of the foremost medieval theologians, whose restraint in this regard was not observed by lesser theologians, who caused such difficulties as the Galileo affair (pp. 9–150). The involvement of astronomy with religion is also illustrated by means of his discussion of the astronomical themes used by such literary figures as Dante, Milton, Klopstock, and Chateaubriand (pp. 152–204). Ortolan then turns to the pluralist question, claiming that the "romancers of astronomy" have seriously overstated the possibility of extensive extraterrestrial life. Double stars and variable stars, for example, would scarcely support habitable planets, and in our solar system it is only for Mars that Ortolan holds out some hope of habitability (pp. 206–82). That scientific rather than religious reasons were fundamental to his rejection of the extreme pluralist position is suggested by Ortolan's urgings that if the pluralist position were proved, it could easily be reconciled with Christianity. As he states: "Our venerable dogmas accommodate very well the teachings of modern science on the habitation of stars, for it teaches nothing on the subject. . . ." (p. 286) In support of his position, Ortolan analyzes such scriptural passages as those on the one lost sheep and on the many mansions to show that a pluralist interpretation would be possible (pp. 294–7). He also claims that various passages in the writings of such Church fathers as Origen and Saint Basil may contain a "pressentiment" of the pluralist doctrine. He immediately adds: "Like contemporary theologians, [the Church fathers] would find no opposition between it and our faith. The Congregation of the Index, consulted on this point, has officially responded that there is not." (p. 311)

Concerning the question of why Christ came only to our planet, Ortolan offers a number of responses, including that the inhabitants of other planets may benefit from this action or may not need redemption (pp. 320–1). He provides a lengthy critique of the eschatological and pluralist ideas of Reynaud, Figuier, and Flammarion, who created "poetic utopias of the future life." (p. 327) Noting that Reynaud's *Terre et ciel* had been placed on the Index on December 19, 1865 (p. 329) and that Figuier's *Le lendemain de la mort* had joined it on March 1, 1873 (p. 333), he describes them as only reasserting centuries later "the false mysteries of the Orient." (p. 330) Flammarion is presented as "scarcely having written a chapter in his numerous works without mixing in it the most violent and most unjust accusations against the Church . . ." (p. 341) and as "relying solely on the facts of astronomy to found a new religion and to establish a cult of which the object will be the divine Nature, for [to him] Nature is God. . . ." (p. 336) In his final section, Ortolan reveals his own opinions and conclusions. He argues that the end of the earth will probably not entail the destruction of the entire universe and that although most planets are probably bereft of intelligent life, they nonetheless may manifest God's greatness by having their own specific perfections. In this regard, he asks: "Would it be illogical to suppose . . . that certain stars have been created to represent in space, in some cases, the mineral world in a more or less grand perfection, in others, the plant world, still others will have animal life, and others finally intellectual life?" (pp. 386–7) The "globes of gold or diamond" (p. 388) that Ortolan proposes in this context show that the seductive charms of astronomical speculation also acted on this opponent of the "romancers of astronomy." With a final note that although he has attacked the adversaries of his faith, he has respected them as persons, Ortolan concludes his not unimpressive volume, which soon appeared in a German translation. Ortolan returned to the pluralist debate in 1897 with his *Études sur la pluralité des mondes habités et dogme de l'incarnation* (Paris), which appeared in three small volumes and went through nine editions. In 1898, motivated probably by the increasing interest in spiritualism at the end of the nineteenth century, he published *La fausse science contemporaine et les mystères d'outre-tombe* (Paris), which was aimed at Figuier and to a lesser extent at Reynaud and Flammarion.

A review of Ortolan's 1897 book begins with this statement:

For twenty or thirty years, certain malicious minds have joined forces, who have found a method of transforming the gracious, poetic, and perfectly inoffensive hypothesis of the habitation of the celestial bodies into an engine of war against spiritual and Christian doctrines, into a sort of pantheistic and materialistic system based on unlimited evolutionism.[29]

The author of this review was Charles de Kirwan (1829–1917), a French Catholic scientist who, sometimes using the pen name Jean d'Estienne, made a number of forays against the "machine de guerre" of the pluralists. For example, in 1891 he published a response to a pluralist paper of the Potsdam astronomer J. Scheiner, urging among other points that

Scheiner's pluralist conclusions were excessive.[30] During the same decade he published favorable reviews of both Ortolan's *Astronomie et théologie* and his *Études,* as well as of Boiteux's third edition.[31] Although usually writing in a theological context, de Kirwan formulated most of his arguments in scientific terms and generally adopted a more antipluralist position than that of Ortolan. He continued to contribute to the debate during the early years of the twentieth century, publishing, for example, a short book in 1902 in which he no doubt developed the antipluralist arguments set out in his earlier periodical publications.[32]

Three books published in France in the 1892–1900 period illustrate a remarkable feature of the pluralist debate: the frequency with which otherwise almost unknown individuals, usually situated far from the centers of culture, became consumed with a compulsion to contribute not an essay, not a modest memoir, but an immense tome treating, and often, in their eyes, at last resolving, the question of extraterrestrial life. Invariably with boundless energy and enthusiasm and usually with a deficiency of sense and sophistication, they offered their vast volumes to the public, which – almost invariably – paid scant attention. Appearing in both pluralist and antipluralist strains, this mania led to books sometimes strange in their conclusions and always bizarre in some of their arguments, and frequently of little value to their times but interesting to the historian concerned to observe the pluralist debate at its fringes. These symptoms were seen before in such authors as Nares, Maxwell, Knight, Simon, Hequembourg, Williams, Girard, and perhaps Boiteux, but in few cases were the symptoms more striking than in the writings of G. Prigent, R. M. Jouan, and F. X. Burque.

From the far northwestern tip of France (Kerlouan in Finistère), Gabriel Prigent surveyed the universe and in the 456 pages of his *De l'habitabilité des astres* (1892) pronounced it pluralist. Accurately admitting to being a "writer unknown and without scientific or literary credit"[33] and profusely apologizing for his limitations, Prigent, amidst dozens of digressions and oversized quotations frequently decades out of date, urges that the considerations brought forth in his book concerning the planets make it "almost impossible for the reason of man to refuse to admit the existence of human creatures on the surface of these distant bodies. . . ." (p. 389) Not a Martian canal convert to the pluralist cause, and apparently no more aware of Schiaparelli's 1877 "observation" than of Hall's simultaneous discovery of the moons of Mars, Prigent cites as his ultimate authority, as his "last and supreme witness," the "eminent astronomer" Camille Flammarion. It is characteristic of Prigent, who laments that few thinkers have studied this subject (p. 380), that the only volume by Flammarion he cites is his *Merveilles célestes,* also decades out of date. Although most of Prigent's arguments are garbed in the language

of science, one suspects that his confidence in and enthusiasm for pluralism had religious roots. His book certainly contained little of value for
the scientifically sophisticated reader.

Eight years after Prigent's ponderous publication, an even larger book
(478 pages) appeared from the pen of R. M. Jouan. Entitled *La question
de l'habitabilité des mondes étudiée au point de vue de l'histoire, de la
science, de la raison et de la foi*, it was published "chez l'auteur" from
Saint-Ilan (Côtes du Nord), a village perhaps fifty miles less remote from
Paris than Prigent's Kerlouan. Yet the efforts expended by the later
Breton were far greater, and his sources more up to date, even if a
number of his conclusions were no less curious. Jouan, who according to
his title page was a former professor, presents his arguments in the four
sections suggested in his title. The aim of the historical sections is to show
that "The hypothesis of a plurality of worlds . . . has been universally
accepted by the elite of humanity of all times and all places [and] consequently ought to enjoy all the privileges attached to the consent of all
peoples."[34] Although his research for this section took him "to all the
libraries of Paris" (p. 56) and involved scrutiny of numerous obscure
sources, Jouan's judgments are not infrequently extravagant, as when he
tries to claim Adam and various other antediluvians (pp. 8–22), as well
as Saint Augustine (pp. 71–2), for the pluralist camp. His scientific section suffers from the same tendency: Although one cannot but admire his
dedication to delving into the then current astronomical literature, one is
disturbed by his failure to see that an extended discussion (pp. 215–42)
of the possibility of life on the sun could not in 1900 be considered sober
science. His third section, that written "from the point of view of reason," consists largely of metaphysical and theological arguments for his
extreme pluralist position. In his fourth section, Jouan presses two main
points. First, he attempts to show that pluralism can be reconciled with
Catholic doctrine (pp. 305–92), citing many of the French clerics already
considered. Second, he develops the controversial hypothesis that Christ
became incarnate on numerous planets (pp. 393–429). No doubt many
readers concurred with the judgment of de Kirwan, who in reviewing
Jouan's book found his first thesis commendable and his second "im-
proper," "dangerous," and "inadmissable."[35]

The third French book, *Pluralité des mondes habités considérée au
point de vue négatif* (Montreal), came in 1898 from the Canadian curé of
Fort Kent, Maine, Abbé François Xavier Burque (1851–1923), whose
commitment to his cause kept him at the composition of his 407-page
treatise despite incessant interruptions of his parishioners (p. vi). Although Burque had been bothered as early as the 1870s by the acceptance
of pluralism among Canadian seminary teachers, it was his disappointment in Ortolan's *Astronomie et théologie* that finally set his pen in

motion (pp. v–vii). The first half of his book, which treats the astronomical aspects of the question, reveals an author of wide if unsystematic reading, deeply committed to overcoming what he labels the "materialist" position in the pluralist debate, as well as "the grand leader at the present time of the materialists, Mr. Camille Flammarion." (p. 9) Treating first the objects in our solar system, he argues that in no case are the seven conditions (pp. 23–9) requisite for intelligent life simultaneously present. Fullest by far is his discussion of Mars, where he first attacks arguments that it is inhabited and then that it is habitable (pp. 94–119). Although frequently exposing the excessive claims of Flammarion, Burque himself at times falls into such extreme assertions as that the "human form . . . is the only one that can adequately serve an intelligence." (p. 111) Claims for the habitability of planets orbiting other suns he combats by arguments against spontaneous generation, especially against the materialist view that where matter exists in appropriate forms, living beings must arise (pp. 142–58). On the level of language, he faults Schiaparelli not only for describing the lines he saw as "water courses, veritable channels" but also for poetically naming a certain Martian elevation *"Fons Juventutis."* (pp. 70–1)

In the theological half of his book, Burque admits that God could have placed intelligent beings elsewhere in the universe, but this should be seen as unlikely because of the scarcity of suitable habitats. He argues that neither scriptural authors nor the Church fathers discussed, let alone endorsed, the pluralist position (pp. 186–216), Origen, for example, having treated only the question of successive "worlds" on the earth (pp. 208–10). To the pluralists' teleological argument, Burque responds that the celestial bodies provide us with heat, light, and methods of distance and time reckoning as well as delighting and instructing us, especially in regard to God's glory (pp. 227–45). For Burque, the fundamental problem with pluralism for the Christian is that "only with extreme difficulty" (p. 246) can it be reconciled with belief in Christ's incarnation and redemption. If extraterrestrials exist, they must almost certainly have sinned, which entails postulating a multicrucified Christ, in contradiction to Hebrews 9:16, or otherwise explaining how Christ's terrestrial atonement could bring salvation to other worlds (pp. 246–61). Burque concludes: "it appears absolutely impossible to extend to sidereal persons the benefits of this redemption, impossible to imagine how the Divine Blood which has flowed on Calvary could be . . . of some utility and efficacy for their justification." (p. 261) In this context, he speculates that the primary aim of the redemption was to expiate the sin of the fallen angels, which would seem to require that God take on a material-spiritual nature such as that of man rather than the unitary nature of angelic existence, because only in this way could Christ experience death without thereby being annihilated (pp. 262–302). Burque states that even were proofs of pluralism provided, Catholic doctrine would not be affected, because pluralism is not central to it, nor has that church ever spoken authoritatively on the topic. Such a discovery would present new mysteries for Christians, but they already know that God's ways are not their ways (pp. 308–14). Late in this book, Burque refers to the "fact" that supporters of evolutionary theory are "precisely the same men who with the greatest ardor preach the suspect doctrine of the plurality of worlds." (pp. 341–2) This was not a fact; for example, in the 1890s, Boiteux and de Kirwan supported Darwinian doctrines but opposed pluralism.[36]

Prigent, Jouan, and Burque must have been very disappointed at the public's reaction to their vast volumes, which disappeared almost as suddenly as they had appeared, none attaining a second edition. Jouan and Burque deserved a better fate, marred though their books were by excessive claims. Among the few later references to Burque's book is one in a 1932 survey of the pluralist debate in which it is stated that "most writers of [Burque's] creed have to the present day firmly opposed the doctrine of other inhabited worlds."[37] That this claim is probably wrong is indicated by the fact that of the fifteen Catholic authors discussed in this section, only Filachou, Boiteux, Ortolan, de Kirwan, and Burque opposed the pluralist position.

That the situation in Spain may not have been greatly different from that prevailing in France is suggested by a perusal of the long book that Niceto Alonso Perujo (1841–90), a prominent Spanish priest, contributed to the pluralist debate in 1877.[38] Like many French Catholic authors, his chief concerns were to refute a number of the irreligious claims of Flammarion and also to assure Catholics that pluralism was not inimical to religion.

The openness of Catholics in France and elsewhere during the 1860-to-1900 period to pluralism is in at least one sense surprising. Certainly in the judgment of most, the Roman church turned at that time in an increasingly conservative direction in intellectual matters. This trend can be seen in the "Syllabus of Errors" of 1864, the decree on papal infallibility of 1870, the encyclical on Thomistic philosophy of 1879, and that on biblical inerrancy of 1893. A partial explanation of Rome's restraint regarding pluralism may be that it was championed for over two decades by the most prolific pre-1881 astronomer, Rev. Angelo Secchi (1818–78),[39] the director of the Roman College Observatory. As early as 1856, Secchi made known his belief in pluralism: "it is with a sweet sentiment that man thinks of these worlds without number, where each star is a sun which, as minister of the divine bounty, distributes life and goodness to the other innumerable beings, blessed by the hand of the Omnipotent."[40]

In the 1870s, Secchi drew together his important spectroscopic researches in two books, *Les étoiles* and *Le soleil*, both of which contain pluralist passages, these being repeatedly quoted by Catholic pluralists in a number of nations. For example, in *Les étoiles* Secchi states that

. . . the creation, contemplated by the astronomer, is not a simple mass of incandescent matter; it is a wonderful organism where when the incandescence ceases, life commences. Even though it may not be accessible to our telescopes, nonetheless, by analogy with our globe, we are able to conclude that it exists in the others. The atmospheric constitution of the other planets which, in certain points, is so similar to ours as that of the stars is similar to that of the sun, persuades us that these bodies are in a state similar to that of our system, or are traversing one of the periods that we have already traversed or that we will traverse some day.[41]

These and other passages show Secchi to have been a convinced pluralist, ironically ready to overlook the vast differences in spectral types among the stars, which differences he probably knew better than any of his contemporaries. Moreover, it is a little known fact that in 1859 Secchi introduced a crucial term into the Martian literature when he described two fine lines he had seen on Mars as "canali."[42] One may conjecture that Secchi's eminence, eloquence, and closeness to Pope Pius IX, who was his patron and friend, acted to cut off any attempts to condemn pluralism that may have arisen in Rome.

2. German religious writings: "Heathens, Christians, atheists . . . hand in hand" in the pluralist cause?

German publications on religious aspects of the plurality of worlds debate in the latter half of the nineteenth century came largely from materialists or from Christians. Late in the century, one participant in the debate claimed that a consensus for pluralism had been attained: "A Proctor and a Secchi judge not otherwise than a Pythagoras and Epicurus. . . . Heathens, Christians, atheists of all varieties go dispassionately hand in hand, although each strives to utilize [pluralism] in his own sense and to set it in harmony with his own worldview."[43] Is this characterization correct? The thesis that is suggested in the following discussion, first of materialist authors and then of Christian writers, is that it is not; in fact, the publications discussed point to the conclusion that even within each camp (materialist and Christian), major disagreements were present.

In his *History of Materialism*, F. A. Lange stated concerning *Kraft und Stoff*, the famous presentation of materialism published in 1855 by Ludwig Büchner (1824–99), that it "perhaps created a greater sensation, and was at all events more bitterly condemned, than any other book of the kind."[44] Büchner's book was both controversial and popular: It cost

him his position on the medical faculty at Tübingen, but established his
literary reputation by going through numerous editions, as well as translations into seventeen languages. Influenced by earlier materialists such
as Feuerbach, Moleschott, and Vogt, Büchner sets forth the materialist
program in these words:

Starting from the recognition of the indissoluble relation that exists between
force and matter as an indestructible basis, the view of nature resting upon
empirical philosophy must result in definitely relegating every form of supernatu-
ralism or idealism from what may be called the hermeneutics of natural facts, and
in looking upon these facts as wholly independent of the influence of any external
power dissociated from matter.[45]

Despite repeatedly advocating an empiricist methodology, Büchner does
not hesitate to discuss the question of extraterrestrial life or to draw far-
reaching conclusions from his position on that question. His conclusions
were, of course, tied to his materialist (or monist, as he preferred to call
it) system, but in a very different way from that adopted by most materialists.

    In his chapter on "The Heavens," Büchner asserts that in the cosmos of
modern astronomy the old notion that we are destined to "go to heaven"
is replaced by the recognition that "we are already in this dreamed of
heaven, surrounded by countless worlds. . . ." (p. 132) In this context,
he attacks traditional teleology on the basis of his belief not in the abundance of inhabited planets but in their paucity. He states:

    If, as must be assumed according to the teleological idea of the world, a
personal creative power, guided by definite aims, meant to create worlds as
dwelling-places for intelligent thinking beings worshipping his omnipotence, why
should there be these huge, vacant and useless tracts of space, in which but here
and there isolated suns and earths swim as almost imperceptible dots . . . ?
Why then are not the other planets of our solar system (with perhaps the solitary
exception of the planet Mars) adapted to be likewise inhabited by men or by
man-like beings? (pp. 138–9)

His critique of teleology continues as he claims that the sun "is constantly
squandering uselessly huge quantities of light and heat in the cold realms
of space [because] the whole of the planets . . . are benefited to only the
230-millionth part of this enormous waste of force." (p. 14) Those who
argue for a divine designer of the universe are also challenged in this
book, written two years after Whewell published his geological argument
against pluralism, to explain why the sun and stars shone on the earth
"during those untold ages of the past in which no creature existed on its
surface to turn these glorious arrangements to account. . . . (p. 141)
Finally, this philosopher, prone to pessimism, laments that the solar
system, "must perish in time, and with it all that is great, all that man has
ever accomplished or ever done on earth, must subside again into the
chaos of eternal oblivion." (p. 142)

David Friedrich Strauss (1808–74), notorious since 1835 for his *Das Leben Jesu,* created another sensation in 1872 by the extreme materialism advocated in his *Der alte und der neue Glaube,* which was denounced by figures as different as Gladstone and Nietzsche, but went through six editions in its first six months. The book is structured around four questions:

   I.   Are We Still Christians?
  II.   Have We Still a Religion?
 III.   What Is Our Conception of the Universe?
 IV.   What Is Our Rule of Life?[46]

In his first section, he asserts that "if we would speak as honest, upright men, we must acknowledge we are no longer Christians." (p. 107) Scientific thought, Strauss urges, is a major reason for this, pluralism being cited as opposing the idea of a personal God. In this regard, he claims that when it came to be believed that the stars are surrounded by inhabited planets, "the ancient personal God was, as it were, dispossessed of his habitation." (p. 124) Pluralism also made problems for belief in the general resurrection by raising the difficulty that the planets, being already inhabited, could not become residences of the redeemed (p. 152). In response to the question "What Is Our Conception of the Universe?" Strauss champions a pluralist, evolving universe, citing Darwin and Kant as authorities (p. 174). He also praises Kant for his identification of nebulae with island universes, finding no difficulty for this in the spectroscopic determination that some nebulae are gaseous. In fact, he states that this result proves that "space contains not only completed worlds, but also such as are only in process of formation. . . ." (p. 188) After recounting Kant's speculation that the inhabitants of the more distant planets of our solar system are the most developed, he comments: "is it not amusing that we must be on our guard lest we be led into extravagant fancies by him who was destined to write the *Critique of Pure Reason?"* (p. 192) Championing Darwinian evolution later in this section, Strauss seeks to destroy the old faith so as to open the way for his version of materialism. Judged a poorly worked out book even by friends of Strauss, his *Der alte und der neue Glaube* nonetheless won a wide audience and was more typical of the materialists' treatment of pluralism than was Büchner's book.

Difficulties stand in the way of knowing how to classify by period and language a manuscript written between 1872 and 1882 but first published in 1927 and containing such sentences as this: "Wenn Coulomb von particles of electricity spricht, which repel each other inversely as the square of the distance, so nimmt Thomson das ruhig hin als bewiesen," but there can be no question that Frederick Engels's *Dialecticts of Nature* is a work in the materialist tradition and its author one of the most prominent materialists of his day.[47] In that book, Engels (1820–95), the Barmen-born, Berlin-educated, and Manchester-based collaborator with Marx, sets out the scientific aspects of their program. His belief in extra-terrestrial life is conveyed first in a discussion (pp. 14–18) of the development through purely natural processes of organic life, then in such statements as "the eternally repeated succession of worlds in infinite time is only the logical complement of the co-existence of innumerable worlds in infinite space. . . ."[48] This is part of the "same iron necessity," according to which, just as nature "will exterminate on the earth its highest creation, the thinking mind, it must somewhere else and at another time again produce it." (p. 25) Engels's pluralism also appears in his *Socialism: Utopian and Scientific:* "Modern materialism embraces the more recent discoveries of natural science, according to which Nature also has its history in time, the celestial bodies, like the organic species that, under favourable conditions, people them, being born and perishing."[49]

Because Baron Carl Du Prel (1839–99) employed a number of approaches favored by materialists, he has sometimes been classified among them, although at least in his later years he would have designated himself a spiritualist.[50] Among Du Prel's earliest books, that most relevant to astronomy was *Der Kampf ums Dasein am Himmel: Versuch einer Philosophie der Astronomie* (1873), in which he sought to apply the Darwinian survival of the fittest doctrine to the material universe, proposing, for example, that planets that moved in very elliptical orbits were destroyed by falling into the sun, whereas planets with stable, nearly circular orbits survived. His support for pluralism, already evident in his 1873 book, was especially apparent when in 1880 he published *Der Planetenbewohner und die Nebularhypothese: Neue Studien zur Entwicklungsgeschichte des Weltalls* (Leipzig). Pluralist themes also occur in his *Die Philosophie der Mystik* (1885).

The first three chapters of Du Prel's *Planetenbewohner* are a presentation of his form of the nebular hypothesis, the sources of which are evident in his statement "We must thus complete Kant and Laplace by means of Darwin." (p. 29) His fourth chapter contains his most focused discussion of the possibility of life on the planets of our solar system. Urging the Proctorian point that Jupiter generates much internal heat and lacks a solid crust, he denies life to it, and for similar reasons to the three planets beyond it. They will, he states, eventually attain life, although it will last less long than on the smaller planets nearer the sun (pp. 62–5). Life may exist on the moon, but not on comets (pp. 66–7). He supplies no arguments for life on the four inner planets, nor does he describe the conditions of life on them. However, in his *Philosophy of Mysticism*, he claims that because Mars cooled earlier and has a more favorable distribution of land and water than on earth, it may be inhabited by beings more advanced than we are.[51]

The last two chapters of Du Prel's *Planetenbewohner* contain an interesting discussion, based on epistemology and psychology, of the nature of extraterrestrials. Noting that human organs of sight and hearing respond only to certain wavelengths, he urges that other senses are possible. Speculations about magnetic, electric, and chemical senses occur as well, leading him to conclude that man's "sense organs are equipped only for the perception of a fraction of reality" and that "Not all that our senses perceive is present in reality." (p. 139) These two "laws" are cited to support his claim that beings with different senses than our own will perceive reality very differently from us; as he succinctly and dramatically states:

> If we consider the immeasurable differences of the celestial bodies in astronomical and physical respects, and that the life forms carried by them must generally be adapted to their conditions, we are led to the recognition: Other worlds, other beings. But the complementary epistemological investigation, which has been carried through in the preceding, allows us to recognize the not less doubtless truth: Other beings, other worlds! (p. 171)

Du Prel's book concludes with comments on the possibility of communication with extraterrestrials (pp. 174–5), an idea that is developed more fully in his *Philosophy of Mysticism* and that forms one of the links between his spiritualism and pluralism. In his 1885 book, he suggests that the invention of an apparatus comparable to but more advanced than the spectroscope, or possibly some spiritualist mode of perception, may make such communication attainable and thereby allow "the history of man . . . to debouch into the general stream of cosmic history." (p. 269) On the other hand, Du Prel's analysis as summarized in his statement "Other beings, other worlds!" points to the problem that Martian messages might elude our sensory equipment or possess an intelligibility foreign to our modes of thought. In the same book, Du Prel deploys his idea of other or more developed senses to attack the materialism of such authors as Büchner, who, he claims, fail to see that evolution may lead once again to new modes of access to the world (pp. 266–8).

Another materialist who supported pluralism was Ernst Haeckel (1834–1919), whose doctrines were drawn together in his *Die Welträthsel* (1899), in which he advocates a pantheistic, Darwinian monism that most labeled materialism. Much of one chapter of that book is devoted to discussing pluralism, which he notes is a doctrine of Flammarion, whose writings are "equally distinguished by exuberant imagination and brilliant style, and by a deplorable lack of critical judgment and biological knowledge."[52] Drawing on his extensive knowledge of biology and on Darwinian doctrines, of which he was a leading champion in

Germany, Haeckel speculates on the bodily forms of extraterrestrials, suggesting that lower forms of life elsewhere are probably similar to terrestrial forms, but that higher forms probably differ; for example, they may not be vertebrates. He adds: "perhaps from some higher animal stem, which is superior to the vertebrate in formation, higher beings have arisen [on other planets] who far transcend us earthly men in intelligence." (p. 371)

Looked at broadly, these five authors can be seen to have shared an enthusiasm for an evolutionary, mechanistic view of the universe and a hostility to traditional Christian concepts of the cosmos. These similarities should not obscure the fact that the forms of pluralism they adopted differed significantly, ranging from the extreme pluralism of Strauss through the Proctorian position of Du Prel to the antipluralism of Büchner. The Christian authors to whom we now turn were no less divided among themselves.

In the 1860s, two German theologians spoke out against pluralism. Johann Ebrard (1818–88), a theologian of the Reformed church teaching at Erlangen University, did so in his *Der Glaube an die heilige Schrift und die Ergebnisse der Naturforschung* (1861) and in his *Apologetik* (1874–5).[53] In responding to the pluralist objection to the atonement, he first cites Scripture against the idea of multiple incarnations, adding that a pluralist system not entailing repeated redemptions would not conflict with Christianity (pp. 354–6). What does make extraterrestrials improbable for him is astronomy, which indicates that all planets beyond the earth, except possibly Mars, receive too little heat from the sun, whereas the interior planets receive an excess. Moreover, in some cases the length of the orbital period, the inclination of the planet's axis of rotation (Uranus), or the extreme height of its mountains (Venus!) present further obstacles (pp. 357–63). That Ebrard's antipluralism may have been influenced by Whewell is suggested by the Erlangen theologian's reliance on the double star argument against stable planetary systems (pp. 363–4). However, this possibility is diminished by Ebrard's statement that the stars were designed for "personal creatures, who are in no danger of coming to a resolve to rebel against God, or who no longer are in such danger, or who have already overcome it, who therefore are not in need of a redemption, nor of any alternation of seasons or conditions." (p. 364) Whether Ebrard's inhabitants of the stars are angels, the souls of earthlings, or some other, possibly material species he does not make explicit. Another source that might be proposed for Ebrard's antipluralism is Hegel, but this is decisively excluded by his turning the "astronomical argument" against Hegel and his Absolute: "It does indeed seem strange that the Absolute . . . 'should come itself' only upon our small planet . . . and have been brought to a conclusion in the beginning of the

nineteenth century by Professor Hegel in Berlin. . . ." (p. 355) A number of Ebrard's arguments were employed in 1864 against pluralism by
Christoph Ernst Luthardt (1823–1902), a Lutheran theologian at Leipzig University, in his *Apologetische Vorträge*.[54] Luthardt also drew arguments from J. H. Kurtz (discussed previously), while avoiding the speculations into which both Ebrard and Kurtz had fallen.

The German Protestant theologian of this period most active in the
pluralist debate was Otto Zöckler (1833–1906), a Lutheran professor at
Greifswald, who throughout his life wrote widely on the interactions of
science and religion. Zöckler's first relevant publication was an 1866
paper on the history of the pluralist debate, which paper, although based
to some extent on Flammarion's writings, reveals an author possessing a
more critical spirit and a deeper knowledge of German sources than the
French astronomer. Concentrating on the nineteenth century, Zöckler
discusses the Whewell debate, comparing Whewell's antipluralism to
that of such German philosophers as Hegel.[55] German readers learning
from Zöckler of the Whewell debate found Powell's book praised as
"spirited, learned, and adroit." (p. 365) Flammarion's writings are also
discussed, as well as the transmigrational teachings of Reynaud, Pezzani,
and others, Zöckler commenting that German thinkers recently engaged
in the pluralist debate "have abstained entirely from these and similar
extravagances and caprices and their suppositions [concerning extrater-
restrials] manifest no forms transgressing the biblical norm . . . or the
stand of empirical natural science." (p. 371) A pluralist himself, although
open to a solar system with life confined to the earth, Zöckler shows
special interest in Chalmers's ideas. His concluding paragraph stresses
that Christianity can accommodate either a pluralist or antipluralist system, the decision between them resting with science.

In an 1877 paper focused on comparing the ideas of Flammarion and
Proctor, Zöckler states that the pluralist position of the "Leibniz, Kant,
Herder, and Herschel" tradition has won out over the antipluralism of
the "pantheist natural philosophy" of Hegel, Schelling, and their disciples. Even Whewell's protest "died away rather soon among the numerous vigorous replies of distinguished colleagues. . . ."[56] Zöckler's analysis divides pluralists into two groups, the first holding that although in
our solar system life is probably confined to the earth, it exists on planets
of other systems, the second group maintaining that nearly all planets
everywhere possess life. Friedrich Pfaff and Johannes Huber are cited as
proponents of the former position,[57] Flammarion of the latter. That astronomer's *Les terres du ciel* is analyzed, Zöckler repeatedly quoting its
excessive claims as justification for his characterization of it as filled with
"extravagantly daring and in part Baroque expositions." (p. 645) Flammarion's rhapsodic description of the sun leads Zöckler to accuse him of

fostering a new "Cult of the Sun," connected with "Bruno's infinite reproduction of suns and worlds and likewise [involving] countless genies, demons, and spirits of all kinds and orders, to animate and populate the immeasurable plurality of worlds." (pp. 645–6) Turning to Proctor, Zöckler notes his shift toward a Whewellite position and detects in Proctor the presence of "Spencerian-Darwinian evolutionism." (p. 649) Overall, Proctor is portrayed as a pluralist, albeit in a qualified sense, and as a more scientific author than Flammarion. In concluding his paper, Zöckler again urges the reconcilability of Christianity with nearly any form of pluralism that astronomers may ultimately establish as correct.

Among historical studies of the relations between science and theology, none surpasses in size, and few in quality of scholarship, the two-volume treatise Zöckler published in 1877–8, two sections of it being devoted to the pluralist debate.[58] The first section, which treats the pre-1800 period, is especially rich in references to eighteenth-century German authors. In the second section, Zöckler, whose knowledge of the Whewell debate had increased since his 1866 paper, describes that debate as having led to a more rigorous analysis of pluralism, an effect intensified by the development shortly thereafter of evolutionary theory and of astrophysical modes of investigation (p. 435). Zöckler's treatment of transmigrational teachings as related to pluralism includes discussion of the ideas of H. Baumgartner, J. P. Lange, and C. H. Weisse and shows that by the 1870s such doctrines were attracting attention in Germany. However, the disdain that most people felt for such theories, Zöckler suggests, produced an aversion to pluralist speculations and also gave support to relatively restricted forms of the pluralist hypothesis (pp. 437–8).

Zöckler's contributions to the extraterrestrial life debate include informing German readers of the ideas of a nearly worldwide array of authors, setting their studies within an illuminating historical context, and critically evaluating some of their systems. It is ironic that in his study of the historical relations of theology to pluralism Zöckler was above all concerned to claim that Christian theology need have no relation to pluralism, being able to accommodate almost any pluralist or nonpluralist system.

Germany, which produced neither a Proctor nor a Flammarion, possessed a prolific Catholic priest who during the 1880s, while teaching in four countries, published a two-volume book and an assortment of essays, two nearly of book length, on the question of other worlds.[59] Born in Germany, educated in Italy, Rev. Joseph Pohle (1852–1922) had taught in Switzerland, England, and Prussia by the time of his 1889 call to the founding faculty of the Catholic University of America. From there, in 1894, he proceeded first to Münster and finally to Breslau

University, where he wrote a twelve-volume *Lehrbuch der Dogmatik* (1902–5). A generation or more of German and American seminary students knew him as the author of those volumes, but he was earlier known in Germany primarily as a proponent of the idea of a plurality of worlds.

Pohle's first book was a biography of Angelo Secchi, who had instructed him in astronomy and who no doubt instilled in him that enthusiasm for pluralism present in Pohle's *Die Sternenwelten und ihre Bewohner* (1884–5).[60] Stating early in that book that "a broad, unleapable chasm" separates us "from our brothers who perhaps live on Mars, Venus, the satellites of Sirius . . ." (p. 18), Pohle advocates an approach combining science with "teleological consideration based on metaphysics and theodicy." Although differing from Proctor in methodology, Pohle adopts a similar universe:

. . . on general teleological and cosmological grounds we must accept that in consideration of their different masses and temperatures, not all worlds succeed simultaneously to readiness for life. Each celestial body has its definite period of life which is enclosed . . . between unequally long epochs of death. (p. 29)

After a capable historical survey of the extraterrestrial life debate, in which he urges that "Heathens, Christians, atheists, of all varieties" have joined together in accepting pluralism, he turns to specific topics, the first being meteorites. These he treats at length, rejecting Hahn's fantastic claims. His discussion of astrophotography and spectrum analysis reveals an author with adequate expertise to write effectively on these advances and with sufficient candor to admit that neither proves the pluralist position (p. 128).

In discussing the sun, Pohle considers the views of Carl Goetze and William Preyer (1841–97), both of whom had recently advocated its habitability.[61] He rejects their conclusion, although accepting as a possibility the proposal of the Jesuit astronomer Carl Braun (1831–1907) that the sun may eventually cool sufficiently to become habitable.[62] In his survey of the sidereal realm, Pohle marshals an array of arguments, observational and theoretical, on behalf of the stars being surrounded by planets. Spectroscopy, gravitational theory, including Bessel's "astronomy of the invisible," and various forms of the nebular hypothesis are used as sources. Pohle admits that these point only to the habitability, not necessarily to the inhabitation, of these planets, but this does not deter him from asserting: "From the habitability of a celestial body follows philosophically its actual habitation." (p. 257)

Scientifically, the most interesting of Pohle's chapters is that in which he treats in turn the habitability of each solar system object. His discussion of Mars, which he labels "A Second Earth" (p. 275), is based on extensive readings in the then current American, British, French, and

German authorities. Although aware of the opposing arguments, he accepts Schiaparelli's observations of the Martian canals and their doubling, which observations had since the first edition of his book "kept the astronomical world continuously in agitation and out of breath." (p. 298) In fact, Pohle asserts:

. . . the existence and organized arrangement of the canal system with almost irresistible power forces the acceptance of an intelligent origin for them and consequently of thinking beings on our neighboring planet. All other hypotheses . . . are destroyed by their too great arbitrariness and internal improbability. [Thus most probably] intelligent, person-like creatures, either directly or indirectly, have created the astronomical miracle. . . .[63]

Although admitting the obstacles to observing Venus, he labels it "The Twin Sister of the Earth" (p. 304) and pronounces it habitable. Mercury he designates as at present an undecidable case (p. 314), but Jupiter he judges is still "in planetary youth," with nothing more than "the first sea-monsters and fish in its warm seas." (p. 319) Expressing caution about catching the "population fever of a Flammarion" (p. 321), who had proposed that "aereal beings" live in Saturn's atmosphere (p. 326), Pohle rejects both that planet and its rings as being presently habitable by intelligent beings. "Arctic" Uranus and Neptune are also denied habitability (p. 331). Pohle concludes his discussion of the planets by a long quotation from Carl Braun, who held essentially identical views (p. 331). The planetoids Pohle presents as "dead rocky masses," but leaves as possible exceptions such objects as Vesta, on which signs of an atmosphere had been reported (p. 337). A measure of caution characterizes his analysis of the likelihood of life on the planetary satellites, but does not deter him from stating: "Since Saturn is still probably self-illuminating and like Jupiter represents a secondary sun, so must we declare the habitability of [Saturn's] eight moons as rather possible." (p. 343) The moon he leaves lifeless, but adds: "perhaps no argument goes against the supposition that the moon, in a long passed period when our earth still in a chaotic, molten condition . . . represented for its satellite a sort of nearby sun, was filled with corresponding life forms." (p. 356) Comets and nebulae are treated together, the former being found uninhabitable by intelligent forms of life (p. 375). Nebulae he divides into two classes, the extragalactic and those within the Milky Way. The latter, he suggests, are solar systems in the process of formation and will eventually contain inhabited planets (p. 390).

Pohle's final two chapters contain his philosophical and theological arguments for extraterrestrial life. In the first of these chapters, he contrasts the Christian world view with that of such authors as Du Prel and Flammarion, Pohle suggesting that the crucial problem with the materialist viewpoint is that it contradicts established science by assuming spontaneous generation. In place of the materialists' arguments for pluralism, Pohle proposes four arguments derived more or less from Scholastic philosophy. His first argument, the "Metaphysical-teleological Argument from the Highest Purpose of the World," rests on the assumption that God made the universe for his own glorification. It also invokes a distinction between "objective" and "formal" glorification, objective glorification being that arising from the unconscious physical creation, whereas formal glorification comes only from intelligent beings. The latter being the nobler, Pohle concludes that "it seems suitable to the highest purpose of the world that the habitable heavenly bodies be populated with creatures that utilize the physical splendors of their worlds for the glorification of the Creator. . . ." (p. 415) Concerning this and Pohle's other arguments, it should be kept in mind that they are not presented as direct metaphysical proofs for extraterrestrials; such proofs he considers impossible (p. 395). His methodology consists rather in joining metaphysical arguments to the empirical evidences of habitability presented in the astronomical sections of his book. His second argument, that from the "Perfection of the Universe,"[64] begins with the "observation" that the universe manifests a high degree of perfection. Using Aquinas and Secchi as sources, he asserts that a universe widely populated with intelligent beings is more perfect than one filled with "unadorned deserted wastelands." (p. 416) His third argument, that based on the "Omnipotence and Wisdom of God," is that from the ingenuity and power shown by God in populating the earth with the most diverse creatures, we can infer that God has acted similarly on the other planets. Pohle's fourth argument is that so much evil exists in mankind that it seems highly probable that God would populate other worlds with beings more disposed to glorify him (pp. 427–8). After dealing with objections to his arguments (pp. 428–49), Pohle concludes this very metaphysical chapter.

In his final chapter, "The Plurality of Inhabited Worlds before the Tribunal of Christianity," Pohle points out that the French priests Félix and Moigno had urged the reconcilability of pluralism with Christianity. He also discusses St. Augustine's treatment of the question of inhabitants of the antipodes, as well as the condemnation of Vergilius in 748 for holding what to some appeared to be, but actually was not, a pluralist position (pp. 453–7). Surprisingly, he gives little attention to the question how Christ's incarnation and redemption are to be viewed in a pluralist universe. Most of what he says is contained in the following statement from his final paragraph:

Concerning the dogma of the Redemption of fallen men through the God-man Christ, it is not necessary to assume as probable also the fall of species on other celestial bodies. No reason . . . obliges us to think others as evil as ourselves.

However even if the evil of sin had gained its pernicious entry into those worlds, so would it not follow from it that also there an Incarnation and Redemption would have to take place. God has at his disposal many other means to remit a sin that weighs either on an individual or on an entire species. . . . (pp. 457–8)

Pohle's *Sternenwelten* is in many ways an impressive presentation. Rich in references to works in half a dozen languages, filled with illustrations and interesting astronomical information, it served simultaneously as an introduction to astronomy and to a more or less Proctorian pluralism, including some of its theological aspects. However, not all of his contemporaries were in agreement with it. For example, Canon Julius Zucht of Ulm published a theological critique of some of the claims Pohle made in his first edition, Pohle responding at length.[65] In discussing Pohle's second edition, the Jesuit astronomer Adolf Müller urged in 1900 that Pohle had overestimated what astronomy teaches concerning extraterrestrial life and also what reason tells us of the ways of God.[66] Ludwig Günther (1846–1910), in reviewing the fourth edition, described it as overly philosophical, and Pohle's acceptance of the Mars of Schiaparelli and Lowell as somewhat naive.[67] On the other hand, Paul Schanz (1841–1905), a Catholic theologian at Tübingen, cited Pohle in support of his statement in his *Apologie des Christentums* that

. . . when it is said that Christ died for all men, it means after all the men on earth and no other. . . . the Schoolmen taught in opposition to Anselm, that the Incarnation was not an absolute necessity. Some, however, held that even apart from sin the Incarnation formed part of God's eternal plan; nor did several Incarnations seem to them impossible. . . . Why not [then] admit other possibilities for rational beings in other planets? Perhaps they did not fall in their progenitor and head; maybe they fell and were redeemed by an Incarnation of their own or in some other way. . . .[68]

Furthermore, Pohle's book and other pluralist writings constitute the most extensive study ever made by a Catholic theologian of comparable prominence of the question of extraterrestrial life. That the public appreciated his book is shown by the fact that it went through seven editions, frequently with substantial revisions. In fact, it was last published in 1922, the year in which Pohle died, having shortly before once again revised it.

Pohle's claim (cited at the beginning of this section) that unbelievers and Christians had reached agreement in favor of pluralism is certainly excessive. As this discussion of the extraterrestrial life debate in German religious writings of the late nineteenth century shows, neither camp had attained a consensus even within itself. Among materialists, Büchner's antipluralist universe was strikingly different from that of Strauss, Du Prel, and Haeckel. Among Christians, the Lutheran Zöckler was similarly opposed to the antipluralist views of his confessional colleague Luthardt. Catholics were scarcely less divided; whereas Braun, Pohle, and Schanz endorsed pluralism, Zucht and Müller expressed major reservations. Once again the pluralist hypothesis had shown its remarkable flexibility and appeal, while those who discussed it revealed the difficulties man has experienced in attaining any sort of evidence – scientific, philosophical, or religious – on which even persons of similar perspectives could concur.

3. British religious writings: "What links are ours with orbs that are/So resolutely far"?

Although French and German publications from the final four decades of the nineteenth century on religious aspects of pluralism fit in each case into two broad categories, the corresponding British publications are of six types, all of earlier origin. These are the anti-Christian, the highly speculative, the Swedenborgian, the natural theological, the Whewellian, and those by literary figures. Beyond these groupings, one overarching tendency can be detected; it will be discussed at the conclusion of this section.

Victorian religious sensitivities received two shocks in the early 1870s, the first being administered in 1872 by the physician, African explorer, and literary hopeful Winwood Reade (1838–75) in his *The Martyrdom of Man*. Begun as a world history, Reade's book became a treatise on comparative religion, featuring such themes as: "Supernatural Christianity is false. God-worship is idolatry. Prayer is useless. The soul is not immortal."[69] Reade offers consolations; by means of science,

Disease will be extirpated . . . immortality will be invented. And then, the earth being small, mankind will migrate into space, and will cross the airless Saharas which separate planet from planet, and sun from sun. . . . Finally, men will master the forces of nature; they will become themselves architects of systems, manufacturers of worlds. (p. 515)

The *Saturday Review* of October 12, 1872, described Reade's book as "wild, mischievous, . . . blasphemous," and it found during the rest of the century not a single favorable review – but thousands of readers (pp. xxxv–xxxvi).

A greater shock came in 1874 at the Belfast meeting of the British Association for the Advancement of Science, where some claimed to detect the smell of brimstone as John Tyndall (1820–93), superintendent of the Royal Institution, delivered his presidential address. It contains such claims as: "The impregnable position of science may be described in a few words. We claim, and we shall wrest from theology, the entire domain of cosmological theory."[70] Tyndall's cosmology was that of a materialist and pluralist; in his address, he asks if the time has not now come

> . . . to close to some extent . . . with Bruno, when he declares that Matter is not "that mere empty *capacity* which philosophers have pictured her to be, but the universal mother who brings forth all things . . .?" Believing, as I do, in the continuity of nature, . . . I cross the boundary of the experimental evidence, and discern in . . . Matter . . . the promise and potency of all terrestrial Life. (pp. 203–4)

And he might have added extraterrestrial life, for he had earlier in his address described Bruno's plurality of worlds doctrine as a "sublime generalization." (p. 165) Moreover, in the draft of an 1867 paper attacking miracles, Tyndall had criticized the Judeo-Christian concept of God by stating:

> Transferring our thoughts from this little sand-grain of an earth to the immeasurable heavens, where countless worlds with freights of life probably revolve unseen . . . and bringing these reflections face to face with the idea of the Builder and Sustainer of it all showing Himself in a burning bush, exhibiting His hinder parts, or behaving in other familiar ways ascribed to Him in Jewish Scripture, the Incongruity must appear.[71]

In a recent study of Victorian science and belief battles, Frank M. Turner stated concerning Tyndall's Belfast address: "Probably no single incident in the conflict of religion and science raised so much furor."[72] Responses to it were numerous, one of the most widely read coming from the prominent physicists Balfour Stewart (1828–87) and Peter Guthrie Tait (1831–1901), who in 1875 issued (at first anonymously) *The Unseen Universe, or Physical Speculations on a Future State*. In the preface to their book, which some reviewers compared to Isaac Taylor's *Physical Theory of Another Life*,[73] Stewart and Tait describe it as designed "to show that the presumed incompatibility of Science and Religion does not exist [and moreover] that immortality is strictly in accordance with the principle of Continuity. . . ."[74] To this end, they argue that recent scientific findings support the idea that an invisible universe exists in association with our visible universe, interacting with it by means of energy exchanges in the ether. After death, we pass into this invisible universe by means of processes that strictly obey the "principle of Continuity" (or uniformity of nature), this occurring in accordance with their new law of conservation of energy, which stipulates that the total energy of the universe (visible and invisible) remains constant. Pluralist views are discussed periodically in their book. They reject, for example, the doctrines of Swedenborg and those of the spiritualists (pp. 36–40), as well as traditional transmigrational teachings about human souls ascending to higher forms of life in the visible universe, asserting that "man and beings at least analogous to man represent the highest order of living things connected with the present visible universe." (p. 140) But they do leave open the possibility that as the visible universe moves through collisions to its final thermodynamic degradation, solar systems "on a far grander scale" may be formed (p. 86). They also discuss William Thomson's theory that life may spread throughout the visible universe by germs carried on meteorites, but they maintain that life must originally have arisen from the invisible universe (pp. 173–5). They even speculate that the energy that seems wastefully to pass into space from the stars and planets may not in fact be lost, but may instead supply a memory for denizens of the invisible universe of events in the visible universe (pp. 146–7, 188–9). Their attempt to apply science to some of man's spiritual concerns ensured their book brisk sales; by 1883 it was in its tenth edition, and there was a French translation. But it also drew a number of unfavorable reviews,[75] and it has come to be seen in the science-religion debate more as a curiosity than as a contribution.

The Unseen Universe illustrates the tendency, evident throughout the history of the pluralist debate, for authors to advance highly speculative systems of the universe. Few such pluralist systems are more fantastic than that presented in 1863 by the schoolmaster Nicholas Odgers (1839–89) in *The Mystery of Being; or, Are Ultimate Atoms Inhabited Worlds?* (London). In his preface, Odgers states that readers of his earlier *Glance at the Universe* had objected to the suggestion made in it that in a speck of dust "there may exist more suns, and stars, and planets, than our eyes behold in the heavens above us. And upon them may exist inhabitants, surpassing far in strength of mind the most intellectual of all earthly men." (p. iv) To counter this objection, Odgers felt compelled to compose his 1863 treatise, which draws heavily on teleological considerations. His main argument consists in asking, concerning the cosmos, whether God's "creative energies [were] exhausted at the moment this sprang into being?" (p. 144) Such a prospect appearing as impious and as impossible as the idea that God would waste those portions of creation below human visibility, Odgers feels that he has made his case for atom-sized worlds. But he does not stop at that; as the logic of his position dictates, he also claims that "the entire universe visible to human beings [forms] a single atom of a larger and more extended economy." (p. 14) All this is set amidst urgings that a human must not measure "the ability of God by the capacities of his own mind." (p. 159) Odgers cannot be given a place of distinction in the pluralist debate, but he secured preeminence in a history written by Augustus De Morgan, who in his *Budget of Paradoxes* memorialized Odgers as the author of a book that "as a paradox, beats quadrature, duplication, trisection, philosopher's stone,

perpetual motion, magic, astrology, mesmerism, clairvoyance, spiritualism, homeopathy, hydropathy, kinesipathy, Essays and Reviews, and
Bishop Colenso, all put togehter."[76]

Odgers's system attracted few, if any, adherents, but that presented a
century earlier by Swedenborg had a significant following in Victorian
Britain. R. A. Proctor's 1876 essay criticizing the Swedish seer's system
provoked the English Swedenborgian Rev. Augustus Clissold (1797?–
1882) to publish in response *The Divine Order of the Universe*.[77] Among
other charges made by Proctor was that Swedenborg's visions contain no
mention of Uranus, Neptune, or the asteroids, which objects, although
discovered after Swedenborg's death, should have been known to the
spirits who conversed with him.[78] Clissold responds that Swedenborg
had asserted that the spirits to whom he talked, having departed from
their physical bodies, had lost their physical sight and could supply information about their planetary lives only from memory (pp. 70–2). Clissold's book is directed against Proctor, but even more against Whewell.
In reacting to Whewell's book and Proctor's post-1870 Whewellite tendencies, Clissold draws on Brewster, John Herschel, and even Proctor's
early propluralist writings (pp. 107–14).

The Swedenborgian position in the pluralist debate was also advocated
in 1899 by Rev. John E. Bowers in *Suns and Worlds in the Universe
According to the Philosophy of Emanuel Swedenborg* (London). It consists of an elementary presentation of astronomy embellished by a bevy
of quotations from Swedenborg. Bowers, like Clissold before him, bases
many of his astronomical statements on sources decades out of date.
Moreover, again like Clissold, he champions life on the moon, asserting
that "it is absurd to suppose that a planetary body more than one-fourth
the diameter of the earth, exists for the sole purpose of reflecting its light,
for the benefits of the inhabitants of the earth. Enlightened reason . . .
perceives it to be an indubitable fact that the moon is inhabited." (p. 89)
"Enlightened reason," of course, claimed just the opposite, as many
readers must have realized. Moreover, such claims probably acted at the
end of the nineteenth century to discredit the Swedenborgian system.
That this was the case is indicated by a story recounted by Theodore F.
Wright, editor of the Swedenborgian *New-Church Review*. Writing in
1897, Wright related that as a clergyman and young lady watched the
starry sky, and in particular one planet, "she was heard to say to her
companion, 'Don't you think it must be inhabited?' 'No, indeed,' was the
quick reply, 'that is Swedenborgian nonsense.' "[79] Wright also admitted:
"Much ridicule has been heaped on the New Church [because of its
pluralism] as if this single fact . . . were enough to discredit all else."
Nonetheless, Wright devoted his paper to presenting Swedenborg's ideas
of extraterrestrial life.

The death knell for traditional natural theology may have been sounded in 1859 by Darwin's *Origin of Species,* but by no means all pluralists chose to hear it. Because he was both well informed and not uncritical, Rev. Josiah Crampton, an Irish Anglican clergyman and astronomical author, illustrates this point with particular effectiveness. Crampton's critical spirit can be seen in the fact that to the 1863 fourth edition of his *Lunar World* he added a chapter devoted to refuting Brewster's extravagant claims for lunar life. There Crampton states concerning the question of a lunar atmosphere: "It is wonderful . . . how small an amount of air the advocates for the moon's atmosphere are content with. Give them as much air as will remain under an exhausted receiver, and straightaway our satellite is clothed with verdure and thickly populated. . . ."[80] Nonetheless, Crampton retains the teleological perspective seen previously in his propluralist contribution to the Whewell debate; this is indicated by his pointing out in his 1863 book that the moon, although lifeless, serves a purpose, "viz., the enlightenment and regulating the climactical arrangements" of the earth (p. 95). His continuing attachment to the natural theological tradition and to a strongly pluralist position is even more evident in his *The Three Heavens* (1871), especially in his discussion of Mercury. After admitting that unless modifying conditions exist on that planet to alleviate the intensity of the sun's rays, it must be as uninhabitable as a "red-hot ball," he adds: "But this *very circumstance* . . . is the *strongest proof possible* that there is a *modifying* cause at work there for the purpose of cooling the climate [for otherwise how] completely indeed the argument for *design* . . . would be annihilated . . . !"[81] Such comments suggest that Crampton did not pay adequate attention to his own reminder that God "has not admitted us to His creative counsels. . . ." (p. 349)

Another example of the persistence of a natural theological perspective is *The Starry Hosts* (1875), by Joseph Hamilton, who argues that "the planets must serve some higher purpose than the lighting up of the earth" and that the "purpose most worthy of their Creator, and most in harmony with His revealed operations, is that of sentient and intellectual habitation."[82] Part of the explanation of how Hamilton came to repeat in 1875 this old and much attacked argument and also to argue for lunar life (pp. 98–9) lies in his statement "that in arranging my thoughts on the subject before us, I have avoided as much as possible coming in contact with what others may have said or written upon it before me." (p. 9) This lamentable practice, followed by many pluralists, although rarely so readily admitted by them, ensured that Hamilton's book would contain little of value. In 1904, he returned to the pluralist debate with *Our Own and Other Worlds* (New York), an equally unimpressive performance.

Hamilton's position may have been due to lack of learning, but no such charge could be brought against Sir Edwin Arnold (1832–1904), an Oxford-educated journalist, poet, and educator. As late as 1894, Arnold published an essay attacking astronomers for "rashly" and "foolishly" denying life to the moon and planets and for failing to see that "there may be creatures on the sun which thrive upon incandescent hydrogen. . . ."[83] What provided Arnold with confidence in such claims was his belief in the principle of plenitude. By century's end, however, such claims were becoming far less common; in fact, that, in a sense, was precisely what had angered Arnold.

The fourth feature of this period is that some authors advocated what was essentially Whewell's position. For example, writing with both the scientific credibility of a Cambridge graduate with first-place finishes in both the Wrangler and Smith Prize competitions and with the religious renown of being bishop of Edinburgh, Henry Cotterill (1812–86) published in 1883 a book on science and religion with a chapter devoted to the question of "Life on Other Worlds."[84] Describing Whewell as "unquestionably the physicist of the most philosophical mind and of the largest and most attainments of all scientific men at the beginning of this half-century . . ." (p. 137), Cotterill claims that subsequent advances in science have strengthened Whewell's position. He cites, for example, the new energy laws and spectroscopy, which "confirm the similarity, if not the identity, of nature throughout the universe," making it appear that "the requirements of life [elsewhere] cannot very greatly vary" from those on earth (pp. 137–8). He also draws on Proctor's Whewellite writings, but with the caution that those rare planets that are at a stage that can support life may not in fact have higher forms of life. Proctor reviewed Cotterill's book, describing his antipluralist position as "doubtless possible, however unlikely," and claiming that Cotterill's reasons for advocating it were "not scientific, but depend on that particular form of religious belief which is his."[85] Proctor was probably correct about the origin of Cotterill's claims, but their validity was another matter.

Although it would be excessive to claim that Richard Holt Hutton (1826–97), a well-known essayist and editor, became a latter-day disciple of Whewell, it is clear that four essays he published in the *Spectator* in the 1882–92 period belong, broadly speaking, to the Whewell tradition. This is all the more interesting because his position in those essays differs dramatically from the position he adopted when in 1855 he vigorously attacked Whewell's book.[86] Hutton's changed conception of the cosmos may be linked to the fact that whereas he wrote in the 1850s as a Unitarian, by the 1880s he had become a High Church Anglican. In any case, Hutton in an 1882 essay criticizes the use of analogical argumentation as a basis for inferring the existence of beings comparable to man on other planets, concluding that "a great deal too much is made [of it], when the facts on which we reason are a mere infinitesimal fraction of the facts which would be wanted in order to draw any certain inference."[87] In 1888, Hutton returned to the debate in an effort to combat the charge made by Frederic Harrison (1831–1923), the leading English proponent of Comtean positivism, that "With a geocentric astronomy . . . , the anthropomorphic Creator, the celestial resurrection, and the Divine Atonement, were natural and homogeneous ideas. . . . But with a science where this planet shrinks into an unconsidered atom . . . , the Augustan Theology goes overboard."[88] Hutton's response to Harrison's in one sense Whewellian formulation of the alternatives includes arguments that the pluralist position is conjectural and that both Old and New Testament authors were already sensitive to the vast size of the universe. In his 1889 "The Humility of Science," Hutton discusses Aubrey de Vere's poem "The Death of Copernicus," drawing from it the lesson that man should show humility in his cosmological speculations.[89] The fourth Hutton essay, that of 1892, criticizes Galton's proposal for sending signals to Mars. As noted previously, Hutton argues in it that such a proposal makes little sense, because we have no satisfactory evidence of the existence of Martians capable of such communication.[90]

British literary figures, especially poets, retained an active interest in the pluralist debate as it developed in the latter half of the nineteenth century. In their writings, a greater diversity of views is evident concerning extraterrestrial life than in earlier eras. As a first example, consider Arthur High Clough (1819–61), who in his "Uranus" writes

> Of suns and stars, by hypothetic men
> Of other frame than ours inhabited,
> Of lunar seas and lunar craters huge.
> And was there atmosphere or was there not?
> And without oxygen could life subsist?[91]

But Clough's recommendation in the poem is "Mind not the stars, mind thou thy Mind and God." Robert Browning (1812–89), in *The Ring and the Book,* has a seventeenth-century pope describe his position in society in these lines:

> I it is who have been appointed here
> To represent Thee, in my turn, on earth,
> Just as, if new philosophy know aught,
> This one earth, out of all the multitude
> Of peopled worlds, as stars are now supposed, —
> Was chosen, and no sun-star of the swarm,
> For stage and scene of Thy transcendent act. . . .[92]

A more interesting instance of the invocation of pluralism occurs in "Meditation under Stars," published in 1888 by George Meredith (1828–1909), who in it asks whether on other planets "Life climbs the self-same Tree" and also "What links are ours with orbs that are/So resolutely far. . . ."[93] His answer to the latter question refers to rays from other celestial bodies:

> We who reflect those rays, though low our place
> To them are lastingly allied.
> So may we read, and little find them cold:
> Not frosty lamps illumining dead space,
> Not distant aliens, not senseless Powers.
> The fire is in them whereof we are born;
> The music of their motion may be ours.
> Spirit shall deem them beckoning Earth and voiced
> Sisterly to her, in her beams rejoiced
> Of love, the grand impulsion, we behold
> The love that lends her grace
> Among the starry fold. (p. 455)

Meredith's vision of a love-filled universe, although possessing similarities to that seen by Alyosha Karamazov in a climatic scene of Dostoevsky's *Brothers Karamazov* (1879–80),[94] differs dramatically from that in the post-1860 writings of Alfred Lord Tennyson, who (as noted previously) had responded negatively to Whewell's antipluralist book. Nonetheless, echoes of Whewell's ideas and of Proctor's Whewellite phase occur in Tennyson's late poetry. For example, his "Vastness" (1889) opens with the lines

> Many a hearth upon our dark globe sighs after many a vanish'd face,
> Many a planet by many a sun may roll with the dust of a vanish'd race.[95]

In his "Locksley Hall Sixty Years After," he sees the moon as lifeless – "Dead the new astronomy calls her" – but he speculates that Venus and Mars may have life. In doing this, he must have been aware of Whewell's geological objection; in fact, Tennyson writes:

> Many an aeon moulded earth before her highest, man, was born,
> Many an aeon too may pass when earth is manless and forlorn.[96]

A few months after Tennyson's death in 1892, his nephew's wife, Agnes Grace Weld, published some of his remarks on extraterrestrial life. One of these recalls the Psalms and also endorses an idea espoused by Kant:

> When I think of the immensity of the universe, I am filled with the sense of my own utter insignificance, and am ready to exclaim with David: "What is man that Thou art mindful of him!" The freedom of the human will and the starry heavens are the two greatest marvels that come under our observation, and when I think of all the mighty worlds around us, to which ours is but a speck, I feel what poor little worms we are, and ask myself, What is greatness? I do not like such a word as *design* to be applied to the Creator of all these worlds, it makes Him seem a mere artificer. A certain amount of anthropomorphism must, however, necessarily enter into our conception of God, because, though there may be infinitely higher beings than ourselves in the worlds beyond ours, yet to our conception man is the highest form of being.[97]

Another passage shows that pluralism was not only a recurrent element in Tennyson's poetry but also a fundamental tenet of his view of life and the afterlife:

We shall have much to learn in a future world, and I think we shall all be children to begin with when we get to heaven, whatever our age when we die, and shall grow on there from childhood to the prime of life, at which we shall remain forever. My idea of heaven is to be engaged in perpetual ministry to souls in this and other worlds.[98]

The pessimistic perspective present in the cosmos of Tennyson's late poetry is even more evident in the universe described by Thomas Hardy (1840–1928) in his novel *Two on a Tower* (1882). It is the tale of a young "scientific Adonis," Swithin St. Cleeve, whose passion for making astronomical observations from his tower observatory comes into tension with his attraction for Lady Constantine, the victim of an unfortunate marriage. In the background of the novel are Hardy's reading of Proctor's *Essay on Astronomy,* as well as the efforts of the novelist to familiarize himself with an observatory setting by visiting Greenwich in 1881.[99] In an 1895 preface to the novel, Hardy revealed that his artistic goal in it was "to set the emotional history of two infinitesimal lives against the stupendous background of the stellar universe, and to impart to readers the sentiment that of these contrasting magnitudes the smaller might be the greater to them as men." (p. 29) It is crucial in this context to note that St. Cleeve beheld in his telescope not the poetically pleasing universe of earlier pluralists but an impersonal, law-dominated cosmos concerning which he felt a fascination fettered to fear. Early in the novel, he responds to Lady Constantine's questions as to how many stars can be seen in a good telescope by stating: "Twenty million. So that, whatever the stars were made for, they were not made to please our eyes. It is just the same in everything; nothing is made for man." (p. 56) To Lady Constantine's remark that such information "annihilates" her, St. Cleeve replies: "If it annihilates your ladyship, . . . think how it must annihilate me to be, as it were, in constant suspension amid them night after night." (p. 56) Urging that "the actual sky is a horror" and that "horrid monsters lie up there," he goes on to elaborate:

Impersonal monsters, namely, Immensities. Until a person has thought out the stars and their interspaces, he has hardly learnt that there are things much more terrible than monsters of shape, namely, monsters of magnitude without known shape. Such monsters are the voids and waste places of the sky. Look, for instance, at those pieces of darkness in the Milky Way. . . . In these our sight plunges quite beyond any twinkler we have yet visited. Those are deep wells for the human mind to let itself down into, leave alone the human body! (p. 57)

Portions of Proctor's evolving universe are also seen by Swithin:

And to add a new weirdness to what the sky possesses in its size and formlessness, there is involved the quality of decay. For all the wonder of these everlasting stars, eternal spheres, and what not, they are not everlasting, they are not eternal, they burn out like candles. You see that dying one in the body of the Greater Bear? Two centuries ago it was as bright as the others. . . . Imagine them all extinguished, and your mind feeling its way through a heaven of total darkness, occasionally striking against the black, invisible cinders of those stars. . . . If you are cheerful, and wish to remain so, leave the study of astronomy alone. Of all the sciences, it alone deserves the character of the terrible. (p. 58)

The terrors of such a universe may help explain the intensity of the attachment of religious pluralists to their God of plenitude and of Whewell to the primacy of his God-redeemed earth. These options, which Hardy seems to have rejected, left him without a way to exorcise the monsters of his heavens.

Three Catholic poets, Coventry Patmore (1823–96), Aubrey de Vere (1814–1902), and Alice Meynell (1847–1922), also addressed the religious issues raised by a pluralist universe. In his 1866 "The Two Deserts," Patmore pleads: "Put by the Telescope!" In explanation, he writes:

> Not greatly moved with awe am I
> To learn that we may spy
> Five thousand firmaments beyond our own.

Moreover, astronomy has revealed the sun as "Too horrible for hell" and the moon to be "A corpse on Night's highway, naked, fire-scarr'd, accurst. . . ."[100] Patmore's negative reaction contrasts with that conveyed by Aubrey de Vere in his "The Death of Copernicus," published in 1889. In it the dying Copernicus is described as pondering whether or not to publish his book, fearing that it will raise theological problems. In reference to pluralism and the idea that it discredits Christianity, the astronomer reflects:

> 'Tis Faith and Hope that spread delighted hands
> To such belief; no formal proof attests it.
> Concede them peopled; can the sophist prove
> Their habitants are fallen? That too admitted,
> Who told him that redeeming foot divine
> Ne'er trod those spheres?

And he adds:

> Judaea was one country, one alone:
> Not less Who died there died for all. The Cross
> Brought help to vanished nations: Time opposed
> No bar to Love: why then should Space oppose one?[101]

Some of the same questions were on the mind of Alice Meynell when she wrote "Christ in the Universe." Its final four verses are:

No planet knows that this
Our wayside planet, carrying land and wave,
Love and life multiplied, and pain and bliss,
Bears, as chief treasure, one forsaken grave.

Nor, in our little day,
May His devices with the heavens be guessed,
His pilgrimage to tread the Milky Way,
Or His bestowals there be manifest.

But, in the eternities,
Doubtless we shall compare together, hear
A million alien Gospels, in what guise
He trod the Pleiades, the Lyre, the Bear.

O, be prepared, my soul!
To read the inconceivable, to scan
The million forms of God those stars unroll
When, in our turn, we show to them a Man.[102]

Having surveyed these six late-nineteenth-century British approaches
to the religious aspects of pluralism, we can turn to the question whether
or not any broader patterns or convergences can be detected. In doing
this it is well to keep in mind the danger of precipitous generalization.
This danger can be illustrated by citing two generalizations about this
period, each made by an author possessing expertise in the history of the
pluralist debate. The first generalization is: "The completion of the secu-
larization of the plurality [of worlds doctrine] by Proctor was verified by
the swift decline of Christian interest after the 1870's. The one Christian
author interested was George M. Searle."[103] The second generalization,
which occurs in a discussion of Ortolan's 1897 book on pluralism, is:
"At the end of the 19th century, . . . the question of a 'plurality of
inhabited worlds' presented itself with more acuteness than in the past.
This work [Ortolan's] is without doubt the first to treat the problem from
the theological point of view."[104] These claims are mutually contradictory, and they are both patently wrong, as can be seen even from the
authors treated in this section on the British.

With this caution as background, it may be suggested that Proctorian
pluralism was winning an increasing following in Britain as the century
waned. This point, developed more fully in the concluding section of this
chapter, can be seen in the declining confidence in the habitability of the
objects of the solar system evident in the writings of such authors as
Cotterill, Hutton, and Tennyson. It is also indicated by the fact that even
though the spiritualist movement grew in strength in Britain during this
period, it eschewed the association with pluralism that had been fostered
in France by Reynaud, Pezzani, Flammarion, and Figuier. Moreover,
those who championed extreme forms of pluralism were authors on the

fringes of science such as Arnold, Crampton, and Hamilton, or Sweden-borgians such as Bowers, Clissold, and Wright.

4. American religious writings: "The world! H'm! there's billions of them!"

Before surveying American Protestant and then Catholic reactions to the idea of a plurality of worlds in the final four decades of the nineteenth century, let us look at the two leading American literary figures of that period, Walt Whitman and Mark Twain. Both employed pluralism in their writings, doing this in contexts somewhat distant from orthodox Christianity.

Readers of *Leaves of Grass,* the classic work Walt Whitman (1819–92) published in 1855, can be misled about his interest in astronomy by these lines:

> When I heard the learn'd astronomer
> When the proofs, the figures, were ranged in columns before me,
> . . . .
> How soon unaccountable I became tired and sick. . . .[105]

An array of scholarly studies[106] and the more than two hundred astro-nomical allusions in *Leaves of Grass* show Whitman's fascination with astronomy, as well as the seriousness with which he took up, in compos-ing that volume and in revising and supplementing it, a challenge he had set for himself: "Modern science and democracy seem'd to be throwing out their challenge to poetry to put them in its statements in contradis-tinction to the songs and myths of the past."[107]

Whitman's pluralism was that of a pantheist. When asked in 1890: "Are you still as firmly pantheistic as you were in the earlier poems?" Whitman replied: "Yes indeed . . . if anything more and more so!"[108] Probably because of his pantheism, he saw the universe as somehow unified – a conviction set out in his "On the Beach at Night Alone":

> As I watch the bright stars shining, I think a thought of the clef of the universe and of the future.
> A vast similitude interlocks all,
> All spheres, grown, ungrown, small, large, suns, moons, planets,
> All distances of place however wide,
> All distances of time, all inanimate forms,
> All souls, all living bodies, though they be ever so different, or in different worlds . . .
> All identities that have existed or may exist on this globe, or any globe
> . . . .
> This vast similitude spans them, and always has spann'd
> And shall forever span them and completely hold and enclose them.
> (pp. 260–1)

His pluralism was also linked to his interest in transmigrational doctrines; both appear in the opening lines of *Leaves of Grass*:

> Come, said my Soul,
> Such verses for my Body let us write, (for we are one,)
> That should I after death invisibly return,
> Or, long, long hence, in other spheres,
> There to some group of mates the chants resuming
> (Tallying Earth's soil, trees, winds, tumultuous waves,)
> Ever with pleas'd smile I may keep on,
> Ever and ever yet the verses owning – as, first, I here and now,
> Signing for Soul and Body, set to them my name,
> Walt Whitman[109]

Pluralist evolutionary ideas occur in his "Night on the Prairies":

> I was thinking the day most splendid till I saw what the not-day exhibited,
> I was thinking this globe enough till there sprang out so noiseless around me
> myriads of other globes.

> Now while the great thoughts of space and eternity fill me I will measure
> myself by them,
> And now touch'd with the lives of other globes arrived as far along as those
> of the earth,
> Or waiting to arrive, or pass'd on farther than those of the earth,
> I henceforth no more ignore them. . . . (p. 452)

Whence came Whitman's pluralism? Scholars have suggested Thomas Dick or Ormsby Mitchel as source of the following lines from his "Song of Myself":

> My sun has his sun and round him obediently wheels,
> He joins with his partners a group of superior circuit,
> And greater sets follow, making specks of the greatest inside them.[110]

Whether or not either attribution is acceptable, it is possible if not probable that Whitman's introduction to pluralism came from Thomas Paine, who had been a friend of his father and whose *Age of Reason* was a prized possession in his father's household.[111] Paine, whom Whitman eulogized in 1877,[112] is widely regarded as an important source of Whitman's strongly democratic and anticlerical sentiments, the latter being expressed in the preface to *Leaves of Grass* by such comments as: "There will soon be no more priests. Their work is done. . . . the gangs of kosmos and prophets en masse shall take their place."[113] Nonetheless, Whitman never seems to have invoked Paine's pluralist objection to Christianity. Although he wrote to Emerson "The churches are one vast lie,"[114] he told another correspondent that in speaking with a minister of the Dutch Reformed church: "I not only assured him of my retaining faith in that sect, but that I had perfect faith in all sects, and was not inclined to reject a single one. . . ."[115] "Cosmic optimism" has been used by Howard Mumford Jones to describe Whitman's overall philosophy; this phrase not only captures the spirit of Whitman's pluralism but also suggests how strikingly he differed from his great contemporary, also a pluralist, Mark Twain, whom Jones with equal aptness designated a proponent of "cosmic pessimism."[116]

That Mark Twain was intrigued by astronomy was amply demonstrated in a 1937 article by H. H. Waggoner, who discussed whether Twain's anti-Christian and pessimistic world view had its source primarily in science, as P. H. Boynton and S. T. Williams had maintained, or in Twain's readings of eighteenth-century philosophers, especially Thomas Paine, as M. M. Brashear had urged.[117] When this question is approached from the perspective of the pluralist debate, Waggoner's dichotomy between the influence of science and of Paine (himself strongly influenced by astronomy) appears unnecessarily sharp. In fact, at a number of points in Twain's life, pluralist ideas played a major role, certainly in the formulation, possibly in the formation, of some of his most fundamental convictions. Around 1858, Twain read Paine's *Age of Reason* with an intensity revealed in his recollection that "I . . . read it with fear and hesitation, but marveling at its fearlessness and wonderful power."[118] This may explain why on three occasions in the 1870s Twain attacked Christianity on the basis of Paine's pluralist objection.

In an 1870 letter to his future wife, Twain describes an astronomy volume he had been reading and in that context reveals his religious doubts:

How insignificant we are, with our pigmy little world! – an atom glinting with uncounted myriads of other atom worlds . . . & yet prating complacently of our speck as the Great World, & regarding the other specks as pretty trifles made to steer our schooners by & inspire the reveries of "puppy" lovers. Did Christ live 33 years in each of the millions & millions of worlds that hold their majestic courses above our heads? Or was *our* small globe the favored one of all?[119]

Later in 1870, Twain composed an essay contrasting ancient and modern notions of Divinity. After noting the vastness of the modern universe, he states:

The difference between that universe and the modern one revealed by science is as the difference between a dust-flecked ray in a barn and the sublime arch of the Milky Way in the skies. Its God was strictly proportioned to its dimensions. His sole solicitude was about a handful of truculent nomads. He worried and fretted over them in a peculiarly and distractingly human way. . . . He sulked, he cursed, he raged, he grieved . . . , but all to no purpose . . . he could not govern them.[120]

No doubt aware that such passages would offend many readers, Twain withheld that essay from publication and did the same – for over three decades – with the manuscript finally published in 1907 as his *Extract from Captain Stormfield's Visit to Heaven*.[121] He had begun work on it in the late 1860s, inspired by a tale told him by his friend Captain Ned Wakeman, revised it in 1873, and put it into final form in 1878.[122] Writing partly to satirize the "little ten-cent" heaven he found in E. S. Phelps's *The Gates Ajar,*[123] Twain was intent to present, as Brashear put it, "a burlesque of the Christian heaven."[124] Such an interpretation accounts both for the relish with which he wrote it[125] and for his strong temptation in 1906 (which he overcame) to burn it.[126] More than a satirization of heaven, however, it includes a raucous restatement of Paine's pluralist objection to Christianity. Arriving at *a* heaven, which he mistakes for *the* heaven, Stormfield is asked by the gatekeepers where he is from. Trying in succession the replies "San Francisco," "California," and "America," he is mystified by such responses as "There ain't no such orb." (p. 17) To the gatekeepers' insistent "Come, come, what world?" Stormfield replies: "Why, *the* world of course." Their patience at an end, they reply: "*The* world! H'm! there's billions of them! . . . Next!" (pp. 17–18) Stormfield goes off to ponder the problem, finally finding what seems the ideal identification of his world: "it's the one the Saviour saved." This fares no better, the gatekeepers telling him: "The worlds He has saved are like to the gates of heaven in number – none can count them." (pp. 19–20) After a few days of searching on a giant map, they locate Stormfield's planet, which they first fear may be a flyspeck, but which under microscopic examination proves to be our planet, known in heaven as "the Wart." (p. 24) The denunciations that Twain had feared would follow his *Stormfield* were not forthcoming, probably because its humorous character led readers to miss its anti-Christian intent. Moreover, Twain's increasingly bitter hostility to Christianity and his cynical pessimism, both of which are evident in a number of his post-1900 manuscripts, were not yet widely recognized. As these have become known, the true intent of his *Stormfield* has become all the more clear.

Twain's interest in the idea of a plurality of worlds continued into the twentieth century, appearing even in some of his latest writings. For example, he wrote a review of *The Cities of the Sun,* which Colonel George Woodward Warder (1848–1907) published in 1901 to prove "that the suns are not hot, nor burning gaseous spheres, but are the self-luminous perfected worlds of the universe and the future abode of man . . . and the planets are the hatcheries of human souls, and the suns the places of their development and growth to perfection."[127] Twain was not impressed by Warder's "New Jerusalem," to which he had "added the modern improvements."[128] His reading in 1903 of A. R. Wallace's *Man's Place in the Universe* provoked Twain to write "Was the World Made for Man?" which was first published in 1962. This short essay uses the late appearance of man on the earth to argue the negative side of the question in its title.[129] In his 1909 "Letters from the Earth," which was his last long composition before his death, Twain again employed a version of Paine's pluralist objection to Christianity. While discussing the statement in Genesis that God made the universe in six days, he adds in regard to the earth that God "constructed it in five days – and then? It took him only *one* day to make *twenty million suns and eighty million planets!*"[130]

American Protestants of the 1860-to-1900 period, although deeply divided about Darwinian theory, seem to have seen pluralism as largely unproblematic. Indeed (as indicated previously), it had become doctrine for the Latter-day Saints, the Seventh-day Adventists, the Swedenborgians, and the followers of Thomas Lake Harris. Although pluralism did not attain comparable prominence in the mainline Protestant denominations, members of a number of groups endorsed it. As evidence of this generalization, publications by a Baptist, Methodist, Congregationalist, and two Presbyterians can be cited.

In 1871, the *Baptist Quarterly* carried an essay by Rev. Edwin T. Winkler (1823–83), whose opening sentence and overall message was Edward Young's claim: "An undevout astronomer is mad."[131] After presenting reasons why astronomy should inspire devotion, Winkler turns to the "one [astronomical] speculation . . . which has been made the ground of skeptical objection against Christianity . . ." (p. 70), that is, the idea of a plurality of worlds, which he admits has been rejected by Ebrard, Hegel, and Whewell. Nonetheless, Winkler supports it, citing scriptural and scientific arguments on its behalf. To relieve the tension some saw between it and Christianity, Winkler employs a basically Chalmersian approach, suggesting that although the incarnation of Christ is unique to our planet, its benefits are universal:

As a battle may be fought at some grey pass of Marathon . . . that shall change the fortunes of a world for a thousand years, so here, on this small world, a triumph may have been achieved by the Son of God, that distributes its spoils to all systems, through all times; and for the temptation and anguish of Jesus of Nazareth, the sweet influences of the Pleiades may be fuller of vernal promise . . . and seasons of salvation may have befallen all the signs of the zodiac. . . . (p. 73)

Winkler's essay is also embellished by the Brewsterian eschatological speculation that on some celestial orb, "my loved and lost are gathered. There the wasted cheeks are flushed with immortal beauty, and the glazed eyes with immortal lustre, and the voice that was once so dear, that was once so frozen by the chills of death, is melted into immortal melodies." (p. 72)

The Chicago Methodist preacher and homeopathic physician Rev. Adam Miller (1810–1901) was no less enthusiastic than Winkler in championing pluralism. This Miller did in 1878 in his *Life in Other*

Worlds (Chicago), which includes three pluralist sermons by the Chicago Methodist Rev. Hiram W. Thomas, as well as numerous quotations from Adam Clarke and T. L. Harris. From Brewster, Miller borrows the idea that after the resurrection we shall live on other planets (pp. 168–9). The most novel feature of his book is a new theory of solar heat, with which he had challenged Proctor when the latter lectured in Chicago in 1873 (p. 83). The level of Miller's sophistication is indicated by his populating the planetoids (p. 110), while his skill in exposition is suggested by the fact that his book attained no later editions.[132]

After graduating in 1839 from Yale as class orator, Enoch Fitch Burr (1818–1907) may have considered a career in science; in fact, he studied science for six years and published in the 1840s two technical papers in astronomy. However, in 1850, he was ordained as the minister of a Congregational church in Hamburg, Connecticut, serving in that position for the remainder of his life. His scientific studies were at times put to use, for example, between 1868 and 1874 when he lectured on the "scientific evidences of religion" at Amherst, and in a number of books bridging science and religion. Among these was his *Ecce Coelum; or, Parish Astronomy* (1867), which went through at least twenty editions, assuring its readers that "An undevout astronomer is mad" and telling them of the "peopled heavens."[133] Burr's enthusiasm for pluralism is seen most clearly in his "Are the Heavens Inhabited?" which he published in 1885 both in the *Presbyterian Review* and as a chapter of his *Celestial Empires*.[134] So propluralist was Burr that he states: "I should be disappointed if, in landing on Mars, I should not find the equivalents of *men*. Almost as much can reasonably be said as to the planet Venus." (p. 258) Later, after stressing God's omnipotence, he denies "that any physical differences . . . among the Heavenly Bodies interpose any shadow of objection to their being, every one of them, the home of intelligent beings of even as high an order as man." (p. 261) Although admitting that some celestial bodies may not yet be inhabited and that stars may never be, Burr informs his readers that the Scriptures reveal that "God, his holy angels . . . , the spirits of saved men, . . . Satan, the evil angels, and the lost souls . . . have their proper homes on glorious materialisms *somewhere* out yonder in the profound of space." (pp. 263–4) All this is associated with various religious injunctions, the last of which is, ironically, "With what modesty, then, should we . . . venture to pronounce on the ways of Divine Providence. . . ." (p. 267)

Two Presbyterian authors wrote books that included arguments for pluralism. In 1862, *God's Glory in the Heavens* was published by Rev. William Leitch (1818–64), a Glasgow graduate who at one time had served as J. P. Nichol's observatory assistant and who in 1860 came to Canada as principal of Queen's College, Kingston.[135] Designed "to present a survey of recent astronomical discovery and speculation, in connexion with the religious questions to which they give rise" (p. v), this book reveals an author of both scientific and theological sophistication. A reader of Laplace, whose nebular hypothesis he endorses, and of Leverrier, whose recent prediction of Vulcan he chronicles, Leitch also shows his familiarity with the religio-scientific ideas of Kant, Comte, Chalmers, Whewell, and Brewster. Pluralist passages occur frequently, special attention being given to the moon, concerning which he admits that there is little observational evidence of lunar life. He does, however, discuss Hansen's hidden-hemisphere hypothesis, using this as a means not of proving lunar life but rather of refuting those who attack natural theology on the grounds of the uselessness of the apparently dead moon.

Leitch's final chapter, devoted to the plurality of worlds question, shows not only unusual analytical ability but also a readiness to occupy a middle position in the pluralist debate. He classifies propluralist arguments under four categories: (1) *a priori* or theological, (2) metaphysical, (3) scriptural, and (4) astronomical or analogical (pp. 302–16). Each of these he carefully critiques, suggesting specific flaws in many traditional arguments. Scripture, he urges, cannot be used to prove pluralism, although it can be seen to be in harmony with it. Analogical arguments are vitiated when variations are excessively large. The influence of Whewell is evident in his analysis, but Whewell's arguments are in turn dissected and judged less than satisfactory (pp. 317–22). From these analyses emerges Leitch's conclusion that "astronomy does afford a probability, but only a probability, that some of the planets have a life epoch in their history." (p. 316) The cautions and qualifications contained in his statement are each developed in detail by Leitch, who within this perspective considers Paine's pluralist attack on Christianity as well as some of the responses to it (pp. 322–30). Leitch rules out multiple incarnations of Christ, because these would contradict "Scripture, which declares that He will forever bear His human nature." On the same basis he rejects Brewster's argument that Christ's crucifixion should secure redemption for extraterrestrials. These claims combine to leave Leitch in effect asserting that redemption is required only by earthlings: "The universe is a great harp, and each orb a string of that harp; but one string, at least is untuned. . . . One great end of redemption is to re-adjust this jarring string of our world." (p. 329) Thus, we see from revelation that although the earth is not the "*material* centre of the universe . . . it is still the *spiritual* centre." (p. 329) Leitch deserves credit for a sober and sophisticated contribution to the debate.

Jermain Gildersleeve Porter (1853–1933) was also a Presbyterian and had in fact studied for two years at Auburn (N.Y.) Theological Seminary. However, the career on which he settled was astronomy, which he studied with C. H. F. Peters at Hamilton College and also in Berlin. In 1884, Porter began his forty-six years as director of the Cincinnati Observatory, but his preference for the observing platform over the pulpit did not preclude writing on religious concerns; this he did in *Our Celestial Home: An Astronomer's View of Heaven* (London, 1888). His view of heaven turns out to be that it is a material entity, as are angels (pp. 15–23). Porter even proclaims that the "whole universe constitutes heaven" (p. 108) and that the astronomer, in studying the night sky, observes "the glittering lamps of the many mansions" mentioned in Scripture and "may even catch a glimpse of the celestial city itself. . . ." (pp. 115–16) In a chapter on "The Habitability of the Celestial Worlds," Porter argues that the earth is probably the only inhabited planet in our solar system (pp. 43–55), but he salvages his teleologically grounded pluralism by the thoughts that some planets of other systems are probably now inhabited and that others will come to be (pp. 54–9). In this way, Porter, like the other Protestant authors discussed in this section, sought to save astronomy and pluralism from charges that they were irreligious. That this tension lingered even into the final decade of the century is shown by the anti-Christian polemics employed by William Fretts in an 1892 pamphlet entitled *Inhabited Worlds Is the Universal Law of Nature* (Washington, D.C.).

Whereas these five Protestant authors wrote largely in isolation from each other, four Catholic priests in the final two decades of the century became involved in what was in effect a debate. Rev. George Mary Searle (1839–1918), whose brother Arthur had participated in the Whewell debate, was the earliest, best informed, and most active participant, having published three essays. After graduating from Harvard, where he had been a Unitarian, George Searle held a number of positions before becoming in 1866 an assistant at Harvard Observatory. In 1862, he became a Catholic and in 1868 began studies for the priesthood as a Paulist.[136] In 1883, he published a paper arguing that astronomy gives no conclusive evidence of extraterrestrials. He admits that life may exist on Mars or on planets (if such exist) circling other suns,[137] but he attacks the pluralists' teleological argument by drawing a sharp dichotomy between the spiritual and material (pp. 55–6).

In 1889, at the founding of the Catholic University of America, Searle was chosen observatory director and in 1890 presented a lecture entitled "Are the Planets Habitable?" Possibly the influence of Rev. Joseph Pohle, who taught apologetics at Catholic University from 1889 to 1894, led Searle to advocate, albeit cautiously, a much more pluralist position in this lecture than in his 1883 paper. For example, he shows that although the sun's light is nine hundred times dimmer on Neptune than on the earth, it is nonetheless seven hundred times brighter than moonlight, making it suitable even for reading.[138] Despite not feeling "the craving for the plurality of worlds . . . which seems so general" (p. 174), Searle suggests that the giant planets of our system and possibly our sun may eventually become habitable and that Mars may be ending its period of habitability (pp. 173–5). Having questioned the nebular hypothesis in 1883, he remarks that "it would be strange" were other suns not to "have attendant planets like our own." (p. 177)

Searle's third paper (1892) contains the fullest statement of his views. Suggesting that the problem of possible conflicts between science and religion should be approached piecemeal, he isolates two difficulties that, it is said, modern astronomy creates for Christianity. These are that our planet appears to be an "insignificant little speck" in space and the related problem of other inhabited worlds, in particular, why God chose to become incarnate on our planet.[139] He dismisses the first problem by urging that for the astronomer, distance and size are merely relative. Turning to the pluralist issue and arguing against the habitability of the sun and giant outer planets because they have not sufficiently cooled, he concludes that we are "left about one ten-thousandth part of the whole surface of our solar system . . . as an admissible habitation for life." (p. 869) Mercury and Venus are described as at present not sufficiently developed for life, leaving only Mars and some of the larger satellites as possibilities. Concerning Mars, he notes that the conditions that some see as suggesting life on its surface were present on the earth long before humans appeared. Repeatedly he stresses that astronomy can give evidence only of habitability, not of actual habitation, and that even such claims are at most matters of probability. In fact, he asserts that "the majority of astronomers do not really believe in the existence of intelligent inhabitants of the planets . . ." of our solar system (p. 872). Turning to the sidereal realm, Searle points out that multiple star systems would probably not support habitable planets and that the nebular hypothesis is a conjecture beset with technical problems. He summarizes his views concerning habitable planets by stating that "there may be one like ours here and there. The checks, balances, and adjustments which we have are not the natural or unavoidable result of the celestial mechanism; they are an extraordinary—perhaps a very extraordinary or almost unique—occurrence." (p. 876) He stresses, however, that astronomy does not preclude the possibility that God created worlds inhabited by man-like beings, but that this by no means implies that divine incarnations occurred elsewhere.

Shortly after Searle's 1883 paper, Januarius De Concilio (1836–98), priest, professor, pastor, and even playwright, arose to play Brewster opposite Searle's Whewell. Never mentioning Searle, but making much of Father Secchi's propluralist statements,[140] De Concilio states that his

chief arguments "will be both theological and metaphysical, and especially drawn from the most fundamental principles of St Thomas. . . ."
(p. 196) His first argument is that "the number of species to be created
was determined by the place which each one holds in the scale of being or
perfection; those species holding a higher place being created in much
larger number than those which occupy a lower grade." (p. 205) This
idea is also involved in his second argument:

. . . as there is an immense distance between the highest intellect of mankind and
that of the lowest of the pure spiritual substances [the angels], the cosmological
law of affinity demands that there should be some intermediate species to soften
down the immense contrast, and thus to exhibit and represent a most beautiful,
harmonious order in the universe. (p. 206)

These intermediate species between man and angels are extraterrestrials.
In presenting his third argument, which is that uninhabited planets
would be a wastage of God's powers, De Concilio makes a statement
about how God "must" have designed the universe, which statement
suggests how easily speculative philosophers and theologians can fall into
an excessive dogmatism. He writes that God "must, if He would follow
the requirements of wisdom, draw from the given forces to be created all
the possible good in view of the end; in other words, He must . . . use
the least possible means to attain an object, and follow the fundamental
law of wisdom laid down by St. Thomas." (p. 207) It is interesting in this
context to note, as De Concilio does not, that Thomas Aquinas, rather
than arguing for pluralism, had rejected it.[141] De Concilio also discusses
its relation to Christianity, stating that extraterrestrials were created "in
and through Christ" and through Christ must attain their eternal end.
Christ's incarnation and redemption, although unique to the earth, must
extend to all other "species of incorporated intelligences [which] may
have fallen, and very likely did fall. . . ." (p. 215)

Later in 1884, Thomas Hughes (1849–1939), a Jesuit teaching at St.
Louis University, published a response to De Concilio, faulting him not
only for taking a "utilitarian" view of the universe but also for being
narrow in this approach, seeking to specify the use of minor matters.
How, Hughes urges, can one label an uninhabited planet useless if one
cannot explain "why the drop of rain in *our* mid-ocean is not utterly
wasted?"[142] In a passage that echoes God's final response to Job, Hughes
reminds his contemporary that the ways of God are beyond man's knowing. Hughes admits that in a broad sense we can say that the heavens
serve the function of supplying man with a lofty object of study and
contemplation (pp. 458–9). Hughes reponds to a number of De Concilio's specific arguments, urging, for example, that De Concilio's claim
that extraterrestrials exist to fill the gap between the material and the
spiritual breaks down because man himself is a sufficient mean between

those extremes. Hughes's fundamental objection is that his opponent overvalues the material but undervalues man and his contemplative nature: "an earth for man's body, a universe for his soul." (p. 467) In no respect scientific, more poetic than philosophical, Hughes's critique questions De Concilio's arguments more than his conclusions; nonetheless, looked at broadly, it is an antipluralist plea made by a man more inclined to the God of Job than the God of Aquinas.

That De Concilio's confidence in his pluralist arguments was not shaken by Hughes's critique is shown by the five chapters devoted to pluralism in De Concilio's *Harmony between Science and Revelation* (1889). After citing propluralist statements by such Continental clerics as Félix, Gratry, Moigno, and Secchi, De Concilio admits that no conclusive evidence of extraterrestrial life is presently available.[143] He then summarizes the astronomical evidence for life elsewhere, drawing the conclusion that "the planets, and the sun even, present very little difficulties in the way of their being inhabited . . ." (p. 214), and restates his main pluralist arguments, making no mention of Hughes's criticisms of them. To his earlier treatment of the incarnation and redemption, he adds only the conviction that "when Christ died and paid the ransom of our redemption, He included [extraterrestrials] also in that ransom, the value of which was infinite and capable of redeeming innumerable worlds." (p. 232) He again leaves as an open question how Christ may have bestowed "knowledge of Himself, and His Church" on extraterrestrials (p. 233). De Concilio, whose *Harmony* was primarily aimed at attacking evolutionary doctrines, may have been pleased that his pluralist chapters could be read as showing his openness even to radical and fascinating scientific ideas.

Between De Concilio's enthusiastic pluralism and Searle's somewhat Whewellite approach, a compromise position may seem inconceivable. Yet such was put forward in late 1892 by the superior general of the Paulists, Rev. Augustine F. Hewit (1820–97), who, having been selected with Searle and Pohle for the founding faculty of Catholic University, may have worked out his ideas over coffee with them. In his paper, which appeared in *Catholic World* just a month after Searle's, Hewit endorses Searle's claims but proposes to supplement them by a theological conjecture of his own devising. In particular, Hewit labels the belief in the past or present existence of other worlds inhabited by intelligent beings as "unphilosophical, untheological, and unscriptural," but the future existence of such worlds he describes as "possible, and even probable."[144] Interpreting various New Testament texts as indicating that the incarnation and redemption must be unique to the earth and that redeemed mankind holds the highest place next to angels in heaven, Hewit urges that God will populate the planets only after the last judgment. Ruled by the resurrected of earth, these planetarians will experience no probation but will nonetheless be immortal and enjoy a form of felicity below that available to mankind. Although claiming that his position solves many difficulties, Hewit admits that it "cannot be demonstrated scientifically or philosophically [nor] can we pretend that it is explicitly revealed." (p. 26)

Having surveyed religious writings dealing with pluralism in both America and Europe in the period from 1860 to 1900, we may turn to the question of what broad patterns are suggested. For example, can correlations be found between authors' denominational affiliations and the positions they advocated in the pluralist debate? Twenty-five identifiably Protestant authors have been discussed; of these, nineteen supported and four opposed pluralism, with two defying classification.[145] Among the twenty-seven Catholics considered, fourteen favored and ten opposed pluralism, with four being unclassified. Combining these figures, thirty-three Christians endorsed and fourteen denied the pluralist doctrine during these decades. Of the anti- or non-Christian authors whose views have been examined in these sections, all except Büchner were pluralists. It may seem a plausible conjecture (to Burque it seemed a "fact") that proponents of pluralism were also advocates of evolutionary ideas. In the present context, this conjecture can to some extent be tested, as well as its opposite: that anti-pluralists opposed Darwinian doctrines. From various sources it has been possible to determine the positions taken in the Darwinian debates by ten of the Christian authors discussed in these sections. This sample, although so limited that no strong inferences should be drawn from it, does suggest that neither conjecture is sound. For example, whereas Burque and Hughes opposed both evolutionary and pluralist thought, Burr, De Concilio, and Schanz rejected Darwinian doctrines while accepting pluralism. Moreover, whereas Boiteux, Hewit, Kirwan, Pohle, and Searle sided with evolutionary theorists, only Pohle endorsed pluralism, the rest rejecting it. In short, the extraterrestrial life debate in its late-nineteenth-century religious phase retained to some extent that feature found in the Whewell debate: It continued to be a night fight in which enemies and allies could not easily be distinguished until close conflicts commenced.

5. Scientific writings: the prevalence of "Proctorian pluralism"

In concluding this chapter and setting the stage for the next, it is useful to survey an array of late-nineteenth-century British, American, and Continental publications that do not easily fit the categories employed up to this point. These publications were in nearly every case written from a scientific rather than a religious perspective, but were not focused on such specific topics as lunar life, extraterrestrial signaling, or meteorites. The thesis that emerges from this survey is that by century's end, pluralism had taken on an attenuated form, at least if compared with the pluralism of midcentury; in particular, pluralists had relinquished claims for extraterrestrials on nearly every planet of our solar system, holding out hopes, however, for the remote past or distant future or for planets of other systems. Even the last were placed in jeopardy as habitations by the fact that analogy with our solar system now pointed not to widespread planetary life but rather to life as a development on comparatively few, although possibly numerically many, planets. This attenuated or "Proctorian pluralism" formed the background for the dramatic efforts (discussed in the next chapter) to salvage at least Mars as an abode of life. In calling this gradually emerging position "Proctorian pluralism," it is not necessarily implied that Proctor's writings were its only or even chief source; in fact, a major portion of Proctor's importance was due to the fact that he had taken Whewell's astronomical arguments more seriously than most other astronomers and developed them in a new direction. Nor is it suggested that all astronomers accepted a form of Proctorian pluralism; proponents of a Whewellite position continued to appear at times, and at the other extreme, some authors championed a traditional pluralism with extensively populated planetary systems.

Let us begin with Britain and with Rev. James O. Bevan (d. 1930), who in an 1880 essay advocated a Proctorian position. Admitting his inclination to both the nebular hypothesis and the "principle of development" in biology,[146] Bevan urges that some planets of the solar system are now inhabited, whereas others have been or will be (p. 172). At one point he compares extreme pluralists to those medieval philosophers who speculated on the number of angels who could dance on the head of a needle (p. 169). This comparison is apt; in fact, its force is increased by a recent study indicating that medievals indulged in no such speculation about angels.[147]

A much more impressive analysis appeared in 1883 from the pen of an Edinburgh solicitor, William Miller, who should not be confused with Huggins's collaborator, William A. Miller. Entitled *The Heavenly Bodies: Their Nature and Habitability* (London), Miller's book is rich in systematic analyses based on extensive readings in two centuries of pluralist literature. Despite this, and although receiving at least sixteen generally favorable reviews,[148] it secured but a single edition, probably because Miller's position was that of a Whewellite. Miller's first two chapters, devoted to the sun, of which he had made a special study, set the stage for his overall argument by building an effective case against solar life. Miller then devotes forty pages to the history of the pluralist debate, providing thereby the most detailed history of it then available in English. Although he avoids revealing in it his own antipluralist position, his frequent criticisms of pluralists certainly aided his overall purpose. Miller also lists nine philosophical and theological arguments employed by pluralists, including the arguments from Scripture, God's wisdom, final causes, and the fullness of terrestrial life (pp. 134–9). His basic response to these arguments is that they unjustifiably assume that we can know God's designs in creating the universe.

After returning briefly to his arguments against life on the sun and thereby on other stars, Miller proceeds to the question of planetary life. Because he accepts the nebular hypothesis, he is forced to stress the obstacles to life on planets orbiting stars in multiple star systems (p. 160). Although well read on both sides of the evolutionary debate in biology, he opposes the Darwinian position, urging that "the Earth was . . . specially made for man. . . ." (p. 216) To this a reader could justifiably object that one cannot jettison the knowability of final causes in one chapter and accept it in another. Turning to the moon, he rejects selenites, who by 1883 had few partisans, against whom Miller marshals a mass of authorities. Drawing on Zöllner's photometric researches to oppose life on Mercury, Miller disposes of life on Venus by citing arguments from Proctor, such as that the Venusian atmosphere may intensify rather than alleviate the excessive heat of that planet. Miller makes his case against Martians largely by means of arguments presented in 1882 by E. W. Maunder and cites the discrediting of Hahn against pluralists employing meteorites (p. 281). Miller's thoroughness and the source and nature of his arguments against life on the planets beyond Mars are both evident in his remark that Proctor's views of Jupiter "are scattered through several of his works; I have extracted them from six. . . ." (p. 303) Despite his heavy reliance on Proctor, Miller passes over that author's position on the possible habitability of the Jovian and Saturnian satellites with only the remark that they are probably as lifeless as our moon (p. 330). Nor does Miller seriously consider Proctor's suggestion that the giant planets may eventually attain life. The irony involved in this situation must be evident: Although Miller's position was that of a Whewellite rather than that of a Proctorian pluralist, and although Miller's book was dismissed (because of its Whewellite position) in a review that appeared in Proctor's journal *Knowledge* and that was probably written by Proctor himself,[149] many if not most of Miller's arguments were in fact derived from Proctor's writings.

In his concluding chapter, Miller explicitly states his extreme antipluralist position: "we can hold *life to exist only upon the Earth*." (p. 343) Aware that this conclusion removes the "pleasantly beguiling . . . supposition of a plurality of worlds," Miller asks: "have we no bright and noble thought to take its place?" (p. 343) His answer suggests what probably led him not necessarily to his precise position but to writing his large book:

But when we find that this world alone . . . is fitted for life, and has been prepared for . . . MAN, are we not justly led . . . to feel . . . that there must be drawn to it . . . the boundless love of our heavenly Father. . . . Or when this being, Man, . . . sinned,. . . can we not better see why it was that the Eternal Son . . . humbled Himself to take upon Him our nature . . . that Man might . . . be restored to the privileges he had lost. . . . (pp. 344–6)

Although flawed by digressions, by neglect of some opposing arguments, and at times by excessive claims, Miller's book deserved a wider readership than it secured. Part of the reason for its neglect was that whereas Whewell was well informed, systematic, highly original, and already famous when he wrote, this Scottish solicitor was only well informed and systematic. Nonetheless, the chief reason was probably the message it carried.

Among the legacies left by Proctor at his death in 1888 was an audience interested in astronomy. Of those who subsequently published works aimed at this audience, two of the most successful were the Irishmen R. S. Ball and J. E. Gore. Both enhanced the attractiveness of their books by sprinkling them with comments on the plurality of worlds. Sir Robert Stawell Ball (1840–1913) was admirably suited to succeed Proctor as the leading British astronomical popularizer. Attracted to astronomy at an early age by Ormsby Mitchel's *Orbs of Heaven,* Ball studied at Trinity College, Dublin, and then gained observational experience at Lord Rosse's observatory.[150] In 1874, he became Royal Astronomer of Ireland and Andrews Professor of Astronomy at Trinity and in 1892 Lowndean Professor of Astronomy and Geometry at Cambridge and director of the University Observatory. Ball followed in Proctor's footsteps both by publishing thirteen popular-level astronomy books and by giving widely attended lecture tours, including three in the United States and Canada. A contemporary remarked: "There is no important town in England, Scotland, Ireland or Wales . . . where Sir Robert Ball has not lectured. . . . At a very moderate estimate over one million people have heard him lecture."[151] Even in the tiny English town of Goole, over a thousand heard Ball lecture on "Other Worlds."

Ball's pluralism was influenced not only by astronomy but also by his acceptance of an evolutionary perspective. On reading Darwin's *Origin of Species* while in college, he became an "instantaneous convert"[152] and was equally enthused about the nebular hypothesis, which he endorsed and developed in his *The Earth's Beginning* (1901). Ball's evolutionary propluralist position is evident in his most detailed presentation of astronomy, his *Story of the Heavens* (1885). After urging in it that the proliferation of life on earth indicates that life is widespread elsewhere, even though "It does not seem probable that a man could live for one hour on any body . . . except the earth," Ball adds: "Could we obtain a closer view of some of the celestial bodies, we should probably find that they, too, teem with life, but with life specially adapted to the environment. Life in forms strange and weird, . . . stranger than ever Dante described, or Doré drew."[153] However, he was no extreme pluralist; lunar life he rejects (pp. 76–80), and accepting a Proctorian view of Jupiter and Saturn as generating great quantities of internal heat, he presents them as comparable to our earth "countless ages ago and hence uninhabitable." (pp. 216, 221, 494) He also mentions the question of life on Mercury and Mars, leaving it open (pp. 136, 190). He was aware of Schiaparelli's reports of Martian canals, commenting in his *In Starry Realms:* "For the present, all we can say is, that the 'canals' present problems of a very mysterious nature, which have not yet been solved." (p. 167)

His fullest pluralist presentations occur in his 1893 *In the High Heavens* and in an 1894 paper. In the former, his perspective is indicated by his statement that although the earth has reached maturity, some planets "exhibit different phases of progress. Some will appear as worlds which are to be regarded in extreme old age, while others again seem to be in an imperfect or immature condition."[154] In this book, he also argues against lunar life, adding the argument, probably derived from his friend G. J. Stoney, that the moon's gravitational field is too weak to retain the high-velocity molecules of most atmospheric elements (p. 131). His evolutionary approach is especially evident in his comment concerning intelligent life on Mars that although he does "not see how anyone can deny the possibility" of such life, the brief period of terrestrial intelligent life points to the conclusion that "the occupancy of any other world by intelligent beings might be only a very minute fraction in the span of the planet's history. . . . I should therefore judge that . . . the laws of probability pronounce against the supposition that there is such life there at this moment." (pp. 145–6) Jupiter's internal heat makes it too hot for life, whereas distant Neptune is too cold. The later assertion is curious, because in his 1894 paper he rejects life on all four giant outer planets on account of their supposed excessive internal heat.[155] It is also odd that this paper, aimed at surveying the pluralist question in the decades after Brewster's book (p. 147), concludes with the assertion that "the tendency of modern research has been in favour of the supposition that there may be life on some of the other globes." (p. 56) What is strange is that although (as he notes) spectrum analysis had given indications that other planets contain chemicals necessary for life on earth (p. 149), other evidences had gone against Brewster's belief in life on the sun, moon, and giant outer planets, as well as leaving Mercury and Venus as undecidable cases (pp. 152–4). Even Mars was beset with difficulties, as Ball admits. The significance of Ball is that he was a successor not only to Proctor's audience but also to his message. Somewhat more cautious than Proctor and even more sparing in religious remarks, Ball with his impressive credentials championed Proctor's evolving pluralist universe as well as his internally heated and uninhabitable (as yet) Jupiter and Saturn.

Although John Ellard Gore (1845–1910) was, like Ball, born in Ireland and educated at Trinity College, Dublin, his early career was quite different. In 1868, Gore left Ireland to work on engineering projects in India, returning in the late 1870s with a modest pension and such intense interest in astronomy that he devoted the last three decades of his life to expounding it in a dozen books.[156] Gore, whose translation of Flammarion's *Astronomie populaire* appeared in 1894, also influenced the pluralist debate by bringing out in the same year his *Worlds of Space* (London), in which he included as the lead items three articles he had published on life beyond the earth. Among a number of fascinating features in these essays, the most striking is the case Gore makes for the possibility of life on Mercury. Accepting Schiaparelli's 1889 report of a rotation period for Mercury equal to its period of revolution, and adapting to Mercury John Herschel's suggestion that life might be possible on the moon near the intersection of its dark and illuminated sections, Gore suggests that Mercury may be able to support life in its comparable regions (pp. 7–8). In doing this, Gore had to pass over Zöllner's evidence against an atmosphere on Mercury, but having accepted Herschel's twilight zone habitability, this may have caused him no difficulty. Venus, he suggests, if inhabited, is inhabited only in its polar regions, whereas Mars may have life in its equatorial zone. He adds, however, that "I am disposed to think . . . that if life ever existed on the surface of Mars it has now become extinct." (p. 13–14) He rules out life on the planets beyond Mars because of their supposed internal heat, Neptune being considered a possible exception. The first essay concludes with the suggestion that as the sun and planets cool down, "Venus will probably form the theatre of life. . . . Later still . . . Mercury will become cool enough – even at the centre of its sun-lit side – to be inhabited by animal life." (p. 15) His second essay picks up Proctor's suggestion that Jupiter and Saturn may form quasi suns for their satellites, making some of them habitable (pp. 29–30). Gore's third essay deals with the possibility of planets orbiting stars. Limiting his analysis to stars with spectra similar to that of the sun, he presents four arguments for such planets. First, because such stars have similar spectra, they are probably also similar in having planets. Second, the great distances of stars place their planets beyond the range of our telescopes. Third, because the stars do not shine for us, "They must . . . have been formed for some other purpose." Fourth, the nebular hypothesis makes such planets probable. These arguments possess special interest as representing the case for extrasolar planets as it was made in the 1890s by a leading expert on sidereal astronomy. After admitting that life in our solar system is probably presently confined to the earth, Gore suggests that "very possibly a planet resembling our earth may revolve round each of these distant suns. I say *a* planet, for evidently there would be only *one* distance from the central luminary . . . at which the temperature necessary for the support of life would exist . . . over the *whole* of the planet's surface." (p. 39) Using this adaptation of Whewell's habitable zone idea, Gore estimates the number of habitable planets. Taking 100,000,000 as the number of visible stars and assuming that one-tenth of these have solar-type spectra and one-tenth of the latter are of appropriate size, he reports "1,000,000 worlds in the *visible* universe fitted for the support of animal life." (p. 40) Overall, this suggests that Gore, even more than Ball, was in the tradition of Proctor.

During the final decades of the nineteenth century, the most prominent American astronomer was Simon Newcomb (1835–1909), superintendent of the Nautical Almanac Office and recipient of honorary degrees from six American and eleven European universities. First introduced to some areas of science and possibly to the pluralist debate by his father's gift of Lardner's *Popular Lectures on Science and Art,*[157] Newcomb in the course of his career contributed to that debate on many occasions. Between his 1868 refutation (noted previously) of Hansen's hypothesis and the attack (considered in the next chapter) he made in the last years of his life on Lowell's canal observations, Newcomb published at least five discussions of the overall question of extraterrestrial life. The most influential of these appeared in his book *Popular Astronomy* (1878), which went through many editions as well as translations into Czech, Dutch, German, Japanese, Norwegian, and Russian. That volume concludes with a section entitled "The Plurality of Worlds,"[158] in which he asserts: "Many thinking people regard [it] as the great ultimate object of telescope research." Nonetheless, the "extremely disappointing" conclusion he advances is that "attainment of any direct evidence of such life seems entirely hopeless. . . ." Although urging that astronomers know "no more on the subject than anyone else," he sets out such astronomical information as may "guide and limit our speculations." In regard to the nebular hypothesis, described earlier in his book as far from certain, he states that it is "quite possible that retinues of planets revolving in circular orbits may be rare exceptions . . . among the stars." He also offers an analysis of the conditions necessary for life elsewhere, urging that intelligent life is possible only in very special circumstances. The similarity of his position to Proctor's is particularly evident in his claim that because "the brevity of civilization" on earth is such that a check made every ten thousand years for intelligent life on the earth would have come out negative "a thousand times or more," we should expect a comparable result from a check of a thousand planets. Moreover, and again like Proctor, he adds that although only one in a thousand planets may be inhabited, the existence of "hundreds of millions" of planets entails that "this small fraction may be really a very large number, and . . . many may be peopled by beings much higher than ourselves. . . ." Newcomb presented basically the same position in papers published in 1897, 1902, 1904, and 1905.[159]

Charles A. Young (1834–1908) of Princeton rivaled Newcomb in prestige among American astronomers and surpassed him in the success of his astronomical texts. Among these were his *Elements of Astronomy,* *General Astronomy,* and *Lessons on Astronomy,* none of which contains a plurality of worlds section, although he treats such topics as the habitability of Mars. Young's avoidance of the extraterrestrial life issue may be attributed to his belief, expressed in an 1882 paper, that it is a "barren question."[160] Another astronomically astute contemporary of Newcomb was Charles S. Peirce (1839–1914), whose father taught astronomy at Harvard and who himself worked at the Harvard Observatory from 1868 to 1875. Peirce, who possessed the most sophisticated understanding of scientific method of any American of the period, could have made a major contribution to the pluralist debate, but he seems to have confined himself to a single comment, that occurring in an analysis of whether or not all scientific questions will ultimately be resolved. His 1885 remark is: "we take it as certain that other intellectual races exist on other planets, – if not of our solar system, then of others; and also that innumerable new intellectual races have yet to be developed; so that on the whole, it may be regarded as most certain that intellectual life in the universe will never finally cease."[161] An interesting irony is evident in Peirce's attempt to shed light on whether or not all scientific questions are answerable by assuming the answer to a question that still awaits solution. Also in 1885, Herbert Alonzo Howe (1858–1928), a University of Denver astronomer, argued that although at present the earth is probably the only inhabited planet in our system, the nebular hypothesis and evolutionary theory combine to suggest that "each of the planets may have been in the past the abode of man-like creatures, or may be in the future."[162]

Howe mentions a recent book presenting a similar point of view; this was *World Life or Comparative Geology,* by the University of Michigan geologist Alexander Winchell (1824–91). A skillful organizer sometimes described as the father of the Geological Society of America, Winchell was also a prolific writer, concerned about the relations of religion and science. In his preface, he describes his book as an attempt to blend cosmology, especially the nebular hypothesis, with geology to construct "a thoughtful view of the process of world formation, world growth and world decadence."[163] In his chapter on "Habitability of Other Worlds," Winchell surveys pluralist literature, proceeding then to claim that life may exist in very diverse forms and under conditions drastically different from those on earth. Extraterrestrials may possess additional senses or, on Neptune, pupils "as large as dinner plates." He even asks: "Why might not psychic natures be enshrined in indestructible flint and platinum?" (p. 499) Turning to the more tractable topic of whether or not other planets can be inhabited by beings similar to earthlings, he concludes that the earth is "in the middle of the habitable zone of the solar system. . . . On either side, the rigor of the physical conditions seems to proclaim our system a voiceless and lifeless desert." (p. 507) However, like Proctor, he claims that most planets and even the sun will be or have been habitable.[164] Moreover, from the premise that every star has a habitable zone, he concludes that the number of habitable worlds must be "countless."

Whereas Winchell presented his evolutionary cosmogony, including the nebular hypothesis, as reconcilable with Christianity, John Fiske (1842–1901) of Harvard, who espoused similar ideas in his *Outlines of Cosmic Philosophy Based on the Doctrine of Evolution* (1874), viewed himself as an "infidel," although not an atheist.[165] Written to champion Spencerian philosophy, Fiske's *Outlines* includes an analysis of planetary evolution, on the basis of which he asserts that "it is a fair inference from the theory of natural selection, that upon a small planet, there is likely to be a slower and less rich and varied evolution of life than upon a large planet." And he adds: "The moon would thus appear to be not merely an extinct world, but a partially aborted world, and the still smaller asteroids are perhaps totally aborted worlds."[166] Fiske was more favorably disposed to Martian life, suggesting that "the moderate temperature and habitable aspect of Mars, are alike deducible from the nebular hypothesis." (p. 384) Fiske's enthusiasm for the nebular hypothesis and evolutionary theory in general and his openness to life on Mars are of particular interest because of the possibility that these ideas may have influenced Percival Lowell, whose undergraduate years at Harvard culminated in 1876 when he contributed to his commencement exercises by an address entitled "The Nebular Hypothesis,"[167] and who later became a foremost advocate of an evolutionary cosmogony and of Martian life.

Various American biologists also became involved in the late-nineteenth-century pluralist debate. In an 1883 paper, John Pratt argued that vast quantities of energy are expended without benefiting life and that the conditions necessary for life are relatively restrictive. The basis for his antipluralist position is encapsulated in his claim that "No thinker is so loosely hinged now as to imagine life without a certain degree of heat, light, and without oxygen, hydrogen, carbon, and all the chemical elements. . . ."[168] Three months later, Charles Morris, without mentioning Pratt's views, published a paper in the *American Naturalist* suggesting that extraterrestrial life would be possible without the formation of protoplasm or even without the presence of carbon. In particular, Morris maintained: "it is possible, and even probable, that in other spheres whose atmospheric constituents may consist of simple chemical compounds analogous to, but not identical with, those of our atmosphere, a like process of decompounding and recompounding into complex and unstable molecules may be active, and organic forms exist."[169] In opposing the necessity of protoplasm, Morris was seconding arguments made by the prominent editor of the *American Naturalist,* Edward D. Cope (1840–97).[170] A major cause of this dispute was that scientists as yet possessed only a relatively elementary understanding of the biochemical features of life. This situation gradually improved; for example, in 1899, the botanist Daniel T. MacDougal (1865–1958) pointed out that because the color of a plant is a function of the wavelength at which it absorbs energy, Martian vegetation, if such exists, would not necessarily be green.[171]

In urging in an 1894 paper in the *Astronomical Society of the Pacific Publications* that "in explaining the phenomena outside of our earth, we must not assume the existence of new and unknown forces and properties of matter,"[172] the mining engineer Carl A. Stetefeldt (1838–96) was pressing for a sort of cosmic uniformitarianism, as well as laying the groundwork for his claim that although extraterrestrials may exist on planets orbiting various stars, the earth is the only inhabited planet in the solar system. His paper ends with the comment, for which he had supplied no adequate justification, that "we cannot help admiring the inductive acumen of the theologians who considered the Earth the most important of the planets, and the center of creation. Although their opinions were not based upon scientific facts, they arrived at the truth nevertheless." A few months later, a rejoinder appeared in the form of "A letter from a citizen of Mars," who argued that solar system life is confined to Mars. The earth, it is asserted, has too dense an atmosphere, too warm a climate, and too strong a gravitational field to be inhabited by anything more than "some vile crawling creatures" with at most "five or six senses." Although this interplanetary parody is presented as having come by meteorite, readers no doubt recognized its true origin from the note that it was "Communicated by M. Camille Flammarion."[173]

In 1898, Edwin C. Mason published a paper similar to Stetefeldt's, arguing that life in our solar system is confined to the earth. However, some of his arguments are weak; for example, he describes the four outermost planets as incandescent,[174] even though it had been known for at least a decade and a half that when Jupiter's satellites enter the shadow of Jupiter, they become dark.[175] Mason allows extraterrestrials on planets of other systems, but suggests that they may be similar to "ants and dragon flies." His concluding remark is also reminiscent of Stetefeldt: "far from being the merest speck in the universe, our world may be the apple of the Creator's eye, and man, in all verity, the very image of God." It is noteworthy that both Stetefeldt's and Mason's papers appeared in astronomical rather than religious journals.

In an 1896 paper, Charles Etler reflected on the comment of an unidentified scientist that "The whole vast universe . . . looks to the eye of science like a huge blunder," Etler adding that the stars are "purposelessly . . . draining their hot life-blood" into space and finally succumbing.[176] Repelled by the awful economy of this eventual heat death, he cites problems in the nebular hypothesis and in the theory that solar heat results from gradual gravitational collapse of the sun as a basis for questioning the belief "that our sun is a raging furnace of fire, and therefore totally destructive of every conceivable form of life." (p. 13) Some might have agreed that the *fin de siècle* cosmos of thermodynamics and evolutionary theory was a scene, as Etler puts it, of "the chance interaction of blind and blundering forces, in which a murderous struggle for a worthless existence is a fitting prelude to that universal suicide so confidently predicted by the prophet of the most 'advanced' type of thought" (p. 14), but few sought solace in solarians.

Whereas Etler saw the cosmos in pessimistic terms, Lester Frank Ward (1841–1913), a paleobotanist turned sociologist, urged in a cosmological chapter of his *Outlines of Sociology* that both optimistic and pessimistic views of the universe should be jettisoned in favor of "meliorism," a straightforward and empirically grounded acceptance of the nature of the universe combined with vigorous efforts to improve man and society.[177] Claiming that the universe should be seen "as in a certain sense fortuitous," he states that terrestrial life is "merely an accident, or rather the convergence of a number of accidents." And he adds that of all the solar system planets, probably only the earth is inhabited (pp. 30–1). If extraterrestrials exist on another planet, they will be "entirely different from [beings] that inhabit this earth. The plan of structure of organic forms depends entirely on the first . . . wholly fortuitous . . . initiative which first launched each type on its career." (p. 37) Although some might see in this fortuitousness grounds for pessimism, Ward claims that this "fortuity is laden with the highest hopes for mankind. . . ." (p. 35) Both the optimist and the pessimist are paralyzed for action, the one because he views the universe as benevolent, the other because he assumes malevolence, whereas the meliorist realizes that in a fortuitous universe, "Whatever 'turns up' must be turned up." (p. 36) Ward's efforts to provide a cosmological backdrop for his sociological theories also led him to become involved in the canals of Mars controversy.[178]

One of the most singular papers in the entire pluralist debate appeared in 1897, but it was not an atypical production of its author, Dr. Thomas Jefferson Jackson See (1866–1962). The prospects of the Missouri-born See must have appeared exceptional in 1893 when with a Berlin doctorate he began teaching at the University of Chicago. Two decades later, in what may be the most effusive biography ever written about an astronomer, W. L. Webb recounted See's "Unparalleled Discoveries," which in the words of Champ Clark, speaker of the U.S. House of Representatives, had established him as "The American Herschel, the greatest astronomer now living."[179] The truth was drastically different, as noted shortly after See's death by Joseph Ashbrook, who urged sympathy for the "painful fate of a man who saw his career collapse, and was trapped for more than half a century in its ruins."[180] The cause of that collapse, whether due to drugs, as F. R. Moulton speculated, or due to See's being "largely devoid of moral principles," as George D. Purinton of the University of Missouri warned President Harper of Chicago, or due to his being "mentally imbalanced," as W. A. Cogshall suggested,[181] need not be discussed here, where the intent is only to sketch one early scene in the tragedy of See's career.

See's 1897 paper, written from Lowell Observatory, where he worked from 1896 to 1899, announces remarkable results: He describes his new theory of the formation and motion of double stars in highly elliptical orbits, concludes from it that our solar system with its nearly circular orbits is "unique among the thousands of known systems,"[182] and claims that he has mathematically proved that in the double star system F. 70 Ophiuchi "there is some dark body or other cause disturbing the regularity of its elliptical motion. . . ." (p. 489) Moreover, he announces a sensational discovery:

Our observations during 1896–97 have certainly disclosed stars more difficult than any which astronomers had seen before. Among these obscure objects about half a dozen . . . seem to be dark, almost black in color, and apparently are shining by a dull reflected light. . . . If they should turn out dark bodies in fact, shining only by the reflected light of the stars around which they revolve, we should have the first case of planets – dark bodies – noticed among the fixed stars. (p. 491)

See's claims provoked a number of responses, of which three may be mentioned. His argument for a dark object in the F. 70 Ophiuchi system, which he had previously presented in the *Astronomical Journal*, was demolished by F. R. Moulton in an 1899 paper in that journal.[183] In the same year, H. H. Turner published an anonymous paper in *Observatory,* in which, without mentioning See's name, he provides a solar-sized suggestion as to the identity of his adversary. Turner proposes a law relating four variables, of which the first, *T,* is defined as the measure of "the scientific reputation that may be established by any means whatever, on the condition that it be of perfectly gaseous nature." *J* represents the "value which the investigator places upon popular newspaper or magazine notoriety," whereas J_1 is "sound reputation." Lastly, *C* is taken as the asymptotically increasing measure of egotism. The resulting equation is "$T = (J/J_1)C$."[184] The third attack came from Garrett Putnam Serviss (1851–1929), who criticized See's supposed sighting of dark objects orbiting certain stars by showing that his six objects "apparently . . . shining by a dull reflected light" and labeled "planets" would have to be far beyond planetary size to be visible at all. Even if giant Jupiter, he notes, were moved to the distance of the nearest star, it would be one hundred times too dim to be seen by the most powerful telescope.[185] In rejecting See's claims, Serviss stresses that he is not denying that planets may orbit stars.

The 1897 paper by Serviss was not his only contribution to the pluralist debate; in fact, because he became for a time the chief American successor to Proctor's audience and message, he can be seen as an ideal concluding figure for this discussion of the pluralist debate in late-nineteenth-century America. Educated in science at Cornell,[186] Serviss pursued a career as a journalist, writing hundreds of newspaper columns on astronomy. He also wrote various works of science fiction, publishing, for example, in 1898 a sequel to H. G. Well's *War of the Worlds.* Moreover, he became a successful popular lecturer on astronomy whose Proctorian approach is apparent in a number of his astronomy books, especially in his *Other Worlds* (1901). Similar in title to Proctor's *Other Worlds than Ours,* it reveals an author comparably well informed and commendably cautious in his speculations. The solar system presented in it lacks life on Mercury and the planets beyond Mars, Serviss providing a careful analysis of all the arguments, pro and con, for Venusian and Martian life. Other contributions to the pluralist debate followed, including essays on the Martian canal controversy, and in 1928 an edition of Swedenborg's *Earths in Our Solar System.* Serviss is consequently another illustration of the fact that the position developed by Proctor before his death in 1888 continued to be attactive even as the new century began.

English or American readers coming to Flammarion-influenced France found a more plentifully populated solar system. Possibly at Flammarion's urging, and certainly to that author's delight, Charles Delaunay

(1816–72), shortly before he became director of the Paris Observatory in 1870, stated:

The examination of the conditions in which the planets find themselves and of the circumstances that their surfaces present shows that the planets can be inhabited as well as the earth.

Moreover the stars . . . are nothing else but suns of diverse dimensions, and among them our sun is certainly not the most grand. It is extremely probable that each of these suns is accompanied by a retinue of planets . . . ; and it is entirely natural to admit that these planets can be inhabited. . . .[187]

Ferdinand Hoefer (1811–78) was no less supportive of pluralism when in 1873 he surveyed its history in the final chapter of his *Histoire de l'astronomie* (Paris). However, his mistaken claim (p. 621) that Huygens's *Cosmotheoros* (1698) "gave birth" to Fontenelle's *Entretiens* (1686) suggests that he was more concerned to defend that doctrine than to delineate its history. The intensity of his attachment to pluralism is also evident when after quoting pluralist passages from Lambert and Kant and listing some books such as Flammarion's *La pluralité,* he concludes his book by stating:

In rendering account of [Flammarion's] volume (in *Cosmos,* 1864) we said: "The sky itself seems to solicit man no longer to believe himself the alpha and the omega of creation. The light of astronomy, said Kepler, comes to us from Mars. It was, in effect, by the observations of this planet that the immortal astronomer attained . . . the laws which led Newton to the idea of universal gravitation. Ah well! The observation of the spots of Mars will serve perhaps one day as point of departure for the integration of vital force in the infinite continuity." Our prediction appears to be on the way to accomplishment. (p. 624)

Although from the perspective of the present Hoefer must be pronounced a poor prophet, in the short run he may have appeared far otherwise: Just four years after his book, Schiaparelli announced his "discovery" of the Martian "canali."

Hoefer's book was far less widely read than the popularizations of astronomy published by Amédée Guillemin (1826–93), whose *Le ciel* (1864) had gone through five French and seven English editions by 1877. Guillemin's propluralist sentiments are evident in the 1872 English edition, where without much discussion he accepts life on Mercury, Mars, Jupiter, and Saturn.[188] The question of solar (p. 53) and lunar (p. 145) life he leaves largely open, citing Arago in support of the former, but adding that such questions "will remain eternally . . . in the domain of the probable." (p. 53) This claim contrasts with Guillemin's assertion in his *Le soleil* (1869) that "Unless we fall back . . . to the superstitious ravings of times gone by, and believe in the existence of certain imaginary animals capable of living in fire, we cannot do otherwise than to consider the Sun as a globe upon or in which life is absolutely impossible."[189] In the same work he rebukes the French astronomer E. Liais for his "gratuitious hypothesis" of "a low temperature at the surface of the solar globe" (p. 295) and complains about those pluralists "who will, at any risk, people both the largest and the smallest celestial globes . . . comets and nebulae, sun and planets by imagining that matter in these distant regions possesses different properties to those which it reveals to us in this world. . . ." (p. 296) This is much closer to the Guillemin who in his *Les comètes* (1875) excoriated Fontenelle, Lambert, and Andrew Oliver as writers of "physical romances" and as "partisans of a preconceived idea of the habitability" of comets,[190] and who in 1892 (as noted earlier) faulted those proposing that signals be sent to the moon or Mars. Guillemin's last book carried a pluralist title, *Autres mondes,* but its discussion of extraterrestrial life in its two concluding chapters was above all cautious. In his chapter on Mars, he accepts the Martian canals, suggesting that they were constructed by inhabitants "much more advanced than us in technical and industrial procedures."[191] His final chapter, although claiming that the probability of life elsewhere "is equivalent to a certitude" (p. 261), avoids attributing life to any planet or satellite of our system except earth and Mars.

The enthusiasm for pluralism manifested by many French intellectuals, including the religious writers discussed earlier in this chapter, makes all the more striking the fact that in 1874 an antipluralist came forth from among the most famous French astronomers of the period. This was Hervé Faye (1814–1902) who had recently succeeded Delaunay as professor of astronomy at the École polytechnique and who in 1876 would become president of the Bureau des longitudes. In the 1874 *Annuaire* of the latter institution, Faye published a monograph on the physical constitution of the sun, the last section of which consists of an analysis of the conditions requisite for extraterrestrial life. After urging that we should view the "conditions of organic existence on our globe as entirely applicable to other globes . . . ,"[192] Faye stresses the crucial importance of thermal considerations for life. He maintains that such considerations cast serious doubt on Thomson's theory of the transmission of life by meteorites and also place in jeopardy the possibility of inhabited planets orbiting many types of stars, including variables, small stars, stars in closely packed clusters, and those lacking certain radiations. Pointing out that life on a celestial body requires the simultaneous presence of a number of specific geological, atmospheric, and climatic conditions, as well as carbon and oxygen, he concludes that in our solar system only earth and possibly Mars and Venus can be inhabited (pp. 485–90).

In 1884, Faye proposed a new form of the nebular hypothesis in his *Sur l'origine du monde,* in the last chapter of which he again analyses the requirements for life on other worlds. His version of the nebular hypothesis entailed not only condensation of the solar system from meteorites and the idea that the outer planets were formed last but also the conclusion that the formation of solar systems comparable to our own must be a rare event. In regard to the last point, which undercut a central tenet of the pluralists' position, he claims that among the primitive chaotic collections of materials, some "have led to an isolated star, without planets circulating around it; others to a central star encompassed by very small bodies circulating in every sense as our comets; others to double or triple stars in excentric movements; others, finally, but in very particular cases, have led to a star encompassed by planets moving in orbits almost circular."[193] Among the numerous competing variations of the nebular hypothesis, Faye's remained for some years among the more prominent and as such made difficulties for pluralists. Faye's analysis also includes a reassertion of one of his earlier arguments:

> One sees . . . how many are the conditions, simultaneously multiple and delicate, of life. . . . If it were possible to make a complete enumeration of these conditions which, in the majority of cases, are independent of each other, one would see that there are indeed few chances that they would be found united on any globe. Nature has consequently had to form a great number of worlds for that one habitable milieu to be produced, here or there, by a fortunate concourse of favorable circumstances. (p. 305)

Faye's antipluralist arguments attracted substantial attention, Courbet, de Kirwan, and Ortolan being among those who discussed them. Nonetheless, in the French phase of the pluralist debate, the influence of Flammarion, not Faye, was paramount.

A major source of Flammarion's influence was *Astronomie*, the journal that he founded in 1882 and that functioned as a forum for pluralists. The first volume, for example, contains a paper by the scientist and scientific editor Dr. Louis Olivier (1854–1910), who presents evidences of the adaptability of life to the depths of the oceans and to regions of attenuated atmosphere, as well as the the changing geological conditions that arose in the evolution of the earth. Olivier's aim is to argue that the laws of evolution should apply elsewhere in the universe, especially if Thomson's theory of the interplanetary transmission of life is accepted, and thus to imply that extraterrestrial life can develop on diverse planets.[194] The adaptability of life to a variety of environments is also stressed in an 1891 essay in *Astronomie* by Dr. Julius Scheiner (1858–1913), a prominent Potsdam astrophysicist, who first published the essay in *Himmel und Erde*. Flammarion's fondness for it led him to publish it in both French and English, as well as to annotate it so as to enhance its pluralist message.[195] After a historical introduction, Scheiner asserts that three hypotheses have been proposed to explain the origin of terrestrial life: (1) a special act of creation, (2) spontaneous generation, and (3) importation from space. Urging that observation does not permit a decision among these hypotheses, Scheiner states that the second and third entail extra-terrestrial life, whereas the first leaves the question open (p. 218). Turning directly to that issue, he asserts that "three conditions ought to be held as essential: water, an atmosphere containing oxygen and carbonic acid, and a temperature . . ." between 0°C and 50°C.[196] These necessary conditions he mistakes for jointly sufficient conditions and urges the habitability of Mercury, Venus, and Mars, and possibly Jupiter, Saturn, and Uranus. On behalf of life on Mercury, Scheiner cites Schiaparelli's report that Mercury keeps the same side toward the sun, thus providing a small habitable zone at the intersection of its dark and bright regions. He rejects lunar life, Flammarion rescuing it in a footnote. Silicon, Scheiner states, is similar to carbon and may fulfill its functions on some planets. Finally, he claims that because the sun is orbited by at least three inhabited planets, each star should have at least one, making the inhabited planets as numerous as the stars.[197]

Flammarion himself contributed essays to almost every issue of *Astronomie*. Most of these are not specifically on pluralism, but many treat Mars or other matters relating to what his 1892 "Hommes et femmes planétaires" calls "the supreme question posed to the human spirit by the spectacle of the universe [i.e.] are other worlds inhabited, and if they are, do their citizens resemble us?"[198] Richer in exhortation than information, this paper warns against the mistake of a fish who believes life possible only in its own environment. Oxygen, Flammarion asserts, is not only not essential to life, it is destructive of anaerobic organisms. Scheiner's suggestion of silicon-based life is also advocated by the French astronomer, who urges that extraterrestrials may not resemble men and women at all. They may, for example, have additional senses – electric, magnetic, infrared, and so forth – or other sexes, or even be incombustible and/or immortal (pp. 247–9).

Written three decades after his *La pluralité,* this paper shows that Flammarion's fervor for pluralism had not faded in the face of the tendency at century's end to reject the robust pluralism of midcentury. This tendency, less typical of France than of the English-speaking countries, nonetheless was making inroads among the French. For example, Gaspard Bovier-Lapierre left the question of extraterrestrial life open in his *Astronomie pour tous* (1891), quoting Faye on one side and Lescoeur and Secchi on the other.[199] Moreover, Guillemin's gradual abandonment of the extreme pluralism of his early writings and the antipluralism of Faye, de Kirwan, and others show that even Flammarion could not prevent the spread of an attenuated pluralism.

Among Italian scientific writers, the pluralist position was no doubt treated with the respect due a doctrine advocated by the two leading

Italian astronomers of the latter half of the nineteenth century. The confident pluralism of Angelo Secchi (discussed previously) was still fresh in the minds of his contemporaries when just a year before Secchi's death in 1878, Giovanni Schiaparelli announced his discovery of the Martian "canali." Schiaparelli's extensive pluralist writings, which centered on but were not confined to Martian life, are most conveniently discussed in the next chapter.

In Germany, many forms of pluralism were debated, as was evident in the discussion of Rev. Joseph Pohle and his fellow German contributors to the pluralist debate. Because of the extent and especially the message of Pohle's publications, he may be described as a "German Proctor." Was there also a "German Flammarion"? Desiderius Papp answered that question affirmatively when in a 1931 book, after mentioning some pluralist publications by Flammarion, he commented: "M. Wilhelm Meyer, the 'German Flammarion,' travels similar paths of thought in his conversations on other earth-stars [in his] *Die Königen des Tages und ihre Familie* Wien, 1885."[200] Max Wilhelm Meyer (1853–1910) advocated pluralism not only in that book but also in various other publications, most notably his *Bewohnte Welten* (Leipzig, 1909). Moreover, he was the founder of the Urania Gesellschaft, a society in Berlin that operated an observatory and provided public lectures on science, and the editor of its journal, *Himmel und Erde*.[201] The important role of that journal in the pluralist debate is indicated by the fact that Scheiner's 1891 paper first appeared in it. Moreover, it carried an 1889 paper by Schiaparelli as well as 1893 and 1896 papers by Meyer himself written in support of life on Mars. Meyer is also known for a theory of the origin of terrestrial life, according to which a planet exploded, propelling organisms from its oceans into space, where they became dormant until falling to the earth.[202]

The pluralism of Meyer may be compared to the Proctorian position presented by Wilhelm Schur (1849–1901), director of the Göttingen Observatory, in an 1899 paper prepared for the general public. Schur begins by reviewing the claims made by Fontenelle for life on all the planets.[203] Readers must have been struck by the contrast between the lighthearted, imaginative, and optimistic assessment of that seventeenth-century amateur and the sober and pessimistic analysis of this late-nineteenth-century professional. Schur argues against life on both inner planets and all the giant outer planets, Jupiter, for example, being presented as possibly glowing and certainly without a solid crust. Denying life to our moon, Schur nonetheless holds out for the possibility of life on some of Jupiter's satellites. Mars he views as the most favorable locale for life, although he cites the width of its canals as evidence of their formation by natural processes. Spectral analysis, Schur suggests, rules out life on the sun and stars, but he concludes by noting that planets of other suns may in some cases be abodes of life.

Schur's analysis, when considered in conjunction with those of Du Prel, Zöckler, and Pohle, points to the acceptance among many late-nineteenth-century Germans of a limited or "Proctorian pluralism." The shift to this form of pluralism from that dominant in the first half of the century is more noticeable in Germany than in France, but is above all evident in Britain and the United States. The reasons for this change are numerous. Among those of astronomical nature is that planets beyond the solar system had been placed in jeopardy both by objections to the nebular hypothesis and by the evidence against the earlier view of nebulae as island universes. Even more important, claims for widespread life in our solar system were increasingly perceived as problematic. This does not, however, represent the whole story; as the confident pluralism of midcentury came gradually to be seen as excessive, a dramatically new pluralism centered on the planet Mars kept, in Pohle's words, "the astronomical world continuously in agitation and out of breath."[204] Originating in Italy with Schiaparelli, spreading most rapidly in Flammarion's France and in the United States, where Lowell championed it, this new pluralism claimed observational evidence of an advanced civilization on Mars. To that story we now turn.

Figure 10.1 *Giovanni Schiaparelli.*

Figure 10.2. *Camille Flammarion (left) and Percival Lowell (right) at Flammarion's Juvisy Observatory in* 1908. *(Courtesy of the Lowell Observatory)*

Figure 10.3. Edward Walter Maunder.

Figure 10.4. William Wallace Campbell.

Figure 10.5. Eugène Michael Antoniadi. (Courtesy of the Royal Astronomical Society)

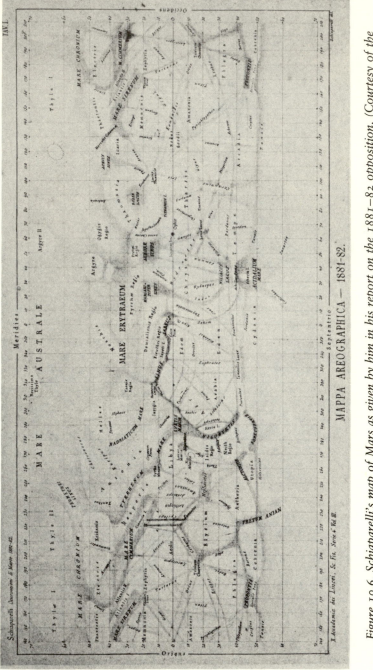
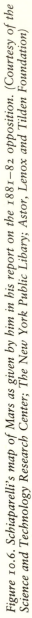

Figure 10.6. Schiaparelli's map of Mars as given by him in his report on the 1881–82 opposition. (Courtesy of the Science and Technology Research Center; The New York Public Library; Astor, Lenox and Tilden Foundation)

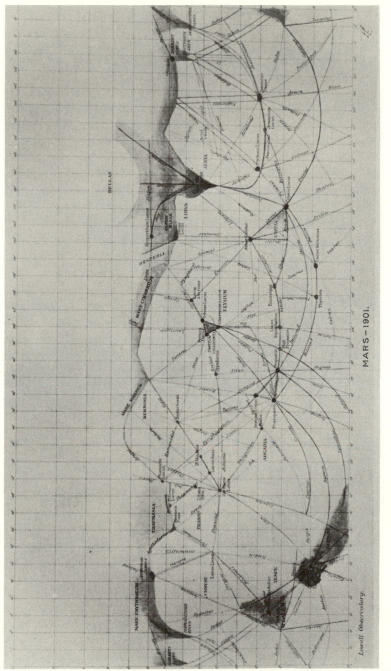

Figure 10.7. Lowell's map of Mars as drawn in 1901 and published in Annals of the Lowell Observatory, 3 (1905), 144.

479

10

The battle over the planet of war

1. The commencement of the canal controversy: enter
Giovanni Schiaparelli: "a gazer gifted with that supreme power of
brain-directed vision"

The continuing erosion near the end of the nineteenth century of the confidence, widespread at midcentury, in the existence of intelligent life on most or all solar system planets was offset in the hearts of some by an increasing conviction that compelling evidence for life on Mars was becoming available. Although so small in apparent diameter, even when nearest the earth, that a teacup half a mile distant from the eye will cover it, the planet Mars, named for the god of war, became in 1877 the center of an international controversy. Dozens of books, hundreds of telescopes, thousands of articles, and millions of people focused on whether or not intelligent beings, possibly desperately trying to survive, conceivably seeking to signal us, roamed its surface.

The history of ideas concerning Mars during the 1877-to-1910 period, which is the subject of the present chapter, exhibits a number of periodicities determined by the relations of the Martian motions to those of the earth. The most important of these periodicities is that approximately every 780 days Mars comes especially near our planet, making optimal observation possible. The effect is dramatic: At its nearest, Mars subtends an angle of more than 25 seconds, whereas at its farthest its diameter subtends less than 4 seconds. Superimposed on this is another periodicity of approximately 15 years, arising from the relatively large ellipticity of the Martian orbit. Because of this second factor, Mars, when at opposition (i.e., when situated on the opposite side of the earth from the sun), may be nearer to us than 35 million miles or more distant than 63 million miles. Other factors also influence observation of Mars, such as whether the planet is high or low in the sky when at opposition and whether its north or south pole is visible, but the crucial factors are the first two mentioned. They combined to make the oppositions of 1877, 1892, and 1909 the finest in the interval of interest for this study.

Although before 1877 over a thousand drawings (mostly unpublished) of Mars had been made,[1] astronomers had not as yet reached a consensus concerning the form of the planet's surface. The lack of agreement is illustrated by the fact that in 1877 a leading British astronomical journal carried a paper suggesting that the redness of Mars may be due to the planet glowing with a "dull red heat."[2] As the excellent 1877 opposition approached, many astronomers waited with high expectations, but none foresaw the remarkable results soon reported from Washington, D.C., and Milan, Italy.

In Washington, Asaph Hall (1829–1907) used the giant 26-inch-aperture refractor recently installed at the Naval Observatory to show that Tennyson's "Moonless Mars" is orbited by two miniature moons. Whether or not Hall was motivated to search for Martian moons by the repeated speculations of pluralists that Mars should have one or more moons[3] must remain a matter of conjecture or future research, but that Hall had pluralist inclinations is doubly evident from his paper announcing that discovery. After estimating the diameter of the outer satellite as 0.031 second, he points out that this "would correspond to a distance of 187 feet on the Moon's surface." He then adds that from this "it appears that the proposition of a German astronomer to establish on the plains of Siberia a system of fire signals for communicating with the inhabitants of the Moon is by no means a chimerical project."[4] He also devotes a long paragraph to observing how the moons must appear to the "Martian astronomers." Various attempts have been made to explain why these moons were not discovered earlier, the most bizarre of these being one put forward around 1960. The Russian astrophysicist I. S. Shklovskii proposed in 1959, on the basis of indirect evidence, that the Martian moons may be artificial satellites. Another Russian author then seized on this idea to explain the failure of pre-1877 searches by suggesting that the moons had been launched into orbit shortly before 1877.[5]

The second remarkable 1877 report came from the Brera Observatory in Milan, where Giovanni Schiaparelli (1835–1910) (Figure 10.1), despite having access to a refractor of only eight inches aperture, announced his discovery of an extensive system of "canali" on Mars. Before discussing his report, it will be useful to survey Schiaparelli's career, which by the time of his death in 1910 was so rich in accomplishments that astronomers praised him in obituary notices as "Italy's greatest scientist" and as "not only the greatest astronomer of Italy, but one of the greatest astronomers of our times."[6] Even the two leading critics of his canal observations, E. W. Maunder and E. M. Antoniadi, wrote of him in such terms as "the most distinguished astronomer on the continent of Europe" and "the premier planetary astronomer of modern times."[7] He was held in high esteem not only for the magnitude of his researches (over 250 publications) but also for the cautious manner in which he presented his conclusions. For example, one author stated concerning his canal observations: "Schiaparelli always guarded himself carefully from countenancing any theoretical conclusions as to the nature of these markings."[8] In this regard he was seen as practicing a professionalism rarely ascribed to Flammarion or Lowell, who also wrote on the canals. In the materials that follow, however, various facts will be brought forward to suggest that the great Italian astronomer was not quite so detached from the pluralist position as has been believed, at least in the English-speaking world.

After training at Turin University in civil engineering, Schiaparelli studied astronomy at Berlin with J. F. Encke and at Pulkowa with Otto Struve. In 1860, he joined the staff of the Brera Observatory, becoming director in 1862. The origins of Schiaparelli's pluralist convictions are unknown; possibly they stem from Encke, who had strongly endorsed the pluralist position,[9] from the writings of Boscovich, who had been director of the Brera Observatory in the eighteenth century, or from Angelo Secchi, whom Schiaparelli quoted in support of pluralism in an 1889 essay on the inhabitants of other worlds.[10] Despite the modest instrumentation available to him, Schiaparelli soon established a solid reputation, especially for his studies of the relations of meteors and comets, for which he was awarded in 1872 a gold medal by England's Royal Astronomical Society. Ever anxious to improve the telescopic equipment at his observatory, Schiaparelli in 1878 found a way to do this, the details of which came to light only in 1963 when a large number of his letters were first published. Among these is a letter to Otto Struve recounting that in April 1878 he went to Rome to make a presentation of his Martian discoveries at a session of the Academy of the Lynxes attended by various government officials. Seizing this opportunity, Schiaparelli stressed that with a telescope comparable to that in Washington, he could attain even better results. His lecture was so warmly received that a few days later he was asked to appear at the Quirinal Palace, where, as he told Struve, he explained to the king and queen "that Mars appears to be a world little different from our own; and by employing a little the Flammarionesque style, I managed the affair rather well."[11] With the support of the Academy of the Lynxes, Schiaparelli's request for a large telescope came before the Chamber of Deputies, where, after he had addressed the members, it was overwhelmingly endorsed, with the Senate and king subsequently adding their approval. As a result of this "very exciting phantasmagory," as Schiaparelli described it, his observatory in 1886 acquired an 18-inch refractor. The enthusiasm he had generated for his request was no doubt due to many factors, including government concern for pure science, nationalistic sentiments, and Schiaparelli's skill in adapting the "Flammarionesque style" so as to highlight the pluralist aspects of his observations of Mars.

The high esteem in which Schiaparelli was held in the 1890s was due not only to his meteor and Martian studies but also to his determination during the 1880s of the rotational periods of Mercury and Venus. Although both his determinations differed markedly from earlier results, they were generally accepted, until shown in the mid-1960s to be seriously in error. Let us examine his study of Mercury (the case of Venus being largely analogous). This will lay the groundwork for the conjecture that, in a manner not previously recognized, Schiaparelli's determinations of the rotation periods of Mercury and Venus were, like his Mars observations, involved with extraterrestrial life ideas. The small size of Mercury and its closeness to the sun were widely recognized in the nineteenth century as inhibiting observation of its surface features and consequently as impeding attempts to determine its rotation period. Early in the century, Schröter had made observations from which he inferred that Mercury's rotation period was slightly over 24 hours. Although his figure may have been influenced by terrestrial analogies, if not pluralist convictions, it was still generally accepted in the 1880s when Schiaparelli took up the matter. On the basis of about 150 mainly daytime drawings made during that decade, Schiaparelli concluded that although Mercury presents the same features over two, three, or four days, its rotation "lasts not a day . . . but rather occurs more slowly."[12] This was a correct conclusion, but Schiaparelli went on to assert that Mercury's rotation period is equal to its period of revolution, which is approximately 88 days. This meant that Mercury, like the moon, always presents the same side to its primary. G. H. Darwin had shortly before explained how tidal forces could account for the moon's behaving in this way, and a similar argument might apply for Mercury. Moreover, Schiaparelli attributed a 47° libration movement to Mercury; that is, Mercury, again like the moon, does not keep precisely the same face toward its primary (the sun), but rocks back and forth through an angle of 47° to each side. Observations made by an array of later astronomers, such as Lowell, Jarry-Desloges, Antoniadi, McEwen, Lyot, and Dollfus, were in agreement with the rotation period assigned by Schiaparelli,[13] but in 1965, radar techniques revealed that Mercury's rotation period is about 59 days.[14] Schiaparelli's claim that the rotation period of Venus is also equal to its period of revolution was also shown to be erroneous.[15]

How did Schiaparelli slip into such a sizable error in regard to Mercury? Part of the explanation is linked to the facts (1) that Mercury's rotation period (58.65 days) is exactly equal to two-thirds of its sidereal period (87.97 days) and (2) that six of its rotation periods equal nearly three synodic periods (115.88 days).[16] Because of these factors, Mercury sometimes presents features reconcilable with an 88-day rotation period. In fact, some of the observations of the astronomers who later claimed to have confirmed Schiaparelli's rotation period can be explained in this manner, but this is less true of Schiaparelli's observations.[17] Part of the explanation of Schiaparelli's error, it may be suggested, emerges from the perspective provided by this history, which contains numerous instances of astronomers being misled not only in theories but also in observations by their pluralist proclivities. Evidence exists that Schiaparelli's observations of Mercury may have been influenced in this regard in two ways. The first of these is that he dismissed some discordant observations by attributing them to Mercury having a cloudy atmosphere.[18] That he described the presence of an atmosphere on Mercury as "almost a certainty" [19] is surprising, not only because Mercury lacks an atmosphere but also because Zöllner's analysis of Mercury's albedo more than a decade earlier had made the existence of such an atmosphere improbable.

Schiaparelli may have been misled by his pluralist inclinations in a second way. He must have realized that his 88-day rotation period was the only period that would save part of Mercury from repeated fluctuations between the frigid darkness of prolonged night and a solar radiation seven times more intense than that reaching earth. In fact, his rotation period, when combined with his 47° libration, provides, as he notes, one-fourth of Mercury's surface (the libration zone) with a sun that rises and sets every 88 days on the same horizon, but never inflicts the intensity of vertical or nearly vertical rays. Although he does not press this point, he says enough to make this feature obvious. Having implied that Mercury has a region of moderate seasons, he proceeds to provide seas for the planet, arranged in a manner that earthlings "might envy." His argument, which refers to certain dark spots he reported on the planet, is yet another illustration of the penchant of pluralists for double negatives: "if any one, taking into account the fact that there exists an atmosphere upon Mercury capable of condensation and perhaps also of precipitation, should hold the opinion that there was something in those dark spots analogous to our seas, I do not think that a conclusive argument to the contrary could be advanced."[20] Later he adds that Mercury is a world that

. . . receives light and heat from the sun, not only in a greater amount but in a different manner than the earth; and where life, if so be life exists there, finds conditions so different from those to which we are accustomed that we can scarcely imagine them. The perpetual presence of the sun almost vertically above certain regions, and its perpetual absence from other regions . . . should produce an atmospheric circulation which is at the same time stronger, more rapid, and more regular than that which sows the elements of life on the earth; and that on this account it may come about that an equilibrium of temperature is produced quite as complete as ours, and possibly even more so. (p. 141)

That Schiaparelli's rotation period did not simply emerge from his observations over the 1880s is further suggested by the fact that as early as 1882 he had revealed his 88-day period to the astronomer François Terby.[21]

These ideas, which Schiaparelli also presented before the king and queen, demonstrate his disposition toward life on Mercury. The conjecture that this led him to give primacy to some and to downplay other observations discordant with an 88-day period so as to attain an attractive and self-consistent view of Mercury by no means implies that in this difficult determination he was in any way deceitful, nor should it detract from the credit due him for his discovery that Mercury rotates more slowly than others had believed. To put this conjecture differently, it may be suggested that a double meaning, unintended by Ellen M. Clerke, may be found in her description of Schiaparelli as "a gazer gifted with that supreme power of brain-directed vision. . . ."[22]

2. Mars oppositions from 1877 to 1884: Schiaparelli's "curious drawings" and the responses of Green and Maunder

In 1878, Schiaparelli presented his "canali" in a nearly book-length memoir[23] that begins with a new determination of the axis of rotation of Mars. This was based on his micrometric determinations of the locations of sixty-two features on Mars's surface. These also served as the foundation for his first map of Mars for which he developed a new nomenclature based on ancient geography and mythology. His nomenclature created controversy, because in 1877 most astronomers used a system employing the names of prominent observers of Mars that R. A. Proctor had introduced in 1867, that Nathaniel Green had expanded, and that Camille Flammarion had revised so as to reduce Proctor's excessive use of names of British astronomers.[24] For example, what Proctor and Green called "Lockyer Land," Flammarion named "Terre de Secchi" and Schiaparelli "Hellas"; what the two Englishmen designated "De La Rue Ocean" was for Flammarion "Océan Kepler" and for Schiaparelli "Mare Erythraeum" (Indian Ocean); and the "sea" that Proctor and Flammarion named for Phillips, Green named after Maunder, whereas Schiaparelli proposed "Promethei Sinus" (Bay of Prometheus). Although in his memoir Schiaparelli cautions that use of such terms as "mare" and "canale" should not be allowed to prejudice discussions of the nature of the surface features, and although he urges that his nomenclature should be considered only as "a simple artifice to aid the memory and to abbreviate the descriptions" (p. 60), it is clear that in the ensuing controversy too little attention went to whether or not such geographical terms as "sea" and "continent" were justified. Schiaparelli did not himself introduce the term "canale"; rather, he took it from an 1859 publication by Angelo Secchi.[25] Moreover, the Italian word "canale," which primarily means "channel," but which can also mean "canal," was somewhat less theory-laden than its usual English translation as "canal," the difference being that channels are natural features, whereas canals are constructed. Nor did Schiaparelli view himself as the discoverer of the "canali," only as discoverer of the extensive system of "canali." This point and Schiaparelli's confidence in the aquatic aspects of Mars are evident in this statement:

> One sees that the hypothesis of a marine and continental constitution of the surface of Mars is endowed with a great probability. But this becomes almost elevated to a certainty, if one succeeds in affirming in an indubitable manner the real disappearance of the eastern drain of Solis Lacus. . . . This canal, which had been seen by Maedler in 1830, by Kaiser, Lockyer, Rosse, and Lassell in 1862, by Kaiser and Dawes in 1864 had in 1877 become entirely invisible to an instrument certainly not inferior in this way . . . to any employed previously. . . . If this variation is verified in the future, I believe it will be difficult to find an interpretation more simple and more natural than that of a change in the hydraulic regime of this region, similar perhaps to that which within our memory changed in China the course of the Yellow river. (pp. 158–9)

Although much of Schiaparelli's memoir consists of discussions of technical matters unrelated to the canal system, it was above all this feature – vividly clear in his map even to those innocent of Italian – that gave substance to Flammarion's remark in an 1882 letter to Schiaparelli: "The entire world is interested in your excellent observations of Mars."[26]

Unlike some astronomers, Flammarion was not slow to accept Schiaparelli's observations nor to see in them proof of extraterrestrial life. In an 1879–80 paper, the French astronomer cited recent Mars observations as well as studies of meteorites and spectra as resolving the pluralist debate. Referring to Mars researches, although not directly to Schiaparelli, Flammarion reveals that "sea shores there, as here, receive the tribute of aquatic canals. . . ."[27] After mentioning the low gravitational force on Mars, he adds: "It is almost certain that the *inhabitants of Mars are of a different form from us, and fly in its atmosphere.*" (p. 159) He also populates the moons of Mars "with reasoning microscopic mites." (p. 159) For the 1882 volume of his journal *Astronomie*, Flammarion wrote a three-part essay on Mars in which he endorses the canals and provides a new map of Mars showing many canals.[28] A translation of a Mars paper by Schiaparelli appears in the same volume, but the Italian astronomer does not seem to have held his French contemporary in high esteem. In an 1878 letter to Otto Struve, Schiaparelli stated concerning an earlier Flammarion Mars map: "This map, as many others produced by this literary astronomer, is a corrupt imitation of previous works."[29] Likewise, Schiaparelli probably was not enthused about the 1879 book written by the German poet, philosopher, linguist, and spiritualist Jakob Heinrich Schmick (1824–1905) to introduce German readers to the new Mars. Although Schiaparelli was himself involved in spiritualism,[30] he was no doubt distressed by some of the claims made in Schmick's book, the title and message of which was *Der Planet Mars: Eine zweite Erde, nach Schiaparelli* (Leipzig).

The early reception of Schiaparelli's canal observations was neither "universal scepticism," as Lowell later claimed,[31] nor widespread acceptance. One of Schiaparelli's earliest supporters was the Belgian Mars expert François Terby (1846–1911), who in 1880 published a memoir urging that canals had been seen by Knott and Schmidt in 1862, by Secchi in 1864, by Gledhill, Lehardelay, Vogel, and Lohse in 1871, by Knobel, Lohse, and Trouvelot in 1873, and by Cruls and Niesten in 1877.[32] In discussing his observations of Mars in 1879, Louis Niesten (1844–1920) claimed to have "recognized traces corresponding to [twelve canals] and many others, often more as delimitations of large spots of different tints, than as detached lines."[33]

Among British astronomers, Schiaparelli's canal observations received mixed responses, a result influenced by the fact that the chief rival map produced from the 1877 opposition was drawn by Nathaniel E. Green (1823–99), an English artist who in 1880 was chosen to teach painting to Queen Victoria and other members of the royal family. After spending August and September of 1877 observing Mars with a 13-inch reflector on the island of Madeira, Green presented his delicately shaded map and accompanying memoir for publication by the Royal Astronomical Society (R.A.S.).[34] The English-speaking world may first have learned of Schiaparelli's Mars researches when *Nature* reported that at the April 12, 1878, R.A.S. meeting Green "read a letter from Prof. Schiaparelli on Mars . . . and showed some curious drawings."[35] Green, recognizing that his Mars differed drastically from Schiaparelli's, proposed at that meeting three alternative explanations of these differences. First, Green suggested the possibility that the "several dark channels" seen by Schiaparelli, whose observations had commenced around the time of the September 5, 1877, opposition and extended to March 1878 (i.e., later than Green's), were "great physical changes . . . which were not visible till some considerable time after the opposition." Second, artist Green proposed that Schiaparelli's "hard and sharp lines" might have resulted less from his observations than from his drawing techniques. Third, Green mentioned the possibility of "a tendency, either in the object glass, or the eyepiece, or in the eye of the professor, to elongate and develope dark points seen against a light background."[36] At the December 13, 1878, R.A.S. meeting, Green offered yet another explanation: "that the atmospheric vibrations have a tendency to cause a series of points in a line to appear connected together. . . . Green also reported that the Franco-American astronomer Étienne Trouvelot (1827–95), then at Harvard, had written him: "about the canals observed by Signor Schiaparelli, my observations give them little support." Herbert Sadler added that he doubted that Schiaparelli could observe such fine detail with a telescope with only an 8-inch aperture.[37] In a March 1880 paper, Green proposed other explanations of Schiaparelli's lines: "some . . . may be the boundaries of faint tones of shade . . . or they may be spaces between veil-like masses of atmosphere. . . ."[38]

By the 1879–80 opposition, British support for Schiaparelli was beginning to build. In a note in which Green reported that Schiaparelli had written him concerning the canals that "It is as impossible to doubt their existence, as that of the Rhine," Green added that the Irish astronomer Charles E. Burton (1846–82) had seen traces of the canals.[39] Burton's published reports of canal observations began to appear in 1880, containing such statements as "Dr. Schiaparelli has accepted the drawings of these objects made here, as representing the objects named by him as 'canals,' and most valuable support has been given to my seeing by the drawings obtained by Mr. J. L. E. Dreyer. . . ."[40] Green graciously responded to Burton's report by stating: "after Mr. Burton's exact description of these forms . . . it would be impossible to doubt their existence."[41] Burton's published map contains more than a dozen canals, and in his report on it he states that in 1879 he had "independently detected" many of the canals that had been seen earlier by Schiaparelli.[42] Terby was also emerging as a champion of the canal observations; in an 1880 letter to *Observatory,* he attacked Green's suggestion that the canal observations were due to differences of tint, adding that Trouvelot had reported seeing nine canals.[43] Thomas William Webb (1807–84), writing in late 1879, compared the Green and Schiaparelli maps, stating that Green had produced a "picture," whereas Schiaparelli's map resembled a "plan." Webb also mentioned that Schiaparelli was "inconvenienced by colour blindness."[44] When in 1881 Webb issued the fourth edition of his *Celestial Objects for Common Telescopes,* he selected Burton's map for Mars, a decision probably influenced by his belief, also expressed in that book, that life exists on Mars.[45] Burton was not the only R.A.S. member to see canals; Edward B. Knobel announced in 1882 that in 1879 he had seen some canals.[46]

Schiaparelli carefully observed Mars during the 1879–80 opposition, producing another lengthy report, as well as a map showing canals, although the sharpness at the edges of features was substantially reduced. At that opposition, he also sighted a new phenomenon that he named "gemination," which, after observing additional instances in 1882, he revealed to the public, describing it as follows: "To the right or left of a pre-existent line, without any change in its course or prior position, another line is produced equal and parallel to the first. . . . Between the lines thus generated the distance varies from 12° to 16° in circular measure (350 to 700 kilometers). . . ."[47] In a similar paper published in Flammarion's journal *Astronomie* in 1882, he states that by that year he had seen as many as sixty canals and no less than twenty geminations or doublings.[48] In that paper, he again warns against speculating about the nature of the canals and stresses that the canals and their doublings are not optical illusions, against which he took precautions. He concludes by stating: "I am absolutely sure of what I have observed."

Word of Schiaparelli's new discovery spread rapidly. T. W. Webb announced it in the London *Times* for April 10, 1882, and discussed it more fully in an article in the May 4 issue of *Nature*. A few days later, *Scientific American* noted it, citing the *London Telegraph* as its source. Webb's letter to the *Times* stimulated R. A. Proctor to write that newspaper, noting that Dawes's drawings of Mars, made in the 1860s, show some canals, but that the possibility of these being optical illusions, as Green had proposed, should not be ruled out. Proctor added that the lower gravity of Mars may lead to the inference that "engineering works on a much greater scale than any which exist on our globe have been carried on, upon the surface of Mars." This rash speculation is followed by the qualification that "It would be rash, however, at present to speculate in this way."[49] Proctor's letter was mentioned in the discussion of the canals that broke out at the April 14, 1882, R.A.S. meeting. At that session Green claimed that because individual canals were sometimes differently seen by different observers (e.g., Dawes and Schiaparelli), "great caution" should be exercised before accepting the canals. E. W. Maunder of Greenwich Observatory then commented that he had seen a number of canals, but that their positions had sometimes shifted from night to night. After some comments by other members, Maunder and Green reasserted that the "canals" may only represent differences in shading between adjacent areas.[50] On May 16, Burton, up to then a leading advocate of the canals, wrote a letter stating in part that in regard to the fainter streaks: "I strongly incline to the opinion expressed by Messrs. Green and Maunder, that they are boundaries of differently tinted districts. . . ."[51] Whether or not this signified Burton's break from the canal camp soon became irrelevant, because he died a few months later, followed in 1884 by Webb. Terby, however, remained a faithful and forceful proponent of the canals, sending off a paper in June

1882 to the R.A.S. asserting that the Dawes, Burton, and Schiaparelli drawings showed a high level of agreement in their canal positions.[52]

By the time the poor 1884 opposition had passed and the somewhat better opposition of 1886 was approaching, a number of astronomers, including Burton, Dreyer, Flammarion, Knobel, Niesten, Terby, and Trouvelot, had provided some support for Schiaparelli's canals, although not as yet for their doubling. This and the Italian astronomer's excellent reputation were probably the reasons that led Agnes Clerke in the 1885 first edition of her *Popular History of Astronomy during the Nineteenth Century* to state that Schiaparelli's canals had been "fully substantiated. . . . The 'canals' of Mars are an actually existent and permanent phenomenon."[53] However, a measure of opposition had emerged, centered in England, where Green had raised objections, some of which were developed by others. Among these persons was E. W. Maunder, who before long became the most active adversary of the canal claims.

After studies at King's College, London, Edward Walter Maunder (1851–1928) (Figure 10.3) joined the staff of Greenwich Observatory as photographic and spectroscopic assistant, remaining there for over four decades and establishing a solid reputation, especially in solar research. In addition, he edited *Observatory* from 1881 to 1887 and in 1890 founded the British Astronomical Association, serving as one of its first presidents and for a decade as editor of its journal.[54] That Maunder became a critic of the canals is doubly surprising. First, as Maunder himself stated: "I had recorded some of the markings now familiar to us as 'canals' and 'oases' even before Schiaparelli had published his results. . . ."[55] Second, in 1877, Maunder had found confirmation of Huggins's (spurious) spectroscopic detection of water vapor on Mars,[56] and he continued to accept this result, indicative of water on Mars, throughout the 1880s. A clue to the origin of Maunder's skepticisms is contained in a series of Mars articles he published in 1882 in the *Sunday Magazine,* in which, after arguing on scientific grounds against life on Mars, he concludes that

. . . it was no mere lucky chance that brought together so many qualifications, each of them essential to our welfare, in this planet of ours, whilst larger, brighter orbs possess no such fitness for our use; and as we look, on the one hand, at the indescribable magnificence of the starry heavens, suns without number, planets untold . . . , and on the other at the infinite care with which this earth has been fashioned . . . , we are compelled . . . to unite in the outburst of the Psalmist . . . and exclaim –

> "When I consider Thy heavens, the works of Thy fingers,
> The moon and the stars, which Thou has ordained,
> What is man, that Thou art mindful of him,
> And the son of Man, that Thou visitest him"[57]

Thirty years later, in a paper for the Victoria Institute, Maunder again assigned man a special place in the universe.[58] This suggests that Maunder, the son of a Wesleyan minister and himself a very active member of a small pentecostal and adventist denomination known as the (Irvingite) Catholic Apostolic church,[59] may have been led, as Whewell before him, to question the pluralist position on the basis of the "Christian" objection to it. He never in print argued against pluralism on this basis, no doubt realizing that this would be seen as sectarian and as discrediting his numerous scientific arguments. In fact, in his *Are the Planets Inhabited?* (1913), he explicitly urges in his first chapter that although religious factors were involved in the Whewell-Brewster debate, he views religion and the doctrine of a plurality of worlds as irrelevant to each other. Nonetheless, in the concluding chapter of this antipluralist book, he suggests that the conditions of life are so complex that life on other planets must be very rare; in fact, "it is even conceivable that this Earth of ours may be unique." He then asks whether astronomy can say anything about life not subject to these conditions. The answer he offers is that although astronomy must be mute, the Christian, by accepting Christ's message, may proclaim: "I LOOK FOR THE RESURRECTION OF THE DEAD AND THE LIFE OF THE WORLD TO COME."[60]

3. The oppositions from 1886 to 1892: Schiaparelli supported and Mars masked with "grotesque polygonations and geminations"

Although during the relatively poor 1886 opposition Schiaparelli sighted only a single gemination,[61] that year provided him with some sources of satisfaction, not least the erection of his new 18-inch refractor. Moreover, he learned in that year that Henri Perrotin (1845–1904), working with Louis Thollon (1829–87) at the Nice Observatory with a 15-inch refractor, had observed many canals and some doubles in positions agreeing with Schiaparelli's 1882 map.[62] A delighted Schiaparelli wrote Terby of this "grand news," adding: "I attach very great importance to this confirmation for people will hereafter cease to scoff at me in certain places. The geminations are very difficult to explain, but it is indeed necessary to admit their existence."[63] The author of an article in the June 3, 1886, issue of *Nature* was no less impressed, asserting that Perrotin's observations had "fully demonstrated" Schiaparelli's claims.[64] The English astronomer William F. Denning (1848–1931), writing in the same issue of *Nature,* presented his recent Mars observations as also supportive of Schiaparelli.[65] In addition, the American astronomer Herbert C. Wilson (1858–1940) succeeded in seeing some canals in 1886.[66] That the popular press was becoming more interested in Mars is illustrated by an

article in *Chambers's Journal*, the anonymous author of which rejects life
on all the other planets, but concludes for "large engineering works" on
Mars and, without mentioning Schiaparelli, suggests that "perhaps there
may be some truth in the story of the Italian astronomer who says he has
lately detected lights on the planet moving about in such a way as seems
to indicate a deliberate intention to open communication with the
earth."[67]

Schiaparelli awaited the 1888 opposition with heightened hopes; not
only was his new telescope now operational, but also Perrotin had in
1887 acquired a 30-inch telescope, the largest refractor in Europe, exceeded in size only by the 36-inch Lick refractor, which was completed in
1887 amidst claims that it would resolve the canal controversy.[68] Perrotin, for example, created a sensation when in May 1888 he announced
concerning the Martian "continent" called "Libya" by Schiaparelli:
"Clearly visible two years ago, it no longer exists today. The nearby sea
(if sea it is) has totally inundated it."[69] The disappearance of Martian
Libya, larger than France itself, was only one of a number of spectacular
changes he reported, including, later that year, the partial reappearance
of Libya.[70] The disappearance of Libya drew widespread press coverage,
with some authors claiming that Schiaparelli and Terby agreed with
Perrotin,[71] but some saying that they did not.[72] Schiaparelli's detailed
memoir on the 1888 opposition did not appear until 1899, but in late
1888 he published a long, semipopular account in German of his Martian studies, in which he stated that he had seen Libya gradually dimin-
ishing in the early 1880s, and for the most part supported Perrotin.[73]
Although Schiaparelli had specifically avoided any tendency to "*flamma-
rionize*"[74] the issues in this presentation, Flammarion decided to publish
a translation of it in *Astronomie* for 1889.
    Disconcerting news came from the Naval Observatory and from Lick
Observatory. After Asaph Hall reported that his June 1888 efforts to
observe the canals with the large Naval Observatory telescope had been
"without success,"[75] Edward S. Holden, the Lick director, published
drawings of Mars he had made in the late 1870s while at the Naval
Observatory, which drawings showed a number of canals.[76] Possibly
these papers contributed to the rumor then circulating that the lens of the
Naval Observatory telescope had deteriorated.[77] In a September 1888
report from Lick, Holden discussed forty-two Martian drawings he and
his colleagues James E. Keeler and John M. Schaeberle had made. Although they did not commence observing until July 16, they observed
some canals, but no geminations or changes in Libya.[78]
    Flammarion was deeply moved by the Lick observations: "These ob-
servations have made us despair. The more one dedicates time, study,

and care to the analysis of the numerous and varied observations made of this mysterious planet, the more one is obstructed from arriving at a definite opinion." And he adds:

> The most powerful telescope in the world has been applied to the study of the globe of Mars. . . . Yet it is necessary to admit that the drawings made by Messrs. Holden, Schaeberle, and Keeler . . . correspond neither with those of Monsieur Schiaparelli of Milan, nor with those of Monsieur Perrotin of Nice. Has each astronomer, then, in physical as in moral matters, a "manner of seeing?"[79]

He also calls attention to the lack of agreement in drawings made only moments apart, even by the same Lick observer. This point is repeated by Flammarion in his *La planète Mars* (1892), where he also notes that the drawings made in 1888 by Niesten and by the German astronomer O. Lohse were in poor accord with the Mars of Perrotin and Schiaparelli.[80] Flammarion's dismay at these discordances had its origin in the fact that they jeopardized his claim, made as early as 1876,[81] that actual changes had occurred on Mars. Although he believed that evidences of change were the strongest proofs of Martian life, he was also enthusiastic about Schiaparelli's observations, stating in 1888: "we dare to recall here that we have been the first – and for a long time the only one in France – to declare our absolute confidence in the competency and certainty of the observations of Monsieur Schiaparelli. . . ."[82] In the same year, Flammarion suggested that some of the canals are rivers.[83]

Richard Proctor died in September 1888, but not before having his say on Mars in three publications. An April paper repeats the suggestion he had made four years earlier that the double canals are neither "objective realities" nor "optical *illusions*," but rather "Diffraction-images of Martian rivers at times when mist hangs over the river-beds. . . ."[84] He also pronounces negatively on Schiaparelli's maps: "No one who has ever seen *Mars* through a good telescope can accept the hard and unnatural configurations depicted by Schiaparelli." Later in 1888, Proctor issued revised Martian maps showing numerous rivers.[85] That Proctor realized that a canal or river on Mars would have to be "fifteen or twenty miles broad" to be seen from earth is evident from his discussion of Mars in his posthumous *Old and New Astronomy*, where he develops his diffraction theory in more detail.[86]

William H. Pickering entered the debate in 1888 with a critique of various canal theories.[87] That labeled "least probable" is the hypothesis that they are "water canals." Perrotin, he notes, had reported a canal crossing the northern ocean. "If this observation is correct, it is clear that the ocean is not ocean, or the canals are not canals." Pickering adds: "Think of the labor involved in covering over, and then re-opening, a canal, say, sixty miles wide by three thousand miles long. . . ." He also rejects Proctor's mists and the theory of the physicist Hippolyte Fizeau that the lines are analogous to glacial crevasses, citing, in response to Fizeau, Flammarion's arguments that ice is not red and that, judging by the melting of the polar caps, Mars's temperature is too high for widespread glaciation.[88] Pickering's preference turns out to be for the view that "the stripes are due to differences in vegetation," which position gradually won increasing acceptance.

Maunder, in a review of some of the 1888 writings, rejects Proctor's mist-covered rivers, but suggests that his idea that the double canals "are 'optical *products*' may prove more satisfactory."[89] He also states that some astronomers had opposed the canals not simply on account of their failure to see them but rather because "they had seen other markings which they could not reconcile with them." The most striking passage in his paper consists of a quotation from a letter Schiaparelli had written to Terby:

But that which is most extraordinary and unexpected are the changes . . . in the Boréosyrte and the surrounding regions. . . . What strange confusion! What can all this mean? Evidently the planet has some fixed geographic details, similar to those of the earth. . . . Comes a certain moment, all this disappears to be replaced by *grotesque polygonations and geminations* which, evidently, attach themselves to represent approximately the previous state, but it is a gross mask, and I say almost ridiculous.[90]

The developments associated with the 1888 opposition illustrate an important generalization made by Maunder: "Before 1877 the study of planetary markings was left almost entirely to the desultory labours of amateurs. . . . [S]ince 1877, the most powerful telescopes . . . have turned upon Mars, and the most skillful and experienced of professional astronomers have not been ashamed to devote their time to it."[91] That this transformation was well under way by 1888 is illustrated by Flammarion's remark to Schiaparelli: "Your observations have made Mars the most interesting point for us in the entire heavens."[92] That this transformation was intimately tied to the topic of Mars's habitability is suggested by an 1889 statement of the Princeton astronomer Charles A. Young: "Probably there is no astronomical subject concerning which opposite opinions are so positively and even passionately held. . . ."[93] As the 1890s began, the intensity of the debate once again increased.

The opposition of June 1890 brought Mars nearer than it had been in a decade, but its proximity to the horizon hindered observation. Nonetheless, the number of persons reporting canals increased. Although Hall in Washington again met with no success, despite his 26-inch telescope,[94] Rev. Giovanni Giovannozzi at Florence, with a 4-inch refractor, Walter Wislicenus at Strasbourg, with a 6-inch telescope, and J. Guillaume at

Péronne, with an 8-inch reflector, all succeeded in seeing canals.[95] Most striking were the results attained by England's Arthur Stanley Williams (1861–1930), who reported that with his 6.5-inch reflector "as many as 43 of Schiaparelli's canals were seen with certainty. . . . Seven were also seen distinctly double." Williams added the interesting comment: "On most nights Schiaparelli's map was referred to during the observation. . . . But on several of the best nights reference to the map was carefully avoided until the end of the observation."[96]

Schiaparelli must have been pleased when from the Holden, Keeler, and Schaeberle group at Lick came a report that said in part:

E. S. H. and J. E. K. always saw the *canals* as dark, broad, somewhat diffuse bands. In bad vision they were drawn in this way by J. M. S. also. Under good conditions, however, the latter observer described them as narrow lines, a second of arc or so in width.

On April 12th, J. M. S. saw two of the canals doubled. It may, therefore, be said that the observations of Professor SCHIAPARELLI have been verified by this observer. The position of most of Professor SCHIAPARELLI'S canals have been verified by some one of us.[97]

Schiaparelli sent Terby the exciting news that Solis Lacus had acquired a new set of canals and moreover "has not been able to escape the principle of doubling which tyrannizes the entire planet: *it is cut crosswise by a yellow band which divides it into two parts of unequal extension.*"[98] These discoveries were rapidly imparted to the public, in a substantially embellished form, by Flammarion.[99] W. H. Pickering provided Flammarion with the dramatic news that he had photographically detected a Martian snowstorm, which, in the words of the French author, "in twenty-four hours covered a territory larger than the United States."[100]

The canal controversy became especially lively at this time in the British Astronomical Association (B.A.A.), founded in 1890 as an organization to which both professionals and amateurs, including women, could belong. Over subsequent decades, its *Journal, Memoirs,* and meetings frequently served as fora in the Mars debate.[101] At its December 31, 1890, meeting, the artist-astronomer Green sparked a discussion by criticizing Schiaparelli's drawing technique. In particular, he asserted that the Milan astronomer and those who drew canals "have not *drawn* what they have *seen,* or, in other words, have turned soft and indefinite pieces of shading into clear, sharp lines."[102] Moreover, Green maintained that comparisons among drawings of a particular region of Mars made by himself in 1877, by Maunder in 1879, and by Boeddicker in 1882 showed more agreement than could be found in Schiaparelli's drawings from the same oppositions. In the ensuing discussion, Maunder and Captain William Noble (1828–1904), then president of B.A.A., agreed with Green, whereas Sadler sided with Schiaparelli. A. S. Williams continued the controversy by publishing a paper stressing that many astronomers had seen canals.[103]

During 1891, a dispute concerning the doubling of canals erupted in the Belgian journal *Ciel et terre* and spilled over into the *English Mechanic*. Adolphe de Boë began it by claiming that a thin dark line would under certain circumstances, such as eye fatigue, be seen as double. Terby responded on behalf of Schiaparelli, with H. Schleusner supporting the optical-psychological theory of gemination.[104] Schiaparelli was no doubt disturbed by these challenges to his observations, including the fact stressed by John Ritchie that astronomers of proven ability and excellent instrumentation had not seen the canals.[105] However, what began to trouble him most at this time may have been his deteriorating eyesight; in 1909, he revealed to Antoniadi that

. . . *fully satisfying* observations, made in good atmosphere and with the full power of my left eye (the other is rather defective in some respects) have ceased with 1890. During the 1892-94-97-99 oppositions, I still observed, but I no longer made drawings. . . . Towards 1894 (or 1896) a deteriorization in my eye was becoming more evident: the field of vision became more and more darkened. Finally, in 1900, I verified, with great sorrow, that the images began also to be deformed. . . . I have also decided *not to publish* my observations of Mars made after 1890.[106]

Schiaparelli continued to write about Mars, but because by then a dozen or more astronomers had seen the canals, the case for them no longer rested exclusively on his shoulders.

The August 1892 opposition was excellent. The best indication of the flood of papers flowing from it and from the next four oppositions is the fact that whereas Flammarion devoted the more than six hundred pages of the first volume of his *La planète Mars* to all pre-1892 Mars publications, his comparably sized second volume covered only those five oppositions. The appearance of the first volume of Flammarion's *Mars* was itself among the most important events of 1892. Although Flammarion admitted that his pluralist convictions had motivated him to write it,[107] it is for the most part free of the flamboyant prose and excessive claims characteristic of most of his writings. For example, he shows an openness to explaining the doubling of the canals as "due either to mists . . . or especially to the double refraction in the Martian atmosphere." (p. 588) He accepts the canal observations, suggesting: "The canals [canaux] may be due to superficial fissures produced by geological forces or perhaps even to the rectification of old rivers by the inhabitants for the purpose of the general distribution of water on the surface of the continents." (p. 591) As to Martian life, his conclusion is that "the actual habitation of Mars by a race superior to our own is in our opinion very probable."

(p. 591) Favorably reviewed in a number of journals,[108] it has remained a standard source for the history of Martian studies.

Possibly a hundred persons in 1892–3 published papers on Mars or made observations discussed in print by others. Among the most curious papers was that of Asaph Hall, who had repeatedly reported his failures at earlier oppositions to see canals. His 1892 report, however, contains this enigmatic remark: "the usual markings on the planet could be seen, but no duplication of the so-called canals could be made out. . . ."[109] Holden, at Lick, distressed by the departures of Keeler and Burnham from his staff and by garbled newspaper reports on Lick's Mars observations,[110] announced that his staff had seen many canals and that on August 17, Professors Schaeberle, Campbell (Keeler's replacement), and Hussey (visiting for the summer from Stanford) "made three entirely independent drawings each of which shows the canal . . . Ganges . . . to be distinctly double."[111] Later that year, Holden reported: "Crossing the darker areas are still darker streaks. . . .,"[112] which some interpreted as canals crossing the Martian seas. Holden had frequently expressed doubts that the dark areas on Mars correspond to water, urging that as yet no evidence had even refuted Brett's earlier idea of a glowing Mars.[113] Schaeberle proposed that were the dark areas interpreted as land and the light areas as seas, this would fit better with the variations in tone seen in the dark areas and allow the "canals" to be considered partly submerged mountain ranges, parallel ridges of which would be seen as double "canals."[114] Schaeberle's claim drew a response from Schiaparelli, to which Schaeberle in turn responded.[115] Edward Emerson Barnard (1857–1923), of Lick, famed for the sensitivity of his vision, and by 1894 an opponent of the canals, reported seeing some canals but no doubles.[116] Moreover, in 1893 Barnard visited Schiaparelli, asking the Italian astronomer: "In your published drawings of Mars the canals are shown very strongly marked. The drawings in your notebook do not show these lines so heavy. Is it an accident of the reproduction that they are so heavy and dark in the engraving?" Schiaparelli made this interesting response: "The canals can have a different aspect in different times. They may disappear wholly, or be nebulous or indistinct, or be so strongly marked as a pen line. The reproductions of my drawings unfortunately can mislead the reader. I cannot find artists who reproduce them well."[117] Keeler, who had left Lick to direct the Allegheny Observatory, presented his 1892 observations as more supportive of Green than of Schiaparelli, with Princeton's Young also publicly favoring Green's representation of Mars.[118]

Early in 1892, Edward C. Pickering, the Harvard director, sent his brother, William H. Pickering, to Arequipa, Peru, to set up an observing station for photographic studies of southern stellar and nebulous objects. Soon tensions developed as the younger Pickering not only greatly exceeded his budget but also, when Mars came into opposition, jettisoned his assigned project for visual planetary studies, concerning which he issued sensational telegrams containing such statements as *"Mars* has two mountain ranges near the south pole. Melted snow has collected between them before flowing northward. In the equatorial mountain range, to the north of the gray regions, snow fell on the two summits on August 5 and melted on August 7."[119] E. C. Pickering, who had privately warned his brother against such excessive claims, must have been disturbed when Holden publicly stated that Pickering's telegrams were received "with a kind of amazement" at Lick, where such questions were asked as "How does he know that the flow will be northwards?"[120] Referring to such telegrams and newspaper reports, the *British Astronomical Association Journal* wryly commented: "our American colleagues suffered many things of many reporters, and it is to be feared (or rather hoped) they were grievously slandered by them."[121]

Pickering's four reports in *Astronomy and Astro-Physics* on his Arequipa Mars observations show scarcely more restraint than his telegrams. An excited tone pervades them, as when he writes in the third paper that the canals "can now be observed readily any evening," adding: "Changes were now coming thick and fast . . . we never knew what we should see next."[122] The third paper also contains the announcement: "Some very developed canals cross the oceans. If these are really water canals and water oceans there would seem to be some incongruity here." In the fourth paper, Pickering sets out his "definite conclusions" from the 373 drawings of Mars made by his colleague A. E. Douglass and himself. These include statements that "clouds undoubtedly exist upon the planet," that the "so-called canals exist upon the planet, substantially as drawn by Professor Schiaparelli," and that "we have found a large number of minute black points [which] occur almost without exception at the junctions of the canals with one another. . . . Called "lakes" by Pickering, these points became Lowell's "oases." Perhaps the most dramatic of his conclusions is that "Excepting the two very dark regions . . . , all of the shaded regions . . . have at times a greenish tint."[123] In his third paper he had attributed these greenish tints to life forms. Pickering's four papers, if accepted uncritically, seemed very impressive; writing in 1893, George Searle described their author as the "principal source of information as to Mars this last year."[124] E. C. Pickering was not so impressed; exasperated by his brother's indiscretions, he recalled him from Peru and placed Solon Bailey in charge.

In England, Maunder, the first director of the B.A.A. Mars section, issued guidelines for members intending to observe Mars during the opposition. Their central, albeit implicit, theme was to seek observations that would settle the Green-versus-Schiaparelli controversy.[125] In his report on the drawings submitted by over two dozen observers, Maunder reluctantly used the term "canal," stressing that it was being employed "in a purely technical and not in a geographical sense" and that " 'canals' in the sense of being artificial productions [do not exist and it] is difficult indeed to understand how so preposterous an idea obtained currency. . . ."[126] Nonetheless, Maunder reported on the numerous "canal" observations he received and in his composite map showed over thirty canals, including some doubles (p. 196). The brief report published by the R.A.S. contains the lament that the lowness of Mars on the horizon retarded observation and the suggestion that the changes seen on the planet jeopardize "the usually accepted theory that the brighter and darker markings indicate respectively land and sea."[127]

In France, Perrotin again saw canals in 1892, but concentrated on watching for "bright projections" on Mars, seeing three in the summer of 1892. Because the Guzman prize had recently been announced and similar phenomena had been reported from Lick, these observations attracted widespread attention.[128] Terby, writing Schiaparelli in 1891, warned that Flammarion's support for the canals was "wavering," but by 1892 he reported: "Flammarion is now entirely reconciled with the canals and I am pleased for I had viewed with sorrow that he allowed himself to be won over a little by Schiaparellophobomania. . . ."[129] Whether or not Terby's fears for Flammarion were justified, the Frenchman's confidence in the canals must have been increased by the confirmations that came in 1892 from his own Juvisy Observatory, where Léon Guiot used a 9-inch refractor to observe ten canals on a single night, and from the Flammarion Observatory in Bogotá, Colombia, where J. M. Gonzalez detected many canals with a 4-inch telescope, but in such a fleeting way as to defy drawing.[130]

From Belgium, Terby reported that he had earlier "confirmed the existence . . . of a great number of canals, and the doubling at least of Phison."[131] He also revealed his method of seeing the canals:

We took inspiration from the principle announced by some great observers: "Often," they say, "one can see well what one especially seeks." . . . we have sought the canals . . . where we knew that M. Schiaparelli had proved them to exist . . . and, map in hand, we have patiently and obstinately pursued these very difficult details. It is to this method . . . we owe our partial success. (p. 479)

As is obvious, this passage admits quite a different interpretation than that intended by Terby.

Not only observations but also theories of the canals attracted international attention. T. W. Kingsmill's tidal theory, as published in the *Shanghai Mercury,* was picked up by both English and American journals, while S. E. Peal of India presented a geological theory before astronomical groups in both England and Canada.[132] Readers of the American journal *Science* in 1892 could select among two geological theories and one optical theory of the canals,[133] while in France, Stanislas Meunier hypothesized that the doubling of the canals was due to the Martian atmosphere, illustrating his theory by producing double lines on a marked sphere by suspending muslin above it. Whatever its merits, Meunier's theory was soon discussed in English and American journals.[134]

Three of the most important theoretical papers came from England. Maunder, writing on "The Climate of Mars," urged that the low density of the Martian atmosphere and the remoteness of Mars from the sun might suggest "that the mean temperature even of the equator of Mars lies below the freezing-point," were we not aware of Mars's moderate climate from the observed melting of the Martian polar caps and the presence of water vapor in the Martian atmosphere, the latter result having been discovered spectroscopically by Huggins and twice confirmed by Maunder himself. To alleviate this tension, Maunder suggested that Mars, more effectively than our planet, retains the heat received from the sun, partly because less sunlight is reflected away by the less dense Martian atmosphere and partly because clouds may form during the Martian night, preserving the heat accumulated during the day.[135] J. Norman Lockyer, early observer of Mars, prominent astrophysicist, and editor of *Nature,* wrote a wide-ranging essay in which he interpreted his observations of Mars from the 1860s as supporting more recent views of Mars as wet and cloudy. Protesting the translation of Schiaparelli's "canali" as "canals," while endorsing the Italian astronomer's "channel" observations, he urged that they are "true water channels" that sometimes appear double because of overhanging cloud banks.[136] Sir Robert Ball called attention to Stoney's analysis of planetary atmospheres based on the kinetic theory of gases, stating that this theory sanctions the retention of both oxygen and water vapor in Mars's atmosphere. Like Lockyer, he supported Schiaparelli's canal observations. Nonetheless, he maintained that the dissimilarities between Mars and the earth make it improbable that intelligent life now exists on our neighbor planet.[137]

The question of life on Mars underlay many if not most of these publications and was made explicit by some writers. Mary Proctor, the daughter of Richard Proctor, wrote that during the Mars opposition, "The astronomer . . . can scarcely . . . avoid the thought that contests such as have raged upon our earth for the possesion of various regions . . . may be in progress out yonder in space."[138] Ellen Clerke, like her sister Agnes, endorsed Schiaparelli's observations; moreover, she accepted the possibility of life on Mars, although describing the idea of sending signals to Mars as "outside the bounds of rational speculation."[139]

Probably the most widely read Mars essay of this period was a presentation by Schiaparelli that appeared first in Italian and then, in the form of an abridgement and translation by W. H. Pickering, in an array of American and English journals.[140] Some features of this paper merit discussion.

1. Because Pickering omitted without comment two sections of this paper, in which sections Schiaparelli presents a historically developed argument for extraterrestrial life and for Mars as the last, best hope for its discovery, this paper perpetuated Schiaparelli's image in the English-speaking world as a technical astronomer wary of pluralist ideas.

2. Persons puzzled by the widespread acceptance of the Schiaparellian canals may find an answer in this paper: They appear, first of all, not as controversial new observational entities but as deductive consequences of almost universally accepted observations and assumptions that Schiaparelli did not initiate, but at most refined. His argument begins with a description of the polar caps and of their seasonal contractions and expansions. First seen by William Herschel, the polar caps had subsequently been observed by dozens of astronomers. Schiaparelli's statement that in 1892 the southern polar cap shrank from 1,200 to 180 miles in diameter was untraditional only in its precise quantification. Difficulties enter with his identification of the caps as composed of snow and ice. This seemingly natural step, although made after A. C. Ranyard had suggested (correctly, as is now known) that carbon dioxide may be the main constituent of the caps,[141] is the fatal flaw in Schiaparelli's argument, because from it he concludes that their melting produces "gigantic inundations" (p. 637) that in turn cause the seas, lakes, and canals of Mars. The melting of the snowy caps, moreover, convinced Schiaparelli, as it had also convinced Maunder, that Mars possesses more or less terrestrial temperatures. Schiaparelli's main argument is, in short, that the conclusion "that the lines called canals are truly great furrows or depressions . . . destined for the passage of liquid mass, and constituting for [Mars] a true hydrographic system, is demonstrated by the phenomena which are observed during the melting of the northern snows." (p. 719) The cautions characteristic of his 1878 memoir are absent from this paper: "We conclude therefore that the canals ["canali"] are such in fact, and not only in name." (p. 719) In support of this inferential chain, he cites his observations of the canals, including their seasonal alterations. But he provides scant explanation as to how they could be as wide as 180 miles or why "only a few are visible at once." (p. 718)

3. As to the origin of the canals, he suggests that their network "was probably determined in its origin in the geological state of the planet, and has come to be slowly elaborated in the course of centuries." (p. 719) Relying on the pluralists' double negative, he adds that their geminations and geometrical character have "led some to see in them the work of

intelligent beings. . . . I will indeed abstain from combating this supposition, which includes nothing impossible." His pluralism is also evident in his remark that the canals are "the principal mechanism . . . by which water (and with it organic life) may be diffused over the arid surface of the planet." (p. 639)

4. Concerning the geminations, he states that they begin around the time of the Martian vernal equinox, that they appear and disappear within a few days or even hours, that both canals and lakes geminate, that the second canal appears at distances varying from 30 to 360 miles from the original canal, that they have "been seen and described at eight or ten observatories," and that although unsatisfied with all published explanations of geminations, he has nothing better to offer (pp. 720–3). Following that admission, and after ruling out all explanations in terms of "forces pertaining to organic nature" as excessively hypothetical, he adds: "Changes in vegetation over a vast area, and the production of animals, also very small but in enormous magnitudes, may well be rendered visible at such a distance." (p. 723)

Schiaparelli's paper has among its strengths a noteworthy internal consistency, observational evidence collected over sixteen years by a well-equipped expert, and an exciting subtheme – Martian life. These factors go far to explain its appeal. Another factor was its appearance shortly after one splendid Mars opposition and just before another.

4. The canal controversy in 1894: enter Percival Lowell, who "has taken the popular side of the most popular scientific question afloat"

The opposition of October 1894 rivaled that of 1892 in quality and surpassed it in quantity of controversy. Although less near the earth, Mars rose higher above the horizon, easing observational difficulties. Three astronomers emerged in 1894 as figures of crucial and continuing importance; although each observed Mars in 1894, the chief contribution of the first came from "observing" with a spectroscope, that of the second from observing the sun and from observing other observers, whereas the third made his observations from an observatory that had not existed when the year began. These astronomers were W. W. Campbell, E. W. Maunder, and Percival Lowell.

Raised by his widowed mother on an Ohio farm, trained in civil engineering at the University of Michigan, where Newcomb's *Popular Astronomy* had interested him in astronomy, William Wallace Campbell (1862–1938) (Figure 10.4) in 1891 joined the staff of the Lick Observatory, where a decade later he began his three-decade directorship of that institution. Presidencies of the University of California and of the National Academy of Sciences crowned his scientific career, which, one obituary noted, was characterized by an "admiration amounting to a worshipful passion for observational results of the greatest possible precision. . . ."[142] Although much was uncertain about Mars in 1894, Huggins's 1867 report of water vapor in the Martian atmosphere seemed almost beyond dispute, having been confirmed by Janssen, Secchi, Vogel, and even Maunder, who in an 1892 paper loaded with evidence difficult to reconcile with Huggins's delicate determination had remarked: "I have repeated it myself on two occasions, and have little doubt as to its accuracy."[143] Convinced of the correctness of this result, which earlier observers had attained "under conditions *extremely unfavorable*," Campbell in the summer of 1894 undertook the "simple and easy matter," as he viewed it, of reconfirming it, aided by the advantages of an "improved spectroscopic apparatus" harnessed to a superior telescope located at high altitude in dry summer air.[144] His technique, which involved eliminating the spectral lines due to the earth's atmosphere by comparing the Martian spectrum with that of the airless moon, led him to conclude that his observations, repeated ten times, "furnish no evidence whatever of a Martian atmosphere containing aqueous vapor," even though a "Martian atmosphere one-fourth as extensive as our own ought to be detected by the method employed." (p. 236) His paper, which may have been seen by many as the report of a research failure of a neophyte in spectroscopy, concludes with a qualifying comment reflecting his continuing confidence in the overall belief from which he had begun: "the polar caps on *Mars* are conclusive evidence of an atmosphere and aqueous vapor. . . ."

   This last statement, from which Campbell soon wished he had omitted the word "aqueous,"[145] shows how slowly he recognized the revolutionary ramifications of his result. Probably aided by a letter to Holden from Dr. Henry H. Bates, who maintained that Campbell's "aqueous vapor" claim based on the polar caps lacked sufficient support because the caps could consist of carbon dioxide or some other salt, Campbell composed in November 1894 a second paper, urging that many of the leading features of Mars admit explanation without recourse to a relatively dense atmosphere containing substantial amounts of aqueous vapor.[146] For example, Mars's redness, rather than being an atmospheric effect, may be due to the color of the Martian soil, thereby also explaining why the poles appear white instead of red. The brightness of the caps, he suggests, is difficult to reconcile with the assumption that rays reflected from them have twice passed through a relatively dense atmosphere. Were the caps composed not of snow but of frozen crystals of carbonic acid, this would fit with the colder climate expected on Mars because of its remoteness from the sun. Citing R. S. Ball's 1892 discussion of Stoney's kinetic theory analysis of planetary atmospheres, Campbell suggests that Mars may have lost much of its atmosphere (pp. 280–1). Finally he adds that the albedo of Mars, measured by Zöllner as comparable to that of the moon, is easily reconciled with a Mars on which both dark and light regions are land. In short, whereas in June 1894 the closest analogue of Campbell's Mars had been the earth, by November it had become the moon. Holden, among the few who publicly supported Campbell, dramatically summarized his colleague's claims by stating that "the lakes, oceans, canals (of water), the snowstorms, inundations, inhabitants (like ourselves) and the signals they were making us, etc., etc., etc., have all vanished with the aqueous vapor."[147] Campbell, seemingly to his surprise, found himself surrounded by controversy.

In October 1894, Huggins composed a letter reaffirming the "substantial accuracy" of his 1867 observation and claiming that Campbell had inadequately described the precautions taken in it.[148] After Campbell further explained his position, Huggins softened his criticism, but on repeating his spectroscopic determination, Huggins again reported a positive result.[149] Janssen, who had not previously published a full account of his 1867 detection of water vapor, proceeded to do so, affirming his confidence in it.[150] Hermann Carl Vogel (1841–1907), director of the Potsdam Astrophysical Observatory, also defended his 1873 observations of a Martian atmosphere and carried out a new, likewise positive, determination.[151] A response that first appeared in the *Martha's Vineyard Herald* and later in astronomical journals noted:

> Marseania has struck the fellows who are running the big Lick telescope. They have it bad.
> According to them, Mars has no atmosphere at all, and consequently no life exists on that planet. This is symptom No. 1.
> Mars has no more atmosphere than the Moon. It is a dead world, of no more value than the state of Nevada which only supports two Republican senators. Symptom No. 2.
> If Mars has an atmosphere it is not one-quarter as extensive as that of the Earth. Symptom No. 3.
> No. 4 is, the atmosphere of Mars is so thin a man from the Earth would be discouraged if he tried to breathe it for a steady occupation.
> Well then, who dug "them" canals on Mars?
> Who set up "them" signals on Mars and lighted 'em?[152]

To each of these objections (except the last), Campbell responded in his 1895 "A Review of the Spectroscopic Observations of Mars." George Ellery Hale, concerned about international goodwill among astronomers, urged restraint on Campbell,[153] but the young spectroscopist, with a confidence fully vindicated only decades later, pressed on ahead. Few, at first, followed.

It is intriguing to ask how the Mars controversy might have been different had Campbell not come to Lick. Such speculation is legitimate, because Holden had first offered the position to Maunder.[154] Had Maunder accepted it, not only might Campbell never have undertaken this research, but moreover a second crucial attack on the Mars of tradition might not have been made. This critique, also appearing in 1894, was composed by Maunder himself; although entirely different from Campbell's, it was no less revolutionary.

In 1894, Maunder became president of the British Astronomical Association, having resigned the directorship of its Mars Section in 1893. The latter position was filled for two years by Bernard E. Cammell, who in his report on the 1894 opposition admitted that he was "comparatively a beginner in the study of Mars."[155] As Cammell prepared his report on the 156 drawings of Mars sent him by twenty-two observers, he must have felt confused by the lack of agreement in the materials submitted to him. For example, whereas in 1894 Maunder reported seeing only one canal with the 28-inch Greenwich refractor, A. S. Williams, with a reflector of less than one-fourth the aperture, saw sixty canals. Whether it was because Cammell himself saw canals or because the majority (thirteen) of the observers saw at least some, Cammell concluded for a Schiaparellian Mars: "The present apparition of Mars seems to have fully demonstrated the accuracy of Schiaparelli's observations and maps. . . ." (p. 111)

Maunder was no doubt disappointed at Cammell's claim, especially because in an 1894 paper he had presented the first fully formulated theory of the canals as optical illusions. After admitting the excellence of the visual evidence for the "canals," Maunder mentions some anomalies that suggest caution, such as the "great divergency between the descriptions of different observers" and the "greatness and suddenness of the changes remarked in the 'canal' system."[156] He also examines the problem of seeing objects at the limit of visibility. Having discovered while viewing the sun in 1891 that a group of sunspots, separately invisible, were jointly seen as a line or, in effect, as a "canal,"[157] he conducted experiments in which he found that he could see dots on white paper only if their widths subtended in excess of 30 seconds of angle and could see them clearly only for widths equivalent to over 40 seconds. However, a line equivalent to 7 or 8 seconds could be seen, and "a pair of lines, each only 4″ in breadth, and the pair separated by say 20″, was visible as a faint single line. . . ." He adds that "a chain of dots, each of 20″, irregularly disposed along a straight line, the average interval between any two dots being three times the diameter of a dot, was easily seen as a continuous straight line. . . ." Because the canals drawn by Schiaparelli and others lie below these limits, he suggests that the canals, their doublings, and some of the other observed changes may be illusory. His conclusion consists in the important warning: "We cannot assume that what we are able to discern is really the ultimate structure of the body we are examining."

In 1895, Maunder published a comparative study of drawings of the "Lacus Solis" or "Oculus" region of Mars made over many years by a variety of observers. Although attributing some differences among these drawings to "imperfect seeing and imperfect drawing,"[158] he concludes that others represent actual changes on the planet. To account for these changes, he ascribes to that region a "very level surface intersected by shallow channels, most of them probably too narrow to be separately defined by any telescopic power we possess. . . ." And he adds: "it seems . . . probable that throughout the planet . . . the difference of level between land and sea-bed is but slight . . . so that the difference in the sea-level consequent on the melting of the winter snows, has a vast effect on the extent of the areas submerged." (p. 58) Maunder's 1895 interpretation of Mars is doubly ironic. First, because the changes he accepted as real are probably illusory, it appears that he fell into the very trap he had warned against in 1894. Second, Maunder's retention of a watery Mars shows that he had failed to accept Campbell's results, which in effect pointed to the conclusion that spectroscopic evidence of Martian water vapor could be as illusory as visual observations of its canals.

Maunder was in the president's chair at an early 1895 B.A.A. meeting when an anticanal paper by J. Orr, a Scottish member, was read. Orr's argument is that if we assume that the canals are 33 miles wide and 70 feet deep, it follows that they "contain about 1,634,000 of our Suez Canals, and would require an army of 200 millions of men working for 1,000 of our years for their construction." This being improbable, Orr opts for geological fissures as the source of the canal observations. In the ensuing discussion, Maunder praised the paper as "one more nail in the coffin of a very absurd idea . . . that the canals could possibly be the work of human agents."[159]

Maunder's 1894 critique of the canals received even less attention than Campbell's paper. One reason was the relatively unsophisticated under-standing of the nature and limits of observation prevalent in that period among astronomers as well as psychologists and philosophers, not to mention the public. Another reason was that while Campbell and Maunder wrote, the world's attention was turning to a dynamic new champion of Schiaparelli's Mars.

Whereas in mid-1893 he had been conceptually and geographically thousands of miles distant from the Mars debate, by mid-1895 Percival Lowell (1855–1916) (Figure 10.2) had positioned himself at its very center. Fame was nothing new to Lowell, the oldest offspring of parents so prominent that their lineages had been memorialized in Massachusetts town names: Lawrence and Lowell. The maternal family name was also carried by Lowell's distinguished brother, A. Lawrence Lowell, president of Harvard from 1909 to 1933. When, after early studies that included two years in France, Lowell entered Harvard, it was not to prepare for a career in astronomy, which had been a childhood hobby. It is true that at Harvard Lowell won the praises of Professor Benjamin Peirce for his mathematical abilities and that when chosen to give one of the commencement addresses he took the title "The Nebular Hypothesis," but his commitment to an astronomical career came only in 1893 after seventeen years devoted to family business concerns and three extended stays in the Orient. The latter led to his four books on Eastern thought and culture, the best known of which is *The Soul of the Far East* (1888). The most intriguing question about this period is what motivated Lowell to leave Japan in autumn 1893 to devote the majority of his energies to studying the planets, especially Mars. Despite a number of biographical studies of Lowell, this question has remained largely unanswered.[160] The most frequently advanced explanation is that Schiaparelli's failing eyesight led Lowell to take up the Italian astronomer's promising Martian researches. However, this theory is negated by the fact that in 1893 Schiaparelli had not yet realized, let alone revealed, that his vision was impaired.[161] A less well known explanation is that Flammarion's *Mars* (1892) functioned as the decisive factor. After Lowell had visited Flammarion in the winter of 1895–6 to show Flammarion his own recently completed volume on Mars, Flammarion stated that Lowell had told him: "It was your work upon the planet Mars which gave us the impetus. . . ."[162] This explanation is difficult to reconcile with at least two facts: In 1891, a year before the publication of Flammarion's book, Lowell took a 6-inch telescope to Japan and moreover told W. L. Putnam of his plan to write a work on the philosophy of the cosmos.[163] A third theory, which may appeal to those attracted to psychoanalytic interpretations, was proposed in 1964 by the psychiatrist C. K. Hofling, who suggested that Lowell's Martian work was "heavily influenced by unconscious forces, taking the final form of incompletely sublimated voyeuristic impulses [arising from] unresolved oedipal conflicts."[164] Despite the difficulties in these explanations, it is clear that Lowell's interest in astronomy and commitment to an evolutionary cosmology date from his Harvard days, if not earlier. The latter perspective he may have acquired from Harvard's John Fiske, who, as noted previously, published his *Outlines of Cosmic Philosophy* in 1874 while Lowell was a student at Harvard. Another possible source is R. A. Proctor, five of whose books are preserved in Lowell's personal library, two (*Saturn and Its System* and *The Sun*) having been given to Lowell by his mother for Christmas in

1872.[165] So intense was Lowell's commitment to an evolutionary perspective that he included it even in some of his writings on the Orient.[166] On this basis it seems probable that by 1891 Lowell had resolved to publish his evolutionary cosmic philosophy and soon realized that Martian studies would provide an optimal context for this.

Whatever its origin, Lowell's fascination with Mars led him in early 1894 to contact Harvard, proposing a jointly sponsored observing station in Arizona. Harvard authorities, no doubt delighted by the prospect of having their own James Lick or Charles Yerkes, may have underestimated what their fellow Bostonian had in mind. Lowell, whose pen was scarcely less powerful when dashing off popular magazine articles than when applied to his checkbook, envisioned himself not only as funder but also as founder, chief observer and theoretician, publicist, and director of the new observatory. The tensions that soon developed between Lowell and E. C. Pickering led them to jettison the joint venture. That Harvard president Charles W. Eliot found no fault with his observatory director's decision on this matter is suggested by Eliot's statement in a November 22, 1894, letter to Pickering:

But Mr. Percival Lowell is undoubtedly an intensely egoistic and unreasonable person, and in my opinion his frame of mind towards the Observatory is a hopeless one. Fortunately he is generally regarded in Boston among his contemporaries as a man without good judgment. So strong was this feeling a few years ago that it was really impossible for him to live in Boston with any comfort.[167]

Nonetheless, Harvard leased a 12-inch telescope to Lowell and gave unpaid leaves to W. H. Pickering and his Arequipa assistant, Andrew Ellicott Douglass (1867–1962), to permit them to work with Lowell.[168] Lowell's association with Pickering at this period seems unfortunate. As a neophyte in astronomical science, Lowell scarcely needed counsel from a person whose Martian mania and sensational pronouncements had contributed to his recent recall from Arequipa.

After dispatching Douglass to Arizona, where Flagstaff was selected as the observatory site, and after borrowing from J. A. Brashear an 18-inch refractor, Lowell began to publicize his "Lowell Observatory." Addressing the Boston Scientific Society on May 22, 1894, he explained that the "main object" of his observatory would be "the study of our own solar systems [i.e.] an investigation into the conditions of life in other worlds. . . ." And he added: "there is strong reason to believe that we are on the eve of pretty definite discovery in the matter."[169] Espousing a Proctorian solar system and a Schiaparellian Mars, Lowell urged concerning the canals that they were probably "the work of some sort of intelligent beings." On June 1, just as Lowell made his first Flagstaff observations, Holden, at Lick, completed a paper labeling Lowell's Boston address "very misleading and unfortunate" and warning its author of the dangers of "overstatement," of mixing fact with conjecture, of holding out "hopes . . . unlikely to be realized . . . during this century," and of telling "half-truths [which] the whole human race would be delighted to have verified."[170] In the same paper, Holden (as noted previously) excoriated Pickering for issuing sensational claims from Arequipa. The battle between the "millionaire telescopes" had begun and did not cease for over two decades.

Lowell frequently stressed that observatories should be located "where they may see rather than be seen"[171] and that Mars ought be observed not sporadically but night after night. Although Lowell observed for only about eighty days during the first ten months of his observatory's operation, Pickering, present until late November, and Douglass, who temporarily closed down the observatory in April 1895, made nearly nightly observations of Mars.[172] Their efforts produced striking results; in his *Mars* (1895), Lowell listed 184 canals as sighted, more than double the 79 shown by Schiaparelli, although less than the 700 eventually seen from Flagstaff. Only 8 doubles were detected in 1894–5, but hundreds of diameter determinations and thousands of drawings of Mars were made.[173] According to W. G. Hoyt, Lowell proceeded to "formalize his thinking into what he considered to be a full-blown scientific theory . . . late in July of 1894 . . .," and throughout the remainder of his life, "neither subsequent observations nor mounting criticisms . . . ever caused him to abandon or alter any of its essential points."[174]

Put briefly, Lowell's theory is that the snowy polar caps alternately melt, causing water to flow toward and beyond the equatorial regions through numerous constructed canals, which, although not themselves seen, are evident from the vegetation adjacent to and fertilized by them. Vegetation also furnished an explanation for the observed changes of color in the dark regions, at least after Lowell accepted the sightings beginning in July 1894 by Douglass and Pickering of canals in the dark regions and concluded from them that the dark regions could not be bodies of water.[175] The view of Mars as having few large bodies of water was also supported by Pickering's observations with a polariscope, which led him in July 1894 to urge that "the permanent water area upon Mars, if it exists at all, is extremely limited in its dimensions."[176] The small circular spots at the intersections of canals, which Pickering in 1892 had seen as lakes, Lowell considered to be "oases," tracts of vegetation that become visible with the "wave of darkening" resulting from the flow of water away from the melting polar caps. Arguing from such matters as the small size and mass of Mars, Lowell proposes a Martian atmosphere only one-seventh as dense as the earth's.[177] This harmonizes with his theory, because a more dense atmosphere would imply precipitation, making canals unnecessary, whereas too thin an atmosphere would not return water to the poles. Lowell's theory includes some clouds, these being used to account for the projections that Campbell attributed to reflections from Martian mountains.[178] In his *Mars,* Lowell deals colorfully if cavalierly with the problems presented for intelligent life by a thin Martian atmosphere:

That beings constituted physically as we are would find [Mars] a most uncomfortable habitat is pretty certain. But lungs are not wedded to logic, as public speeches show, and there is nothing in the world or beyond it to prevent, so far as we know, a being with gills, for example, from being a most superior person. A fish doubtless imagines life out of water to be impossible; and similarly to argue that life of an order as high as our own, or higher, is impossible because of less air to breathe than that to which we are accustomed, is, as Flammarion happily expresses it, to argue, not as a philosopher, but as a fish. (pp. 74–5)

Lowell's arguments for the constructed character of the canals rely heavily on the geometrical form of the canal network, a feature less easily accounted for in the geological explanations offered by others.

In at least two senses, Lowell's theory could be seen as strong. As one wit remarked, Lowell "is certain that the canals are artificial. And nobody can contradict him."[179] Moreover, by suggesting "giant" Martians "physically, fifty-fold more efficient than man," endowed with a "highly intelligent mind," and possibly "possessed of inventions of which we have not dreamed,"[180] Lowell *could* explain how the canals had been constructed. But this mode of explanation left Lowell open to a serious objection that the Nobel-Prize-winning physical chemist Svante Arrhenius phrased as follows:

The theory that intelligent men exist on Mars is very popular. With its help everything may be explained, particularly if we attribute an intelligence vastly superior to our own to these beings, so that we not always are able to fathom the wisdom with which their canals are constructed. . . . The trouble with these "explanations" is that they explain anything, and therefore in fact nothing.[181]

To put this point differently, Lowell's idea of superhuman Martians constructing canals suffered from the identical difficulties that beset the medieval theory that an angel produces Mars's motions. On the other hand, Lowell's Martians were well suited to such science fiction writers as H. G. Wells, Kurd Lasswitz, Mark Wicks, and Edgar Rice Burroughs.[182]

Lowell was not slow to publish his observations and theories. In 1894–5, he blitzed both *Astronomy and Astro-Physics* and *Atlantic Monthly* with four articles each, while *Popular Astronomy* received six.[183] He also proceeded to the speaker's platform for four February lectures before a packed Huntington Hall audience in Boston. Rev. Edward Everett Hale used the same powerful pen that produced *Man without a Country* to praise Lowell's lectures in an often reprinted piece that noted: "his humor, his ready wit, his complete knowledge of the subject with which he deals are such as one has no right to expect in the same speaker."[184] What, above all, gave Lowell's theory its celebrity was his *Mars,* published in December 1895, as Lowell set off for Europe to meet Flammarion, Schiaparelli, and other Martian enthusiasts and to seek in the Sahara a still clearer atmosphere for Martian observations. His *Mars,* free of the detail in the volumes of his lavishly printed *Annals,* sold very well. In fact, Lowell later argued for a tax exemption for his observatory because its products were "marketable."[185]

Numerous reviews of Lowell's *Mars* appeared, some by prominent astronomers. The review solicited by *Science* from Campbell accuses Lowell of sensationalism, preconceived ideas, and failure to give adequate credit to earlier astronomers. Campbell labels Lowell's claim that mankind tends to be "shy" of extraterrestrials "certainly wrong," adding: "In my opinion, he has taken the popular side of the most popular scientific question afloat. The world at large is anxious for the discovery of intelligent life on Mars, and every advocate gets an instant and large audience."[186] After quoting from Lowell's May 22, 1894, Boston address, Campbell charges: "Mr. Lowell went direct from the lecture hall to his observatory in Arizona, and how well his observations established his pre-observational views is told in this book." And Campbell noted the anomaly of this occurring in an author who "has written vigorously . . . of the dangers of bias [from] preconceived notions. . . ." (p. 232) Campbell also labels Lowell's theories "mostly old ones, suggested by Schiaparelli, Pickering and others, many of them having been elaborated by Flammarion and others. . . ." (p. 238)

Campbell's views were shared by a number of astronomers, including Holden, who in 1895 borrowed a phase from Kipling to describe Lowell as:

> Hanging like a reckless seraphim,
> On the reins of red-maned Mars.[187]

Barnard's disagreement with Lowell was on a more fundamental level than those of Campbell, who basically accepted Schiaparelli's and Lowell's canal observations. Writing in 1896, Barnard stated concerning his 1894 Lick observations that he had not seen any "straight hard sharp lines," but had observed Mars with "a wonderful clearness and amount of detail [which] was so intricate, small, and abundant, that it baffled all attempts to properly delineate it. . . ."[188] Nonetheless, when Douglass, using the title "The Lick Review of 'Mars,' " responded to Campbell's review, Campbell reported that he had written it without consulting with his Lick colleagues. On the other hand, as W. G. Hoyt has recently shown, Douglass's response was largely written by Lowell.[189] Lowell was not without some support in the astronomical community; for example, W. W. Payne and H. C. Wilson, the editors of *Popular Astronomy*, republished the Lowell-Douglass response without even a reference to Campbell's review and rejoinder.[190] Furthermore, Payne favored Lowell's position in two papers he published in 1896,[191] and he continued for years to publish Lowell's papers. Lowell may have needed this, because George Ellery Hale, founding director of Yerkes Observatory and co-founder of the *Astrophysical Journal*, rejected some of the papers Lowell submitted to his journal.[192] Moreover, in his dedicatory address at the opening of Yerkes, Hale pointedly stressed his staff's concern "that the work to be done here shall acquire a reputation for thorough reliability. We mean to do all we can to discourage sensationalism, the evils of which have been only too apparent in recent astronomical literature."[193] In the same year, Simon Newcomb expressed similar sentiments: "While every astronomer has entertained the highest admiration for the energy and enthusiasm shown by Mr. Percival Lowell . . . , they cannot lose sight of the fact that the ablest and most experienced observers are liable to error when they attempt to delineate the features of a body 50 to 100 million miles away through such a disturbing medium as our atmosphere."[194] Newcomb also warned: "The astronomer cannot afford to waste his energies on hopeless speculations about matters of which he cannot learn anything. . . ." Young, of Princeton, in this period linked Lowell with Flammarion by remarking that the "extreme popular interest in the question of the habitability of 'other worlds' . . . has . . . been greatly intensified by the rather sensational speculations and deliverances of FLAMMARION, LOWELL, and others. . . ."[195]

Two other supporters of Lowell were the New York journalist Garrett Serviss and the astronomer T. J. J. See. Reviewing *Mars* for *Harper's Weekly*, Serviss praised Lowell's theory as "the most complete that has yet been advanced" and its author as possessing "ample means, an active imagination, an enduring enthusiasm, a proper regard for facts, and a clear literary style."[196] In the July 16, 1896, *Dial*, See commended Lowell for having "remarkable literary skill and a clear grasp of subject matter," adding that his canal theory "at least demands more respectful attention than any interpretation that has ever been made before."[197] Whatever weight See's effusive praises carried with the public, their effect among astronomers probably diminished as word spread that See, almost simultaneously with the publication of his review, had become a member of Lowell's staff.

The published reaction to Lowell's *Mars* was somewhat more favorable in Britain than in the United States. Both the astrophysicist W. J. S. Lockyer (1868–1936), reviewing *Mars* for his father's *Nature*, and Agnes Clerke, writing in the *Edinburgh Review*, found much to recommend in it. Lockyer praises its style, logical character, and observational evidence, accepts the canals, and predicts that the vegetation hypothesis will be "hard to oppose."[198] Clerke discusses Lowell's book along with Flammarion's *Mars* and Schiaparelli's 1894 *Astronomy and Astro-Physics* paper. Concentrating on the first but favoring the last, she labels Lowell a "purely Martian product" who appeared after the "great Mars boom" of 1892 "possessed of dollars at discretion."[199] Despite describing Lowell as a "gifted observer" with a "buoyant and original" style, she urges that as "a contribution to science, [his book] can scarcely be taken quite seriously." Her preference is for Schiaparelli, that "miraculous observer" whose "canals have . . . gradually triumphed . . . and . . . rank among the least questionable although perhaps the very strangest of planetary phenomena." (p. 374) Although alluding to Campbell's spectroscopic work and to Douglass's dark area canals, she opts for the traditional Schiaparellian Mars with seas, oceans, and an atmosphere with aqueous vapor. This she supports by calculating the quantity of water contained in the snowy caps, finding it inadequate as the sole source for irrigation. Adopting Schiaparelli's supposed conservatism concerning conjectures about extraterrestrials at a time when the Italian astronomer had already broken from its rigors, she cautions against Lowell's Martian life speculations:

The Alexander's sword of cosmical intelligence, so freely wielded by Mr. Lowell, is not a scientific weapon. In physical investigations knots have to be untied, not cut. . . . Irrigation hypotheses . . . are superfluous and . . . inadmissible. Not that they are, in all shapes, demonstrably false, but that they open the door to pure license in theorising. (pp. 383–4).

Too cautious, too critical is the complaint in the *Spectator's* review of Clerke's review. So important is the search for life elsewhere that "a generation of millionaires or a mountain of gold might worthily and rightfully be expended."[200] Convinced that Mars "is habitable by corporeal beings but slightly different from ourselves," this reviewer welcomes Lowell's book and predicts that further observations "may produce results so enlightening that the chance of attaining them is well worth the devotion of millions of treasures. . . ." The Mars boom was still booming.

Flammarion was among the earliest enthusiasts for Lowell, publishing one of Lowell's papers in the September 1894 issue of *Astronomie* and praising the "distinguished American" for founding "an observatory, inspired, as that at Juvisy, with the dominant idea of studying the conditions of life on the surface of the planets of our system."[201] Flammarion, no less concerned than Lowell about the hydrography of Mars, published a paper in 1895 on "The Circulation of Water in the Atmosphere of Mars." Flammarion's tact may be seen in the version he published in *Knowledge,* where he states: "Mr. Maunder was perfectly right in thinking that 'we cannot assume that what we are able to discern is really the ultimate structure of the body we are examining.' " Flammarion nonetheless asserts that it has been "ascertained indubitably" that water from the melted snows flows through the canals.[202] In the French version, however, the mention of Maunder is omitted, but Lowell, along with five of his drawings, is included.[203] In the winter of 1895–6, Lowell dined with Flammarion in Paris. As he wrote his father, "There were fourteen of us, and all that could sat in chairs of the zodiac, under a ceiling of a pale blue sky, appropriately dotted with fleecy clouds. . . . Flammarion is nothing if not astronomical."[204] Early in 1896, Flammarion published an account of their conversation, in which Lowell informed him that the Schiaparellian seas were in fact prairies and the lakes oases.[205] Although at first opposed to Lowell's prairies, Flammarion by May 1896 had settled on the compromise of making the seas "very shallow." In fact, he states: "In a great number of places they are hardly more than marshes, sometimes dry and sometimes overflowed." Not to be outdone by Lowell, Flammarion adds his own touch by urging that Martians "may have received the privilege of flight."[206]

During his 1895–6 trip to Europe, Lowell also visited his "cher Maître Martien," as he later called Schiaparelli in correspondence.[207] In their conversation, they no doubt discussed an 1895 Italian paper by Schiaparelli on Martian life that reveals the speculative inclinations of the Milan astronomer more clearly than any of his other writings. This aspect of his intellect, so little known in the English-speaking world, must also have been recognized in France, at least after 1898, when Flammarion published a translation of part of this paper.[208] Schiaparelli's paper begins modestly enough with an endorsement of the idea that the canals constitute the irrigation system of Mars. Although retaining his lakes and oceans, which were rapidly evaporating from the English-language Mars under the Pickering-Lowell-Campbell critique, he nonetheless advocates the idea that what one sees is not the waterway but its surrounding vegetation, perhaps lining the slopes of a shallow natural valley. That Schiaparelli possessed pluralist speculative impulses as strong as (although more restrained than) those of Lowell and Flammarion emerges from his explanation of the "mysterious geminations." Urging that the idea that these "may be due to intelligent beings ought not be rejected as an absurdity . . ." (p. 427), he speculates that the Martian engineers may have built dikes at various levels along the slopes of the shallow valleys through which the waterways run. When the spring inundations begin, the "Minister of Agriculture orders the opening of the most elevated sluices and fills the upper canals with water. . . . The irrigation then spreads to the two (lower) lateral-zones . . . , the valley changes color in these two lateral zones, and the terrestrial astronomer perceives a gemination." (p. 428) Gradually the water is released into the lower portions of the valley, fertilizing the lowest region of the valley and producing a single "canal" appearance. Most remarkably, he adds:

The institution of a collective socialism ought indeed result from a parallel community of interests and of a universal solidarity among the citizens, a veritable phalanstery which can be considered a paradise of societies. One may also imagine a great federation of humanity in which each valley constitutes an independent state. The interests of all are not distinguished from the other; the mathematical sciences, meteorology, physics, hydrography, and the art of construction are certainly developed to a high degree of perfection; international conflicts and wars are unknown; all the intellectual efforts which, among the insane inhabitants of a neighboring world are consumed in mutually destroying each other, are [on Mars] unanimously directed against the common enemy, the difficulty which penurious nature opposes at each step. (p. 429)

Although Fourier would have been jubilant at Schiaparelli's socialist Mars, Lowell was probably distressed. In a 1911 address, this aristocratic Bostonian with social Darwinist inclinations turned Mars to his own very different political purposes, presenting it as a benevolent oligarchy in which "the fittest only have survived."[209] That Schiaparelli felt he had exposed in this paper a side of himself best kept concealed is suggested by his concluding remark: "I now leave to the reader the need to continue these considerations, and, as for myself, I descend from the hippogryph."[210] Moreover, across the top of the copy he sent Flammarion, he wrote: "Semel in anno licet insanire," which means "Once a year it is permissible to act like a madman."[211] Schiaparelli's speculations in this paper reached German readers in 1896, when Max Wilhelm Meyer translated a major portion of it for inclusion in his survey of that year on recent Martian researches. Meyer, impressed by Lowell's work, selected most of his illustrations from Lowell's writings.[212]

Schiaparelli's speculations in this paper are surprising in a number of ways. For examples, in late 1894 he told Terby that he would confine himself to publishing Martian observations. His reason is implied in his rhetorical question: "Is there a self-respecting man who still risks publicly mentioning this unfortunate planet [Mars, which has] become the field of action for all the charlatans of the world; which (according to what *Punch* of London says) will in the future supersede the great sea serpent and other similar enticements to the curiosity of the star-crazed?"[213] Moreover, he repeatedly responded to inquires concerning the nature of the canals by such statements as: "I do not know!"[214] Also, Schiaparelli, by 1899, if not earlier, had adopted *"in pectore,"* as he wrote Otto Struve, the view that the canals are atmospheric phenomena, positioned as a result of Martian magnetic forces.[215]

After leaving Milan, Lowell traveled to the town of Lussinpiccolo on an island in the Adriatic to meet a man whose canal claims soon rivaled and conflicted with those flowing from Flagstaff. This self-proclaimed astronomer, almost simultaneously with Lowell, had erected an observatory devoted to planetary astronomy and located at a site (Lussinpiccolo) chosen for its superior viewing conditions. Moreover, he had fabricated for himself the pseudonym "Leo Brenner" and shortly thereafter began fabricating (as then seemed possible and now is certain) observations as well. Born Spiridion Gopčević in Trieste in 1855 of a father who by 1861 had lost his fortune and taken his own life, Lowell's host had by the 1890s acquired something of a reputation based on various political writings, including a best-selling narrative of military conflict in which he (falsely) implied he had participated. He had also secured a wealthy wife and a subsidy from the Austrian government.[216] Using a 7-inch refractor as his chief instrument, Brenner by late 1894 was publishing papers containing extravagant claims concerning his Mars observations.[217] By 1895, he had announced a rotation period for Venus of 23 hours, 57 minutes, and 36.2396 (!) seconds, which he corrected (!) the next year by adding 0.1377 second. Also in 1896 he announced new rotation periods for Mercury ($33\frac{1}{4}$ hours) and Uranus (8 hours, 17 minutes). His 1896–7 Mars observations were so successful, he reported, that he was able "to see not only all 88 Schiaparellian canals and 12 Lowellian canals, but also to discover 68 new canals, 12 seas, and 4 bridges."[218] His immensely detailed Mars map, published and praised by Pohle, Flammarion, and others, as well as his Mars articles scattered in a half dozen journals in three languages, helped establish him as a sort of German-language Lowell. An apostle of pluralism, he championed extraterrestrial life ideas in most or all of his four books, especially in his *Die Bewohnbarkeit der Welten* (1905). His message was also carried in *Astronomische Rundschau,* a journal he founded in 1899 that provided an outlet for his writings when such periodicals as *Astronomische Nachrichten* refused to publish any more of his contributions.[219] During the eleven years his journal survived, he filled it with his astronomical essays, his incessant and vituperative polemics, and papers (some probably pirated) by other astronomers. In its final March 1909 issue, Doctor (which he was not) Professor (which he was not) Brenner revealed to his readers that he was in fact Count (which he was not) Spiridion Gopčević and that he had decided to forsake astronomy (which he did). After some political publications, he disappeared so totally from history that his recent biographer was unable to determine which, if any, of the three death dates and death locations ascribed to him is in fact correct.[220]

In summary, the opposition of 1894 had brought with it an enlarged cast of characters and an intensification of the drama that had been developing. Support for Schiaparelli's astounding observations had grown, Lowell being only the most vocal of those who championed the canaled Mars of the Italian astronomer. Even Campbell, whose spectro-scopic determination doomed the canals in the eyes of Holden, continued to accept the canal observations. Campbell and Maunder, allies in ways they failed to perceive, saw only dimly the revolutionary ramifications of the papers they had published. The public saw still less. However, over the next decade and a half the positions each glimpsed in 1894–5 began to be visible to others. In hindsight, one may discern the beginning of this process in the two oppositions after 1894.

5. The last oppositions of the century: why Schiaparelli was finding Mars a "frightful and almost disgusting subject"

The marginal-quality Mars oppositions centered on December 1896 and January 1899 represented the last, best hope that nineteenth-century man might number the detection of extraterrestrial life among the achieve-ments of his century. Writing in 1895, Samuel Phelps Leland, professor of astronomy at Charles City College in Iowa, confidently attributed to Mars "a civilization as high, if not higher, than our own. Is it possible to know this of a certainty? Certainly." Moreover, because the 40-inch Yerkes refractor was nearing completion, he believed the discovery of Martian life to be imminent; with that telescope, "it will be possible to see cities on Mars, to detect navies in his harbors, and the smoke of great manufacturing cities and towns."[221] Alas, Leland's prophecy went unful-filled, but by century's end some significant progress had been made in the canal controversy. Most notably, Maunder's criticisms of the canal observations and Campbell's evidence against water vapor in the Mar-tian atmosphere began to win support. Two European astronomers, E. M. Antoniadi and V. Cerulli, were central to the first development, while J. E. Keeler and G. J. Stoney contributed to the second. Simulta-neously, the Lick-Lowell phase of the fray subsided for a few years after 1897, in which year Holden resigned his troubled Lick directorship, and Lowell, afflicted with nervous exhaustion, gave up astronomy until 1901.

The level of energy Lowell displayed in 1895–6 was no less than it had been in 1894. Convinced by early 1895 that both his borrowed telescope and its Flagstaff location were unsatisfactory, he contracted for a 24-inch Clark refractor that by July 23, 1896, was in operation at Flagstaff and by year's end had been erected temporarily in Mexico for better viewing of the December opposition. Increasing his staff by hiring three assis-tants, including T. J. J. See, who in a number of ways may be seen as a successor to the recently departed Pickering, Lowell opened war on a second front in late 1896 by claiming that he had "unmistakably" con-firmed Schiaparelli's still disputed 225-day Venusian rotation period by means of his discovery of a system of "distinct and definite . . . broadish lines" on Venus.[222] Lowell's paper, which was met with widespread skepticism, contains the confident claim concerning these lines that "seeing must be distinctly bad to have the more prominent among them not discernible." This assertion is all the more striking in light of the fact, uncovered by Lowell's most recent biographer, that "Not even his assistants saw them as he did. . . ."[223] Although Lowell stressed that there was "nothing about the system bearing resemblance to the noticeable artificiality of the markings of Mars" (pp. 284–5), some critics saw this report as discrediting Lowell's canal observations. In fact, S. M. B. Gemmill took it to be evidence of "diseased imitative faculties" and protested: "We are being treated to canals everywhere and it would have created no surprise had we been assured (preferably from the Lowell Observatory), that the 'spurious discs' of the stars exhibited a canal network. . . ."[224] In 1902, Lowell publicly retracted his report of "spoke-like markings" on Venus, but subsequently he regained confidence in them and defended them in his later years.[225]

Although Lowell's four-year affliction with neurasthenia kept him from his observatory, Douglass continued its observing program, edited its *Annals,* and published various papers. Gradually, however, he grew concerned about the accuracy of many observations of Mars and about Lowell's methodology. In fact, he publicly stated his conclusion that doubled canals were probably a "subjective effect"[226] and in a January 9, 1901, letter to the psychologist Joseph Jastrow revealed: "I would have written you long before but for Mr. Lowell's indifference to taking up the psychological question. . . . I have made some experiments . . . by means of artificial planets . . . placed at a distance of nearly a mile from the telescope. . . . I found at once that some well known planetary appearances could, at least in part, be regarded as very doubtful. . . ."[227] Jastrow did not answer, but Douglass did receive, much to his distress, a form of response from his March 12, 1901, confidential letter to W. L. Putnam, Lowell's brother-in-law, who oversaw the observatory during Lowell's illness. In his letter, Douglass stressed his loyalty to Lowell, but lamented: "His work is not credited among astronomers because he devotes his energy to hunting up a few facts in support of some speculation instead of perseveringly hunting innumerable facts and then limiting himself to publishing the unavoidable conclusions. . . . I fear it will not be possible to turn him into a scientific man."[228] Putnam showed the letter to Lowell, who responded by discharging Douglass. Later Douglass published a paper urging that "illusions of vision" were involved in the canal observations.[229]

The drama of Douglass's defection from the supporters of the canal observations was all the more striking because he had spent seven years at an observatory dedicated to the question of planetary life and had himself reported many canals. Even more noteworthy is the parallel case of Eugène M. Antoniadi (1870–1944) (Figure 10.5), who despite his extended association with Flammarion's Juvisy observatory, his long-standing advocacy of extraterrestrial life,[230] and his many canal sightings, eventually emerged as one of the three most important (with Campbell and Maunder) opponents of the Mars of Schiaparelli, Flammarion, and Lowell. Born in Constantinople to Greek parents, Antoniadi was by the early 1890s sending impressive Mars drawings to Flammarion for *Astronomie* and to Maunder for his B.A.A. report on the 1892 opposition. After moving to Paris in 1893 and becoming associated with Flammarion's observatory, he was in 1896 named director of the B.A.A. Mars Section, preparing in that role highly professional reports on each of the next ten oppositions. The experience gained thereby led to his classic *La planète Mars* (1930).[231]

Antoniadi, who in 1894 had seen 42 canals and at least one double, no doubt awaited the 1896 opposition with Schiaparellian expectations. These must have been shaken by the differences among the reports he received from members of the Mars Section. His developing critical capacity may be seen in his altered assessment of the observations of C. Roberts of Aberdeen. At first describing these as "incontestably the most detailed views" of Mars he had seen, he later rejected Roberts's "astounding results" as probably caused "by some sort of illusion."[232] Antoniadi's 1897 analysis in *Knowledge* of the great differences among the drawings made since 1659 of the "most easily recognizable [Martian] features . . . , the 'Hourglass Sea,' " leads to the comment: "These displacements of 'seas' and 'lakes' (or 'forests' and 'oases') . . . are simply familiar occurrences to the areographer." In explanation, he offers "the over bold and almost absurd assumption that what we are witnessing on Mars is the work of rational beings . . . capable of dealing with thousands and thousands of square miles of grey and yellow material with more ease than we can cultivate or destroy vegetation in a garden one acre in extent."[233] Possibly Maunder, then astronomy editor of *Knowledge* and editor of both the B.A.A. *Journal* and *Memoirs*, suggested to Antoniadi that these changes were more satisfactorily explained as illusions, but whatever the reason, no such Flammarionesque statements appear in his report on the 1896 opposition, which is a model of sobriety. Moreover, before it appeared, he presented a paper to the B.A.A. urging that the doubling of the canals is an optical illusion arising from telescopic "focussing errors."[234] Antoniadi soon broadened this theory, which he was told others had advanced earlier,[235] by urging that the perception of geminations also involves illusions of the eye itself, in particular, "diplopic oscillations arising from an intermittent and uneven action of the muscles of accommodation."[236] Moreover, in a number of other 1898–1900 publications, he championed his illusion interpretation of the geminations, which Schiaparelli and Stanley Williams opposed, but which received support from Abbé Théophile Moreux, W. H. Pickering, and Edwin Holmes.[237] However, in 1901 he backed off from his explanation of gemination as due to erroneous telescope focusing, although he continued to argue that the geminations were illusory.[238]

This dispute led some to question even the canals themselves; for example, Williams in 1899 expressed his "belief that if we could approach Mars to within a few miles, the appearance presented by the so-called 'canals' would be so changed that we would not recognize them at all."[239] What Williams had in mind in this remark, striking because he was possibly the leading British observer of the canals, was revealed when in 1900 he stated that he agreed with Maunder in considering the canals "to be probably formed, in a great many cases at least, of more or less detached spots, streaks, and stipplings which, at the distance and under the conditions in which they are viewed from the earth, run together to form apparent lines or streaks."[240]

Although Antoniadi's B.A.A. report on the 1896 opposition does not directly dispute the reality of the "canals," which had in fact been seen by all members, it does include the remark of Captain P. B. Molesworth (1867–1908), who from Ceylon saw more canals (96) than any other member, that the canals he saw lacked "the hard line-like appearance and angularity with which they are drawn by Schiaparelli."[241] Moreover, Antoniadi, who saw 46 canals, stresses the difficulty in detecting them: "had it not been for Prof. Schiaparelli's wonderful discoveries, and the foreknowledge that 'the canals are there,' [I] would have missed three-quarters at least of those seen now." (p. 63) He also attributes some apparent changes to "nothing more than barbarous representations of a detail too minutely complex to be accessible to our means" (p. 90) and reasserts his illusion theory of gemination (p. 102). Concerning life on Mars, he recommends vegetation as the best explanation of certain reported regional color changes, but advocates a geological rather than an artificial origin of the canals (pp. 100–2). Antoniadi's increasing skepticism concerning the "canals," at least as fine lines, is evident in his report on the 1899 opposition, especially in his stress on "the detection, made by the Rev. P. H. Kempthorne, that a considerable number of [canals] are merely the edges of diffused shadings," or, in other words, that some "of the so-called 'canals' coincide with the boundaries of adjacent areas of different *albedos*."[242] He links this point with reports from Henry Corder, Arthur Mee, and Rev. T. E. R. Phillips that they had seen diffuse streaks, rather than canals (p. 68), in order to attack directly the reliability of the canal observations. This Antoniadi report was also prophetic in its advocacy of both a Martian atmosphere low in aqueous vapor and polar caps formed of carbonic acid. Stoney, rather than Campbell, is cited in support of these conclusions (p. 75).

In the same years that Antoniadi began to question the trustworthiness of the canal observations, the Italian astronomer Vincenzo Cerulli (1859–1927) independently launched a critique of the canals. Cerulli, who had studied physics in Rome and astronomy in Berlin and Bonn, established an observatory in 1890 at Teramo, Italy, and equipped it with a 15.5-inch Cooke refractor.[243] After observing Mars with this instrument during the 1896 opposition, he formulated three main arguments for an optical origin of the canals and presented them in a number of papers and in his book on his opposition observations. His first argument is that if the moon is observed with a low-power opera glass, it "appears streaked with numerous narrow lines . . . having the greatest similarity to the canals of Mars. . . ." This points to the conclusions that "the canals of Mars are themselves only simple alignments of spots similar to those which the telescope shows us on the moon" and that when more powerful telescopes become available, "the canals of Mars will lose that linear form which presently makes them so mysterious and so interesting. . . ."[244] His second argument is that on January 4, 1897, he succeeded in resolving the canal Lethes "into a complex and indecipherable system of the most minute distinct spots."[245] Third, he cites the anomaly, irreconcilable with the reality of the canals, that although from July to December 1896 as Mars drew closer and the angle subtended by its diameter increased from 7 seconds to 17 seconds, he could detect no increase in the visibility (width) of the canals during this period.[246] In short, Cerulli explained the canals as illusions arising from the tendency of the eye to link together discrete points so as to give the appearance of lines. Only in 1909 did he learn that Maunder had advanced this theory in 1894.[247]

Cerulli's claims must have been especially disturbing to Schiaparelli, who had come to view Cerulli as his most promising successor. For example, in an 1899 letter to Struve, he described his own work, and then remarked:

> I had once hoped that a similar work would have been able to be done by Mr. Percival Lowell, but he is very ill; moreover he is more a literary man than an astronomer; he is much attracted to theatrical matters and sensational news. Mr. Cerulli at Teramo is a more serious man; perhaps he will come to do something solid especially if he does not surrender to the tendency that he has shown to judge observations by the theories in his head.[248]

Cerulli, however, persisted in his theory, developing it further in 1900 in his *Nuove osservazioni di Marte (1898–1899): Saggio di una interpretazione ottica delle sensazioni areoscopiche* (Collurania). He encountered opposition from Schiaparelli, who critically reviewed his first book, and from Flammarion, who published vast numbers of naked eye drawings of the moon in an effort to refute Cerulli's "canals" on the moon claim.[249] Although Cerulli's publications were not translated into English, Miss M. A. Orr favorably presented his optical theory in a widely reprinted essay in *Knowledge* in which she noted its similarity to Maunder's theory and suggested that Cerulli should have given more attention to the influence of drawing techniques and of "unconscious imitation."[250] Cerulli's theory received support in France from Gaston Millochau, who reported in 1901 that using the 32.7-inch Meudon refractor at the last two oppositions, he had seen "the canals as a sort of string of small, dark, irregular masses."[251] In the same year, the Spanish astronomer José Comas Solá and the German astronomer Adolf Müller also commented favorably on Cerulli's theory.[252] Two years later, Antoniadi publicly quoted from a letter sent him by Molesworth, who had concluded: "After careful consideration I am inclined to believe with Signor Cerulli, that the canals are not true continuous lines at all. I think an increase of power, if attainable, would show many of them as chains of discontinuous irregular markings, giving the idea of straight hard lines, by their combined impression."[253]

During the 1896 opposition, both W. W. Campbell of Lick and J. E. Keeler of Allegheny Observatory used spectrography to check Campbell's 1894 evidence against a substantial quantity of aqueous vapor in Mars's atmosphere. As Campbell reported in 1897, "Neither Professor Keeler nor I [detected] the slightest difference between the spectrum of Mars and that of the Moon."[254] Moreover, Campbell concluded his paper by recommending: "Astronomers would wisely turn the question of life [on Mars] over to the physiologists for solution; and possibly the latter would wisely hand it over to the domain of pure speculation for the present." Keeler not only was confident of his own result but also urged that Campbell's methods were superior to his own.[255] Campbell then dropped Martian research, returning to it in 1908 with a greatly enhanced reputation. In the interim, his results continued to be an anomaly that at times attracted attention, as in 1899, when Everett I. Yowell favorably reviewed them and noted that they agreed with researches of G. J. Stoney.[256]

As noted previously, George Johnstone Stoney, during the late 1860s, used the kinetic theory of gases and the concept of escape velocity to analyze the compositions of the atmospheres of the planets and their satellites, but he did not at that time publish his detailed results. They were, however, known to such astronomers as R. S. Ball, who had reported favorably on them in his 1893 "Mars" paper. In 1898, Stoney, heartened by the information that the recently isolated gas helium is (in agreement with his theory) very sparsely present in our atmosphere, finally published his analysis. Although making no mention of Campbell's spectroscopic research, he concludes for a Mars bereft of water vapor and capped with carbon dioxide. As he states in reference to his unpublished 1870 discourse: "what was before probable is now certain — that water cannot in any of its forms be present on Mars."[257] Stoney's assumed initial conditions and necessarily hypothetical approach to this complicated problem were called into question by both C. R. Cook of the University of Nebraska and George Hartley Bryan of England. Their analyses, although worked out independently, concurred in permitting water vapor on Mars. In the ensuing controversy, Stoney, Cook, and Bryan all held their ground. Thus, as the new century began, Mars's atmosphere was as beset with riddles as its surface.[258]

Important as the publications of Antoniadi, Cerulli, Campbell, and Stoney were for the future, most Mars papers from the 1895–1900 period were traditional in character. Observations of the canals, as well as theories of their structure, continued to proliferate. For example, when in 1897 Perrotin reported on his recent Mars observations with the 32.7-inch refractor at Meudon, his results differed only in detail from those he had earlier attained at Nice.[259] Brenner and Flammarion were especially active in issuing observational papers, the latter being as energetic in championing a changing Mars as Lowell had been on behalf of the canals. Edwin Holmes faulted the French astronomer for this tendency, stating: "M. Flammarion says 'we must not take differences of drawing for real changes upon the planet,' but this is what he actually does continually."[260] In 1898, Brenner, who by then claimed sightings of 165 canals, reported on the achievement of his observatory during 1897, asserting that in that year alone he had published 17 scientific papers, 52 newspaper articles, and a 408-page book.[261] In the same year, Brenner presented a theory of Mars as a much eroded planet, on the nearly flat surface of which the Martians had erected low dikes, thus forming the canals. Moreover, he claimed that Lowell's "oases" were actually "reservoirs" and that "we see but the largest canals; the millions of lesser ones remain invisible."[262] This theory and his criticisms of observations of Lowell and Pickering involved him in extended and heated controversies with an array of adversaries.[263] Moreover, Brenner's dike theory had to compete not only with all the earlier canal theories but also with new ones, such as those of John Joly of Ireland and M. Teoperberg of The Hague, both of whom argued for the canals being mountain ranges, and that of Abbé Moreux, who viewed the canals as cracks resulting from surface contractions.[264]

What is most noteworthy about this array of elaborate theories concerning the origin and structure of the canals is that by 1900 they were being challenged by sophisticated theories of the illusory character of the canals. On the acceptability of the latter theories rested not only the reputations of many astronomers but also the question whether or not the century's most extensively developed evidences for extraterrestrial life would ultimately carry conviction. That the new illusion theories were placing all this in jeopardy may help explain why in 1900 Schiaparelli confessed to Antoniadi that Mars "has become for me a frightful and almost disgusting subject."[265]

6. The early oppositions of our century and the disappearance of the "marvellous legend of the canals of Mars"

Although the post-1900 portion of the canals of Mars controversy falls beyond the scope of this book, sufficient should be said of it to show the plausibility of two claims. The first was made shortly after the excellent 1909 opposition by José Comas Solá, who asserted that "the marvellous legend of the canals of Mars has disappeared . . . after this memorable opposition."[266] The second claim is that Antoniadi, Campbell, and Maunder played the dominant roles in this dénouement. The first claim needs to be qualified by the admission that some proponents of a canaled Mars appeared in later decades. Among these was Wells Alan Webb, who in a 1956 book cited Edison Pettit, Earl Slipher, and Robert Trumpler as among the post-1920 supporters of the canal observations. However, it is significant that Webb described the challenge they faced by stating: "Antoniadi . . . convinced the world that the Martian network was non-existent."[267] The second claim also needs qualification; the argument developed in this section is not that Antoniadi, Campbell, and Maunder single-handedly demolished the canals but that they were the leaders of a wrecking crew that included Cerulli, Douglass, Newcomb, Stoney, and others and that their efforts were aided by such factors as the deaths of Schiaparelli in 1910 and Lowell in 1916.

In 1901, Lowell, with restored health, returned to his observatory, where Schiaparelli, having retired in 1900 from his, wrote him in 1904 with the encouraging comment: "your theory of vegetation becomes more and more probable."[268] Flammarion also remained active in the Mars debate, which he described in 1904 as having spread to "Buenos Ayres, Mexico, Caracas, . . . Paris, St. Petersburg, Budapest, and Stockholm" and "as having become as much a subject of conversation as politics or art."[269] Although Flammarion, Lowell, and Schiaparelli retained their prominence, other astronomers were emerging as Martian experts, Antoniadi being described in 1904 by one author as the "highest living authority" on Mars.[270] Antoniadi summarized his position in a 1902 paper in which he accepts some canals as well as vegetation on

Mars, but concentrates on the two main theories of what he had called the *"canaliform illusion"*: (1) the Green-Kempthorne theory that the canals are seen at "the boundaries of half-tones" and (2) Maunder's theory that "dark points on the planet's surface, too small to be appreciated individually, produce on the retina the idea of diffused lines."[271] Antoniadi's paper led B. W. Lane to make drawings of Mars, showing the dark areas, but no canals, and to ask persons positioned at certain distances from them to draw what they saw. He found that such persons frequently included canals in their sketches. Maunder, in a note appended to Lane's paper, explained this effect by stating: "in 1882 . . . I noted [that] many of the Schiaparellian canals were prolongations of what other observers had drawn as indentations on the coast lines of Mars."[272] Maunder also summarized his 1894 paper and revealed that in collaboration with J. E. Evans, headmaster of the Royal Hospital School at Greenwich, he was undertaking experiments in which canal-free drawings of Mars would be shown to schoolboys to test whether or not they would draw canals. The extensive experiments he carried out showed, as Maunder reported to the R.A.S. in June 1903, that canals were drawn where none existed on the Martian drawings, pointing to the conclusion that canals "can be seen by perfectly unbiassed and keen-sighted observers upon objects where no marking of such a character actually exists."[273] This result, he claimed, supported his theory that "the most fruitful source of the canal-like impression is the tendency to join together minute dot-like markings." Although stressing that the astronomers who drew canals "have drawn, and drawn truthfully, that which they saw . . . ," he urges against the reality of the canals: "It seems a thousand pities that all those maginificent theories of human habitation, canal construction . . . , and the like are based on lines which our experiments compel us to declare non-existent. . . ." Shortly after presenting this paper to the R.A.S., Maunder summarized its contents for the B.A.A. and also recounted additional experiments of a similar nature carried out with his wife Annie, who was herself an astronomer.[274]

Both the B.A.A. and the R.A.S. discussions of Maunder's papers were quite favorable. However, at the B.A.A. meeting, one member, the Lowellian Mark Wicks, dissented, becoming so upset that eight years later he attacked the opponents of the canals in a work of science fiction entitled *To Mars via the Moon*.[275] At the R.A.S. meeting, Maunder secured an extremely important American ally, Simon Newcomb, who happened to be present. Although in a December 1902 publication Newcomb had accepted the "canal" observations,[276] at that meeting, according to the report, "Newcomb, the great American astronomer, strongly supported [Maunder's] paper. . . ."[277] Four years later, Newcomb published an influential paper (discussed later) arguing for an illusion theory

of the canal observations. By late 1903, Maunder had restated his arguments in both *Observatory* and *Knowledge,* declaring in the former journal that Antoniadi's inclusion in his B.A.A. report on the 1901 opposition of a chart free of canals should be seen as "marking an epoch"
because of it being "the first chart of Mars since 1878 in which the canal
element has not been prominent."[278] After admitting that Antoniadi had
also included some maps showing canal-like features, Maunder concluded that "the great network [of canals] has been discredited, and with
it has gone the evidence . . . of the inhabitants of Mars." After a similar
statement in his *Knowledge* paper, he added: "To have been set free from
the grotesque in observation is to have been freed also from the grotesque
in speculation."[279]

Maunder must have been pleased by an erudite analysis of the reliability of the canal observations that Rev. Edmund Ledger (1841–1913)
published in May 1903. Drawing together the numerous anomalies concerning the canal observations that had emerged since 1877, and also
surveying the ideas of such illusion theorists as Antoniadi, Lane, Maunder, and Pickering, Ledger urged that with some exceptions "the so-
called canals may not really exist" and that at future oppositions "nerve
specialists and oculists [should] work in conjunction with practised observers," at least until photography settles the issue.[280] Maunder, however, must have been disappointed that as yet Antoniadi only partially
accepted the illusory nature of the canal observations. In a July 1903
paper, he was still insisting on the "incontestable objective reality" of at
least some canals.[281] Although mentioning Maunder's experiments in a
number of his 1903 publications, Antoniadi preferred a "contrast of
shading" explanation of some canals and, in conjunction with eye strain,
of all geminations.[282] That Antoniadi harbored suspicions about
Schiaparelli himself is suggested by his remark in one of these papers that
"when the Milan observer spoke of 'robust dikes' and 'inundations' of
these 'canals,' the care of opening whose locks would have been committed to the Martian Secretary of State for Agriculture himself, a distinct
shade was cast on the seriousness of the term in question."[283] Antoniadi's
determination to separate the subjective from the objective in canal observation may be seen in his B.A.A. report on the 1901 opposition. In it,
he included the map (mentioned by Maunder) that was designed to summarize "only the solid data bequeathed to us by the labours of the last
125 years."[284] This map, which showed only a few canals, and these as
broad, diffuse streaks, contrasted with his canaled general reference map
on which Antoniadi bestowed the enigmatic heading: "Chart of Mars.
Of all Authenticated, but not necessarily Objective, Details." Stressing
that "exactly one half of the 'canals' seen in 1900–1901 . . . correspond
to edges of indefinite half tones" (p. 91), Antoniadi also included many

quotations from Molesworth, who had written him: "I am inclined to believe with Signor Cerulli, that the canals [are] chains of discontinuous irregular markings. . . ."[285] Antoniadi's continuing attachment to some traditional ideas of Mars is revealed by his rejection of Stoney's dry Mars capped with carbon dioxide, which he felt was refuted by the spectroscopic observations of Huggins and Maunder (p. 93). The crisis reflected in this report may explain why Antoniadi, who previously had published his reports within two years of the opposition analyzed, delayed his 1903 report until 1910.

That the crisis emerging in Britain had not yet crossed the Atlantic is suggested by the claim, made in a review in *Popular Astronomy* of Antoniadi's report for 1901, that American astronomers, although dubious about Lowell's theories, had not questioned his observations.[286] Even if this questionable claim is correct, Americans were learning at this time of the aforementioned illusion publications from reprints in American periodicals or from discussions of them by W. W. Payne and by W. H. Pickering.[287] Lowell, whose observations had been directly attacked in some of these publications, responded in various papers, as well as by initiating efforts to photograph the canals.[288] By 1905, the Maunder-Evans paper had been discussed in at least one Russian and three German journals, as well as being reprinted in a somewhat free translation in a French journal, in which Flammarion inserted a brief statement that when he repeated their experiments, essentially no canals were drawn.[289] Although Flammarion provided no details on his experiments, Lowell referred to them as constituting a refutation of Maunder's "Small Boy Theory."[290] Cerulli also discussed his own illusion theory of the canals in a French publication in which he asserted: "All modern areography has come to seem to me a marvellous chapter of physiology. . . ."[291]

By early 1905, Maunder may have felt that after two decades of effort, he had finally won a major victory. S. A. Saunder, president of the B.A.A. when Maunder presented his June 1903 paper, characterized his achievement by stating: "The onus of proof now lay upon those who thought the canals were there."[292] Maunder had won a battle, but not the war; his taste of victory, however sweet, was short. From Flagstaff in mid-1905 flowed reports of such conclusive evidence that the new B.A.A. president, A. C. D. Crommelin, took it as decisive proof of the canals, and an American author stated that the "hot controversy [over the reality of the canal observations] may now be considered definitely settled. . . ."[293]

This dramatic development was Lowell's announcement that on May 11, 1905, one of his staff, Carl Otto Lampland (1873–1951), had succeeded in photographing canals on Mars. While Lowell circulated this news in numerous papers and personally carried it to Europe in late summer of 1905, an excited Schiaparelli wrote him "I should never have believed it possible," and Mark Wicks publicly predicted that the illusion theory of the canals "will doubtless now be relegated to obscurity."[294] In Lowell's late 1906 *Mars and Its Canals,* he reported that in various photographs "thirty-eight canals were counted . . . and one double. . . ."[295] Anxious for even better images from the 1907 opposition, Lowell funded an expedition to the Andes led by the Amherst astronomer David Todd (1855–1939), accompanied by Earl C. Slipher of Lowell's staff. Slipher's three months at this Chilean observatory netted 13,000 new images, while Lampland and Lowell, photographing from Flagstaff, secured 3,000. In these, Lampland, Lowell, and Slipher sighted numerous canals. Todd, however, at first saw none, but by late 1907 he announced "Almost all the photographs exhibit Martian canals, so called . . . ," and in 1908 he claimed "our plates ought to silence forever the optical-illusion theorists."[296] Lampland's Martian photographs, although awarded a medal by the Royal Photographic Society, neither silenced the illusion theorists nor fully satisfied most Mars experts. Less than one-fourth inch in diameter, these photographs would not withstand reproduction without loss of their fine detail, promoting a practice of authors publishing them along with drawings showing the unreproducible detail they claimed to see in the originals. In at least one case, a journal labeled the drawings photographs, causing further confusion.[297] Moreover, when experts examined the negatives, many saw no canals or only diffuse markings or different figures in photos of the same region.[298] In short, the old problems had reappeared, raising the possibility that Maunder's message of 1894 might need rephrasing as: "We cannot assume that what we are able to discern [even in photographs] is really the ultimate structure of the body we are examining."[299]

The ambiguity of photography emerged in a new form in 1908 when Vesto Slipher (1875–1969), who had joined Lowell's staff in 1901, secured spectrographs of the previously little studied red end of the spectrum. These, Lowell and Slipher asserted, showed enhancement of the "little a" spectral line, indicative of Martian water vapor. Coming shortly after even Huggins had given up his claim for the detection of water vapor,[300] Slipher's paper, not surprisingly, sparked a sustained controversy that brought W. W. Campbell back to the fray and eventually to the three-mile summit of Mount Whitney, where his spectrographs yielded once again no evidence of water vapor on Mars.[301] Of the few dozen papers that by 1911 had fueled this controversy, little need be said to supplement two recent analyses of it showing that in the same Slipher spectrographs in which Slipher and Lowell saw enhancement and Frank W. Very (1852–1927) reported measuring it, Campbell, Hale, and Newcomb saw nothing significant.[302] Concerning Slipher and the astrophysicist Very, it may be relevant that both were pluralists; in fact, Very was a Swedenborgian.[303] Concerning Campbell, it is noteworthy that when in 1925 W. S. Adams and C. E. St. John published new spectrographic evidence of Martian water vapor, but in a sparse quantity reconcilable with Campbell's earlier results, he accepted their determination. Ironically, it was also illusory, being excessive by a factor of about forty-five.[304] In regard to Lowell, critics may agree with Hale's complaint to Campbell that "What I particularly dislike is [Lowell's] absolutely unscientific method of dealing with the material and of stating his case,"[305] but they should also recognize that Lowell was the first to publish the "velocity-shift" method of investigating planetary spectra and that his pluralist theories later led him to urge Slipher to study the spectra of certain spiral nebulae, from which observations Slipher in 1913–14 discovered the high velocities of these objects.[306] Lastly, it is very significant that in this controversy, Campbell's cause was taken up by Stoney, who from kinetic theory had earlier reached conclusions comparable to Campbell's, but who had not until then urged that their results were mutually supportive.[307]

Among the most prominent of Lowell's American supporters were the scientists George R. Agassiz and Edgar S. Morse and the sociologist Lester Frank Ward, all of whom had succeeded in seeing canals. In 1906, Morse, director of the Peabody Museum, recounted his Flagstaff observations of Mars in his *Mars and Its Mystery,* and in 1907 Ward published his "Mars and Its Lesson," which, he told Lowell, "has been copied by scores, perhaps hundreds, of papers all over the country. . . ."[308] In Britain, Lowell had the support of three astrophysicists: Sir Norman Lockyer, his son W. J. S. Lockyer, and J. H. Worthington.[309] Lowell's other British allies included E. H. Hankin and C. E. Housden. In a paper in *Nature,* Hankin praised Lowell's "brilliant researches" and proposed, with complete seriousness: "Perhaps on Mars there is only one living being, a gigantic vegetable the branches or pseudopodia of which embrace the planet like the arms of an octopus, suck water from the melting polar caps, . . . and are visible to us as the Martian canals."[310] The contribution of Charles Edward Housden, a hydraulic engineer, consisted of a long paper and short book detailing a pumping system (Figure 10.8) for the Martian canals, followed by another book in which he worked out the hydraulics of Venus.[311] The Hankin and Housden hypotheses were not the most bizarre schemes attributed to Mars in the first fifteen years of our century. The credit for those go to German authors: (1) Ludwig Kann, who maintained that Mars is largely covered by seaweed, the canals being separations caused by rapid ocean currents, (2) Adrian Baumann, who proposed that the dark Martian regions are land, covered with vegetation and possibly animal life, whereas the yellow regions consist of ice, on top of which lie ashes from volcanoes,

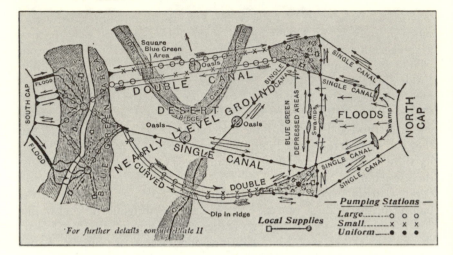

Figure 10.8. C. E. Housden's "PLAN ILLUSTRATING THE TRANSPORT OF WATER ACROSS MARS" (from his Riddle of Mars).

which also cause the cracks seen as canals, and (3) Philip Fauth and Hanns Hörbiger, who championed a cracked Martian ice mantle a thousand miles thick.[312]

With allies such as Hankin and Housden, Lowell scarcely needed opponents – yet these he had, including the energetic octogenarian Alfred Russel Wallace (1823–1913), codiscoverer of the theory of evolution by natural selection. While surveying nineteenth-century science for his *The Wonderful Century* (1898), Wallace discerned that the growing evidence against life elsewhere in the solar system, when combined with recent researches in sidereal astronomy, offered possibilities for a fresh analysis of the extraterrestrial life question. This he sketched in a 1903 essay entitled "Man's Place in the Universe" and elaborated in a book of the same title.[313] Wallace's paper contains some shocking claims, including that recent astronomical discoveries pointed to the probability that man's place in the universe "is special and probably unique" and that "the supreme end and purpose of this vast universe was the production and development of the living soul in the perishable body of man." (p. 396) He summarized his argument by stating:

The three startling facts – that we *are* in the centre of a cluster of suns, and that that cluster *is* situated not only precisely in the *plane* of the Galaxy, but also *centrally* in that plane, can hardly now be looked upon as chance coincidences without any significance in relation to the culminating fact that the planet so situated *has* developed humanity. (p. 411)

Although Wallace's first two "facts" were supported to a surprising extent by the astronomy of his day,[314] his paper and book created an international controversy involving a number of the figures discussed in this and the two preceding chapters.[315] Although Wallace referred in his paper (p. 396) to the Christian objection to pluralism, it is doubtful that this played a role, as it had for Whewell and others, in the origin of his antipluralism.[316] An analysis of Wallace's position and the controversy it created[317] is beyond the scope of this book, but mention may be made of an important appendix that Wallace in 1904 added to his book. This appendix, which has recently attracted attention,[318] consists of a critique of those who on the basis of evolutionary theory "consider it absurd . . . to suppose that man, or some being equally well organised and intelligent, has not been developed many times over in many of the worlds they assume must exist." In response to such persons, the cofounder of evolutionary theory suggested that they do not

. . . give any indication of having carefully weighted the evidence as to the number of very complex and antecedently improbable conditions which are absolutely essential for the development of higher forms of life from the elements that exist upon the earth or are known to exist in the universe. Neither does any one of them take account of the enormous rate at which improbability increases with each additional condition which is itself improbable. (pp. 326–7)

In developing his argument, Wallace stressed not only the number of conditions necessary for the evolution of intelligent life and the improbability of each condition but also that they must simultaneously be present and act over vast periods of time. So compelling did Wallace consider this argument, especially when joined to his other arguments, that he stated: "the total chances against the evolution of man, or an equivalent moral and intellectual being . . . will be represented by a hundred million of millions to one." (p. 335)

In 1907, Wallace responded to Lowell's *Mars and Its Canals* by publishing *Is Mars Habitable?* in which the great biologist drew not only on the astrophysical studies of Campbell and Stoney but also on those of J. H. Poynting, whose researches Wallace used against a mid-1907 paper by Lowell arguing for a moderate Martian temperature.[319] Their contemporaries cannot have missed the irony when Lowell urged that Wallace's book be dismissed as having been written by "a man, however eminent in one branch [of learning], wandering into another not his own."[320] Despite the fact that Wallace contributed a much needed biological perspective to the debate, and although Carl Sagan has admitted to being "astounded" by the ingenuity and foresight shown in Wallace's book,[321] it suffered from a serious flaw evident in Wallace's remark on Lowell's canal observations: "I myself accept unreservedly the substantial accuracy of the whole series." (p. 14)

By 1907, Lowell and his allies had survived an array of attacks, but they remained vulnerable on the most fundamental level: the objectivity of their canal observations. In 1907–8, a veritable blitz of illusion theory papers began, the earliest coming from Simon Newcomb, whose eminence assured that his paper would attract attention and whose ingenious discussion of the "Optical and Psychological Principles" involved in the canal observations repaid that attention. In his introductory optical section, Newcomb urges that the effects of aberrations and diffraction associated with refractors make a thin line seen on Mars appear broader than its actual width, just as stars seem wider than they are.[322] Turning to psychological factors, he introduces the term *"visual inference,"* defined as "the act by which the mind unconsciously draws conclusions as to an observed object from the image formed by its light on the retina." (p. 8) As an example of this process, Newcomb notes that although a very thin bright line must cast an image 2 or 3 minutes in breadth on the retina, we perceive it as a geometrical line. The important role of visual inference in observing objects at the limits of visibility was illustrated by experiments in which subjects, including four astronomers, observed disks on which lines *broken* at regular intervals had been drawn, these drawings being seen by transmitted light. In agreement with Maunder's earlier results, these observers saw *continuous* lines when suitably distanced from the drawings. After remarking favorably on the work of Lowell and his staff, Newcomb combines the two portions of his paper to urge that on the assumption that the albedo of the Lowellian canals is about half that of the surrounding areas, it seems probable that each canal could be seen only if about 10 miles in width. Moreover, he states: "Adding the border of 20 miles on each side necessarily produced by aberration, diffraction, and softening, the apparent breadth . . . would be 50 miles. . . ." (p. 15) Taking the conservative figure of 40 miles as the probable minimum apparent width of the canals, and using Lowell's claim that at least 400 canals averaging about 2,000 miles in length exist on Mars, Newcomb calculates that the total apparent surface area of the canals should be approximately 60 percent of Mars's surface! (pp. 15–16) Because nothing like this was seen, Newcomb concludes that the objectivity of the canal system is far from established. Lowell responded to Newcomb's critique, but with limited effect; the master had spoken, and by late 1907 a dozen or more astronomical journals had published reports of his results.[323]

The impact of Newcomb's illusion arguments was increased when in a 1908 *Harper's Weekly* essay, he re-presented them along with a report on new physical studies of Mars. Citing improved estimates of the Martian surface temperature and Campbell's spectrographic results, Newcomb urges: "For snow-fall substitute frost-fall; instead of feet or inches say fractions of a millimetre, and instead of storms or wind substitute little motions of an air thinner than that on the top of the Himalayas, and we have a general description of Martian meteorology."[324] Concerning life on Mars, Newcomb concludes that the conditions on that planet are "unfavorable to any form of life unless of the very lowest order. . . ." Newcomb's tactful praise for portions of Lowell's planetary work becomes excessive when he states: "What is especially remarkable is that Mr. Lowell was almost, if not quite, the first astronomer to test the limits of vision, and lay down exact rules for deciding between what could be seen and what could not be seen upon a planet." Although Lowell's efforts in this area were not insignificant, the person most deserving of credit for developing such criteria was E. W. Maunder, whose 1903 R.A.S. address, it is important to note, Newcomb had heard.

Newcomb repeated his analysis of the illusion issue, along with a summary of Campbell's and of Poynting's astrophysical studies of Mars, when shortly before his death in 1909 he wrote the Mars article for the eleventh edition (1910–11) of the *Encyclopaedia Britannica*. Newcomb had earlier proposed that the editors ask Maunder to write the article on this controversial planet, leading Lowell to protest to Newcomb and to send the editors Slipher's spectrographs.[325] When the article appeared, it included a long footnote so hostile to Newcomb's Mars that one suspects that Lowell wrote it.

Almost simultaneously with Newcomb's 1907 publication, A. E. Douglass, by then professor of astronomy at the University of Arizona, argued in two papers that because halo and ray phenomena distort our views of Mars, we should doubt the reality of all but the most conspicuous canals.[326] Later in 1907, Solon I. Bailey (1854–1931), of Harvard, published an illusionist interpretation of Lowell's canal photographs, some of which had recently been displayed at the Massachusetts Institute of Technology, which university had appointed Lowell to a nonresident professorship in 1902.[327] In 1908, Harold Jacoby (1865–1932), professor of astronomy at Columbia University, attacked the reliability of both Lowell's photographs and his canal observations, citing in support of the latter point some of Douglass's arguments, as well as Lowell's own disputed Venus observations.[328]

The 1907–8 period also included a number of illusionist essays published by Europeans. In Britain, both E. W. Maunder and his wife published restatements of their views,[329] but the most elaborate analysis of the limits of vision issue came from G. J. Stoney, whose three-part "Telescopic Vision" warns that for objects as distant as Mars, astronomers can have little certainty that what they see in their telescopes corresponds to the actual surface features of the planet; in fact, he urges that telescopes transform the details of an object such as Mars "into something unlike itself."[330] Stoney describes the problem as so severe that it is erroneous to suppose that present-day astronomers "have even so much knowledge of what exists upon Mars, as [pretelescopic] astronomers . . . could have of what exists upon the Moon." (p. 809) In Italy, Cerulli continued to present his point of view, publishing papers in both 1907 and 1908. In the latter year, he favorably discussed Newcomb's illusion arguments and argued against Lowell's photographs as evidence for canals.[331] Henri Dierckx drew on the theories of both Cerulli and Newcomb when in a 1908 essay in a new Belgian astronomical journal he set out to refute claims made by Flammarion that Lowell's photographs proved the accuracy of the canal observations.[332] In short, the Mars of Flammarion, Lowell, and Schiaparelli was under international attack.

Nonetheless, the array of 1907–8 illusion papers served more to rally the revisionists than to rout those who accepted the canal observations. Lowell continued to carry his message far and wide. Londoners in 1908 saw the honeymooning Lowell ascend by balloon above Hyde Park to compare the sight of the park paths to his canals. The multilingual Lowell lectured at the Sorbonne in the same year, amidst such exclamations as: "Why, he is even clever in French!"[333] Other Frenchmen knew of him from Flammarion's excited articles on Lowell's Martian photographs[334] or saw him during visits to his friend Flammarion or heard him address the Société astronomique de France, which in 1904, probably at Flammarion's urging, had awarded him its Janssen medal. Lowell was winning converts in Scotland, where in 1908 the *Edinburgh Review,* so negative toward him in 1896, now asserted:

Would-be destructive criticisms have been showered upon Flagstaff – suggestions of instrumental defects, of diplopic vision, of optical interference, of obscure physiological laws seemingly devised by Providence for the deception of planet-observers, of hypnotic suggestion. All, except perhaps the last, have been received seriously if not patiently; have been tested in ingenious ways, killed to the satisfaction of Mr. Lowell and his assistants. . . .[335]

So famous had Lowell become that his South American expedition needed no more specific address than "Lowell expedition to Andes."[336] Moreover, the complex attacks of the illusionists seem to have carried little weight with a public aware that the "canals" *had been seen* and convinced that observation forms the bedrock of science. What the illusionists consequently needed was *observational* evidence against the canals. This they soon secured.

The excellent 1909 Mars opposition brought forth a flood of publications, about ninety in 1909 and a similar number in 1910. Antoniadi authored two dozen or more papers, many showing the influence of the transforming experience he had when from September 20 to November

27, 1909, he observed Mars with the 32.7-inch Meudon refractor, seeing many of the so-called canals dissolve into fine detail. In his paper dated December 23, 1909, in which he draws together his observations with those of other observers, he asserts that the disappearance of the canals under the high resolution of large telescopes justifies the conclusions that

I.   *The true appearance of [Mars] is . . . comparable to that of the Earth and of the Moon;*
II. *Under good seeing there is no trace whatever of a geometrical network; and*
III. *The "continental regions" of the planet are variegated with innumerable dusky spots of very irregular outline and intensity, whose sporadic groupings give rise, in small telescopes, to the "canal" system of Schiaparelli.*[337]

And he adds: "No doubt we have never seen a *single* genuine canal on Mars. . . ." His reference in point III to small telescopes was probably directed against those authors who from the mid-1880s had been urging that large telescopes are inferior to those of more modest aperture for planetary observation.[338] This debate was also in the background of a famous telegram (mentioned by Antoniadi) that Yerkes director E. B. Frost sent to R. Jonckheere, a French canal enthusiast, to respond to the latter's telegraphed inquiry as to whether or not canals were seen with the Yerkes telescope. Frost's laconic reply was: "Forty-inch telescope too big to show canals."[339] Antoniadi, initially incensed at what he took to be Frost's flippancy, soon saw the significance of his telegram – and proceeded to claim priority over it, urging that he had announced the resolution of the canals in a Greek publication eight days before Frost's telegram.[340] In his December 23 paper, Antoniadi also cites letters from E. E. Barnard and A. S. Williams to show that both agreed with his view of the canals as congeries of diffuse detail (pp. 137–8). Moreover, Millochau, he reports, had seen Mars in the same way while observing with the Meudon telescope in 1899, 1901, and 1903. Antoniadi then turns to Maunder, quoting with approval his 1894 denial that our telescopes necessarily "reveal to us the ultimate structure of the surface of the planet" and praising him as "the man who by a masterly interpretation of facts, saw clearly into the 'canal' deadlock at a time when everything was darkness to all." After mentioning the Maunder-Evans experiments and noting that they had recently been endorsed by Charles André, director of the observatory at Lyon,[341] the cosmopolitan Greek astronomer provides the fullest exposition of Cerulli's views available in English up to that time and also summarizes Newcomb's 1907 paper. In his concluding remarks, Antoniadi adds that the canal network "vanished when the planet was practically at its closest approach to earth. . . . And the fact that no straight lines could be held steadily when much more delicate detail was continually visible constitutes a fatal objection to their crumbling existence." Never before had such a broadly based and well-documented critique of the canals been prepared. The effect of Antoniadi's analysis was enhanced when on December 29, 1909, Maunder presented it at a B.A.A. meeting that propitiously began with a report on some new photographs of Mars made by G. E. Hale with the 60-inch Mount Wilson reflector. These were judged superior to Lampland's and moreover agreed with Barnard's description of Mars. Antoniadi's paper received a warm welcome, especially from Maunder, who described it as bringing members "to a most satisfactory conclusion as to the 'canal system' of Mars" and thereby freeing them from "the idea that there were miraculous engineers at work on Mars. . . ."[342]

Antoniadi continued his attack in a January 1910 paper that again shows his command of the international Mars literature. Calling Cerulli and Maunder "the ablest theorists in the 'canal' question," he cites observations of Millochau, Molesworth, and C. A. Young (in 1892) as supportive of a canalless Mars. His most impressive evidence consists of a recent letter from Hale, who from Mount Wilson wrote that he had seen "a vast amount of intricate detail" on Mars but "no trace of narrow straight lines. . . . I may add that observations of Mars with the 60-in. reflector were also made by Messrs. Abbot, Adams, Babcock, Douglass (formerly of the Lowell Observatory), Ellerman, Fath, Seares, and St. John, all of whom agree with me regarding the character of the details shown."[343] Antoniadi's other 1910 publications include his long B.A.A. reports on the oppositions of 1903, 1905, and 1907 that were in a number of cases supportive of his fully matured conception of the planet. For example, in his report for 1903, he replaced the term "canal" by "streak," and in that for 1907, he backed off from a changing Martian surface in favor of one sometimes obscured by clouds.[344] His strongest statements appear in his report on the 1909 opposition, where he asserts that on first viewing Mars on September 20, 1909, with the Meudon telescope, he

. . . thought he was dreaming and scanning Mars from his outer satellite. The planet revealed a prodigious and bewildering amount of sharp or diffused natural, irregular, detail, all held steadily; and it was at once obvious that the geometrical network of single and double canals discovered by Schiaparelli was a gross illusion.[345]

Such observations convinced him of the correctness of "Mr. Maunder's theory of 1894–1895 [that] the lines, which are glimpsed severally, are merely a summation of complex details. The true theory of the 'canal' fallacy is thus due to Mr. Maunder, and to him alone." (p. 44) His report on the 1909 opposition and his subsequent book on Mars contain dozens of drawings in which he contrasts Mars as he had seen it at Meudon with Schiaparelli's representations (Figure 10.9).

In late 1909, Maunder, perhaps sensing that the decisive battle in the

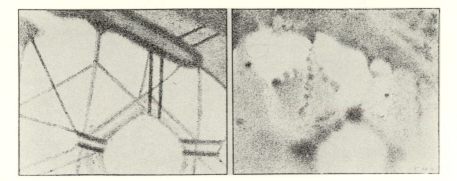

Figure 10.9. A region of Mars as represented by Schiaparelli (left) and by Antoniadi (right). (From Antoniadi, La planète Mars; *courtesy of Hermann éditeurs des sciences et des arts)*

thirty-year war against the Martians and their canals was at hand, roused himself to action, publishing in November 1909 a paper arguing that the temperature of Mars would not support life. Unfortunately, neither he nor Antoniadi as yet had accepted Campbell's conclusions concerning the Martian atmosphere; in fact, Antoniadi continued to describe Mars as a "living and probably still inhabited world."[346] In 1910, Maunder drew together in a long paper three decades of criticisms of the canal observations, citing Antoniadi's recent work as at last settling the controversy.[347]

A feeling of high drama must have been in the air at the March 30, 1910, B.A.A. meeting when Maunder and other members arrived to find Percival Lowell present. In London to lecture at the Royal Institution, Lowell had learned of the meeting less than two hours before; nonetheless, at the invitation of H. P. Hollis, then president of the B.A.A., he addressed the association that for two decades had been a major source of criticisms of the canals. As reported in the B.A.A. *Journal,* Lowell's remarks revealed the charisma of the American astronomer, whom Hollis praised for providing the most "thrilling half hour" he could recall at any B.A.A. meeting.[348] Civilities, more than substance, characterized the discussion following Lowell's lecture, but the controversy continued in the next issue of the B.A.A. *Journal,* where Maunder noted that although canals could be seen in small photographs of his 1903 test diagrams, these disappeared under magnification. Antoniadi, responding to a letter from Lowell, stressed that his Meudon drawings agreed far better with Hale's photographs than did recent Flagstaff drawings, and M. E. J. Gheury proposed that Lowell go to Meudon and Antoniadi to Flagstaff.[349] Shortly after the B.A.A. meeting, Lowell addressed the Société astronomique de France and the Royal Astronomical Society; not surprisingly, he received a warmer welcome from the former group.[350]

By early 1910, many publications had presented the claims (1) that experiments had shown that canalless drawings containing irregular, mottled markings could be seen as covered with canals and (2) that astronomers such as Antoniadi, Barnard, Cerulli, Hale, and Millochau had succeeded in resolving the canals into diffuse detail. Antoniadi had abandoned his contrast of shading explanation of the canal observations, placing his authority squarely behind Maunder. In short, it was becoming increasingly clear that astronomers must take very seriously Maunder's 1894 statement that "we cannot assume that what we are able to discern is really the ultimate structure of the body we are examining." Soon an international array of astronomers allied themselves with the anticanal position. J. Comas Solá of Spain concluded an essay by asserting that "the marvellous legend of the canals of Mars has disappeared. . . ."[351] Lucien Libert approvingly quoted Comas Solá's claim in a 1910 review of recent Mars research, in which he also asserted that the "geometrical network of canals is a pure optical illusion" and that "Mars is not a world overflowing with life as it has previously been portrayed; and the hour is not far distant when it will be a desert of red sand. . . ."[352] Abbé J. Belpaire, writing in 1911 for the astronomical society of Antwerp, stated: "The supposition that one can see straight lines on the surface of Mars is ever more losing credibility. . . ."[353] Abbé T. Moreux advocated similar conclusions in his *Les autres mondes sont-ils habités?* (1912), which Charles de Kirwan reviewed and cited in support of the claim: "Mars appears to be a deserted planet . . . where life, if it exists, ought no longer manifest itself except by the lowest types of cryptograms."[354]

In 1910, Sweden's Svante Arrhenius attacked Lowell's Mars in a pair of German publications.[355] Although aware of the illusionist ideas of Antoniadi, Cerulli, and Maunder, Arrhenius for the most part accepted the canal observations, explaining the canals in terms of chemical changes in areas adjacent to cracks caused by earthquakes. He also drew on Campbell's spectroscopic researches and various physical considerations to urge that Mars is too cold and barren to be inhabited. His opposition to Lowell's theories is striking, because he favored life on Venus and was championing during this period the theory that life on earth arose from spores of extraterrestrial origin.[356]

In the United States, criticism of the canal observations became especially intense around 1910, in which year Henry Paradyne, writing on "The Mythical Canals of Mars," declared that the December 29, 1909, B.A.A. meeting marked the decisive turning point, although he admitted that "it is probable that the great Martian myth will persist . . . [as] one of the large stock of fallacies which science has annexed, to its great embarrassment. . . . But for those who prefer the cold precision of fact to the warm, nebulous glow of fancy, there will be no canals on Mars now or for all time to come."[357] Paradyne could have made more of his term "myth," used by him only to mean "bogus," had he elaborated on Lowell's efforts to enhance his canal observations by developing a dramatic myth of brilliant Martian engineers desperately endeavoring to irrigate their dying planet. E. B. Frost, the witty Yerkes director, succinctly expressed the views of many American astronomers when a reporter wired: "Send us three hundred words expressing your ideas of the habitability of Mars." In response, Frost telegraphed: "Three hundred words unnecessary – three enough – no one knows."[358] That E. E. Barnard of Yerkes not only rejected the canal observations but also supported Maunder's illusion theory of them was revealed by Antoniadi, who informed American, English, and French readers that Barnard had written him: "I quite agree with you in respect to Mr. Maunder's work in trying to clear up the tangle about the canals of Mars. I think he has thrown much light on the subject."[359]

The most comprehensive critique of the canals made by an American at this time came from Robert G. Aitken, of Lick, whose 1910 review of recent Martian research includes the information that Lowell had reported sighting two newly constructed canals, numbered 659 and 660 by Lowell.[360] Aitken also mentions that Jonckheere had reported verifying every Lowell canal and *adds twenty-three new ones!*" On the other hand, Aitken asserts that at Lick "we do not see the canals," nor, he notes, do Antoniadi, Barnard, Comas Solá, Hale, and Williams accept them. What gives Aitken's essay special importance is that he endorses Campbell's work, thereby becoming the first author after Newcomb's 1908 paper to accept both Campbell's Mars and the illusionists' Mars. That the two were mutually supportive is clear from Aitken's concluding comment: "It is difficult to understand how so small an amount of water can keep a geometrical canal system on *Mars* in active operation." Soon, other authors recognized the correctness of the conjunction of these two views of Mars.

Some advocates of the Mars of Flammarion, Lowell, and Schiaparelli appeared after 1910, and even into the 1960s, but the evidence adduced in this chapter points to the conclusion that by around 1912, astronomers had reached a consensus that the canal observations had been discredited. Although Antoniadi, Campbell, and Maunder deserve the most credit for this, other factors contributed as well. The deaths of Schiaparelli in 1910 and of Lowell in 1916, the poor quality of the five post-1910 oppositions, the onset of World War I, and the increasing interest in stellar astronomy were also factors. Moreover, the widespread rejection of the nebular hypothesis in the period from about 1920 to the mid-1940s in favor of encounter theories of planetary formation,[361] the latter (unlike the former) entailing that few stars are orbited by planets, combined with the abandonment of belief in life elsewhere in our solar system to make the pluralist position less attractive than at any time during the previous two centuries. Two British astronomers may be cited to illustrate the antipluralism of this period. In 1928, Arthur Eddington stated: "I do not think that the whole purpose of the creation had been staked on the [earth]. . . . But I feel inclined to claim that at the present time our race is supreme; and not one of the profusions of stars . . . looks down on scenes comparable to those which are passing beneath the rays of the sun."[362] A year later, James Jeans asserted that because "planets are very rare . . . , life must be limited to a tiny fraction of the universe."[363]

7. Conclusion: "the canal fallacy . . . relegated into the myths of the past"

To conclude this chapter, let us look for a final time at the six figures central to the canal controversy: Antoniadi, Campbell, Flammarion, Lowell, Maunder, and Schiaparelli. In 1913, Antoniadi wrote the epitaph of the canals: "Ponderous volumes will still be written to record the discovery of new canals. But the astronomer of the future will sneer at these wonders; and the canal fallacy, after retarding progress for a third of a century, is doomed to be relegated into the myths of the past."[364] Although resigning the directorship of the B.A.A. Mars Section in 1917, Antoniadi continued to write about that planet, publishing in 1930 his classic *La planète Mars,* in which he presented it as in a state of "advanced decrepitude," as a "vast red solitude" on which remnants of water and vegetation may persist, but almost certainly no higher forms of life.[365] When this amateur who had taught so much to professionals died in 1944, P. M. Ryves described him in the B.A.A. *Journal* as "the world's greatest authority on the Planet Mars." Ironically, Antoniadi's obituary was followed by a review, also written by Ryves, of a new book devoted to reviving Lowell's theories.[366]

W. W. Campbell returned briefly to the canal controversy in 1918 with a paper supportive of the illusionist interpretation of the canals. Juxtaposing discordant passages from W. H. Pickering's and Lowell's observations of Mars, Campbell concluded: "If two observers so advantageously situated, so capable and enthusiastic as Pickering and Lowell cannot agree better. . . , what hope is there for. . . ordinary observers. . . ?"[367]

In 1937, a year before his death, Campbell was still distressed because, as he wrote H. N. Russell, "there are many astonomers . . . who are sceptical of the dependability of my [Mars] results."[368] Campbell need not have been concerned; subsequent studies fully vindicated his claims.

Flammarion's death in 1925 prevented him from finishing the manuscript for the third volume of his treatise on Mars. That he was championing in it the same Mars to which he had been devoted for six decades is indicated by a message he cabled the *New York Times* eighteen months before he died: "Mars is as full of life as the earth. We do not see the inhabitants, but we can observe the upheavals on the surface and must deduce cause from effect."[369]

Lowell, shortly before his death in 1916, stated that "since the theory of intelligent life [on Mars] was first enunciated twenty-one years ago, every new fact discovered has been found to be accordant to it."[370] Moreover, on March 1, 1916, he assured the editor of *Scientific American* that "among astronomers qualified to judge there is no difference of opinion as to the existence of the 'canals' of Mars, opposition arising solely from those who without experience find it hard to believe or from lack of suitable conditions find it impossible to see." As W. G. Hoyt remarked when making this letter public, Lowell's statement is "not only self-serving but slanderous."[371] In response to Antoniadi's 1913 paper, Lowell wrote the editor to urge that Antoniadi's claim "that Schiaparelli is responsible for the theory of [the canals's] artificiality should be corrected. The blame rests wholly with me."[372] The assessment of "blame" being among the tasks of critical history, it may be noted that because Lowell's writings contained at times sensationalistic and biased statements and even instances of deception and slander, he deserves his share of censure. Nonetheless, Antoniadi, who in his 1913 essay had referred to Schiaparelli's "Martian minister of agriculture," was correct in claiming priority for the Italian astronomer. Moreover, Flammarion in his obituary on Lowell reasserted that Lowell "had been inspired, according to what he himself recounted . . . , to make discoveries on the world of Mars . . . by the publication of my work *La planète Mars* in 1892."[373] In this context, mention may also be made of the probable influence of W. H. Pickering, whose close association with Lowell came precisely when Lowell was formulating his Martian theories. In fact, Pickering, who had assumed the Martian mantle years before Lowell, reclaimed it between Lowell's death and his own demise in 1938, making declarations during those decades as excessive as any ever ascribed to Lowell.[374] Among Lowell's legacies, not the least was the magnificent observatory he founded and endowed, which is now a major center for astronomical research. Recently, however, Clyde Tombaugh, who joined the Lowell staff in 1929 and in 1930 discovered the planet Pluto, suggested another result of Lowell's Mars controversies; this was that they left his observatory for a number of years "virtually an outcast in professional astronomical circles" and its older staff members under "ostracism from the astronomical community."[375]

E. W. Maunder retired from Greenwich in 1913, becoming for a time secretary of the Victoria Institute, a society devoted to reconciling reason and religion. In 1912, Maunder had addressed this society on the question of extraterrestrial life. After recounting his anticanal and anti–Martian life arguments, he urged that because of the restrictive conditions necessary for life and the failure of at least multiple star systems to provide such conditions, the probability of life elsewhere is small. Consequently, the earth must be among the "small portion, at best" of inhabited planets in the universe.[376] Asking what this may indicate about the "purpose and design" of earthly life, he suggested: "The Wisdom of God Who was with Him 'when He prepared the heavens, . . . when He appointed the foundations of the earth,' desired that, as 'the Word made flesh,' He might 'rejoice in the habitable part of His earth, and have His delights with the Sons of men.' " (p. 94) Maunder must have been disappointed that his antipluralist arguments received no warm welcome from his audience, one member characteristically complaining:

I came to this meeting hoping that Mr. Maunder would tell us something about life upon other worlds, and I have been much disappointed. . . . Surely all the millions of stars . . . were not created without some purpose . . . ? And of the planets in the solar system, may there not be forms of life quite unlike those with which we are familiar which would flourish under such conditions as they offer?[377]

That the physicotheological pluralism of the Whewell era had survived into the new century was also evident in the reactions when in 1913 Maunder published his *Are the Planets Inhabited?* For example, William Dean Howells, the editor of *Harper's Monthly,* demanded to know, concerning Mars and Jupiter, whether God could not "invent kinds of life which should not only be possible, but comfortable in both?"[378] History has dealt harshly with Maunder. Unnoticed in England's *Dictionary of National Biography,* his solar researches described as "undistinguished" in the *Dictionary of Scientific Biography,* where his Martian studies are scarcely mentioned, he is now nearly forgotten. An evaluation of his accomplishments as Greenwich solar spectroscopist, founder of the B.A.A. and editor for extended periods of both its *Journal* and of *Observatory,* and author of an array of respected astronomical books and papers cannot now be attempted. But in future histories of astronomy, Maunder deserves mention as the central figure in freeing astronomy from perhaps its most famous fallacy, and thereby sensitizing astronomers to the need for a more sophisticated understanding of observation.

In assessing Maunder's contribution to the Mars debate, it is relevant to ask whether subsequent information, including that derived from recent close approach photographs of Mars, has verified the Maunder-Cerulli theory that the canal observations result from an unconscious tendency in observers to see discrete details on Mars's surface as integrated into lines. Two recent studies shed light on this question. In concluding their 1975 comparison of classic canal maps with Mariner photographs, Carl Sagan and Paul Fox state: "while a small subset of the classical Lowellian canals corresponds to topographic or albedo features on Mars, the bulk of the canals do not. Indeed there are many canals where there are no real surface features, and there are many surface features where there are no canals."[379] Although this conclusion is difficult to reconcile with Maunder's specific theory, it does not negate his broader 1894 caution against assuming that observation reveals actual planetary features. In fact, Sagan and Fox, in a sense, restate Maunder's view: "The vast majority of the canals appear to be largely self-generated by the visual observers of the canal school, and stand as monuments to the imprecision of the human eye-brain-hand system under difficult observing conditions." (p. 609) The second, more recent study is that of R. A. Wells, who in his *Geophysics of Mars* (1979) reports a substantially higher correlation between specific Martian surface detail and the classic canal drawings. Wells summarizes his analysis by stating: "While there is no evidence supporting Lowell's extensive network of some 800 canals, Mariner 9 data do support a much smaller basic framework more on the order of that originally observed by Schiaparelli,"[380] Moreover, Wells asserts that whereas Sagan and Fox had found from about 6 to "no more than 10 or 20" acceptable correlations, his studies indicate that between 20 and 100 alignments are detectable (pp. 436–7). Because Wells's analysis, although more sophisticated than Maunder's, is based on the Maunder-Cerulli theory, it may be taken as evidence of the correctness of that theory, but with the qualification that purely subjective factors play an even larger role than Maunder expected. This point continues to be of relevance, because in the future astronomers may need to perform additional exorcisms of the demons that lie in wait for those making observations relevant to the question of extraterrestrial life. In fact, as recently as 1973, Bruce C. Murray suggested:

So this hopeful view of Mars is not just a popular thing. It affects science very deeply, I think. I am not sure that we are out of it yet. My own personal view is that we are all so captive to Edgar Rice Burroughs and Lowell that the observations are going to have to beat us over the head and tell us the answer in spite of ourselves.[381]

Schiaparelli's canal observations, as well as the view he took of them late in his life, are discussed by Professor Wells, who has formulated an interesting conjecture on the basis of an extract of a letter from Schiaparelli to Cerulli that Wells located in M. Maggini's *Il pianeta Marte* (Milan, 1939). Although Wells states that a search for Schiaparelli's original letter in the relevant archival sources proved unsuccessful and that consequently the letter cannot be dated more precisely than Maggini's assignment of it to sometime in 1907,[382] he sees the surviving extract as evidence for the conclusion that "it is highly probable that the stigma associated with 'Lowell's Legacy' might have been eradicated almost before it began." (p. 422) Wells's specific suggestion is that because Schiaparelli seems in this extract to accept Cerulli's theory, it is conceivable that Lowell, had he known of Schiaparelli's change of mind, might have been led by his esteem for his "cher maître Martien" to forsake the canal observations as well. In fact, Wells suggests: "To have learned [Schiaparelli's view] in a public manner might have dealt [Lowell] a mortal blow!" (p. 426)

My researches have led to materials that shed light on this letter and conjecture. In particular, a fuller extract of the letter was published in 1908 by Cerulli, who stated that he received it in July 1907.[383] Thus, the letter was made public during Lowell's lifetime, and moreover its significance was summarized in a German journal.[384] Furthermore, I have been able to reconstruct Schiaparelli's views from the last two years of his life. On August 29, 1909, he wrote Antoniadi admitting that the Flagstaff photographs that Lowell claimed showed canals might be deceptive.[385] However, by December 15, 1909, when he again wrote Antoniadi, his views seem to have changed: "The polygonations and geminations of which you show such horror (and with you indeed others also) are an established fact against which it is useless to protest. Dr. Cerulli was convinced some weeks ago. I have shown him a series of fine photographs obtained by Mr. Lowell in July, 1907. . . ."[386] In 1910, Schiaparelli twice published letters defending the canal observations. In a February 10 letter to Hale, who had sent him some Mount Wilson photographs of Mars as evidence for Antoniadi's arguments, Schiaparelli urged: "The experience of Prof. Lowell in 1907 has demonstrated that photography is capable of distinctly representing the geometrical lines and polygonations, about which it pleases some still to doubt; what has been possible at Flagstaff will certainly be possible also at Mount Wilson."[387] On May 19, 1910, Schiaparelli responded to Arrhenius's article in the German journal *Kosmos* by writing:

So far as I am concerned, I have still not succeeded in forming an organic whole of reasonable and credible thoughts on the phenomena of Mars which are perhaps more complicated than Arrhenius assumes. However I am myself entirely clear on one point concerning which I am in full agreement with him, namely, that one must take into consideration the geological structure of the planet. . . .

I also believe with Arrhenius that the lines and streaks of Mars (the name "canal" should be avoided) may be entirely explained through the action of physical-chemical forces, excepting certain periodical color changes which can perhaps be explained as the result of organic formations of large extent, as on the earth the flowering of steppes and similar phenomena. I am likewise of the opinion that the geometrical and regular lines (the existence of which is still disputed by many persons) teach us nothing at all at present in regard to the probability or improbability of intelligent beings on this planet. However I would judge it to be good were someone to collect everything . . . that can reasonably be advanced on behalf of their existence. And from this viewpoint, I hold in high esteem the noble-minded endeavors of Mr. Lowell and the expenditures in gold and work made by him for this purpose as well as his very sagacious arguments on the matter.[388]

This final defense of his canal observations suggests that when six weeks later Schiaparelli died, he was still astride his hippogryph.

Leaving aside the question of the influence of the canal controversy on popular thought, what was its effect on astronomy? A number of authors have endorsed the assessment offered in 1938 by Reginald Waterfield:

Now the story of the "canals" is a long and sad one, fraught with back-bitings and slanders; and many would have preferred that the whole theory of them had never been invented. Yet whatever harm was done was more than outweighed by the tremendous stimulus the theory gave to the study of Mars, and indirectly of the planets in general. Whether in a postive way to champion it, or in a negative way to oppose it, it attracted many able observers who otherwise might never have taken an interest in the planets. . . . So the pistol which Schiaparelli had so unwittingly let off, though it shocked the finer feelings of many, had undoubtedly been the starting signal of that race for discovery which the planetary astronomers are still successfully pursuing.[389]

It is possible, however, that Waterfield's assessment was influenced by the facts that he had "no doubt" as to the existence of some Martian canals and believed Martian vegetation "extremely probable."[390] Writing in 1966, Carl Sagan expressed a radically different view of the canal controversy:

It became so bitter and seemed to many scientists so profitless, that it led to a general exodus from planetary to steller astronomy, abetted in large part by the great scientific opportunities then developing in the application of physics to stellar problems. The present shortage of planetary astronomers can be largely attributed to these two factors.[391]

This historical study finds support both for Waterfield's and for Sagan's claims, but the majority of evidence favors Sagan's assessment. Many benefits came to astronomy from the canal controversy, but the price paid by the astronomical community in loss of credibility, internal discord, methodological misconceptions, and substantive errors, as well as the efforts wasted on the observation of ambiguous detail, was far too high. If history can teach a lesson on this matter, it is that should a comparable controversy arise, astronomers may wish to emulate the actions of Antoniadi, Campbell, and Maunder. They were the heroes of the canal controversy, whereas Schiaparelli, Flammarion, and Lowell were its chief causes – and victims.

11

Some conclusions concerning the unconcluded debate

1. Extent and character of the extraterrestrial life debate before 1917

The historical materials in these chapters suggest many conclusions, two of which concern the extent and character of the extraterrestrial life debate in the pre-1917 period. First, they provide a massive refutation of the widespread assumption that the debate began in this century and that finally in recent years the "long-standing belief that the only intelligent life in the universe exists on our planet, Earth, is gradually disappearing."[1] We have seen that the debate in fact arose in antiquity, continued in almost every subsequent century, and (as documented in the Appendix) had by 1916 produced over one hundred forty books on this topic. The great majority of these, moreover, as well as most of the thousands of pre-1917 essays, papers, and reviews in the debate, advocated extraterrestrial life. Although some authors of these publications possessed only modest learning, a large number were persons of indisputable prominence. About three-fourths of the most prolific astronomers[2] and nearly half of the most prominent intellectuals of the eighteenth and nineteenth centuries contributed to the debate.[3]

Extensive as the pluralist debate has been, brilliant as many of its participants were, it differs from most debates in the history of science by the fact that it remains unresolved. Centuries of searching for evidence of extraterrestrials have produced hundreds of claims, thousands of publications, and millions of believers, but not as yet a single solid proof. The vast extent but inconclusive character of the debate suggest two questions, treated in the next three sections of this chapter. First, how has pluralism retained its attractiveness even though its arguments have in many cases proved defective? Second, what persistent fallacies can be detected in pre-1917 pluralist writings?

2. The unfalsifiability, flexibility, and richness in explanatory power of many pluralist theories

The attractiveness of some pluralist theories may be attributed to their unfalsifiability, flexibility, and richness in explanatory power. The generalized pluralist position is clearly unfalsifiable. If the moon be found bereft of life, Mars remains. If Mars be barren, Venus is available. If the solar system is shown to lack life except on the earth, emphasis can be placed on planets orbiting other stars. If no evidence for such planets is found, attention can turn to other galaxies. If this fails, the past and future offer other opportunities. Even specific pluralist theories, for example, that of lunar life, have shown remarkable resistance to falsification. When the near side of the moon appeared lifeless, pluralists populated the far side. When this ploy became impossible, attention turned to lungless lunarians, the past of the moon, or to insects and later to subsurface microorganisms. Moreover, the assumption of an omnipotent God unconstrained by laws of terrestrial life expanded almost endlessly the repertoire of rescue techniques available to pluralists. Pluralism has also shown remarkable flexibility, having been adjusted to fit extremely diverse astronomical, religious, philosophical, and literary contexts. The moon with or without an atmosphere, Mars as moist or dry, Jupiter as fluid or solid – these and more have been accommodated by pluralist astronomers. Writers on religious subjects have been no less ingenious in adapting that doctrine. Atheists and evangelicals, fundamentalists and physicotheologians, spiritualists and Swedenborgians – all have found pluralism pliable enough for their needs. Philosophers also: Empiricists and idealists, positivists and the speculative, optimists and pessimists – all at times have used pluralism for their purposes. Science fiction writers feeling constrained by characters limited to only one head, two sexes, or five senses, by plots tied to merely terrestrial cataclysms, or by scenery showing but a single moon have been no less ready to draw on pluralism. Extraterrestrial life ideas are also extraordinarily rich in explanatory power. Flashes on the moon, lines on Mars, radio signals of unknown origin, and even odd images in ancient statuary can be and have been explained by postulating appropriate extraterrestrials. Pluralism has proved scarcely less attractive to religious writers anxious to explain where heaven and hell may be, why so much evil can be present on earth, what certain Scriptures mean, or how man may be immortal.

The claim that the unfalsifiability, flexibility, and richness in explanatory power of pluralist theories have contributed to their attractiveness needs to be supplemented by an examination of how these features can be present in theories that have now generally been discarded. The explanation lies in the fact that these features are now recognized as poor criteria for assessing theories. As illustration, consider the theory, sometimes called vitalism, that organic matter is animated by an indwelling nonphysical entity. This theory is unfalsifiable; for example, the failure of scientific tests to detect the vitalist entity is to be expected, because the entity is nonphysical, Moreover, the vitalist theory is highly flexible and rich in explanatory power; if organic objects behave in a certain way, one can claim that the vitalist entity directed them to behave in that way. If they behave in another way, the theory can be adjusted by supposing that in that instance the entity caused actions of another sort. Despite these apparent strengths, vitalism, at least in the oversimplified version presented here, is rejected by scientists, who stress, among other points, that it is not a scientific theory because it cannot be tested by scientific means. The vitalist theory can be contrasted with Newton's mechanics, which is falsifiable, inflexible, and in one sense poor in explanatory power. Were any planet not to have the orbit or speed specified by Newton's theory, his theory would be falsified. Moreover, because his theory applies only to areas governed by inverse-square fields of attraction, it cannot be adjusted to fit or used to explain other areas.

Some present-day philosophers of science warn against unfalsifiability, flexibility, and richness of explanatory power as tests for theories. For example, Karl Popper points out that many theories that are unfalsifiable and rich in explanatory power are known to be wrong.[4] He even presents falsifiability as the best test of scientific theories. Although Popper developed his views after the events narrated in this book, cases may be cited in which early opponents of pluralism made objections that fit with his analysis. For example, one wit complained in 1905: "Prof. Percival Lowell is certain that the canals on Mars are artificial. And nobody can contradict him."[5] Svante Arrhenius urged in the next decade that Lowell's theories "explain everything, and therefore in fact nothing."[6] In 1925, Reginald Waterfield warned that although Lowell's canal theory "seems to explain everything," it nonetheless "must be remembered *that a theory that postulates intelligent existence will*, ipso facto, *account for practically any phenomenon*."[7] Within this context, Schröter, Gruithuisen, Weinland, Lowell, Tesla, and others may be described as the vitalists of the history of astronomy.

3. The importance of empirical evidence

Pluralists as well as their opponents have repeatedly stressed the importance of empirical evidence in judging extraterrestrial life hypotheses. Yet it is an illuminating fact that, with the exceptions of Whewell, Antoniadi, and Proctor, almost no figures in the pluralist debate significantly changed their positions. William Herschel's failures to detect evidences of extraterrestrial life only led him to postulate new locales for it. Although evidence against life elsewhere in the solar system continually increased from 1850 to 1900, the pluralist camp suffered few defections. No more striking example of the willingness of pluralists to disregard empirical evidence can be cited than the readiness with which they accepted life on the sun, an idea favored by Bodmer, Boscovich, G. Knight, Bode, both Herschels, Arago, Gauss, Brewster, Read, Schimko, Phipson, Ponton, Liagre, Coyteux, Preyer, Goetze, Etler, Warder, and others. In short, the pattern that is apparent is that most individuals took new empirical evidence to be an occasion for adjusting rather than abjuring their theories.

In one sense, however, observational claims have played an overly large role in the debate, this being due to the prevalence, not only in the public but possibly even among astronomers, of an insufficiently sophisticated understanding of the role and reliability of observation. That this problem may persist is suggested by an examination of the methodology of astronomy advocated by the astronomer Edwin Hubble (1887–1953) in his classic *Realm of the Nebulae* (1936) and in other writings.[8] The selection of Hubble as a representative of the position to be analyzed is justified both by his numerous contributions to astronomy and by the fact that Allan Sandage in his preface to the 1958 reprinting of Hubble's *Realm of the Nebulae* stated that although minor modifications of Hubble's approach may be necessary, these changes are "not in fundamental philosophy or direction of attack. Hubble's original approach to observational cosmology remains." (p. vii) Sandage also asserted that Hubble's book remains "a source of inspiration in the scientific method" of astronomy. Hubble's position rests on a sharp dichotomy between observation and theory. This dichotomy and the primacy Hubble assigns to observation are both conveyed in his claim: "The observations and the laws which express their relations are permanent contributions to the body of knowledge; the interpretations and the theories change with the spreading background." (p. 4) Hubble stresses the same points when at the conclusion of his book he asserts: "Not until the empirical results are exhausted need we pass on to the dreamy realms of speculation." (p. 202)

In 1955, the British cosmologist Hermann Bondi, without explicitly attacking Hubble, presented a strikingly different position. Bondi wrote:

There is undoubtedly a widespread opinion that astronomical theory is, by and large, airy speculation, that it changes from day to day in its most fundamental tenets, . . . and that its reliability should accordingly be assessed to be low. By contrast the results of observational work are, according to this opinion, solid incontrovertible facts, permanent and precise achievements, that will never change and whose reliability is accordingly high.[9]

In support of his belief that "these opinions are unfounded and false, and their prevalence does great harm to the progress of astronomy," Bondi cites many historical examples of the unreliability of observations, claiming even that "errors in theories are if anything less frequent than in observational work." (p. 159) For example, he states that in 1931 "HUBBLE and HUMASON . . . inferred from their data that the constant of red-shift of the nebulae was 4.967 ± 0.012 and soon after, from almost the same data, that it was 4.707 ± 0.016. . . ." As a partial explanation of the tendency for observational claims to be incorrect, Bondi notes that they "are frequently obtained at the very limit of the power of the instruments used. . . ." (p. 158)

Bondi's evidence for his position need not be repeated in a history rich in illustrations of the tendency to place excessive trust in observation. Before citing specific cases, it is crucial to state that these are not instances of fraudulent observations or of interpretations of observations, but simply observations, reports of what the astronomer, in a reasonably straightforward meaning of the word, *saw*. Thus, when during the 1840s Lord Rosse of Ireland and W. C. Bond at Harvard reported resolving Orion into stars, they were not asserting a theory but stating what they believed they had observed. Reports of spectra indicative of water vapor in Mars's atmosphere or of changes, figures, or flashes on the moon were also presented not as theoretical but as observational claims. Schiaparelli's determination of Mercury's rotation period, repeatedly confirmed until the 1960s, and See's observation of distant nonluminous bodies orbiting stars, never confirmed, are other examples of observational reports that have proved ephemeral. The most striking case in the present context is the linear network seen on Mars by dozens of astronomers. The tragically large expenditure of talent and time involved in the Mars case testifies to the need for a critical approach to observation such as that taken by Maunder, Antoniadi, Cerulli, Newcomb, and others. The more sophisticated understanding that they developed of the Martian observations was crucial in resolving the debate about the Martian canals. Moreover, it supplied astronomers with analytical techniques unavailable in the philosophical or psychological literature of the period, and still perhaps not sufficiently appreciated. It is only in the last few decades that such philosophers of science as N. R. Hanson, Thomas Kuhn, and Michael Polanyi, in some cases building on Kantian, Wittgensteinian, or Gestaltist views, have developed theories of science that recognize and respond to the problematic character of observation itself. Their writings have shown that observations, rather than being totally separate from theory, may in many instances be theory-laden or, to put the point more generally, that the sharp dichotomy assumed by Hubble between observation and theory can rarely be detected in actual scientific developments. In short, what may with irony be called the "message from Mars" should be kept in mind as astronomers use ever more powerful techniques for attaining observations that may, or may not, contain evidence of extraterrestrials.

4. Recurrent fallacies and linguistic abuses

Over seven centuries ago, Albertus Magnus described the issue of a plurality of worlds as "one of the most wondrous and noble questions in Nature." Studies of that question, however, have repeatedly led to wild speculations and acrimonious quarrels that some have seen as the chief scandal of astronomy. The frequency with which pluralists have fallen into logical fallacies and linguistic abuses has contributed to this problem.

Among the logical and/or methodological fallacies in the debate, one of the most pervasive is mistaking necessary conditions for sufficient conditions. According to most analyses, air is a necessary condition for intelligent life, but it is only one of numerous necessary conditions, all of which must be present for a celestial object to be habitable. All too often, evidence of an atmosphere has been taken as proof of a planet's habitation, whereas such evidence is conclusive only when joined to other facts that with it constitute sufficient conditions for life. Moreover, until we learn the necessary and sufficient conditions for life to arise, we shall not be able to infer actual habitation of any planet from proofs of its habitability. An effective statement of this overall point was made a century ago by Faye, who justifiably objected that pluralists rarely recognize the multitude of delicate conditions simultaneously necessary for life.

The complexities involved in analogical arguments are such that present-day logicians still dispute their forcefulness. A penetrating analysis of their logic was presented a century ago by C. S. Peirce, who urged: "There is no greater nor more frequent mistake in practical logic than to suppose that things that resemble one another strongly in some respects are any more likely for that to be alike in others."[10] Citing a logician who illustrated the fallacious nature of much analogical argumentation by pretending to prove that Napoleon was "an impersonation of the sun," Peirce added: "any two things resemble one another just as strongly as any two others, if recondite resemblances be admitted." Peirce's analysis, which raises problems for claims that the similarities of a particular planet to the earth prove its habitability, received no more attention from pluralists than Mill's earlier analysis. Neither Mill nor Peirce would have denied all usefulness to analogical argumentation: both accepted such approaches as methods of discovery, but not (except under special conditions) as methods of proof. Pluralists, however, rarely confined their claims to the former context.

A third widespread fallacy involves misuse of large numbers. Typically, pluralists, explicitly or implicitly, have relied on such assertions as this: Among the countless planets orbiting the billions of stars in our galaxy, some at least must be inhabited. Frank W. Cousins, in 1972, demonstrated the error in pluralist arguments of this form, using as illustration the well-known conjecture that were a monkey to type long enough, it would eventually hammer out *Hamlet*. Making reasonable assumptions about typing speed, and so forth, he showed that this would occur once in $10^{460,000}$ seconds, which is effectively never. To see this, consider a universe containing a billion galaxies, each made up of a billion stars, around each of which a hundred planets revolve. Place a billion monkeys on each of these planets and set them typing for fifteen billion years (the approximate age of the universe). This would produce only 10^{46} seconds of typing time toward the $10^{460,000}$ seconds needed.[11]

A fourth type of fallacy, not infrequently associated with large number arguments, involves the failure to realize that most probabilistic arguments used by pluralists must take the form not of inductive probabilities but of theoretical probabilities. Ernan McMullin has effectively discussed the applicability of this distinction to extraterrestrial life arguments.[12] In attempting to predict the probability of heads coming up when a coin is tossed, we may either count the occurrences of heads in a large number of tosses or apply considerations of mechanics, such as the laws of fall, in order to make predictions. The former method is that of inductive probability, the latter that of theoretical probability. This distinction, rarely drawn in probabilistic discussions of extraterrestrial life, is important because such discussions in most cases rest only on theoretical probabilities. The data provided by astronomy on such problems as the existence of planetary systems for other stars are far too limited for the assigning of meaningful inductive probabilities. Consequently, such discussions must derive from theoretical considerations. The difficulty in this is that present-day theories of, for example, planetary formation are not well established. Nonetheless, some pluralists have proceeded to make probabilistic assertions about this matter.

A fifth source of fallacies is the failure to realize that most pluralist arguments are hypothetico-deductive in nature, which is to say that proving them takes the form not of inductive inferences gradually culminating in a generalization but of testing whether or not the deductive consequences of the hypothesis in fact occur. For example, the methodological relationship between flashes seen on Mars and the claim that Mars is inhabited depends on the possibility of inferring such flashes from assuming Martian life. If such an inference seems reasonable, then the observation of flashes counts as support for life on Mars. The difficulty is that other hypotheses, for example, the supposition that Martian mountains reflect sunlight, also explain the flashes. Strictly speaking, only when all hypotheses except that of habitation have been disproved can that hypothesis be considered established.

A sixth category of fallacies derives from misunderstandings about evolutionary theory. Despite the rich explanatory power of Darwinian theory and its strong empirical support, many leading theorists of evolution and philosophers of science agree that Darwinian theory is not, except in the broadest sense, predictive. Consequently, it cannot generate detailed predictions of the direction evolution will take in a given population, even of terrestrial animals. One reason is that evolutionary change works from chance variations that are not themselves predictable. Loren Eiseley has drawn a striking conclusion from this:

. . . nowhere in all space or on a thousand worlds will there be men to share our loneliness. There may be wisdom; there may be power; somewhere across space great instruments, handled by strange, manipulative organs, may stare vainly at our floating cloud wrack, their owners yearning as we yearn. Nevertheless, in the nature of life and in the principles of evolution we have had our answer. Of men elsewhere, and beyond, there will be none forever.[13]

Eiseley's elegant prose should not be understood as indicating that evolutionists generally support the existence of extraterrestrial intelligent life in some form. In fact, both George Gaylord Simpson and Theodosius Dobzhansky have, as A. R. Wallace earlier, presented strong antipluralist arguments based on evolutionary theory.[14]

The seventh and final fallacy is the assumption that the question of other worlds is purely a scientific issue. Although the claim that extraterrestrials exist is, in principle, verifiable, and despite the fact that scientific considerations enter most discussions of it, it is not, as noted before, empirically falsifiable. Furthermore, Arthur Lovejoy's claim that metaphysical considerations influenced seventeenth- and eighteenth-century pluralists may be extended to later authors as well. It is hardly surprising that men's conceptions of matter, nature, God, life, goodness, and so forth, affect their responses to such a large question as that of extraterrestrial life. That this pattern continues in contemporary discussions is illustrated, for example, by the stress Carl Sagan and others place on the "assumption of mediocrity," the doctrine that "our surroundings are more or less typical of any other region of the universe."[15] Another example is R. B. Lee's assertion that we possess three powerful tools for analyzing the possibilities of life elsewhere in the universe: "evolutionism, historical materialism . . . pioneered by Marx and Engels . . . , and uniformitarianism."[16] The point to be stressed is not that metaphysical ideas must not enter the debate but that authors recognize and candidly admit their presence, rather than claiming that the issues can be analyzed solely in scientific categories.

Abuses of language are also evident in a number of pluralist writings, one of the three most common being the use in observational reports of terms unjustifiably carrying connotations of evidence of extraterrestrial life. The distortions that derived from calling the lines seen on Mars "canals" serve as a warning of the dangers of such theory-laden terms. A second form of linguistic legerdemain is the deployment of double negatives. An excellent example is Flammarion's claim concerning supposed changes on the lunar surface that "we cannot affirm that . . . there are not some changes which can be due to the vegetable kingdom or even the animal kingdom, or — who knows? — to some living formations which are neither vegetable nor animals." This deceptive form of expression, for which contemporary pluralists have also been criticized,[17] is no more acceptable in scientific discourse than are such statements in the political realm as "It is not impossible that, were I elected, we would have fifty percent reduction in taxes" or "I am not able to disprove the allegation that my opponent is a crook." The third abuse (this list is not exhaustive) involves appeals to the prejudices and passions of the public. W. W. Campbell, in his review of Lowell's *Mars,* was especially critical of that author's propensity for taking "the popular side of the most popular scientific question afloat." The need for accurate, engaging, and understandable public presentations of science is indisputable, but these and comparable linguistic abuses may be taken as marks of authors more interested in influencing than informing, in converting than clarifying.

5. The place of the idea of a plurality of worlds in the history of astronomy

One of the most remarkable paradoxes encountered in this history is that the idea of extraterrestrial life, long a delight of poets and a dogma of physicotheologians, entered at many points into the evolution of scientific astronomy and into the careers of some of its most distinguished practitioners. Figures as influential as the elder Herschel, Lalande, Flammarion, Barnard, Brashear, Newcomb, and Lowell were all first attracted by pluralist writings to the field that they eventually so enriched. Moreover, evidence has been presented indicating that pluralist concerns played a major role in the establishment of the Plumian professorship and the observatory at Cambridge University, in the improved instrumentation acquired by William Herschel and by Schiaparelli, and in the founding of the Juvisy, Lick, Lowell, and possibly other observatories. Reports from the 1860s and 1870s of changes on the moon may have been spurious, but their effect was real: They resulted in a renaissance in selenography. Schiaparelli's "canali" have also proved ephemeral, but they contributed crucially to interest in Mars. It is also noteworthy that all four pioneers of sidereal astronomy – Wright, Kant, Lambert, and Herschel – were deeply involved with pluralism. Moreover, solid evidence indicates that the majority of the authors most successful at interesting the public in astronomy did so in good measure by embellishing their books with extraterrestrial life ideas. Both Flammarion and Proctor, the most widely read nineteenth-century astronomical authors, launched their careers by pluralist publications and sustained them by scattering essays on extraterrestrials in their later volumes. That they were not alone in employing this strategy is shown by the popularity of such authors as Fontenelle, Ferguson, Bode, Chalmers, Dick, and Lowell.

On another level, pluralist and antipluralist convictions influenced both the formulation and acceptance of various theories and promoted and sometimes permeated numerous astronomical observations. Stanley Jaki has recently developed the thesis that pluralist preferences led many astronomers to favor cosmogonies supplying planets orbiting other stars and to neglect evidence conflicting with such cosmogonies.[18] That William Herschel's theory of the sun originated less in observation than in his penchant for proposing locations for extraterrestrial life seems beyond dispute. The attraction felt by later astronomers for Herschel's solar model may largely be explained in similar terms. The lunar buildings and roads described by Gruithuisen and the spurious lunar changes announced by a number of astronomers, as well as the seas, canals, and oases repeatedly reported on Mars, were surely in good part due to the pluralist sentiments of their observers. On the other hand, the antipluralist convictions of Whewell contributed to his successful search in the 1850s for flaws in the then dominant view of nebulae.

Numerous as the interactions between astronomy and pluralism were, beneficial as some of them may have been, it cannot be denied that in many cases disputes over extraterrestrial life distracted astronomers from concerns more central to their science. The excessive claims made by Schröter, Gruithuisen, Dick, Flammarion, W. H. Pickering, Lowell, Brenner, and others not only distorted the truths of astronomy but also detracted from the credibility of the astronomical community. Astronomy popularized frequently became astronomy "pluralized," while the success of popularizers tended at times to be a function less of the accuracy of their expositions than of the audacity of their assertions. Whatever the overall effect of the interactions between astronomy and pluralism may have been, this much is certain: They continue into the present. Today's astronomers, aware that their discipline heavily depends on public confidence and support, need, no less than their predecessors, to be vigilant about the credibility of their community as long as the question of extraterrestrial life remains the most emotionally and ideologically charged issue associated with astronomy.

6. Interactions between extraterrestrial life ideas and religion

Interactions between pluralism and religion have been numerous and varied. In fact, an important conclusion repeatedly documented in this book is that in numerous pre-1900 publications in the debate, no sharp line can be drawn between an author's astronomical and religious perspectives. Although before 1700 pluralism was frequently opposed on religious grounds, by the eighteenth century many authors were appropriating it for religious purposes. This was especially true in the natural theology and deist traditions, where extraterrestrial life was taken as evidence of God's benevolence and omnipotence. The attractiveness and flexibility of pluralism are also illustrated by the enthusiastic response accorded it by proponents of the doctrine of transmigration of souls. Moreover, despite Wesley's reservations, evangelicals, led by Chalmers, incorporated pluralism into their publications and preaching. Not only did pluralism become associated with the most diverse religious and philosophical systems, it was made a fundamental feature of three major new church groups: the Swedenborgians, the Latter-day Saints, and the Seventh-day Adventists. So extensive were pre-1900 religious discussions, including those by mainline Protestants and Catholics concerned to reconcile it with their creeds, that such publications number in the hundreds.

The readiness with which extraterrestrial life was accepted by an array of religious figures should not obscure the fact, repeatedly illustrated in this history, that many persons continued to maintain that the Christian belief that God became incarnate on this planet and died to redeem its sinful inhabitants could not easily be freed of tension with pluralism. So powerfully felt was this tension that Paine, Shelley, Emerson, Flammarion, Harrison, Twain, and others deemed it sufficient reason for rejecting Christianity. Moreover, some persons, no less convinced of the reality of this tension, perceived it as casting doubt on pluralism. In fact, evidence adduced in this book indicates that both Whewell and Maunder began to seek the important scientific arguments they marshaled against pluralism precisely because of their Christian convictions. This tension sets off the pluralist debate from many other conflicts in the science-religion area; for example, in the Darwin debate, theism was thought to be directly challenged, but Christianity only indirectly, whereas in the Whewell debate, theism was accepted, but the reconcilability of pluralism with Christianity was disputed.

Among the most remarkable features of the religious phase of the pluralist controversy is the degree to which it may be characterized as a night fight in which participants could not distinguish friend from foe until close combat commenced. Allies in a dozen conflicts, authors agreeing on a hundred issues, disagreed on extraterrestrial life. Anglicans argued against Anglicans, Catholics against Catholics, materialists against materialists, the most striking case being the Whewell controversy, in which denominational divisions provide few clues to the participants' positions. The fact that pairs as dissimilar as Hume and Wesley, Paine and Whewell, Maunder and Wallace agreed on fundamental issues in the debate, whereas pairs as close in overall philosophies as Whewell and Sedgwick, Büchner and Strauss, Pohle and Searle opposed each other, must make historians cautious in claiming correlations.

Another relation between religion and pluralism is suggested by Carl Sagan's comment that "over the last few centuries science has systematically expropriated areas which are the traditional concern of religion." As an example, Sagan cites the tendency of some UFO advocates to expect "that we are going to be saved from ourselves by some miraculous interstellar intervention. . . ."[19] Sagan's assertion is no doubt correct and may be further illustrated by aspects of the pluralist debate. Persons skeptical of traditional Christian conceptions of heaven or the afterlife have imagined planetary paradises populated by angelic extraterrestrials. Moreover, such persons as James Stephen and John Herschel, concerned about the overall goodness of a universe in which sinful and suffering men exist, took solace in such beliefs. Frank J. Tipler has recently illustrated the quasi-religious character of some pluralist writings by citing statements made by Sagan himself as well as by his Cornell colleague, the radio astronomer Frank Drake. Sagan, although disparaging the messianic motives of some flying saucer enthusiasts, has suggested that the mere detection of an extraterrestrial radio signal would provide "an invaluable piece of knowledge: that it is possible to avoid the dangers of the period through which we are now passing. . . ." Furthermore, according to Sagan, "it is possible that among the first contents of such a message may be detailed prescriptions for the avoidance of technological disaster. . . ."[20] Drake has written in even more enthusiastic terms:

I fear we have been making a dreadful mistake by not focusing all searches . . . on the detection of the signals of the immortals. For it is the immortals we will most likely discover. . . . An immortal civilization's best assurance of safety would be to make other societies immortal like themselves, rather than risk hazardous military adventures. Thus we could expect them to spread actively the secrets of their immortality among the young, technically developing civilizations.[21]

Such passages support the thesis, advanced by Karl S. Guthke in his study of the extraterrestrial life debate, that pluralism, once deemed a heresy, has become "the myth of modern times" and a "religion or alternate religion."[22]

7. A concluding comment

What final suggestion toward resolution, or at least more irenic analyses of the issues in this debate, emerges from these historical materials? What is the chief lesson to be learned from, for example, the conflicts between Paine and Dick, Whewell and the Brewsterites, Maunder and Lowellians? The point that seems to stand out most clearly is the need for a greater humility in dealing with the philosophical, religious, and scientific issues so central in this debate. Many of our finest minds have faced these issues; directly or indirectly, the message their writings seem most dramatically to demonstrate is that the ways of the universe and of God are more difficult to discern than most inhabitants of our planet have been willing to recognize.

Notes

1 Essentially all astronomers are convinced that because of the strong evidence against the possibility of life elsewhere in the solar system and because of the vast distances to even the nearest stars, contact, if it occurs at all, will probably be by signals sent through space. Moreover, that we probably would be the "barbarians" in such a contact is indicated by the following line of argument. Our capability for radio transmission is less than a century old. Among civilizations in the universe that also possess such a capability, we must be among the youngest. Consequently, most such civilizations, having developed radio transmission earlier than we, also will have attained higher levels of technology than we now possess. A comparable analysis explains why young children who have just learned to talk find nearly everyone with whom they converse more advanced than they are.

2 Although many of these ten authors have published more than one study relevant to this topic, I have listed here only the single publication of each that has contributed most to my research: (1) Ralph V. Chamberlin, "Life in Other Worlds: A Study in the History of Opinion," *University of Utah Bulletin*, 22 (1932), 1–52; (2) Steven J. Dick, *Plurality of Worlds: The Origins of the Extraterrestrial Life Debate from Democritus to Kant* (Cambridge, England, 1982); (3) Sylvia L. Engdahl, *Planet-Girded Suns* (New York, 1974); (4) Camille Flammarion, *La pluralité de mondes habités*, 33rd ed. (Paris, ca. 1885); (5) Karl S. Guthke, *Der Mythos der Neuzeit: Das Thema der Mehrheit der Welten in der Literatur- und Geistesgeschichte von der kopernikanischen Wende bis zur Science Fiction* (Bern, 1983); (6) William G. Hoyt, *Lowell and Mars* (Tucson, 1976); (7) Stanley L. Jaki, *Planets and Planetarians: A History of Theories of the Origin of Planetary Systems* (Edinburgh, 1978); (8) Arthur O. Lovejoy, *The Great Chain of Being: A Study of the History of an Idea* (New York, 1960 reprint of the 1936 original); (9) Grant McColley, "The Seventeenth Century Doctrine of a Plurality of Worlds," *Annals of Science*, 1 (1936), 385–430; (10) Marjorie Nicolson, *Voyages to the Moon* (New York, 1960 reprint of the 1948 original). This listing, which can also function as an introductory bibliography on this area, needs to be qualified in one way: Guthke's book reached me only after my manuscript had been completely drafted, making it difficult for me to exploit the richness of his research except in my revisions.

1 Steven J. Dick, *Plurality of Worlds: The Origins of the Extraterrestrial Life Debate from Democritus to Kant* (Cambridge, England, 1982). Use has also been made of the following publications: (1) Ralph V. Chamberlin, "Life in Other Worlds: A Study in the History of Opinion," *University of Utah Bulletin*, 22 (1932), 1–52; (2) Pierre Duhem, "La pluralité des mondes," in Duhem's *Le système du monde*, vol. IX (Paris, 1958), pp. 363–430; see also Duhem, *Système*, vol. X (Paris, 1959), pp. 94–5, 111, 116–17, 145–6, 319–24, 437–9, and Duhem, *Études sur Léonard de Vinci*, vol. II (Paris, 1904), pp. 57–96, 408–32; (3) Camille Flammarion, *Les mondes imaginaires et les mondes rèels*, 20th ed. (Paris, 1882); (4) Camille Flammarion, *La pluaralité des*

mondes habités, 33rd ed. (Pairs, ca. 1885); (5) Karl S. Guthke, *Der Mythos der Neuzeit: Das Thema der Mehrheit der Welten in der Literatur- und Geistesgeschichte von der kopernikanischen Wende bis zur Science Fiction* (Bern, 1983); (6) Stanley L. Jaki, *Planets and Planetarians: A History of Theories of the Origin of Planetary Systems* (Edinburgh, 1978); (7) Arthur Lovejoy, *The Great Chain of Being: A Study of the History of an Idea* (New York, 1960 reprinting of the 1936 original); see especially Chapter 4; (8) Grant McColley, "The Seventeenth Century Doctrine of a Plurality of Worlds," *Annals of Science,* 1 (1936), 385–430; (9) Charles Mugler, *Deux thèmes de la cosmologie grecque: devenir cyclique et pluralité des mondes* (Paris, 1953); (10) Milton Munitz, "One Universe or Many?" *Roots of Scientific Thought,* ed. Aaron Noland and Philip P. Wiener (New York, 1957), pp. 593–617; (11) Paolo Rossi, "Nobility of Man and Plurality of Worlds," *Science, Medicine and Society in the Renaissance,* ed. Allen Debus, vol. II (New York, 1972), pp. 131–62; (12) Frank J. Tipler, "A Brief History of the Extraterrestrial Intelligence Concept," *Royal Astronomical Society Quarterly Journal,* 22 (1981), 133–45.

2 Epicurus, "Letter to Herodotus," trans. C. Bailey, in *The Stoic and Epicurean Philosophers,* ed. Whitney J. Oates (New York, 1957), pp. 3–15:5.

3 Epicurus, "Herodotus," p. 12.

4 At least one qualification to this generalization may be necessary. Grant McColley, in his "Plurality," p. 388, quotes Eusebius as stating: "Heraclitus and the Pythagoreans consider that every star is a world in an infinite aether, and encompasseth air, earth, and aether. This opinion is current among the disciples of Orpheus, for they suppose that each of the stars does make a world."

5 As quoted from Simplicius in F. M. Cornford, "Innumerable Worlds in Presocratic Philosophy," *Classical Quarterly,* 28 (1934), 1–16:13.

6 Lovejoy, *Chain,* p. 52.

7 Lucretius, *The Nature of the Universe,* trans. R. E. Latham (Baltimore, 1951), p. 91. Emphasis added by the translator.

8 Lucretius, *Nature,* p. 92. Emphasis added by the translator.

9 In their *Presocratic Philosophers* (Cambridge, England, 1964), p. 412, G. S. Kirk and J. E. Raven state that Leucippus and Democritus "are the first to whom we can with absolute certainty attribute the odd concept of innumerable worlds. . . ."

10 Plato, *Timaeus,* 31a and 33a.

11 For Aristotle, see his *On the Heavens,* bk. I, chs. 8 and 9, and *Metaphysics,* bk. XII. Dick's discussion in his *Plurality,* pp. 14–19, is especially thorough.

12 As quoted from Pseudo-Plutarch's *Placita,* in McColley, "Plurality," pp. 386–7.

13 On Lucian, see Roger Lancelyn Green, *Into Other Worlds: Space Flight in Fiction from Lucian to Lewis* (New York, 1975), and Marjorie Hope Nicolson, *Voyages to the Moon* (New York, 1960).

14 For Cicero's rejection of Epicurean pluralism, see his *De natura deorum and Academia,* trans. H. Rackham (London, 1933), pp. 629–30. For his comments on lunar life, see p. 625. In his "Dream of Scipio," he accepted successive worlds; for a discussion of Cicero and of Stoic thought in general in regard to other worlds, see Stanley L. Jaki, *Science and Creation* (New York, 1974), pp. 114ff.

15 See McColley, "Plurality," p. 393. No thorough study of the interactions of pluralism and Christianity during the first millennium of Christian history has been written. When such a task is attempted, McColley's "Plurality" will provide useful leads.

16 Saint Augustine, *The City of God,* trans. Marcus Dods (New York, 1950), bk. XI, ch. 5. See also bk. XII, chs. 11–15, 19, and bk. XIII, ch. 16.

17 As translated by and quoted in Dick, *Plurality,* p. 23.

18 As quoted by Dick in *Plurality,* p. 28, whose analysis of thirteenth-century developments parallels that of Duhem and is closely followed in the present account.

19 For a detailed study of Ockham, see Mary Anne Pernoud, "Tradition and Innovation in Ockham's Theory of the Possibility of Other Worlds," *Antonianum,* 48 (1973), 209–33.

20 Dick, *Plurality,* pp. 35–6.

21 Nicole Oresme, *Le livre du ciel et du monde,* trans. Albert D. Menut (Madison, 1968), pp. 177, 179.

22 See Dick, *Plurality,* p. 43, and Duhem, *Système,* vol. X, p. 324.

23 Thomas Aquinas, *Summa Theologica,* Part I, Q. 47, Art. 3.

24 Nicolas Cusanus, *Of Learned Ignorance,* trans. Germain Heron (London, 1954), pp. 114–15.

25 Duhem, *Système,* vol. X, p. 324.

26 As translated by and quoted in Grant McColley and H. W. Miller, "Saint Bonaventure, Francis Mayron, William Vorilong, and the Doctrine of a Plurality of Worlds," *Speculum,* 12 (1937), 386–9:388. In a footnote (p. 387), the authors state so far as they know, Vorilong was the first to raise this topic explicitly. However, they add: "Others doubtless preceded him."

27 Lovejoy, *Chain,* pp. 99, 111.

28 Lovejoy, *Chain,* pp. 116–17, 124–5, 130.

29 Dick, *Plurality,* p. 3. Dick is by no means an advocate of the extreme position that astronomy was the sole source of pluralism in the 1500–1750 period; moreover, he is well aware of the strengths of Lovejoy's position. However, at least implicitly in his book and in his earlier analysis, "The Origins of the Extraterrestrial Life Debate and Its Relation to the Scientific Revolution," *Journal of the History of Ideas,* 41 (1980), 3–27, he criticizes Lovejoy's stress on the ties of pluralism to philosophy.

30 For example, Frank J. Tipler has suggested that "The principle of plenitude has become the principle of mediocrity in twentieth-century discussions of extraterrestrial intelligence." See Tipler, "History," p. 133. See also my review of three current books in the extraterrestrial life debate in *Physics Today,* 35 (January 1982), 71–3.

31 Curiously, Lovejoy (*Chain,* p. 121) errs and weakens his own argument for the primacy of considerations of plenitude over observation by asserting that Brahe, Galileo, and Kepler were all proponents of life throughout the solar system. In what follows, I attempt to show that Brahe rejected pluralism and that the evidence for Galileo's pluralism is at most inconclusive and certainly does not justify Lovejoy's claim that Galileo "inclined to the Brunonian view." Moreover, Kepler's pluralism was so restricted that he held the earth to be the primary planet in the universe.

32 Among the numerous sources on Bruno, see especially Frances A. Yates, *Giordano Bruno and the Hermetic Tradition* (New York, 1964), and her "Bruno," in the *Dictionary of Scientific Biography,* as well as Paul-Henri Michel, *The Cosmology of Giordano Bruno,* trans. R. E. W. Maddison (Ithaca, 1973).

33 *Kepler's Conversation with Galileo's Sidereal Messenger,* trans. Edward Rosen (New York, 1965), pp. 36–7.

34 *Kepler's Conversation,* p. 42. On the same page he also refers to the inhabitants of Saturn. He makes no comment about life on Mercury, Venus, or Mars, although he speculates (p. 47) whether or not Venus and Mars have moons.

35 Dick, *Plurality,* pp. 73–4, 77, 204–5.

36 Dick, *Plurality,* pp. 70–2.

37 *Kepler's Conversation,* p. 45.

38 Johannes Kepler, *Epitome of Copernican Astronomy,* bks. IV and V, trans. Charles Glenn Wallis, *Great Books of the Western World,* vol. XVI (Chicago, 1952), p. 873; see, in general, pp. 854–87 for Kepler's arguments. Kepler goes to substantial lengths to argue that the sun is far larger than any other star (p. 886).

39 Alexandre Koyré, *From the Closed World to the Infinite Universe* (New York, 1958), p. 99.

40 Galileo, "Letter on Sunspots," in *Discoveries and Opinions of Galileo,* trans. Stillman Drake (Garden City, N.Y., 1957), p. 137. For other passages, see *Galileo's Early Notebooks: The Physical Questions,* trans. William A. Wallace (Notre Dame, Ind., 1977), p. 44; Galileo, *Dialogue Concerning the Two Chief World Systems – Ptolemaic and Copernican,* trans. Stillman Drake (Berkeley, 1953), pp. 61, 99–100; and Galileo's letters to Prince Cesi (1/25/1613) and to Giacomo Muti (2/28/1616), as quoted in J. J. Fahie, *Galileo: His Life and Work* (London, 1903), pp. 134–6.

41 As translated by and quoted in Dick, *Plurality,* p. 89.

42 Lambert Daneau's *Physica Christiana* (1576), which reached English readers as *The Wonderfull Woorkmanship of the World* (London, 1578); see ch. 11.
43 Guthke, *Mythos,* pp. 65–6.
44 Yates, *Bruno,* ch. XX.
45 Tommaso Campanella, "The Defense of Galileo," trans. Grant McColley, *Smith College Studies in History,* 22, nos. 3 and 4 (1937), 66–7.
46 Marjorie Hope Nicolson, *Science and Imagination* (Ithaca, 1956), p. 25.
47 John Wilkins, *Discovery of a New World in the Moone,* in Wilkins, *Mathematical and Philosophical Works,* vol. I (London, 1970 reprinting of the London 1802 edition), p. 102. The last of these quotations did not appear in the 1638 first edition.
48 For the continuing dispute, compare Jaki, *Planets,* pp. 55–6, with Dick, *Plurality,* especially p. 105.
49 Marjorie Hope Nicolson, "English Almanacs and the 'New Astronomy'," *Annals of Science,* 4 (1939), 1–33:18.
50 Wilkins, *Works,* I, p. 190. See, on this point, Lovejoy, *Chain,* pp. 101–2.
51 For the South and Newcastle incidents, see Barbara Shapiro, *John Wilkins* (Berkeley, 1969), pp. 262, 264; for Butler's poem, see Grant McColley (ed.), *Literature and Science* (Chicago, 1940), pp. 86–91.
52 Among those classifying him as antipluralist are Dick, in *Plurality,* pp. 85–6, and Carl Sagan, in *Cosmos* (New York, 1980), p. 146. Professor Sagan quotes from the *Anatomy of Melancholy,* mentioning the book but attributing the quotation taken from it not to Robert Burton but to Robert Merton. For Burton as propluralist, see Richard G. Barlow, "Infinite Worlds: Robert Burton's Cosmic Voyage," *Journal of the History of Ideas,* 34 (1973), 291–301. See also Robert M. Browne, "Robert Burton and the New Cosmology," *Modern Language Quarterly,* 13 (1952), 131–48.
53 Dick, *Plurality,* pp. 93–5.
54 Lovejoy, *Chain,* p. 126.
55 *Pascal's Pensées,* intro, T. S. Eliot (New York, 1958); see #206 and 207.
56 Brunschveicg's #206 and 207 are Lafuma's #201 and 42. See Pascal, *Oeuvres complètes,* ed. Louis Lafuma (Paris, 1963). As Pascal remarked (Brunschveicg #23, Lafuma #784), "meanings differently arranged have a different effect."
57 Lovejoy, *Chain,* p. 129.
58 In the Brunschveicg ordering, see #72, 194, 266, 348, 692, and 792, which correspond to Lafuma's #199, 427, 782, 113, 269, and 499.
59 Dick, *Plurality,* p. 111.
60 *Oeuvres des Descartes,* ed. Charles Adams and Paul Tannery, vol. V (Paris, 1903), pp. 54–5.
61 See Lovejoy, *Chain,* pp. 124–5, and Dick, *Plurality,* ch. 5.
62 See Dick, *Plurality,* pp. 112ff, for a full discussion.
63 Henry More, *Democritus Platonissans,* as reprinted by the Augustan Reprint Society, with an introduction by P. E. Stanwood (Los Angeles, 1968), Preface. On More's thought, see Stanwood's introduction and the discussions in M. H. Nicolson, *Mountain Gloom and Mountain Glory* (New York, 1963), and her *Breaking of the Circle,* rev. ed. (New York, 1962).
64 [Henry More], *Divine Dialogues,* vol. I (London, 1668), pp. 523–36.
65 Ralph Cudworth, *True Intellectual System of the Universe* (London, 1678); see pp. 675 and 882–3.
66 Quoted in and translated from Flammarion, *Pluralité,* p. 341. Compare his concern with that of Gabriel Naudé, as expressed to the astronomer Boulliau in a 1640 letter: "I fear that the old theological heresies are nothing compared to the new ones that the astronomers want to introduce with their worlds, or rather lunar and celestial earths. For the consequences of these latter heresies will be much more dangerous than that of the preceding ones and will introduce yet stranger revolutions." As quoted in Rossi, "Plurality," p. 131. For further evidence of antipluralist views among French Catholics of this period, see Henri Busson, *La pensée religieuse Française de Charon à Pascal* (Paris, 1933), ch. VI.
67 For a detailed discussion of his analysis, see Dick, *Plurality,* pp. 53–60.

68 See Walter Charleton, *Physiologia Epicuro-Gassendo-Charltoniana* (London, 1654). For background, see Robert Hugh Kargon, *Atomism in England from Hariot to Newton* (Oxford, 1966).

69 P. Chabbert, "Pierre Borel," *Revue d'histoire des sciences*, 21 (1968), 303–43:336. Chabbert indicates that the same characterization is applicable to most of Borel's writings.

70 On Borel, see Rossi, "Plurality," pp. 146–50; Dick, *Plurality*, pp. 117–20; and Marie-Rose Carré, "A Man between Two Worlds: Pierre Borel and his *Discours nouveau prouvant la pluralité des mondes* of 1657," *Isis*, 65 (1974), 322–35.

71 The preview is given in Alexandre Calame's "Introduction," pp. vii–viii, to his critical edition of Fontenelle's *Entretiens sur la pluralité des mondes* (Paris, 1966).

72 This point, as well as an excellent overview of Fontenelle, is included in Edward John Kearns, *Ideas in Seventeenth-Century France* (New York, 1979), pp. 161–76:165.

73 Calame, "Introduction," pp. xxxix–lx. Calame suggests that it was primarily the mechanistic approach advocated in the book that upset church authorities. For information on the theologians at the Sorbonne, see Dick, *Plurality*, pp. 213–14, fn. 50.

74 Despite their thoroughness, Calame in his critical edition and Robert Shackleton in his earlier (Oxford, 1955) critical edition note only five translations. Dick (*Plurality*, pp. 136–7) lists a sixth, and a seventh was made by Richard A. Proctor and published in the journal *Knowledge* in 1884 and 1885.

75 As quoted in George R. Havens, *The Age of Ideas* (New York, 1965), p. 73.

76 To preserve the flavor of seventeenth-century prose, these passages are taken from John Glanvill's 1688 translation of Fontenelle's *A Plurality of Worlds*, as reprinted in facsimile in Leonard M. Marsak (ed.), *The Achievement of Bernard le Bovier de Fontenelle* (New York, 1970), pp. 96, 121.

77 Fontenelle, *Plurality* (Glanvill), pp. 122, 14.

78 Such is the consensus of scholars; see, for example, Suzanne Delorme's article on Fontenelle in the *Dictionary of Scientific Biography* and Shackleton's introduction (pp. 11–20) in his critical edition. Calame (pp. xxi–xxvii) expresses reservations.

79 Calame, *Pluralité*, p. 103.

80 For a succinct analysis of the very limited extent of Fontenelle's Cartesianism, see Kearns, *Ideas*, pp. 167–8.

81 Fontenelle, *Entretiens* (Calame, ed.), p. 161.

82 Alexander von Humboldt, *Cosmos*, trans. E. C. Otté, vol. III (London, 1851), p. 22.

83 For Fontenelle's first and fourth arguments, see Christiaan Huygens, *The Celestial Worlds Discover'd: or, Conjectures Concerning the Inhabitants, Plants and Productions of the Worlds in the Planets* (London, 1968 facsimile reprinting of the London 1698 edition), pp. 17–19; for Fontenelle's second, p. 21; for his third, pp. 10 and 22; for his fifth, p. 18.

84 This point is developed in David Knight, "Celestial Worlds Discover'd," *Durham University Journal*, 58 (1965), pp. 23–9.

85 For Huygens's manuscripts prepared prior to and in preparation for his *Cosmotheoros*, see *Oeuvres complètes de Christiaan Huygens*, vol. XXI (The Hague, 1944), pp. 345–71, 529–68.

86 On this, see the article on Plume in the *Dictionary of National Biography* and J. Edleston, *Correspondence of Sir Isaac Newton and Professor Cotes* (London, 1969 reprinting of the 1850 original), pp. lxxiv–lxxv.

87 As quoted in R. C. Hebb, *Bentley* (New York, n.d.), p. 20.

88 *Four Letters from Sir Isaac Newton to Doctor Bentley* . . . (London, 1756), as reprinted in facsimile in *Isaac Newton's Papers and Letters on Natural Philosophy*, ed. I. Bernard Cohen (Cambridge, Mass., 1958), p. 302.

89 *Letters . . . to . . . Bentley*, p. 282.

90 Richard Bentley, *A Confutation of Atheism from the Origin and Frame of the World* (London, 1693), in *Newton's Papers and Letters*, p. 326.

91 *Newton's Papers and Letters*, p. 356.

92 *Newton's Papers and Letters*, p. 356.

93 *Newton's Papers and Letters*, p. 357–8.

94 *Newton's Papers and Letters*, p. 358.

95 *Newton's Papers and Letters*, p. 359.

96 *Newton's Papers and Letters*, p. 368.

97 Isaac Newton, *Opticks*, with a foreword by Albert Einstein, introduction by Sir Edmund Whittaker, and preface by I. Bernard Cohen (New York, 1952), p. 400.

98 Newton, *Opticks*, pp. 403–4.

99 Richard S. Westfall, *Never at Rest: A Biography of Isaac Newton* (Cambridge, England, 1980), pp. 648–9, 699.

100 Isaac Newton, *Mathematical Principles of Natural Philosophy and His System of the World*, Andrew Motte translation revised by Florian Cajori (Berkeley, 1947), p. 544.

101 These points were supplied to me in private correspondence by Professor R. S. Westfall, who notes that the originals are preserved at King's College, Cambridge.

102 David Brewster, *Memoirs of Life, Writings, and Discoveries of Sir Isaac Newton*, vol. II (Edinburgh, 1855), p. 354.

103 This information was also supplied by Professor Westfall.

104 Coduitt's record, which was called to my attention by Professor I. Bernard Cohen, appears in Edmund Turnor, *Collections for the History of the Town and Soke of Grantham Containing Authentic Memoirs of Sir Isaac Newton* (London, 1806), pp. 172–3. For an interesting discussion of some features of the Conduitt conversation, see David Kubrin, "Newton and the Cyclical Cosmos: Providence and the Mechanical Philosophy," *Journal of the History of Ideas*, 28 (1967), 325–46.

105 On Derham, see A. D. Atkinson, "William Derham, F.R.S. (1657–1735)," *Annals of Science*, 8 (1952), 368–92, and James Moseley, "Derham's Astro-Theology," *Journal of the British Interplanetary Society*, 32 (1979), 396–400.

106 William Derham, *Astro-Theology*, 2nd ed. (London, 1715), p. xli.

107 For the pluralist statements of Burnet, Ray, and Grew, see Thomas Burnet, *Sacred Theory of the Earth*, 2nd ed. (Carbondale, Ill., reprinting of the London 1691 edition), pp. 128, 218–25, 367–8; John Ray, *The Wisdom of God Manifested in the Works of Creation*, 5th ed. (London, 1709), pp. 71, 75–6; and Nehemiah Grew, *Cosmologica Sacra: or A Discourse of the Universe As It Is the Creature and Kingdom of God* (London, 1701), pp. 10–11.

108 John Locke, *Essay Concerning Human Understanding*, bk. IV, ch. III, #23. This passage was called to my attention by Ralph Kenat. Locke's endorsement of pluralism is no doubt linked to his enthusiasm for the great chain of being, expressed in *Essay*, bk. III, ch. VI, #12.

109 John Locke, *Elements of Natural Philosophy*, in *The Works of John Locke*, vol. III, new ed. (Darmstadt, 1963 reprinting of the London 1823 edition), ch. III.

110 *The Works of George Berkeley, Bishop of Cloyne*, ed. A. A. Luce and T. E. Jessop, vol. II (London, 1949), p. 211.

111 *Works of Berkeley*, vol. III (London, 1950), p. 172. On Berkeley, see also Sylvia Louise Engdahl, *Planet-Girded Suns* (New York, 1974), p. 60.

112 The historiography of Leibniz's position in the pluralist debate seems beset with anomalies. Despite his numerous discussions of "possible worlds," Leibniz is rarely mentioned in historical surveys of ideas of extraterrestrials. Even in Lovejoy's *Great Chain of Being*, where Leibniz is extensively discussed, his name never occurs in the chapter on the plurality of worlds doctrine. Moreover, his name is absent from the pluralist treatises written by Kant in 1755 and Lambert in 1761, even though both authors must have known of Leibniz's ideas. The historian also faces the problem that although Leibniz on possible worlds has been discussed by a number of contemporary philosophers, these presentations rarely contain historical analyses. A partial exception is Nicholas Rescher, *Leibniz's Metaphysics of Nature* (Dordrecht, 1981).

113 See Rescher, *Leibniz's Metaphysics of Nature*, p. 87, and *Leibnizens mathematischen Schriften*, ed. C. I. Gerhardt, vol. III (Halle, 1855), pp. 545–54.

114 G. W. Leibniz, *Theodicy: Essays upon the Goodness of God, the Freeedom of Man, and the Origin of Evil*, trans. E. M. Huggard (New Haven, 1952), p. 127.

115 G. W. Leibniz, *New Essays on Human Understanding*, trans. Peter Remnant and Jonathan Bennett (Cambridge, England, 1981), p. 307.

116 Christian Wolff, *Elementa matheseos universae*, vol. III (Hildesheim, 1968 reprinting of the 1735 original), pp. 576–7. Other pluralist passages occur in Wolff's *Vernünfftige Gedancken von den Würckungen der Natur* (Hildesheim, 1981 reprinting of the Magdeburg 1723 original); see pp. 125–226; and Wolff's *Vernüfftige Gedancken von den Ansichten der natürlichen Dinge,* 2nd ed. (Hildesheim, 1981 reprinting of the Frankfurt and Leipzig 1726 edition); see pp. 55ff, 98ff, 113–14, 136ff.

117 Katherine Brownell Collier, *Cosmogonies of Our Fathers* (New York, 1968 reprinting of the 1934 original), p. 143. Moncharville's book is extremely rare; locating it is also complicated by the fact that the *National Union Catalog* lists the one known U.S. copy under Brodeau, whereas it appears in the *Catalogue général des livres imprimés de la bibliothèque nationale* under the name Châtres.

118 David Gregory, *Elements of Physical and Geometrical Astronomy,* 2nd ed., 2 vols. (New York, 1972 reprint of the 1726 London edition), pp. 3, 810–13; John Keill, *Introduction to the True Astronomy,* 4th ed. (London, 1748), p. 40.

119 Edmond Halley, *Miscellanea curiosa,* 3rd ed., vol. I (London, 1726), pp. 55–9. This collection of papers by Halley first appeared in 1708; the paper in question was first published in 1692.

120 Jaki, *Planets and Planetarians,* p. 94.

121 William Whiston, *Astronomical Principles of Religion* (London, 1717), pp. 91–7.

122 William Whiston, *Praelectiones astronomiae* (1707), translated as *Astronomical Lectures,* 2nd ed. (London, 1727); see pp. 38–42.

123 Marjorie Nicolson and G. S. Rousseau, *"This Long Disease, My Life": Alexander Pope and the Sciences* (Princeton, 1968); see "Part Three: Pope and Astronomy," pp. 131–235.

124 One example of the popularity of Pope's *Essay on Man* is the fact that by the end of the eighteenth century it had been translated into French ten times. See Richard Gilbert Knapp, "The Fortunes of Pope's *Essay on Man* in 18th Century France," *Studies on Voltaire and the Eighteenth Century,* 82 (1971), 1–156:17.

125 Alexander Pope, *Essay on Man,* Epistle I, lines 23–8.

126 This point is developed by Joseph Anthony Mazzeo in his *Nature and the Cosmos* (Oceanside, N.Y., 1977), pp. 97–8.

127 Nicholson and Rousseau, *Pope,* pp. 150–6.

128 *The Spectator,* ed. Donald F. Bond, vol. IV (Oxford, 1965), p. 346. As Bond points out (vol. III, p. 576), Addison owned a copy of the 1707 edition of Fontenelle's *Entretiens.*

129 For additional Addisonian discussions of pluralism, see *Spectator* 420 and 580.

130 As quoted in Albert Rosenberg, *Sir Richard Blackmore* (Lincoln, Neb., 1953), p. 104.

131 Richard Blackmore, *The Creation,* in *Minor English Poets 1660–1780,* ed. David P. French, vol. III (New York, 1967), pp. 27–71:39.

132 As quoted in Linda Ruth Thornton, *The Influence of Bernard de Fontenelle upon English Writers of the Eighteenth Century* (1977 doctoral dissertation, University of Oklahoma), p. 133. Chapter III of this dissertation discusses the English influences of Fontenelle's *Entretiens.*

133 The books in chronological order are (1) Hareneus Geierbrand [Andreas Ehrenberg], *Curiöse und wohlgegründete Gedancken von mehr als einer bewohnten Welt* (Jena, 1711); (2) Daniel Sturmy, *A Theological Theory of a Plurality of Worlds* (London, 1711); (3) William Derham, *Astro-Theology* (London, 1714); (4) Johann Wilhelm Weinreich, *Disputatio philosophica de pluralitate mundorum* (Torun, 1715); (5) Andreas Ehrenberg, *Die noch unümgestossene Vielheit der Welt-Kugeln, oder: Dass die Planeten Welt-Kugeln seyn . . .* (Jena, 1717); (6) Johann Jacob Schudt, *De probabili mundorum pluralitate* (Frankfurt, 1721); (7) William Arntzen, *Dissertatio astronomico-physica de luna habitabili* (Utrecht, 1726); (8) Johann Heinrich Herttenstein, *Dissertatio mathematica, sistens similtudinem inter terram et planetas intercedentem* (Strasbourg, 1732); (9) John Peter Biester, *An Enquiry into the Probability of the Planets Being Inhabited* (London, 1736); (10) Johann Christoph Hennings, *Specimen planetographiae physicae inquirens praecipue an planetae sint habitabiles* (Kiel, 1738); (11) Eric Engman, *Dissertatio astronomico-physica de luna non habitabili*

(Upsala, 1740); (12) Isacus Svanstedt, *Dissertatio philosophica, de pluralitate mundorum* (Upsala, 1743); and (13) D.G.S. [David Gottfried Schöber], *Gedanken von denen vernünftig freyen Einwohnern derer Planeten* (Liegnitz, 1748). Derham has already been discussed. I know Ehrenberg's books only from secondary sources and received a microfilm of Weinreich's book only after this volume had gone to press. Weinreich's book is a dissertation directed by Reinhold Fryderyk Bornmann, to whom it is misleadingly attributed in Jérôme de la Lande, *Bibliographie astronomique* (Paris, 1803), p. 362.

134 Dick, *Plurality*, p. 181.

135 On Engman, see the thorough discussion in Dick, *Plurality*, pp. 181–3.

136 Otto Zöckler, *Geschichte der Beziehungungen zwischen Theologie und Naturwissenschaft*, 2nd ed., vol. II (Gütersloh, 1879), pp. 62–3. Zöckler, writing two years after the first observation of the moons of Mars, was very impressed by the correctness of the predictions made by Ehrenberg and Schudt.

137 This point is stressed in Guthke, *Mythos*, pp. 183–5.

138 Zöckler, *Geschichte*, vol. II, pp. 62, 248.

139 Sturmy, *Theory*, p. 25.

140 [Thomas Baker], *Reflections upon Learning* (London, 1699), pp. 97–8.

141 Robert Jenkin, *The Reasonableness and Certainty of the Christian Religion*, bk. II (London, 1700), pp. 218–23:222.

142 For Watts's poem, see *Minor English Poets 1660–1780*, ed. David P. French, vol. III (New York, 1967), p. 635. For interpretation of these lines as antipluralist, see A. J. Meadows, *The High Firmament: A Survey of Astronomy in English Literature* (Leicester, 1969), p. 131; and Vincent Cronin, *The View from Planet Earth* (New York, 1981), p. 150.

143 Isaac Watts, *Knowledge of the Heavens and the Earth Made Easy; or, The First Principles of Astronomy and Geography*, 2nd ed. (London, 1728), p. 103.

144 The nine supporters of pluralism were Ehrenberg, Schöber, and Schudt (already discussed), J. H. Becker (1698–1772), Joachim Böldicke, J. Carpov (1699–1768), J. F. W. Jerusalem (1709–89), J. E. Reinbeck (1682–1741), and A. F. W. Sack (1703–82); the opponents were J. F. Buddeus (1667–1729), V. E. Löscher (1674–1747), and Hermann Witsius (1632–1708). For information on Jerusalem and Sack, see Gerhard Kaiser, *Klopstock: Religion und Dichtung* (Gütersloh, 1963), pp. 46ff. For Witsius, see his *Sacred Dissertation on What Is Commonly Called the Apostles' Creed*, trans. Donald Fraser, vol. I (Glasgow, 1823), pp. 213–21. For the others, see Zöckler, *Geschichte*, vol. II, pp. 62–4.

145 As Bonamy admitted, many of his references were from the bibliography of ancient pluralist writings supplied by Johann Albert Fabricius in his *Bibliotheca Graeca*, vol. I (Hamburg, 1708), pp. 131–6. Fabricius also contributed to the pluralist debate by his German translation of Derham's *Astro-Theology*.

146 Pierre Bonamy, "Sentiments des anciens philosophes sur la plualité des mondes," *Histoire de l'academie des inscriptions et belles lettres*, 9 (1736), 1–19. On Bonamy's essay, see the comments in Flammarion, *Les mondes*, pp. 455–6.

147 John Dillenberger, *Protestant Thought and Natural Science: A Historical Survey* (London, 1961), p. 135.

148 See Collier, *Cosmogonies*, pp. 179–82, and D. P. Walker, *The Decline of Hell* (Chicago, 1964), pp. 39–40.

149 Jacob Ilive, *The Oration . . . Proving I. The Plurality of Worlds. II. That this Earth is Hell. III. That the Souls of Men are the Apostate Angels. IV. That the Fire which will punish those who shall be confined to this Globe after the Day of Judgement will be Immaterial . . .* (London, 1733; 2nd ed. 1736).

CHAPTER 2

1 This story is told in detail in Stanley Jaki's "Wright's Wrong" in his *The Milky Way* (New York, 1972), pp. 183–220.

2 Augustus De Morgan, "An Account of the Speculations of Thomas Wright of Durham," *Philosophical Magazine*, 3rd ser., 32 (1848), 241–52:252.

3 The first two of these phrases are contained in Wetherill's "Dedication" (p. 2), and the third is from Rafinesque's "Preface" (p. 7) to their reprinting of Wright's *Original Theory* under the title *The Universe and the Stars*, edited with notes by C. S. Rafinesque, printed for Charles Wetherill (Philadelphia, 1837).

4 As quoted in Jaki, *Milky Way*, p. 210. For a typical statement of Paneth's views, see his "Thomas Wright and Immanuel Kant," in F. A. Paneth, *Chemistry and Beyond* (New York, 1964), pp. 91–119.

5 See Hoskin's "Introduction" to Wright's *An Original Theory or New Hypothesis of the Universe* (New York, 1971 reprinting of the London 1750 original), as well as M. Hoskin, "The Cosmology of Thomas Wright of Durham," *Journal for the History of Astronomy*, 1 (1970), 44–52.

6 As quoted in Edward Hughes, "The Early Journal of Thomas Wright of Durham," *Annals of Science*, 7 (1951), 1–24:4.

7 See Thomas Wright, "The Elements of Existence or A Theory of the Universe," as edited by M. Hoskin for Wright, *Theory* (1971 ed.), pp. 1–15:3.

8 Thomas Wright, *Clavis coelestis* (London, 1967 reprinting of the London 1742 original), p. 75. See also p. 49, where Wright expresses doubts about the habitability of comets.

9 See Edward Young, *Night Thoughts,* edited with notes by George Gilfillan (Edinburgh, 1853), "Night Ninth," lines 748–9.

10 As quoted in Wright, *Theory*, p. 36. This passage is from Addison's *Spectator* of July 9, 1714.

11 As quoted in Wright, *Theory*, p. 47. See Young, *Night Thoughts,* "Night Ninth," lines 772–3.

12 In his identification of nebulae with other universes, he was anticipated by both Pierre Gassendi and by Christopher Wren. See Jaki, *Milky Way*, pp. 122, 130.

13 Thomas Wright, *Second or Singular Thoughts upon the Theory of the Universe,* edited from the manuscript by M. A. Hoskin (London, 1968), p. 79.

14 Rafinesque, in Wright, *Universe and the Stars*, pp. 155, 6.

15 See W. Hastie's "Translator's Introduction," pp. xvi, xcviii, lxxxv, and ix in his *Kant's Cosmogony* (Glasgow, 1900). Although Hastie stated (p. cvi) that he had translated all of Kant's book, he excluded his translation of Chapter VIII of Part II and all of Part III from this edition. This omission was perpetuated when between 1968 and 1970 three separate republications of Hastie's translation, all with new introductions, appeared. See (1) *Kant's Cosmogony*, ed. Willy Ley (New York, 1968); (2) I. Kant, *Universal Natural History and Theory of the Heavens*, new introduction by Milton K. Munitz (Ann Arbor, 1969); and (3) *Kant's Cosmogony*, new introduction by Gerald J. Whitrow (New York, 1970). Recently, Stanley L. Jaki has completed a new translation of the entire book; see his I. Kant, *Universal Natural History and Theory of the Heavens*, translated with an introduction and notes by Stanley L. Jaki (Edinburgh, 1981). References to Kant's book are to the Jaki translation.

16 See Svante Arrhenius, "Emanuel Swedenborg as a Cosmologist," in Swedenborg's *Opera quaedem*, vol. II (Stockholm, 1908), pp. xxx–xxxi; and C. V. L. Charlier, "On the Structure of the Universe," *Astronomical Society of the Pacific Publications*, 37 (1925), 53–76:63.

17 I. I. Polonoff, "Force, Cosmos, Monads and Other Themes of Kant's Early Thought," *Kantstudien*, 107 (1973), 1–214:2; S. L. Jaki, *Planets and Planetarians* (Edinburgh, 1978), pp. 111–20.

18 For a reprinting of this review, see Kant, *Allgemeine Naturgeschichte und Theorie des Himmels*, ed. Fritz Kraft (Munich, 1971), pp. 199–211.

19 For a listing and discussion of these and other comparable works, see Wolfgang Philipp, "Physicotheology in the Age of Enlightenment: Appearance and History," *Studies on Voltaire and the Eighteenth Century*, 57 (1967), 1233–67.

20 Kant, *Theory* (Jaki), p. 81. Emphasis added.

21 Kant, *Theory* (Jaki), p. 84. Emphasis added.

22 Kant, *Theory* (Jaki), pp. 89–90. Kant derived his optical theory of the Milky Way from Wright, but not the disk theory, which Wright did not have, nor his theory of the nebulae as independent systems, which the review did not mention.

23 Kenneth Glyn Jones, "The Observational Basis of Kant's *Cosmogony:* A Critical Analysis," *Journal for the History of Astronomy,* 2 (1971), 29–34:33.

24 For Jaki's analysis, see his *Planets and Planetarians,* pp. 111–20.

25 W. Hastie, "Translator's Introduction," p. lvii.

26 Immanuel Kant, *The One Possible Basis for a Demonstration of the Existence of God,* trans. Gordon Treash (New York, 1979), p. 117. For the summary of his earlier exposition of his theory of the universe, see pp. 187–215.

27 As quoted in F. A. Paneth, *Chemistry and Beyond* (New York, 1964), pp. 110–11.

28 Immanuel Kant, *Critique of Pure Reason,* trans. J. M. D. Meiklejohn (London, 1956), p. 468. I am indebted to Dr. Curtis Wilson for bringing this passage to my attention.

29 Immanuel Kant, *On History,* ed. Lewis White Beck (Indianapolis, 1963); see pp. 36–7 and 18.

30 Immanuel Kant, *Critique of Practical Reason and Other Writings in Moral Philosophy,* trans. Lewis White Beck (Chicago, 1949), pp. 258–9. For an interesting analysis of this passage and its background in Kant's thought, see Rudolf Unger, " 'Der bestirnte Himmel über mir . . .' Zur geistesgeschichtlichen Deutung eines Kant-Wortes," in Ungar, *Gesammelte Studien,* vol. II (Darmstadt, 1966 reprinting of the Berlin 1929 edition), pp. 40–66.

31 Immanuel Kant, *Critique of Judgement,* trans. James Creed Meredith, in *Great Books of the Western World,* vol. XLII (Chicago, 1952), p. 604; see also pp. 507 and 591.

32 Immanuel Kant, *Anthropology from a Pragmatic Point of View,* trans. Mary J. Gregor (The Hague, 1974), p. 48.

33 Immanuel Kant, *Lectures on Philosophical Theology,* trans. Allen W. Wood and Gertrude M. Clark (Ithaca, 1978), p. 138. For Kant on the moon's habitability, see p. 121.

34 Kant, *Basis,* p. 51.

35 See Kant's December 31, 1765, letter to Lambert as translated in Arnulf Zweig, *Kant's Philosophical Correspondence 1759–1799* (Chicago, 1967), p. 47.

36 Lambert's letter of November 13, 1765, to Kant, as translated in Zweig, *Kant's Philosophical Correspondence,* p. 46.

37 Johann Heinrich Lambert, *Cosmological Letters on the Arrangement of the World-Edifice,* translated with an introduction and notes by Stanley L. Jaki (New York, 1976), p. 111. Quotations from Lambert, except where noted, are from this translation. The original is Lambert, *Cosmologische Briefe über die Einrichtung des Weltbaues* (Augsburg, 1761).

38 Although the passage previously quoted from Kant's foreword to his 1763 book implies that Lambert held this view, Kant's April 19, 1791, letter to J. F. Gensichen contains a specific denial that Lambert identified nebulae as other Milky Ways. See Zweig, *Correspondence,* p. 171. Perhaps the safest brief statement is that Lambert did endorse the existence of other Milky Ways and speculated on whether or not they might be telescopically detectable, suggesting Orion as a possible case. See Lambert, *Letters,* pp. 119, 125, 132, 160, 166.

39 Wright's *Second Thoughts* has cosmogonical features, but these do not appear in his *Theory* of 1750. See Simon Schaffer, "The Phoenix of Nature: Fire and Evolutionary Cosmology in Wright and Kant," *Journal for the History of Astronomy,* 9 (1978), 180–200.

40 See Jaki, "Introduction" to Lambert, *Letters,* pp. 24, 27.

41 Lambert, *Briefe,* p. XXV.

42 Roger Jaquel, *Le Savant et philosophe Mulhousien Jean-Henri Lambert* (Paris, 1977), p. 61.

43 Lambert, *Letters,* p. 92. For a discussion of Lambert's teleological approach by a prominent astronomer, see Karl Schwarzschild, "Ueber Lambert's Kosmologische Briefe," *Nachrichten von der Königlichen Gesellschaft der Wissenschaften zu Göttingen, Geschäftliche Mitteilungen* (1907), 88–102.

44 For William Herschel's published statement on Lambert, see Herschel's "On the Direction and Velocity of the Motion of the Sun, and Solar System," in his *Scientific Papers,* ed. J. L. E. Dreyer, vol. II (London, 1912), p. 318. See the discussion of Herschel later in this book for more details on Herschel's view of Lambert.

45 The restraint and rigor evident in Lambert's book are especially stressed in S. L. Jaki, "Lambert and the Watershed of Cosmology," *Scientia,* 72 (1978), 75–95.

46 As quoted with approval in Roland Stromberg, *An Intellectual History of Modern Europe,* 2nd ed. (Englewood Cliffs, N.J., 1975) p. 124.

47 James Ferguson, *Astronomy Explained upon Sir Isaac Newton's Principles,* 2nd ed. (London, 1757), p. 1.

48 *The Newtonian System of Philosophy Adapted to the Capacities of Young Gentlemen and Ladies . . . Being the Substance of Six Lectures . . . by Tom Telescope, A. M., and Collected and Methodized . . . by . . . Mr. Newbery* (London, 1758). This work, which contains pluralist remarks in its second lecture, is attributed to John Newbery in Sylvia Engdahl, *The Planet-Girded Suns* (New York, 1974), p. 66, whereas Oliver Goldsmith is proposed in William Cushing, *Initials and Pseudonyms,* 2nd ser. (New York, 1888), p. 144. Goldsmith knew of the pluralist position; in fact, he satirized members of the Royal Society in his *Citizen of the World* by suggesting that they indulged in calculations as to "How many geese are bred in Saturn's ring;/ How soon a snipe in Mercury would roast. . . ." As quoted in William Powell Jones, *The Rhetoric of Science* (Berkeley, 1966), p. 195.

49 As quoted in Engdahl, *Suns,* p. 67.

50 Edwin Hubble, *Realm of the Nebulae* (New York, 1958 reprinting of the 1936 original), p. 25.

51 Despite his prominence, no paper on Herschel's philosophical or religious ideas has ever been published, and his name does not appear in the indexes of such respected surveys of Enlightenment thought as those by Ernst Cassirer, Peter Gay, and Leslie Stephen.

52 The leading Herschel scholar is Michael A. Hoskin. A full listing of his writings on Herschel would be very long, but mention may be made of his *William Herschel and the Construction of the Heavens* (London, 1963). His former student Simon Schaffer has enriched Herschel scholarship by a number of papers, including " 'The Great Laboratories of the Universe': William Herschel on Matter Theory and Planetary Life," *Journal for the History of Astronomy,* 11 (1980), 81–111. Another Hoskin student, J. A. Bennett, catalogued the materials in the Herschel archives and arranged them for reproduction on microfilm.

53 Herschel's biographers are numerous; I have drawn upon (1) Angus Armitage, *William Herschel* (London, 1962), (2) Günther Buttmann, *Wilhelm Herschel* (Stuttgart, 1961), (3) Agnes M. Clerke, *The Herschels and Modern Astronomy* (New York, 1895), (4) Edward S. Holden, *Sir William Herschel* (New York, 1881), (5) Constance Lubbock, *The Herschel Chronicle* (Cambridge, England, 1933), (6) J. B. Sidgwick, *William Herschel* (London, 1953), and (7) James Sime, *William Herschel and His Work* (New York, 1900), as well as other works.

54 Another possible source of Herschel's pluralism is Benoît de Maillet's *Telliamed.* This book is listed along with about thirty others on p. 83 of Herschel's unpublished "Commonplace Book," which seems to have been written during the mid-1770s. This item is owned by the Linda Hall Library in Kansas City, Missouri, where I was allowed to examine it.

55 As recorded by his sister Caroline; see Mrs. John Herschel, *Memoir and Correspondence of Caroline Herschel* (London, 1879), p. 35.

56 A. J. Turner, in his *Science and Music in Eighteenth Century Bath* (Bath, 1977), refers to Ebenezer Henderson's *Life of James Ferguson* (London, 1867); Turner states (p. 53): "A note cited by Henderson (pp. xxxv–vi) claims that Herschel attended Ferguson's lecture courses at Bath in 1767. Although not unlikely, there is no evidence available to confirm this. . . ."

57 *The Scientific Papers of William Herschel,* ed. J. L. E. Dreyer, vol. I (London, 1912),

p. 5. All subsequent references to Herschel's published writings are to the two volumes of this work.

58 Microfilm (reel 17) of the Royal Astronomical Society Herschel MSS, W. 3/1.4, pp. 1–2. Mark M. Moes provided helpful assistance in searching for pluralist materials in Herschel's manuscripts.

59 Ibid., p. 4.

60 Ibid., pp. 8–10.

61 Ibid., p. 17.

62 Ibid., p. 17.

63 Ibid., pp. 65–8.

64 Ibid., pp. 65, 69.

65 Ibid., pp. 70–1. See also p. 75 for a 1793 observation of a lunar twilight.

66 See reel 12 of the Royal Astronomical Society Herschel MSS, W. 1/1, pp. 66–7.

67 See Lubbock, *Herschel Chronicle*, pp. 99, 103, and Simon Schaffer, "Herschel in Bedlam: Natural History and Stellar Astronomy," *British Journal for the History of Science*, 13 (1980), 211–39.

68 Herschel, *Papers*, vol. I, pp. xxxiii–xxxiv.

69 As quoted in Lubbock, *Herschel Chronicle*, p. 170.

70 See *Gentleman's Magazine*, 57 (1787), 636.

71 Herschel, *Papers*, vol. I, p. 479. R. A. Proctor, in his *Sun* (London, 1871), quotes this statement by Herschel, describing it (p. 185) as "that remarkable passage which every student of astronomy knows by heart. . . ." For additional information on Herschel's theory of the inhabitability of the sun, see Steven Kawaler and J. Veverka, "The Habitable Sun: One of William Herschel's Stranger Ideas," *Royal Astronomical Society of Canada Journal*, 75 (1981), 46–55.

72 Holden, *Herschel*, p. 149.

73 Agnes M. Clerke, *Popular History of Astronomy during the Nineteenth Century* (Edinburgh, 1885), p. 71; A. J. Meadows, *Early Solar Physics* (Oxford, 1970), p. 6.

74 For Herschel's notes on Lambert's book, see the microfilm (reel 24) of the Royal Astronomical Society Herschel MSS, W. 7/2, pp. 17–22. See also Michael Hoskin, "Lambert and Herschel," *Journal for the History of Astronomy*, 9 (1978), 140–2.

75 M. E. W. Williams, "Was There Such a Thing as Stellar Astronomy in the Eighteenth Century?" *History of Astronomy*, 21 (1983), 369–85.

76 Simon Schaffer, "Uranus and the Establishment of Herschel's Astronomy," *Journal for the History of Astronomy*, 12 (1981), 11–26.

77 Ida Macalpine and Richard Hunter, *George III and the Mad-Business* (New York, 1969), p. 41.

78 Schaffer, "Planetary Life," p. 101.

79 Clerke, *Popular History*, p. 288.

80 Johann Schröter, *Selenotopographische Fragmente, zur genauern Kenntniss der Mondfläche*, vol. I (Göttingen, 1791), p. 670.

81 John Jerome Schroeter, "Observations on the Atmospheres of Venus and the Moon . . . ," *Philosophical Transactions of the Royal Society*, 82 (1792), 309–61:337. For an excellent analysis of Schröter's Venusian claims, see Richard Baum, *The Planets: Some Myths and Realities* (New York, 1973), pp. 48–83.

82 William Herschel, "Observations on the Planet Venus," in Herschel, *Papers*, vol. I, pp. 441–51:442.

83 J. J. Schroeter, "New Observations in Further Proof of the Mountainous Inequalities, Rotation, Atmosphere, and Twilight, of the Planet Venus," *Philosophical Transactions of the Royal Society*, 85 (1795), 117–76:169.

84 J. H. Schröter, *Aphroditographische Fragmente, zur genauern Kenntniss des Planeten Venus* (Helmstedt, 1796), pp. 193–4.

85 J. Ashbrook, "Schröter and the Rings of Saturn," *Sky and Telescope*, 36 (1968), 230–1:231.

86 Robert Grant, *History of Physical Astronomy* (London, 1852), p. 233. Schröter was not the first to assign approximately 24-hour rotation periods to the inner planets, but his advocacy of these periods helped perpetuate them in the literature.

87 J. Ashbrook, "Schröter's Observations of Mars," *Sky and Telescope,* 14 (1955), 140.
88 Ashbrook, "Mars," p. 140.
89 J. E. Bode, "Gedanken über die Natur der Sonne und Entstehung ihrer Flekken," *Beschäftigungen der Berlinischen Gesellschaft Naturforschender Freunde,* 2 (1776), 225–52:233.
90 Stanley L. Jaki, *Planets and Planetarians* (Edinburgh, 1978), p. 121.
91 Roger Jaquel, *Le savant et philosophe Mulhousien Jean-Henri Lambert (1728–1777)* (Paris, 1977), p. 51.
92 J. E. Bode, *Kurzgefasste Erläuterung der Sternkunde und den* [sic] *dazu gehörigen Wissenschaften* (Berlin, 1778), p. 375.
93 J. E. Bode, "Die Betrachtung des Weltgebäudes," in Bode, *Betrachtung der Gestirne und des Weltgebäudes* (Berlin, 1816), pp. 325–413:355.
94 Stanley L. Jaki, in his "Introduction" (p. 39) of his translation of Immanuel Kant, *Universal Natural History and Theory of the Heavens* (Edinburgh, 1981).
95 On Laplace's theory and its history, see Stanley L. Jaki, *Planets and Planetarians: A History of Theories of the Origin of Planetary Systems* (Edinburgh, 1978), and Ronald L. Numbers, *Creation by Natural Law: Laplace's Nebular Hypothesis in American Thought* (Seattle, 1977).
96 Pierre Simon Laplace, *The System of the World,* trans. John Pond from the first French edition, vol. II (London, 1809), p. 355. Jaki (*Planets,* pp. 122–34) has traced the substantial changes in Laplace's presentation of the nebular hypothesis in the five editions of his *Exposition* that appeared between 1796 and 1824.
97 Vincent Cronin, *The View from Planet Earth* (New York, 1981), p. 164; Emil Ludwig, *Napoleon,* trans. Eden and Cedar Paul (New York, 1926), p. 120.
98 As quoted in and translated from Hervé Faye, *Sur l'origine du monde,* 2nd ed. (Paris, 1885) p. 131.
99 G. Sarton, "Laplace's Religion," *Isis,* 33 (1941), 309–12. See also Jean Pelseneer, "La religion de Laplace," *Isis,* 36 (1946), 158–60.
100 R. Hahn, "Laplace's Religious Views," *Archives internationales d'histoire des sciences,* 8 (1955), 38–40:40. See also Hahn's "Laplace and the Vanishing Role of God in the Physical Universe," in *The Analytical Spirit,* ed. Harry Woolf (Ithaca, 1981), pp. 85–95.
101 As quoted in Constance Lubbock, *The Herschel Chronicle* (New York, 1933), p. 310.
102 Jérôme Lalande, *Astronomie,* vol. III (Paris, 1792), p. 353.
103 Bernard de Fontenelle, *Conversations on the Plurality of Worlds,* trans. Elizabeth Gunning, with notes by Jerome de la Lande (London, 1803), pp. iv–v.
104 Hélène Monod-Cassidy, "Un astronome-philosophe, Jérôme de Lalande," *Studies on Voltaire and the Eighteenth Century,* 56 (1967), 907–30:928.
105 As quoted in Monod-Cassidy, "Lalande," p. 909.
106 As quoted in Ludwig Büchner, *Force and Matter,* reprint of the 4th English edition (New York, 1920), pp. 105–6. Possibly this statement comes from the newspaper articles or public lectures that Lalande during the last years of his life devoted to advocating atheism. See Hahn, "Laplace and the Vanishing Role of God," p. 87.
107 Blaise Pascal, "Pensées," #781, in Pascal, *Oeuvres complètes,* ed. Louis Lafuma (Paris, 1963), p. 599.

CHAPTER 3

1 Franklin L. Baumer, *Modern European Thought* (New York, 1977), p. 141.
2 Sheridan Gilley, "Christianity and Enlightenment: An Historical Survey," *History of European Ideas,* 1 (1981), 103–21; Donald Greene, "Augustinianism and Empiricism: A Note on Eighteenth-Century English Intellectual History," *Eighteenth Century Studies,* 1 (1967–8), 33–68; Carl L. Becker, *The Heavenly City of the Eighteenth-Century Philosophers* (New Haven, 1932).
3 David Mallet, "The Excursion," in *Minor English Poets 1660–1780,* vol. V, ed. David P. French (New York, 1967), pp. 17–24:21.

4 James Thomson, *The Seasons* and *The Castle of Indolence,* ed. James Sambrook (Oxford, 1972), p. 40.

5 Thomson, *Seasons,* p. 84. On this influence, see Alan Dugald McKillop, *The Background of Thomson's Seasons* (Minneapolis, 1942), pp. 67–8.

6 As quoted in William Powell Jones, *The Rhetoric of Science* (Berkeley, 1966). p. 103.

7 As quoted in Hoxie Neale Fairchild, *Religious Trends in English Poetry,* vol. I (New York, 1931), p. 464. On the date of this poem, see G. R. Potter, "Henry Baker, F.R.S. (1698–1774)," *Modern Philology,* 29 (1932), 301–21.

8 Henry Brooke, "Universal Beauty," in *Minor English Poets 1660–1780,* ed. David P. French, vol. VI (New York, 1967), pp. 591–619:593.

9 For a discussion of Gray's "Luna habitabilis," see Marjorie Hope Nicolson, *Voyages to the Moon* (New York, 1960), pp. 127–9.

10 As quoted in Jones, *Science,* p. 134.

11 As quoted in Fairchild, *Religious Trends,* vol. I, p. 392.

12 Fairchild, *Religious Trends,* vol. I, p. 508.

13 Henry C. Shelley, *The Life and Letters of Edward Young* (London, 1914), p. 198. A more satisfactory biography is Isabel St. John Bliss, *Edward Young* (New York, 1969). See also Fairchild, *Religious Trends,* vol. II, pp. 131–49; and Jones, *Science,* pp. 153–9.

14 Edward Young, *Night Thoughts,* ed. George Gilfillan (Edinburgh, 1853), Night IX, lines 772–5. Future references are from this text and in this format.

15 Among Hay's contemporaries, Wesley and Swedenborg, for example, rejected this position. For a carefully considered contemporary analysis of the concept of Christ as related to a pluralistic universe, see Charles Davis, "The Place of Christ," *Clergy Review* (London), 45 (1960), 706–18.

16 As quoted by James Boswell in his *Life of Johnson,* ed. George Birkbeck Hill, vol. I (Oxford, 1934), p. 268.

17 Henry St. John, Lord Viscount Bolingbroke, *Philosophical Works,* 5 vols., vol. II (London, 1754), p. 144.

18 Hume and Burke, as quoted in Ernest Campbell Mossner, "Bolingbroke," *Encyclopedia of Philosophy,* ed. Paul Edwards, vol. I (New York, 1967), p. 332.

19 Samuel Pye, M.D., *Moses and Bolingbroke; A Dialogue in the Manner of the Right Honourable *******, Author of Dialogues of the Dead* (London, 1765). The author of *Dialogues of the Dead* was Fontenelle.

20 Pye, *Moses and Bolingbroke,* pp. 60–3. The four additional lights for Jupiter are its four Galilean moons. The fifteen days assigned to the creation of Jupiter correspond roughly to the seven days of earth's creation, because Jupiter completes a rotation in about ten terrestrial hours.

21 These and Pye's other publications were insufficient to secure his inclusion in the *Dictionary of National Biography.* The only subsequent references to him I have encountered are brief comments by Wesley in his *Journal* and in the book by Nares discussed subsequently. There may, however, remain a tale to be told about Pye, for a Samuel Pye was a member of the Bath Philosophical Society at the same time as William Herschel. See A. J. Turner, *Science and Music in Eighteenth Century Bath* (Bath, 1977), p. 83.

22 Frank J. Tipler, "A Brief History of the Extraterrestrial Intelligence Concept," *Royal Astronomical Society Quarterly Journal,* 22 (1981), 133–45:139.

23 Robert E. Schofield, *Mechanism and Materialism* (Princeton, 1970), p. 122. Other publications on this group include (1) Albert J. Kuhn, "Glory or Gravity: Hutchinson versus Newton," *Journal of the History of Ideas,* 22 (1961), 303–22; (2) G. N. Cantor, "Revelation and the Cyclical Cosmos of John Hutchinson," in *Images of the Earth,* ed. L. J. Jordanova and Roy S. Porter (Chafont St. Giles, 1979), pp. 3–22; and (3) C. B. Wilde, "Hutchinsonianism, Natural Philosophy and Religious Controversy in Eighteenth Century Britain," *History of Science,* 18 (1980), 1–24. None of these publications mentions the antipluralist features of the Hutchinsonian movement, for which the best source is Alexander Maxwell, *Plurality of Worlds,* 2nd ed. (London, 1820).

24 Duncan Forbes, *Reflexions on the Sources of Incredulity with Regard to Religion* (Edinburgh, 1752), pp. 1–2.

25 As quoted by Maxwell, *Plurality*, p. 19.

26 As quoted in Edward Nares, ΕΙΣ ΘΕΟΣ, ΕΙΣ ΜΕΣΙΤΗΣ; *or An Attempt to Show How Far the Philosophical Notion of a Plurality of Worlds Is Consistent, or Not So, with the Language of the Holy Scriptures* (London, 1801), p. 14.

27 On the younger Catcott, see Michael Neve and Roy Porter, "Alexander Catcott: Glory and Geology," *British Journal for the History of Science*, 10 (1977), 37–60.

28 Robert Clayton's book is *A Vindication of the Histories of the Old and New Testament. Part I . . .* (Dublin, 1752); *Part II Wherein the Mosaical History of the Creation and Deluge Is Philosophically Explained; . . . Together with Some Remarks on the Plurality of Worlds* (Dublin, 1754); I have consulted the Dublin 1758 edition, where Clayton's pluralist comments occur on pp. 172–96. Clayton's pluralism extended to populating the moon with "Pigmy"-sized creatures living in a "perpetual Serenity of Weather." (p. 182) Catcott's response is *Remarks on the Second Part of The Lord Bishop of Clogher's Vindication of the Histories of the Old and New Testament . . .* (London, 1756); for Catcott's antipluralist comments, see especially pp. 20–34.

29 As quoted in Catcott, *Remarks*, p. 31.

30 Maxwell, *Plurality*, p. 199.

31 References to Wesley's writings are to *The Works of John Wesley*, 3rd ed., 14 vols. (Grand Rapids, Mich., 1978 reprinting of the London 1878 edition). For this incident, which is recorded in Wesley's *Journal*, see his *Works*, vol. II, p. 515.

32 See Bernard Semmel, *The Methodist Revolution* (New York, 1973), p. 5, where Semmel discusses the present state of Wesley historiography and indicates his intention to argue for the latter description.

33 See Frank W. Collier, *John Wesley among the Scientists* (New York, 1928), and Robert E. Schofield, "John Wesley and Science in the 18th Century," *Isis*, 44 (1953), 331–40.

34 Twenty-eight comments on Hutchinson, Catcott, Forbes, Horne, and Jones are listed in the index to Wesley's works; none of these directly relates to their antipluralism. For evidence of Wesley's concern in the 1750s with Hutchinsonian thought, see, for example, his *Works*, vol. II, pp. 353, 388, 441, 454. Wesley was at most ambivalent to the Hutchinsonian system; as a first approximation it may be suggested that he rejected their exegetical doctrines, but was sufficiently impressed by their scientific ideas that he devoted a section to them in his *Survey*. Moreover, in 1765 he described Jones's *Essay on the Principles of Natural Philosophy* as "ingenious," and added: "He seems to have totally overthrown the Newtonian principles; but whether he can establish the Hutchinsonian is another matter." See *Works*, vol. II, pp. 237–8.

35 John Wesley, *A Survey of the Wisdom of God in the Creation*, vol. II (Bristol, 1763), p. 143.

36 See "What Is Man?" in Wesley, *Works*, vol. VII, pp. 167–74. My tentative conclusion that this sermon was from 1788 is based largely, but not exclusively, on L. Tyerman, *The Life and Times of the Rev. John Wesley, M.A.*, vol. III, 6th ed. (London, 1890), p. 563.

37 See Wesley, *Survey*, vol. II (1763), pp. 156–9, for these passages. For James Hervey's "Contemplations on the Starry Heavens," see Hervey's *Meditations and Contemplations*, vol. II (New York, 1845), pp. 209–77. On Hervey, see Alan D. McKillop, "Nature and Science in the Works of James Hervey," University of Texas *Studies in English*, 28 (1949), 124–38. Hervey's materials were put into poetry by Rev. Thomas Newcomb (1682?–1765), who issued *Mr. Hervey's Meditations and Contemplations, Attempted in Blank Verse*, 2 vols. (London, 1764). Newcomb is discussed and some of his pluralist verses are quoted in Jones, *Science*, pp. 173–5.

38 For Wesley's interest in Bonnet and Dutens, see Collier, *Wesley*, pp. 77–81.

39 For Wesley's editing of Young, see Wesley, *Works*, vol. III, p. 350; vol. XIV, pp. 336–8.

40 This is most conveniently available in Wesley, *Works,* vol. XIII, pp. 488–99. See also his sermon on the same theme in *Works,* vol. VI, pp. 337–50.

41 John Dillenberger, *Protestant Thought and Natural Science* (London, 1961), p. 156.

42 Wesley, *Survey,* Mayo ed., vol. II, pp. 112, 122, 143–8.

43 Ernest Campbell Mossner, *The Life of David Hume,* 2nd ed. (Oxford, 1980), pp. 391, 487, 588; Wesley, *Works,* vol. III, p. 462.

44 David Hume, "The Sceptic," in Hume's *Essays Moral, Political, and Literary,* ed. T. H. Green and J. H. Grose, vol. I (1964 reprinting of the London 1882 edition), pp. 213–31:226–7. For Hume's other remark, see p. 227.

45 Wesley, *Works,* vol. VII, pp. 336, 342.

46 David Hume, *Dialogues Concerning Natural Religion,* edited and introduced by Norman Kemp Smith (Indianapolis, 1947), p. 148.

47 Kant labeled Swedenborg "der Erzphantast unter allen Phantasten" in his 1766 *Träume eines Geisterseher, erläutet durch Träume der Metaphysik,* 2nd pt., 1st section. The "ghost-seeker" whose "dreams" are discussed in this work is Swedenborg. For Wesley's references to Swedenborg as a "madman," see Wesley, *Works,* vol. III, p. 387; vol. IV, p. 150; for his 1782 "Thoughts on the Writings of Baron Swedenborg," see Wesley, *Works,* vol. XIII, pp. 425–48.

48 For biographical information on Swedenborg, see (1) George Trobridge, *Swedenborg: Life and Teaching,* 4th ed. (London, 1945); (2) Signe Toksvig, *Emanuel Swedenborg: Scientist and Mystic* (New Haven, 1948); (3) Cyriel Odhner Sigstedt, *The Swedenborg Epic* (New York, 1952); and (4) Inge Jonsson, *Emanuel Swedenborg* (New York, 1971). Jonsson also contributed articles on Swedenborg to the *Encyclopedia of Philosophy* and the *Encyclopaedia Britannica,* 15th ed.

49 For recent discussion, see Stanley L. Jaki, *The Milky Way* (New York, 1972), pp. 168–72, and also his *Planets and Planetarians* (Edinburgh, 1978), pp. 77–9.

50 Emanuel Swedenborg, *Posthumous Theological Works,* edited and translated by Rev. John Whitehead, vol. I (New York, 1954), p. 7.

51 Emanuel Swedenborg, *Arcana coelestia,* vol. I, revised and edited by J. F. Potts (New York, 1956), #5. Swedenborg himself translated this work.

52 I have used the 1970 reprinting of the 1894 third edition of this work. Presumably the translation is that of Clowes in a slightly revised form.

53 Katherine Brownell Collier, *Cosmogonies of Our Fathers* (New York, 1968 reprinting of the 1934 original), p. 185.

54 Svante Arrhenius, *The Life of the Universe,* trans. H. Borns, vol. I (London, 1909), p. 119.

55 Marjorie Hope Nicolson, *Voyages to the Moon* (New York, 1960), p. 63.

56 Theodore F. Wright, "The Planets Inhabited," *New Church Review,* 4 (1897), 117–21:117.

57 Soame Jenyns, "An Essay on Virtue," in *Modern English Poets 1660–1780,* vol. VII, ed. David P. French (New York, 1967), pp. 296–8:296.

58 For a selection, see *Science and Religious Belief 1600–1900: A Selection of Primary Sources,* ed. D. C. Goodman (Dorchester, 1973), pp. 297–302, especially p. 302.

59 Newcome Cappe, "On the Glory of God, as Displayed by the Heavenly Luminaries," in his *Discourses Chiefly on Devotional Subjects* (York, 1805), pp. 320–54:324.

60 Reid endorsed pluralism in his 1785 *Essays on the Intellectual Powers of Man.* In his discussion of the uses of analogy in argumentation, he cites as his first and, to his mind, fully acceptable illustration of this method the analogical argument for life on other planets. See Essay I, ch. IV.

61 James Beattie, *Evidences of the Christian Religion* (Annapolis, 1812), p. 179.

62 Beilby Porteus, "On the Christian Doctrine of Redemption," in Porteus's *Works,* vol. III (London, 1811), pp. 59–86:78.

63 Porteus, "Redemption," p. 79. The scriptural references given by Porteus are Colossians 1:16, 20 and Ephesians 1:10.

64 Edward King, *Hymns to the Supreme Being. In Imitation of the Eastern Songs,* new ed. (London, 1798), p. 7.

65 Edward King, *Morsels of Criticism,* 2nd ed., vol. I (London, 1800), p. 108.

66 [Richard Gough], "[Review of] *Morsels of Criticism.* . . . By Edward King," *Gentleman's Magazine,* 63, pt. 1 (1788), 141–5:145. This review is attributed to Gough in the *Dictionary of National Biography* article on King.

67 Roger Long, *Astronomy in Five Books,* vol. II (Cambridge, England, 1764), p. 646.

68 Long, *Astronomy,* vol. II, p. 646; the reference is to John Milton's *Paradise Lost,* bk. VIII, lines 175–6.

69 George Adams, *Lectures on Natural and Experimental Philosophy,* vol. IV (London, 1794), p. 241. Emphasis added.

70 Adams, *Lectures,* vol. IV, p. 244. After Adams's death in 1795, a second edition of his *Lectures* was published in 1799 "with considerable corrections and additions" by William Jones, mentioned earlier as a Hutchinsonian. In that edition the cited paragraph was deleted, although all the other passages quoted from Adams remained.

71 Olinthus Gregory, *Lessons Astronomical and Philosophical,* 2nd ed. (London, 1799), p. 74.

72 Cotton Mather, *The Christian Philosopher* (Gainesville, 1968 facsimile reprinting of the London 1721 original), p. 2.

73 For Mather's relations to these and other such authors, see Theodore Hornberger, "The Date, the Source, and the Significance of Cotton Mather's Interest in Science," *American Literature,* 6 (1935), 413–20.

74 Among Mather's readers were Benjamin Franklin and John Winthrop IV. On their favorable reactions, see George H. Daniels, *Science in American Society* (New York, 1971), p. 83. Daniels also states (p. 82) that Mather's *Christian Philosopher* "was the first general Newtonian approach published by an American."

75 For these facts and for a compact discussion of pluralism in eighteenth-century America, see Herbert Leventhal, *In the Shadow of the Enlightenment; Occultism and Renaissance Science in Eighteenth-Century America* (New York, 1976), pp. 242–7.

76 L. W. Labaree et al. (eds.), *Papers of Benjamin Franklin,* vol. I (New Haven, 1959), p. 102. Subsequent references are to this edition of his papers.

77 The possibility of Newton's influence on Franklin in this regard was first set out by James Parton in his *Life and Times of Benjamin Franklin,* vol. I (New York, 1864), p. 175, and has been accepted in I. Bernard Cohen, *Franklin and Newton* (Philadelphia, 1956), p. 209, and also in Alfred Owen Aldridge, *Benjamin Franklin and Nature's God* (Durham, N.C., 1967), pp. 29–30.

78 James Burgh (1714–75) was an English dissenting schoolmaster. See Franklin, *Papers,* vol. IV, p. 404n.

79 Franklin's almanacs are reprinted in facsimile in *The Complete Poor Richard Almanacks,* 2 vols. (Barre, Mass., 1970).

80 On Ames, see Marion Barber Stowell, *Early American Almanacs: The Colonial Weekday Bible* (New York, 1977), pp. 72–6, 164–5.

81 As quoted in Leventhal, *Enlightenment,* p. 247.

82 As quoted in Stowell, *Almanacs,* p. 165.

83 As quoted in Silvio A. Bedini, *The Life of Benjamin Banneker* (New York, 1972), p. 76.

84 As quoted in Bedini, *Banneker,* p. 137.

85 *The Monmouth Almanac, for the Year M,DXX,XCV* (Middletown Point, N.J., printed and sold by Philip Freneau), p. 7. Freneau's almanac is thoroughly discussed by Richard Waldron in "The Artist as Scientific Educator: Philip Freneau's *Monmouth Almanac for 1795,*" which Mr. Waldron presented at the 1979 meeting of the History of Science Society in New York and of which he generously supplied me a copy.

86 This judgment needs qualification in two ways First, my reading of his second essay does not permit me to exclude the possibility that it may have been written, as a number of his almanac essays certainly were, with satiric intent. Second, as noted later, one of Freneau's chief sources for the strong pluralism of his first essay was the leading American astronomer of the period, David Rittenhouse, from whose *Oration* Freneau lifted at least two very pluralist passages.

87 Philip Freneau, *Poems Written and Published during the American Revolutionary War* (Delmar, N.Y., 1976 reprinting of the 1809 original), pp. 221–2:221.

88 On Freneau's religious thought, see especially Nelson P. Adkins, *Philip Freneau and the Cosmic Enigma* (New York, 1949).

89 Freneau, in a short piece in the cosmic voyage genre, bestowed this title on Rittenhouse in 1793. See *A Freneau Sampler,* ed. Philip M. Marsh (New York, 1963), pp. 292–6:293.

90 For a detailed analysis of Rittenhouse's *Oration,* see Brooke Hindle, *David Rittenhouse* (Princeton, 1964), pp. 112–22. Although fully aware of Professor Hindle's well-justified reputation as *the* Rittenhouse expert, my reading of Rittenhouse's *Oration* has led me to question his conclusion that it is the work of a man whose "beliefs in many respects could be tagged as deistic." (p. 117) Partial support for my view of it as expressing a basically Christian perspective is provided in my analysis; see also William Barton, *Memoirs of the Life of David Rittenhouse* (Philadelphia, 1813), pp. 501ff.

91 David Rittenhouse, *An Oration* (Philadelphia, 1775), pp. 7–8. This is most conveniently consulted in its facsimile reprinting in Brooke Hindle (ed.), *The Scientific Writings of David Rittenhouse* (New York, 1980).

92 Rittenhouse, *Oration,* pp. 13–14. Rittenhouse's paragraph on Orion, with very minor alterations, was included (without acknowledgment) in Freneau, *Almanac,* p. 17. Some of his statements on comets were also borrowed by Freneau.

93 John Winthrop, *Two Lectures on Comets* (Boston, 1759), pp. 39–40. See also John C. Greene, "Some Aspects of American Astronomy 1750–1815," *Isis,* 45 (1954), 339–58.

94 Hugh Williamson, "An ESSAY on the Use of COMETS, and an Account of Their Luminous Appearance; Together with Some Conjectures Concerning the Origin of HEAT," *American Philosophical Society Transactions,* 1 (1770), Appendix, 27–36. References are to the reprinting in John C. Burnham (ed.), *Science in America: Historical Selections* (New York, 1971), pp. 29–37:31–2.

95 Brooke Hindle, *The Pursuit of Science in Revolutionary America* (Chapel Hill, 1956), p. 172.

96 Professor Burnham, in his *Science in America* (p. 29), titles his republication of Williamson's paper: "A Well Read Physician's Facts and Fancies as Respectable Science."

97 Andrew Oliver, *An Essay on Comets* (Salem, 1772), pp. 15–30.

98 Greene, "Aspects," p. 348, notes the approval of Oliver's essay in Morse's *American Universal Geography.* Morse had earlier endorsed the pluralist position on the basis of the principle of plenitude in his *The American Geography* (New York, 1970 facsimile reprinting of the Elizabeth Town 1789 original), p. 3.

99 James Bowdoin, "A Philosophical Discourse," *American Academy of Arts and Sciences Memoirs,* 1 (1785), 1–20:20.

100 James Bowdoin, "Observations Tending to Prove, by Phaenomena and Scripture, the Existence of an Orb, Which Surrounds the Whole Visible Material System . . . ," *American Academy . . . Memoirs,* 1 (1785), 208–33.

101 *Diary and Autobiography of John Adams,* ed. L. H. Butterfield, vol. I (New York, 1964), p. 22. See also Adams's reflections for April 25 and 30 and May 22, 23, and 27.

102 Page Smith, *John Adams,* vol. II (Garden City, N.Y., 1962), pp. 676–7.

103 Sylvia Engdahl, *Planet-Girded Suns* (New York, 1974), p. 70. For Jefferson's interest in astronomy, see Henry Raphael, "Thomas Jefferson, Astronomer," *Astronomical Society of the Pacific Leaflet,* #174 (August, 1943).

104 *The Works of John Adams,* vol. X (Boston, 1856), p. 415.

105 For Stiles, see Leventhal, *Enlightenment,* p. 245.

106 William Darlington, *Memorials of John Bartram and Humphrey Marshall* (Philadelphia, 1849), p. 399.

107 Leventhal, *Enlightenment,* p. 244.

108 The main source for information on Johnson is Herbert and Carol Schneider, *Samuel*

Johnson: President of King's College: His Career and Writings, 4 vols. (New York, 1929). See Johnson's "A Sermon on the Creation of the World," in Schneiders, *Johnson,* vol. III, pp. 422–34:423–4, for a cautious endorsement of pluralism. See also vol. I, pp. 495–526, for Johnson's readings between 1719 and 1756. During this period, Johnson read not only Derham's *Astro-Theology,* Thomson's *Seasons,* Young's *Night Thoughts,* and Hervey's *Meditations* but also Whiston's *Astronomical Principles of Religion* and Fontenelle's *Dialogues* twice each, as well as Huygens's *Cosmotheoros* three times!

109 M. Gerard, "Of the Philosophers Who Have Believed in a Plurality of Worlds . . . ," *American Museum, or Universal Magazine,* 12 (1792), 241–7. This survey is of uneven quality, but especially good on Leibniz.

110 This and subsequent references to Voltaire's writings, except where noted, are to *Oeuvres complètes de Voltaire,* 97 vols. (Paris: Baudouin Frères, 1825–34). This passage is from vol. XXXV, pp. 102–3.

111 This contention is supported by comparison of a celestial vision passage in the first edition of Voltaire's 1723 *Henriade* with the revision he gave it for its 1730 edition. Although altering the cosmos from Cartesian to Newtonian, Voltaire presented a pluralistic universe in both. See Harcourt Brown, *Science and the Human Comedy: Natural Philosophy in French Literature from Rabelais to Maupertuis* (Toronto, 1976), pp. 136–7.

112 Voltaire, *Oeuvres,* vol. LXVIII, pp. 81–4. The present analysis of Voltaire's pluralism has profited in many ways from two earlier publications broadly focused on his *Micromégas.* See Ira O. Wade, *Voltaire's Micromégas: A Study of the Fusion of Science, Myth, and Art* (Princeton, 1950), and W. H. Barber, "Voltaire's Astronauts," *French Studies,* 30 (1976), 28–52.

113 As translated from the passage quoted in Wade, *Micromégas,* pp. 43, 153. Wade notes (p. 153) that this passage was published in 1738 and that essentially the same idea was expressed three decades later in Voltaire's article "Sensations," in his *Dictionnaire philosophique.*

114 Voltaire, *The Elements of Sir Isaac Newton's Philosophy,* trans. John Hanna (London, 1967 facsimile reprinting of the London 1738 original), pp. 334–5. See also Voltaire, *Élémens de la philosophie Neuton* (Amsterdam, 1738), p. 375. I am indebted to Donald Beaver for calling this passage to my attention. For information on the Dutch publication of this book, see Theodore Besterman, *Voltaire* (New York, 1969), pp. 192–4.

115 Wade, *Micromégas,* pp. 37–40, 52–9.

116 For Voltaire's views on Leibniz, see Richard A. Brooks, *Voltaire and Leibniz* (Geneva, 1964), and W. H. Barber, *Leibniz in France from Arnauld to Voltaire* (Oxford, 1955).

117 The reference here is to Pascal, who according to his sister discovered thirty-two Euclidean propositions. Wade, in his edition of *Micromégas,* has analyzed the many sources of this story, as well as urging that it was for the most part composed in 1739. This conclusion has to some extent been questioned by W. H. Barber in his "The Genesis of Voltaire's Micromégas," *French Studies,* 11 (1957), 1–15. George R. Havens, in his "Voltaire's *Micromégas* (1739–1752): Composition and Publication," *Modern Language Quarterly,* 33 (1972), 113–18, reexamined the evidence, siding for the most part with Barber's contention that important post-1739 revisions occurred.

118 Voltaire, *Oeuvres,* vol. LIX, pp. 181–2. For other comments made by Voltaire on Fontenelle during this period, see *Oeuvres,* vol. XXV, p. 132, where he praises Fontenelle's *Pluralité des mondes* as a "work unique in its genre," and vol. XXVI, p. 358, where he laments its Cartesian basis.

119 Although the moons of Mars were first discovered in 1877, they appear frequently in pre-1877 pluralist writings. Voltaire's two moons probably came not only from pluralism but also from Jonathan Swift, whose *Gulliver's Travels* supplied a number of elements in *Micromégas.*

120 Voltaire, "Poème sur la dèsastre de Lisbonne," in *Oeuvres,* vol. XV, p. 240.

121 Jean-Jacques Rousseau, *Confessions,* edited and revised by Lester G. Crocker (New York, 1957), pp. 221–2.

122   R. A. Leigh, "Rousseau's Letter to Voltaire on Optimism" (containing a critical edition of the letter as sent to Voltaire), *Studies on Voltaire and the Eighteenth Century*, 30 (1964), 247–308:284. See also R. A. Leigh, "From the *Inégalité* to *Candide:* Notes on a Desultory Dialogue between Rousseau and Voltaire (1755–1759)," in *The Age of the Enlightenment: Studies Presented to Theodore Besterman,* ed. W. H. Barber et al. (Edinburgh, 1967), pp. 66–92. I am indebted to Stanley Jaki for calling Rousseau's letter to my attention.

123   Rousseau, *Confessions,* p. 222. R. A. Leigh, "Dialogue," pp. 91–2, has made the interesting suggestion that the publication in September 1759 of Rousseau's letter to Voltaire may have been intended by someone as a response, in turn, to *Candide!*

124   Wade, *Micromégas,* pp. 44–5, takes it seriously, as does Theodore Besterman in his *Voltaire* (New York, 1969), p. 490. But Besterman goes on to describe *Tout en Dieu* as "a prolonged exercise in irony." (p. 491)

125   Payton Richter and Ilona Ricardo, *Voltaire* (Boston, 1980), p. 165.

126   As quoted in Arthur M. Wilson, "Encyclopédie," in *Encyclopedia of Philosophy,* ed. Paul Edwards, vol. II (New York, 1967), pp. 505–8:506.

127   Jacques Roger, in his *Les sciences de la vie dans la pensée Française du XVIIIe siecle,* 2nd ed. (Paris, 1971), p. 664, states that "the deepest source of the thought of Diderot and of his vision of the world is still and always the philosophy of Epicurus and of Lucretius."

128   Denis Diderot, "Épicureisme," in *Oeuvres complètes,* vol. XIV, ed. J. Assezat (Paris, 1876), p. 515.

129   The influence of de Maillet on Diderot's thought in the late 1740s is discussed in Aram Vartanian, "From Deist to Atheist: Diderot's Philosophical Orientation 1746–1749," *Diderot Studies,* 1 (1949), 46–63:59–60. Moreover, de Maillet may have been a source of Diderot's pluralism; his *Telliamed* contains a number of pluralist passages.

130   Denis Diderot, "Letter on the Blind for the Use of Those Who See," in *Diderot's Early Philosophical Works,* trans. Margaret Jourdain (New York, 1972 reprinting of the London 1916 original), pp. 68–141:113.

131   Jean Le Rond d'Alembert, "Cosmologie," *Encyclopédie ou dictionnaire raisonné des sciences, des arts et des métiers,* vol. IV (Paris, 1754), pp. 294–7:294. Articles written by d'Alembert were signed "O".

132   [d'Alembert], "Étoile," *Encyclopédie,* vol. VI (Paris, 1756), pp. 60–4:64.

133   Three sentences of minor significance were given to the topic under the article "Pluralité."

134   [d'Alembert], "Monde," *Encyclopédie,* vol. X (Neufchastel, 1765), pp. 640–1:640.

135   [d'Alembert], "Monde," pp. 640–1.

136   [d'Alembert], "Planète," *Encyclopédie,* vol. XII (Neufchastel, 1765), pp. 703–8:705.

137   [d'Alembert], "Planète," p. 705.

138   Alan Charles Kors, *D'Holbach's Coterie: An Enlightenment in Paris* (Princeton, 1976), p. 87.

139   Baron d'Holbach, *The System of Nature,* trans. H. D. Robinson (New York, 1970 reprinting of the 1868 edition), pp. 44–5.

140   Pierre Maupertuis, *Essai de cosmologie,* in Maupertuis, *Oeuvres,* vol. I (Lyon, 1758), pp. 55–6.

141   Pierre Brunet, *Maupertuis: Étude biographique* (Paris, 1929), p. 128.

142   Maupertuis, *Oeuvres,* vol. III (Hildesheim, 1965 reprinting of the Lyon 1768 edition), p. 251.

143   Leonhard Euler, *Letters of Euler on Different Subjects in Natural Philosophy Addressed to a German Princess,* trans. Henry Hunter, vol. I (New York, 1975 reprinting of the New York 1833 original), p. 207.

144   As quoted in Raymond Savioz, *La philosophie de Charles Bonnet de Geneva* (Paris, 1948), p. 49.

145   Charles Bonnet, *Contemplation de la nature,* vol. I (Amsterdam, 1764), pp. 8–9.

146   Charles Bonnet, *La palingénésie philosophie,* vol. I (Geneva, 1769), pp. 203–4.

147 Arthur O. Lovejoy, *The Great Chain of Being* (New York, 1960 edition of the 1936 original), p. 283.

148 For Bonnet's influence, see, in addition to Savioz's book, Jacques Marx, "Charles Bonnet contre les Lumières, 1738–1850," *Studies on Voltaire and the Eighteenth Century*, 156–157 (1976), 1–782.

149 Robert R. Palmer, *Catholics and Unbelievers in Eighteenth Century France* (Princeton, N.J., 1939), pp. 112, 116–17.

150 For an analysis of Buffon's theory that the planets were formed by a comet having forced material out of the sun, see Stanley L. Jaki, *Planets and Planetarians* (Edinburgh, 1978), pp. 96–106.

151 Comte de Buffon, "Partie hypothétique," *Histoire naturelle,* vol. II of *Supplément* (Paris, 1775), pp. 361–564.

152 As quoted in Jacques Roger, *Les sciences de la vie dans la pensée Française du XVIIIe siècle,* 2nd ed. (Paris, 1971), p. 580.

153 Roger, *Sciences,* p. 580.

154 Buffon, "Partie hypothétique," p. 513.

155 Buffon, "Partie hypothétique," pp. 513–15.

156 Otis E. Fellows and Stephen F. Milliken, *Buffon* (New York, 1972), pp. 168–9.

157 Comte de Buffon, *Histoire naturelle,* vol. I (Paris, 1749), p. 182.

158 Published anonymously from Liège in 1771, this book went through two later editions. Feller published the 1778 edition under the pseudonym Flexier de Revel, whereas the 1788 edition carried his own name. References are to the 1771 edition.

159 Feller, *Observations,* pp. 178–9. Feller claims that Abbé Noël-Antoine Pluche (1688–1761) had criticized the pluralist position in vol. IV of his *Spectacle de la nature,* p. 496. An examination of pp. 496–504 of vol. IV of the Paris 1749 edition shows that in this extremely popular work Pluche attempted to combat pluralism as an objection to religion. The nature of his approach is effectively summarized in his statement in his *History of the Heavens,* trans. J. B. de Freval, vol. II (London, 1740), p. 235: "Let us not presume to speak of what God has made in other places; since we have no manner of knowledge thereof." In his mention of Dulard, Feller is probably referring to Paul-Alexandre Dulard, who in 1749 published *La grandeur de Dieu dans les merveilles de la nature.*

160 Mention of this controversy occurs in the article on Feller in *Biographie universelle,* vol. III (Lyon, 1860), pp. 541–6:544. Because the founding editor of the *Biographie universelle* was Feller himself, this article should possibly be read with caution.

161 F. X. de Feller's *Catéchisme philosophique* first appeared in 1773 in French and later was translated into Spanish and English. For his antipluralist views, see vol. I of the 1820 Liège fifth edition, p. 139.

162 The evidence for Terrasson's authorship is set out in Francisque Bouillier, *Histoire de la philosophie Cartésienne,* 3rd ed., vol. II (Paris, 1868), pp. 608–16. Bouillier's arguments are accepted in Palmer, *Catholics,* pp. 110–12, in Aram Vartanian, *Diderot and Descartes* (Princeton, 1953), pp. 73–5, 97–8, and by André Robinet in *Oeuvres complétes de Malebranche,* vol. XX (Paris, 1967), pp. 321–6. Robinet indicates that three variant printings of the *Traité* were made in 1769, two of which have different paginations and somewhat different texts. The only edition of the work to have been printed after 1769 seems to be that edited in 1915 by Emile Lafuma, who attributed it to Malebranche. Lafuma's text is strongly criticized by Robinet. I have used [Jean Terrasson], *Traité de l'infini créé* (Amsterdam, 1769), which corresponds to Robinet's edition A with the *Traité* occurring on pp. 1–156.

163 Bouillier, *Histoire,* vol. II, pp. 611–12, and Vartanian, *Diderot and Descartes,* p. 73.

164 Robinet, in Malebranche, *Oeuvres,* vol. XX, p. 322, mentions this manuscript, of which neither Bouillier nor Vartanian seems to have been aware.

165 "[Review of] *Traité de l'infini créé*," *Journal encyclopédique* (January 1770), 147–8:147.

166 "[Review of] *Traité de l'infini créé*," *Journal encyclopédique* (March 1770), 180–94:194.

167 As quoted by Robinet in Malebranche, *Oeuvres*, vol. XX, p. 323. See *Oeuvres de M. le Chancelier d'Aguesseau*, vol. XII (Paris, 1783), pp. 162–4, for his full statement.

168 Étienne Bonnot de Condillac, *La logique*, in *Oeuvres philosophiques de Condillac*, ed. Georges Le Roy, vol. II (Paris, 1948), pp. 369–416:412.

169 The discussion extends over about ten pages at the end of Chapter 30 and the beginning of Chapter 31. It can be consulted in English in Jean Jacques Barthélemy, *Voyage of Anacharsis the Younger in Greece during the Middle of the Fourth Century before the Christian Era*, 5th ed., vol. III (London, 1817), pp. 96–104.

170 Louis de Fontanes, "Essai sur l'astronomie," in Fontanes, *Oeuvres*, vol. I (Paris, 1839), pp. 14–25:21–2. For Fontanes and for the interactions in general of French poetry and science, see Casimir Alexandre Fusil, *La poésie scientifique de 1750 à nos jours* (Paris, 1917).

171 Walter Schatzberg, *Scientific Themes in the Popular Literature and the Poetry of the German Enlightenment 1720–1760* (Bern, 1973), and Karl S. Guthke, "Die Mehrheit der Welten: Geistesgeschichtliche Perspektiven auf ein literarisches Thema im 18 Jahrhundert," *Zeitschrift für deutsche Philologie*, 97 (1978), 481–512. Guthke expanded this paper in his *Das Abenteuer der Literatur* (Bern, 1981), pp. 159–86. See also his *Der Mythos der Neuzeit: Das Thema der Mehrheit der Welten in der Literatur- und Geistesgeschichte von der kopernikanische Wende bis zur Science Fiction* (Bern, 1983), especially ch. IV.

172 For information on these translations, see Schatzberg, *Scientific Themes*, pp. 21–46, and for his analysis of these compendia, see pp. 46–63.

173 Johann Christoph Gottsched, *Erste Gründe der gesammten Weltweisheit* (Frankfurt, 1965 reprinting of the Leipzig 1731 original), frontispiece.

174 Gottsched, *Weltweisheit*, pt. I, pp. 384–7.

175 On Gottsched's career, see Gustav Waniek, *Gottsched und die deutsche Litteratur seiner Zeit* (Leipzig, 1897), and G. L. Jones, "Johann Christoph Gottsched," *German Men of Letters*, 6 (1972), 45–69. On his contribution to spreading knowledge of science, see Walter Schatzberg, "Gottsched as a Popularizer of Science," *Modern Language Notes*, 83 (1968), 752–70.

176 Johann Christoph Gottsched, "Als der Verfasser sein funfzigsten Jahr zurücklegte," in Gottsched, *Ausgewählte Werke*, ed. Joachim Birke, vol. I (Berlin, 1968), pp. 224–37:227.

177 Schatzberg, *Scientific Themes*, p. 178.

178 Barthold Brockes, "Das, durch die Betrachtung der grösse Gottes, verherrlichte Richts der Menschen," in Brockes, *Irdisches Vergnügen in Gott*, vol. I (Bern, 1970 reprinting of the Hamberg 1737 edition), pp. 423–57:431.

179 Brockes, "Traum-Gesicht," *Vergnügen*, vol. IV, pp. 192–9, and "Zum Traum-Gesicht," *Vergnügen*, vol. VI, pp. 291–5. See also his "Inseln," "Vier Welte," and the analyses by Guthke and Schatzberg.

180 Guthke, "Mehrheit," pp. 508–10.

181 Albrecht von Haller, "Ueber die Ursprung des Uebels," in Haller, *Gedichte* (Bern, 1969 reprinting of the Göttingen 1762 edition), pp. 161–95:193. This translation is from Arthur O. Lovejoy, *Great Chain of Being* (New York, 1960 reprinting of the 1936 original), p. 200.

182 Haller, *Gedichte*, pp. 205–11:207.

183 Christlob Mylius, *Vermischte Schriften* (Berlin, 1754), p. 355. For a full discussion of Mylius's numerous pluralist writings, see Schatzberg, *Scientific Themes*, pp. 93ff and 233–43, on which the present discussion is largely based.

184 As quoted in Schatzberg, *Scientific Themes*, p. 110.

185 Schatzberg, *Scientific Themes*, p. 114.

186 See, for example, Henry E. Allison, *Lessing and the Enlightenment* (Ann Arbor, 1966), pp. 50–1. One suspects that Mylius may also have been an early source of Lessing's advocacy of metempsychosis, a doctrine sometimes associated with pluralism, as in Bonnet, whose *Palingénésie* Lessing had read.

187 Gotthold Lessing, "Die Planetenbewohner," in *Lessings Werke*, vol. I, ed. Georg Witkowski (Leipzig, 1911), pp. 67–8.

188 *Lessings Werke*, vol. I, pp. 106–8.

189 As quoted in Brian Keith-Smith, "Friedrich von Hagedorn," *German Men of Letters*, 6 (1972), 149–67:150.

190 Friedrich von Hagedorn, "Die Glückseligkeit," in Hagedorn, *Sämmtliche poetische Werke*, pt. I (Darmstadt, 1968 reprinting of the Hamburg 1757 original), pp. 14–27:17.

191 Ewald von Kleist, "Priase of the Godhead," in *The Poetry of Germany*, trans. Alfred Baskerville (Leipzig, 1854), pp. 6–9:7. See also Kleist's "Unzufriedenkeit des Menschen."

192 In addition to Klopstock, Bodmer, and Wieland, who are subsequently discussed, these authors are Georg Heinrich Behr (1708–61), Friedrich von Creuz (1724–70), Johann Friedrich von Cronegk (1731–58), Johann Jakob Dusch (1725–87), Johann Siegmund Leinker (1724–88), Michael Richey (1678–1761), Christian Benjamin Schubert, Johann Peter Uz (1720–96), and Justus F. W. Zachariä (1726–77). For discussions of their pluralist writings, see the publications by Guthke and Schatzberg cited earlier.

193 On the initial reception of Klopstock's *Der Messias*, see Frederick Henry Adler, *Herder and Klopstock* (New York, 1914), pp. 18ff.

194 One indication of Klopstock's stature is that in *German Poetry of the Enlightenment* (University Park, Pa., 1978), Robert M. Browning devotes one-third of his book to Klopstock and scarcely considers *Der Messias*.

195 This is the title of Albrecht Ritschl's chapter on the eighteenth century in his *Critical History of the Christian Doctrine of Justification and Reconciliation*, trans. John S. Black (Edinburgh, 1872), p. 320. The contrast between the conceptual problems faced by Milton and Klopstock in writing Christian epics is thoroughly discussed in Gerhard Kaiser, *Klopstock: Religion und Dichtung* (Gütersloh, 1963).

196 For comments by Klopstock's contemporaries C. F. Cramer and A. G. Kästner on his extensive use of astronomy in *Messias*, see Kaiser, *Klopstock*, p. 51.

197 Extensive evidence for this aspect of Klopstock's thought is provided in Kaiser, *Klopstock*, pp. 104–22.

198 Friedrich Klopstock, *Der Messias*, in Klopstock's *Ausgewählte Werke*, ed. K. A. Schleiden (Munich, 1962), canto I, lines 195–6.

199 Hans Wöhlert, *Das Weltbild in Klopstocks Messias* (Halle, 1915), pp. 36–7.

200 Klopstock, *Messias*, XX, 578–9. See also Kaiser, *Klopstock*, pp. 238–40.

201 For Klopstock's use of the chain of being idea, see Kaiser, *Klopstock*, pp. 54–6. For his attraction to Young's *Night Thoughts*, see Lawrence Marsden Price, "English Literature in German," *University of California Publications in Modern Philology*, 37 (1953), 1–548:114.

202 As quoted in Adler, *Herder and Klopstock*, p. 78.

203 Friedrich Klopstock, "Die Genesung," in *Werke*, p. 77. Subsequent references to odes, except where noted, are to his *Werke*.

204 Friedrich Klopstock, "Psalm," in *Klopstocks Werke*, pt. III, ed. R. Hamel, Deutsche National-Litteratur edition (Berlin, n.d.), pp. 178–9.

205 On Bodmer's use of Klopstockian materials in *Der Noah*, see C. H. Ibershoff, "Bodmer's Indebtedness to Klopstock," *Publications of the Modern Language Association*, 41 (1926), 151–60, and Ibershoff's "Bodmer and Klopstock Once More," *Journal of English and Germanic Philology*, 26 (1927), 112–23.

206 As translated from Schatzberg, *Scientific Themes*, p. 164. This discussion of Bodmer is based largely on Schatzberg, pp. 163–9, and Guthke, "Mehrheit," p. 502.

207 On Wieland's life, see W. E. Yuill, "Christoph Martin Wieland," *German Men of Letters*, 6 (1972), 93–119. For his pluralism, see Schatzberg, *Scientific Themes*, pp. 287–93, and Guthke, "Mehrheit," pp. 502–3.

208 As translated from Alexander Gode-von Aesch, *Natural Science in German Romanticism* (New York, 1941), p. 125. This volume also contains a useful discussion of G. F. Meier.

209 As translated from Guthke, "Mehrheit," p. 502.

210 Schatzberg, *Scientific Themes*, p. 293.

211 As translated from Yuill, "Wieland," p. 78.

212 As quoted in Gode-von Aesch, *Romanticism*, pp. 131–2. For Gellert's "Die Ehre Gottes aus der Natur," see Browning, *German Poetry*, pp. 64–5.

213 Christian Fürchtegott Gellert, "Moralische Vorlesungen," in Gellert's *Sämmtlich Schriften*, vol. VII (Hildesheim, 1968 reprint of Leipzig 1770 original), pp. 396–7.

214 Christoph Christian Sturm, *Reflections on the Works of God and His Providence throughout All Nature* (Philadelphia, 1832), p. 50. Sturm's book was also translated into French and Swedish.

215 The fullest discussion of Herder's involvement with the sciences is H. B. Nisbet, *Herder and the Philosophy and History of Science* (Cambridge, England, 1970).

216 As translated from Nisbet, *Herder*, p. 237. On this manuscript, see also pp. 141–3.

217 As translated from Nisbet, *Herder*, p. 234, whose detailed discussion of Herder's complex and changing views on palingenesis can only be sketched here.

218 Johann Gottfried Herder, *Ueber die Seelenwanderung: Drei Gespräche*, in Herder, *Sämtliche Werke*, ed. Bernhard Suphan, vol. XV (Hildesheim, 1967 reprinting of the Berlin 1888 original), p. 272. All subsequent references, except where noted, are to Suphan's edition of Herder's works.

219 Herder, *Seelenwanderung*, p. 276.

220 Herder, *Werke*, vol. XIII, pp. 19–20. This and other passages in Herder's *Ideen* provoked Kant in 1785 to write a very critical review of Herder's work. See *Kants gesammelte Werke*, vol. VIII (Berlin, 1923), pp. 45–55.

221 Nisbet, *Herder*, p. 234.

222 Herder, *Werke*, vol. XXIII, p. 534. See, in general, pp. 526–35.

223 See Herder, *Werke*, vol. XXIV, p. 317, and Johann Bode, *Betrachtung der Gestirne und des Weltgebäudes* (Berlin, 1816), p. 409.

224 Friedrich Schiller, "To Astronomers," as translated in *Schiller's Works*, ed. J. G. Fischer, vol. I (Philadelphia, 1883), p. 136. See Guthke, "Mehrheit," pp. 511–12. Schiller's "An die Astronomen" is one of his later poems. A number of his earlier poems, such as his "Laura" poems, contain pluralist references. For Schiller's changing uses of cosmic images, see Wolfgang Düsing, "Kosmos und Natur in Schillers Lyrik," *Jahrbuch der deutschen Schillergesellschaft*, 13 (1969), 196–220.

225 As translated from Guthke, "Mehrheit," p. 512. The quotation is from Goethe's *Rede über Winckelmann*.

226 Camille Flammarion, in his *Pluralité des mondes habités*, 33rd ed. (Paris, ca. 1885), p. 43, states that Goethe accepted pluralism but cites no source. Desiderius Papp, in his *Was lebt auf den Sternen?* (Zurich, 1931), p. 13, quotes Goethe as saying "Das leben wohnt im jedem Sterne," but again gives no source. On the other hand, no pluralist passages occur in Goethe's *Faust*, nor is any definite mention of pluralism provided in Carl Hammer, Jr., "Goethe's Astronomical Pursuits," *Studies by Members of SCMLA*, 30 (1970), 197–200.

227 As quoted in Gode-von Aesch, *Romanticism*, p. 126. For the original, see *Deutsch National-Litteratur, Historisch-Critische Ausgabe*, ed. Joseph Kürschner, vol. L, pt. 2 (Berlin, 1893), pp. 18–27:24.

228 Ernst Gottfried Fischer, "Etwas aus der transcendenten Astronomie," *Astronomisches Jahrbuch* (1792), 222–32:222.

229 Pluralist themes appear also in a number of Jean Paul's later writings; see, for example, the "Traum über das All" section of his *Der Komet* and his *Selina oder über die unsterblichkeit der Seele*.

230 Jean Paul, "The Moon," as translated by John Oxenford and C. A. Feiling in their *Tales from the German* (London, 1844), pp. 261–8:262.

231 On this experience, see Dorothea Berger, *Jean Paul Friedrich Richter* (New York, 1972), pp. 22–3, and J. W. Smeed, *Jean Paul's Dreams* (London, 1966), pp. 4–5.

232 Jean Paul, *Flower, Fruit, and Thorn Pieces; or, The Wedded Life, Death, and Marriage of Firmian Stanislaus Siebenkaes*, trans. Alexander Ewing (London, 1895), pp. 262–3.

233 Smeed, *Dreams*, pp. 81–91.

234 As translated from the Hugo quotation in Smeed, *Dreams*, p. 84.

235 In his edition of Fontenelle's *Entretiens sur la pluralité des mondes* (Paris, 1966), Alexandre Calame (p. 189) lists translations by Bernardino Vestrini (1711), possibly Annibal Antonini (1748), Vincenzo Garzia (1765), and Galiani (1780).

236 Giovanni Cadonici, *Confutazione teologica-fisica del sistema di Guglielmo Derham inglese, che vuole tutti i planeti da creature ragionevoli, come la terra, abitati* (Brescia, 1760), p. 21.

237 See the article "Planet" in Ephiam Chambers's unpaginated *Cyclopaedia: or, An Universal Dictionary of Arts and Sciences,* vol. II (London, 1741).

238 For biographical information on Boscovich, see the article on him in the *Dictionary of Scientific Biography* and Lancelot Law Whyte (ed.), *Roger Joseph Boscovich* (London, 1961).

239 On Boscovich's *De lunae atmosphaera,* see Joseph Ashbrook, "Roger Boscovich and the Moon's Atmosphere," *Sky and Telescope,* 16 (1957), 378, and on the subsequent studies, e.g., that of Du Sejour, see Robert Grant, *History of Physical Astronomy* (London, 1852), pp. 230–3.

240 Roger Joseph Boscovich, *A Theory of Natural Philosophy,* trans. J. M. Child (Cambridge, Mass., 1966), p. 166.

241 As quoted in Zeljko Marković, "Boscovich's *Theoria,*" in Whyte, *Boscovich,* pp. 127–52:150.

242 Valentin Boss, *Newton and Russia: The early Influences 1698–1796* (Cambridge, Mass., 1972), pp. 50ff. See also Alexander Vucinich, *Science in Russian Culture: A History to 1860* (Stanford, 1963), p. 56.

243 Boss, *Newton and Russia,* p. 61.

244 As quoted in Boss, *Newton and Russia,* p. 64.

245 Boss, *Newton and Russia,* pp. 64–5.

246 Boss, *Newton and Russia,* pp. 116–27.

247 Boss, *Newton and Russia,* p. 222.

248 For details on Lomonosov's life, see Boris Menshutkin, *Russia's Lomonosov: Chemist – Cortier – Physicist – Poet,* trans. J. E. Thal, E. J. Webster, and W. C. Huntington (Princeton, 1952). For his reading of Fontenelle, see Walter Sullivan, *We Are Not Alone,* rev. ed. (New York, 1966), p. 35. For his astronomical work, see Otto Struve, "Lomonosov," *Sky and Telescope,* 13 (1954), 118–20.

249 Leo Wiener, *Anthology of Russian Literature* (New York, 1902), pp. 253–4.

250 Struve, "Lomonosov," p. 119.

251 As quoted in Struve, "Lomonosov," p. 120.

252 As quoted in Menshutkin, *Lomonosov,* p. 149. See also pp. 82–3.

253 Menshutkin, *Lomonosov,* pp. 147–8. This paper was not printed during Lomonosov's lifetime, but is included in his collected works.

254 Baumer, *European Thought,* p. 141.

255 Becker, *Heavenly City,* p. 31.

256 Jean Milet, *God or Christ?,* trans. John Bowden (New York, 1981), p. xi.

257 As quoted in Ralph C. Roper, "Thomas Paine: Scientist-Religionist," *Scientific Monthly,* 58 (1944), 101–11:102.

258 I have used the reprinting of *The Age of Reason* in Thomas Paine, *Representative Selections,* edited with an introduction, bibliography, and notes by Harry Hayden Clark (New York, 1961). In his long introduction, Clark stresses (pp. xv–xxvii, cxli–cxlii) that Paine's view of science was a much more powerful influence than Quakerism in his thought.

259 Paine, *Age of Reason,* p. 276. The extent of Paine's knowledge of astronomy should not be exaggerated, as was done by Roper, who claimed that he anticipated Herschel in some of his discoveries. A better assessment is provided by Joseph V. Metzger, "The Cosmology of Thomas Paine," *Illinois Quarterly,* 37 (1974), 47–63. Metzger rightly points out (p. 57) that although Paine wrote his *Age of Reason* a dozen years after Herschel's discovery of Uranus, Paine makes mention of only the six traditional planets.

260 Marjorie Nicolson, "Thomas Paine, Edward Nares, and Mrs. Piozzi's Marginalia," *Huntington Library Bulletin,* 10 (1936), 103–33:107.

261 Nicolson, "Paine," p. 108.

262 On its reception, see Clark's comments in his Paine, *Selections,* p. 428.

263 See Nicolson, "Paine," p. 114, and Michael L. Lassar, "In Response to *The Age of Reason, 1794–1799,*" *Bulletin of Bibliography,* 25 (1967), 41–3.

264 For information on a number of the replies dealing with pluralism, see Nicolson, "Paine," pp. 115ff.

265 As quoted in "Walpoliana . . . Number IV," *Monthly Magazine,* 6 (1798), 116. Walpole's reading of Fontenelle probably occurred in 1772 or earlier; see *Horace Walpole's Correspondence,* ed. W. S. Lewis, vol. XXVIII (New Haven, 1955), p. 161.

CHAPTER 4

1 The publications on which this characterization of the period around 1800 rests are (1) William Wordsworth's "Preface" to the second edition of *Lyrical Ballads* (1800); (2) Madame de Staël's *De la litterature considérée dans ses rapports avec les institutions sociales* (1800) and her *De l'Allegmagne* (1810); (3) Thomas Young's "On the Theory of Light and Colours" (1802); (4) Humphry Davy's "Essay on Heat, Light, and the Combinations of Light" (1799); (5) Erasmus Darwin's *Zoonomia* (1794–6); (6) Robert Malthus's *Essay on the Principles of Population* (1798); (7) Friedrich Schleiermacher's *Reden über Religion* (1799); (8) Timothy Dwight's *Theology Explained and Defended* (1818), based on sermons he delivered at Yale between 1795 and 1817; and (9) Comte Joseph de Maistre's *Considérations sur la France* (1796).

2 The endorsements of pluralism in the writings of Wordsworth, de Staël, Davy, E. Darwin, Dwight, and Maistre are subsequently discussed. For the statements of Young, see *A Course of Lectures on Natural Philosophy,* vol. I (London, 1845), p. 468, and *Miscellaneous Works of Thomas Young,* ed. George Peacock, vol. I (London, 1855), p. 417; for Malthus, see *Essay on the Principles of Population,* ed. Philip Appleman (New York, 1976), p. 125; for Schleiermacher, see his *On Religion: Speeches to Its Cultured Despisers,* trans. John Oman (New York, 1958), pp. 35, 106, 142.

3 Thomas Thomson, *A System of Chemistry,* 4th ed., vol. I (Edinburgh, 1810), p. 583.

4 Young, *Course of Lectures,* vol. I, p. 399.

5 James Mitchell, *On the Plurality of Worlds: A Lecture in Proof of the Universe Being Inhabited. Read in the Mathematical Society, London* (London, 1813), pp. 16–18. This was reprinted in 1820 as an appendix to Mitchell's *Elements of Astronomy.*

6 Adam Walker, *A System of Familiar Philosophy: In Twelve Lectures: Being the Course Usually Read by Mr. A. Walker,* new ed., vol. II (London, 1802).

7 William Wordsworth, *The Poetical Works,* ed. Thomas Hutchinson (London, 1910), pp. 236–7.

8 Wordsworth, "To the Moon," *Works,* pp. 460–1. See also E. E. Marwick, *Astronomy in Wordsworth's Poetry* (Dublin, 1918).

9 As quoted in M. K. Joseph, *Byron the Poet* (London, 1964), p. 336. See pp. 311–12 for Joseph's reasons for describing these "Detached Thoughts" as Byron's "credo."

10 Manfred Eimer, "Byron und der Kosmos," *Anglistische Forschungen,* 34 (1912), 1–234.

11 *The Works of Lord Byron, Letters and Journals,* ed. R. E. Prothero, vol. II (London, 1903), pp. 221–2.

12 For Walker's influence on Shelley, see Richard Holmes, *Shelley: The Pursuit* (New York, 1975), pp. 16–17; Carl Grabo, *A Newton among Poets: Shelley's Use of Science in Prometheus Bound* (Chapel Hill, 1930), pp. 4–6; and Kenneth Neill Cameron, *Shelley: The Golden Years* (Cambridge, Mass., 1974), pp. 296, 544–5, 549, 551.

13 Holmes, *Shelley,* p. 26, and Grabo, *A Newton among Poets,* p. 6.

14 Grabo, *A Newton among Poets,* pp. 30ff.

15 As quoted in Grabo, *A Newton among Poets,* p. 43.

16 As quoted in Grabo, *A Newton among Poets,* p. 169.

17 Grabo, *A Newton among Poets,* p. 45.

18 Percy Bysshe Shelley, *The Complete Poetical Works,* ed. Neville Rogers, vol. I (Oxford, 1972), p. 296.
19 Ifor Evans, *Literature and Science* (London, 1954), p. 69.
20 Cameron, *Shelley,* p. 602.
21 Percy Bysshe Shelley, "On the Devil, and Devils," *Works,* ed. Harry Buxton Forman, vol. VI (London, 1880), pp. 383–406:390.
22 Strictly speaking, this line was written by Robert Southey, who translated Coleridge's 1793 "A Greek Ode on Astronomy" into English. The original of Coleridge's poem is now lost, according to James Dykes Campbell, *Samuel Taylor Coleridge: A Narrative of the Events of His Life,* 2nd ed. (London, 1896), p. 23n. For Southey's translation, see *Coleridge's Poems,* ed. J. B. Beer (London, 1963), pp. 13–16:15.
23 Samuel Taylor Coleridge, *Inquiring Spirit: A New Presentation of Coleridge from His Published and Unpublished Prose Writings,* ed. Kathleen Coburn (London, 1951), p. 257.
24 Samuel Taylor Coleridge, *Complete Works,* ed. Professor Shedd, vol. VI (New York, 1853), p. 502–3.
25 Andrew Fuller, *The Gospel Worthy of All Acceptation and The Gospel Its Own Witness* (1961 Sovereign Grace reprinting), pp. 270–83.
26 George Gilfillan, *Gallery of Literary Portraits,* vol. I (Edinburgh, 1845), p. 113.
27 Paine is never mentioned in Nares's book, but Professor Nicolson's implicit judgment that Nares wrote largely in response to Paine seems correct. See Marjorie Nicolson, "Thomas Paine, Edward Nares, and Mrs. Piozzi's Marginalia," *Huntington Library Bulletin,* 10 (1936), 103–33. That Nares knew of Paine's *Age of Reason* is evident from his *A View of the Evidences of Christianity at the Close of the Pretended Age of Reason* (Oxford, 1805). That Nares's literary virtues did not include conciseness is indicated by Macaulay's review of Nares's three-volume biography of William Cecil, Lord Burghley. Wrote Macaulay: "We cannot sum up the merits of this stupendous mass of paper . . . better than by saying, that it consists of about two thousand closely printed pages, that it occupies fifteen hundred inches, cubic measure, and that it weighs sixty pounds avoirdupois. Such a book might, before the deluge, have been considered as light reading by Hilpa and Shalum." As quoted in Nicolson, "Nares," pp. 105–6.
28 According to G. Cecil White, *A Versatile Professor: Reminiscences of the Rev. Edward Nares, D.D.* (London, 1903), p. 148, Nares's book first appeared anonymously, but in response to the many inquiries the publisher received, a new title page bearing Nares's name was added.
29 Nares, *Attempt,* pp. 27, 50, 52, 77–9, 171, 209, 214, etc.
30 I have used Adam Clarke, *The Holy Bible . . . with a Commentary and Critical Notes,* 6 vols. (New York, 1817–25). It is difficult to describe adequately the size of this unpaginated collection; suffice it to say, using the criteria Macaulay applied to Nares's *Burghley* volumes, that Clarke's *Commentary* occupies over 1,600 cubic inches and weighs 35 pounds.
31 Because Clarke's *Commentary* is unpaginated, all references to it are given in terms of the scriptural passage discussed.
32 For other passages in Clarke, *Commentary,* see his discussions of Nehemiah 9:6, Psalms 8:3, and Amos 9:6. For his use of pluralist ideas in his sermons, see Adam Clarke, *Discourses on Various Subjects Relative to the Being and Attributes of God* (New York, 1829), pp. 20, 25–6.
33 John Wesley, *Works,* 3rd ed., 14 vols., vol. XII (Grand Rapids, Mich., 1978 reprinting of the London 1878 edition), p. 398.
34 Merle Curti, *The Growth of American Thought,* 3rd ed. (New York, 1964), p. 153.
35 Russell Blaine Nye, *The Cultural Life of the New Nation* (New York, 1960), p. 205.
36 For Harvard and Yale, see Curti, *American Thought,* p. 192.
37 Nye, *Cultural Life,* p. 215.
38 As quoted in Nye, *Cultural Life,* p. 212.
39 Timothy Dwight, *Theology Explained and Defended in a Series of Sermons,* 5 vols. (Middletown, Conn., 1818). On Dwight's career, see Stephen E. Berk, *Calvinism*

versus Democracy (Hamdem, Conn., 1974), Charles E. Cunningham, *Timothy Dwight* (New York, 1952), and Kenneth Silverman, *Timothy Dwight* (New York, 1969). See also Roland Bainton, *Yale and the Ministry* (New York, 1957).

40 A hint of an answer can be found in Dwight, *Theology*, vol. II, p. 385.

41 Cunningham, *Dwight*, p. 330, and Bainton, *Yale*, p. 77.

42 Pierre Samuel Du Pont de Nemours, *Philosophie de l'univers* (Paris, 1796), p. 236. For his discussion of pluralism, see especially pp. 119–46.

43 Jacques Necker, *Cours de la morale religieuse*, in Necker, *Oeuvres complètes*, ed. Baron de Staël, vol. XII (Paris, 1821), p. 48.

44 Necker, *Cours*, p. 49. See also pp. 50, 57–61. For commentary on Necker's attachment to the great chain of being as a form of "cosmic Toryism," see Henri Grange, *Les idées de Necker* (Paris, 1974), pp. 589–611.

45 Madame de Staël, *Corinne, ou l'Italie*, new ed. (New York, 1953), p. 140.

46 In his "Bernardin de Saint-Pierre and the Idea of 'Harmony'," in *Stanford French Studies*, 2 (1978), 209–21, Basil Guy asserts (pp. 217–18) that in the period before Humboldt's *Kosmos* of 1845, Bernardin's *Harmonies de la nature* "was the most popular effort of its kind that Western Europe had ever known."

47 Jacques Henri Bernardin de Saint-Pierre, *Harmonies of Nature*, 3 vols., vol. III, trans. W. Meeston (London, 1815), pp. 226–365.

48 Louis Cousin-Despréaux, *Leçons de la nature*, vol. IV (Lyon, 1836), p. 248. For his advocacy of pluralism, see especially Chapters CCCXXI–CCCXXV.

49 Paul Gudin de la Brenellerie, *L'astronomie. Poëme en quatre chants*, new ed. (Paris, 1810), p. 193. For Gudin's discussion of pluralism, see pp. 193–211.

50 Edmond Grégoire, *L'astronomie dans l'oeuvre de Victor Hugo* (Paris, 1933), p. 201. In discussing Chateaubriand (pp. 191–205) as a source for Hugo's interest in astronomical imagery, Grégoire states: "Repeatedly Chateaubriand shows the same distant preoccupations which were those of Hugo: the habitability of worlds, the plurality of inhabited worlds, science, the two infinities." Jean d'Estienne (pseudonym of Charles de Kirwan) also ascribed pluralism to Chateaubriand; see *Revue des questions scientifiques*, 36 (1894), 317.

51 François-René de Chateaubriand, *Les martyrs*, in his *Oeuvres*, vol. III (Paris, 1837), p. 641; Grégoire, *Hugo*, p. 195.

52 Chateaubriand, *Oeuvres*, vol. III, p. 37.

53 Comte Joseph de Maistre, *Soirées de Saint-Pétersbourg*, vol. II (Paris, n.d.), pp. 318–19. Maistre cites Cardinal Gerdil as holding this view; this was probably Cardinal Hyacinthe Gerdil (1718–1802).

54 Edward Hitchcock, "Introductory Notice to the American Edition" of [William Whewell], *The Plurality of Worlds* (Boston, 1854), p. x.

55 The chief source of biographical information on Chalmers is William Hanna, *Memoirs of the Life and Writings of Thomas Chalmers*, 4 vols. (Edinburgh, 1849–52). I have also consulted (1) Hugh Watt, *Thomas Chalmers and the Disruption* (London, 1943); (2) W. P. Blaikie, *Thomas Chalmers* (New York, [1896]); (3) David Cairns, "Thomas Chalmers's Astronomical Discourses: A Study in Natural Theology," *Science and Religious Belief*, ed. C. A. Russell (Bungay, Suffolk, England, 1973), pp. 195–204; (4) Daniel F. Rice, "Natural Theology and the Scottish Philosophy in the Thought of Thomas Chalmers," *Scottish Journal of Theology*, 24 (1971), 23–46; and (5) Stewart J. Brown, *Thomas Chalmers and the Godly Commonwealth of Scotland* (Oxford, 1982).

56 As quoted in Hanna, *Chalmers*, vol. I, p. 33.

57 As quoted in Blaikie, *Chalmers*, p. 12.

58 On Chalmers's relation to Scottish philosophy, see especially Rice, "Chalmers."

59 As quoted in Hanna, *Chalmers*, vol. I, pp. 437–8.

60 As quoted in Hanna, *Chalmers*, vol. I, p. 449.

61 [Samuel Warren], "Speculators among the Stars," *Blackwood's Edinburgh Magazine*, 76 (1854), 288–300; 370–403:373. For evidence of Warren's authorship, see Isaac Todhunter, *William Whewell*, vol. I (London, 1876), pp. 199–200.

62 Hanna, *Chalmers*, vol. II, p. 88.

63 Hanna, *Chalmers,* vol. II, p. 89.
64 As quoted in Hanna, *Chalmers,* vol. II, p. 89.
65 Hanna, *Chalmers,* vol. II, pp. 89–90.
66 Hanna, *Chalmers,* vol. II, p. 92.
67 Translated by von Reinecke, it appeared under the title: *Reden über die christliche Offenbarung.*
68 Watt, *Chalmers,* p. 51.
69 Thomas Chalmers, *A Series of Discourses on the Christian Revelation Viewed in Connection with the Modern Astronomy* (New York, 1918), p. 3.
70 It may be helpful in attempting to understand the attractiveness to nineteenth-century readers of Chalmers's presentation to recall that in the twentieth century, C. S. Lewis used a number of the same images and ideas in his very popular *Out of the Silent Planet* and *Perelandra.*
71 Cairns, "Chalmers," p. 197.
72 Cairns, "Chalmers," p. 196.
73 On the Bridgewater Treatises, see (1) D. W. Gundry, "The Bridgewater Treatises and Their Authors," *History,* 31 (1946), 140–52; (2) F. H. Amphlett, "Thomas Chalmers and the Bridgewater Treatises," *History,* 32 (1847), 122–3; and (3) W. H. Brock, "The Selection of the Authors of the Bridgewater Treatises," *Notes and Records of the Royal Society,* 21 (1966), 162–79.
74 Thomas Chalmers, *On the Wisdom, Power, and Goodness of God, As Manifested in the Adaption of External Nature to the Moral and Intellectual Constitution of Man* (Philadelphia, 1836).
75 Amphlett, "Chalmers," p. 122.
76 *British Review,* 10 (August 1817), 1–30:7.
77 *Christian Observer,* 16 (September 1817), 588–609:593.
78 *Evangelical Magazine,* 25 (1817), 267–70:267.
79 "On the Pulpit Eloquence of Scotland," *Blackwood's Edinburgh Magazine,* 2 (November 1817), 131–40:139. The bases for suggesting that Robert Hall wrote this article are that (1) it is signed "R. H." and (2) Robert Hall (Hanna, *Chalmers,* vol. II, p. 107) in a letter expresses sentiments similar to those found in the article.
80 [John Foster], *Eclectic Review,* N.S., 8 (1817), 205–19, 354–66, 466–76:209. For evidence of Foster's authorship, see Hanna, *Chalmers,* vol. II, pp. 91–2.
81 [Foster], *Eclectic Review,* 8 (1817), 212. Foster's review was more judicious in many of its other statements; in fact, it was held in high esteem by Chalmers, as is noted in Hanna, *Chalmers,* vol. II, p. 92.
82 Besides the discussion by "R. H.", a review appeared in *Blackwood's,* 1 (April 1817), 73–5, written by the noted Edinburgh literary critic John Wilson, who wrote under the pseudonym (but not in this review) "Christopher North." For evidence of Wilson's authorship, see Blaikie, *Chalmers,* p. 84.
83 *Monthly Review,* 84 (1817), 68–73:72.
84 *Blackwood's,* 1 (April 1817), 73.
85 *British Review,* 10 (August 1817), 19; but see also pp. 28–9.
86 As quoted in Blaikie, *Chalmers,* pp. 58–9.
87 [Henry Fergus], *An Examination of Some of the Astronomical and Theological Opinions of Dr. Chalmers, As Exhibited in a Series of Discourses on the Christian Revelation, Viewed in Connection with Modern Astronomy. With Some Remarks on the History of Pulpit-Eloquence in Scotland* (Edinburgh, 1818). Both the *National Union Catalog* and the University of Edinburgh Library attribute this work of Fergus, although the records at Edinburgh do not explain why this was done (private correspondence). Hanna, *Chalmers,* vol. II, p. 501, mentions that rumor had it that Bishop Gleig of Sterling was its author. The *National Union Catalog* also attributes to Fergus a work entitled *A Review of Dr. Chalmers' Astronomical Discourses: An Article Not to Be Found in Any Number, Hitherto Published, of the Edinburgh or Quarterly Review* (Glasgow, 1818). Although having more pages (66 as compared with 42), this work probably expresses similar views. The other two treatises are (1) Anonymous, *A Free Critique on Dr. Chalmers's Discourses on Astronomy, or, An English Attempt to*

"Grapple It" with Scotch Sublimity (London, 1817), and (2) John Overton, *Strictures on Dr. Chalmers' Astronomical Discourses* (Deptford, Kent, 1817). The first item concentrates almost entirely on criticizing Chalmers's style, whereas Overton faults him above all for not having adopted a peculiar system of biblical chronology that Overton had earlier advocated.

88 Augustus De Morgan, *Budget of Paradoxes,* 2nd ed., ed. D. E. Smith (Chicago, 1915), p. 102.

89 I have used the second edition enlarged (London, 1820). The first edition appeared in 1817.

90 This makes Maxwell's book especially useful for the historian; most pluralist works contain few quotations and fewer references.

91 De Morgan, *Budget,* p. 102.

92 The correct value is about 300 miles. Maxwell does not seem to realize that these were not parallax measurements, but rather were based on measurements of the arc width of Pallas.

93 According to the British Museum catalog, portions of Maxwell's treatise were re-printed in 1872.

94 *Gentleman's Magazine,* 124 (November 1818), 441–3; *Evangelical Magazine,* 26 (1818), 517–19, 562–3; *Monthly Review,* 87 (1818), 108–11.

95 Charles D. Cleveland, *English Literature of the Nineteenth Century* (Philadelphia, 1860), p. 575. For biographical information, see also (1) Anonymous, "Thomas Dick, LL. D.," *Living Age,* 61 (April 16, 1859), 131–6; (2) John A. Brashear, "A Visit to the Home of Dr. Thomas Dick," *Royal Astronomical Society of Canada Journal,* 7 (1913), 19–30; (3) Davide Gavine, "Thomas Dick, LL. D., 1774–1857," *British Astronomical Association Journal,* 84 (1974), 345–50; (4) Edward T. Jones, "The Theology of Thomas Dick and Its Possible Relationship to That of Joseph Smith," a 1969 M.A. thesis at Brigham Young University; and (5) J. V. Smith, "Reason, Revelation and Reform: Thomas Dick of Methven and the 'Improvement of Society by the Diffusion of Knowledge,' " *History of Education,* 12 (1983), 255–70.

96 I have used the printing in *The Works of Thomas Dick* (Hartford, 1844).

97 Thomas Dick, *Philosophy of Religion,* in *Works* (Hartford, 1844), p. 65.

98 Thomas Dick, *Philosophy of a Future State,* in *Works* (Hartford, 1844), p. 89.

99 Thomas Dick, *On the Improvement of Society by the Diffusion of Knowledge,* in *Works* (Hartford, 1844), p. 101.

100 Thomas Dick, *On the Mental Illumination and Moral Improvement of Mankind* in *Works* (Hartford, 1848), pp. 108–16.

101 I have used the edition in Dick, *Works* (Hartford, 1848).

102 I have used the edition in Dick, *Works* (Hartford, 1848).

103 An interesting example of his knowledge of Laplace occurs in the section (p. 111) in which he discusses Laplace's idea that there may exist celestial bodies with gravitational fields so strong that no light escapes from them, i.e., what are now known as black holes.

104 Dick recounts one such criticism in an unpublished October 12, 1840, letter to the amateur astronomer Rev. B. W. S. Vallack. This letter was called to my attention by Dr. David W. Dewhirst of the Institute of Astronomy of Cambridge University, where the letter is preserved in Vallack's correspondence. See also Dewhirst's "The Correspondence of Rev. B. W. S. Vallack," *Royal Astronomical Society Quarterly Journal,* 23 (1982), 552–5.

105 Brashear, "Visit," p. 24, and Vincent Cronin, *The View from Planet Earth* (New York, 1982), pp. 181, 320.

106 See Robert Hardie, "The Early Life of E. E. Barnard," *Astronomical Society of the Pacific Leaflets,* #415 (January 1964), 4–5; for Brashear, see his "Visit," pp. 19ff; for Livingstone, see Gavine, "Dick," p. 347.

107 The list appears in J. C. Houzeau and A. Lancaster, *Bibliographie générale de l'astronomie jusqu' en 1880,* vol. II (London, 1964 reprint of the Bruxelles 1882 original), p. lxxiv. Gruithuisen edited *Analekten für Erd- und Himmels-Kunde* (1828–31), *Neue Analekten . . .* (1832–6), and *Astronomisches Jahrbuch* (1839–50). On Gruit-

huisen, see Dieter B. Herrmann, "Franz von Paula Gruithuisen and seine 'Analekten für Erd- und Himmelskunde'," *Sterne,* 44 (1968), 120–5, and Siegmund Gunther, *Kosmo- und geophysikalische Anschauungen eines vergessenen bayerischen Gelehrten* (Munich, 1914).

108 Franz von Paula Gruithuisen, "Selenognostische Fragmente," *Nova Acta Leopoldina (Physico-Medica),* 10, pt. 2 (1821), 636–92:651; 11, pt. 2 (1823), 584–602.

109 F. von P. Gruithuisen, "Entdeckung vieler deutlichen Spuren der Mondebewohner, besonders eines collossalen Kunstgebäudes derselben," *Archiv für die gesammte Naturlehre,* 1 (1824), 129–71, 257–322.

110 See the summary of the first part of it published in *Bullétin des sciences mathematiques, astronomiques, physiques, et chimiques,* 2 (1824), 282–4. See also *Observatory,* 42 (1919), 327, for information on its reception in Vienna and Britain. The anonymous report on it under the title "The Moon" in the *Edinburgh Journal of Science,* 11 (October 1824), 211, was the source for the discussion of Gruithuisen that Dick added as an appendix to later editions of his *Christian Philosopher.*

111 T. W. Webb, "Gruithuisen's City in the Moon," *Intellectual Observer,* 12 (1868), 214–22:217.

112 F. von P. Gruithuisen, "Kann man denn gar nichts Gewisses von den Bewohnern anderer Weltkörper wissen?" *Neue Analekten für Erd- und Himmels-Kunde,* 1, 3 (1833), 30–46; 1, 4 (1833), 40–55; 2, 1 (1835), 40–6; 2, 2 (1835), 60–7.

113 Camille Flammarion, in his *Dreams of an Astronomer* (New York, 1923), pp. 191–2, quoted extensively from it without indicating its source, as did Willy Ley in his *Rockets, Missiles, and Men* (New York, 1968), pp. 32–3, and his *Watchers of the Skies* (London, 1963), pp. 211–12.

114 Gruithuisen, "Bewohnern," 1, 4 (1833), 52–3.

115 As quoted without reference in G. Waldo Dunnington, *Carl Friedrich Gauss: Titan of Science* (New York, 1955), p. 253.

116 C. Schilling, *Wilhelm Olbers: Sein Leben und seine Werke,* vol. II (Berlin, 1909), pp. 470, 321.

117 J. J. von Littrow, *Die Wunder des Himmels,* 2nd ed. (Stuttgart, 1842), p. 384.

118 For typical versions of this story, see Willy Ley, *Rockets,* pp. 29–30; Ley, *Mariner IV to Mars* (New York, 1966), pp. 29–30; Ley, *Watchers,* p. 502; Isaac Asimov, *Extraterrestrial Civilizations* (New York, 1979), p. 250; and Walter Sullivan, *We Are Not Alone,* rev. ed. (New York, 1966), p. 178.

119 The late Willy Ley, who was extremely widely read in pluralist literature, admitted in his *Watchers* (p. 502) that he had been unable to locate Gauss's proposal in his writings. He also supplied no source for the Littrow version. Nonetheless, he recounted the story in at least three of his books.

120 Littrow, *Wunder,* p. 387n.

121 F. E. Plisson, *Les mondes* (Paris, 1847), pp. 72–5.

122 Patrick Scott, *Love in the Moon* (London, 1853), p. 19.

123 *Christian Remembrancer,* 29 (1855), 66.

124 Asaph Hall, "The Discovery of the Satellites of Mars," *Royal Astronomical Society Monthly Notices,* 38 (1878), 205–9:207; J. Norman Lockyer, "The Opposition of Mars," *Nature,* 46 (September 8, 1892), 443–8:443.

125 Anonymous, "Life in Mars," *Chambers's Journal,* 63 (June 12, 1886), 369–71:371.

126 Simon Newcomb, "Are Other Worlds Inhabited?" *Youth's Companion,* 76 (December 11, 1902), 639–40:639. The astronomer referred to by Newcomb is probably Francis Xavier von Zach (1754–1832).

127 Anonymous, "The Moon and Its Inhabitants," *Edinburgh New Philosophical Journal,* 1 (October 1826), 389–90.

128 "The Moon and Its Inhabitants," *Annals of Philosophy,* 12 (December 1826), 469–70:470.

129 C. Schilling, *Wilhelm Olbers: Sein Leben und seine Werke,* vol. I (Berlin, 1894), p. 663. One caution needs to be expressed concerning this letter; Schilling cites as his source for this letter the publication of it by Gruithuisen in 1841 in the *Astronomishes*

Jahrbuch, which Gruithuisen edited. The fact that Schilling published only the excerpt previously published by Gruithuisen suggests that Schilling did not see the original.

130 Wilhelm Olbers, "Ueber die Durchsichtigkeit des Weltraums" as reprinted in S. L. Jaki, *The Paradox of Olbers' Paradox* (New York, 1969), pp. 256–64:257.

131 Schilling, *Olbers*, vol. II, pp. 321.

132 Schilling, *Olbers*, vol. II, p. 457, 470.

133 Littrow, *Wunder*, pp. 357–61, 382–8.

134 Schilling, *Olbers*, vol. II, p. 180.

135 As quoted without reference in Wolfgang Sartorius von Waltershausen, *Carl Friedrich Gauss: A Memorial,* trans. Helen W. Gauss (Colorado Springs, 1966), p. 41. Sartorius's biography was first published in 1856. In his *Gauss: Titan of Science,* Dunnington stated that the Göttingen *Almanach* attributed this statement to Gauss (p. 122).

136 See Dunnington, *Gauss,* p. 253, for evidence of this visit.

137 Kurt R. Biermann (ed.), *Briefwechsel zwischen Alexander von Humboldt und Carl Friedrich Gauss* (Berlin, 1977), pp. 115–16. Whewell's book, which is subsequently analyzed, does not explicitly make this claim, although Humboldt's interpretation of it was not uncommon.

138 Biermann, *Briefwechsel,* p. 118.

139 For a discussion of these conversations, see Dunnington, *Gauss,* pp. 301–10. Wagner's report of these conversations, which was suppressed for many years, has recently been published: Rudolf Wagner, "Gespräche mit Carl Friedrich Gauss in den letzten Monaten seines Lebens," edited by Heinrich Rubner, *Nachrichten der Akademie der Wissenschaften in Göttingen, philologisch-historische Klasse,* nr. 6 (Göttingen, 1975), pp. 141–171. Rubner refers to Gauss's "kosmische Religiosität." (p. 149)

140 Sartorius, *Gauss,* p. 73.

141 Friedrich Wilhelm Bessel, "Ueber die physische Beschaffenheit der Himmelskörper," in Bessel, *Populäre Vorlesungen über wissenschaftliche Gegenstände,* ed. H. C. Schumacher (Hamburg, 1848), pp. 68–93.

142 In his "Ueber den Monde," an 1838 lecture included in his *Vorlesungen,* Bessel specified that the lunar atmosphere could be no more than one five-hundredth of the density of the earth's atmosphere (p. 620).

143 As quoted in Joseph Ashbrook, "Lohrmann's Atlas of the Moon," *Sky and Telescope,* 15 (December 1955), 62–3.

144 As quoted in H. Percy Wilkins and Patrick Moore, *The Moon* (New York, n.d.), p. 39. In his *Populäre Astronomie* (Berlin, 1846), Mädler expressed reservations about lunar life, noting that pluralists frequently relied more on religious than on astronomical considerations (p. 196).

145 Edmund Neison, *The Moon* (London, 1876), p. 104.

146 In discussing the report that appeared in the *Sun,* I have used the reprinting of it in William N. Griggs, *The Celebrated "Moon Story," Its Origin and Incidents with a Memoir of Its Author* (New York, 1852). All references are to this edition. My interpretation relies heavily on Griggs's analysis, which differs substantially from other presentations. Among these are (1) Edgar Allan Poe, "Richard Adams Locke," *Complete Works of Edgar Allan Poe,* vol. XV, ed. James A. Harrison (New York, 1902), pp. 126–37; (2) Anonymous, "Locke among the Moonlings," *Southern Quarterly Review,* 24 (1853), 501–14; (3) Richard A. Proctor, "The Lunar Hoax," in Proctor's *Myths and Marvels of Astronomy* (London, 1880), pp. 242–67; (4) Frank M. O'Brien, *The Story of the Sun,* new ed. (New York, 1928), pp. 37–57; (5) William H. Barton, Jr., "The Moon Hoax: The Greatest Scientific Fraud Ever Perpetrated," *Sky,* 1 (February 1937), 6–10, 22; (March 1937), 10–11, 23–5; (April 1937), 10–11, 22–4, 28; (6) Gibson Reaves, "The Great Moon Hoax of 1835," *Griffith Observer* (November 1954), 126–34; (7) Willy Ley, *Watchers of the Skies* (London, 1963), pp. 268–75; (8) Richard Adams Locke, *The Moon Hoax,* with an introduction by Ormond Seavey (Boston, 1975); (9) Donald Fernie, *The Whisper and the Vision* (Toronto, 1976), pp. 91–102; and (10) David S. Evans, "The Great Moon

Hoax," *Sky and Telescope,* 62 (1981), 196–8, 308–11. See also M. J. Crowe, "New Light on the 'Moon Hoax'," *Sky and Telescope,* 62 (1981), 428–9. Items 5 and 8 contain a full text of the articles that appeared in the New York *Sun.*

147   On the number of copies of the reprint, see Ley, *Watchers,* p. 273; on the lithographs, see O'Brien, *Story,* p. 53.

148   These five quotations, which are frequently cited in accounts of the moon hoax, are all given in an appendix to a reprinting of the *Sun* articles: Richard Adams Locke, *A Discovery That the Moon Has a Vast Population of Human Beings,* introduction by William Gowans (New York, 1859), pp. 61–2. This appendix itself consists of a reprinting of an article attributed to the September 1, 1835, issue of the *Sun.* This point may not be without importance; the tone of the appendix is such that one may suspect that some or all of these quotations may themselves be fabrications!

149   Sidney P. Moss, *Poe's Literary Battles* (Durham, N.C., 1963), p. 87, and Josiah Crampton, *The Lunar World* (Edinburgh, 1863), p. 84.

150   This claim from "Locke among the Moonlings" should not perhaps be taken seriously; the tone of the article suggests that satire may have been mixed with factual information in it.

151   "Locke among the Moonlings," p. 502.

152   Ley, *Watchers,* p. 273. David F. Musto disputes this in his "Yale Astronomy in the Nineteenth Century," *Ventures,* 8 (1968), 7–18:9.

153   Poe, "Locke," p. 134.

154   O'Brien, *Story,* pp. 54–5.

155   This sketch of Locke is a composite from various sources, but is based primarily on Griggs, "Moon Story," pp. 40–5. Different authors dispute various features of Locke's life, for example, whether he was born in England or the United States. I have been unable to confirm that he studied at Cambridge.

156   Poe, "Locke," p. 136.

157   As quoted in Griggs, "Moon Story," p. 30.

158   Griggs, "Moon Story," p. 8.

159   Griggs, "Moon Story," pp. 8–9.

160   The account given by Griggs, "Moon Story," pp. 4–18, of how Locke came to satirize Dick seems largely correct, but is flawed in some details. For example, I can find no evidence to support Griggs's statement that Dick wrote the 1826 *Edinburgh New Philosophical Journal* article; in fact, Dick was critical of it in his *Celestial Scenery.* Moreover, Griggs's contention that Locke satirized Dick's *Celestial Scenery* is surely wrong; although it contains much that could be satirized, it was published two years after Locke's 1835 lunar articles. A more plausible conjecture as to how Locke came to satirize Dick is that Locke may have come upon one of the countless copies of Dick's *Christian Philosopher,* which by 1827 had been given a pair of appendixes discussing the July 1824 report in the *Edinburgh Philosophical Journal* of Gruithuisen's and Schröter's supposed sighting of lunar buildings. Although critical of Schröter and Gruithuisen (whom Dick mistook for Fraunhofer), Dick does approvingly discuss in these appendixes the possibility of finding *"direct proofs"* of lunar life by telescopic means. These appendixes and probably Dick's *Philosophy of a Future State* inspired Locke to his satire.

161   Griggs, "Moon Story," p. 33.

162   For details on Herschel's reception, see Griggs, "Moon Story," pp. 37–40. Concerning the inquiries he received, see David S. Evans et al. (eds.), *Herschel at the Cape* (Austin, 1969), p. 282.

163   As quoted in Evans, "Hoax," p. 196.

164   "Locke among the Moonlings," p. 502.

165   Dick, *Celestial Scenery,* in *Works* (Hartford, Conn., 1848), p. 121.

166   The *New York Herald,* in its September 2, 1835, issue, published: "A Better Story – Most Wonderful and Astounding Discoveries by Herschell, the Grandson, L.L.D., F.R.S., R.F.L., P.Q.R., &c.&c.&c." This is reprinted in Seavey, *Hoax,* pp. 65–7. Also, the London publisher B. D. Cousins brought out, probably in 1836, not only a partial reprint of Locke's moon articles but also a sixteen-page booklet entitled *The*

History of the Sun, which purports to be a report of John Herschel's observations of solar civilizations! This not unskillful but almost unknown work was brought to my attention in 1982 by Mrs. Enid Lake, who was at that time librarian at the Royal Astronomical Society.

CHAPTER 5

1 For biographical information, see Günther Buttmann, *The Shadow of the Telescope,* trans. B. E. J. Pagel (New York, 1970).
2 John Herschel, *A Treatise on Astronomy* (London, 1833), section #1; John Herschel, *Outlines of Astronomy,* 3rd ed. (London, 1850), section #1.
3 Herschel, *Treatise,* #2; *Outlines,* #2.
4 Herschel, *Treatise,* #592; *Outlines,* #819.
5 John Herschel's comments on Whewell's book are discussed subsequently.
6 Herschel, *Treatise,* #332; *Outlines,* #389.
7 Herschel, *Treatise* #334; *Outlines,* #396. This was slightly amended in 1858.
8 Herschel, *Treatise,* #364. In his *Outlines,* this passage is slightly altered to read: "The moon has no clouds, nor any decisive indications of an atmosphere." (#431)
9 Herschel, *Treatise,* #363; *Outlines,* #430.
10 Herschel, *Treatise,* #364; also *Outlines,* #431, but weakened by the qualification: "this process . . . must . . . be confined within very narrow limits."
11 Patrick Scott, *Love in the Moon* (London, 1853), pp. 10–11.
12 Herschel, *Treatise,* #365; *Outlines,* #433.
13 Herschel, *Treatise,* #368; *Outlines,* #436, where this forms the next to last paragraph of the chapter.
14 Willy Ley, *Rockets, Missiles, and Men in Space* (New York, 1968), p. 31.
15 Simon Newcomb, "On Hansen's Theory of the Physical Constitution of the Moon," *American Association for the Advancement of Science Proceedings,* 17 (1868), 167–71:171.
16 Herschel, *Outlines,* 5th ed. (London, 1858), #436a and b. Herschel's discussion was described as "one of the most remarkable additions" to that edition in the review of it in *Eclectic Review,* 47 (1859), 33–9:36.
17 Herschel, *Treatise,* #435; *Outlines,* #508.
18 Herschel, *Treatise,* #436; *Outlines,* #506.
19 Herschel, *Treatise,* #437; *Outlines,* #510.
20 Herschel, *Treatise,* #437; *Outlines,* #510, where this is softened to "some probability."
21 Herschel, *Treatise,* #446; *Outlines,* #533.
22 Herschel, *Treatise,* #448; *Outlines,* #525, where the remark is preserved, but the hazy appearance is described as probably an illusion.
23 Herschel, *Treatise,* #448; *Outlines,* #525.
24 Herschel, *Treatise,* #609; *Outlines,* #847.
25 Herschel, *Treatise,* #610; *Outlines,* #851.
26 John Herschel, "The Sun," in Herschel's *Familiar Lectures on Scientific Subjects* (New York, 1871), p. 84. This volume was first published in 1868, but a footnote (p. 79) indicates that this lecture was delivered in late 1861. The lecture was, according to the preface, published in the journal *Good Words.* See C. F. Bartholomew, "The Discovery of the Solar Granulation," *Royal Astronomical Society Quarterly Journal,* 17 (1976), 263–89.
27 As quoted in William Hanna Thomson, *Some Wonders of Biology* (New York, 1909), p. 176.
28 As quoted in Robert Perceval Graves, *Life of Sir William Rowan Hamilton,* vol. II (Dublin 1885), p. 383. See also vol. I (Dublin, 1882), p. 95, for a pluralist poem by Hamilton and vol. III (Dublin, 1889) for Hamilton's view of the Whewell debate.
29 Jane Marcet, *Conversations on Natural Philosophy* (Philadelphia, 1826), pp. 85–6. I am indebted to Donald Beaver for calling this to my attention.
30 David Milne [Home], *Essay on Comets* (Edinburgh, 1828), pp. 142–4.

31 The question of life on comets was also seriously debated in issues from 1829 and 1830 of the *Revue britannique*.

32 Humphry Davy, *Consolations in Travel, or The Last Days of a Philosopher* (London, 1830), p. 54. Pluralist themes also occur in Davy's poetry; see J. Z. Fullmer, "The Poetry of Humphry Davy," *Chymia*, 6 (1960), 102–26:120,123.

33 John Davy, *Memoirs of the Life of Sir Humphry Davy, Bart.,* vol. II (London, 1836), pp. 384–5.

34 As quoted in Fullmer, "Poetry of Davy," p. 125.

35 The narrator in the second dialogue of *Consolations* (p. 67) states that "the most important parts of the vision really occurred to me in sleep, particularly that in which I seemed to leave the earth. . . ." One of the companions then (p. 68) compares him, presumably in jest, to Swedenborg. J. Davy, in *Memoirs*, vol. II, pp. 377–80, presents compelling evidence that Davy experienced dreams in 1819 and 1821 that supplied materials for this vision.

36 As quoted in Harold Hartley, *Humphry Davy* (Yorkshire, 1972), p. 148.

37 Mrs. Mary Somerville, *The Connexion of the Physical Sciences* (London, 1834), p. 225.

38 Charles Lyell, "Presidential Address," *Geological Society of London Proceedings*, 2 (1838), 520. Phillip R. Sloan called this passage to my attention.

39 Charles Darwin, *The Foundation of the Origin of Species: Two Essays Written in 1842 and 1844*, ed. Francis Darwin (Cambridge, England, 1909), p. 254.

40 For Chambers and the response to his book, see Milton Millhauser, *Just before Darwin: Robert Chambers and Vestiges* (Middletown, Conn., 1959). For Chambers and the nebular hypothesis, see Marilyn B. Oglivie, "Robert Chambers and the Nebular Hypothesis," *British Journal for the History of Science*, 8 (1975), 214–32, and Ronald Numbers, *Creation by Natural Law: Laplace's Nebular Hypothesis and American Thought* (Seattle, 1977), pp. 28–34.

41 [Robert Chambers], *Vestiges of the Natural History of Creation* (London, 1844, 1st ed., as reprinted with an introduction by G. de Beer, Leicester, 1969), pp. 2–4.

42 Millhauser, *Chambers*, pp. 87, 147.

43 [Chambers], *Vestiges*, 10th ed. (London, 1853), p. v.

44 John Hedley Brooke has urged that William Whewell's conversion to the antipluralist position resulted from reading Chambers. See his "Natural Theology and the Plurality of Worlds: Observations on the Brewster-Whewell Debate," *Annals of Science*, 34 (1977), 221–86, especially pp. 264–8.

45 Richard Owen, *On the Nature of Limbs* (London, 1849), p. 83.

46 George Gilfillan, *A Gallery of Literary Portraits*, vol. II (Edinburgh 1855), pp. 254–5.

47 Numbers, *Creation by Natural Law*, p. 21.

48 J. P. Nichol, *Views of the Architecture of the Heavens*, 4th ed. (Edinburgh, 1843), p. 76.

49 J. P. Nichol, *The Phenomena and Order of the Solar System* (Edinburgh, 1838), pp. 170–1.

50 Anonymous, "[Review of] *The Steam Engine* by Dionysius Lardner," *Athenaeum*, no. 624 (December 5, 1840), 962.

51 On Lardner, see J. N. Hays, "The Rise and Fall of Dionysius Lardner," *Annals of Science*, 38 (1981), 527–42, and Morse Peckham, "Dr. Lardner's Cabinet Cyclopaedia," *Papers of the Bibliographical Society of America*, 45 (1951), 37–58.

52 "Publisher's Advertisement," in Dionysius Lardner, *Popular Lectures on Science and Art; Delivered in the Principal Cities and Towns of the United States*, vol. I, 12th ed. (New York, 1850), p. 5. Further information on the tour is supplied in Lardner's "Preface" to this volume, pp. 7–21.

53 Samuel Noble, *An Appeal in Behalf of the . . . New [Jerusalem] Church*, 10th ed. (London, 1881), p. vi.

54 Samuel Noble, *The Astronomical Doctrine of a Plurality of Worlds Irreconcilable with the Popular Systems of Theology, but in Perfect Harmony with the True Christian Religion* (London, 1838), pp. 1, 25.

55 This position is expounded in Noble, *Plurality,* pp. 33–48. For a good summary, see his note L, pp. 63–4.

56 Sharon Turner, *Sacred History of the World, as Displayed in the Creation and Subsequent Events to the Deluge,* 3rd ed., vol. I (London, 1833), p. 513.

57 Alexander Copland, *The Existence of Other Worlds, Peopled with Living and Intelligent Beings, Deduced from the Nature of the Universe* (London, 1834). For Copland as "Advocate," see his title page. For the views of his father, who was probably Patrick Copland, professor of natural philosophy at Marischal College, Aberdeen, see pp. 1 and 77–80.

58 Edward Walsh, "Are There More Inhabited Worlds than Our Globe?" *Amulet,* 5 (1830), 65–100, of which only the last eight pages are devoted to the question in his essay's title. Although antipluralist, Walsh admits that various planets "may be now in preparation for inhabitants. . . ." (p. 99)

59 Isaac Taylor, *Physical Theory of Another Life* (London, 1858), pp. 1–3. All subsequent references are to this edition.

60 J. S. Mill, *System of Logic,* in Mill, *Collected Works,* vol. VII, ed. J. M. Robson (Toronto, 1973), p. 555. This section in Mill was called to my attention by Mary Hesse.

61 Hallam Lord Tennyson, *Alfred Lord Tennyson: A Memoir,* vol. I (New York, 1897), p. 20. For Tennyson and astronomy, see (1) Sir Norman Lockyer and Winifred Lockyer, *Tennyson as a Student and Poet of Nature* (London, 1910), and (2) Jacob Korg, "Astronomical Imagery in Victorian Poetry," *Victorian Science and Victorian Values: Literary Perspectives,* ed. J. Paradis and T. Postlewait, which appears as vol. 360 (1981) of the *Annals of the New York Academy of Sciences;* see pp. 137–58.

62 Philip James Bailey, *Festus: A Poem,* 1st. American ed. (Boston, 1845), p. 25; see also p. 326.

63 Whewell's initial adoption but eventual rejection of pluralism is discussed subsequently. Tennyson's use of the nebular hypothesis in *The Princess* (bk. III) may very well have had its source in Whewell.

64 For "Timbuctoo," see Alfred Lord Tennyson, *Poems,* ed. Hallam Lord Tennyson, vol. I (London, 1908), pp. 317–25:320. See also Tennyson, *The Devil and the Lady and Unpublished Early Poems,* ed. Charles Tennyson (Bloomington, 1964), p. 24, and for "Armageddon," p. 12 of part II.

65 *The Poems and Plays of Alfred Lord Tennyson* (New York, 1938), p. 60.

66 Ibid., p. 430.

67 An excellent example of the derivative character of American pluralism is Duncan Bradford's *Wonders of the Heavens* (New York, 1843), which first appeared in 1837. Although possibly the most sumptuous astronomy text then available, a number of its pluralist passages were plagiarized from Chalmers, Dick, and John Herschel. Compare Bradford, pp. 29 and 173, with Herschel, *Treatise,* #592 and #448; Bradford, pp. 144–5, with Dick, *Christian Philosopher,* pp. 84 and 150–2; and Bradford, pp. 218–20, with Chalmers, *Discourse,* ch. I.

68 Ormsby MacKnight Mitchel, *The Orbs of Heaven,* 4th ed. (London, 1854); see pp. iv and 221–3.

69 As quoted in Russell McCormmach, "Ormsby MacKnight Mitchel's *Sidereal Messenger,* 1846–1848," *Proceedings of the American Philosophical Society,* 110 (February 1966), 35–47:39.

70 Two weeks before he died, Webster wrote an inscription for his tombstone, which said in part: "Philosophical argument, especially that drawn from the vastness of the universe in comparison with the apparent insignificance of this Globe, has sometimes shaken my reason for the faith that is in me; but my heart has assured, and reassured me, that the Gospel of Jesus Christ must be a Divine Reality." As quoted in Irving H. Bartlett, *Daniel Webster* (New York, 1978), p. 291.

71 Ralph Waldo Emerson, "Nature," *Selected Prose and Poetry,* ed. Reginald L. Cook (New York, 1959), p. 3. This essay was originally published anonymously as a booklet. For Emerson's interest in astronomy, see pp. 230–7 of Harry Hayden Clark, "Emerson and Science," *Philological Quarterly,* 10 (July, 1931), 225–60.

72 David Gavine, "Thomas Dick, LL. D., 1774–1857," *British Astronomical Association Journal,* 84 (1974), 345–50:346; *The Journals and Miscellaneous Notebooks of Ralph Waldo Emerson,* vol. IV, ed. Alfred R. Ferguson (Cambridge, Mass., 1952), p. 170.

73 *Journals,* vol. IV, p. 25.

74 Sherman Paul, *Emerson's Angle of Vision* (Cambridge, Mass., 1952), p. 95. For quotations in support of this point, see especially pp. 82–4 and 95–8.

75 Jonathan Bishop, *Emerson on the Soul* (Cambridge, Mass., 1964), p. 53.

76 As quoted in *Young Emerson Speaks: Unpublished Discourses on Many Subjects,* ed. A. C. McGiffert, Jr. (Boston, 1938), p. 253.

77 Edward Emerson believed that this topic was first broached in the week following June 2, 1832. See *Journals,* vol. IV, p. 27.

78 Ralph Waldo Emerson, "Astronomy," *Young Emerson Speaks,* pp. 170–9. I have been unable to uncover from the vast number of writings on Emerson a careful analysis of the role this sermon played in the evolution of his thought. Two possibilities may be suggested: (1) The sermon may simply have borrowed arguments, perhaps from Paine's *Age of Reason,* which arguments Emerson used to justify his shrinking belief in Christianity to his congregation. (2) The sermon may, however, have presented arguments that were as crucial for Emerson as they were for Paine in forcing the rejection of revealed religion. The fact that this sermon immediately preceded his decision to give up the pastorate points toward the latter possibility. H. H. Clark, in his "Emerson and Science," pp. 236–7, suggested that astronomy may have been an important factor in Emerson's decision, but Clark, writing in 1931, before this sermon was published, could rely only on the journals.

79 In a journal entry for May 26, 1832, Emerson cites Laplace as someone who in this way became an infidel. See Emerson, *Journal,* vol. IV, p. 26.

80 Emerson, *Journal,* vol. IV, p. 24.

81 Emerson, *Selected Prose and Poetry,* pp. 75–6.

82 Henry David Thoreau, *Walden and Civil Disobedience,* ed. Owen Thomas (New York, 1966), p. 6.

83 Thoreau, *Walden,* p. 89; see also p. 59. Other relevant passages are cited in Mary I. Kaiser, " 'Conversing with the Sky;' The Imagery of Celestial Bodies in Thoreau's Poetry," *Thoreau Journal Quarterly,* 9 (1977), 15–28.

84 Ralph Waldo Emerson, "Swedenborg; or, The Mystic," in *The Works of Ralph Waldo Emerson* (New York, n.d.), pp. 456, 472. See also Clarence Hotson, "Emerson and the Swedenborgians," *Studies in Philology,* 27 (1930), 517–45.

85 Marguerite Beck Block, *The New Church in the New World* (New York, 1968 reprinting of the 1930 original), p. 173.

86 Thomas Lake Harris, *An Epic of the Starry Heaven* (New York, 1854), p. 26. On Harris, see Arthur A. Cuthbert, *The Life and World-Work of Thomas Lake Harris* (Glasgow, 1909), and Herbert W. Schneider and George Lawton, *A Prophet and a Pilgrim, Being the Incredible History of Thomas Lake Harris and Laurence Oliphant* (New York, 1942).

87 T. L. Harris, *Arcana of Christianity,* vol. I, pts. 1 and 2 (New York, 1976 reprinting of the New York 1858 original). In 1867, Harris published part 3, being a commentary on the Apocalypse. Pluralist comments occur less frequently in it.

88 As quoted in Schneider and Lawton, *Prophet,* p. xiv.

89 Sidney Ahlstrom, *A Religious History of the American People* (New York, 1972), p. 488.

90 My discussion of Seventh-day Adventist views is largely based on information received from Orville R. Butler, especially on two of his unpublished papers: "The Concept of Extraterrestrial Life in Early Adventist Thought" and "Heaven and the Second Coming: Popular Astronomy in Millerite and Early Seventh-day Adventist Thought." Other sources include (1) J. N. Loughborough, *The Great Second Advent Movement: Its Rise and Progress* (New York, 1972 reprinting of the Washington, D.C., 1905 edition); (2) Ronald Numbers, *Prophetess of Health: A Study of Ellen G. White* (New York, 1976); and (3) M. Ellsworth Olsen, *A History of the Rise and*

Progress of the Seventh-day Adventists (New York, 1972 reprinting of the Washington, D.C., 1925 edition).

91 As quoted in Joseph Bates, *The Opening Heavens* (New Bedford, Mass., 1846), p. 7. See also *The Autobiography of Elder Joseph Bates* (Battle Creek, Mich., 1868), p. 151.

92 As quoted in Loughborough, *Advent Movement,* p. 260.

93 As quoted in Loughborough, *Advent Movement,* p. 258. The astronomer referred to is William Parsons, the third Earl of Rosse, who was an expert on nebulae.

94 Ellen G. White, *Early Writings* (Washington, D.C., 1945), pp. 40–1. This first appeared in the Adventist journal, *The Present Truth,* 1, no. 3 (August 1849), 23–4.

95 Numbers, *White,* p. 44.

96 Ellen G. White, *The Story of Patriarchs and Prophets* (Mountain View, Calif., 1948), pp. 69–70.

97 White, *Story,* p. 154.

98 Ellen G. White, *The Great Controversy between Christ and Satan,* 11th ed., 72nd thousand (New York, 1889), pp. 667–78.

99 See, for example, Donald R. McAdams, "Shifting Views of Inspiration: Ellen G. White Studies in the 1970's," *Spectrum: A Quarterly Journal of the Association of Adventist Forums,* 10 (March 1980), 27–41.

100 The literature on Joseph Smith and the Mormons is very extensive. Two of the best known biographies of Smith are Donna Hill, *Joseph Smith: The First Mormon* (Garden City, N.Y., 1977), and Fawn Brodie, *No Man Knows My History: The Life of Joseph Smith* (New York, 1945). The former is friendly, the latter hostile to Smith. For a critique of Brodie's biography, see Hugh W. Nibley, *No Ma'am, That's Not History* (Salt Lake City, 1946). See also John A. Widtsoe, *Joseph Smith as Scientist* (Salt Lake City, 1964 reprinting of the 1908 original). A standard history of the Mormon church is Brigham H. Roberts, *A Comprehensive History of the Church of Jesus Christ of Latter-day Saints,* 6 vols. (Provo, Utah, 1965). I have also consulted Leonard J. Arrington and Davis Bitton, *The Mormon Experience: A History of the Latter-day Saints* (New York, 1979); Klaus J. Hansen, *Mormonism and the American Experience* (Chicago, 1981); and Thomas F. O'Dea, *The Mormons* (Chicago, 1959). The drafts of my discussion of Mormon ideas have been greatly improved by the criticism of them made by Phillip R. Sloan of Notre Dame University and E. Robert Paul of Dickinson College.

101 I have used *The Doctrine and Covenants of the Church of Jesus Christ of Latter-day Saints, Containing Revelations Given to Joseph Smith, The Prophet* (Salt Lake City, 1957), and *The Pearl of Great Price: A Selection from the Revelations, Translations, and Narrations of Joseph Smith* (Salt Lake City, 1957). An earlier and shorter version of the former book was printed in 1833 as *Books of Commandments.*

102 Roberts, *History,* vol. II, p. 394. See also Widtsoe, *Smith as Scientist,* p. 49, where it is noted that Smith taught the plurality of worlds as an "Absolute truth." Widtsoe adds that although others had discussed this doctrine earlier, "Probably no other philosopher had gone quite that far."

103 See, for example, Roberts, *History,* vol. II, p. 387.

104 Robert J. Matthews, *"A Plainer Translation:" Joseph Smith's Translation of the Bible: A History and Commentary* (Provo, Utah, 1975), p. 52. For information on the composition, compilation, and revision of the *Pearl of Great price,* see Matthews, ch. 11.

105 For its date of first publication, see Matthews, *Translation,* p. 47. For its republication in the 1851 edition of *Pearl,* see Matthews, p. 220. This book is not accepted as canonical by the Reorganized Church of Jesus Christ of Latter Day Saints, a group centered in Independence, Missouri.

106 It was long believed that these papyri had been lost. However, portions of them were discovered in the 1960s. When examined by experts, it was learned that they contain funeral rites, rather than corresponding to what was published as the "Book of Abraham." See Hill, *Smith,* p. 194.

107 As quoted in O'Dea, *Mormons,* p. 126.

108 *Discourses of Brigham Young,* selected and arranged by John A. Widtsoe (Salt Lake City, 1925), p. 31.
109 Phelps is so described in Roberts, *History,* vol. II, p. 387. The hymn is given on p. 388.
110 See, for example, Chapter 13 of Frank B. Salisbury, *Truth by Reason and by Revelation* (Salt Lake City, 1965); R. Grant Athay, "Worlds without Number: The Astronomy of Enoch, Abraham, and Moses," *Brigham Young University Studies,* 8 (1968), 255–69; and Hollis R. Johnson, "Civilizations out in Space," *BYU Studies,* 11 (1970), 1–12.
111 Brodie, *Smith,* pp. 171–2. See also Hansen, *Mormonism,* pp. 79–80.
112 Edward T. Jones, "The Theology of Thomas Dick and Its Possible Relationship to That of Joseph Smith," which is a 1969 Brigham Young University M.A. thesis. Also see E. Robert Paul, "Joseph Smith and American Culture: The Emergence of Astronomical Pluralism, 1820–1836." Professor Paul graciously showed me a draft of this essay, which is scheduled for publication in *Dialogue: A Journal of Mormon Thought.*
113 Thomas Dick, *Philosophy of a Future State,* 2nd ed. (Brookfield, Mass., 1830), p. 249.
114 François Arago, *Astronomie populaire,* vols. I–IV, ed. J. A. Barral (Paris, 1854–7).
115 For the Elliot-Boydell incident, see the discussion of W. Herschel in the second section of Chapter 2.
116 This mistake seems to have occurred in [William Williams], *The Universe No Desert, The Earth No Monopoly,* vol. II (Boston, 1855), p. 164, where it is stated that "Arago . . . speaks of 'the inhabitants of the moon.' " Arago may somewhere in his writings have advocated selenites; however, he does not seem to have done this in his *Astronomie populaire.*
117 See Flammarion's dedication in his *Popular Astronomy,* trans. J. E. Gore (New York, 1931).
118 F. E. Plisson, *Les mondes ou essai philosophique sur les conditions d'existence des êtres organisés dans notre système planétaire* (Paris, 1847). For Plisson's presence at Arago's lectures, see, for example, p. 75.
119 Samuel Rogers, *Balzac and the Novel* (New York, 1969), p. 16. See also Pauline Bernheim, *Balzac und Swedenborg* (Berlin, 1914), and Theodore Wright, "Balzac and Swedenborg," *New-Church Review,* 3 (1896), 481–503.
120 Honoré de Balzac, *Seraphita and Other Stories,* trans. Clara Bell (Philadelphia, 1900), p. 48.
121 Philippe Bertault, *Balzac and the Human Comedy,* trans. Richard Monges (New York, 1963), p. 87. Balzac was a Roman Catholic.
122 Although not unsympathetic to Fourier, Nicholas V. Riasanovsky, in his *The Teaching of Charles Fourier* (Berkeley, 1969), frequently labels Fourier's ideas "mad." See, for example, pp. ix, 38, and 86. Frank E. Manuel, in his *Prophets of Paris* (Cambridge, Mass., 1962), on the other hand, suggests that Fourier used his cosmological ideas to serve as a "mask" to convince critics "that he was a visionary, and nobody prosecuted visionaries. . . ." (p. 244) On Fourier's cosmology, see Hélène Tuzet, "Deux types de cosmogonies vitalistes: 2. Charles Fourier, hygieniste du cosmos," *Revue des sciences humaines,* no. 101 (January-March 1961), 37–53.
123 As translated in Riasanovsky, *Fourier,* p. 37, from Charles Fourier, *Oeuvres complètes,* vol. I (Paris, 1966), pp. 29–30.
124 Manuel, *Prophets,* p. 205, and Riasanovsky, *Fourier,* pp. 38–9.
125 As translated in Riasanovksy, *Fourier,* p. 87n.
126 Fourier, "Cosmogonie," in *Oeuvres,* vol. XII (Paris, 1968), p. 13.
127 As translated in Riasanovsky, *Fourier,* p. 88, from Fourier, *Oeuvres,* vol. XI, p. 326.
128 Riasanovsky, *Fourier,* p. 89.
129 Riasanovsky, *Fourier,* pp. 93–5.
130 For information on Grandville, see *Bizarreries and Fantasies of Grandville,* with introduction and commentary by Stanley Appelbaum (New York, 1974), which includes his illustrations from *Un autre monde.*
131 See, for example, Carl Sagan, *Other Worlds* (New York, 1975), and *The Cosmic*

Connection (New York, 1973), and especially Ronald N. Bracewell, *The Galactic Club* (San Fransicso, 1974).

132 Auguste Comte, *Cours de philosophie positive,* vol. II, in *Oeuvres de Auguste Comte,* vol. II (Paris, 1968), p. 1.

133 Comte, *Cours,* vol. II, p. 10. For criticism of these Comtean doctrines, see Stanley L. Jaki, *Relevance of Physics* (Chicago, 1966), pp. 470–2.

134 Auguste Comte, *Système de politique positive,* vol. I, in *Oeuvres,* vol. VII (Paris, 1964), p. 511.

135 Comte, *Cours,* vol. II, p. 24. The context makes clear that by "theology," Comte meant primarily Christian theology.

136 Comte, *Cours,* vol. II, p. 130.

137 From a selection in Denis Mack Smith (ed.), *The Making of Italy 1796–1870* (London, 1968), p. 179. Thomas Kselman brought this passage to my attention.

138 Giacomo Leopardi, *Storia dell'astronomia dalla sua origine fino all'anno MDCCCXI,* in Leopardi, *Le Poesie et le prose,* vol. II, in *Tutti le opere,* ed. F. Flora (Milan, 1937), pp. 723–1069. On this book, see Maurice A. Finocchiaro, "A Curious History of Astronomy: Leopardi's *Storia dell'astronomia,*" *Isis,* 65 (1974), 517–19.

139 Leopardi, *Storia,* pp. 809–18.

140 G. Leopardi, "Dialogue between the Earth and the Moon," trans. Norman A. Shapiro, in *Leopardi: Poems and Prose,* ed. Angel Flores (Bloomington, Ind., 1966), pp. 196–204. See also Leopardi, "Copernicus: A Dialogue," trans. Rufus Suter, *American Scientist,* 54 (1966), 119–27, especially p. 125.

141 H. C. Oersted, *The Soul in Nature,* trans. L. and J. B. Horner (London, 1966 reprint of the London 1852 edition).

142 Per F. Dahl, *Ludwig Colding and the Conservation of Energy Principle* (New York, 1972), pp. 124–7.

143 The best source on the antipluralist leanings of various German philosophers of this period is Johannes Huber, *Zur Philosophie der Astronomie* (Munich, 1878), pp. 53–60. See also John Henry Kurtz, *The Bible and Astronomy,* translated from the third German edition by T. D. Simonton (Philadelphia, 1857).

144 Part Two of Hegel's *Enzyklopädie* appears in English as *Hegel's Philosophy of Nature,* trans. A. V. Miller (Oxford, 1970), p. 62. This quotation and the other passages quoted are from the sections called "Zusätze," which were added by Michelet, who took them in most cases, from notes made by himself or by others who attended Hegel's lectures.

145 C. L. Michlet, *Vorlesungen über die Persönlichkeit Gottes und Unsterblichkeit der Seele oder die ewige Persönlichkeit des Geistes* (Bruxelles, 1968 reprinting of the Berlin 1841 original), p. 227.

146 Ludwig Feuerbach, *Thoughts on Death and Immortality,* trans. James A. Massey (Berkeley, 1980), p. 62.

147 Ludwig Feuerbach, *The Essence of Christianity,* trans. George Eliot (New York, 1957), p. 12.

148 Huber, *Philosophie,* pp. 57–8, and Kurtz, *Bible and Astronomy,* p. 53.

149 Henrich Steffens, *Christliche Religionsphilosophie* (Breslau, 1839), pp. 205–6.

150 Arthur Schopenhauer, *Parerga and Paralipomena,* trans. E. F. J. Payne, vol. II (Oxford, 1974), p. 299.

151 Schopenhauer, *Parerga,* vol. II, p. 34; see also pp. 143–4.

152 James Branch Cabell, *The Silver Stallion* (New York, 1926), p. 129.

153 Johann Heinrich Kurtz, *Die Astronomie und die Bibel* (Mitau, 1842). I have used J. H. Kurtz, *The Bible and Astronomy: An Exposition of the Biblical Cosmology, and Its Relations to Natural Science,* translated from the third German edition by T. D. Simonton (Philadelphia, 1857).

154 Otto Zöckler described Schubert as having "developed similarly extreme pluralist views as those of Bode and Herschel, especially also concerning the supposed habitability of the dark core of the sun. . . ." See Zöckler's *Geschichte der Beziehungen zwischen Theologie und Naturwissenschaft,* 2nd ed., vol. II (Gütersloh, 1879), p. 427.

Judging from Kurtz, *Bible*, pp. 438–9, Schubert later adopted a spiritual conception of the cosmos in which the earth is assigned a primacy.

155 Heinrich Heine, *Werke und Briefe* (Berlin, 1961), pp. 207–8.

156 *Heine's Poetry and Prose*, introduction by Ernest Rhys (London, 1934), p. 330.

157 *Heine's Poetry and Prose*, p. 360.

158 As quoted in Hallam Lord Tennyson, *Alfred Lord Tennyson, A Memoir*, vol. I (New York, 1897), p. 38.

CHAPTER 6

1 The two major biographies of Whewell are Isaac Todhunter, *William Whewell*, 2 vols. (London, 1876), and Mrs. Stair Douglas, *The Life of William Whewell*, 2nd ed. (London, 1882).

2 Todhunter, *Whewell*, vol. I, pp. 70, 188.

3 Otto Zöckler, *Geschichte der Beziehungen zwischen Theologie und Naturwissenschaft*, 2nd ed., vol. II (Gütersloh, 1879), p. 432; Zöckler, "Der Streit über die Einheit und Vielheit der Welten," *Der Beweis des Glaubens*, 2 (1866), 352–76:361–3.

4 As quoted in Todhunter, *Whewell*, vol. II, p. 353.

5 F. B. Burnham, "Religion and the Extraterrestrial Intelligent Life Debate in the Nineteenth Century," a paper presented at the December 30, 1977, meeting of the History of Science Society.

6 J. H. Brooke, "Natural Theology and the Plurality of Worlds: Observations on the Brewster-Whewell Debate," *Annals of Science*, 34 (1977), 221–86:267.

7 On this question, see, for example, Silvestro Marcucci, "William Whewell: Kantianism or Platonism?" *Physis*, 12 (1970), 69–72.

8 Todhunter, *Whewell*, vol. I, pp. 187–8.

9 Douglas, *Whewell*, p. 74. Internal evidence makes it nearly certain that this letter was written in 1822.

10 As quoted in Todhunter, *Whewell*, vol. I, p. 326.

11 As quoted in Todhunter, *Whewell*, vol. I, p. 327.

12 As quoted in Todhunter, *Whewell*, vol. I, p. 327.

13 Todhunter, *Whewell*, vol. I, p. 327.

14 Douglas, *Whewell*, pp. 161, 199.

15 Todhunter, *Whewell*, vol. I, p. 67.

16 William Paley, *Natural Theology*, 15th ed. (London, 1815), pp. 378–9.

17 William Whewell, *Astronomy and General Physics Considered with Reference to Natural Theology* (Philadelphia, 1833), pp. 7–8.

18 Possibly Whewell derived this position from Pascal's *Pensées*. Whewell discussed Pascal's views on natural theology in Whewell, *Astronomy*, p. 240.

19 As quoted in Whewell, *Astronomy*, p. 261.

20 On this, see Ronald Numbers, *Creation by Natural Law: Laplace's Nebular Hypothesis in American Thought* (Seattle, 1977), pp. 20–1, where Numbers notes that Whewell in his 1833 book seems to have been the first to use the term "nebular hypothesis."

21 Todhunter, *Whewell*, vol. I, pp. 67–74.

22 For Brewster's review, see *Edinburgh Review*, 58 (January 1834), 422–57. For evidence of Brewster's authorship, see Todhunter, *Whewell*, vol. I, p. 72. Whewell responded to Brewster in *British Magazine*, 5 (1834), 263–8.

23 Charles Darwin, "D Notebook: Transmutation of Species," in Howard D. Gruber and Paul H. Barrett, *Darwin on Man* (New York, 1974), p. 455.

24 The third edition of Whewell's *Philosophy of the Inductive Sciences* appeared in three parts: (1) *History of Scientific Ideas*, 2 vols. (London, 1858), (2) *Novum Organum Renovatum* (London, 1858), and (3) *Philosophy of Discovery* (London, 1860). This quotation is from p. 335 of the last named work.

25 John Herschel, "The Reverend William Whewell, D. D.," *Royal Society of London Proceedings*, 16 (1868), li–lxi:liii.

26 Harvey Carlisle, "William Whewell," *Macmillan's Magazine*, 45 (December 1881), 138–44:141.

27 William Whewell, *History of the Inductive Sciences*, 3 vols., vol. III (London, 1837), p. 457; see also pp. 467–8.
28 Whewell, *History*, vol. III, pp. 469–70.
29 William Whewell, *Philosophy of the Inductive Sciences*, 2 vols., vol. I (New York, 1967 reprinting of the 1847 2nd ed.), p. 620.
30 Whewell, *Philosophy*, vol. I, pp. 623–5.
31 Whewell, *Philosophy*, vol. I, p. 628.
32 Whewell, *History*, vol. III, pp. 571–3.
33 Whewell, *History*, vol. III, p. 574.
34 Whewell, *History*, vol. III, p. 574.
35 As quoted in Milton Milhauser, *Just Before Darwin: Robert Chambers and Vestiges* (Middletown, Conn., 1959), p. 122.
36 Millhauser, *Chambers*, pp. 122–4.
37 Todhunter, *Whewell*, vol. I, p. 135.
38 References are to Whewell, *Indications of the Creator*, 2nd ed. (London, 1846).
39 Todhunter, *Whewell*, vol. II, p. 327.
40 As quoted in Douglas, *Whewell*, p. 318. See also Whewell, *Indications*, p. 8.
41 Peter J. Bowler, "Darwinism and the Argument from Design: Suggestions for a Reevaluation," *Journal of the History of Biology*, 10 (1977), 29–43.
42 Whewell, *Indications*, p. 36.
43 William Whewell, "Second Memoir on the Fundamental Antithesis of Philosophy," *Cambridge Philosophical Society Transactions*, 8 (1848), 614–20:614.
44 Todhunter, *Whewell*, vol. I, p. 317; see also p. 188.
45 William Whewell, "Astronomy and Religion, Dialogue I," p. 1, in the Whewell collection at Trinity College, Cambridge, as R. 6. 13[25].
46 This letter is preserved in the Whewell correspondence at Trinity College, Cambridge; see Add. Ms. a. 216[135].
47 As quoted in Todhunter, *Whewell*, vol. II, p. 392; see also p. 393.
48 As quoted in [Samuel Warren], "Speculators among the Stars, Part II," *Blackwood's Edinburgh Magazine*, 76 (October 1854), 370–403:372.
49 [William Whewell], *Of the Plurality of Worlds: An Essay* (London, 1853). The English editions appeared in 1853, summer 1854, autumn 1854, 1855, and 1859. The American editions appeared in 1854 and 1861. Future references are, except where noted, to the London 1853 first edition. For details on the English editions, see Todhunter, *Whewell*, vol. I, pp. 184ff.
50 [William Whewell], *Dialogue on the Plurality of Worlds, Being a Supplement to the Essay on That Subject* (London, 1854). This work was reprinted in the *Essay* in later editions. My references are to the reprinting in the fifth edition (London, 1859) of the *Essay*.
51 Todhunter, *Whewell*, vol. I, p. 182.
52 A copy of these proof sheets has been preserved at Trinity College, Cambridge. Hereafter it is referred to as [Whewell], *Essay* (proof sheets).
53 As quoted in Todhunter, *Whewell*, vol. II, p. 398.
54 As quoted in Todhunter, *Whewell*, vol. II, p. 380. Later in the letter, Whewell discusses whether a new edition of Pascal is better than a previous edition, suggesting that Whewell may have been reading Pascal shortly before.
55 As quoted in [Whewell], *Essay*, p. 43. Whewell erroneously attributed this view to p. 31 of Bessel's *Vorlesungen über wissenschaftliche Gegenstände*; the page number should have been 81.
56 As quoted in Todhunter, *Whewell*, vol. I, p. 187, where Todhunter also rejects Herschel's speculation.
57 For a good summary, see [Whewell], *Essay*, pp. 99–100.
58 Michael Hoskin, "Apparatus and Ideas in Mid-Nineteenth Century Cosmology," *Vistas in Astronomy*, 9 (1967), 79–85; Stanley L. Jaki, *The Milky Way* (New York, 1972).
59 Agnes Clerke, *A Popular History of Astronomy during the Nineteenth Century* (Edinburgh, 1885), p. 436.

60 Richard A. Proctor, "Varied Life in Other Worlds," *Open Court*, 1 (1888), 595–600:595.

61 [Whewell], *Essay*, pp. 143–4. John Herschel's *Outlines of Astronomy* is cited in support of the latter conclusion.

62 In fact, Algol's companion is now known to be a star.

63 As quoted in [Whewell], *Essay*, pp. 161–2.

64 Todhunter, *Whewell*, vol. I, pp. 205–10.

65 [Whewell], *Essay*, p. 253. In the proof sheet edition of the *Essay*, this chapter is entitled "The Argument from Law." It consists of a detailed presentation of the argument for God's existence from the laws and archetypes that man sees in nature.

66 [Whewell], *Essay*, p. 252. This point was more fully developed in ch. XIII, "The Omnipresence of the Deity," in Whewell, *Essay* (proof sheets).

67 [Whewell], *Essay*, p. 258. These points are more fully developed in ch. XIV, "Man's Intellectual Task," and in ch. XV, "Man's Moral Trial," in [Whewell], *Essay* (proof sheets).

68 [Whewell], *Essay*, p. 259. This section was taken from ch. XVI, "The Design of Animal Springs of Action," in [Whewell], *Essay* (proof sheets).

69 Whewell's letters to Stephen were published in Todhunter, *Whewell*, vol. II, pp. 379–96. Stephen's letters are preserved in the Whewell correspondence at Trinity College, Cambridge.

70 Stephen to Whewell (September 20, 1853), in the Whewell papers at Trinity College, Cambridge (Add. Ms. a. 216^{118}).

71 Todhunter, *Whewell*, vol. II, p. 382. Either the Stephen letter (Add. Ms. a. 216^{121} at Trinity College) is misdated as September 26 or Todhunter misdated Whewell's response, for he dated it as September 25.

72 Todhunter, *Whewell*, vol. II, p. 382.

73 Stephen to Whewell (October 6, 1853) (Add. Ms. a. 216^{124}).

74 Stephen to Whewell (October 8, 1853) (Add. Ms. a. 216^{126}).

75 Stephen to Whewell (October 10, 1853) (Add. Ms. a. 216^{127}).

76 Stephen to Whewell (October 12, 1853) (Add. Ms. a. 216^{128}).

77 Stephen to Whewell (October 13, 1853) (Add. Ms. a. 216^{129}).

78 Whewell to Stephen (October 14, 1853), in Todhunter, *Whewell*, vol. II, p. 389.

79 Stephen to Whewell (October 15, 1853) (Add. Ms. a. 216^{130}).

80 Whewell to Stephen (October 16, 1853), in Todhunter, *Whewell*, vol. II, p. 390.

81 Stephen to Whewell (October 18, 1853) (Add. Ms. a. 216^{131}).

82 Stephen to Whewell (October 24, 1853) (Add. Ms. a. 216^{134}).

83 Stephen to Whewell (October 31, 1853) (Add. Ms. a. 216^{137}).

84 Stephen to Whewell (November 10, 1853) (Add. Ms. a. 216^{142}).

85 Whewell to Stephen (November 14, 1853), in Todhunter, *Whewell*, vol. II, p. 395.

86 Whewell to Stephen (October 9, 1853), in Todhunter, *Whewell*, vol. II, p. 386.

87 Stephen to Whewell (November 10, 1853) (Add Ms. a. 216^{142}).

88 Whewell to Stephen (November 14, 1853), in Todhunter, *Whewell*, vol. II, p. 396. For Whewell's earlier contacts with Birks, see Todhunter, *Whewell*, vol. I, p. 90, and vol. II, p. 175.

89 Thomas Rawson Birks, *Modern Astronomy* (London, 1850), pp. 25–7.

90 Birks, *Modern Astronomy*, pp. 61–2. The earth-Bethlehem comparison also appears in contemporary writings on "exochristology"; see Andrew J. Burgess, "Earth Chauvinism," *Christian Century* (December 8, 1976), 1098–1102.

91 In 1846, Hugh Miller published an article in the *Witness*, setting out a form of the geological argument. Miller republished his ideas in his *First Impressions of England*, 2nd ed. (London, 1848); see pp. 335–40. Miller's pluralist publications are discussed in the next chapter.

92 Todhunter, *Whewell*, vol. I, p. 377.

93 See Todhunter, *Whewell*, vol. I, pp. 376–406, for an account of and generous quotations from this manuscript.

94 As quoted in Leslie Stephen, "William Whewell," *Dictionary of National Biography*, vol. XX (London, 1921–2), p. 1370.

95 Brooke, "Brewster-Whewell Debate," passim.
96 Paul R. Thagard, "Darwin and Whewell," *Studies in the History and Philosophy of Science*, 8 (1977), 353–6; Michael Ruse, "Darwin's Debt to Philosophy," *Studies in the History and Philosophy of Science*, 6 (1975), 159–81.
97 James R. Moore, *The Post-Darwinian Controversies* (Cambridge, England, 1981), p. 328.
98 [Whewell], *Essay*, 5th ed., p. 401.
99 Todhunter, *Whewell*, vol. I, p. 100.

CHAPTER 7

1 The starting points for research on the Whewell debate are the discussion of it in Isaac Todhunter, *William Whewell*, vol. I (London, 1876), pp. 184–210, and the materials collected by Whewell himself and preserved at Trinity College, Cambridge. Other sources of information and interpretation of the debate are (1) Otto Zöckler, *Geschichte der Beziehungen zwischen Theologie und Naturwissenschaft*, 2nd ed., vol. II (Gütersloh, 1879), pp. 432–6; (2) Camille Flammarion, *Les mondes imaginaires et les mondes réels*, 20th ed. (Paris, ca. 1882), passim; (3) William Miller, *The Heavenly Bodies: Their Nature and Habitability* (London, 1883), pp. 120–6; (4) Camille Flammarion, *La pluralité des mondes habités*, 33rd ed. (Paris, ca. 1885), passim; (5) Michael Ruse, "The Relationship between Science and Religion in Britain, 1830–1870," *Church History*, 44 (1975), 505–22:516–18; (6) John Hedley Brooke, "Natural Theology and the Plurality of Worlds: Observations on the Brewster-Whewell Debate," *Annals of Science*, 34 (1977), 221–86; and (7) Richard Yeo, "William Whewell, Natural Theology and the Philosophy of Science in Mid Nineteenth Century Britain," *Annals of Science*, 36 (1979), 493–516, especially 505–11. I have also used Frederic B. Burnham's unpublished paper, "The Teleology and Plurality of Worlds: William Whewell versus David Brewster."
2 Edgar W. Morse, *Natural Philosophy, Hypotheses, and Impiety: Sir David Brewster Confronts the Undulatory Theory of Light* (a 1972 doctoral dissertation at the University of California, Berkeley).
3 Whewell to Murchison (May 30, 1854), in Whewell papers at Trinity College, Cambridge (O. 15. 47[311]).
4 Mrs. Gordon, *The Home Life of Sir David Brewster*, 2nd ed. (Edinburgh, 1970), p. 312.
5 Gordon, *Home Life*, pp. 177–8.
6 [David Brewster], "[Review of] *Astronomy and General Physics* . . . By the Rev. William Whewell," *Edinburgh Review*, 58 (1834), 422–57:427.
7 Morse, *Brewster*, passim. For Morse's discussion of Brewster's review of Whewell's *Astronomy*, see pp. 58–60 and 221–30.
8 [David Brewster], "[Review of] *Philosophy of the Inductive Sciences* . . . By Rev. William Whewell," *Edinburgh Review*, 74 (1842), 265–306:266.
9 [David Brewster], "The Revelations of Astronomy," *North British Review*, 6 (1847), 206–55; see, for example, pp. 215, 224–5, 241–3, and 254–5.
10 David Brewster, "Presidential Address," *British Association for the Advancement of Science Report for 1850* (London, 1851), xxxi–xlii:xxxiii.
11 As quoted in Gordon, *Home Life*, p. 247.
12 I have used the following edition: David Brewster, *More Worlds than One* (London, 1870). The review is item 31 in the appendix to this chapter. Subsequent references to it and to the other articles listed in the appendix are given in the form $(x, p. y)$, where x indicates the number of the item in the appendix and y is the page in that article.
13 As quoted in Gordon, *Home Life*, p. 250. Whewell's book is clearly not aimed at Brewster, who is never criticized in it.
14 Gordon, *Home Life*, p. 249.
15 Gordon, *Home Life*, p. 248.
16 Charles S. Peirce, *Essays in the Philosophy of Science*, ed. Vincent Tomas (Indianapolis, 1957), p. 217.

17 Brooke, "Brewster-Whewell Debate," pp. 259–60.
18 [William Whewell], *Of the Plurality of Worlds: An Essay*, 5th ed. (London, 1859), pp. 392–3.
19 [Whewell], *Essay* (5th ed.), p. 392.
20 [David Brewster], "William Whewell," *Royal Society of Edinburgh Proceedings*, 6 (1866–7), 29–32:31.
21 Gordon, *Home Life*, pp. 314–18.
22 The *National Union Catalog* lists an 1870 edition carrying the words "Ninth Thousand." I have secured references to five later editions.
23 Owen Chadwick, *The Victorian Church*, Part I (New York, 1966), pp. 553–4.
24 Baden Powell, "On the Study of the Evidences of Christianity," in *Essays and Reviews* (London, 1860), p. 139.
25 Alec Vidler, *The Church in an Age of Revolution* (Baltimore, 1965), p. 123.
26 As quoted without reference in Milton Millhauser, *Just before Darwin: Robert Chambers and Vestiges* (Middletown, Conn., 1959), p. 134.
27 Millhauser, *Just before Darwin*, p. 134.
28 The best source on Powell is William Tuckwell, *Pre-Tractarian Oxford* (London, 1909), pp. 165–225. See also David M. Knight, "Professor Baden Powell and the Inductive Philosophy," *Durham University Journal*, 60 (1968), 81–7.
29 As quoted in Todhunter, *Whewell*, vol. II, p. 399. Herschel's copy of Whewell's *Essay* is now in the extensive collection of books from the library of William and John Herschel that is owned by Professor Sydney Ross, of Rensselaer Polytechnic, who graciously allowed me to examine it for marginalia. Although not extensive, they further confirm Herschel's negative reaction.
30 Herschel to Whewell (undated), in the Whewell materials at Trinity College, Cambridge (Add. Ms. a. 207⁹⁰).
31 For Whewell's response, see [Whewell], *Essay* (5th ed.), pp. 357–8.
32 For Whewell's response, see [Whewell], *Essay* (5th ed.), pp. 382–3.
33 Lord Rosse to Whewell (February 28, 1854), in Whewell papers at Trinity College, Cambridge (Add. Ms. a. 216¹¹⁴).
34 [Whewell], *Essay* (5th ed.), pp. 343–53. Rosse seems to have arrived at his changed view of nebulae independent of Whewell.
35 Sabine to Whewell (March 4, 1854), in Whewell papers at Trinity College, Cambridge (Add. Ms. a. 216⁹⁰).
36 William Stephen Jacob, *A Few More Words on the Plurality of Worlds* (London, 1855), p. 3.
37 For Whewell's response to Jacob's probabilistic arguments, see [Whewell], *Essay* (5th ed.), pp. 407–8.
38 James Breen, *The Planetary Worlds: The Topography and Telescopic Appearances of the Sun, Planets, Moon, and Comets* (London, 1854), preface.
39 For evidence that he wrote the review under discussion, see Margaret Harwood, "Arthur Searle," *Popular Astronomy*, 29 (1921), 377–81.
40 As quoted in Phebe Mitchell Kendall, *Maria Mitchell: Life, Letters and Journals* (Boston, 1896), p. 121.
41 As quoted in Helen Wright, *Sweeper of the Sky: The Life of Maria Mitchell* (New York, 1949), p. 112.
42 As quoted in Kendall, *Mitchell*, p. 163.
43 For a discussion of Olmsted's paper and evidence that Olmsted had for years presented pluralism in his writings, lectures at Yale, and public addresses, see Gary Lee Schoepflin, *Denison Olmsted (1791–1859), Scientist, Teacher, and Christian* (a 1977 doctoral dissertation at Oregon State University), pp. 292–345.
44 De Morgan to Whewell (January 24, 1854), in Whewell papers at Trinity College, Cambridge (Add. Ms. a. 202¹²⁶). For Whewell's response, see [Whewell], *Essay* (5th ed.), p. 340.
45 De Morgan to Whewell (May 21, 1854), in Whewell papers at Trinity College, Cambridge (Add. Ms. a. 202¹²⁷).

46 De Morgan to Whewell (August 11, 1863), in Whewell papers at Trinity College, Cambridge (Add. Ms. a. 202^{151}).

47 Todhunter, *Whewell*, vol. I, pp. 202–3.

48 Todhunter, *Whewell*, vol. I, p. 192.

49 Todhunter, *Whewell*, vol. I, p. 193.

50 [Whewell], *Essay* (5th ed.), pp. 344–5.

51 [Whewell], *Essay* (5th ed.), pp. 412–13.

52 Appendix item 7 carries the initials "T. H." This fact, combined with Hill's Unitarianism and mathematical background, makes it probable that he wrote it.

53 As quoted in John W. Clark and Thomas M. Hughes, *Life and Letters of Rev. Adam Sedgwick*, vol. II (Cambridge, 1890), p. 269. Evidence of Sedgwick's negative reaction to Whewell's book may also be seen in the marginalia in his copy of the *Essay*, which is preserved at the Sedgwick Museum at Cambridge.

54 For this letter, see Mrs. Stair Douglas, *The Life of William Whewell*, 2nd ed. (London, 1882), pp. 434–5. Another clue to Sedgwick's views may be found in Whewell's *Dialogue*, where an argument of one of the interlocutors is marked in Whewell's personal copy with the letters "A. S.." See [Whewell], *Essay* (5th ed.), p. 353, line 13.

55 Leonard G. Wilson (ed.), *Sir Charles Lyell's Scientific Journals on the Species Question* (New Haven, 1970), pp. 99, 156, 177.

56 Whewell to Murchison (December 23, 1853), in Whewell papers at Trinity College, Cambridge (O. 15. 47^{310}).

57 Murchison to Whewell (January 15, 1854), in Whewell papers at Trinity College, Cambridge (Add. Ms. a. 216^{94}).

58 [Whewell], *Essay* (5th ed.), p. 339.

59 Whewell to Murchison (May 30, 1854), in Whewell papers at Trinity College, Cambridge (O. 15. 47^{311}).

60 As quoted in Todhunter, *Whewell*, vol. II, p. 398.

61 Forbes to Whewell (November 7, 1853), in Whewell papers at Trinity College, Cambridge (Add. Ms. a. 214^{109}).

62 Forbes to Whewell (December 26, 1853), in Whewell papers at Trinity College, Cambridge (Add. Ms. a. 214^{110}).

63 Forbes to Whewell (February 16, 1854), in Whewell papers at Trinity College, Cambridge (Add. Ms. a. 216^{88}).

64 As quoted in Todhunter, *Whewell*, vol. II, pp. 400–1. In evaluating Whewell's statements concerning the nebular hypothesis, it is wise to keep in mind that there were numerous forms of that hypothesis; see Brooke, "Brewster-Whewell Debate," pp. 268–73.

65 [Whewell], *Essay* (5th ed.), pp. 377–80.

66 This point is made in [Whewell], *Essay* (5th ed.), p. 377, and in Todhunter, *Whewell*, vol. I, pp. 194–5.

67 [Whewell], *Essay* (5th ed.), p. 358.

68 For Whewell's response, see [Whewell], *Essay* (5th ed.), pp. 362–6.

69 Todhunter, *Whewell*, vol. I, pp. 194–5.

70 Hugh Miller, *Geology versus Astronomy: or, The Conditions and the Periods; Being a View of the Modifying Effects of Geologic Discovery on the Old Astronomic Inferences Respecting the Plurality of Inhabited Worlds* (Glasgow, [1855]).

71 William Samuel Symonds, *Geology As It Affects a Plurality of Worlds* (London, 1856). The identity between the book and the article establishes Symonds's authorship of the latter.

72 Symonds derived this characterization from an anonymous review of the books by Brewster and Whewell that appeared as "Other Worlds in Space," *Times* (London) (December 26, 1855), p. 4. I saw this review only after this book was in press.

73 Dirk J. Struik, *Yankee Science in the Making*, rev. ed. (New York, 1962), p. 381.

74 Edward Hitchcock, "Introductory Notice" to [William Whewell], *The Plurality of Worlds* (Boston, 1854), pp. ix–xvi.

75 As translated from the French original given in Todhunter, *Whewell*, vol. I, pp. 204–5. For the views of another German who wrote on geology, see Oscar Peschel, "Ueber

die Pluralität der Welten," in Peschel's *Abhandlungen zur Erd- und Völkerkunde,* neue Folge (Leipzig, 1878), pp. 187–202; reprinted from the journal *Ausland* for the year 1855.

76 Humboldt to Gauss (March 6, 1854), in Kurt R. Biermann (ed.), *Briefwechsel zwischen Alexander von Humboldt und Carl Friedrich Gauss* (Berlin, 1977), pp. 115–16.

77 Michael Ruse, "William Whewell and the Argument from Design," *Monist,* 60 (1977), 244–68:263. For direct evidence that Darwin read Whewell's *Essay* in May 1854, as well as Powell's *Unity of Worlds* in January 1856, see Peter J. Vorzimmer, "The Darwin Reading Notebooks (1838–1860)," *Journal for the History of Biology,* 10 (1977), 107–53:144, 149.

78 Walter E. Houghton (ed.), *The Wellesley Index to Victorian Periodicals,* vol. III (Toronto, 1978), pp. 620–2. See also Leonard Huxley, *Life and Letters of Thomas Henry Huxley,* vol. I (London, 1900), p. 85.

79 (48, p. 314) See [Whewell], *Essay* (5th ed.), pp. 366–74, for Whewell's response to the latter charge and to other objections raised in the *Westminster Review*.

80 (49, p. 245) For the original, see [Whewell], *Essay* (5th ed.), p. 374.

81 Thomas H. Huxley, *Science and Christian Tradition* (New York, 1896), p. 39. I am indebted to W. Paul Fayter for calling this passage to my attention. In a letter sent while this book was in press, Mr. Fayter stated that he has located an unpublished April 1, 1854, letter from Richard Owen to R. Dockray in which Owen mentions reading Whewell's *Essay* "with a sense of dissatisfaction, and of pain." Moreover, Owen viewed Whewell's book as a piece of "special pleading." See Royal College of Surgeons MSS., *Correspondence of Richard Owen, 1826–1889,* 3.438–9.

82 For evidence of Mann's authorship, see Walter E. Houghton (ed.), *The Wellesley Index to Victorian Periodicals,* vol. I (Toronto, 1966), p. 505.

83 This was reprinted in [Whewell], *Essay* (5th ed.), pp. 404–16.

84 Wilson's letter is addressed to "The Author of the Plurality of Worlds." It is preserved in the Whewell papers at Trinity College, Cambridge (Add. Ms. a. 216[103]).

85 As quoted from *Vestiges,* in Wilson, *Chemistry of the Stars,* p. 29.

86 Pages 46–50 seem reconcilable even with the idea of transmigration of souls to other planets, but other passages, too numerous and ambiguous to quote, suggest that his view was that life exists only on the earth.

87 William Miller, *The Heavenly Bodies: Their Nature and Habitability* (London, 1883), p. 124.

88 Brodie to Whewell (January 5, 1854), in Whewell papers at Trinity College, Cambridge (Add. Ms. a. 216[84]). Internal evidence makes it probable that the letter was from the elder Brodie rather than his son of the same name.

89 For Holland's objections, see [Whewell], *Essay* (5th ed.), pp. 345–8 (on nebulae as galaxies), 350–3 (on spiral nebulae), 354–6 (on whether double stars are rare, as Holland thought, or frequent, as Whewell, supported by Struve, argued, and whether or not they have planetary systems), 358 (on Venus), and 374 (on the probability of Whewell's position).

90 As quoted in Todhunter, *Whewell,* vol. I, p. 188. See also Sir Henry Holland, *Recollections of Past Life* (London, 1872), pp. 240–1.

91 Todhunter, *Whewell,* vol. I, p. 197.

92 No papers by Phillips or Simon are listed in the *Royal Society Catalogue of Scientific Papers*. Simon did publish a few books in the areas of science, religion, and philosophy.

93 In an effort to check Simon's claim concerning Cullen and/or to locate the manuscript referred to by him, I have consulted a number of experts on Cullen. Neither Rev. Peadar MacSuibhne, who authored a five-volume biography of Cullen, nor Monsignor Patrick Corish was able to support Simon's report.

94 [Whewell], *Essay,* 4th ed. (London, 1855), p. xi.

95 In associating denominations with journals referred to in this section, I have used the listings in Alvar Ellegård, *Darwin and the General Reader* (Göteborg, 1958), pp. 368–84, for British journals, and in Ronald Numbers, *Creation by Natural Law* (Seattle, 1977), pp. 172–3, for American journals.

96 This letter, which mentions an earlier letter presumably also to the *Standard,* is preserved in the Whewell review volume at Trinity College, Cambridge. Symonds (16, p. 39) quotes from another Croly letter that appeared in the *Morning Herald.*

97 Musgrave to Whewell (December 14, 1854), in Whewell papers at Trinity College, Cambridge (Add. Ms. a. 216⁹⁵).

98 Rev. Josiah Crampton, *Testimony of the Heavens to Their Creator: A Lecture to the Enniskillen Young Men's Christian Association* (Dublin, 1857).

99 Whewell himself, in his 1855 Paris Exhibition sermon, chose as his theme: *Heavenly Mansions,* as the sermon was titled when published. In doing this, he may have been suggesting that Chalmersian themes could be invoked without tying them to the material "heavenly mansions" of the pluralists.

100 Peat to Whewell (December 26, 1855), in the Whewell review volume at Trinity College, Cambridge.

101 As quoted in Hallam Lord Tennyson, *Alfred Lord Tennyson: A Memoir,* vol. I (New York, 1897), p. 379.

102 Carus to Whewell (February 4, 1854), in Whewell papers at Trinity College, Cambridge (Add. Ms. a. 216⁸⁵).

103 Carden to Whewell (April 2, 1859), in Whewell papers at Trinity College, Cambridge (Add. Ms. a. 216⁸⁶).

104 I have used the second edition, which contains a biographical introduction by his son, i.e., Robert Knight, *The Plurality of Worlds: An Essay* (London, 1878). The texts of the two editions appear to be identical.

105 E. C. Brewer, *Theology in Science* (London, n.d.), pp. 326–34. The *National Union Catalog* lists an 1860 second edition; presumably the first edition appeared shortly before 1860. Internal evidence shows that it appeared after Whewell's *Essay.*

106 Walter E. Houghton (ed.), *The Wellesley Index to Victorian Periodicals,* vol. II (Toronto, 1972), pp. 762–3.

107 (Liverpool, 1855). Tarbet is listed in the *British Museum Catalogue* as a minister of the Catholic Apostolic church. For Tarbet's November 23, 1855, letter, see the Whewell review volume at Trinity College, Cambridge (Adv. C. 16. 35).

108 According to Houghton, *Wellesley Index,* vol. III, p. 147, this review was probably written by the joint editor of the *National Review,* Richard Holt Hutton. If correct, this would be interesting, because decades later (as shown subsequently) Hutton adopted a position closer to that of Whewell.

109 The *British Museum Catalogue* lists an 1858 edition, as does the *National Union Catalog.* However, the latter source refers to the existence of an 1857 edition. I have used the Philadelphia 1867 second edition.

110 Morison's authorship is known from his November 17, 1854, letter to Whewell in the Whewell review volume at Trinity College, Cambridge (Adv. C. 16. 35).

111 Robert A. Wilson and Elizabeth S. Wilson, *George Gilfillan: Letters and Journals, with Memoir* (London, 1892), p. 34. For evidence that Gilfillan wrote this review, see this biography, p. 203.

112 George Gilfillan, *Gallery of Literary Portraits,* vol. I (Edinburgh, 1845), pp. 105–23, especially pp. 115–19; see also Gilfillan, *The Christian Bearings of Astronomy* (London, 1848), especially pp. 18–28.

113 Gilfillan to "The Author of 'Of the Plurality of Worlds'," in Whewell papers at Trinity College, Cambridge (Add. Ms. a. 216⁹¹).

114 For evidence of Leavitt's authorship of the first and third reviews, see Todhunter, *Whewell,* vol. I, p. 201. The table of contents of the *Methodist Quarterly Review* reveals him as the author of the second.

115 According to Frank Luther Mott, *A History of American Magazines 1850–1865* (Cambridge, Mass., 1938), p. 75, Lord wrote most of the materials in his journal. Lord was probably also the author of the article "Christ the Saviour Only of Mankind" that opposed Brewster's views of the relations of Christianity to pluralism; see *Theological and Literary Journal,* 11 (October 1858), 177–96.

116 On the title page of the Library of Congress copy of this book, the words "by William

Williams" appear, and it is so cataloged there. I have been unable to secure biographical information on Williams.

117   William Swan Plumer (1802–80) was an American Presbyterian minister. In his *The Bible True and Infidelity Wicked* (New York, 1848?), pp. 43–7, he objected to those presenting pluralism as more than conjecture.

118   This sentence is the result of a number of inferences. That the author of the sixteen-page letter, dated October 8, 1858, preserved in Whewell's papers, was J. J. Larit is based on an analysis of its not perfectly legible signature. I have not, however, been able to find this name in standard reference sources. That the author of this letter wrote *Rêveries et vérités* is certain from comparing the contents of the letter and book. That the author of the book was a Protestant is indicated by the statement on its title page that it was sponsored by "La société centrale d'évangélisation," which seems to have been a Protestant organization. Todhunter (*Whewell*, vol. I, p. 204) discussed the letter, but without identifying its author or mentioning the title of his book. This very rare book is in the library of Trinity College, Cambridge. This is almost certainly the copy sent to Whewell.

119   For evidence that Christians have felt a tension concerning whether Christ or God should have primacy, see Jean Milet, *God or Christ?* (New York, 1981).

120   A check of dictionaries of pseudonyms and/or initialisms has produced no promising leads as to the identities of the six debaters.

121   Proof of Warren's authorship is provided in Todhunter, *Whewell*, vol. I, pp. 199–200. Warren published an expanded form of his review in his *Miscellanies* of 1854, pp. 437–94.

122   Warren's letters are preserved at Trinity College, Cambridge (Add. Ms. a. 216$^{105-112}$). Whewell's responses do not seem to have been preserved.

123   The eight-column review, headed "More Worlds than One," discussed in Todhunter, *Whewell*, vol. I, p. 197, and preserved as the sixth item in the Whewell review volume at Trinity College, Cambridge (Adv. C. 16. 35), corresponds, with minor changes, to pp. 14–22 of Hugh Miller's *Geology versus Astronomy* (Glasgow, [1855]). Almost certainly this was first published in the *Witness*. The four-column review of Whewell's *Essay,* discussed in Todhunter, *Whewell,* vol. I, p. 222, and placed first in the Whewell review volume, corresponds to item 49 in the appendix to this chapter.

124   G. W. Featherstonhaugh to Whewell (September 5, 1854), in Whewell papers at Trinity College, Cambridge (Add. Ms. a. 216^{90}).

125   Egerton to [Whewell] (March 15, 1854?), in Whewell papers at Trinity College, Cambridge (Add. Ms. a. 216^{87}).

126   Hallam to Whewell (January 9, 1854), in Whewell papers at Trinity College, Cambridge (Add. Ms. a. 216^{92}).

127   Henslow to [Whewell] (February 8, 1854?), in Whewell papers at Trinity College, Cambridge (Add. Ms. a. 216^{93}).

128   Rorison to [Whewell] (August 12, 1854), in Whewell papers at Trinity College, Cambridge (Add. Ms. a. 216^{98}).

129   Anonymous letter to [Whewell] (March 1854), in Whewell papers at Trinity College, Cambridge (Add Ms. a. 216^{104a}).

130   The November *Putnam's* (p. 503) kept the controversy before readers by a short, well-done poem entitled "Plurality of Worlds."

131   Anthony Trollope, *Barchester Towers and The Warden* (New York, 1950), pp. 378–9.

132   Todhunter, *Whewell*, vol. I, pp. 184–210.

133   As quoted in Todhunter, *Whewell*, vol. I, p. 187.

134   [Whewell], *Essay* (5th ed.), p. 374; see also p. 385.

CHAPTER 8

1   Theodor Appel, "Man and the Cosmos," *Mercersburg Review,* 16 (April 1867), 278–306:279. In one sense, an even later review of Whewell's book appeared in *Southern Review,* 8 (October 1870), 369–85. This anonymous review, although purportedly

directed at the books of Brewster and Fontenelle, is in fact a long critique of Whewell's *Essay*.

2 Two excellent examples are Johann Gottlieb Schimko, *Die Planetenbewohner* (Olmütz, 1856), and Hollis Read, *The Palace of the Great King* (Glasgow, 1860 reprinting of the New York 1859 original); see pp. 153–72. Schimko attempts to prove that the farther a planet is from the sun, the higher is the level of intellectual perfection of its inhabitants (p. 16). Read's discussion is in the natural theology tradition; in fact, like Thomas Dick, he even specifies the population of the rings of Saturn (p. 160).

3 Examples of such publications are (1) Anonymous, "Are the Planets Inhabited?" *Eclectic Magazine*, 55 (1862), 327–9; (2) Anonymous, "About the Plurality of Worlds," *Knickerbocker Monthly*, 61 (1863), 395–405; (3) George Leigh, "Are the Planets Inhabited? *Once a Week*, 9 (July 11, 1863), 80–82; (4) Anonymous, "The Seas and Snows of Mars," *Living Age*, 76 (1863), 537–9; reprinted from the *Spectator*; (5) Pierre Samuel Tzaut, "La pluralité des mondes habités," *Bibliothèque universelle et revue suisse*, nouvelle période, 29 (1867), 97–118, 189–209; (6) Robert Hogarth Patterson, "Are There More Worlds than One?" *Belgravia*, 6 (October 1868), 523–30; (7) Georg Holtzhey, "Ueber die Bewohnbarkeit der Weltkörper," *Sirius*, 2 (1869(, 52–3; and (8) H . . . t, "Ueber die Bewohnbarkeit der Welten," *Sirius*, 2 (1869), 91–3.

4 Read, *Palace*, pp. 155, and Schimko, *Planetenbewohner*, pp. 30–2.

5 T. L. Phipson, "Inhabited Planets," *Belgravia*, 3 (October 1867), 63–6:65.

6 J. N. Lockyer, *Elements of Astronomy*, American edition of his *Elementary Lessons on Astronomy* (New York, 1879), p. 69.

7 Mungo Ponton, *The Great Architect as Manifested in the Material Universe*, 2nd ed. (London, 1866), pp. 243, 262–6.

8 J. B. J. Liagre, "Sur la pluralité des mondes," *Bulletin de l'académie de Belgique*, 2nd ser., 8 (1859), 383–416:413.

9 F. Coyteux, *Qu'est-ce que le soleil; peut-il être habité [?]*, (Paris, 1866). The response made by Joseph-Louis Trouessart, "Professeur à la Faculté des Sciences de Poitiers," as well as the rejoinder of Coyteux, are contained in *Rapport fait les 4 décembre 1866 et 8 janvier 1867 par M. Trouessart . . . á la cosiété académique d'agriculture, belleslettres, sciences et arts [de Poitiers] sur un ouvrage intitulé Qu'est-ce que le soleil? peut-il être habité? par M. Coyteux* (Poitiers, 1867).

10 William Huggins, "The New Astronomy: A Personal Retrospect," *Nineteenth Century*, 41 (1897), 907–29.

11 See, for example, William McGucken, *Nineteenth Century Spectroscopy* (Baltimore, 1969); Herbert Dingle, "A Hundred Years of Spectroscopy," *British Journal for the History of Science*, 1 (1963), 199–216; and Donald H. Menzel, "The History of Astronomical Spectroscopy," *Annals of the New York Academy of Sciences*, 198 (1972), 225–44.

12 References to Huggins's papers will be as they appear, without alteration, according to their author (p. viii), in *The Scientific Papers of Sir William Huggins*, ed. Sir William Huggins and Lady Huggins (London, 1909).

13 Huggins, *Papers*, p. 60, but see also p. 493. The degree to which pluralism entered Huggins's writings is striking. A possible explanation is provided by the claim made by Camille Flammarion that Huggins was led to take up spectroscopic study of the planets by reading his *La pluralité des mondes habités*, which first appeared in 1862. For Flammarion's claim, see *La pluralité*, 33rd ed. (Paris, ca. 1885), p. 125, and Flammarion's *Les mondes imaginaires et les mondes réels*, 20th ed. (Paris, 1882?), p. 572.

14 Robert Hunt, "The Physical Phenomena of Other Worlds," *Popular Science Review*, 4 (1865), 311–23:323.

15 William Carter, "On the Plurality of Worlds," *Journal of Science*, 2 (1965), 227–39.

16 H. Schellen, *Spectrum Analysis*, translated from the second German edition by Jane and Caroline Lassell, ed. William Huggins (London, 1872), p. 506. See the note on p.

488, and compare p. 506 of Schellen with Huggins, *Papers*, p. 493, for evidence that Huggins held similar views.

17 As quoted in "Water on the Planets and Stars," *Annual of Scientific Discovery for 1869*, p. 345.

18 Jules Janssen, "Life on the Planets," *Popular Scientific Monthly*, 50 (1897), 812–14:813.

19 For biographical information on the Drapers, see Donald Flemming, *John William Draper and the Religion of Science* (Philadelphia, 1950).

20 J. W. Draper, *History of the Conflict between Religion and Science* (New York, 1897), p. 179.

21 J. W. Draper, *History of the Intellectual Development of Europe*, rev. ed., vol. II (New York, 1876), p. 279; see also p. 292.

22 Draper, *Development*, vol. II, p. 336. See Frederick Engels, *Dialectics of Nature* (New York, 1940), p. 24, where Engels comments that this principle "forced itself even on the anti-theoretical Yankee brain of Draper."

23 Henry Draper, "Are There Other Inhabited Worlds?" *Harper's Magazine*, 33 (June 1866), 45–54.

24 S. P. Langley, *The New Astronomy* (Boston, 1889), p. 14.

25 S. P. Langley, "The First 'Popular Scientific Treatise,' " *Popular Scientific Monthly*, 10 (April 1877), 718–25.

26 Isaac Asimov, *Extraterrestrial Civilizations* (New York, 1979), p. 35.

27 George Johnstone Stoney, "On the Physical Constitution of the Sun and Stars," *Proceedings of the Royal Society*, 17 (1869), 1–57. An abstract of this paper appeared both in the *Proceedings of the Royal Society*, 16 (1868), 25–34, and in the *Philosophical Magazine*, ser. 4, 34 (1867), 304–12.

28 G. J. Stoney, "Of Atmospheres upon Planets and Satellites," *Royal Dublin Society Scientific Transactions*, 6 (1898), 305–28.

29 John James Waterston, "On the Physics of Media That Are Composed of Free and Perfectly Elastic Molecules in a State of Motion," *Royal Society Philosophical Transactions*, 183A (1892), 1–80. For Waterston's discussion of lunar and planetary atmospheres, see pp. 36–8.

30 F. Zöllner, "Photometrische Untersuchungen über die physische Beschaffenheit des Planeten Merkur," *Annalen der Physik und Chemie*, Jubelband (1874), 624–43:639.

31 See Stanley Jaki, *Planets and Planetarians* (Edinburgh, 1978), ch. VI.

32 For an insightful analysis of whether or not the theory of evolution by natural selection supports pluralism, see Alfred Russel Wallace, *Man's Place in the Universe*, 4th ed. (London, 1904), pp. 326–36.

33 *The Times* (London) (September 14, 1888), p. 5.

34 As quoted in Charlotte R. Willard, "Richard A. Proctor," *Popular Astronomy*, 1 (1894), 319–21:319.

35 R. A. Proctor, *The Borderland of Science* (London, 1882), p. v.

36 J. C. Houzeau and A. Lancaster, *Bibliographie générale de l'astronomie jusqu'en 1880*, vol. II (London, 1964 reprinting of the 1882 original), p. lxxiv.

37 On his use of pseudonyms, see Anonymous, "Richard Anthony Proctor," *The Critic*, 13 (September 22, 1888), 134, where the pseudonyms Thomas Foster and Edward Clodd are attributed to him. The former ascription is correct, being accepted in the *British Museum Catalogue* and other sources. The latter ascription is excessive; this is indicated by the fact that the October 1, 1888, issue of *Knowledge* contained an obituary of Proctor – written by Clodd.

38 As quoted in Willard, "Proctor," p. 319.

39 The best source of information on this portion of Proctor's career is his "Autobiographical Notes," *New Science Review*, 1 (April 1895), 393–7.

40 In 1875, Proctor stated that the one thousand copies of it that had been printed still had not sold out. See Proctor, *Science Byways* (London, 1875), p. xiii, and Proctor's letter in *Atlantic Monthly*, 34 (1874), 750–1.

41 Proctor, "Autobiographical Notes," p. 396.

42 As quoted in A. C. R. [Arthur C. Ranyard], "Richard Anthony Proctor," *Royal Astronomical Society Monthly Notices*, 49 (February 1889), 165.
43 As quoted in [Ranyard], "Proctor," p. 165.
44 [A. M. Clerke], "[Review of] *Old and New Astronomy*. By Richard A. Proctor," *Edinburgh Review*, 177 (1893), 544–64:545.
45 John Fraser, as quoted in "Proctor the Astronomer," *English Mechanic*, 18 (December 12, 1873), 322.
46 W. Noble, "Richard A. Proctor," *Observatory*, 11 (October 1888), 366–8:367.
47 [R. A. Proctor], "Life in Other Worlds," *Knowledge*, 11 (August 1, 1888), 230–2:231.
48 R. A. Proctor, "Other Worlds and Other Universes," in *Myths and Marvels of Astronomy*, new ed. (London, 1880), p. 135; see also p. 137.
49 R. A. Proctor, *Other Worlds than Ours* (London, 1870), p. 4. Subsequent references are provided in the text; when cited passages occur in signficantly altered form in Proctor's fourth edition (New York, 1890?), these changes are noted. The last cited sentence, as well as all mention of the Whewell debate, are not present in the introduction to his fourth edition.
50 As quoted, without reference, in L. T. Townsend, *The Stars Not Inhabited* (New York, 1914), pp. 143–5.
51 Proctor, *Other Worlds*, p. 145. This sentence does not appear in Proctor, *Other Worlds*, 4th ed., p. 150.
52 R. A. Proctor, *Saturn and Its System* (London, 1865), pp. 156–85.
53 [R. A. Proctor], "Life in Other Worlds," *Knowledge*, 11 (August 1, 1888), 230.
54 Proctor, *Other Worlds*, p. 242. In Proctor, *Other Worlds*, 4th ed., p. 244, "must" is replaced by "may."
55 [Felix Eberty], *The Stars and the Earth; or, Thoughts upon Space, Time, and Eternity* (London, 1846). Proctor edited an edition of this work, for which Thomas Hill had earlier written a "Recommendatory Letter."
56 R. A. Proctor, *The Orbs around Us*, 2nd ed. (New York, 1899), p. vii.
57 R. A. Proctor, *The Borderland of Science* (London, n.d.), pp. 156–7.
58 R. A. Proctor, *Science Byways* (London, 1875), p. 4.
59 R. A. Proctor, *Our Place among Infinities*, 2nd ed. (London, 1876), p. 67.
60 In the year of his death, he published two essays presenting this point of view; see his "Varied Life in Other Worlds," *Open Court*, 1 (1888), 595–600, and "Life in Other Worlds," *Knowledge*, 11 (August 1, 1888), 230–2.
61 E. Clodd, "In Memoriam. Richard Anthony Proctor," *Knowledge*, 11 (October 1, 1888), 265.
62 Anonymous, "Richard A. Proctor Dead," *New York Times* (September 13, 1888), p. 1.
63 I have examined twenty-eight articles on Proctor's life, chiefly obituary notices. The cited quotations embody essentially all that they contain concerning his Christian convictions.
64 R. A. Proctor, *Myths and Marvels of Astronomy* (New York, 1877), p. 109.
65 R. A. Proctor, *The Poetry of Astronomy* (London, 1882?), p. 148.
66 Proctor, *Poetry*, pp. 250–4; see also the remark in Proctor's *Pleasant Ways in Science* (London, 1893), p. iv.
67 A. F. O'D. Alexander, *The Planet Saturn: A History of Observation, Theory and Discovery* (London, 1962), p. 111. It is noteworthy that Whewell's ideas importantly influenced Proctor's theory of Saturn.
68 Proctor, "Varied Life," p. 599.
69 Anonymous, "WILL SEE MEN ON MARS," *New York Times* (June 21, 1896), p. 22.
70 R. A. Sherard, "Flammarion the Astronomer," *McClure's*, 2 (May 1894), 569–77:569.
71 Simon Newcomb, "A Very Popular Astronomer," *Nation*, 59 (December 20, 1894), 469–70:469.
72 Houzeau and Lancaster, *Bibliographie*, vol. II, p. lxxiv.
73 Camille Flammarion, "How I Became an Astronomer," *North American Review*, 150

(January 1890), 100–5:102. The literature on Flammarion is vast. The most important source is his autobiography, which despite being published in 1911 and running to 556 pages, covers only the first thirty years of his life. See Flammarion, *Mémoires: biographiques et philosophiques d'un astronome* (Paris, 1911). See also (1) A. F. Miller, "Camille Flammarion," *Royal Astronomical Society of Canada Journal,* 19 (1925), 265–86; (2) E. Touchet, "La vie et l'oeuvre de Camille Flammarion," *Astronomie,* 39 (July 1825), 341–65; (3) Hilaire Cuny, *Flammarion* (Vienna, 1964), which contains a selection of his writings as well as biographical information; and (4) A. Duplay, "La vie de Camille Flammarion," *Astronomie,* 89 (December 1975), 405–19.

74 Flammarion, "How I Became an Astronomer," p. 103.

75 As quoted in Sherard, "Flammarion," p. 569.

76 These are listed in Flammarion, *Mémoires,* p. 215.

77 Camille Flammarion, *La pluralité des mondes habités,* 33rd ed. (Paris, ca. 1885), p. 480, lists translations (sometimes with translators and dates of publication) into Danish, English, German, Greek, Italian, Polish, Portuguese, Russian, Spanish (2), and Swedish. In Flammarion, *Mémoires,* pp. 217–18, Arabic, Braille, Chinese, and Czech editions are also listed. This information is at least in part erroneous. For example, despite the fact that Flammarion's reference (*La pluralité,* 33rd ed., p. 480) to an English translation by "Charles Powel . . . Boston . . . 1873" seems so definite as to be beyond question, I have been unable to locate any English translation in the *National Union Catalog* or in the catalogs of numerous American and European libraries. The only satisfactory explanation for the rarity of this item seems to be its nonexistence. Conclusive evidence exists for Danish, Dutch, German, Portuguese, Spanish, and Swedish translations.

78 Flammarion, *Mémoires,* p. 202.

79 Flammarion, *Mémoires,* pp. 168–88.

80 C. Flammarion, *Les mondes imaginaires et les mondes réels,* 20th ed. (Paris, 1882?), p. 566. See also Flammarion, *Mémoires,* pp. 242–3.

81 Flammarion, *Mémoires,* p. 202.

82 Flammarion, *Mémoires,* pp. 203–4.

83 Anonymous, "Camille Flammarion," *The Times* (London) (June 5, 1925), p. 16. See also Flammarion, *Mémoires,* pp. 457–458.

84 C. Flammarion, *La pluralité des mondes habités* (Paris, 1862), p. 7.

85 Flammarion, *Mémoires,* p. 242.

86 As quoted in Flammarion, *Mémoires,* p. 216. Flammarion published two books in 1862, the lesser known being *Les habitants de l'autre monde* (Paris). It is on spiritualism, in particular, table rapping and mediumistic writing.

87 Flammarion, *Mémoires,* p. 242.

88 Flammarion, *Mémoires,* pp. 218–19.

89 Miller, "Flammarion," pp. 272–3.

90 Flammarion, *La pluralité,* 33rd ed., p. 355. The final sentences in each of the two passages quoted from Whewell on pp. 354–5 do not follow the preceding sentences and may not occur anywhere in Whewell's book.

91 From Saint-Beuve's review in the May 22, 1865, issue of *Le constitutionnel,* as republished in Saint-Beuve's *Nouveau Lundis,* vol. X (Paris, 1886), p. 105.

92 Flammarion, *Mémoires,* p. 304.

93 Flammarion, *Mémoires,* p. 431.

94 Camille Flammarion, *The Atmosphere,* trans. C. B. Pitman, ed. James Glaisher (New York, 1874), p. 3.

95 Kenneth Allott, *Jules Verne* (London, n.d.), pp. 108, 152, 215.

96 C. Flammarion, *Les terres du ciel,* 11th ed. (Paris, 1884), p. 7.

97 R. A. Proctor, *The Poetry of Astronomy* (London, 1882?), pp. 250–4.

98 W. H. M. Christie, "[Review of] *Les terres du ciel,*" *Observatory,* 1 (1878), 355–8. See also *Astronomical Register,* 15 (1877), 121–2.

99 John Ellard Gore's "Preface" to Camille Flammarion, *Popular Astronomy,* trans. J. E. Gore (New York, 1931 printing of the 1907 edition), p. vii.

100 Roger Servajean, "Camille Flammarion," *Dictionary of Scientific Biography,* vol. V (New York, 1972), p. 21.
101 C. Flammarion, *Astronomie populaire,* 70th ed. (Paris, 1885), p. 199.
102 Flammarion, *Astronomy,* p. 79. Simon Newcomb noted the same passage; see his "Popular Astronomer," p. 469.
103 Anonymous, "[Review of] *Les étoiles,*" *Observatory,* 5 (1882), 265–7:266.
104 Fuldah LeCocq de Lautreppe, "A Poet Astronomer," *Cosmopolitan,* 17 (1894), 146–50:150.
105 Servajean, "Flammarion," p. 22.
106 See, for example, I. M. Stefan, "Camille Flammarion et la Roumanie," *Astronomie,* 89 (April 1975), 165–8.
107 C. Flammarion, *La planète Mars et ses conditions d'habitabilité,* vol. I (Paris, 1892), p. 592.
108 See also the article by him on Mars in the *New York Times* (March 2, 1924), section IX, p. 3.
109 Edmund Neison, *The Moon* (London, 1876), p. 104.
110 William Huggins and William Miller, "Observations of the Moon and Planets," in Huggins, *Papers,* p. 365.
111 On Stoney's studies in the 1860s, see G. Johnstone Stoney, "Of Atmospheres on Planets and Satellites," *Royal Dublin Society Scientific Transactions,* 6 (1898), 305–28.
112 Edmund Neison, "Hyginus N," *Astronomical Register,* 17 (1879), 199–208:199. See also Neison's "Physical Changes on the Surface of the Moon," *Quarterly Journal of Science,* n.s., 7 (1877), 1–26:5.
113 Patrick Moore, "The Linné Controversy," *British Astronomical Association Journal,* 87 (1977), 363–8:365.
114 For discussions of the controversy that resulted and evidence that Linné had not in fact changed, see Moore, "Linné," pp. 363–8; Joseph Ashbrook, "Linné in Fact and Legend," *Sky and Telescope,* 20 (1960), 87–8; and Richard J. Pike, "The Lunar Crater Linné," *Sky and Telescope,* 46 (1973), 364–6.
115 Houzeau and Lancaster, *Bibliographie,* vol. II, pp. 1283–5, where over eighty discussions of Linné published between 1866 and 1880 are listed.
116 W. R. Birt, "Supposed Changes in the Moon – Letter from Schmidt," *Student and Intellectual Observer,* 2 (1869), 48–50.
117 W. R. Birt, "Report on the Discussion of Observation of Spots on the Surface of the Lunar Crater Plato," *British Association for the Advancement of Science Report 1871)* (London, 1872), 60–97; Birt, "Report on the Discussion of Observations of Streaks on the Surface of the Lunar Crater Plato," *B.A.A.S. Report 1872* (London, 1873), 245–301.
118 Neison, "Changes," pp. 12–16.
119 E. Neison, "The Supposed New Crater on the Moon," *Popular Science Review,* 18 (1879), 138–46.
120 Camille Flammarion, *Astronomie populaire,* 70th ed. (Paris, 1885), p. 196.
121 W. R. Birt, "Lunar Atmosphere and Vegetation," *English Mechanic,* 14 (1871), 248.
122 James Nasmyth and James Carpenter, *The Moon: Considered as a Planet, a World, and a Satellite,* 3rd ed. (London, 1885), p. 186.
123 Edmund Neison, *The Moon* (London, 1876), pp. 19ff.
124 Neison, "Changes," p. 2.
125 R. A. Proctor, *The Moon* (London, 1873), pp. 263, 271–2.
126 R. A. Proctor, "Supposed Changes in the Moon," *Belgravia,* 37 (1878–9), 304–20:318. See also Proctor's "Changes in the Moon's Surface, with Special Reference to Supposed Changes in Linné and Plato," *Quarterly Journal of Science,* 3 (1873), 483–510.
127 W. R. Birt, "Is the Moon Dead?" *English Mechanic,* 25 (July 27, 1877), 484–5; Neison, "Changes," pp. 20–6.
128 Joseph Ashbrook, who had detailed knowledge of the history of lunar studies, did not mention Proctor in his "A Plato Illusion," *Sky and Telescope,* 19 (1959), 92; Patrick

Moore, another master of this area, presented Proctor as accepting changes in Linné; see Moore, "Linné," p. 365.

129 Timothy Harley, *Moon Lore* (Detroit, 1969 reprinting of the London 1885 original), pp. 227–57.

130 E. S. Holden, *Handbook of the Lick Observatory* (San Francisco, 1888), p. 12. See also Dorthy Tye, "When Fantasy Becomes History," *Pacific Historian*, 14 (Fall 1970), 96–102.

131 Simon Newcomb, *Reminiscences of an Astronomer* (London, 1903), p. 315.

132 P. A. Hansen, "Sur la figure de la lune," *Royal Astronomical Society Memoirs*, 24 (1856), 29–90:32.

133 For further information, see Daniel A. Beck, "Life on the Moon? A Short History of the Hansen Hypothesis," *Annals of Science*, 41 (1984), 463–70, and N. T. Roseveare, *Mercury's Perihelion from Le Verrier to Einstein* (Oxford, 1982), pp. 52–7.

134 Hervé Faye, "Remarques sur l'hypothèse de l'atmosphère de la Lune . . . ," *Comptes rendus de l'académie des sciences*, 51 (1860), 445–8.

135 [John Herschel], "Figure of the Moon and of the Earth," *Cornhill Magazine*, 6 (1862), 548–50:549. Internal evidence and Walter E. Houghton (ed.), *Wellesley Index to Victorian Periodicals*, vol. I (Toronto, 1966), p. 332, support ascription of this essay to Herschel.

136 W. Leitch, *God's Glory in the Heavens*, 3rd ed. (London, 1867), ch. III; H. Draper, "Are There Other Inhabited Worlds?" *Harper's Magazine*, 33 (June 1866), 45–54:50. On Verne, see Mark R. Hillegas, "Victorian 'Extraterrestrials,' " in *The Worlds of Victorian Fiction*, ed. Jerome Buckley (Cambridge, Mass., 1975), 391–414:398–9.

137 Watson's lectures are described in "The Moon a Habitable Globe," *English Mechanic*, 9 (1869), 323, and in *Astronomical Register*, 7 (1869), 115–16.

138 R. K. Miller's *The Romance of Astronomy* (London, 1873), as reprinted in *The Humboldt Library of Science*, vol. II (New York, 1881?), pp. 387–439:416.

139 Simon Newcomb, "On Hansen's Theory of the Physical Constitution of the Moon," *American Association for the Advancement of Science Proceedings 1868* (Cambridge, Mass., 1869), 167–71:171.

140 C. Delaunay, "Sur la constitution physique de la Lune," *Comptes rendus de l'académie des sciences*, 70 (1870), 57–61; P. A. Hansen, "Ueber die Bestimmung der Figur des Mondes, in Bezug auf Ausätze der Herren Newcomb und Delaunay darüber," *Berichte über die Verhandlungen (mathematisch-physische Classe) der königlich säch-sische Gessellschaft der Wissenschaften*, 23 (1871), 1–12. On Newcomb's meeting with Hansen in 1870, see Newcomb, *Reminiscences*, pp. 315–18.

141 R. A. Proctor, *Other Worlds than Ours* (London, 1870), p. 181n; see also Proctor's *Saturn and Its System* (London, 1865), pp. 209–12; *The Moon* (London, 1873), 298–302; "Note on Mr. Plummer's Reply," *Royal Astronomical Society Monthly Notices*, 23 (1873), 419–20; and *Borderland of Science* (London, 1882?), p. 229–34.

142 Willy Ley, *Rockets, Missiles, and Men in Space* (New York, 1968), p. 31.

143 William H. Pickering, "Is the Moon a Dead Planet?" *Century* 42 (1902), 90–9:90–1.

144 Ashbrook, "A Plato Illusion," p. 92.

145 As quoted in Bessie Zaban Jones and Lyle Gifford Boyd, *The Harvard College Observatory: The First Four Directorships, 1839–1919* (Cambridge, Mass., 1971), p. 307.

146 W. H. Pickering, "The Canals in the Moon," *Century*, 42 (1902), 189–95:195.

147 Waldemar Kaempffert, "Life on the Moon," *Munsey's Magazine*, 33 (August 1905), 588–92:592.

148 As quoted in Jones and Boyd, *Harvard*, p. 373.

149 W. H. Pickering, "Eratosthenes, No. 6: Migration of the Plats," *Popular Astronomy*, 32 (1924), 393–404:404.

150 W. H. Pickering, "Life on the Moon," *Popular Astronomy*, 45 (1937), 317–19.

151 E. P. Martz, Jr., "Professor William Henry Pickering," *Popular Astronomy*, 46 (1938), 299–310:299.

152 I. S. Shklovskii and Carl Sagan, *Intelligent Life in the Universe* (New York, 1966), p. 308.

153 Patrick Moore, *Survey of the Moon* (London, 1963), p. 185.

154 George H. Leonard, *Somebody Else Is on the Moon* (New York, 1976).

155 Victor Meunier, "La Lune est-elle habitable?" in Meunier's *Science et démocratie*, 2nd ser. (London, 1866), pp. 97–107:107.

156 Howard Sutton, "Charles Cros, the Outsider," *French Review*, 39 (1966), 513–20.

157 Camille Flammarion, *Mémoires: biographiques et philosophiques d'un astronome* (Paris, 1911), pp. 480–1.

158 For a report, see *L'Institut: Journal universelle des sciences*, 37 (Juillet 7, 1869), 209–10.

159 Republished in Charles Cros, *Oeuvres complétes*, ed. Jean-Jacques Pauvert (Paris, 1964), pp. 463–77, with editorial notes (pp. 622–3).

160 Anonymous, "Concerning the Means of Communication with the Planets," republished from *L'Italie, Astronomical Register*, 8 (1870), 166–7.

161 As quoted in Camille Flammarion, "Idée d'une communication entre les mondes," *Astronomie*, 10 (1891), 282–7:282. This essay appeared in English in an expanded form as "Inter-Astral Communication," *New Review*, 6 (January 1892), 106–14. See also "Shall We Talk with Men in the Moon? Probably, Says M. Camille Flammarion," *Review of Reviews*, 5 (February 1892), 90.

162 The first official announcement of the prize seems to have been "Prix Pierre Guzman," *Comptes rendus de l'académie des sciences*, 131 (1900), 1147. For further information, see C. Flammarion, *La planète Mars*, vol. II (Paris, 1909), 500–1, and Frank H. Winter, "The Strange Case of Madame Guzman and the Mars Mystique," *Griffith Observer*, 48 (February 1984), 2–15.

163 Flammarion, "Inter-Astral Communication," p. 107.

164 A. Guillemin, "Communication with the Planets," *Popular Science Monthly*, 40 (January 1892), 361–3:363; republished from *La nature*.

165 [R. H. Hutton], "Telegraphing to Mars," *Spectator*, 69 (August 13, 1892), 218–19:218. For Hutton's authorship, see Robert H. Tener, "R. H. Hutton's Editorial Career," *Victorian Periodicals Newsletter*, 7 (December 1974), 6–13:12.

166 [R. H. Hutton], "De We Need Wider Horizons?" *Spectator*, 69 (August 20, 1892), 253–4. The contents make clear that this was written by the author of the earlier *Spectator* article, i.e., by Hutton.

167 As quoted from the *Pall Mall Gazette* of August 18, 1892, in J. Norman Lockyer, "The Opposition of Mars," *Nature*, 46 (September 8, 1892), 443–8:444.

168 Anonymous, "The Signals from Mars," *Popular Astronomy*, 3 (1895), 47.

169 Francis Galton, "Intelligible Signals between Neighboring Stars," *Fortnightly Review*, N.S., 60 (November 1896), 657–64. This paper is a summary of a sixty-page manuscript preserved in the Galton Archives. See D. W. Forrest, *Francis Galton: The Life and Work of a Victorian Genius* (New York, 1974), p. 238.

170 For quotations, see Forrest, *Galton*, pp. 239–40.

171 Konstantin Tsiolkovskii, "Can the Earth Ever Inform the Inhabitants of Other Planets about the Existence of Intelligent Beings on It?" as translated from the *Kaluga Herald* (*Kaluzkskii Vestnik*) for 1896 (no. 68) and published in N. A. Rynin, *Interplanetary Flight and Communication*, vol. 1, no. 3 (Jerusalem, 1971 translation of the Leningrad 1931 original), pp. 53–5.

172 A. Mercier, *Conférence astronomique sur la planète Mars* (Orléans, 1902).

173 For Douglass, see "The Message from Mars," *Annual Report of the Smithsonian Institution for 1900* (Washington, 1901), 169–71, where Douglass's denial as published in the *Boston Transcript* for February 2, 1901, is quoted. For Lowell, see Percival Lowell, "Explanation of the Supposed Signal from Mars," *Popular Astronomy*, 10 (1902), 185–94.

174 D. E. Parks, as quoted in William Graves Hoyt, *Lowell and Mars* (Tucson, Arizona, 1976), p. 125.

175 Nikola Tesla, "Talking with the Planets," *Collier's Weekly*, 24 (February 9, 1901), 4–5:5.

176 *Colorado Springs Gazette* (March 9, 1901), as quoted in Inez Hunt and Wanetta W.

Draper, *Lightning in His Hand: The Life Story of Nikola Tesla* (Denver, 1964), p. 122.

177 E. S. Holden, "What We Know about Mars," *McClure's,* 16 (1901), 439–44:444.

178 Anonymous, "Nonsense about Mars," *Current Literature,* 30 (1901), 257–8.

179 Sir Robert Ball, "Signalling to Mars," *Living Age,* 229 (1901), 277–84:284.

180 N. Tesla, "That Prospective Communication with Another Planet," *Current Opinion,* 66 (March 1919), 170–1. For Marconi, see "Marconi Sure Mars Flashes Messages." *New York Times* (September 2, 1921), pp. 1, 3.

181 Anonymous, "Signaling to Mars," *Scientific American,* 100 (May 8, 1909), 346.

182 Anonymous, "More about Signaling to Mars," *Scientific American,* 100 (May 15, 1909), 371.

183 George Fleming, "Signaling to Mars with Mirrors," *Scientific American,* 100 (May 29, 1909), 407.

184 Wilfred Griffin, "Signaling to Mars," *Scientific American,* 100 (June 5, 1909), 423.

185 Anonymous, "Prof. David Todd's Plan of Receiving Martian Messages," *Scientific American,* 100 (June 5, 1909), 423.

186 William C. Peckham, "Signaling to Mars," *Scientific American,* 100 (June 26, 1909), 479.

187 W. H. Pickering, "Signaling to Mars," *Scientific American,* 101 (July 17, 1909), 43.

188 As noted in Winthrop Packard, "Signalling to Mars," *Illustrated World,* 12 (December 1909), 393–8:398.

189 Edgar Lucien Larkin, "Signaling to Mars: Its Impossibility by Means of Light," *Scientific American Supplement,* 67 (June 19, 1909), 387.

190 William R. Brooks, "Signaling to Mars," *Collier's,* 44 (September 24, 1909), 27–8.

191 T. C. M., "Communicating with Mars," *Science,* N.S., 30 (July 23, 1909), 117. I am indebted to Professors John Burnham and June Fullmer, who independently suggested to me that T. C. M. should be identified as Thomas Corwin Mendenhall, an American physicist. Their suggestion is supported by the fact that T. C. M. wrote from Dresden, Germany, and it is known that Mendenhall was living in Europe in 1909.

192 Anonymous, "Communicating with Mars," *Independent,* 66 (1909), 1042–3:1043.

193 As cited in Rynin, *Interplanetary Flight,* vol. I, part 3, p. 72.

194 Simon Mitton (ed.), *The Cambridge Encyclopaedia of Astronomy* (New York, 1977), p. 248.

195 For a listing of these and of the pre-1800 and the 1901-to-1950 publications on meteorites, see Harrison Brown (ed.), *A Bibliography of Meteorites* (Chicago, 1953). No definitive history of the study of meteorites has been written, but much useful information is contained in Peter Lancaster Brown, *Comets, Meteorites and Men* (New York, 1974).

196 As quoted in Brown, *Comets,* pp. 205–6.

197 Jöns Jacob Berzelius, "Ueber Meteorsteine," *Annalen der Physik und Chemie,* 2nd ser., 33 (1834), 1–32, 113–48:144.

198 J. J. Berzelius, "On Meteoric Stones," *Philosophical Magazine,* 3rd. ser., 9 (1836), 429–41:440.

199 Friedrich Wöhler, "On the Organic Substance in the Meteoric Stone of Kaba," *Philosophical Magazine,* 4th ser., 18 (1859), 160; Wöhler, "On the Composition of the Cape Meteorite," *Philosophical Magazine,* 18 (1859), 213–18:213.

200 Stanislas Cloëz, "Analyse chimique de la pierre météorique d'Orgueil," *Comptes rendus de l'académie des sciences,* 59 (1864), 37–40:38.

201 Marcellin Berthelot, "Sur la matière charbonneuse des météorites," *Comptes rendus,* 67 (1868), 849.

202 William Thomson, "Inaugural Address," *Nature,* 4 (August 3, 1871), 262–70:269.

203 As quoted in Silvanus P. Thompson, *The Life of William Thomson, Baron Kelvin of Largs,* vol. II (London, 1910), p. 609.

204 As quoted in Agnes Gardner King, *Kelvin the Man* (London, 1925), p. 100.

205 As quoted in Leonard Huxley, *Life and Letters of Sir Joseph Dalton Hooker,* vol. II (London, 1918), pp. 126–7.

206 As quoted in Huxley, *Hooker,* vol. II, p. 126n.

207 Richard A. Proctor, "Comets and Comets' Tails," in Proctor's *The Orbs around Us* (London, 1899), p. 271. Reprinted from the September 1871 issue of *St. Paul's Magazine*.

208 Allen Thomson, "Address," *British Association for the Advancement of Science Report for 1877* (London, 1878), lxxv.

209 Walter Flight, "Meteorites and the Origin of Life," *Popular Science Review*, 16 (1877), 390–401:395–6.

210 As quoted in Thompson, *Thomson*, vol. II, p. 611.

211 As quoted in Thompson, *Thomson*, vol. II, p. 1103.

212 Thompson, *Thomson*, vol. II, p. 1097.

213 Paul Becquerel, "La vie terrestre provient-elle d'un autre monde?" *Bulletin de la société astronomique de France*, 38 (1924), 393–417:399. See also Harmke Kamminga, "Life from Space – A History of Panspermia," *Vistas in Astronomy*, 26 (1982), 67–86.

214 H. E. Richter, "Zur Darwin'schen Lehre," *Schmidts Jahrbücher der in- und ausländischen Medicin*, 126 (1865), 243–9:249.

215 H. von Helmholtz, "On the Use and Abuse of the Deductive Method in Physical Science," trans. Crum Brown, *Nature*, 11 (December 24, 1874), 149–151, and *Nature* 11 (January 14, 1875), 211–12:212.

216 H. von Helmholtz, "The Origin of the Planetary System," in *Selected Writings of Hermann von Helmholtz*, ed. Russell Kahl (Middletown, Conn., 1971), 266–96:294.

217 Johann Zöllner, *Über die Natur der Cometen* (Leipzig, 1872), pp. xxv–xxvi.

218 Helmholtz, "Deductive Method," p. 212.

219 See Becquerel, "Vie," p. 400; Flight, "Meteorites," pp. 400–1; Svante Arrhenius, *Worlds in the Making* (London, 1908), p. 218; and John Farley, *The Spontaneous Generation Controversy from Descartes to Oparin* (Baltimore, 1977), pp. 142–4:142.

220 Arrhenius, *Worlds*, ch. VIII; see also Alphonse Berget, "The Appearance of Life on Worlds and the Hypothesis of Arrhenius," *Annual Report of the Smithsonian Institution for 1912* (Washington, D.C., 1913), 543–51, and Dick Haglund, "Svante Arrhenius och Panspermihypotesen," *Lychnos* (1967–8), 77–104.

221 David Friedrich Weinland, "Korallen in Meteorsteinen," *Das Ausland*, 54 (April 17, 1881), 301–3:301; see also Weinland's "Weiteres über die Tierreste in Meteoriten," *Das Ausland*, 54 (June 27, 1881), 501–8, and his *Ueber die in Meteoriten entdeckten Thierreste* (Esslingen, 1882).

222 Otto Hahn, *Die Meteorite (Chondrite) und ihre Organismen* (Tübingen, 1880), pp. 42–4. This point, as well as a number of others in this section, was called to my attention by Professor John G. Burke, who is completing a book on the history of the study of meteorites.

223 Francis Birgham, "The Discovery of Organic Remains in Meteor Stones," *Popular Science Monthly*, 20 (1881), 83–7:87.

224 As quoted in Joseph Pohle, *Die Sternenwelten und ihre Bewohner*, 2nd. ed. (Cologne, 1899), p. 87.

225 Carl Vogt, "Sur les prétendus organismes des météorites," *Comptes rendus*, 83 (1881), 1166–8. For Meunier, see his "Péridot artificiel produit en présence de la vapeur d'eau, à la pression ordinaire," *Comptes rendus*, 93 (1881), 737–9.

226 Anonymous, "Supposed Organic Remains in Meteorites," *American Journal of Science*, 23 (1882), 156.

227 As quoted in Anonymous, "Organic Remains in Meteorites," *Popular Science Monthly*, 20 (1882), 568–9:569.

228 Harold C. Urey, "Biological Materials in Meteorites: A Review," *Science*, 151 (January 14, 1966), 157–66:157.

229 Louis Figuier, *The Tomorrow of Death*, trans. S. R. Crocker (Boston, 1872), p. 51.

230 F. W. Fitch, H. P. Schwarcz, and E. Anders, "Organic Elements in Carbonaceous Chondrites," *Nature*, 193 (March 24, 1962), 1123–5:1124.

CHAPTER 9

1 L. A. Blanqui, *L'éternité par les astres: hypothèse astronomique* (Paris, 1872), p. 61. For helpful analyses of this book, see Alan B. Spitzer, *The Revolutionary Theories of Louis Auguste Blanqui* (New York, 1957), pp. 34–44, and Stanley L. Jaki, *Science and Creation* (New York, 1974), pp. 314–19. For biographical information, see Samuel Bernstein, *Auguste Blanqui and the Art of Insurrection* (London, 1971).

2 Literature on nineteenth-century transmigrational ideas is sparse; I have derived most assistance from (1) Edouard Bertholet, *La réincarnation* (Neuchâtel, 1949); (2) D. C. Charlton, *Secular Religions in France 1815–1870* (London, 1963); (3) Camille Flammarion, *Les mondes imaginaires et les mondes réels,* 20th ed. (Paris, ca. 1882); and (4) Joseph Head and S. L. Cranston (compilers), *Reincarnation: The Phoenix Fire Mystery* (New York, 1977).

3 Camille Flammarion, *Mémoires: biographiques et philosophiques d'un astronome* (Paris, 1911), p. 243.

4 André Pezzani, *La pluralité des existences de l'âme conforme à la doctrine de la pluralité des mondes,* 6th ed. (Paris, 1872), p. xiii. All subsequent references are to this edition.

5 Flammarion, *Les mondes,* p. 573. Flammarion cites this book as published at Lyon in 1864. In the catalog of the Bibliothèque nationale, no reference to it appears, but Pezzani's *Exposé d'un nouveau système philosophique* (Paris, 1847) is listed as containing a section with this title.

6 Flammarion, *Mémoires,* p. 217.

7 Edmond Grégoire, *L'astronomie dans l'oeuvre de Victor Hugo* (Paris, 1933), p. 173.

8 C. Flammarion, "Victor Hugo astronome," *Société astronomique de France Bulletin,* 16 (1902), 171–5; see also Flammarion, *La pluralité des mondes habités,* 33rd ed. (Paris, ca. 1855), p. 229; Flammarion, *Mémoires,* pp. 216–17.

9 Grégoire, *Hugo,* pp. 189–205.

10 Auguste Viatte, *Victor Hugo et les illuminés de son temps* (Montreal, 1942).

11 Louis Figuier, *The To-morrow of Death; or The Future Life According to Science,* trans. S. R. Crocker (Boston, 1872), p. 3.

12 As quoted in Ida M. Tarbell, "Sketch of Louis Figuier," *Popular Science Monthly,* 51 (1897), 834–41:841.

13 Flammarion, *Les mondes,* p. 590. The Bibliothèque nationale lists an 1876 book by Girard with the title *De la pluralité des mondes habités et des existences de l'âme* (Paris). I have not seen it, but from the fact that it is listed as containing the same number of pages (324) as Girard's *Nouvelles études . . . ,* it seems probable that the books differ only in their title pages.

14 Joseph Félix, *Le progrès par le Christianisme: Conférences de Notre-Dame de Paris – Année 1863,* 2nd ed. (Paris, 1864), pp. 120–1. For the setting of this statement, see Abbé François Moigno, *Les splendeurs de la foi,* vol. II (Paris, 1877), p. 402.

15 Monseigneur de Montignez, "Théorie chrétienne sur la pluralité des mondes," *Archives théologiques,* 9 (1865), 381–404; 10 (1865), 25–46, 102–43, 262–77, 297–313, 369–85; 11 (1866), 57–68, 81–93, 161–80. In this series, the second essay treats scientific aspects of pluralism and is by E. Rauran.

16 (9, p. 400) The quoted phrase is from Philippians 2:7.

17 (9, p. 402) The scriptural reference is to Micah 5:2.

18 As quoted in Flammarion, *La pluralité,* p. 380. I have been unable to determine whether this statement first appeared in the 1862 edition of *Les sources* or in the revised later edition. Gratry entered into astronomical speculations on locations for immortal life in his *De la connaissance de l'âme,* 5th ed., 2 vols. (Paris, 1874), On Gratry's ideas, see Théophile Ortolan, *Astronomie et théologie* (Paris, 1894), pp. 343–59.

19 As quoted in G. Bovier-Lapierre, *L'astronomie pour tous* (Paris, 1891), p. 310.

20 François Moigno, *Les splendeurs de la foi,* vol. II (Paris, 1877), p. 402.

21 Pioger's writings are very rare; I have based my description of these two books on Flammarion, *Les mondes,* pp. 583–4, 588–9.

22 Léger-Marie Pioger, "Introduction: Il y a d'autres mondes que le notre," in Pioger's *Le soleil,* new ed. (Paris, 1893), pp. 1–54. For his quotations from Félix, p. 9; from Frayssinous (1765–1841), pp. 9–10; from Bougaud (1824–88), pp. 14–18; from Monsabré (1827–1907), pp. 31–2.

23 Pioger, *Le soleil,* pp. 293–301; Pioger, *La lune* (Paris, 1883), pp. 290–306.

24 Pioger, *Le soleil,* p. 34.

25 Biographical information is very scarce on Boiteux; this fact is from the title page of what seems to be his only other book: *Notes sur la fonderie de fer* (Frameries, 1903).

26 Jules Boiteux, *Lettres à un matérialiste sur la pluralité des mondes habités et les questions qui s'y rattachent* (Paris, 1876), p. vii. Subsequent references are to the second edition (Paris, 1891).

27 Pierre Courbet, "De la redemption et de la pluralité des mondes habités," *Cosmos,* 4th ser., 28 (May 19, 1894), 208–11; (June 2, 1894), 272–6.

28 Théophile Ortolan, *Astronomie et théologie ou l'erreur géocentrique. La pluralité des mondes habités et le dogme de l'incarnation* (Paris, 1894).

29 Charles de Kirwan, "Les mondes inhabitables et les mondes peut-être habités," *Cosmos* (February 19, 1898), 245–9:245; article continued in *Cosmos* (February 26, 1898), 271–3.

30 Jean d'Estienne, "A propos de habitabilité des astres," *Cosmos* (July 11, 1891), 397–401; (July 18, 1891), 425–7. This was written in opposition to the following paper, which is discussed subsequently: J. Scheiner, "L'habitabilité des mondes," *L'astronomie: Revue mensuelle d'astronomie populaire,* 10 (1891), 221–7. The Bibliothèque nationale also lists Jean d'Estienne [Charles de Kirwan], *Considérations nouvelles sur la pluralité des mondes* (Paris, 1876), but I have been unable to secure information on it.

31 Jean d'Estienne, "[Review of] *Astronomie et théologie,*" *Revue des questions scientifiques,* 36 (1894), 312–20; his review of Ortolan's *Études* is de Kirwan, "Les mondes"; C. de Kirwan, "[Review of] *Lettres à un matérialiste . . .* par Jules Boiteux," *Revue des questions scientifiques,* 44 (1898), 293–5. The content, style, and location of the following article make it probable that it was also written by de Kirwan: "Les astres sont-ils habités?" *Cosmos* (July 24, 1897), 118–22.

32 *The International Catalogue of Scientific Literature* for 1902 lists the following thirty-nine-page book: Charles de Kirwan, *Le véritable concept de la pluralité des mondes* (Louvain, 1902). See also C. de Kirwan, "L'unité de l'universe et de l'homme dans l'univers," *Revue des questions scientifiques,* 64 (1908), 581–601, and C. de Kirwan, "Les mondes présents, passés ou futurs," *Revue des questions scientifiques,* 3rd ser., 23 (1913), 598–614.

33 Gabriel Prigent, *De l'habitabilité des astres* (Landerneau, 1892), p. 379.

34 R. M. Jouan, *La question de l'habitabilité des mondes* (Saint-Ilan, 1900), p. 466.

35 C. de Kirwan, "[Review of] *La question de l'habitabilité des mondes,*" *Revue des questions scientifiques,* 50 (1901), 657–8:658.

36 For Boiteux as supporter of Darwinian doctrines, see Harry W. Paul, "Religion and Darwinism: Varieties of Catholic Reaction," in Thomas F. Glick (ed.), *The Comparative Reception of Darwinism* (Austin, 1974), pp. 403–36:428–9; for de Kirwan, see Harry W. Paul, *The Edge of Contingency: French Catholic Reaction to Scientific Change from Darwin to Duhem* (Gainesville, 1979), pp. 48–52.

37 Ralph V. Chamberlin, "Life in Other Worlds," *Bulletin of the University of Utah, Biological Series,* 1, no 6 (February 1932), 1–52:33.

38 N. A. Perujo, *La pluralidad de mundos habitados ante la fé católica. Estudio en que se examina la habitacion de los astros en relacion con los dogmas católicos, se demeustra su perfecta armonia con estos, y se refutan muchos errores de Mr. Flammarion* (Madrid, 1877).

39 J. C. Houzeau and A. Lancaster, *Bibliographie générale de l'astronomie jusqu'en 1880,* vol. II (London, 1964 reprint of the Bruxelles 1882 original), p. lxxiv. Houzeau and Lancaster list 360 publications by Secchi; actually he published more than double that number.

40 A. Secchi, *Descrizione del nuovo osservatorio del collegio romano* (Rome, 1856), p. 158.

41 A. Secchi, *Les étoiles*, vol. II (Paris, 1879), p. 189. See also pp. 190–1 and Secchi's *Le soleil*, p. 418.

42 Camille Flammarion, *La planète Mars et ses conditions d'habitabilité*, vol. I (Paris, 1892), pp. 135–6.

43 Joseph Pohle, *Die Sternenwelten und ihre Bewohner*, 2nd ed. (Cologne, 1899), pp. 54–6.

44 Frederick Albert Lange, *The History of Materialism*, trans. E. C. Thomas, 3rd ed., vol. II (London, 1957), p. 265.

45 Ludwig Büchner, *Force and Matter*, 4th English ed., translated from the fifteenth German edition (London, 1884), pp. xx–xxi.

46 D. F. Strauss, *The Old Faith and the New*, trans. Mathilde Blind from the sixth German edition (New York, 1874).

47 This passage is quoted from Engels's manuscript by J. B. S. Haldane in his preface to Frederick Engels, *Dialectics of Nature*, trans. Clemen Dutt (New York, 1971), p. ix.

48 Engels, *Dialectics*, p. 24. As indicated in Engels's footnote, he derived this doctrine from J. W. Draper's *History of the Intellectual Development of Europe.*

49 F. Engels, "Socialism: Utopian and Scientific," in *The Marx-Engels Reader*, ed. R. C. Tucker, 2nd ed. (New York, 1978), p. 698.

50 In his *History of Spiritualism*, vol. II (New York, 1975 reprinting of the 1926 original), p. 186, Arthur Conan Doyle described Du Prel's contribution to spiritualism as "probably the greatest yet made by any German."

51 Carl Du Prel, *The Philosophy of Mysticism*, vol. II, trans. C. C. Massey (London, 1889), p. 273; for his pluralist ideas, see pp. 257–91.

52 Ernst Haeckel, *Riddle of the Universe*, trans. Joseph McCabe (New York, 1901), p. 370.

53 My discussion is based on J. Ebrard, *Apologetics; or The Scientific Vindication of Christianity*, trans. W. Stuart and J. MacPherson (Edinburgh, 1886). The relevant section (pp. 353–65) contains, but is not limited to, a long statement evidently quoted from his 1861 volume.

54 I have used C. E. Luthardt, *Apologetic Lectures on the Fundamental Truths of Christianity*, translated from the third edition by Sophia Taylor (Edinburgh, 1865); see pp. 79–87.

55 Otto Zöckler, "Der Streit über Einheit und Vielheit der Welten," *Beweis des Glaubens*, 2 (1866), 353–76:363.

56 Otto Zöckler, "Eine oder viele Welten?" *Beweis des Glaubens*, 13 (1877), 639–51:639. See also Zöckler's "Proctors Theorie des Sonnensystems und des Weltgebaudes," *Beweis des Glaubens*, 7 (1871), 516–25. Also see Kreyer, "Die Vielheit bewohnter Welten im Universum," *Beweis des Glaubens*, 5 (1869), 129–41, 145–51. This article was so enthusiastically pluralist that the editors (Zöckler was one of the four editors) felt the need to append a footnote suggesting that this "somewhat onesided" paper might stimulate discussion.

57 Zöckler, "Welten," pp. 640–1. Pfaff's views, according to Zöckler, appear in his *Schöpfungsgeschichte*, 2nd ed. (1876), p. 203. For Huber, see his *Zur Philosophie der Astronomie* (Munich, 1878), pp. 49–69. Huber's book is especially valuable for information on the antipluralist ideas of Hegel, Schelling, and their followers.

58 Otto Zöckler, *Geschichte der Beziehungen zwischen Theologie und Naturwissenschaft*, 2nd ed., vol. II (Gütersloh, 1879), pp. 55–74, 416–39.

59 The pluralist papers by Pohle include "Ueber das organische Leben auf den Himmelskörpern. Ein Nachtrag zu den Abhandlungen über die Weltanschauung des P. Angelo Secchi," *Katholik*, II Hälfte (1884), 337–81; "Das Problem von der Bewohntheit der Himmelskörper im Lichte des Dogma's," *Katholik*, II Hälfte (1886), 42–70, 113–36, 225–48, 336–58, 449–75, 561–84; and "Neue Untersuchungen über die Vielheit bewohnter Welten mit besonderer Berücksichtigung einiger Schwierigkeiten gegen die Annahme von vernünftigen Astralwesen auf den bewohnbaren

Himmelskörpern," *Natur und Offenbarung*, 33 (1888), 513–33, 595–616; 34 (1888), 139–52, 287–300, 411–26.

60 Joseph Pohle, *Die Sternenwelten und ihre Bewohner*, 2nd ed. (Cologne, 1899), My discussion and references are based on this edition.

61 Carl Goetze, *Die Sonne ist bewohnt. Ein Einblick in die Zustände im Universum* (Berlin, 1898). An examination of this book, which is extremely rare, shows that it was written by an author of very limited scientific sophistication. Preyer's position, as quoted by Pohle, is that the sun itself may be a "glowing organism whose breath may perhaps be shining iron vapor, whose blood may be flowing metal, and whose food may be meteorites." As quoted by Pohle (p. 159), who cites as his source Preyer's *Naturwissenschaftliche Thatsachen und Probleme* (Berlin, 1880), p. 59. The quotation is correct, but occurs on p. 60.

62 Pohle, *Sternenwelten*, p. 161. Braun was a pluralist and a student of Secchi. Pohle frequently cites his views, which are contained in his *Ueber Kosmogonie vom Standpunkt christlicher Wissenschaft*. I have seen the third edition (Münster, 1905).

63 Pohle, *Sternenwelten*, p. 304. Pohle's conviction in this regard seems to have declined between his second edition (1899) and his seventh (1922). Compare, for example, the final paragraph in the first chapter of these editions.

64 The context indicates that perfection is the most suitable translation of Pohle's word "Vollkommenheit," although he uses that term in such a way as to include "completion" or "fulfillment" under its meaning, thus suggesting the idea of a great chain of being.

65 Julius Zucht, "Sind die übrigen kosmischen Körper ausser der Erde bewohnt?" *Katholik*, I Hälfte (1886), 337–60; for Pohle's response, see Pohle, "Das Problem."

66 Adolf Müller, "Die Bewohner der Gestirne," *Stimmen aus Maria-Laach*, 58 (1900), 141–53; 59 (1900), 70–84.

67 Ludwig Günther, "Naturphilosophische Literatur," *Natur und Offenbarung*, 51 (1905), 493–505. This is a joint review of Pohle's book and A. R. Wallace's *Man's Place in the Universe*.

68 Paul Schanz, *A Christian Apology*, trans. M. F. Glancey and V. J. Schobel, 5th ed., vol. I (Ratisbon, 1891), pp. 394–5.

69 Winwood Reade, *The Martyrdom of Man*, 22nd ed., with introduction by F. Leege (London, n.d.), p. 523.

70 John Tyndall, "The Belfast Address," in Tyndall's *Fragments of Science*, vol. II (New York, 1901), p. 210.

71 John Tyndall, "Additional Remarks on Miracles," in Tyndall, *Fragments*, vol. II, pp. 41–2.

72 Frank M. Turner, "The Victorian Conflict between Science and Religion: A Professional Dimension," *Isis*, 69 (1978), 356–76:373.

73 See the anonymous review in *Astronomical Register*, 16 (1878), 98–102:101, and also James Clerk Maxwell's review [*Nature*, 19 (1878), 141–3:142] of the sequel to their book called *Paradoxical Philosophy*.

74 [B. Stewart and P. G. Tait], *The Unseen Universe, or Physical Speculations on a Future State*, 2nd ed. (New York, 1875), pp. ix–x. For an analysis, see P. M. Heimann, "*The Unseen Universe*: Physics and the Philosophy of Nature in Victorian England," *British Journal for the History of Science*, 6 (1972), 73–9.

75 William Kingdon Clifford "savaged" it in the *Fortnightly Review* (June 1875), and it was also criticized in the *Journal of Science*, 11 (1875), 472–86, and by John W. Chadwick in the *Unitarian Review*, 6 (1875), 554–64.

76 Augustus De Morgan, *A Budget of Paradoxes*, 2nd ed., vol. II (New York, 1954 reprinting of the 1915 edition), p. 191.

77 A. Clissold, *The Divine Order of the Universe as Interpreted by Emanuel Swedenborg with Special Relation to Modern Astronomy* (London, 1877).

78 R. A. Proctor, "Swedenborg's Vision of Other Worlds," in Proctor's *Myths and Marvels of Astronomy*, new ed. (London, 1880), pp. 106–34:112.

79 T. F. Wright, "The Planets Inhabited," *New-Church Review*, 4 (1897), 117–21:117.

80 Josiah Crampton, *The Lunar World*, 4th ed. (Edinburgh, 1863), p. 98.

81 Josiah Crampton, *The Three Heavens,* new ed. (London, 1879), p. 190.
82 Joseph Hamilton, *The Starry Hosts: A Plea for the Habitation of the Planets* (London, 1875), p. 38.
83 Sir Edwin Arnold, "Astronomy and Religion," *North American Review,* 159 (1894), 404–15:407–8.
84 Henry Cotterill, *Does Science Aid Faith in Regard to Religion?* 3rd thousand (London, 1886), pp. 130–42.
85 R. A. Proctor, "Science as an Aid to Faith," *Knowledge,* 3 (June 15, 1883), 360–1:361.
86 As noted previously, Hutton probably wrote item 28 in the Whewell debate. For Hutton's thought in the 1870s, see [R. H. Hutton], "The Dog Star and His System," *Spectator* (February 19, 1876), 241–2, and [R. H. Hutton], "A Miniature World," *Spectator* (September 15, 1877), 1144–5. For Hutton's authorship of these and the other astronomical articles in the *Spectator,* see Robert H. Tener, "R. H. Hutton's Editorial Career: Part II. *The Prospective* and *National Review*," *Victorian Periodicals Newsletter,* 7 (December 1974), 6–13:12.
87 [R. H. Hutton], "Other Worlds than Ours," *Spectator,* 55 (December 30, 1882), 1678–80:1679.
88 As quoted in R. H. Hutton, "Astronomy and Theology," reprinted in Hutton's *Criticisms on Contemporary Thought and Thinkers,* vol. I (London, 1894), pp. 288–95:288.
89 R. H. Hutton, "The Humility of Science," reprinted in Hutton's *Aspects of Religious and Scientific Thought,* ed. Elizabeth M. Roscoe (London, 1899), pp. 394–401.
90 [R. H. Hutton], "Telegraphing to Mars," *Spectator,* 69 (August 13, 1892), 218–19.
91 *The Poems of Arthur Hugh Clough,* ed. F. L. Mulhauser, 2nd ed. (Oxford, 1974), p. 193–4:194. This poem was first published in 1869.
92 *The Works of Robert Browning,* with an introduction by Sir F. G. Kenyon, vol. VI (New York, 1966 reprinting of the London 1912 edition), p. 199. These lines were first published in 1869. On Browning and a number of the other poets treated, see Jacob Korg, "Astronomical Imagery in Victorian Poetry," *Victorian Science and Victorian Values: Literary Perspectives,* published as vol. 360 (1981) of *Annals of the New York Academy of Sciences,* pp. 137–58.
93 *The Poems of George Meredith,* ed. Phyllis B. Bartlett, vol. I (New Haven, 1978), pp. 452–5.
94 In a scene in which the weeping Alyosha, disconsolate at the death of Father Zosimov, gazes at the stars, Dostoevsky writes: "There seemed to be threads from all those innumerable worlds of God, linking his soul to them, and it was trembling all over 'in contact with other worlds.' " Feodor Dosteovsky, *Brothers Karamazov,* trans. Constance Garnett, rev. Ralph E. Matlaw (New York, 1976), p. 340.
95 *The Poems and Plays of Alfred Lord Tennyson* (New York, 1938), p. 851.
96 *The Poems . . . of . . . Tennyson,* p. 838. Published in 1886.
97 As quoted in Agnes Grace Weld, "Talks with Tennyson," *Contemporary Review,* 63 (1893), 394–7:395.
98 Weld, "Talks," p. 397.
99 F. B. Pinion, "Introduction" to Thomas Hardy, *Two on a Tower* (London, 1975), p. 13. References are to this edition of the novel.
100 *The Poems of Coventry Patmore,* ed. Frederick Page (London, 1949), pp. 381–2.
101 Aubrey de Vere, "The Death of Copernicus," *Contemporary Review,* 57 (September 1889), 421–30:424.
102 *The Poems of Alice Meynell* (New York, 1923), p. 92. This poem was published in 1913 or possibly earlier.
103 Ivan Lee Zabilka, *Nineteenth Century British and American Perspectives on the Plurality of Worlds: A Consideration of Scientific and Christian Attitudes* (a 1980 doctoral dissertation at the University of Kentucky).
104 Jean-Bruno Renard, "Religion, science-fiction et extraterrestres," *Archives de sciences sociales des religions,* 50 (1980), 143–64:160.
105 Walt Whitman, *Leaves of Grass: Comprehensive Reader's Edition,* ed. Harold W.

Blodgett and Sculley Bradley (New York, 1965), p. 271. All subsequent references are to this edition.

106 These include (1) Alice Lovelace Cooke, "Whitman's Indebtedness to the Scientific Thought of His Day," *Studies in English,* 14 (July 8, 1934), 89–115; (2) Clarence Dugdale, "Whitman's Knowledge of Astronomy," *Studies in English,* 16 (July 8, 1936), 125–37; (3) Joseph Beaver, "Walt Whitman, Star Gazer," *Journal of English and Germanic Philology,* 48 (1949), 307–19; (4) Joseph Beaver, *Walt Whitman – Poet of Science* (Morningside Heights, N.Y., 1951), especially pp. 23–79.

107 Walt Whitman, "A Backward Glance o'er Travel'd Roads," in Whitman, *Leaves,* p. 564.

108 As quoted in Thomas L. Brasher, "A Modest Protest against Viewing Whitman as Pantheist and Reincarnationist," *Walt Whitman Review,* 13 (1967), 92–4:92. On Whitman's pantheistic and reincarnationist beliefs, see Gay Wilson Allen, *Walt Whitman Handbook* (Chicago, 1946), pp. 259–77.

109 Whitman, *Leaves,* page before p. 1.

110 Whitman, *Leaves,* p. 82. Dugdale ("Knowledge," p. 131) suggests as Whitman's source a quotation from Thomas Dick given by Henry Whitall, a friend of Whitman, in his *Treatise on the Principal Stars and Constellations* (1850). Beaver (*Whitman,* p. 159n) questions this, suggesting Mitchel's *A Course of Six Lectures on Astronomy* as the more probable source.

111 Gay Allen Wilson, *A Reader's Guide to Walt Whitman* (New York, 1981), p. 20.

112 Walt Whitman, "In Memory of Thomas Paine," in Whitman, *Prose Works 1892,* vol. I, ed. Floyd Stovall (New York, 1963), pp. 140–2.

113 Whitman, *Leaves,* p. 727. See also Margaret M. Vanderhaar, "Whitman, Paine, and the Religion of Democracy," *Walt Whitman Review,* 16 (1970), 14–22.

114 As quoted in Vanderhaar, "Whitman," p. 16.

115 As quoted in Justin Kaplan, *Walt Whitman: A Life* (New York, 1980), p. 231.

116 Howard Mumford Jones, *Belief and Disbelief in American Literature* (Chicago, 1969), pp. 70, 115.

117 Hyatt Howe Waggoner, "Science in the Thought of Mark Twain," *American Literature,* 8 (1937), 357–70:357–9.

118 As quoted in Minnie M. Brashear, *Mark Twain: Son of Missouri* (Chapel Hill, N.C., 1934), p. 245.

119 As quoted in Dixon Wecter (ed.), *The Love Letters of Mark Twain* (New York, 1949), p. 133.

120 As quoted in Albert Bigelow Paine, *Mark Twain: A Biography,* vol. II (New York, 1929), p. 412. For the dating of this article, see Paine, *Biography,* vol. IV (New York, 1929), p. 1532n.

121 First published in 1907–8 in *Harper's Magazine,* it appeared in 1909 in book form as Mark Twain, *Extract from Captain Stormfield's Visit to Heaven* (New York, 1909). References are to this edition.

122 Dixon Wecter, "Introduction" to Mark Twain, *Report from Paradise* (New York, 1952), pp. ix–xxv:xxi. This edition of *Stormfield* includes passages left out of the 1909 edition. For Wakeman as a model for Stormfield, see Ray B. Browne, "Mark Twain and Captain Wakeman," *American Literature,* 33 (November 1961), 320–9. Browne has also edited an edition of *Stormfield,* providing it with an illuminating introduction. See Mark Twain, *Mark Twain's Quarrel with Heaven,* ed. R. B. Browne (New Haven, Conn., 1970).

123 Robert A. Rees, *"Captain Stormfield's Visit to Heaven* and *The Gates Ajar." English Language Notes,* 7 (March 1970), 197–202.

124 Brashear, *Twain,* p. 208.

125 In the 1880s, Twain remarked that it was the "only book that he had been pertickularly anxious to write. . ." As recorded by his daughter Susy in her youthful biography of Twain and as quoted in Wecter, "Introduction," p. xxi.

126 Wecter, "Introduction," p. xxii.

127 G. W. Warder, *The Cities of the Sun* (New York, 1903), p. 11.

128 For passages from Twain's unpublished review, see Wecter's "Introduction" in Twain, *Report*, pp. ix–x.

129 Mark Twain, "Was the World Made for Man?" in Twain's *What is Man? and Other Philosophical Writings*, ed. Paul Baender (Berkeley, 1973), pp. 101–6.

130 Mark Twain, "Letters from the Earth," in Twain, *Philosophical Writings*, pp. 401–54:413. This essay was first published in 1962. I am indebted to Thomas Werge of the University of Notre Dame for calling both this and the last Twain essay to my attention and for helpful comments on a number of American authors.

131 E. T. Winkler, "Religion and Astronomy," *Baptist Quarterly*, 5 (1871), 58–74:58.

132 That it would be erroneous to conclude from Miller's advocacy of pluralism that all American Methodists were of life mind is indicated by the fact that Luther Tracy Townsend (1838–1922), a prominent Methodist theologian, published an antipluralist book in 1914; see Townsend's *The Stars Not Inhabited* (New York).

133 [Enoch Fitch Burr], *Ecce Coelum; or, Parish Astronomy*, 20th ed. (New York, n.d.), pp. 15, 197; see also p. 165. On Burr's life, see Marc Rothenberg, *The Educational and Intellectual Background of American Astronomers* (a 1974 Bryn Mawr doctoral dissertation), pp. 238–40.

134 E. F. Burr, "Are the Heavens Inhabited?" *Presbyterian Review*, 6 (1885), 265–67. See also Chapter 16 of Burr's *Celestial Empires* (New York, 1885). Burr also championed pluralism in "Astronomy as a Religious Helper," *Homiletic Review*, 23 (1892), 202–10; 24 (1892), 394–402.

135 W. Leitch, *God's Glory in the Heavens*, 3rd ed. (London, 1867). For biographical information on Leitch, see *Dictionary of Canadian Biography*, vol. IX (Toronto, 1976), pp. 461–2.

136 Charles J. Powers, "Father Searle's Distinguished Career," *America*, 19 (1918), 378–80; see also *Catholic World*, 107 (1918), 713–16.

137 G. M. Searle, "The Plurality of Worlds," *Catholic World*, 37 (1883), 49–58:55–6.

138 G. M. Searle, "Are the Planets Habitable?" *Astronomical Society of the Pacific Publications*, 2 (1890), 165–77:169.

139 G. M. Searle, "Is There a Companion World to Our Own?" *Catholic World*, 55 (1892), 860–78:863–4.

140 J. De Concilio, "The Plurality of Worlds," *American Catholic Quarterly Review*, 9 (April 1884), 193–216:196, 211. For biographical information on De Concilio and the other priests in this controversy, see the articles on them in *New Catholic Encyclopedia*.

141 Thomas Aquinas, *Summa Theologica*, Part I, Question 47, Article 3.

142 T. Hughes, "Quid Est Homo? A Query on the Plurality of Worlds," *American Catholic Quarterly Review*, 9 (1884), 452–70:458. This paper was republished in a somewhat shorter form in Hughes, *The Plurality of Worlds and Other Essays*, compiled by M. G. Chadwick (New York. 1927), pp. 1–24.

143 J. De Concilio, *Harmony between Science and Revelation* (New York, 1889), p. 206.

144 A. F. Hewit, "Another Word on Other Worlds," *Catholic World*, 56 (October 1892), 18–26:18–19. On Hewit, see *Catholic World*, 65 (1897), I–XVI.

145 It is difficult if not impossible to provide in this context a precise definition of "pluralist" and of "antipluralist," I have classified as antipluralist those who adopted positions opposed to the pluralist tradition, even if they did not deny all extraterrestrial life.

146 J. O. Bevan, "Arguments for and against the Plurality of Worlds," *Birmingham Philosophical Society Proceedings*, 2 (May 1880), 162–75:171.

147 Edith D. Sylla of North Carolina State University has recently examined a large number of medieval writings without finding any evidence of these speculations.

148 At the end of my copy of Miller's *Heavenly Bodies* appear favorable passages reprinted from sixteen reviews.

149 Anonymous, "The Heavenly Bodies," *Knowledge*, 3 (June 15, 1883), 361. The style is certainly that of Proctor.

150 For biographical information on Ball, see W. Valentine Ball (ed.), *Reminiscences and*

Letters of Sir Robert Ball (London, 1915), and *Royal Astronomical Society Monthly Notices,* 75 (1915), 230–6.

151 As quoted in Hector Macpherson, *Astronomers of To-Day* (London, 1905), p. 102.

152 Robert S. Ball, *In Starry Realms* (Philadelphia, 1892), p. 344.

153 R. S. Ball, *The Story of the Heavens* (London, 1885), p. 80.

154 R. S. Ball, *In the High Heavens,* "new ed." (London, 1894), p. 32.

155 R. S. Ball, "Possibility of Life on Other Worlds," *McClure's Magazine,* 5 (1895), 147–56:152. This paper was also published in *Fortnightly Review,* 62 (1894), 718–27; in *Littell's Living Age,* 203 (1894), 742–50; and in *Eclectic Magazine,* 123 (1894), 818–26.

156 For biographical information, see Hector Macpherson, "John Ellard Gore," *Popular Astronomy,* 18 (November 1910), 519–29, and also Macpherson, *Astronomers,* pp. 145–55. Macpherson's statements (pp. 150–1) on Gore's pluralist views are not fully accurate.

157 Simon Newcomb, *Reminiscences of an Astronomer* (London, 1903), p. 19. See also Arthur Norberg, "Simon Newcomb's Early Astronomical Career," *Isis,* 69 (1978), 209–25.

158 Simon Newcomb, *Popular Astronomy,* 7th edition of the "School Edition" (New York, 1892?), pp. 528–31.

159 Simon Newcomb, "The Problems of Astronomy," *Annual Report of the Smithsonian Institution for 1896* (Washington, 1898), 83–93:93; 639–40; "Are Other Worlds Inhabited?" *Youth's Companion,* 76 (December 11, 1902), 639–40; [Newcomb], "Wallace on Life in the Universe," *Nation* 78 (January 14, 1904), 34–5; "Life in the Universe," *Harper's Magazine,* 111 (August 1905), 404–8.

160 Charles A. Young, "Astronomical Facts and Fancies for Philosophical Thinkers," *Christian Philosophy Quarterly,* 1 (January 1882), 1–22:22.

161 *Collected Papers of Charles Sanders Peirce,* vol. VIII (Cambridge, Mass., 1958), pp. 45–6.

162 H. A. Howe, "The Habitability of Other Worlds," *Sidereal Messenger,* 4 (1885) 294–8:295.

163 A. Winchell, *World Life or Comparative Geology* (Chicago, 1883), p. 5. On Winchell, see F. Garvin Davenport, "Alexander Winchell: Michigan Scientist and Educator," *Michigan History,* 35 (1951), 185–201.

164 Winchell, *World Life,* p. 507. In his earlier *Geology of the Stars* (Boston, 1874), Winchell had also advocated the nebular hypothesis, inferring from it that "every planet, since it must pass through the same succession of states, must attain, at some time, the habitable state." (p. 8).

165 Andrew Oldenquist, "John Fiske," in Paul Edwards (ed.), *Encyclopedia of Philosophy,* vol. III (New York, 1967), pp. 204–5:204.

166 John Fiske, *Outlines of Cosmic Philosophy,* 14th ed., vol. I (Boston, 1892), p. 400n. Fiske's position may have been based in part on Winchell's *The Geology of the Stars,* which Fiske (p. 400n) praises as "lucid and suggestive," although faulting it for a statement concerning the moon that Fiske misunderstood. See Winchell, *World Life,* p. 503.

167 William Graves Hoyt, *Lowell and Mars* (Tucson, 1976), p. 17.

168 John Pratt, "The Cost of Life," *Popular Science Monthly,* 23 (June 1883), 202–7:203. Pratt's paper was marred by some errors of calculation that were pointed out by W. H. Pickering in "Surface Conditions on the Other Planets," *Science,* 2 (July 6, 1883), 10.

169 C. Morris, "The Variability of Protoplasm," *American Naturalist,* 17 (1883), 926–31:930.

170 For E. T. Cope's views, see his "On Archaesthetism," *American Naturalist,* 16 (June 1882), 454–69:464ff, and his "The Evidence for Evolution in the History of the Extinct Mammalia," *Science,* 2 (August 31, 1883), 272–9.

171 D. T. MacDougal, "Life on Other Worlds," *Forum,* 27 (1899), 71–7:76; also printed in *Popular Astronomy,* 7 (1899), 420–6.

172 Carl A. Stetefeldt, "Can Organic Life Exist in the Planetary System outside of the Earth?" *Astronomical Society of the Pacific Publications,* 6 (1897), 91–100:91.

173 [Camille Flammarion], "Can Organic Life Exist in the Solar System Anywhere but on the Planet *Mars?*" *Astronomical Society of the Pacific Publications,* 6 (1894), 214–17.

174 Edwin C. Mason, "Life on Other Worlds," *Popular Astronomy,* 6 (1898), 520–4:522.

175 W. H. Pickering, "Surface Conditions on the Other Planets," *Science,* 2 (July 6, 1883), 10.

176 Charles Etler, "Is the Sun Habitable?" *Dominion Review,* 4 (March 1896), 9–15:9.

177 L. F. Ward, "Relation of Sociology to Cosmology," *Outlines of Sociology* (New York, 1923 reprinting of the 1897 original), pp. 21–42. This chapter appeared under the same title in *American Journal of Sociology,* 1 (1895), 132–45, and in *Dominion Review,* 3 (1898), 141–6, 157–61.

178 For details, see William Graves Hoyt, *Lowell and Mars* (Tucson, 1976).

179 As quoted in W. L. Webb, *Brief Biography and Popular Account of the Unparalleled Discoveries of T. J. J. See* (Lynn, Mass., 1913), p. 257.

180 J. Ashbrook, "The Sage of Mare Island," *Sky and Telescope,* 24 (October 1962), 193–202:202.

181 For these claims, see Hoyt, *Lowell and Mars,* p. 122; Helen Wright, *Explorer of the Universe: A Biography of George Ellery Hale* (New York, 1966), p. 118; and John Lankford, "A Note on T. J. J. See's Observations of Craters on Mercury," *Journal for the History of Astronomy,* 11 (1980), 129–32:130.

182 T. J. J. See, "Recent Discoveries Respecting the Origin of the Universe," *Atlantic Monthly,* 80 (1897), 484–92:491.

183 Forest R. Moulton, "The Limits of Temporary Stability of Satellite Motion, with an Application to the Question of an Unseen Body in the Binary System F. 70 Ophiuchi," *Astronomical Journal,* 20 (May 15, 1899), 33–7. See's intemperate and in fact dishonest reply to Moulton led the editor of the journal, Seth B. Chandler, to point out See's dishonesty and, in effect, to announce that future papers by See would not be welcome. See *Astronomical Journal,* 20 (1899), 56.

184 [Herbert H. Turner], "On the Fundamental Law of Increase of Gaseous Reputation," *Observatory,* 22 (1899), 292.

185 G. P. Serviss, "Are There Planets among the Stars?" *Popular Science Monthly,* 52 (December 1897), 171–6:173.

186 Clyde Fisher, "Garrett P. Serviss," *Popular Astronomy,* 37 (1929), 365–9.

187 C. Delaunay, "Notice sur la constitution de l'univers," *Annuaire du bureau des longitudes pour l'an 1869* (Paris), p. 447. On the possibility of Flammarion's influence, see C. Flammarion, *Les mondes imaginaires et les mondes réels,* 20th ed. (Paris, 1882?), pp. 571–2.

188 A. Guillemin, *The Heavens,* ed. J. Norman Lockyer, revised by Richard Proctor (London, 1872), pp. 66, 186, 205, 227.

189 A. Guillemin, *The Sun,* trans. A. L. Phipson (New York, 1870), p. 297.

190 A. Guillemin, *The World of Comets,* translated and edited by James Glaisher (London, 1877), p. 520.

191 A. Guillemin, *Autres mondes* (Paris, 1892), p. 227.

192 Hervé Faye, "Sur la constitution physique du soleil," *Annuaire du bureau des longitudes pour l'an 1874,* pp. 407–90:478. The final section, "Conditions astronomiques de la vie" (pp. 476–90), was reported upon, in fact, largely translated as "Habitable Worlds," *All the Year Round,* 32 (May 23, 1874), 127–30.

193 Hervé Faye, *Sur l'origine du monde,* 2nd ed. (Paris, 1885), pp. 299–300.

194 Louis Olivier, "La vie et les milieux cosmiques," *Astronomie,* 1 (1882), 379–84.

195 First published as Julius Scheiner, "Die Bewohnbarkeit der Welten," *Himmel und Erde,* 3 (1891), 18–32, 65–78; this paper appeared in French as "L'habitabilité des mondes," *Astronomie,* 10 (1891), 211–27, with annotations by Flammarion. It later appeared, with even more annotations, as Chapter X of Camille Flammarion, *The Dreams of an Astronomer,* trans. E. E. Fournier D'Albe (New York, 1922), pp. 179–222. References are to the French printing.

196 Scheiner, "L'habitabilité," p. 219. Flammarion, in a footnote, urges that the first two conditions are overly restrictive.
197 Scheiner, "L'habitabilité," pp. 226–7. The obvious weakness in this probability argument was noted in an 1891 article in *Cosmos* by Charles de Kirwan, whose article was discussed earlier.
198 C. Flammarion, "Hommes et femmes planétaires," *Astronomie,* 11 (1892), 243–9:243.
199 G. Bovier-Lapierre, *Astronomie pour tous* (Paris, 1891), pp. 308–12.
200 D. Papp, *Was lebt auf den Sternen?* (Zurich, 1931), p. 333.
201 M. Wilhelm Meyer, "The Urania Gesellschaft," *Astronomical Society of the Pacific Publications,* 2 (1890), 143–52.
202 Willy Ley, *Watchers of the Skies* (London, 1964), pp. 490–1. Ley does not identify the publications in which these speculations appeared.
203 Wilhelm Schur, "Welche Planeten können lebenden Wesen bewohnt sein?" *Deutsche Revue,* 24 (1899), 280–6:280–2.
204 Joseph Pohle, *Die Sternenwelten und ihre Bewohner,* 2nd ed. (Cologne, 1899), p. 298.

CHAPTER 10

Abbreviations for journals frequently cited in this chapter

AAP= *Astronomy and Astro-Physics*
AJ= *Astronomical Journal*
ALO= *Annals of the Lowell Observatory*
AN= *Astronomische Nachrichten*
ApJ= *Astrophysical Journal*
AR= *Astronomical Register*
ASPP= *Astronomical Society of the Pacific Publications*
BAAJ= *British Astronomical Association Journal*
BAAM= *British Astronomical Association Memoirs*
CR= *Comptes rendus des séances de l'académie des sciences*
EM= *English Mechanic and World of Science*
PA= *Popular Astronomy*
PSM= *Popular Science Monthly*
RASM= *Royal Astronomical Society Memoirs*
RASMN= *Royal Astronomical Society Monthly Notices*
SA= *Scientific American*
SAFB= *Société astronomique de France bulletin*
SAS= *Scientific American Supplement*
SM= *Sidereal Messenger*

1 According to T. W. Webb, "Mars," *Nature,* 27 (December 28, 1882), 203–5:205, François Terby had collected 1,092 drawings of Mars for his historical study "Étude comparative des observations faites sur l'aspect physique de la planète Mars depuis Fontana (1636) jusqu'à nos jours (1873)," *Mémoires de l'académie royale des sciences de Belgique,* 39 (1875).
2 John Brett, "The Physical Condition of Mars," *RASMN,* 38 (December 1877), 58–61:59.
3 For an early speculation by a pluralist on the possibility of Martian moons, see William Derham, *Astro-Theology,* 2nd ed. (London, 1715), p. 185.
4 Asaph Hall, "The Discovery of the Satellites of Mars," *RASMN,* 38 (February 1878), 205–9:205.
5 I. S. Shklovskii and Carl Sagan, *Intelligent Life in the Universe* (New York, 1966), pp. 373–6. According to a report in *Astronomy,* 5 (January 1977), 5, Shklovskii has now become an opponent of extraterrestrial life claims.
6 Hector Macpherson, Jr., "Giovanni Schiaparelli," *PA,* 18 (1910), 467–74:473, and W. Alfred Parr, "Giovanni Schiaparelli," *Knowledge,* n.s., 7 (November 1910), 466. These statements are not uncharacteristic of claims made in other obituaries, e.g.,

those by Percival Lowell, *PA*, 18 (1910), 457–67; by R. A. S. in *Royal Society Proceedings*, 83 (1911–12), xxxvii–xxxviii; by R. G. Aitken in *ASPP*, 20 (1910), 164–5; by G. Celoria in *AN*, 185 (1910), 193–6; by an anonymous author in *Observatory*, 33 (1910), 311–14; by A. in *ApJ*, 32 (1910), 313–19; by E. B. K. [E. B. Knobel?] in *RASMN*, 71 (1911), 282–7; and by Elia Millosevich in *Gli scienziati Italiani*, 1 (1921), 45–67.

7 E. W. Maunder, *Are the Planets Inhabited?* (London, 1913), pp. 61–2, and E. M. Antoniadi, *La planète Mars* (Paris, 1930), p. 33, respectively.

8 R. A. S., "Schiaparelli," p. xxxviii.

9 For a long propluralist quotation from Encke, see James Breen, *The Planetary Worlds* (London, 1854), pp. 251–2.

10 G. Schiaparelli, "Gli abitanti di altri mondi," in Schiaparelli, *Le opere*, 10 vols., vol. X (New York, 1968 reprinting of the Milan 1930 original), pp. 166–9:168–9.

11 G. Schiaparelli, *Corrispondenza su Marte*, vol. I (Pisa, 1963), pp. 14–18:15.

12 G. Schiaparelli, "Sulla rotazione di Mercurio," in Schiaparelli, *Opere*, vol. V, pp. 323–35:327. This paper first appeared in *AN* in 1889. My discussion is based on it and on his 1889 presentation to the Academy of the Lynxes: "Sulla rotazione e sulla constituzione fisica del pianeta Mercurio," *Opere*, vol. V, pp. 337–45. References to the latter paper are to its English translation: "The Rotation and Physical Constitution of the Planet Mercury," in Edward S. Holden (ed.), *Essays in Astronomy* (New York, 1900), pp. 133–42.

13 Werner Sandner, *The Planet Mercury*, trans. Alex Helm (New York, 1963), pp. 30–1.

14 G. H. Pettengill and R. B. Dyce, "A Radar Determination of the Rotation of the Planet Mercury," *Nature*, 206 (June 19, 1965), 1240.

15 R. B. Dyce, G. H. Pettengill, and I. I. Shapiro, "Radar Determination of the Rotations of Venus and Mercury," *AJ*, 72 (April 1967), 351–9.

16 For details, see Clark R. Chapman, *The Inner Planets* (New York, 1977), pp. 72–7.

17 Dyce, Pettengill, and Shapiro, "Determination," p. 352.

18 Schiaparelli, "Rotazione," pp. 334–5.

19 Schiaparelli, "Rotation," p. 138.

20 Schiaparelli, "Rotation," pp. 140–1.

21 Edward S. Holden, "Announcement of the Discovery of the Rotation Period of Mercury, by M. Schiaparelli," *ASPP*, 2 (1890), 79–82:82.

22 Ellen M. Clerke, "The Planet Mars," *The Month*, 76 (1892), 185–99:188. Ellen Clerke was the sister of the better known Agnes Clerke.

23 G. Schiaparelli, "Osservazioni astronomiche e fisiche sull'asse di rotazione e sulla topografia del pianeta Marte . . . durante l'opposizione del 1877. – MEMORIA PRIMA," in Schiaparelli, *Opere*, vol. I, pp. 11–175. The discussion of this and other pre-1901 publications on Mars is based in part on Camille Flammarion, *La planète Mars et ses conditions d'habitabilité*, vol. I (Paris, 1892) and vol. II (Paris, 1909). This thorough but by no means complete survey of the history of Martian observation up to 1901 includes both historical information on and French versions of many of the most significant writings on Mars known to its author.

24 On the history of Martian nomenclature, see Jurgen Blünck, *Mars and Its Satellites: A Detailed Commentary on the Nomenclature* (Hicksville, N.Y., 1977), and Virginia W. Capen, "History of Martian Nomenclature," *Astronomy*, 3 (April 1975), 20–9.

25 Flammarion, *Mars*, vol. I, pp. 135–6. For evidence that Schiaparelli knew Secchi's publication, see Schiaparelli, "MEMORIA PRIMA," *Opere*, vol. I, pp. 62, 64, 71, 168.

26 Schiaparelli, *Corrispondenza*, vol. I, p. 99.

27 C. Flammarion, "Mars Inhabited, like Our Own Earth," *Kansas City Review*, 3 (1879–80), 86–90, 156–60:157.

28 C. Flammarion, "La planète Mars," *Astronomie*, 1 (1882), 161–75, 206–16, 256–68.

29 G. Schiaparelli, *Corrispondenza*, vol. I, p. 12.

30 Arthur Conan Doyle, *The History of Spiritualism*, vol. II (New York, 1975 reprinting of the 1926 edition), pp. 13, 190–2. It is claimed in an anonymous article, "Commun-

ication with Mars," *Independent,* 66 (1909), 1042–3:1043, that Schiaparelli was a student of the occult and "a believer in the supernatural powers of Eusapio Paladino." In his obituary of Percival Lowell [*SAFB,* 30 (1916), 422–3:423], Flammarion stated: "Like Schiaparelli, like the other observers of Mars that I know . . . Lowell was passionately interested in psychic researches."

31 Percival Lowell, *Mars and Its Canals* (New York, 1911), p. 27.

32 Flammarion, *Mars,* vol. I, pp. 313–14.

33 As quoted in Flammarion, *Mars,* vol. I, p. 316.

34 N. E. Green, "Observations of Mars, at Madeira, in August and September, 1877," *RASM,* 44 (1877–9), 123–40. This memoir actually appeared around September 1879. See *AR,* 17 (1879), 237.

35 *Nature,* 18 (May 9, 1878), 55. For background, see John Burnett, "British Studies of Mars: 1877–1914," *BAAJ,* 89 (1979), 136–43.

36 As reported in *AR,* 16 (May 1878), 122–3.

37 As reported in *AR,* 17 (1879), 15–17. The repeated use in this report of the word "canal" to designate the features seen by Schiaparelli seems to be its first usage in this way in an English-language publication.

38 N. E. Green, "On Some Changes in the Markings of Mars, since the Opposition of 1877," *RASMN,* 40 (March 1880), 331–2:332.

39 N. E. Green, "Mars and the Schiaparelli Canals," *Observatory,* 3 (1879), 252.

40 C. E. Burton, "The Canals of Mars," *AR,* 18 (1880), 116.

41 N. E. Green, "The 'Canals' of Mars," *AR,* 18 (1880), 138.

42 C. E. Burton, "Physical Observations of Mars," *Royal Dublin Society Scientific Transactions,* ser. 2, 1 (1877–83), 151–72:170; see Plate VIII.

43 F. Terby, "The Markings on Mars," *Observatory,* 3 (1880), 416.

44 T. W. Webb, "Mars," *Nature,* 21 (January 1, 1880), 212–13. This claim is supported by an 1879 letter from Schiaparelli to Terby in which Schiaparelli admitted to his eyes "having only a little sensitivity to the nuances of color." Schiaparelli, *Corrispondenza,* vol. I, p. 28.

45 Flammarion, *Mars,* vol. I, p. 351.

46 As reported in *AR,* 20 (1882), 111.

47 G. Schiaparelli, "Osservazioni sulla topografia del pianeta Marte," *Opere,* vol. I, pp. 379–88:385. For an English translation, see *PSM,* 24 (December 1883), 249–53.

48 G. Schiaparelli, "Découvertes nouvelles sur la planète Mars," *Opere,* vol. I, pp. 389–94:392–3.

49 R. A. Proctor, "Canals on the Planet Mars," *Times* (London) (April 13, 1882), p. 12. Proctor republished both Webb's letter and his own letter in his journal *Knowledge,* 1 (April 14, 1882), 519.

50 Accounts of this discussion appeared in *Observatory,* 5 (1882), 135–7, and in *AR,* 20 (1882), 110–11.

51 C. E. Burton, "Canals of Mars," *AR,* 20 (1882), 142.

52 F. Terby, "Remarkques à propos des récentes observations de M. Schiaparelli sur la planète Mars," *RASMN,* 42 (1882), 382–3.

53 A. M. Clerke, *Popular History of Astronomy during the Nineteenth Century* (Edinburgh, 1885), p. 324.

54 On Maunder, see the obituaries by H. P. H. [H. P. Hollis?] in *BAAJ,* 38 (May 1928), 229–33 (also 165–8), and in *RASMN,* 89 (February 1929), 313–18; by E. M. Antoniadi in *SAFB,* 42 (1928), 240–2; by A. C. D. Crommelin in *Observatory,* 51 (May 1928), 157–9; and the anonymous obituary in *Nature,* 121 (April 7, 1928), 545–6.

55 E. W. Maunder, "The 'Canals' of Mars: A Reply to Mr. Story," *Knowledge,* 1, n.s. (May 1904), 87–9:87.

56 George B. Airy, "Physical Observations of Mars, Made at the Royal Observatory, Greenwich," *RASMN,* 38 (1877), 34–6.

57 E. W. Maunder, "Is Mars Habitable?" *Sunday Magazine,* 11, n.s. (1882), 102–4, 170–2:172. See also his "The Red Planet, Mars," *Sunday Magazine,* 11, n.s. (1882), 30–3.

58 E. W. Maunder, "The Conditions of Habitability of a Planet: With Special Reference to the Planet Mars," *Journal of the Transactions of the Victoria Institute,* 44 (1912), 78–94:94.

59 For Maunder's religious affiliation, see H. P. H., "Maunder," *BAAJ,* 38 (May 1928), 229–33:233. The Catholic Apostolic church is a small, somewhat secretive sect wherein pentecostal and adventist elements are combined with High Anglican creed and liturgy. See P. E. Shaw, *The Catholic Apostolic Church, Sometimes Called Irvingite* (Morningside Heights, N.Y., 1946). Maunder's strong religious concerns are also shown by the facts that he published *The Astronomy of the Bible* (London, 1908) and from 1913 to 1918 served as secretary of the Victoria Institute, a society devoted to studying the relations between faith and reason.

60 E. W. Maunder, *Are the Planets Inhabited?* (London, 1913), pp. 161–2.

61 Edmund Ledger, "The Canals of Mars – Are They Real?" *Nineteenth Century,* 53 (1903), 773–85:775. For Schiaparelli's report on his 1886 observations, see his *Opere,* vol. II, pp. 169–229.

62 Henri Perrotin, "Observations des canaux de Mars," *Bulletin astronomique,* 3 (July 1886), 324–9.

63 G. Schiaparelli, *Corrispondenza,* vol. I, p. 153.

64 Anonymous, "The 'Canals' of Mars," *Nature,* 34 (June 3, 1886), 110; reprinted in *SAS,* 22 (July 10, 1886), 8774.

65 W. F. Denning, "The Physical Appearance of Mars in 1886," *Nature,* 34 (June 3, 1886), 104; reprinted in *SAS,* 22 (July 10, 1886), 8774. Also reprinted in *Astronomie,* 5 (September 1886), 321ff, and in Flammarion, *Mars,* vol. I, pp. 387–91, which includes four of Denning's drawings.

66 H. C. Wilson, "The Planets," *SM,* 7 (1888), 400–4. In this paper, Wilson reported seeing three canals in 1886.

67 Anonymous, "Life in Mars," *Chambers's Journal,* 63 (June 12, 1886), 368–70. The mention of the Martians' "large engineering works" was no doubt the reason this article was reprinted in *The American Architect and Building News,* 20 (August 7, 1886), 66–7.

68 Edward S. Holden, "Physical Observations of Mars during the Opposition of 1888, at the Lick Observatory," *AJ,* 8 (September 14, 1888), 97–8:98.

69 H. Perrotin, "Les canaux de Mars. Nouveaux changements observés sur cette planète," *Astronomie,* 7 (1888), 213–15.

70 H. Perrotin, "Sur la planète Mars," *CR,* 106 (1888), 1718–19.

71 Flammarion, in his "Wonders in Mars," in *Littell's Living Age,* 178 (1888), 252–6:253, stated: "The flood has been observed not only by myself but by M. Perrotin . . . by Signor Schiaparelli, and . . . by M. [T]erby."

72 Anonymous, "The Markings of Mars," *Nature,* 38 (October 18, 1888), 601. For other accounts, see Anonymous, "Inundations on the Planet Mars," *Chambers's Journal,* 5th ser., 5 (August 25, 1888), 529–31, and Garrett P. Serviss, "The Strange Markings on Mars," *PSM,* 35 (1889), 41–56.

73 G. Schiaparelli, "Ueber die beobachteten Erscheinungen auf der Oberflaeche des Planeten Mars," *Opere,* vol. II, 1–46:12–13; first appeared in *Himmel und Erde* in 1888.

74 Schiaparelli, *Corrispondenza,* vol. I, p. 226.

75 A. Hall, "The Appearance of Mars, June 1888," *AJ,* 8 (August 14, 1888), 79.

76 Holden, "Mars . . . 1888," p. 97 and Plate II.

77 Simon Newcomb, *Reminiscences of an Astronomer* (London, 1903), pp. 143–4.

78 Holden, "Mars . . . 1888," p. 98.

79 C. Flammarion, "Observations de Mars faites à l'observatoire Lick à l'aide de la plus puissante lunette du mond," *Astronomie,* 8 (1889), 180–4:180.

80 Flammarion, *Mars,* vol. I, pp. 396–7, 411–12.

81 C. Flammarion, "Les inondations de la planète Mars," *Astronomie,* 7 (1888), 241–53:245.

82 C. Flammarion, "Les inondations," p. 252. He also supported Schiaparelli in "Un dernier mot sur la planète Mars," *Astronomie,* 7 (1888), 412–22.

83 C. Flammarion, "Les fleuves de la planète Mars," *SAFB,* 2 (1888), 111–15.

84 R. A. Proctor, "Note on Mars," *RASMN,* 48 (April 1888), 307–8.

85 R. A. Proctor, "Maps and Views of Mars," *SAS,* 26 (October 13, 1888), 10659–60; reprinted from *Knowledge.*

86 R. A. Proctor, *Old and New Astronomy,* completed by A. Cowper Ranyard, new ed. (London, 1895), pp. 544–7. A somewhat similar explanation of the doubling of canals based on refractions in the water vapor above them was put forward by the Halle astronomer Ferdinand Meisel in his "Essai d'une explication optique du dédoublement des canaux de Mars," *Astronomie,* 8 (1889), 461–4. For a critique, see J. Schneider, "Sur l'explication optique de dédoublement des canaux de Mars," *Astronomie,* 9 (1890), 49–50.

87 W. H. Pickering, "The Physical Aspect of the Planet Mars," *Science,* 12 (August 17, 1888), 82–4.

88 H. Fizeau, "Sur les canaux de la planète Mars," *CR,* 106 (1888), 1759–62; for Janssen's comments on this paper, see pp. 1762–4. C. Flammarion, "Les neiges, les glaces et les eaux de la planète Mars," *CR,* 107 (1888), 19–22.

89 E. W. Maunder, "The Canals of Mars," *Observatory,* 11 (September 1888), 345–8:348.

90 As quoted from Schiaparelli's June 8, 1888, letter to Terby, in Maunder, "Canals," pp. 347–8. Emphasis added.

91 E. W. Maunder, *Planets,* p. 62.

92 Schiaparelli, *Corrispondenza,* vol. I, p. 210. This letter is dated July 10, 1888.

93 C. A. Young, "The Planet Mars," *Presbyterian Review,* 10 (1889), 400–14:413.

94 Anonymous, "Observations de Mars à Washington," *Astronomie,* 9 (1890), 410. This note, probably written by Flammarion, concludes with the comment that this is "bien curieuse."

95 Flammarion, *Mars,* vol. I, pp. 477–81.

96 A. S. Williams, "Recent Observations of the Canals and Markings on Mars," *BAAJ,* 1 (November 1890), 82–90:88.

97 E. S. H., J. M. S., and J. E. K., "Note on the Opposition of Mars, 1890," *ASPP,* 2 (1890), 299–300. For letters from this period between Schiaparelli and Holden, as well as other letters in which Schiaparelli comments on the Lick results, see G. V. Schiaparelli, *Corrispondenza su Marte* (1890–1900), vol. II (Pisa, 1976).

98 Schiaparelli's June 12, 1890, letter, as quoted in F. Terby, "Nouvelles découvertes sur Mars," *Astronomie,* 9 (1890), 401–10:408.

99 C. Flammarion, "Variations cértaines observées sur la planète Mars," *SAFB* (Supplement), 4 (1890), 105–19. This paper, with minor changes, appeared as "New Discoveries on the Planet Mars," *Arena,* (1890–1), 275–90.

100 Flammarion, "New Discoveries," p. 286. See also W. H. Pickering, "Photographs of the Surface of Mars," *SM,* 9 (1890), 254–5; [Flammarion], "Observations de Mars faites par M. William Pickering," *Astronomie,* 9 (1890), 410–11; and Flammarion, *Mars,* vol. I, pp. 464–5, which includes two photographs.

101 Norriss S. Hetherington, "Amateur versus Professional: The British Astronomical Association and the Controversy over Canals on Mars," *BAAJ,* 86 (1976), 303–8; see also P. M. Ryves, "Mars Section," in "The History of the British Astronomical Association," *BAAM,* 36, part 2 (December 1948), 86–97.

102 N. Green, "The Canals of Mars," *BAAJ,* 1 (1890–1), 110–13:112. For the discussion, see pp. 113–14.

103 A. S. Williams, "The Canals of Mars," *BAAJ,* 1 (March 1891), 314–16.

104 A. de Boë, "Comment une ligne peut être vue double," *Ciel et terre,* 12 (1891–2), 223–4; letters by H. Schleusner, pp. 257, 307–9, and by F. Terby, p. 285. See also *EM* for July 24, August 7, 14, and 21, and September 4 and 18, 1891. Among these items, those signed "F. R. A. S." should be attributed to William Noble, who according to his obituary notice [*MNRAS,* 65 (1905), 342–3] wrote in *EM* for nearly forty years using these initials.

105 John Ritchie, Jr., "Our Knowledge of Mars," *SM,* 9 (1890), 450–4:454.

106 As quoted in Eugene M. Antoniadi, *La planète Mars* (Paris, 1930), p. 33. On January

27, 1895, Schiaparelli wrote Terby that during the 1890s he had been able to observe very little on Mars, being especially troubled by "a diminution of the sensibility to weak illuminations; I attribute this to the observations of Mercury near the sun carried out from 1882 to 1890. I have entirely abandoned this dangerous kind of observations." Schiaparelli, *Corrispondenza*, vol. II, pp. 166–7. On the other hand, on November 17, 1896, he wrote Lowell: "I have reason to believe that my eye is still tolerably good . . . ," and on November 30, 1896, he wrote Terby of his success in seeing a large number of canals. Schiaparelli, *Corrispondenza*, vol. II, pp. 213, 218–19.

107 Flammarion, *Mars*, vol. I, p. 592.

108 See, for example, the reviews by William J. S. Lockyer in *Nature*, 47 (April 13, 1893), 553–4; by Geoffrey Winterwood in *Good Words*, 34 (1893), 677–85; and by an unidentified author in *BAAJ*, 3 (1892–3), 48.

109 Asaph Hall, "Observations of Mars, 1892," *AJ*, 12 (February 8, 1893), 185–8:187.

110 E. S. Holden, "A Correction," *ASPP*, 4 (1892), 193–4.

111 E. S. Holden, "Note on the Mount Hamilton Observations of Mars, June-August, 1892," *AAP*, 11 (1892), 663–8:667. See also Holden, "Lick Observatory Drawings of *Mars*, 1892," *ASPP*, 5 (June 1893), 133–4.

112 E. S. Holden, "What We Really Know about Mars," *Forum*, 14 (November 1892), 359–68:368.

113 E. S. Holden, "Mr. Brett on the Physical Condition of Mars," *ASPP*, 2 (1890), 17–18, and in *Forum*, 14 (November 1892), 365–6.

114 J. M. Schaeberle, "Preliminary Note on the Observations of the Surface Features of Mars during the Opposition of 1892," *ASPP*, 4 (1892), 196–8.

115 G. Schiaparelli, "Distribution of Land and Water on *Mars*," *ASPP*, 5 (1893), 169–70, and Schaeberle, "Remarks on the Surface Markings of Mars," *ASPP*, 5 (1893), 170–3.

116 E. E. Barnard, "Preliminary Remarks on the Observations of Mars 1892 . . . ," *AAP*, 11 (1892), 680–4; see especially Plate XXXIV.

117 As quoted in John Lankford, "A Meeting of Giants: Milan, 1893," *Strolling Astronomer*, 27 (1979), 217–19:218–19.

118 J. E. Keeler, "Physical Observations of Mars, Made at the Allegheny Observatory in 1892," *RASM*, 51 (1892–5), 45–52; C. A. Young, "Observations of Mars at the Halsted Observatory, Princeton," *AAP*, 11 (1892), 675–8.

119 From Pickering's telegram of September 2, 1892, as quoted in E. S. Holden, "The Lowell Observatory at Arizona," *ASPP*, 6 (1894), 160–9:165.

120 Holden, "Lowell Observatory," p. 166. For Pickering's response and Holden's rejoinder, see *ASPP*, 6 (1894), 221–7. For the relations between the Pickerings in regard to the Arequipa expedition, see Bessie Zaban Jones and Lyle Gifford Boyd, *The Harvard College Observatory* (Cambridge, Mass., 1971), pp. 297–312.

121 Anonymous, "The Opposition of Mars," *BAAJ*, 2 (June 1892), 477.

122 W. H. Pickering, "Mars," *AAP*, 11 (1892), 668–75:669, 672. For his other Mars papers, see pp. 449–53, 545–68, 632, and 849–52.

123 Pickering's perception of green areas on Mars was probably a psychophysiological effect. It is noted in I. S. Shklovskii and Carl Sagan, *Intelligent Life in the Universe* (New York, 1966), p. 273, that when "a neutrally colored area is seen alongside a brightly colored area, it tends to acquire a complementary color. . . . The colors complementary to the red-orange Martian bright areas are greens and blues. . . ."

124 G. M. Searle, "Recent Discoveries in Astronomy," *Catholic World*, 57 (May 1893), 164–81:165.

125 E. W. Maunder, "Mars Section," *BAAJ*, 2 (1891–2), 423–7.

126 E. W. Maunder, "Report of the Mars Section, 1892," *BAAM*, 2 (1895), 157–98:162–3.

127 Anonymous, "The Opposition of Mars," *RASMN*, 53 (February 1893), 281–3.

128 H. Perrotin, "Observations de la planète Mars," *CR*, 115 (1892), 379–81. W. W. Campbell reported on pre-1894 projections, explaining them as reflections from Martian mountains, in his "An Explanation of the Bright Projections Observed on the Terminator of Mars," *ASPP*, 6 (1894), 102–12.

129 Schiaparelli, *Corrispondenza,* vol. II, pp. 66, 97.

130 Flammarion, *Mars,* vol. II, pp. 39, 76.

131 F. Terby, "Physical Observations of Mars," *AAP,* 11 (1892), 478–80:478, 555–7.

132 For Kingsmill, see *Nature,* 47 (December 8, 1892), 133, and *PSM,* 43 (1893), 281–2; for Peal, see *BAAJ,* 3 (1892–3), 223–4; *Canadian Magazine,* 1 (1893), 202–5; and *Science,* 21 (May 5, 1893), 242–3.

133 Hiram M. Stanley, "On the Interpretation of the Markings on Mars," *Science,* 20 (October 21, 1892), 235; Henry W. Parker, "Origins of the Lines of Mars," *Science,* 20 (November 18, 1892), 282–4; C. B. Warring, "The Gemination of the Lines in Mars," *Science,* 20 (September 23, 1892), 177–8, with commentary by W. J. Hussey, *Science,* 20 (October 21, 1892), 235.

134 S. Meunier, "Cause possible de la gémination des canaux de Mars; imitation expéri-mentale du phénomène," *CR,* 115 (1892), 678–80, 901–2. For foreign reports, see *Nature,* 47 (1982), 62, 133, and *SAS,* 35 (March 25, 1893), 14361–2.

135 E. W. Maunder, "The Climate of Mars," *Knowledge,* 15 (September 1, 1892), 167–9.

136 J. N. Lockyer, "The Opposition of Mars," *Nature,* 46 (September 8, 1892), 443–8.

137 R. S. Ball, "Mars," *ASPP,* 5 (1893), 23–6. This essay first appeared in somewhat fuller form in *Goldthwaite's Geographical Magazine* in 1892 and was reprinted in *Annual Report of the Smithsonian Institution for 1900* (Washington, 1901), pp. 157–66.

138 As quoted in Anonymous, "The Planet Mars," *Review of Reviews,* 6 (1892–3), 196–7.

139 E. M. Clerke, "The Planet Mars," *The Month,* 76 (1892), 185–99:199. See also W. J. Baker, "Do People Live on the Planet Mars?" *Chatauquan,* 17 (1893), 443–8.

140 G. Schiaparelli, "Il pianeta Marte," *Opere,* vol. II, pp. 47–74. Pickering's translation appeared first in *AAP,* 13 (1894), 635–40, 714–23. By the end of 1894, it had been reprinted in the *Smithsonian Institution Annual Report* and, with further abridge-ments, in *SA* and in *Nature.* By 1899, a Russian translation had been made. Refer-ences are to the *AAP* version.

141 See A. C. Ranyard's comment in *Knowledge,* 15 (October 1, 1892), 193.

142 Paul W. Merrill, "William Wallace Campbell," *RASMN,* 99 (1938–9), 317–21:321. See also the obituaries by Robert G. Aitken in *ASPP,* 50 (1938), 204–9, and the biography and bibliography by W. H. Wright in *National Academy of Sciences Bio-graphical Memoirs,* 25 (1949), 35–74.

143 Maunder, "Climate," p. 168.

144 W. W. Campbell, "The Spectrum of Mars," *ASPP,* 6 (1894), 228–36:230–1. On the history of this problem, see Campbell's "A Review of the Spectroscopic Observation of Mars," *ApJ,* 2 (1895), 28–44; Campbell, "A Review of the Spectroscopic Observa-tions of Mars," *Lick Observatory Bulletin,* 5, no. 169 (1909), 156–64; and David H. DeVorkin, "W. W. Campbell's Spectroscopic Study of the Martian Atmosphere," *Royal Astronomical Society Quarterly Journal,* 18 (1977), 37–53.

145 Campbell, "Review" (*ApJ*), p. 28.

146 Henry H. Bates, "The Chemical Constitution of *Mars'* Atmosphere," *ASPP,* 6 (1894), 300–2. W. W. Campbell, "Concerning an Atmosphere on *Mars,*" *ASPP,* 6 (1894), 273–83.

147 E. S. Holden, "The Latest News of Mars," *North American Review,* 160 (1895), 636–8:638.

148 William Huggins, "Note on the Spectrum of Mars," *AAP,* 13 (November 1894), 771.

149 W. Huggins, "The Spectrum of Mars," *AAP,* 13 (December 1894), 860, and W. W. Campbell, "On Selecting Suitable Nights for Observing Planetary Spectra," *AAP,* 13 (December 1894), 860–1; W. Huggins, "Note on the Atmospheric Bands in the Spectrum of Mars," *ApJ,* 1 (March 1895), 193–5; see also pp. 207–9.

150 J. Janssen, "Sur la présence de la vapeur d'eau dans l'atmosphère de la planète Mars," *CR,* 121 (1895), 233–7.

151 H. C. Vogel, "Recent Researches on the Spectra of the Planets. I," *ApJ,* 1 (1895), 196–209. See also Lewis E. Jewell, "The Spectrum of Mars," *ApJ,* 1 (1895), 311–17.

152 *Martha's Vineyard Herald,* August 25, 1894, as reprinted in "The Latest News about Mars," *PA,* 2 (October 1894), 92.

153 DeVorkin, "Campbell," pp. 40–1.

154 Hector Macpherson, *Astronomers of To-Day* (London, 1895), p. 195.

155 B. E. Cammell, "Report of the Section for the Observation of Mars," *BAAM,* 4 (1896), 107–37:111. Nathaniel Green condensed this report from the overly long report submitted by Cammell (p. 137).

156 E. W. Maunder, "The Canals of Mars," *Knowledge,* 17 (November 1, 1894), 249–52:250.

157 E. W. Maunder, "The 'Canals' of Mars," *Scientia,* 7 (1910), 253–69:263.

158 E. W. Maunder, "The 'Eye' of Mars," *Knowledge,* 18 (March 1, 1895), 54–9:56.

159 J. Orr, "The Nature of the 'Canals' of Mars," *BAAJ,* 5 (1895), 209, with the discussion following (pp. 209–10). Information on Orr is rare. Identified as Mr. J. Orr in the preceding publication, Orr may have been Miss M. A. Orr, who in 1901 presented an optical theory of the canals in *Knowledge,* 24 (February 1, 1901), 38–9.

160 The most thorough biography of Lowell is William Graves Hoyt, *Lowell and Mars* (Tucson, 1976). Although my discussion of Lowell differs in some ways from Hoyt's, this must not obscure the debt I owe to this careful scholar who has also enriched Lowell research by his *Planets X and Pluto* (Tucson, 1980) and by editing the microfilmed *Early Correspondence of the Lowell Observatory 1894–1916* (Flagstaff, 1973). Other biographical sources are Louise Leonard, *Percival Lowell: An Afterglow* (Boston, 1921); A. Lawrence Lowell, *Biography of Percival Lowell* (New York, 1935); and Ferris Greenslet, *The Lowells and Their Seven Worlds* (Boston, 1946), pp. 345–67.

161 For this explanation, see A. L. Lowell, *Lowell,* p. 60. William Sheehan has pointed out to me that this explanation is contradicted by Schiaparelli's 1909 letter to Antoniadi in which he dated his recognition of his failing eyesight from the late 1890s. For this letter, see E. M. Antoniadi, *La planète Mars* (Paris, 1930), p. 33. See also note 106 in this chapter.

162 As quoted in C. Flammarion, "Recent Observations of Mars," *SA,* 74 (February 29, 1896), 133–4:133. Translated from *L'illustration.* See also C. Flammarion, "Percival Lowell," *SAFB,* 30 (1916), 422–3:423.

163 A. L. Lowell, *Lowell,* p. 60.

164 Charles K. Hofling, "Percival Lowell and the Canals of Mars," *British Journal of Medical Psychology,* 37 (1964), 33–42:42.

165 This information was generously supplied to me by W. G. Hoyt of Lowell Observatory. Lowell also owned Proctor's *Essays on Astronomy* (1872 ed.), *Myths and Marvels of Astronomy* (1877 ed.), and *Poetry of Astronomy* (1881 ed.).

166 Hoyt, *Lowell,* pp. 23–6.

167 As quoted in Bessie Zaban Jones and Lyle Gifford Boyd, *The Harvard College Observatory: The First Four Directorships, 1839–1919* (Cambridge, Mass., 1971), p. 473; see also pp. 325–31.

168 On Douglass, see George E. Webb, *Tree Rings and Telescopes: The Scientific Career of A. E. Douglass* (Tucson, 1983).

169 P. Lowell, "The Lowell Observatory," *Boston Commonwealth* (May 26, 1894), 3–4:3.

170 E. S. Holden, "The Lowell Observatory, in Arizona," *ASPP,* 6 (June 1984), 160–9:160, 162, 165.

171 P. Lowell, *Mars* (Boston, 1895), p. v.

172 For a day-by-day listing of the canal observations during this period by the three observers, see *ALO,* 1 (1898), 101–85.

173 Lowell, *Mars,* pp. 145–8, 219–20, and Hoyt, *Lowell and Mars,* p. 63.

174 Hoyt, *Lowell and Mars,* pp. 68–9.

175 A. E. Douglass, "Canals in the Dark Regions and Terminator Observations," *ALO,* 1 (1898), 253–375; see especially pp. 253–8. Lowell seems to have hesitated to report these observations in his early papers, but the dark region canals are accepted in his *Mars* (December 1895).

176 W. H. Pickering, "The Seas of Mars," *AAP*, 13 (1894), 553–6:554.

177 Lowell, *Mars*, p. 52. Hoyt, *Lowell and Mars*, p. 74, notes that Lowell later revised this to one-twelfth, whereas current estimates are fifteen times lower still.

178 Lowell, *Mars*, pp. 60–75, and W. W. Campbell, "An Explanation of the Bright Projections Observed on the Terminator of Mars," *ASPP*, 6 (1894), 103–12.

179 As quoted in Hoyt, *Lowell and Mars*, p. 89.

180 Lowell, *Mars*, pp. 205–9.

181 Svante Arrhenius, *The Destinies of the Stars*, trans. J. E. Fries (New York, 1918), p. 226.

182 Wells's *War of the Worlds* and Lasswitz's *Auf zwei Planeten* both appeared in 1897. Wicks's *To Mars via the Moon*, dedicated to Lowell, followed in 1911, and in 1912 appeared Burrough's "Under the Moons of Mars." For further information, see Mark R. Hillegas, "The First Invasion from Mars," *Michigan Alumnus Quarterly Review*, 66 (Winter 1960), 107–12, as well as his "Martians and Mythmakers: 1877–1938," in *Challenges in American Culture*, ed. Ray B. Browne et al. (Bowling Green, Ohio, 1970), pp. 159–77, and his "Victorian 'Extraterrestrials,' " in *The Worlds of Victorian Fiction*, ed. Jerome Buckley (Cambridge, Mass., 1975), pp. 391–414. See also Roger Lancelyn Green, *Into Other Worlds*, (London, 1958), chs. 9 and 10, and William B. Johnson and Thomas D. Clareson, "The Interplay of Science and Fiction: The Canals of Mars," *Extrapolation*, 5 (May 1964), 37–48.

183 P. Lowell, "Mars," *AAP*, 13 (1894), 538–53, 645–50, 740, 814–21; *PA*, 2 (1894–5), 1–8, 52–6, 97–100, 154–60, 255–61, 343–8; *Atlantic Monthly*, 75 (1895), 594–603, 749–58; 76 (1895), 106–19, 223–35.

184 E. E. Hale, "Latest News from Mars," *ASPP*, 7 (1895), 116–18:117; reprinted from the March 2, 1895, *SA*, which reprinted it from the *Boston Commonwealth*.

185 As quoted in Hoyt, *Lowell and Mars*, p. 97.

186 W. W. Campbell, "[Review of] Mars. By Percival Lowell," *Science*, n.s., 4 (August 21, 1896), 231–8:232; reprinted in *ASPP*, 8 (1896), 207–20.

187 As quoted in Hoyt, *Lowell and Mars*, p. 90.

188 E. E. Barnard, "Micrometrical Measures of the Ball and Ring System of the Planet Saturn . . . ," *RASMN*, 56 (January 1896), 163–72:166–7.

189 A. E. Douglass, "The Lick Review of 'Mars,' " *Science*, n.s., 4 (September 11, 1896), 358–9. For Campbell's response, see his "Mr. Lowell's Book on 'Mars,' " *Science*, n.s., 4 (September 25, 1896), 455–6. For Lowell's role in the writing of Douglass's response, see Hoyt, *Lowell and Mars*, pp. 91–3.

190 A. E. Douglass, "The Lick Review of 'Mars,' " *PA*, 4 (October 1896), 199–201.

191 William W. Payne, "The Planet Mars," *PA*, 3 (1896), 345–8, 385–90.

192 Helen Wright, *Explorer of the Universe: A Biography of George Ellery Hale* (New York, 1966), p. 116.

193 G. E. Hale, "The Aim of the Yerkes Observatory," *ApJ*, 6 (November 1897), 310–21:320–1. I owe this point and a number of others to the valuable senior thesis completed in 1980 at Harvard by Nancy E. Gittleson; see her "The War of the Worlds: Percival Lowell and His Critics."

194 Simon Newcomb, "The Problems of Astronomy," *Science*, n.s., 5 (May 21, 1897), 777–85:784.

195 C. A. Young, "Is Mars Inhabited?" *ASPP*, 8 (1896), 306–13:306.

196 G. P. Serviss, "Facts and Fancies about Mars," *Harper's Weekly*, 40 (September 19, 1896), 926.

197 T. J. J. See, "The Red Planet Mars," *Dial*, 21 (July 16, 1896), 42–3.

198 William J. S. Lockyer, "Mars as Seen at the Opposition in 1894," *Nature*, 54 (October 29, 1896), 625–7.

199 [Agnes Clerke], "New Views about Mars," *Edinburgh Review*, 184 (1896), 368–85:368, 370. For evidence of Clerke's authorship, see Alfred Russel Wallace, *Is Mars Habitable?* (London, 1907), p. 21.

200 Anonymous, "The 'Edinburgh Review' on Mars," reprinted from the *Spectator*, in *Littell's Living Age*, 211 (1896), 732–5:734.

201 C. Flammarion, "La planète Mars," *Astronomie*, 13 (September 1894), 321–9:328.

202 C. Flammarion, "The Circulation of Water in the Atmosphere of Mars," *Knowledge,* 18 (April 1, 1895), 73–5:74; reprinted in *SAS,* 39 (April 27, 1895), 16112.

203 C. Flammarion, "La circulation de l'eau dans l'atmosphère de Mars," *SAFB,* 9 (1895), 169–76.

204 As quoted in A. L. Lowell, *Lowell,* p. 93.

205 C. Flammarion, "Recent Observations of Mars," *SA,* 74 (February 29, 1896), 133–4.

206 C. Flammarion, "Mars and Its Inhabitants," *North American Review,* 162 (May 1896), 546–57:551, 556.

207 Hoyt, *Lowell and Mars,* p. 79. Flammarion was addressed as "cher ami Martien" (Hoyt, *Lowell and Mars,* p. 329.)

208 G. Schiaparelli, "La vie sur la planète Mars," *SAFB,* 12 (1898), 423–9. For the Italian original, see Schiaparelli, *Opere,* vol. II, pp. 81–95. The commentary that follows is based primarily on the French version.

209 As quoted in Hoyt, *Lowell and Mars,* p. 289. On the political uses to which Mars was put, see Hoyt, pp. 288–90, and Norriss S. Hetherington, "Lowell's Theory of Life on Mars," *Astronomical Society of the Pacific Leaflet,* no. 409 (March 1971).

210 Schiaparelli, "Vie," p. 429. A hippogryph is a mythical flying horse.

211 As quoted by Flammarion in Schiaparelli, "Vie," p. 429. Flammarion's rather curious translation was: "Il est permis de dire des folies deux [!] fois par an."

212 M. W. Meyer, "Die Weltbild des Mars, wie es sich nach den Beobachtungen von 1892 und 1894 darstellt," *Himmel und Erde,* 8 (1896), 15–40. Three years earlier Meyer had published another major survey of Mars researches; see his "Die physische Beschaffenheit des Planeten Mars nach dem Zeugniss seiner hervorragendsten Beobachter," *Himmel und Erde,* 5 (1893), 410–26, 461–72, 505–15, 553–64.

213 Schiaparelli, *Corrispondenza,* vol. II, p. 137.

214 Schiaparelli, *Corrispondenza,* vol. II, p. 192; see also p. 259.

215 Schiaparelli, *Corrispondenza,* vol. II, pp. 299–300, 313; see also p. 72 for an 1892 suggestion that gemination is an atmospheric effect.

216 See Michael Heim, *Spiridion Gopčević: Leben und Werk* (Wiesbaden, 1966), especially ch. 11, and Joseph Ashbrook, "The Curious Career of Leo Brenner," *Sky and Telescope,* 56 (December 1978), 515–16.

217 See, for example, Leo Brenner, "Charts of Mars," *BAAJ,* 4 (1984), 439.

218 Leo Brenner, *Spaziergange durch das Himmelszelt* (Leipzig, 1898), p. 144.

219 Heim, *Gopčević,* p. 133.

220 Heim, *Gopčević,* p. 2–3.

221 S. P. Leland, *World Making* (Chicago, 1895), pp. 68–9. On the title page of the 17th edition (1906) of Leland's book, he is identified as "Emeritus Professor of Astronomy . . . in Charles City College."

222 P. Lowell, "Detection of Venus' Rotation Period and of the Fundamental Physical Features of the Planet's Surface," *PA,* 4 (December 1896), 281–6:282, 284.

223 Hoyt, *Lowell and Mars,* p. 110. Soon, however, both Douglass and See published supportive drawings. See A. E. Douglass, "The Markings of Venus," *RASMN,* 58 (1898), 382–5, and T. J. J. See, "The Study of Planetary Detail," *PA,* 4 (1897), 550–5.

224 S. Maitland Baird Gemmill, "The Martian Canals," *EM,* 67 (May 27, 1898), 333–4. For this controversy, see also the numerous articles and letters in the 1898 issues of *EM,* and E. M. Antoniadi, "Notes on the Rotation Period of Venus," *RASMN,* 58 (1898), 313–20.

225 P. Lowell, "The Markings on Venus," *AN,* 160 (1902), 129–32. For Lowell's later views, see Hoyt, *Lowell and Mars,* pp. 118–21.

226 A. E. Douglass, "Observations of Mars in 1896 and 1897," *ALO,* 2 (1900), 441.

227 As quoted in Hoyt, *Lowell and Mars,* p. 124.

228 As quoted in George E. Webb, *Tree Rings and Telescopes: The Scientific Career of A. E. Douglass* (Tucson, 1983), p. 49.

229 A. E. Douglass, "Illusions of Vision and the Canals of Mars," *PSM,* 70 (1907), 464–74; see also his "Is Mars Inhabited?" *Harvard Illustrated Magazine,* 8 (March 1907), 116–18.

230 E. M. Antoniadi, "La vie dans l'univers," *SAFB*, 52 (1938), 1–14.

231 See the obituary notices on Antoniadi by Fernand Baldet in *SAFB*, 58 (1944), and by P. M. R. [R. M. Ryves?] in *BAAJ*, 55 (September 1945), 163–5.

232 Compare E. M. Antoniadi, "Mars Section (First Preliminary Report)," *BAAJ*, 7 (1896), 54–5, with his "Report of the Mars Section, 1896," *BAAM*, 6 (1898), 55–102:65.

233 E. M. Antoniadi, "The Hourglass Sea on Mars," *Knowledge*, 20 (July 1, 1897), 169–72:172.

234 E. M. Antoniadi, "On the Optical Character of Gemination," *BAAJ*, 8 (1898), 176–8:178.

235 For the priority dispute, see the arguments of Henry Dierckx in *EM*, 67 (1898), 129, 197, 245–6, 311, 362; for Antoniadi's responses, see pp. 175, 219–20, and 333. See also Flammarion, *Mars*, vol. II, pp. 406–24.

236 E. M. Antoniadi, "Further Considerations on Gemination," *BAAJ*, 8 (1898), 308–10:310.

237 E. M. Antoniadi, "L'origine optique des géminations de Mars," *SAFB*, 12 (1898), 170–5; see the same volume, pp. 256 and 313–15, for further discussions by Antoniadi, and, for Schiaparelli's statement, pp. 312–13. For Moreux, see pp. 256 and 315–23, and especially Abbé T. Moreux, "Vues nouvelles sur la planète Mars," *Revue des questions scientifiques*, 44 (1898), 460–87. See also Antoniadi's "On Some Subjective Phenomena (Observed on the Martian Canals)," *BAAJ*, 9 (1899), 269–70. For Pickering, see W. H. Pickering, "Visual Observations of the Moon and Planets," *Annals of the Astronomical Observatory of Harvard College*, 32, pt. II (1900), 155. For the controversy with Williams, see A. S. Williams, "Notes on Mars in 1899," *Observatory*, 12 (1899), 226–9, and his "Considerations on the Double Canals of Mars," *BAAJ*, 10 (1900), 115–29. The dispute that resulted appears in the latter journal; for Antoniadi, see *BAAJ*, 10 (1900), 120–1, 305–6, and *BAAJ*, 11 (1900), 26–30; for Williams, see *BAAJ*, 10 (1900), 211–13, 323–6, and *BAAJ*, 11 (1900), 114–15; for Holmes, see *BAAJ*, 10 (1900), 300–3.

238 E. M. Antoniadi, "Report of the Mars Section, 1898–1899," *BAAM*, 20 (1901), 25–92:45.

239 Williams, "Notes," p. 228.

240 A. S. Williams, "On the Double Canals of Mars," *BAAJ*, 10 (1900), 323–6:324.

241 As quoted in Antoniadi, "Report, 1896," *BAAM*, p. 62.

242 Antoniadi, "Report, 1898–1899," *BAAM*, pp. 71, 105.

243 For biographical information and a partial bibliography of Cerulli's writings, see Mentore Maggini, "Vincenzo Cerulli," *Memorie dela società astronomica Italiana*, 4 (1927), 171–87.

244 V. Cerulli, "Les canaux de Mars et les canaux de la lune," *SAFB*, 12 (June 1898), 270–1; see also his "Mascanäle und Mondcanäle," *AN*, 146 (1898), 155–8.

245 V. Cerulli, *Marte nel 1896–97* (Collurania, Italy, 1898), p. 105.

246 Cerulli, *Marte*, p. 115.

247 E. M. Antoniadi, "Mars Section. Sixth Interim Report for 1909 . . . ," *BAAJ*, 20 (January 1910), 189–92:190.

248 Schiaparelli, *Corrispondenza*, vol. II, p. 307. For other comments by Schiaparelli on Cerulli, see pp. 285 and 316–17; for comments on Lowell, see pp. 206, 286–7, 297, and 317.

249 For Schiaparelli's 1899 review, see his *Opere*, vol. II, pp. 231–44. For Flammarion's moon drawings, see *SAFB*, 14 (1900), 45–50, 93–8, 140–5, 183–8, 227–33, 275–83, 339, 498–506; see also Flammarion, *Mars*, vol. II, pp. 313–37, 460–9.

250 M. A. Orr, "The Canals of Mars," *Knowledge*, 24 (February 1, 1901), 38–9; reprinted in *SAS*, 51 (March 20, 1901), 21108–9, and in *Smithsonian Institution Annual Report for 1900* (1901), 166–9.

251 G. Millochau, "Observations de Mars en 1901," *SAFB*, 15 (1901), 437–8:438.

252 J. Comas Solá, "Les canaux de Mars," *SAFB*, 15 (1901), 122–7:123; A. Müller, "Die Physiologie in der Astronomie," *Die Kultur: Zeitschrift für Wissenschaft, Literatur, und Kunst*, 2 (1901), 280–93:287.

253 As quoted in E. M. Antoniadi, "Report of the Mars Section, 1900–1901," *BAAM*, 11 (1903), 85–142:89.

254 W. W. Campbell, "Recent Observations of the Spectrum of Mars," *ASPP*, 9 (April 1897), 109–12:111.

255 J. E. Keeler, "Spectrographic Observations of Mars in 1896–7," *ApJ*, 5 (1897), 328–31.

256 E. I. Yowell, "Is Aqueous Vapor Present on Mars?" *PA*, 7 (1899), 237–42.

257 G. J. Stoney, "Of Atmospheres upon Planets and Satellites," *Royal Dublin Society Scientific Transactions*, 6 (1898), 305–28:307; see especially pp. 306–7, and 320–2. For commentary on this paper, see Kenneth R. Lang and Owen Gingerich (eds.), *A Source Book in Astronomy and Astrophysics, 1900–1975* (Cambridge, Mass., 1979), pp. 88–9.

258 For Cook's arguments, see especially C. R. Cook, "On the Escape of Gases from Planetary Atmospheres According to the Kinetic Theory," *ApJ*, 11 (1900), 36–43; for Bryan, see especially G. H. Bryan, "The Kinetic Theory of Planetary Atmospheres," *Royal Society of London Philosophical Transactions*, 196A (1901), 1–24. For Stoney's responses, see his "On the Escape of Gases from Planetary Atmospheres According to the Kinetic Theory," *ApJ*, 11 (1900), 251–8, 357–72, and his "Note on Inquiries as to the Escape of Gases from Atmospheres," *Royal Society of London Proceedings*, 67 (1900), 286–91. See also *Nature*, 61 (1900), 501, 515; 62 (1900), 54, 78, 126, 189.

259 H. Perrotin, "Sur la planète Mars," *CR*, 124 (1897), 340–6.

260 E. Holmes, "The Canals of Mars," *BAAJ*, 10 (1900), 300–4:303.

261 Leo Brenner, "Work of the Manora Observatory in 1897," *EM*, 67 (March 4, 1898), 60–1.

262 Leo Brenner, "On the Canals of Mars," *Observatory*, 21 (1898), 296–9.

263 See Leo Brenner, "On the Impossibility of the Martian Hypothesis of Mr. Lowell," *BAAJ*, 9 (1898), 72–5, and Brenner, "Les canaux de Mars," *SAFB*, 13 (1899), 25–33. See also R. du Ligondés and Théophile Moreux, "Les canaux de Mars et l'hypothèse de M. Brenner," *SAFB*, 13 (1899), 174–7; W. H. Pickering, "Les canaux de Mars," *SAFB*, 13 (1899), 170–3; and W. W. Payne, "Current Astronomy," *PA*, 6 (September 1898), 395–8. For Brenner's response, see his "Meine Mars-Hypothese und ihre Gegner," *Astronomische Rundschau*, 2 (1900), 207–12, 234–41.

264 J. Joly, "On the Origin of the Canals of Mars," *Royal Dublin Society Scientific Transactions*, ser. 2, 6 (1898), 249–68; on Teoperberg, see "The Canals of Mars," *Nature*, 55 (January 21, 1897), 280; T. Moreux, "Note on the Physical Constitution of Mars," *BAAJ*, 8 (1898), 278–9.

265 As quoted in Antoniadi, *Mars*, p. 33.

266 José Comas Solá, "Quelques considérations sur la planète Mars," *SAFB*, 24 (1910), 36–7:37.

267 Wells Alan Webb, *Mars, the New Frontier: Lowell's Hypothesis* (San Francisco, 1956), pp. 60–1.

268 As quoted in Hector Macpherson, "The Problem of Mars," *PA*, 29 (1921), 129–37:133.

269 C. Flammarion, "Are the Planets Inhabited?" *Harper's Magazine*, 109 (1904), 840–5:840.

270 C. T. Whitmell, in a letter in *EM*, 78 (August 14, 1904), 12.

271 E. M. Antoniadi, "Recent Observations of Mars," *Knowledge*, 25 (April 1902), 81–4:83. Antoniadi's first use of the phrase "canaliform illusion" seems to be in his "Martian Gemination," *EM*, 67 (July 8, 1898), 474.

272 B. W. Lane, "The Canals of Mars," *Knowledge*, 25 (November 1, 1902), 250–1; see also p. 276. For Maunder's note, see p. 251.

273 J. E. Evans and E. W. Maunder, "Experiments as to the Actuality of the 'Canals' Observed on Mars," *RASMN*, 63 (June 1903), 488–99:497.

274 E. W. Maunder and Annie S. D. Maunder, "Some Experiments on the Limits of Vision for Lines and Spots as Applicable to the Question of the Actuality of the Canals of Mars," *BAAJ*, 13 (1903), 344–51.

275 Norriss S. Hetherington, "Amateur versus Professional: The British Astronomical Association and the Controversy over Canals on Mars," *BAAJ*, 86 (1976), 303–8.
276 In his "Are Other Worlds Inhabited?" *Youth's Companion, 76* (December 11, 1902), 639–40:639, Newcomb accepted "long streaks stretching from point to point on the planet. . . . They must be a hundred miles in breadth. . . ."
277 For the R.A.S. discussion, see *EM*, 77 (June 19, 1903), 407; for the B.A.A. discussion, see *BAAJ*, 13 (1903), 333–8.
278 E. W. Maunder, "A New Chart of Mars," *Observatory*, 26 (1903), 351–6:351, 353.
279 E. W. Maunder, "The Canals of Mars," *Knowledge*, 26 (November 1903), 249–51:251.
280 E. Ledger, "The Canals of Mars – Are They Real?" *Nineteenth Century*, 53 (May 1903), 773–85:784–5.
281 E. M. Antoniadi, "Mars," *EM*, 77 (July 31, 1903), 544–5.
282 For Antoniadi's views in 1903, see his "Considerations on the Planet Mars," *Knowledge*, 26 (November 1903), 246–9, and his "On the Instrumentality of Contrast in 'Duplicating' the Spots of Mars," *AN, 164* (1903), 63–4, as well as his articles and letters in *EM*, 77 (1903), 8, 79, 212–13, 504–5, especially 544–5; 78 (1903), 266–7, especially 285–6, 312–3, 377. Illusion theories were also discussed by a number of other authors in these volumes of *EM*.
283 Antoniadi, "Considerations," p. 246.
284 E. M. Antoniadi, "Report of the Mars Section, 1900–1901," *BAAM*, 11 (1903), 85–142:137; see Plate VIII.
285 As quoted in Antoniadi, "Report, 1900–1901," p. 89. In 1904, Molesworth published a thorough analysis of the canal question that mainly supported Lane, Ledger, and Maunder. See his "The Markings on Mars: A Plea for Moderate Views," *Monthly Review*, 17 (1904), 46–60.
286 Anonymous, "The Study of Mars," *PA*, 11 (1903), 409–10.
287 Lane's experiments were discussed in an anonymous article, "Are Martian Canals a Myth?" *Current Literature*, 34 (1903), 79–80; reprinted from the *New York Mail and Express*. Ledger's paper was reprinted in part in Anonymous, "Evidence of Life on Mars," *Current Literature, 35* (1903), 67–71, and Ledger, "The Canals of Mars," *SAS, 55* (June 27, 1903), 22983–4. Maunder's 1903 *Knowledge* paper was excerpted in "Are the Canals of Mars Illusions?" *SA*, 90 (March 12, 1904), 219. See also W. W. Payne, "The 'Canals' of Mars," *PA*, 12 (1904), 365–75, which included Maunder's "The Canals of Mars: A Reply to Mr. Story," *Knowledge*, n.s., 1 (1904), 87–9. Pickering published two papers partially supportive of Maunder: "Recent Studies of the Martian and Lunar Canals," *PA*, 12 (1904), 77–80, and "An Explanation of the Martian and Lunar Canals," *PA*, 12 (1904), 439–42.
288 See P. Lowell, "Double Canals and the Separative Powers of Glasses," *PA*, 12 (1904), 575–9; "Experiments on the Visibility of Fine Lines in Its Bearing on the Breadth of the 'Canals' of Mars," *Lowell Observatory Bulletin*, no. 2 (1903), 1–2, and in relation to this, his letter to Maunder published in *Observatory*, 27 (January 1904), 49.
289 For information on the Russian and German papers, see the Mars bibliographies for 1903 and 1904 in *Astronomischer Jahresbericht, 5* and 6 (1904 and 1905). J. E. Evans and E. W. Maunder, "Expériences contre la réalité des canaux de Mars," *SAFB*, 19 (1905), 274–83; For Flammarion's note, see p. 283.
290 P. Lowell, *Mars and Its Canals* (New York, 1911 reprint of the 1906 original), p. 202.
291 V. Cerulli, "L'image de Mars," *SAFB*, 19 (1905), 352–8:353.
292 As quoted in *BAAJ*, 13 (1903), 338.
293 Anonymous, "Canals of Mars Photographed," *SA*, 93 (August 5, 1905), 107. On Crommelin, see Hoyt, *Lowell and Mars*, pp. 184–5.
294 Among Lowell's publications, see especially "First Photographs of the Canals of Mars," *Royal Society of London Proceedings, 77A* (1906), 132–5. For Schiaparelli's statement, see P. Lowell, *Mars as the Abode of Life* (New York, 1908), p. 155, and for Wicks's, see his "The 'Canals' of Mars – The End of a Great Delusion," *EM*, 82 (November 3, 1905), 298.
295 P. Lowell, *Mars and Its Canals*, p. 277.

296 D. P. Todd, "The Lowell Expedition to the Andes," *PA*, 15 (1907), 551–3:552, and his "Professor Todd's Own Story of the Mars Expedition," *Cosmopolitan Magazine*, 44 (March 1908), 343–51:349.

297 Anonymous, "The Newsletter," *Sphere*, 30 (September 14, 1907), 237; see also Frank Edward Cane, "The Lowell Photographs of Mars," *EM*, 86 (September 20, 1907), 149.

298 For sample expert analyses, see Walter H. Wesley, "Photographs of Mars," *Observatory*, 28 (1905), 314–15, and E. M. Antoniadi, "Note on Photographic Images of Mars Taken in 1907 by Professor Lowell," *RASMN*, 69 (December 1908), 110–14. See also Anonymous, "The Question of Life on Mars," *Edinburgh Review*, 208 (1908), 74–94, especially 90–2.

299 E. W. Maunder may initially have accepted Lowell's report; see his "Progress of Astronomy in 1906," *PA*, 15 (January 1907), 1–12:5, where he mentioned "Lowell's successful photography of the canals of Mars."

300 Alfred Russel Wallace, *Is Mars Habitable?* (London, 1907), p. 37.

301 V. M. Slipher, "The Spectrum of Mars," *ApJ*, 28 (1908), 397–404, and W. W. Campbell, "The Spectrum of Mars as Observed by the Crocker Expedition to Mt. Whitney," *Lick Observatory Bulletin*, no. 169 (1909), 149–64. On Slipher, see W. G. Hoyt, "Vesto Melvin Slipher," *National Academy of Sciences Biographical Memoirs*, 52 (1980), 410–49.

302 Hoyt, *Lowell and Mars*, pp. 127–50:142–3, and David H. DeVorkin, "W. W. Campbell's Spectoscopic Study of the Martian Atmosphere," *Royal Astronomical Society Quarterly Journal*, 18 (1977), 37–53:43.

303 Hoyt, "Slipher," p. 432. For Very, see the *National Union Catalog*, which lists him as author of three Swedenborgian pamphlets, as well as *An Epitome of Swedenborg's Science*, 2 vols. (Boston, 1927).

304 DeVorkin, "Campbell," pp. 50, 53.

305 As quoted in DeVorkin, "Campbell," p. 43.

306 Hoyt, *Lowell and Mars*, pp. 147–50.

307 See G. J. Stoney's letter in *Nature*, 77 (March 19, 1908), 461–2. See also Charles Lane Poor, *The Solar System* (London, 1908), where Campbell's and Stoney's analyses are presented as mutually supportive.

308 For G. R. Agassiz, see, for example, his "Mars as Seen in the Lowell Refractor," *PSM*, 71 (1907), 275–82; for E. S. Morse, see his *Mars and Its Mystery* (Boston, 1906), and Dorothy G. Wayman, *Edward Sylvester Morse* (Cambridge, Mass., 1942), pp. 392–7; for L. F. Ward, see his "Mars and Its Lesson," *Brown Alumnus Quarterly*, 7 (March 1907), 159–65, and Hoyt, *Lowell and Mars*, pp. 217–18.

309 For Sir Norman Lockyer, see Hoyt, *Lowell and Mars*, passim; for W. J. S. Lockyer, see, for example, his review of A. R. Wallace's *Is Mars Habitable?* in *Nature*, 77 (February 13, 1908), 337–9; for J. S. Worthington, see, for example, his "Markings on Mars," *Nature*, 85 (November 10, 1910), 40.

310 E. H. Hankin, "Life on Mars," *Nature*, 78 (May 7, 1908), 6.

311 C. E. Housden, "Mars and Its Markings," *BAAJ*, 23 (March 1913), 278–90, with further discussion by Antoniadi, pp. 347–8 and 434, and by Housden, p. 395. See also Housden's *The Riddle of Mars* (London, 1914) and his *Is Venus Inhabited?* (London, 1915).

312 Ludwig Kann, *Neue Theorie über den Ursprung der Kohle und die Lösung des Marsrätsels* (Heidelberg, 1901); Adrian Baumann, *Erklärung der Oberfläche des Planeten Mars* (Zurich, 1909); Philip Fauth, *Hörbigers Glazial-Kosmogonie* (Minden, 1912). On these books, see Willy Ley, *Watchers of the Skies* (London, 1962), pp. 295–6, 514–17.

313 A. R. Wallace, "Man's Place in the Universe," *Fortnightly Review*, 73 (March 1, 1903), 395–411, and his *Man's Place in the Universe* (London, 1903).

314 In 1903, the island universe theory was in temporary disrepute, partly because of the objection Whewell had made to it decades earlier. Moreover, astronomers had not yet realized that interstellar obscuring matter was misleading them into believing that the

sun is centrally located in the Milky Way, much as persons in a fog "see" themselves as situated centrally in relation to surrounding objects.

315 I have located over forty published contributions (including translations, but not reprints) to this controversy. The total number must be double that. Many persons wrote more than one paper; among the contributors were A. M. Clerke (2), Flammarion (2), Gore, Günther, Maunder (5), Moye (2), Newcomb, W. H. Pickering, and H. H. Turner (2).

316 Wallace had become an agnostic while a youth; see A. R. Wallace, *My Life: A Record of Events and Opinions,* vol. I (New York, 1905), pp. 226–8. Moreover, during the 1860s he had accepted spiritualism. It is interesting that whereas many authors, e.g., Figuier and Flammarion, accepted both pluralism and spiritualism, Wallace favored only the latter. On his spiritualism, see Malcolm J. Kottler, "Alfred Russel Wallace, the Origin of Man, and Spiritualism," *Isis,* 65 (1974), 144–92.

317 For recent analyses, see James J. Kevin, Jr., *"Man's Place in the Universe:* Alfred Russel Wallace, Teleological Evolution, and the Question of Extraterrestrial Life" (a 1985 MA thesis at the University of Notre Dame), and William C. Heffernan, "The Singularity of Our Inhabited World: William Whewell and A. R. Wallace in Dissent," *Journal of the History of Ideas,* 39 (1978), 81–100.

318 Wallace's arguments in this appendix (pp. 326–36) to his *Man's Place in the Universe,* 4th ed. (London, 1904), are discussed in Frank J. Tipler, "A Brief History of the Extraterrestrial Intelligence Concept," *Royal Astronomical Society Quarterly Journal,* 22 (1981), 133–45:140–1.

319 A. R. Wallace, *Is Mars Habitable?* (London, 1907), pp. 38–77; see also J. H. Poynting, "Radiation in the Universe: Its Effects on Temperature and Its Pressure on Small Bodies," *Royal Society Philosophical Transactions,* 202A (1904), 525–52; P. Lowell, "A General Method for Evaluating the Surface-Temperature of the Planets; with Special Reference to the Temperature of Mars," *Philosophical Magazine,* 6th ser., 14 (July 1907), 161–76; and Poynting's "On Professor Lowell's Method of Evaluating the Surface Temperature of the Planets . . . ," *Philosophical Magazine,* 14 (December 1907), 749–60.

320 P. Lowell, "The Habitability of Mars," *Nature,* 77 (March 19, 1908), 461.

321 Carl Sagan, "Hypothesis," in Ray Bradbury et al., *Mars and the Mind of Man* (New York, 1973), p. 15.

322 Simon Newcomb, "The Optical and Psychological Principles Involved in the Interpretation of the So-Called Canals of Mars," *ApJ,* 26 (July 1907), 1–17:2–8. Newcomb first presented this paper to the Philosophical Society of Washington; for a report, see R. L. Faris, "The Philosophical Society of Washington," *Science,* n.s., 25 (March 1, 1907), 343–4.

323 For the Newcomb-Lowell exchange, see *ApJ,* 26 (1907), 131–40 (Lowell), 141 (Newcomb), and 142 (Lowell). For a listing of 12 reports on Newcomb's paper published in 1907, see *Astronomischer Jahresbericht,* 9 (1908), 407.

324 Simon Newcomb, "Fallacies about Mars," *Harper's Weekly,* 52 (July 25, 1908), 11–12:12.

325 Hoyt, *Lowell and Mars,* pp. 329, 141.

326 A. E. Douglass, "Is Mars Inhabited?" *Harvard Illustrated Magazine,* 8 (March 1907), 116–18, and "Illusions of Vision and the Canals of Mars," *PSM,* 70 (May 1907), 464–74.

327 S. I. Bailey, "The Planet Mars," *Science,* n.s., 26 (December 27, 1907), 910–12.

328 H. Jacoby, "The Case against Mars," *American Magazine,* 65 (1908), 625–8.

329 Mrs. Walter Maunder, "The 'Highways' and the 'Waterways' of Mars," *Knowledge,* n.s., 4 (August 1907), 169–71. For an account of a "communication to the *Daily Graphic"* made by E. W. Maunder, see *EM,* 85 (1907), 534.

330 G. J. Stoney, "Telescopic Vision," *Philosophical Magazine,* 6th ser., 16 (1908), 318–39, 796–811, 950–79:950.

331 V. Cerulli, "L'imagine di Marte," *Rivista di astronomia,* 1 (1907), 93–105; V. Cerulli, "Articoli su Marte di Newcomb e Flammarion. La fotografia dei canali –

risoluzione del Gange" and "Polemica Newcomb-Lowell – Fotografie lunari," *Rivista di astronomia,* 2 (1908), 1–23.

332 H. Dierckx, "Les canaux de Mars existent-ils?" *Gazette astronomique,* 1 (1908), 77–8.

333 As quoted in A. L. Lowell, *Lowell,* p. 149.

334 C. Flammarion, "Photographies de Mars à l'observatoire Lowell," *SAFB,* 21 (1907), 465–8, and *SAFB,* 22 (1908), 153–9.

335 Anonymous, "The Question of Life on Mars," *Edinburgh Review,* 208 (1908), 74–94:74–5. Despite the repeated use of the editorial "we" and the many references to the 1896 review of Lowell's *Mars,* this review of Lowell's *Mars and Its Canals* and of Wallace's *Is Mars Habitable?* cannot have been written by Agnes Clerke (who wrote the earlier review), because she died in early 1907 before Wallace's book was published.

336 Hoyt, *Lowell and Mars,* p. 190.

337 E. M. Antoniadi, "Fifth Interim Report for 1909 . . . ," *BAAJ,* 20 (December 1909), 136–41:136–7.

338 This debate intersected at times with the canal controversy. See, for example, Antoniadi's "On the Advantage of Large over Small Telescopes in Revealing Delicate Planetary Detail," *BAAJ,* 21 (1910), 104–6. For background, see John Lankford, "Amateurs versus Professionals: The Controversy over Telescope Size in Late Victorian Science," *Isis,* 72 (1981), 11–28.

339 Edwin B. Frost, *An Astronomer's Life* (Boston, 1933), pp. 217–18. For evidence that Jonckheere sent the initial telegram, see *BAAJ,* 20 (December 1909), 122.

340 Compare E. M. Antoniadi, "Third Interim Report for 1909 . . . ," *BAAJ,* 20 (October 1909), 25–8:25, with his "Fifth Report," p. 141.

341 Antoniadi, "Fifth Report," pp. 139–40. See also C. André, *Les planètes et leur origines* (Paris, 1909), especially pp. 36–74.

342 "Report of the Meeting of the Association . . . Dec. 29, 1909 . . . ," *BAAJ,* 20 (December 1909), 119–25:123.

343 As quoted in E. M. Antoniadi, "Sixth Interim Report for 1909 . . . " *BAAJ,* 20 (January 1910), 189–92:191–2. See also Antoniadi's "On Some Objections to the Reality of Prof. Lowell's Canal System of Mars," *BAAJ,* 20 (January 1910), 194–7.

344 E. M. Antoniadi, "Report of the Mars Section, 1903," *BAAM,* 16 (1910), 54–104:60, and his "Report of the Mars Section, 1907," *BAAM,* 17 (1910), 65–112:69.

345 E. M. Antoniadi, "Report of the Mars Section, 1909," *BAAM,* 20 (1915), 25–92:32.

346 E. W. Maunder, "Some Facts That We Know about Mars," *BAAJ,* 20 (November 1909), 82–9, and E. M. Antoniadi, "On the Possibility of Explaining on a Geomorphic Basis the Phenomena Presented by the Planet Mars," *BAAJ,* 20 (November 1909), 89–94:94.

347 E. W. Maunder, "The 'Canals' of Mars," *Rivista di scienza,* 7 (1910), 253–69.

348 "Report of the Meeting of the Association Held on March 30, 1910 . . . ," *BAAJ,* 20 (March 1910), 285–94:289.

349 See *BAAJ,* 20 (April 1910), 348–9 for Maunder, 374–7 for Antoniadi, and 385–6 for Gheury.

350 Anonymous, "Observations de la planète Mars à l'observatoire Lowell. Résumé de la conférence faite par M. Lowell à l'assemblée générale annuelle du 6 avril 1910," *SAFB,* 24 (1910), 214–20; "[Report on the] Meeting of the Royal Astronomical Society, Friday, 1910 April 8," *Observatory,* 33 (1910), 192–5.

351 J. Comas Solá, "Quelques considérations sur la planète Mars," *SAFB,* 24 (1910), 36–7.

352 Lucien Libert, "Les progrès récents dans la connaissance de Mars," *Revue scientifique,* 48 (1910), 553–9:559.

353 Abbé J. Belpaire, "Les canaux de Mars," *Gazette astronomique,* 4 (1911), 6–7, 14:7.

354 C. de Kirwan, "Les mondes présents, passés ou futurs," *Revue des questions scientifiques,* 23 (1913), 598–614:611.

355 S. Arrhenius, "Neues vom Mars," *Kosmos,* 7 (1910), 123–8; "Der Planet Mars nach neueren Untersuchungen," *Deutsche Revue,* 35 (1910), 310–24. See also his "Les

conditions physiques sur la planète Mars," *Journal de physique,* 5th ser., 2 (1912), 81–97, and his *Destinies of the Stars,* trans. J. E. Fries (New York, 1918), pp. 180–227.

356   Arrhenius's views of Venus are discussed in Anonymous, "The Limit of Organic Life in Our Solar System," *Current Opinion,* 43 (February 1911), 242–3. For his pan-spermy hypothesis, see S. Arrhenius, "Panspermy: The Transmission of Life from Star to Star," *SA,* 96 (March 2, 1907), 196.

357   H. Paradyne, "The Mythical Canals of Mars," *Harper's Weekly,* 54 (January 15, 1910), 15.

358   Edwin B. Frost, *An Astronomer's Life* (Boston, 1933), p. 217. On this incident, see M. J. Crowe, "Inflation and History: E. B. Frost's Mars Telegram," *Griffith Observer,* 46 (March 1982), 15.

359   As quoted in E. M. Antoniadi, "Considerations on the Physical Appearance of the Planet Mars," *PA,* 21 (1913), 416–24:418; reprinted from *Knowledge* (May 1913); E. M. Antoniadi, "L'aspect physique de la planète Mars," *Ciel et terre,* 32 (1911), 209–22:212.

360   R. G. Aitken, "A Review of the Recent Observations of Mars," *ASPP,* 22 (1910), 78–87:79; P. Lowell, "Mars in 1909 as Seen at the Lowell Observatory," *Nature,* 84 (August 11, 1910), 172–3.

361   Stanley L. Jaki, *Planets and Planetarians* (Edinburgh, 1978), pp. 202ff.

362   A. S. Eddington, *The Nature of the Physical World* (London, 1928), p. 178.

363   James Jeans, *The Universe around Us* (New York, 1929), pp. 320–3.

364   E. M. Antoniadi, "Considerations on the Physical Appearance of the Planet Mars," *PA,* 21 (1913), 416–24:424.

365   E. M. Antoniadi, *La planète Mars* (Paris, 1920), pp. 51–2.

366   P. M. R. [P. M. Ryves], "E. M. Antoniadi," *BAAJ,* 55 (1945), 163–5:165. The book was Donald L. Cyr, *Life on Mars* (El Centro, Calif., 1944).

367   W. W. Campbell, "The Problem of Mars," *ASPP,* 30 (1918), 133–46:146.

368   As quoted in DeVorkin, "Campbell," p. 49.

369   "Flammarion Predicts Talking with Mars," *New York Times* (December 12, 1923), 3.

370   As quoted in H. Macpherson, "The Problem of Mars," *PA,* 29 (1921), 129–37:133.

371   Hoyt, *Lowell and Mars,* p. 297.

372   As quoted in Hoyt, *Lowell and Mars,* p. 294.

373   C. Flammarion, "Percival Lowell," *SAFB,* 30 (1916), 422–3:422.

374   See, for example, W. H. Pickering, "Signals from Mars," *PA,* 32 (1924), 580–1, and Campbell, "Problem of Mars."

375   Clyde W. Tombaugh and Patrick Moore, *Out of the Darkness: The Planet Pluto* (New York, 1980), pp. 82, 99.

376   E. W. Maunder, "Conditions of the Habitability of a Planet: with Special Reference to the Planet Mars," *Journal of the Transactions of the Victoria Institute,* 44 (1912), 78–94:94.

377   For this and other responses, see *Journal of the Transactions of the Victoria Institute,* 44 (1912), 94–102:97.

378   [W. D. Howells], "Editor's Easy Chair," *Harper's Monthly,* 128 (December 1913), 149–51:151.

379   Carl Sagan and Paul Fox, "The Canals of Mars: An Assessment after Mariner 9," *Icarus,* 25 (1975), 602–12:609.

380   R. A. Wells, *Geophysics of Mars* (Amsterdam, 1979), p. 451.

381   Ray Bradbury et al., *Mars and the Mind of Man* (New York, 1973), p. 23. Murray also provides an example from as recently as 1969. For a relevant study of the canal observations, see Lucia Rositani Ronchi and Giorgia Abetti, "Effeti psico-fisiologici nelle osservazioni astronomiche visuali: Il pianeta Marte," *Atti della fondazione Giorgio Ronchi,* 19 (1964), 1–33.

382   Wells, *Mars,* pp. 425–6.

383   V. Cerulli, "Polemica Newcomb-Lowell – Fotografie Lunari," *Rivista di astronomia,* 2 (1908), 13–23:13.

384 *Astronomischer Jahresbericht,* 10 (1909), 445.

385 Antoniadi, *Mars,* pp. 31–2.

386 As quoted in Antoniadi, *Mars,* p. 31.

387 The Hale and Schiaparelli letters appeared as "I canali di Marte," *Rivista di astronomia,* 4 (1910), 113–16:116.

388 "G. Schiaparelli über die Marstheorie von Svante Arrhenius," *Kosmos,* 7 (1910), 303.

389 R. Waterfield, *A Hundred Years of Astronomy* (New York, 1938), pp. 50–1. For endorsements, see Hoyt, *Lowell and Mars,* p. 309, and Otto Struve and Velta Zebergs, *Astronomy of the 20th Century* (New York, 1962), p. 147.

390 Waterfield, *Astronomy,* pp. 424, 429.

391 I. S. Shklovskii and Carl Sagan, *Intelligent Life in the Universe* (New York, 1966), p. 274.

CHAPTER 11

1 Marcia S. Smith, *Possibility of Intelligent Life Elsewhere in the Universe,* rev. ed. (Washington, D.C., 1977), p. xiii.

2 Of the persons in the list of the 23 most prolific astronomers up to 1880 in J. C. Houzeau and A. Lancaster, *Bibliographie générale de l'astronomie,* vol. II (London, 1964 reprint of the Bruxelles 1882 original), p. lxxiv, at least 17 published in the pluralist debate. Among the 8 most prolific, 5 were very active pluralists: Secchi (1), Lalande (2), Flammarion (5), Proctor (7), and Gruithuisen (8).

3 To test this claim, the proper name index in Franklin Baumer's *Modern European Thought* (New York, 1977) has been examined. Leaving aside artists and politicians and defining "prominent persons" as those listed more than once, I have found that for 43% of the eighteenth- and nineteenth-century figures mentioned, I can document their involvement in the pluralist debate. The actual percentage is no doubt higher.

4 Karl Popper, *Conjectures and Refutations* (New York, 1965), pp. 34ff.

5 As quoted in William Graves Hoyt, *Lowell and Mars* (Tucson, 1976), p. 89.

6 S. Arrhenius, *Destinies of the Stars,* trans. J. E. Fries (New York, 1918), p. 226.

7 R. L. Waterfield, "Mars," in *The Splendour of the Heavens,* ed. T. E. R. Philips and W. H. Steavenson, vol. I (New York, 1925), p. 321.

8 Edwin Hubble, *The Realm of the Nebulae* (New York, 1958 reprinting of the 1936 original). See also Hubble's "Points of View: Experiment and Experience," *Huntington Library Quarterly,* 3 (April 1939), 243–50, and his *The Nature of Science and Other Lectures* (San Marino, Calif., 1953).

9 Hermann Bondi, "Fact and Inference in Theory and in Observation," *Vistas in Astronomy,* 1 (1955), 155–62:156. For a recent, very effective presentation of views similar to Bondi's, see Norriss S. Hetherington, "Just How Objective Is Science?" *Nature* 306 (December 22/29, 1983), 727–30.

10 C. S. Peirce, *Essays in the Philosophy of Science,* ed. Vincent Tomas (Indianapolis, 1957), p. 134.

11 F. W. Cousins, *The Solar System* (London, 1972), p. 263.

12 Ernan McMullin, "Persons in the Universe," *Zygon,* 15 (1980), 69–89:81–2. This paper contains a valuable analysis of fallacies frequently found in contemporary discussions of the idea of extraterrestrial life.

13 Loren Eiseley, *The Immense Journey* (New York, 1957), p. 162.

14 G. G. Simpson, "The Nonprevalence of Humanoids," *Science,* 143 (February 21, 1964), 769–75; T. Dobzhansky, "Darwinian Evolution and the Problem of Extraterrestrial Life," *Perspective in Biology and Medicine,* 15 (1972), 157–75; T. Dobzhansky, *Genetic Diversity and Human Equality* (New York, 1973), 97–101.

15 I. S. Shklovskii and Carl Sagan, *Intelligent Life in the Universe* (New York, 1966), p. 357.

16 As quoted in Carl Sagan (ed.), *Communication with Extraterrestrial Intelligence (CETI)* (Cambridge, Mass., 1973), p. 86.

17 David Gelman et al., "Seeking Other Worlds," *Newsweek* (August 15, 1977), 46–

53:51, and Henry S. F. Cooper, Jr., "Profiles: Carl Sagan," *New Yorker* (June 21, 1976), 39–83:46, and (June 28, 1976), 30–61.

18 Stanley Jaki, *Planets and Planetarians: A History of Theories of the Origin of Planetary Systems* (Edinburgh, 1978); see p. 2 for a statement of this thesis.

19 Carl Sagan, "UFO's: The Extraterrestrial and Other Hypotheses," in Carl Sagan and Donald Menzel (eds.), *UFO's – A Scientific Debate* (Ithaca, 1972), pp. 265–75:272.

20 As quoted in Frank J. Tipler, "Additional Remarks on Extraterrestrial Intelligence," *Royal Astronomical Society Quarterly Journal,* 22 (1981), 279–92:288.

21 As quoted in Tipler, "Remarks," p. 289.

22 Karl S. Guthke, *Der Mythos der Neuzeit: Das Thema der Mehrheit der Welten in der Literatur- und Geistesgeschichte von der kopernikanischen Wende bis zur Science Fiction* (Bern, 1983), pp. 9–10. See also William J. O'Malley, "Carl Sagan's Gospel of Scientism," *America,* 144 (February 7, 1981), 95–8, and Robert Short, *The Gospel from Outer Space* (San Francisco, 1982).

Appendix: bibliography of books on the question of a plurality of worlds published before 1917

Criteria for inclusion

Books included in this bibliography are nonfiction works in the chief languages of the Western world that were written primarily to discuss the question of a plurality of worlds (as the topic was referred to throughout most of history) or the question of extraterrestrial life (as it is currently designated). This criterion is applied somewhat loosely for pre-1800 publications, e.g., those of Derham and Kant, but with strictness for books that appeared between 1800 and 1916.

Information provided

1. *Citation:* Wherever possible, the following information is provided: year of publication, author's name, a reasonably full title, place of publication, and approximate number of pages.

2. *Location:* If the book is listed in the *British Museum General Catalogue of Printed Books* (hereafter BMC), the *Catalogue général des livres imprimés de la bibliothèque nationale* (BNC), and/or the *National Union Catalog Pre-1956 Imprints* (NUC), this is noted after the word "In" following the citation for the book. If the book is not listed in any of these catalogs but a location is known, this is noted. If no location is known, a publication referring to the book is noted after the words "Cited in."

3. *Editions:* Following each book, information is provided concerning its subsequent editions and translations. Such information should not be considered exhaustive. If no information is provided, the inference to be drawn is that no later editions or translations are listed in BMC, BNC, and NUC, nor has such a reference been encountered in the research for this book. Because this information has been compiled to aid in assessing

the initial impact of the book, editions and/or translations appearing long after its first publication date are generally not listed.

1584   Giordano Bruno. *De l'infinito universo et mondi*. Venice, 175 pp. In: BMC and NUC.

Editions: In 1716, John Toland published an English trans. of portions of this book. All subsequent editions were after 1800.

1591   Giordano Bruno. *De innumerabilibus immenso et infigurabili; sue de universo et mundis libri octo;* published with *De monade, numero et figura liber consequens quinque de minimo magno et mensura*. Frankfurt, xiv + 655 pp. In: BMC, BNC, and NUC.

Editions: NUC lists a 1614 ed.

1622   Tommaso Campanella. *Apologia pro Galileo*. Frankfurt, 58 pp. In: BMC, BNC, and NUC.

1634   Johannes Kepler. *Somnium, sue opus posthumum de astronomia lunari*. Zagan and Frankfurt, 182 pp. In: BMC, BNC, and NUC.

1638   [John Wilkins]. *The Discovery of a World in the Moone, or A Discourse Tending to Prove That 'tis Probable There May Be Another Habitable World in That Planet*. London, 213 pp. In: BMC, BNC, and NUC.

Editions: Five editions by 1684 with a subsequent printing in 1707; French trans. in 1656; German trans. in 1713.

1646   Henry More. *Democritus Platonissans, or, An Essay upon the Infinity of Worlds*. Cambridge, England, 34 pp. In: BMC and NUC.

1657   Pierre Borel. *Discours nouveau prouvant la pluralité des mondes*. . . . Geneva, 80 pp. In: BMC, BNC, and NUC.

Editions: English trans. in 1658.

1686   [Bernard le Bovier de Fontenelle]. *Entretiens sur la pluralité des mondes*. Paris, 359 pp. In: BMC, BNC, and NUC.

Editions: By Fontenelle's death (1757), it had gone through 33 French printings. By 1800, it had been trans. into English 5 times (1687, 1688, 1688, 1715, and 1760); also trans. into Danish (1748), Dutch (1768), German (twice: 1751, 1780), Greek (1794), Italian (4 times: 1711, 1748, 1765, 1780), Polish (1765), Russian (1740), Spanish (1796), and Swedish.

1698   Christiaan Huygens. Κοσμοθεωρος, *sive de terris coelestis earumque ornatu conjecturae*. The Hague, 144 pp. In: BMC, BNC, and NUC.

Editions: 3 Latin editions by 1704; English trans. in 1698 with at least 4 later editions; Dutch trans. in 1699; French trans. in 1702 with 2 later editions; German trans. in 1703; Russian trans. in 1717.

1711   Hareneus Geierbrand [pseudonym of Andreas Ehrenberg]. *Curiöse und wohlgegründete Gedancken von mehr als einer bewohnten Welt*. Jena. Cited in: Otto Zöckler, *Geschichte der Beziehungen zwischen Theologie und Naturwissenschaft*, 2nd ed., vol. II (Gütersloh, 1879), p. 248.

Editions: 4th ed., Jena, 1718.

1711   Daniel Sturmy. *A Theological Theory of a Plurality of Worlds*. London, 107 pp. In: BMC and NUC.

1714   William Derham. *Astro-Theology; or a Demonstration of the Being and Attributes of God from a Survey of the Heavens*. London, 228 pp. In: BMC, BNC, and NUC.

Editions: 14 English editions by 1777; French trans. in 1729 and German trans. in 1732 that went through 6 editions.

1715   Johann Wilhelm Weinreich. *Disputatio de philosophica pluralitate mundorum*. Torun, 88 pp. Cited in: *Bibliografia Polska*, vol. 32 (Cracow, 1938), p. 323

1717   Andreas Ehrenberg [sometimes Ehrenberger]. *Die noch unümgestossene Vielheit der Welt-Kugeln, oder: Dass die Planeten Welt-Kugeln seyn. . . .* Jena., 87 pp. Cited in: Otto Zöckler, *Geschichte der Beziehungen zwischen Theologie und Naturwissenschaft*, 2nd ed., vol. II (Gütersloh, 1879), p. 248.

1721   Johann Jacob Schudt. *De probabili mundorum pluralitate*. Frankfurt, xii + 82 pp. In: BMC.

1726   William Arntzen. *Dissertatio astronomico-physica de luna habitabili*. Utrecht, 62 pp. In: NUC.

1732   Johann Heinrich Herttenstein [sometimes Hertenstein]. *Dissertatio mathematica, sistens similitudinem inter terram et planetas intercedentem*. Strasbourg, 50 pp. In: BNC.

1736   John Peter Biester. *An Enquiry into the Probability of the Planets Being Inhabited*. London, 24 pp. In: NUC.

1738   Johann Christoph Hennings. *Specimen planetographiae physicae inquirens praecipue an planetae sint habitabiles*. Kiel, 71 pp. In: NUC.

1740   Eric Engman. *Dissertatio astronomico-physica de luna non habitabili*. Upsala, 24 pp. In: NUC (under Anders Celsius).

1743   Isacus Svanstedt. *Dissertatio philosophica, de pluralitate mundorum*. Upsala, 18 pp. In: NUC (under Anders Celsius) and BNC.

1748   D. G. S. [David Gottfried Schöber]. *Gedanken von denen vernünftig freyen Einwohnern derer Planeten*. Liegnitz, 72 pp. In: NUC.

1755   [Immanuel Kant]. *Allgemeine Naturgeschichte und Theorie des Himmels*. Königsberg and Leipzig, 200 pp. In: BMC, BNC, and NUC.

Editions: 4 German editions by 1808.

1758   Emanuel Swedenborg. *De telluribus in mundo nostro solari quae vocantur planetae et telluribus in caelo astrifero*. London, 72 pp. In: BMC, BNC, and NUC.

Editions: German trans. in 1770; English trans. in 1787; French trans. in 1824; Italian trans. in 1886; frequently reprinted.

1760   Giovanni Cadonici. *Confutazione teologico-fisica del sistem di Gugl. Derham, che vuole tutti i pianeti, de creature ragionevoli, come la terra, abitati*. Brescia, xx + 344 pp. In: NUC.

1761   Johann Lambert. *Cosmologische Briefe über die Einrichtung des Weltbaues*. Augsburg, xxviii + 318 pp. In: BNC and NUC.

Editions: French trans. and condensation in 1770; 2nd ed. in 1787; Russian trans. of this condensation in 1797; English trans. of it in 1800; full French trans. in 1801.

1769   [Abbé Jean Terrasson?]. *Traité de l'infini créé, avec explication de la possibilité de la transubstantiation et un petit traité de la confession et de la communion.* Amsterdam, lvii + 156 pp. In: BNC and NUC (under N. Malebranche).

1771   [Abbé François Xavier de Feller]. *Observations philosophiques sur les systêmes de Newton, de Copernic, de la pluralité des mondes, etc., etc., précédées d'une dissertation théologique sur les tremblements de terre. . . .* Liege, 181 pp. In: BNC and NUC.

Editions: 2nd ed. in 1778; 3rd ed. in 1788.

1772   Andrew Oliver. *An Essay on Comets, in Two Parts. Part One . . . and Part Two, Pointing out Some Important Ends for Which These Tails Were Probably Designed: Wherein It Is Shown That Comets May Be Inhabited Worlds . . . and Even Comfortable Habitations.* Salem, New England, vi + 87 pp. In: BMC, BNC, and NUC.

Editions: 2nd ed. in 1811 with two lectures on comets by Winthrop; French trans. in 1777.

1796   Robert Harrington. *A New System of Fire and Planetary Life, Shewing That the Sun and Planets Are Inhabited.* London, 75 pp. In: BMC and NUC.

1801   Johann Elert Bode. *Allgemeine Betrachtungen über das Weltgebäude.* Berlin, 115 pp. In: BMC and NUC.

Editions: 3rd ed. by 1808.

1801   Paul Gudin de la Brenellerie. *L'astronomie, poëme en trois chants.* Auxerre, 68 pp. In: BMC, BNC, and NUC.

Editions: Expanded ed. (223 pp.) in 1810.

1801   Rev. Dr. Edward Nares. ΕΙΣ ΘΕΟΣ, ΕΙΣ ΜΕΣΙΤΗΣ; *or An Attempt to Show How Far the Philosophical Notion of a Plurality of Worlds Is Consistent or Not So, with the Language of the Holy Scriptures.* London, xv + 406 pp. In: BMC and NUC.

1813   James Mitchell. *On the Plurality of Worlds. A Lecture in Proof of the Universe Being Inhabited. Read in the Mathematical Society of London.* London, 25 pp. In: BMC and NUC.

1817   Anonymous. *A Free Critique on Dr. Chalmers' Discourses on Astronomy, or An English Attempt to "Grapple" It with Scotch Sublimity.* London, 42 pp. In: BMC.

1817   Thomas Chalmers. *A Series of Discourses on the Christian Revelation, Viewed in Connexion with Modern Astronomy.* Edinburgh, 275 pp. In: BMC, BNC, and NUC.

Editions: 9 editions in 1817 with frequent British and American reprintings as late as 1871; German trans. by 1841.

1817   [Alexander Maxwell]. *Plurality of Worlds; or, Letters, Notes, and Memoranda, Philosophical and Critical, Occasioned by "A Series of Discourses . . ." by Thomas Chalmers D.D.* London, 221 pp. In: BMC and NUC.

Editions: 2nd enlarged ed. in 1820; abridged ed. in 1872.

1818   [Henry Fergus?]. *An Examination of Some of the Astronomical and Theological Opinions of Dr. Chalmers as Exhibited in a Series of Discourses on the Christian Revelation. . . .* Edinburgh, 42 pp. In: NUC (which attributes it to Fergus).

Editions: 2nd ed. in 1818.

1818   John Overton. *Strictures on Dr. Chalmers' Discourses on Astronomy. . . .* Kent, 27 pp. In: University of Glasgow Library.

1828   Samuel Noble. *The Astronomical Doctrine of a Plurality of Worlds.* London, 64 pp. In: BMC and NUC.

1834   Alexander Copland. *The Existence of Other Worlds, Peopled with Intelligent and Living Beings, Deduced from the Nature of the Universe.* London, 210 pp. In: BMC and NUC.

1836   [Isaac Taylor]. *Physical Theory of Another Life.* London, ix + 321 pp. In: BMC and NUC.

Editions: At least 5 editions with numerous printings by 1871.

1837   Thomas Dick. *Celestial Scenery, or The Wonders of the Heavens Displayed.* London, xvi + 559. In: BMC and NUC.

Editions: At least 6 editions were published. Harper's had printed 6 impressions by 1848; Merriam 2 by 1847; Biddle 7 by 1854; Griffin 4 by 1869; and Claxton printed at least 1. This is in addition to numerous printings of Dick's *Works.* BMC lists the printing of the 5th thousand of the 2nd ed. in 1838.

1840   Thomas Dick. *The Sidereal Heavens, and Other Subjects Connected with Astronomy.* London, xvi + 584 pp. In: BMC and NUC.

Editions: 4 publishing houses repeatedly issued it between 1840 and 1860. Harper's published at least 2 editions and 7 printings; Biddle published 6 printings; W. Collins printed at least 7,000 copies; Thomas Ward and Co. published at least 2 editions totaling 4,000 copies by 1850; and Griffin produced a printing in 1869.

1847   François Édouard Plisson. *Les mondes ou essai philosophique sur les conditions d'existence des êtres organisé dans notre système planétaire.* Paris, vii + 344 pp. In: BMC, BNC, and NUC.

Editions: German trans. in 1851.

1852   George Wilson. *Electricity and the Electric Telegraph to Which Is Added: The Chemistry of the Stars.* London, 77 + 50 pp. In: BMC and NUC.

Editions: New ed. in 1854 with printing in 1856 and revised ed. in 1859 with printing in 1860.

1853   [William Whewell]. *Of the Plurality of Worlds: An Essay.* London, 279 pp. In: BMC and NUC.

Editions: 5 editions by 1859 with another printing in 1867. American printings in 1854, 1855, 1856, 1858, and 1861.

1854   David Brewster. *More Worlds than One: The Creed of the Philosopher and the Hope of the Christian.* London, vii + 262 pp. In: BMC, BNC and NUC.

Editions: 2nd ed. (1854) went through 10 printings by 1871 plus 3 American printings. A 3rd English ed. (1874) was reprinted in 1876 and 1895.

1854   [William Whewell]. *A Dialogue on the Plurality of Worlds.* London, 55 pp. In: NUC.

Editions: Included in the 2nd through 5th editions of Whewell's 1853 book.

1855   Henry Drummond. *On the Future Destinies of the Celestial Bodies.* London, 65 pp. In: BNC and NUC.

1855   Edward Higginson. *Astro-theology; or The Religion of Astronomy: Four Lectures.* London, xiv + 96.pp. In: BMC and NUC.

1855   William Stephen Jacob. *A Few More Words on the Plurality of Worlds.* London, 47 pp. In: BMC.

1855   [Robert Knight]. *The Plurality of Worlds. The Positive Argument from Scripture, with Answers to Some Late Objections from Analogy.* London, 158 pp. In: BMC.

Editions: 2nd ed. in 1878.

1855   Hugh Miller. *Geology versus Astronomy.* Glasgow, 35 pp. In: BMC.

1855   Montagu Lyon Phillips. *Worlds beyond the Earth.* London, vii + 274 pp. In: BMC and NUC.

1855   Rev. Baden Powell. *Essays on the Spirit of the Inductive Philosophy, the Unity of Worlds, and the Philosophy of Creation.* London, xvi + 503 pp. In: BMC and NUC.

Editions: 2nd ed. in 1856.

1855   Thomas Collins Simon. *Scientific Certainties of Planetary Life; or Neptune's Light as Great as Ours.* London, xxiii + 238 pp. In: BMC and NUC.

1855   [William Tarbet]. *Astronomy and Geology as Taught in the Holy Scriptures.* Liverpool, 23 pp. In: BMC.

1855   [William Williams]. *The Universe No Desert, The Earth No Monopoly.* 2 vols. in 1. Boston and Cambridge, Mass., xii + 130 + 239 pp. In: NUC (where it is attributed to Williams).

1856   Rev. John Peat. *Thoughts in Verse on a Plurality of Worlds.* London, 15 pp. In: BMC and NUC.

Editions: A 2nd enlarged ed. in 1856.

1856   Johann Gottlieb Schimko. *Die Planetenbewohner, und ihre aus mathematischen, naturwissenschaftlichen und psychologischen Grunden abgeleitete verschiedene geistige Vollkommenheit.* Olmütz, 68 pp. In: BMC.

1856   William Samuel Symonds. *Geology as It Affects a Plurality of Worlds.* London. In: BMC and NUC.

1857   Rev. Josiah Crampton. *The Testimony of the Heavens to Their Creator.* Dublin, 31 pp. In: BMC.

1858   Frederick W. Cronhelm. *Thoughts on the Controversy as to a Plurality of Worlds.* London, 24 pp. In: BMC and NUC.

1858   [J. J. Larit?]. *Rêvéries et vérités ou de quelques questions astronomiques envisagées sous la rapport réligieux, en résponse à l'ouvrage du Docteur William Whewell sur la pluralité des mondes.* Paris, 318 pp. In: BMC (listed under Whewell).

1859   Charles Louis Hequembourg. *Plan of the Creation; or, Other Worlds, and Who Inhabit Them.* Boston, 398 pp. In: BMC and NUC.

1861   Joseph Emile Filachou. *De la pluralité des mondes.* Montpellier and Paris, 109 pp. In: BMC and BNC.

1862   Camille Flammarion. *La pluralité des mondes habités.* Paris, 54 pp; 2nd ed., Paris, 1864, xx + 550 pp. In: BMC, BNC, and NUC.

Editions: 33 editions by 1880 with reprintings until 1921. German trans. by 1864; Swedish by 1866; Danish by 1868; Spanish by 1875; Portuguese by 1878; Dutch by 1891; Russian by 1896.

1863   Nicholas Odgers. *The Mystery of Being; or Are Ultimate Atoms Inhabited Worlds?* London, vii + 161 pp. In: BMC.

1865   Camille Flammarion. *Les mondes imaginaires et les mondes réels.* Paris, vii + 577 pp. In: BMC, BNC, and NUC.

Editions: 14 editions by 1876; reprintings as late as 1925. Spanish trans. in 1873; Portuguese in 1876; probably also Italian and Russian editions.

1865   J. André Pezzani. *La pluralité des existences de l'âme conforme à la doctrine de la pluralité des mondes.* Paris, xxxv + 432 pp. In: BMC, BNC, and NUC.

Editions: 6 French editions by 1872; Spanish trans. by 1885.

1866   Fernand Coyteux. *Qu'est-ce que le soleil? peut-il être habité?* Paris, 430 pp. In: BMC and BNC

1867   *Rapport fait les 4 décembre 1866 et janvier 1867 par M. Trouessart . . . à la société académique d'agriculture, belles-lettres, sciences et arts [de Poitiers] sur un ouvrage intitulé* **Qu'est-ce que le soleil? peut-il être habité?** *par M. Coyteux. Résponse à ce rapport et notes critiques par F. Coyteux.* Poitiers, 113 pp. In: BNC (under J. Trouessart).

1869   Charles Cros. *Étude sur les moyens de communication avec les planètes.* Paris, 16 pp. In: BMC and BNC.

1869   Maurice-Martin Antonin Macario. *Entretiens populaires sur la formation des mondes.* Paris, 178 pp. In: BNC.

1870   Richard A. Proctor. *Other Worlds than Ours.* London, xiv + 324 pp. In: BMC and NUC.

Editions: At least 7 editions by 1893. NUC lists 29 separate printings by 1909.

1871   Louis Figuier. *Le lendemain de la mort.* Paris, 460 pp. In: BMC, BNC, and NUC.

Editions: 10 French editions by 1894 with printings until 1907. English trans. in 1872 with at least 3 later editions; Portuguese trans. in 1877.

1872   Richard A. Proctor. *The Orbs around Us.* London, 340 pp. In: BMC and NUC.

Editions: 1872 ed. issued in both London and New York. Attained 4th ed. by 1886 with printings as late as 1906.

1873   Camille Flammarion. *Récits de l'infini. Lumen, histoire d'une comète dans l'infini.* Paris, 415 pp. In: BMC, BNC, and NUC.

Editions: 5 editions of *Récits* by 1873 with 12 editions published by 1892; English trans. in 1873; Spanish trans. in 1874. *Lumen* was later published separately, 64,000 copies being printed by 1906.

1873   Richard A. Proctor. *The Borderland of Science.* London, vii + 438 pp. In: BMC and NUC

Editions: NUC lists 3 printings.

1873   Richard A. Proctor. *The Expanse of Heaven.* London, vi + 305 pp. In: BMC and NUC.

Editions: NUC lists 9 printings by 1905.

1874   Abbé Jean Léger-Marie Pioger. *Le dogme chrétien et la pluralité des mondes habités.* Paris, 488 pp. In: BNC.

1875   Abbé Jean Boudon. *Adam à son origine, roi et unique médiateur de tout l'univers planétaire. Question délicate touchant à la pluralité des mondes habités.* Bar-la-Duc, 212 pp. In: BNC and NUC.

Editions: 2nd ed. in 1878; 3rd ed. in 1879.

1875   Joseph Hamilton. *The Starry Hosts: A Plea for the Habitation of the Planets.* London, 114 pp. In: BMC and NUC.

1876   Jules Boiteux. *Lettres à un materialiste sur la pluralité des mondes habités.* Paris, 516 pp. In: BMC and NUC.

Editions: 3rd ed. in 1898.

1876   Jean d'Estienne [Charles de Kirwan]. *Considérations nouvelles sur la pluralité des mondes.* Paris, 31 pp. In: BNC.

1876   Victor Girard. *Nouvelles études sur la pluralité des mondes habités et sur la existence de l'âme.* Paris, 324 pp. In: BMC and NUC.

1877   Augustus Clissold. *The Divine Order of the Universe as Interpreted by Emanuel Swedenborg with Special Relation to Modern Astronomy.* London, 121 pp. In: BMC and NUC.

1877   Camille Flammarion. *Les terres du ciel, description astronomique, physique, climatologique, géographique des planètes qui gravitent avec la terre autour du soleil et de l'état probable de la vie à leur surface.* Paris, 604 pp. In: BMC, BNC, and NUC.

Editions: 10 French editions by 1881; additional ed. in 1884; Spanish trans. by 1877; Italian trans. by 1913.

1877   Niceto Perujo. *La pluralidad de mundos, habitados ante le fé Catolica.* Madrid, 456 pp. In: BMC.

1878   Adam Miller. *Life on Other Worlds.* Chicago, 282 pp. In: NUC.

1879   Jakob H. Schmick. *Der Planet Mars, eine zweite Erde, nach Schiaparelli gemeinverständlich dargestellt.* Leipzig, 64 pp. In: BMC and BNC.

1880   Otto Hahn. *Die Meteorite (Chondrite) und ihre Organismen.* Tübingen, 56 pp. In: BMC, BNC, and NUC.

1880   Karl Du Prel. *Die Planetenbewohner und die Nebularhypothese: Neue Studien zur Entwickelungsgeschichte des Weltalls.* Leipzig, 175 pp. In: BMC, BNC, and NUC.

1881   Richard A. Proctor. *The Poetry of Astronomy.* London, vi + 447 pp. In: BMC, BNC, and NUC.

Editions: NUC lists 4 printings.

1882   David Friedrich Weinland. *Ueber die in Meteoriten entdeckten Thierreste.* Esslingen, 12 pp. In: BMC and NUC.

1883   William Miller. *The Heavenly Bodies: Their Nature and Habitability.* London, 347 pp. In: BMC and NUC.

1883–   Abbé Léger-Marie Pioger. *Dieu dans ses oeuvres. Les splendeurs de*
1884   *l'astronomie, ou il y a d'autres mondes que le nôtre. . . .* 5 vols., Paris. In: BNC.

Editions: Vols. I and II reprinted in 1893.

1884   Manuel Gil y Saenz. *Opuscules sobre mundos habitados ante la Iglesia Catolica.* San Juan, 68 pp. Cited in: *Bibliography of Astronomy 1881–1898* (microfilm).

1884–   Joseph Pohle. *Die Sternwelten und ihre Bewohner.* 2 vols., Cologne. In:
1885   BMC, BNC, and NUC.

Editions: 7 editions by 1922.

1888   Jermain Gildersleeve Porter. *Our Celestial Home: An Astronomer's View of Heaven.* New York, 116 pp. In: BMC and NUC.

Editions: Issued in both New York and London

1892–   Camille Flammarion. *La planète Mars et ses conditions d'habitabilité.* 2
1909   vols., Paris, 608 + 604 pp. In: BMC, BNC, and NUC.

1892   William Fretts. *Inhabitable Worlds Is the Universal Law of Nature as Seen from Material and Spiritual Standpoints* (a lecture delivered before the Liberal League of Jacksonville, Florida, Feb. 14, 1892), Washington, D. C., 15 pp. In: NUC.

1892   Gabriel Prigent. *De l'habitabilité des astres, ou considérations astronomiques, physiques et météorologiques sur l'habitabilité des quelques astres.* Landerneau, 456 pp. In: BMC and BNC.

1894   Théophile Ortolan. *Astronomie et théologie, ou l'erreur géocentrique, la pluralité des mondes habités et le dogme de l'incarnation.* Paris and Lyon, xii + 434 pp. In: BMC and BNC.

1895   Percival Lowell. *Mars.* Boston and New York, viii + 228 pp. In: BMC, BNC, and NUC.

Editions: 2nd ed. in 1896; possible Chinese trans.

1897   Théophile Ortolan. *Étude sur la pluralité des mondes habités et le dogme de l'incarnation.* 3 vols., Paris, 64 + 64 + 64 pp. In: BNC and NUC.

Editions: 9th ed. in 1908.

1898   Abbé François Xavier Burque. *Pluralité des mondes habités considérée au point de vue négatif.* Montreal, viii + 407 pp. In: BMC and NUC.

1898   Carl Goetze. *Die Sonne ist bewohnt. Ein Einblick in die Zustände im Universum*. Berlin, 94 pp. In: BMC.

1898   Théophile Ortolan. *La fausse science contemporaine et les mystères d'outre-tombe*. Paris, 61 pp. In: BNC and NUC.

Editions: 2nd ed. in 1900; also a 1903 ed.

1899   John E. Bowers. *Suns and Worlds of the Universe. Outlines of Astronomy According to the Philosophy of Emanuel Swedenborg*. London, 219 pp. In: BMC and NUC.

1899   Emilia Ferretti. *El dualismo de la eternidad; revelaciones telepáticas*. Valparaiso, 53 pp. In: NUC.

1899   A. Mercier. *Communications avec Mars*. Orléans, 47 pp. In: BMC.

1900   R. M. Jouan. *La question de l'habitabilité du monde étudiée au point de vue de l'histoire, de la science, de la raison et de la foi*. Saint-Ilan, 478 pp. In: BNC.

1901   Otto Dross. *Mars. Eine Welt im Kampf ums Dasein*. Vienna, vii + 171 pp. In: BNC and NUC.

1901   Ludwig Kann. *Neue Theorie über den Ursprung der Kohle und die Lösung des Marsrätsels*. Heidelberg, vi + 96 pp. In: BMC.

1901   Garrett Putnam Serviss. *Other Worlds: Their Nature, Possibilities, and Habitability in the Light of the Latest Discoveries*. New York, xvi + 262 pp. In: BMC and NUC.

Editions: Reprinted in 1928.

1901   George W. Warder. *The Cities of the Sun*. New York, 302 pp. In: BMC and NUC.

Editions: Also a 1901 London printing.

1902   Charles de Kirwan. *Le véritable concept de la pluralité des mondes*. Louvain, 39 pp. Cited in: *International Catalogue of Scientific Literature: Astronomy*, 4 (1905), 78. See *Revue de questions scientifiques*, 51 (1902), 5–39.

1902   A. Mercier. *Conférence astronomique sur la planète Mars . . . Projet d'études sur les moyens practiques d'exécution de signaux lumineux de la Terre à Mars*. Orléans, 48 pp. In: NUC.

1903   Joseph Hamilton. *Our Own and Other Worlds*. New York and Cincinnati, 203 pp. In BMC and NUC.

Editions: NUC lists a 1903 printing in Toronto and a 1904 printing in New York and Cincinnati. BMC lists 1905 and 1917 printings in London.

1903   Alfred Russel Wallace. *Man's Place in the Universe: A Study of the Results of Scientific Research in Relation to the Unity or Plurality of Worlds*. London, xii + 330 pp. In: BMC, BNC, and NUC.

Editions: 7 editions by 1908; another in 1914, trans. into German in 1903 and French in 1907.

1904   Ludwig Zehnder. *Das Leben im Weltall*. Tübingen, 128 pp. In: NUC.

Editions: 2nd ed. in 1910.

1905   Leo Brenner [i.e., Spiridion Gopčević]. *Die Bewohnbarkeit der Welten*. Berlin and Leipzig, 96 pp. In: Arizona State University Library.

1905   William Shuler Harris. *Life in a Thousand Worlds*. Harrisburg, Pa.,
         244 pp. In: NUC.

Editions: 2 printings in 1905.

1906   Svante Arrhenius. *Världarnas utveckling*. Stockholm, 184 pp. In: NUC.

Editions: At least 5 Swedish editions; 1907 German trans. with 7 or more edi-
         tions; 1908 English trans. as *Worlds in the Making*.

1906   Edgar Sylvester Morse. *Mars and Its Mystery*. Boston, xiv + 192 pp. In:
         BNC and NUC.

Editions: Reprinted in 1913.

1907   Svante Arrhenius. *Människan inför världsgåtan*. Stockholm, 181 pp. In:
         NUC.

Editions: Danish trans. in 1908; English trans. in 1909 as *The Life of the Uni-
         verse;* reprinted in 1914.

1907   Percival Lowell. *Mars and Its Canals*. New York and London, 393 pp.
         In: BMC, BNC, and NUC.

Editions: French trans. in 1909.

1907   Alfred Russel Wallace. *Is Mars Habitable?* London, xii + 108 pp. In:
         BMC and NUC.

1908   Percival Lowell. *Mars as an Abode of Life*. New York, xix + 288 pp.
         In: BMC and NUC.

Editions: Reprinted in 1909 and 1910.

1909   Percival Lowell. *The Evolution of Worlds*. New York, xiv + 262 pp. In:
         BMC, BNC, and NUC.

Editions: Reprinted in 1910.

1909   Max Wilhelm Meyer. *Bewohnte Welten*. Leipzig, 94 pp. In: NUC.

1910   Felix Linke. *Ist die Welt bewohnt?* Stuttgart, 110 pp. In: NUC.

1911   Edmond Perrier. *La vie dans les planètes*. 2nd ed., Paris, 126 pp. In:
         NUC.

Editions: Date of 1st ed. not known.

1911   Frank Sewall. *Life on Other Planets as Described by Swedenborg: An
         Address*. Philadelphia, 20 pp. In: BMC and NUC.

1912   Théophile Moreux. *Les autres mondes sont-ils habités?* Paris, 143 pp.
         In: BNC and NUC.

Editions: Published also in 1914; new ed. in 1923; republished in 1950.

1913   Edward Walter Maunder. *Are the Planets Inhabited?* London and New
         York, 166 pp. In: BMC and NUC.

1914   José Comas Solá. *La vida en el planeta Marte según los ultimos datos de
         la ciencia astrónomica*. Barcelona, 102 pp. Cited in: Antonio Palau y
         Dulcet, *Manual de librero Hispanoamericano*. 2nd ed., vol. III (Barce-
         lona and Madrid, 1950), p. 599.

1914   Charles Edward Housden. *The Riddle of the Planet Mars*. New York,
         69 pp. In: BMC and NUC.

Editions: 1914 ed. printed in both London and New York.

1914 Luther Tracy Townsend. *The Stars Not Inhabited: Scientific and Biblical Points of View.* New York and Cincinnati, 254 pp. In: NUC.

1915 Svante Arrhenius. *Stjärnornas öden.* Stockholm, vii + 153 pp. In: NUC.

Editions: English trans. in 1918 as *The Destinies of the Stars;* German trans. 1919; French trans. 1921; Russian trans. 1923.

1915 Charles Edward Housden. *Is Venus Inhabited?* London, 39 pp. In: BMC and NUC.

Editions: 1915 ed. printed in both London and New York.

Name Index

Subject Index

Algol, 66, 287

almanacs, plurality of worlds ideas in, 108–11

analogy: argument for extraterrestrial life from, 43, 50, 52, 59, 63, 66, 67, 73–4, 77, 79, 88-9, 105–6, 115, 126, 128, 137, 153–4, 161, 173, 179, 199, 227, 229–31, 267, 271, 304, 307–10, 315–16, 324, 334, 338, 339, 347, 360–1, 423, 473, 552–3; logic of the argument from, 137, 231–2, 304, 307–8, 315, 343, 440–1, 452, 575n60

Andromeda, 50, 361

angels, 25, 37, 87, 88, 98–100, 107, 130, 136, 148, 161, 172, 187, 198, 238, 243, 244, 261–2, 334, 336, 340, 415, 428, 453, 458

Anglican Church, 103, ch. 6 passim, 332–7, 345, 352, 440–1

asteroids, 227, 288, 432, 438, 465
 advocates of life on: T. Dick, 198–9; T. Hill, 318–9; A. Miller, 451; M. Phillips, 330
 atmosphere of, 66, 220
 origin of, 197, 349
 size of, 194

atheism, 80, 125, 127; and Epicureanism, 3–5, 9

Baptist Church, 190–2, 338, 450

Bible:
 cited in opposition to extraterrestrial life by: Catcott, 92; Kurtz, 262; Leavitt, 342; A. Maxwell, 195; Thomas Aquinas, 4; J. Wesley, 94
 cited in support of extraterrestrial life by: Beattie, 102; Brewster, 303; Burr, 451; T. Dick, 197, 200–1; Ilive, 37; E. King, 104; R. Knight, 336; Lord, 343; Montignez, 412–13; Sturmy, 35

Bridgewater Treatises, 189–90, 268–70, 588n73

Cambridge University, plurality of worlds taught or discussed at, 31, 35, 104–5, 201, ch. 6 passim, 319

canals of Mars: 392, 421, 423, 463, 470, 471, ch. 10 passim, 548, 549, 551, 556, 628n24, 629n37; discovery of system of, 485–6; nomenclature for, 485–6, 498, 500, 536, 545, 555; optical or psychological theories of, 487–8, 489, 494, 495, 497, 500, 505–6, 512, 515, 518, 519–20, 521–2, 523–4, 525–8, 532, 533–4, 535–6, 538, 539, 540, 543, 544, 643n381; origin of the term, 423, 486; photographs of, 526, 527–8, 533–4, 536, 537, 544; physical theories of, 488, 492, 493, 494, 496, 497, 499–500, 501, 506, 509–10, 513, 514–15, 520, 523, 529–30, 538, 544–5; resolved into fine detail, system of, 535–7, 538; see also geminations of Martian canals; Mars

Catholic Apostolic ("Irvingite") Church, 338–9, 491, 630n59

Catholic University of America, plurality of worlds ideas discussed at, 453–4, 456–7

chain of being, great, 29, 32, 35, 52, 57–8, 82, 88–9, 101, 112, 123, 129–30, 138, 139, 142, 145, 151, 172, 178, 197, 260, 455, 621n59

Christianity, 5–9, 111–12, 155, 167, 247, 256, 259, 267, 328, 404, 435, 436, 445, 457, 465, 466, 467, 491, 558, 561n15; see also redemption and incarnation, Christian doctrines of

Church of Jesus Christ of Latter-day Saints, 241–6, 450, 557

Church of the New Jerusalem, 97–101, 228, 237, 250, 438, 450, 548, 557

comets, 68, 82, 102–3, 109, 119, 126, 128, 131, 147, 153, 227, 247, 249, 405, 410, 432, 471, 482, 568n8, 594n31
 advocates of life on: Bernardin de Saint-Pierre, 179; Burgh, 108; Cheyne, 107; Cousin-Despréaux, 180; Davy, 222; T. Dick, 200; Du Pont de Nemours, 178; Ferguson, 60;